Molecular Structures and Dimensions

Vol. 4

Solid State Classes 1–86

Molecular Structures and Dimensions

Vol. 4

Bibliography 1971-72
Organic and Organometallic Crystal Structures

Edited by
Olga Kennard, David G. Watson and William G. Town
University Chemical Laboratory, Cambridge

Springer Science+Business Media, B.V.

Originally published by International Union of Crystallography in 1973.
Softcover reprint of the hardcover 1st edition 1973

Library of Congress catalogue card number 76–133989

ISBN 978-94-017-2343-5 ISBN 978-94-017-2341-1 (eBook)
DOI 10.1007/978-94-017-2341-1

Distributors:

A. Oosthoek, Domstraat 11–13, Utrecht, Netherlands
Polycrystal Book Service, P.O. Box 11567, Pittsburgh, Pennslyvania 15238, USA
Crystallographic Data Centre, University Chemical Laboratory,
Lensfield Road, Cambridge, CB2 1EW, England.

Contents

1	Introduction	page vii
2	Acknowledgements	ix
3	List of Classes in Vol. 4	xi
4	Entries for Classes 1–86	1
5	Cumulative Formula Index for Vols. 1–4	F1
6	Cumulative Transition Metal Index for Vols. 1–4	M1
7	Cumulative Author Index for Vols. 1–4	A1

Introduction

This volume is the fourth classified bibliography of organic and organometallic crystal structures prepared by the Crystallographic Data Centre, University Chemical Laboratory, Cambridge, and published jointly with the International Union of Crystallography.

The first three volumes covered the years 1935–1971. The present volume provides references principally to compounds whose structures were reported in the literature during 1971–1972. A few structures published prior to 1971 and omitted from the previous volumes are also included. The arrangement of entries in the 86 chemical classes is identical with the previous volumes and the reader is referred to the Introduction in Vol. 1 or Vol. 2 for a description of the practical use of the bibliography.

There are three cumulative indexes in the present volume: formula, transition metal and author indexes. All three cover the period 1935–1972 and give references to entries in Vols. 1–4.

The bibliography and indexes were prepared, checked and printed by computer techniques described in the previous volumes. Magnetic tapes of the four volumes are available and anyone interested should contact the Centre for further details.

In the present volume we have attempted to improve the cut-off date by special arrangement with the Centre National de la Recherche Scientifique, Paris, France. Under this arrangement reprints of papers containing crystallographic data are now sent directly to the Crystallographic Data Centre, Cambridge, at the same time as they are sent out to abstractors preparing material for the Bulletin Signalétique. As a result this material is incorporated in our files at an estimated 3–6 months following the publication in the primary journals, and even before the appearance of the abstract in the Bulletin Signalétique.

In addition to the above arrangement, 9 journals, covering approximately 78 % of the crystallographic literature, are scanned directly in Cambridge. The cut-off dates for Volume 4 can be summarised as follows:

Acta Cryst. (B), part 6, 1972
J. Chem. Soc. Dalton, part 13, 1972
J. Chem. Soc. Perkin II, part 9, 1972
J. Chem. Soc. Chem. Comm., part 15, 1972
J. Amer. Chem Soc., part 12, 1972
Acta Chem. Scand., part 3, 1972
Inorg. Chem., part 6, 1972
Tet. Letters, part 27, 1972
Other Journals: complete for 1970
ca. 85% complete for 1971
ca. 70% complete for 1972

Mention should be made of the Conference Proceedings of the American Crystallographic Association, Summer 1971 and Winter 1972, which have been incorporated in Vol. 4.

We would like to draw our readers' attention to the first of the numeric tables which has just been published in this series: Vol. A1 'Interatomic Distances 1960–1965, Organic and Organometallic Crystal Structures'. The new volume is a continuation of the 'Tables of Interatomic Distances and Configuration in Molecules and Ions' (Chemical Society Special Publications No. 11, London 1958; No. 18, London 1965) which covered the literature until the end of 1959. Volume A1 contains numeric data, including bond lengths, bond angles and torsion angles for about 1,300 structures analysed by X-ray and neutron diffraction. Numeric volumes for the later years are also planned.

The work of the Crystallographic Data Centre is supported by the Office for Scientific and Technical Information, Department of Education and Science, as part of the British contribution to international data activities.

We are greatly indebted to readers who have notified us of mistakes and omissions in Vols. 1–3. We have attempted to modify our procedures and are at present considering further changes including changes in the contents of forthcoming volumes. We would be grateful to readers for any suggestions on how these volumes could be further improved.

Cambridge November 1972

Olga Kennard, David G. Watson, William G. Town

Acknowledgements

The production of this bibliography was a collaborative effort by members of the Crystallographic Data Centre: Drs F.H. Allen, N.W. Isaacs, W.D.S. Motherwell, R.C. Pettersen, P.J. Roberts, Miss C.P. Way, Mrs K. Watson and Mrs S. Weeds.

Mrs Weeds was responsible for literature searches, primary abstracting and problems relating to chemical nomenclature. In the latter work she was assisted by Dr R. Huq.

Mrs Watson has been in charge of the encoding of information and the registration and checking of new material. In the secretarial work of documentation she has been assisted by Miss C.P. Way.

Drs Allen, Isaacs, Motherwell, Pettersen and Roberts have contributed to the literature scanning, primary editing and proof reading of the final listings.

The work of the Centre was guided by members of the OSTI Scientific Advisory Committee: Professor J.W. Linnett, FRS, Dr M.F. Lynch, Dr F.W. Matthews, Professor D.C. Phillips, FRS, Professor M.R. Truter and Professor A.J.C. Wilson, FRS (Chairman).

We are grateful to the Medical Research Council for allowing a member of their External Scientific Staff (O. Kennard) to participate in this work.

We thank the University of Cambridge for the provision of accommodation in the University Chemical Laboratory and for the administration of the OSTI grant.

Our task was greatly facilitated by the excellent organisation of the Centre National de la Recherche Scientifique. We are especially grateful to Madam C. Degen of the CNRS who was responsible for the improved literature searches referred to in the Introduction. Dr K. Loening and the Nomenclature Division of the Chemical Abstracts Service have helped with questions of nomenclature.

We have used the IBM 360/44 computer at the Institute of Theoretical Astronomy and we were greatly helped by both the programming staff and operators. We are grateful to INSPEC (Information Service in Physics, Electrotechnology and Computers & Control) and especially to Mr P. Simmons for the use of their computer typesetting programs, which they specially modified for our purposes.

The bibliography was prepared in parallel with the Organic volume of 'Crystal Data' (National Bureau of Standards, Washington D.C., USA). The third edition was published in the summer of 1972 and both publications were strengthened by this collaboration.

List of Classes

1	Aliphatic Carboxylic Acid Derivatives	1
2	Aliphatic Carboxylic Acid Salts (Ammonium, IA, IIA Metals)	4
3	Aliphatic Amines	7
4	Aliphatic (N and S) Compounds	12
5	Aliphatic Miscellaneous	13
6	Enolates (Aliphatic and Aromatic)	15
7	Nitriles (Aliphatic and Aromatic)	16
8	Urea Compounds (Aliphatic and Aromatic)	20
9	Nitrogen-Nitrogen Compounds (Aliphatic and Aromatic)	23
10	Nitrogen-Oxygen Compounds (Aliphatic and Aromatic)	26
11	Sulphur and Selenium Compounds	28
12	Carbonium Ions, Carbanions, Radicals	32
13	Benzoic Acid Derivatives	35
14	Benzoic Acid Salts (Ammonium, IA, IIA Metals)	38
15	Benzene Nitro Compounds	39
16	Anilines	42
17	Phenols and Ethers	47
18	Benzoquinones	51
19	Benzene Miscellaneous	52
20	Monocyclic Hydrocarbons (3, 4, 5-Membered Rings)	55
21	Monocyclic Hydrocarbons (6-Membered Rings)	57
22	Monocyclic Hydrocarbons (7, 8-Membered Rings)	61
23	Monocyclic Hydrocarbons (9- and Higher-Membered Rings)	63
24	Naphthalene Compounds	64
25	Naphthoquinones	67
26	Anthracene Compounds	69
27	Hydrocarbons (2 Fused Rings)	70
28	Hydrocarbons (3 Fused Rings)	73

29 Hydrocarbons (4 Fused Rings) 75
30 Hydrocarbons (5 or More Fused Rings) 77
31 Bridged Ring Hydrocarbons 79
32 Hetero-Nitrogen (3, 4, 5-Membered Monocyclic) 83
33 Hetero-Nitrogen (6-Membered Monocyclic) 87
34 Hetero-Nitrogen (7- and Higher-Membered Monocyclic) 93
35 Hetero-Nitrogen (2 Fused Rings) 94
36 Hetero-Nitrogen (More than 2 Fused Rings) 100
37 Hetero-Nitrogen (Bridged Ring Systems) 106
38 Hetero-Oxygen 109
39 Hetero-Sulphur and Hetero-Selenium 116
40 Hetero-(Nitrogen and Oxygen) 122
41 Hetero-(Nitrogen and Sulphur) 126
42 Hetero-Mixed Miscellaneous 130
43 Barbiturates 132
44 Pyrimidines and Purines 134
45 Carbohydrates 140
46 Phosphates 146
47 Nucleosides and Nucleotides 148
48 Amino-Acids and Peptides 151
49 Porphyrins and Corrins 158
50 Antibiotics 161
51 Steroids 164
52 Monoterpenes 171
53 Sesquiterpenes 172
54 Diterpenes 174
55 Sesterterpenes 177
56 Triterpenes 178
57 Tetraterpenes 179
58 Alkaloids 180
59 Miscellaneous Natural Products 185
60 Molecular Complexes 189
61 Clathrates 198
62 Boron Compounds 199
63 Silicon Compounds 203
64 Phosphorus Compounds 206
65 Arsenic Compounds 212
66 Antimony and Bismuth Compounds 214
67 Groups IA and IIA Compounds 216
68 Group III Compounds 220
69 Germanium, Tin, Lead Compounds 223
70 Tellurium Compounds 228

71	Transition Metal-C Compounds	230
72	Metal π-Complexes (Open-Chain)	242
73	Metal π-Complexes (Cyclopentadiene)	248
74	Metal π-Complexes (Arene)	253
75	Metal π-Complexes (Miscellaneous Ring Systems)	255
76	Metal Complexes (Ethylenediamine)	260
77	Metal Complexes (Acetylacetone)	266
78	Metal Complexes (Salicylic Derivatives)	270
79	Metal Complexes (Thiourea)	273
80	Metal Complexes (Thiocarbamate or Xanthate)	275
81	Metal Complexes (Carboxylic Acid)	278
82	Metal Complexes (Amino-Acid)	286
83	Metal Complexes (Nitrogen Ligand)	289
84	Metal Complexes (Oxygen Ligand)	310
85	Metal Complexes (Sulphur or Selenium Ligand)	316
86	Metal Complexes (P, As, Sb Ligand)	322

ALIPHATIC CARBOXYLIC ACID DERIVATIVES

1.C **Pyridine - N - oxide trichloroacetic acid complex**

$C_2HCl_3O_2 . C_5H_5NO$

For complete entry see 60.4

1.C **Potassium hydrogen bis(trifluoroacetate) (neutron study)**

$C_2HF_3O_2 , C_2F_3O_2^- , K^+$

For complete entry see 2.2

1.C **Potassium deuterium bis(trifluoroacetate) (neutron study)**

$C_2DF_3O_2 , C_2F_3O_2^- , K^+$

For complete entry see 2.3

1.C **Ammonium hydrogen di - chloroacetate (paraelectric form)**

$C_2H_2ClO_2^- , C_2H_3ClO_2 , H_4N^+$

For complete entry see 2.7

1.C **Urea - oxalic acid**

$C_2H_2O_4 , 2CH_4N_2O$

For complete entry see 60.3

1.1 **Monofluoroacetic acid**

$C_2H_3FO_2$

J.A.Kanters, J.Kroon *Acta Cryst. (B),* **28,** 1946, 1972

1.C **Potassium hydrogen diacetate**

$C_2H_4O_2 , C_2H_3O_2^- , K^+$

For complete entry see 2.8

1.C **Deoxycholic acid - acetic acid complex**

$C_2H_4O_2 , C_{24}H_{40}O_4$

For complete entry see 60.42

1.2 **Acetamide (rhombohedral form)**

C_2H_5NO

W.A.Denne, R.W.H.Small *Acta Cryst. (B),* **27,** 1094, 1971

1.3 **Difluoromalonic acid**

$C_3H_2F_2O_4$

J.A.Kanters, J.Kroon *Acta Cryst. (B),* **28,** 1345, 1972

1.4 **Dichloromalonamide**
$C_3H_4Cl_2N_2O_2$
J.A.Lerbscher, K.V.Krishna Rao, J.Trotter *Curr. Sci.*, **39**, 560, 1970

1.C **L - α,γ - Diaminobutyric acid monohydrochloride**
$C_4H_{11}N_2O_2^+$, Cl^-
For complete entry see 48.13

1.C **L - α,γ - Diaminobutyric acid monohydrochloride**
$C_4H_{11}N_2O_2^+$, Cl^-
For complete entry see 48.14

1.5 **Glutarimide**
$C_5H_7NO_2$
C.S.Petersen *Acta Chem. Scand.*, **25**, 379, 1971

1.6 **Muconyl chloride**
$C_6H_4Cl_2O_2$
J.Leser, D.Rabinovich *Israel J. Chem.*, **9**, II, 1971

1.7 **Citric acid monohydrate**
$C_6H_8O_7$, H_2O
G.Roelofsen, J.A.Kanters *Cryst. Struct. Comm.*, **1**, 23, 1972

1.8 **3,3 - Dimethylglutaric acid**
$C_7H_{12}O_4$
E.Benedetti, R.Claverini, C.Pedone *Cryst. Struct. Comm.*, **1**, 27, 1972

1.C **Dilactophorbic acid chloride (absolute configuration)**
(5R,8R) 8 - Carboxy - 2,6 dioxo - 1,7 - dioxo(4,4) - spirononane
$C_8H_7ClO_5$
For complete entry see 38.7

1.C **DL - 6 - Thioctic acid**
α - DL - Lipoic acid
$C_8H_{14}O_2S_2$
For complete entry see 39.9

1.9 **Tetramethyl - β - oxoglutaric acid**
$C_9H_{14}O_5$
G.Avitabile, P.Ganis, U.Lepore *Macromolecules,* **4**, 239, 1971

1.10 **Ethylenediamine tetra - acetic acid**
$C_{10}H_{16}N_2O_8$
M.Cotrait *Acta Cryst. (B),* **28**, 781, 1972
Also classified in 3, 48

1.11 **meso - 2,4,6 - Trimethyl - pimelic acid (high melting form)**
$C_{10}H_{18}O_4$
M.Brufani, W.Fedeli *Helv. Chim. Acta,* **54**, 51, 1971

1.12 **N,N - Dimethyl - p - bromocinnamamide**

$C_{11}H_{12}BrNO$
M.A.M.Meester, H.Schenk *Rec. Trav. Chim. Pays-Bas,* **90,** 508, 1971
Also classified in 19

1.13 **Succinylcholine perchlorate**

$C_{14}H_{30}N_2O_4^{2+}$, $2ClO_4^-$
B.Jensen *Acta Chem. Scand.,* **25,** 3388, 1971
Residue 1 also classified in 3

1.14 **Glycerol - 1,2 - di(11 - bromoundecanoate) - 3 - p - toluenesulfonate**

$C_{32}H_{52}Br_2O_7S$
P.H.Watts Junior, W.A.Pangborn, A.Hybl
Amer. Cryst. Assoc., Abstr. Papers (Summer Meeting), 74, 1971
Also classified in 11

ALIPHATIC CARBOXYLIC ACID SALTS (AMMONIUM, IA, IIA METALS)

2.1 **Strontium formate dihydrate**
$2CHO_2^-$, Sr^{2+}, $2H_2O$
J.L.Galigne *Acta Cryst. (B),* **27,** 2429, 1971

2.2 **Potassium hydrogen bis(trifluoroacetate) (neutron study)**
$C_2F_3O_2^-$, $C_2HF_3O_2$, K^+
A.L.Macdonald, J.C.Speakman, D.Hadzi *J. C. S. Perkin ii,* 825, 1972
Residue 2 classified in 1

2.3 **Potassium deuterium bis(trifluoroacetate) (neutron study)**
$C_2F_3O_2^-$, $C_2DF_3O_2$, K^+
A.L.Macdonald, J.C.Speakman, D.Hadzi *J. C. S. Perkin ii,* 825, 1972
Residue 2 classified in 1

2.4 **Ammonium oxalate monoperhydrate**
$C_2O_4^{2-}$, $2H_4N^+$, H_2O_2
B.F.Pedersen *Acta Cryst. (B),* **28,** 746, 1972

2.5 **Potassium hydrogen oxalate (neutron study)**
$C_2HO_4^-$, K^+
F.H.Moore, L.F.Power *Inorg. Nucl. Chem. Letters,* **7,** 873, 1971

2.6 **Sodium hydrogen oxalate monohydrate**
$C_2HO_4^-$, Na^+, H_2O
H.Follner, J.A.Kanters, J.Kroon *Z. Anorg. Allg. Chem.,* **379,** 225, 1970

2.7 **Ammonium hydrogen di - chloroacetate (paraelectric form)**
$C_2H_3ClO_2$, $C_2H_2ClO_2^-$, H_4N^+
M.Ichikawa *Acta Cryst. (B),* **28,** 755, 1972
Residue 2 classified in 1

2.8 **Potassium hydrogen diacetate**
$C_2H_3O_2^-$, $C_2H_4O_2$, K^+
M.Currie *J. C. S. Perkin ii,* 832, 1972
Residue 2 classified in 1

2.9 **Diaquo - barium oxalate**
$C_2H_4BaO_6$
J.-C.Mutin, A.Courtois, G.Bertrand, J.Protas, G.Watelle-Marion
C. R. Acad. Sci., Fr., C, **273,** 1512, 1971

2.10 **Sodium hydrogen maleate trihydrate**

$C_4H_3O_4^-$, Na^+, $3H_2O$
M.P.Gupta, S.M.Prasad, B.Yadav *Cryst. Struct. Comm.*, **1**, 211, 1972

2.C **DL - Bromopheniramine hydrogen maleate**

1 - (p - Bromophenyl) - 1 - (2′ - pyridyl) - 3 - N,N - dimethylpropylamine hydrogen maleate
$C_4H_3O_4^-$, $C_{16}H_{20}BrN_2^+$
For complete entry see 33.38

2.11 **Tetra - aquo copper(ii) hydrogen maleate**

$2C_4H_3O_4^-$, $H_8CuO_4^{2+}$
C.K.Prout, J.R.Carruthers, F.J.C.Rossotti *J. Chem. Soc. (A)*, 3342, 1971

2.12 **Cesium hydrogen succinate monohydrate**

$C_4H_5O_4^-$, Cs^+, H_2O
A.McAdam, J.C.Speakman *J. Chem. Soc. (A)*, 1997, 1971

2.13 **Potassium hydrogen succinate**

$C_4H_5O_4^-$, K^+
A.McAdam, M.Currie, J.C.Speakman *J. Chem. Soc. (A)*, 1994, 1971

2.14 **Potassium hydrogen succinate (neutron study)**

$C_4H_5O_4^-$, K^+
A.McAdam, M.Currie, J.C.Speakman *J. Chem. Soc. (A)*, 1994, 1971

2.15 **Potassium hydrogen mesotartrate (at $-160\,°C$)**

$C_4H_5O_6^-$, K^+
J.Kroon, J.A.Kanters *Acta Cryst. (B)*, **28**, 714, 1972

2.16 **Triaquo calcium fumarate**

$C_4H_8CaO_7$
M.P.Gupta, S.M.Prasad, R.G.Sahu, B.N.Sahu
Acta Cryst. (B), **28**, 135, 1972

2.17 **Potassium hydrogen glutarate**

$C_5H_7O_4^-$, K^+
A.L.Macdonald, J.C.Speakman *J. C. S. Perkin ii*, 942, 1972

2.18 **Rubidium hydrogen glutarate**

$C_5H_7O_4^-$, Rb^+
A.L.Macdonald, J.C.Speakman *J. Cryst. Mol. Struct.*, **1**, 189, 1971

2.19 **Ammonium hydrogen glutarate**

$C_5H_7O_4^-$, H_4N^+
A.L.Macdonald, J.C.Speakman *J. Cryst. Mol. Struct.*, **1**, 189, 1971

2.20 **Dipotassium cis - aconitate**

$C_6H_4O_6^{2-}$, $2K^+$
J.P.Glusker, W.Orehowsky Junior, C.A.Casciato, H.L.Carrell
Acta Cryst. (B), **28**, 419, 1972

2.21 Garcinia acid calcium salt tetrahydrate (absolute configuration)
(–) - Hydroxycitric acid lactone calcium salt tetrahydrate
$C_6H_4O_7^{2-}$, Ca^{2+}, $4H_2O$
J.P.Glusker, J.A.Minkin, C.A.Casciato *Acta Cryst. (B),* **27,** 1284, 1971

2.22 Potassium dihydrogen trans - aconitate
$C_6H_5O_6^{-}$, K^{+}
J.M.Dargay, H.M.Berman, H.L.Carrell, J.P.Glusker
Acta Cryst. (B), **28,** 1533, 1972

2.23 Calcium nitrilotriacetate dihydrate
$C_6H_7NO_6^{2-}$, Ca^{2+}, $2H_2O$
S.H.Whitlow
Amer. Cryst. Assoc., Abstr. Papers (Summer Meeting), 99, 1971

2.24 Sodium phenoxyacetate hemihydrate
$C_8H_7O_3^{-}$, Na^{+}, $0.5H_2O$
C.K.Prout, R.M.Dunn, O.J.R.Hodder, F.J.C.Rossotti
J. Chem. Soc. (A), 1986, 1971
Residue 1 also classified in 17

ALIPHATIC AMINES

3.1 **tris(Dimethylammonium) chloride tetrachlorocuprate(ii)**
$3C_2H_8N^+ , Cl^- , Cl_4Cu^{2-}$
R.D.Willett, M.L.Larsen *Inorg. Chim. Acta,* **5,** 175, 1971

3.2 **Ethylenediamine (at –60 ° C)**
$C_2H_8N_2$
S.Jamet-Delcroix, H.Gillier-Pandraud
C. R. Acad. Sci., Fr., C, **274,** 771, 1972

3.3 **Tri - ethylenediamine lithium iodide**
$3C_2H_8N_2 , Li^+ , I^-$
H.Gillier-Pandraud, S.Jamet-Delcroix *Acta Cryst. (B),* **27,** 2476, 1971

3.4 **bis(Ethylenediammonium - monobromide) tetrabromocuprate(ii)**
$2C_2H_{10}N_2^{2+} , 2Br^- , Br_4Cu^{2-}$
D.N.Anderson, R.D.Willett *Inorg. Chim. Acta,* **5,** 41, 1971

3.5 **Ethylenediamine - bis(borane)**
$C_2H_{14}B_2N_2$
H.-Y.Ting, W.H.Watson, H.C.Kelly *Inorg. Chem.,* **11,** 374, 1972
Also classified in 62

3.C **Trimethylamino - boron tribromide**
$C_3H_9BBr_3N$
For complete entry see 62.2

3.C **Trimethylamino - boron trichloride**
$C_3H_9BCl_3N$
For complete entry see 62.3

3.C **Trimethylamino - boron tri - iodide**
$C_3H_9BI_3N$
For complete entry see 62.4

3.6 **Trimethylammonionitramidate**
$C_3H_9N_3O_2$
A.F.Cameron, N.J.Hair, D.G.Morris *J. C. S. Perkin ii,* 1071, 1972
Also classified in 9

3.C **L - α,γ - Diaminobutyric acid monohydrochloride**
$C_4H_{11}N_2O_2^+$, Cl^-
For complete entry see 48.13

3.C **L - α,γ - Diaminobutyric acid monohydrochloride**
$C_4H_{11}N_2O_2^+$, Cl^-
For complete entry see 48.14

3.C **Tetramethylammonium tri(acetato) - diphenyl - plumbate(iv)**
$C_4H_{12}N^+$, $C_{18}H_{19}O_6Pb^-$
For complete entry see 69.30

3.7 **bis(Tetramethylammonium) hexachloro(dodeca - μ - chloro - hexaniobate)**
$2C_4H_{12}N^+$, $Cl_{18}Nb_6^{2-}$
F.W.Koknat, R.E.McCarley *Inorg. Chem.*, **11,** 812, 1972

3.8 **Tetramethylammonium 1,6,8 - trichloroheptahydro - closo - decaborate**
$2C_4H_{12}N^+$, $H_7B_{10}Cl_3^{2-}$
F.E.Scarbrough, W.N.Lipscomb *Inorg. Chem.*, **11,** 369, 1972

3.9 **Tetramethylammonium 7,7′ - commo - bis(dodecahydro - 7 - nickela - nido - undecaborate)**
$2C_4H_{12}N^+$, $H_{24}B_{20}Ni^{2-}$
L.J.Guggenberger *J. Amer. Chem. Soc.*, **94,** 114, 1972

3.10 **Diethylammonium tetracyanopalladate**
$2C_4H_{12}N^+$, $C_4N_4Pd^{2-}$
S.Jerome-Lerutte *Acta Cryst. (B),* **27,** 1624, 1971

3.C **Tetramethylammonium carbido(hexadeca - carbonyl)hexa - iron**
$2C_4H_{12}N^+$, $C_{17}Fe_6O_{16}^{2-}$
For complete entry see 71.35

3.11 **bis(2 - Ammonium - ethyl)ammonium chloride tetrachlorocuprate(ii)**
$C_4H_{16}N_3^{3+}$, Cl^- , Cl_4Cu^{2-}
G.L.Ferguson, B.Zaslow *Acta Cryst. (B),* **27,** 849, 1971

3.C **Histamine tetrachlorocobaltate(ii)**
$C_5H_{11}N_3^{2+}$, Cl_4Co^{2-}
For complete entry see 32.8

3.C **Choline chloride**
$C_5H_{14}NO^+$, Cl^-
For complete entry see 59.1

3.C **O - (- L - α - Glyceryl - phosphoryl) - ethanolamine monohydrate (absolute configuration)**
$C_5H_{14}NO_6P$, H_2O
For complete entry see 46.4

3.C **Triethylammonium phenylthioarsenate**
$C_6H_{16}N^+ , C_6H_6AsO_2S^-$
For complete entry see 65.5

3.C **Triethylammonium tris(o - phenylenedioxy) phosphate**
$C_6H_{16}N^+ , C_{18}H_{12}O_6P^-$
For complete entry see 46.7

3.C **6 - Hydroxydopamine hydrochloride**
3,4,6 - Trihydroxyphenylethylamine hydrochloride
$C_8H_{12}NO_3^+ . Cl^-$
For complete entry see 59.3

3.12 **Tetraethylammonium tetracarbonyl - tribromo - tungstate(ii)**
$C_8H_{20}N^+ , C_4Br_3O_4W^-$
M.G.B.Drew, A.P.Wolters *J. C. S. Chem. Comm.*, 457, 1972

3.C **Tetraethylammonium 2,2′ - commo - bis(nonahydrodicarba - 2 - cobalta - closo - decaborate) (tetragonal form)**
$C_8H_{20}N^+ , C_4H_{18}B_{14}Co^-$
For complete entry see 62.9

3.13 **Tetraethylammonium dibromo - bis(tetracarbonyl - cobalt) indate(iii)**
$C_8H_{20}N^+ , C_8Br_2Co_2InO_8^-$
P.D.Cradwick *J. Organometal. Chem.*, **27,** 251, 1971

3.C **Tetraethylammonium tris(ethylxanthato) lead(ii)**
$C_8H_{20}N^+ , C_9H_{15}O_3PbS_6^-$
For complete entry see 69.15

3.C **Compound B**
$C_8H_{20}N^+ , C_9H_{26}B_{17}CoN^-$
For complete entry see 62.13

3.14 **bis(Tetraethylammonium) oxopentachloro - protactinate(v)**
$2C_8H_{20}N^+ . Cl_5OPa^{2-}$
D.Brown, C.T.Reynolds, P.T.Moseley *J. C. S. Dalton,* 857, 1972

3.15 **Tetraethylammonium molybdenum nickel carbonyl**
$2C_8H_{20}N^+ . C_{14}Mo_2Ni_4O_{14}^{2-}$
J.K.Ruff, R.P.White Junior, L.F.Dahl *J. Amer. Chem. Soc.*, **93,** 2159, 1971

3.16 **Tetraethylammonium tungsten nickel carbonyl**
$2C_8H_{20}N^+ . C_{16}Ni_3O_{16}W_2^{2-}$
J.K.Ruff, R.P.White Junior, L.F.Dahl *J. Amer. Chem. Soc.*, **93,** 2159, 1971

3.C **tris(Tetraethylammonium) tris - (1,2 - dicyanoethylene - 1,2 - dithiolato) indate(iii)**
$3C_8H_{20}N^+ . C_{12}InN_6S_6^{3-}$
For complete entry see 68.7

3.17 **Tetra - ethylammonium octa(thiocyanato) uranate(iv)**

$4C_8H_{20}N^+ , C_8N_8S_8U^{4-}$

R.Countryman, W.S.McDonald *J. Inorg. Nucl. Chem.*, **33**, 2213, 1971

3.18 **Adrenalone hydrochloride monohydrate**

$C_9H_{12}NO_3^+ , Cl^- , H_2O$

R.Bergin *Acta Cryst. (B)*, **27**, 2139, 1971

Residue 1 also classified in 17

3.19 **Amphetamine sulfate**

$2C_9H_{14}N^+ , O_4S^{2-}$

R.Bergin, D.Carlstrom *Acta Cryst. (B)*, **27**, 2146, 1971

3.20 **Trimethylammoniobenzamidate**

$C_{10}H_{14}N_2O$

A.F.Cameron, N.J.Hair, D.G.Morris *J. C. S. Perkin ii*, 1071, 1972

Also classified in 9

3.C **Ethylenediamine tetra - acetic acid**

$C_{10}H_{16}N_2O_8$

For complete entry see 1.10

3.21 **Tetramethylene - bis(trimethylammonium) dibromide dihydrate**

$C_{10}H_{26}N_2^{2+} , 2Br^- , 2H_2O$

Y.Barrans *Acta Cryst. (B)*, **28**, 651, 1972

3.22 **2 - Dimethylaminoethyl selenol - benzoate hydrochloride**

$C_{11}H_{16}NOSe^+ , Cl^-$

D.D.Dexter *Acta Cryst. (B)*, **28**, 49, 1972

Residue 1 also classified in 13, 11

3.C **Neostigmine bromide**

$C_{12}H_{19}N_2O_2^+ , Br^-$

For complete entry see 16.11

3.C **3 - (2 - Diethylammoniumethoxy) - 1,2 - benzisothiazole tetrachlorocuprate**

$2C_{13}H_{19}N_2OS^+ , Cl_4Cu^{2-}$

For complete entry see 41.22

3.23 **DL - N - t - Butyl - 2 - (4 - hydroxy - 3 - hydroxymethylphenyl) - 2 - hydroxyethylamine**

Salbutamol

$C_{13}H_{21}NO_3$

J.P.Beale, C.T.Grainger *Cryst. Struct. Comm.*, **1**, 71, 1972

Also classified in 17

3.24 **Procaine hydrochloride**

2 - Diethylaminoethyl p - aminobenzoate hydrochloride

$C_{13}H_{21}N_2O_2^+ , Cl^-$

D.D.Dexter *Acta Cryst. (B)*, **28**, 77, 1972

Residue 1 also classified in 13, 16

3.C **2 - ((2) - Dimethylaminoethyl - 2 - thienylamino)pyridine hydrochloride**
$C_{14}H_{20}N_3S^+ , Cl^-$
For complete entry see 33.34

3.C **Lidocaine hydrohexafluoroarsenate**
2 - Diethylamino - 2′,6′ - acetoxylidide hydrohexafluoroarsenate
$C_{14}H_{23}N_2O^+ , AsF_6^-$
For complete entry see 16.15

3.C **Succinylcholine perchlorate**
$C_{14}H_{30}N_2O_4^{2+} , 2ClO_4^-$
For complete entry see 1.13

3.C **3(a) - Dimethylamino - 2(a) - acetoxy - trans - decalin methiodide**
$C_{15}H_{28}NO_2^+ , I^-$
For complete entry see 27.13

3.C **3 - Chloro - 10 - (3′ - dimethylaminopropyl) - phenothiazine hydrochloride (high temp. form)**
$C_{17}H_{20}ClN_2S^+ , Cl^-$
For complete entry see 41.28

3.25 **DL - N - (2 - (4 - Hydroxyphenyl) - 1 - methyl)ethyl - (2′ - (3,5 - dihydroxyphenyl) - 2′ - hydroxy)ethylamine hydrobromide**
$C_{17}H_{22}NO_4^+ , Br^-$
J.P.Beale *Cryst. Struct. Comm.*, **1**, 223, 1972
Residue 1 also classified in 17

3.26 **DL - N - (2 - (4 - Hydroxyphenyl) - 1 - methyl - ethyl) - 2 - (3,5 - dihydroxyphenyl) - 2 - hydroxy - ethylamine hydrobromide**
$C_{17}H_{22}NO_4^+ , Br^-$
J.P.Beale *Cryst. Struct. Comm.*, **1**, 67, 1972
Residue 1 also classified in 17

3.C **p - Methylbenzyl - dimethyl - benzoylmethylene - ammonium ylide monohydrate**
$C_{18}H_{21}NO . H_2O$
For complete entry see 12.14

3.C **9,10 - Dihydro - 4 - (3 - dimethylamino - propylidene) - 4H - benzo(4,5)cyclohepta(1,2 - b)thiophene hydrochloride**
$C_{18}H_{22}NS^+ . Cl^-$
For complete entry see 39.35

3.C **3 - Methoxy - 10 - (3′ - dimethylaminopropyl) - phenothiazine hydrogen maleate**
$C_{18}H_{23}N_2OS^+ , C_4H_3O_4^-$
For complete entry see 41.30

ALIPHATIC (N AND S) COMPOUNDS

4.1 **Thioacetamide - S - oxide**

C_2H_5NOS

W.Walter, J.Holst, J.Eck *J. Molec. Struct.*, **9**, 151, 1971

4.C α,α' **- Dithio - bis(formamidinium) dichloride**

$C_2H_8N_4S_2^{2+}$, $2Cl^-$

For complete entry see 8.6

4.C **2 - Guanido - ethane sulfinic acid dihydrate**

$C_3H_9N_3O_2S$. $2H_2O$

For complete entry see 8.8

4.2 **syn - S - Methyl - O - (N - methylcarbamoyl)acetothiohydroximate**

$C_5H_{10}N_2O_2S$

M.G.Waite, G.A.Sim *J. Chem. Soc. (B)*, 752, 1971

Also classified in 10

4.3 **S,S - Diethyl - N - dichloroacetylsulphilimine**

$C_6H_{11}Cl_2NOS$

A.Kalman, K.Sasvari, A.Kucsman *J. Chem. Soc. (D)*, 1447, 1971

4.4 **bis(Sulfur dioxide) N,N,N',N' - tetramethylethylenediamine complex**

$C_6H_{16}N_2O_4S_2$

E.L.Enwall, D.van der Helm, R.Lehr, S.D.Christian

Amer. Cryst. Assoc., Abstr. Papers (Summer Meeting), 107, 1971

4.5 **syn - S - Cyanoethyl - O - (N - methylcarbamoyl)acetothiohydroximate**

$C_7H_{11}N_3O_2S$

M.G.Waite, G.A.Sim *J. Chem. Soc. (B)*, 752, 1971

Also classified in 10

ALIPHATIC MISCELLANEOUS

5.1 **Bromoform (at –20 ° C)**
$CHBr_3$
T.Kawaguchi, K.Takashina, T.Tanaka, T.Watanabe
Acta Cryst. (B), **28**, 967, 1972

5.C **Bromoform - hexamethylenetetramine complex (at –35 ° C)**
$2CHBr_3, C_6H_{12}N_4$
For complete entry see 60.1

5.2 **Hexabromoethane**
C_2Br_6
G.Mandel, J.Donohue *Acta Cryst. (B)*, **28**, 1313, 1972

5.3 **Methylchloroform (at –60 ° C)**
1,1,1 - Trichloroethane
$C_2H_3Cl_3$
L.Silver, R.Rudman
Amer. Cryst. Assoc., Abstr. Papers (Summer Meeting), 72, 1971

5.4 **cis,cis - 1,2,3,4 - Tetrachlorobutadiene**
$C_4H_2Cl_4$
Y.Otaka *Acta Cryst. (B)*, **28**, 342, 1972

5.5 **Hexa - 2,4 - diyne - 1,6 - diol**
$C_6H_6O_2$
E.Hadicke, K.Penzien, H.W.Schnell *Angew. Chem.*, **83**, 1024, 1972

5.C **4 - p - Hydroxyphenyl - 2,2,4 - trimethylchroman - n - heptanol**
Dianin's compound - n - heptanol
$C_7H_{16}O, 6C_{18}H_{20}O_2$
For complete entry see 61.2

5.6 **2,5 - Dimethyl - 2,5 - hexanediol tetrahydrate**
$C_8H_{18}O_2, 4H_2O$
G.A.Jeffrey, M.S.Shen
Amer. Cryst. Assoc., Abstr. Papers (Winter Meeting), 88, 1972

5.C **4 - p - Hydroxyphenyl - 2,2,4 - trimethylthiochroman 2,5,5 - trimethylhex - 3 - yn - 2 - ol**
$C_9H_{16}O, 6C_{18}H_{20}OS$
For complete entry see 61.1

5.7 **p - Bromobenzylideneacetone**

$C_{10}H_9BrO$

K.Sugiyama, H.Shimanouchi, Y.Sasada

Bull. Chem. Soc. Jap., **43,** 3624, 1970

Also classified in 19

5.8 **11 - Bromoundecanol (low melting form)**

$C_{11}H_{23}BrO$

L.Rosen, A.Hybl *Acta Cryst. (B),* **28,** 610, 1972

5.C **1,1 - Dimethyl - 2,5 - diphenyl - 1 - silacyclopentadiene diphenylacetylene complex**

$C_{14}H_{10}$, $C_{18}H_{18}Si$

For complete entry see 60.36

5.C **Dibenzoylacetylene**

$C_{16}H_{10}O_2$

For complete entry see 13.19

5.9 **Pentaerythritol tetracinnamate**

$C_{41}H_{36}O_8$

G.Tieghi, M.Zocchi *Cryst. Struct. Comm.*, **1,** 167, 1972

Also classified in 19

ENOLATES (ALIPHATIC AND AROMATIC)

6.C **Potassium 4 - hydroxy - 5,7 - dinitrobenzfurazan monohydrate**
$C_6HN_4O_6^-$, K^+, H_2O
For complete entry see 40.6

6.C **Potassium picrate**
$C_6H_2N_3O_7^-$, K^+
For complete entry see 15.1

6.C **Serotonin picrate monohydrate**
$C_6H_2N_3O_7^-$, $C_{10}H_{13}N_2O^+$, H_2O
For complete entry see 35.11

6.C **tetrakis(Thiourea) lead(ii) picrate**
$2C_6H_2N_3O_7^-$, $C_4H_{16}N_8PbS_4^{2+}$
For complete entry see 69.8

6.C **Isonitroso - acetophenone potassium o - nitrophenolate complex**
$C_8H_4NO_3^-$, $C_8H_7NO_2$, K^+
For complete entry see 10.2

6.1 **Benzoylacetone**
$C_{10}H_{10}O_2$
D.Semmingsen *Acta Chem. Scand.*, **26,** 143, 1972

6.2 **ω - (p - Toluoyl) - acetophenone enol**
$C_{16}H_{14}O_2$
K.Kato *Acta Cryst. (B),* **27,** 2028, 1971

NITRILES (ALIPHATIC AND AROMATIC)

7.1 **Bromo - tricyanomethane**
C_4BrN_3
J.R.Witt, D.Britton, C.Mahon *Acta Cryst. (B),* **28,** 950, 1972

7.2 **Chloro - tricyanomethane (hexagonal form)**
C_4ClN_3
J.R.Witt, D.Britton, C.Mahon *Acta Cryst. (B),* **28,** 950, 1972

7.3 **Potassium tricyanomethanide**
$C_4N_3^-$, K^+
J.R.Witt, D.Britton *Acta Cryst. (B),* **27,** 1835, 1971
Residue 1 also classified in 12

7.4 **1,1,1 - Tricyanoethane**
$C_5H_3N_3$
J.R.Witt, D.Britton, C.Mahon *Acta Cryst. (B),* **28,** 950, 1972

7.5 **Tetracyanoethylene (cubic form)**
C_6N_4
R.G.Little, D.Pautler, P.Coppens *Acta Cryst. (B),* **27,** 1493, 1971

7.C **(3.3)Paracyclophane - tetracyanoethylene complex**
C_6N_4 . $C_{18}H_{20}$
For complete entry see 60.37

7.C **Tetracyanoethylene oxide**
C_6N_4O
For complete entry see 38.2

7.C **Tetracyanoethylene oxide (neutron study)**
C_6N_4O
For complete entry see 38.3

7.C **Tetracyanoethylene oxide (discussion of results of Matthews et al., J.Amer.Chem.Soc.,93,5945,1971)**
C_6N_4O
For complete entry see 38.4

7.6 **Potassium 1,1,3 - tricyanopropan - 2 - one**
$C_6H_2N_3O^-, K^+$
B.Klewe *Acta Chem. Scand.*, **25**, 1988. 1971
Residue 1 also classified in 12

7.C **4 - Cyanopyridine**
$C_6H_4N_2$
For complete entry see 33.13

7.7 **2 - Amino - 1,1,3 - tricyanopropene (form i)**
$C_6H_4N_4$
B.Klewe *Acta Chem. Scand.*, **25**, 1999. 1971

7.8 **2 - Amino - 1,1,3 - tricyanopropene (form ii)**
$C_6H_4N_4$
B.Klewe *Acta Chem. Scand.*, **26**, 317, 1972

7.9 **2,4,6 - Tribromobenzonitrile**
$C_7H_2Br_3N$
V.B.Carter. D.Britton *Acta Cryst. (B)*, **28**, 945. 1972

7.10 **2,4,6 - Trichlorobenzonitrile**
$C_7H_2Cl_3N$
V.B.Carter. D.Britton *Acta Cryst. (B)*, **28**, 945. 1972

7.C **o - Bromobenzene - anti - diazocyanide**
$C_7H_4BrN_3$
For complete entry see 9.4

7.C **Azo - bis(isobutyronitrile)**
$C_8H_{12}N_4$
For complete entry see 9.8

7.11 **Potassium 2 - cyanomethyl - 1,1,3,3 - tetracyanopropene**
$C_9H_2N_5^-, K^+$
B.Klewe *Acta Chem. Scand.*, **25**, 1975. 1971
Residue 1 also classified in 12

7.C **Dipotassium cis - hexacyanobutenedi - ide**
$C_{10}N_6^{2-}, 2K^+$
For complete entry see 12.9

7.C **Quinolinium 2 - dicyanomethylene - 1,1,3,3 - tetracyanopropanedi - ide**
$C_{10}N_6^{2-}, 2C_9H_8N^+$
For complete entry see 60.20

7.C **Anthracene - 1,2,4,5 - tetracyanobenzene**
$C_{10}H_2N_4, C_{14}H_{10}$
For complete entry see 60.32

7.12 **Benzylidene malononitrile**
α - Cyanocinnamonitrile
$C_{10}H_6N_2$
P.Auvray, F.Genet *Acta Cryst. (B),* **27,** 2424, 1971

7.13 **Benzylidenemalononitrile**
α - Cyanocinnamonitrile
$C_{10}H_6N_2$
D.A.Wright, D.A.Williams *J. Cryst. Mol. Struct.,* **2,** 31, 1972

7.C **7,7,8,8 - Tetracyanoquinodimethane - phenazine complex**
$C_{12}H_4N_4$, $C_{12}H_8N_2$
For complete entry see 60.23

7.C **7,7,8,8 - Tetracyanoquinodimethane - N - methylphenothiazine complex**
$C_{12}H_4N_4$, $C_{13}H_{11}NS$
For complete entry see 60.24

7.C **Rubidium 7,7,8,8 - tetracyanoquinodimethane (at −160 ° C)**
$C_{12}H_4N_4^-$, Rb^+
For complete entry see 12.10

7.C **N,N' - Dimethyl - benzimidazolium tetracyanoquinodimethane**
$C_{12}H_4N_4^-$, $C_9H_{11}N_2^+$
For complete entry see 35.9

7.C **N - n - Propylquinolinium bis(7,7,8,8 - tetracyanoquinodimethane)**
$C_{12}H_4N_4^-$, $C_{12}H_{14}N^+$, $C_{12}H_4N_4$
For complete entry see 60.25

7.C **Methyl - triphenyl - arsonium bis($\alpha,\alpha,\alpha',\alpha'$ - tetracyanoquinodimethanide)**
$C_{12}H_4N_4^-$, $C_{19}H_{18}As^+$, $C_{12}H_4N_4$
For complete entry see 65.9

7.C **Methyl - triphenyl - phosphonium bis($\alpha,\alpha,\alpha',\alpha'$ - tetracyanoquinodimethanide)**
$C_{12}H_4N_4^-$, $C_{19}H_{18}P^+$, $C_{12}H_4N_4$
For complete entry see 64.24

7.C **3,3 - Diethylthiacarbocyanine - tetracyanoquinodimethane complex**
$C_{12}H_4N_4^-$, $C_{21}H_{21}N_2S_2^+$, $C_{12}H_4N_4$
For complete entry see 60.40

7.C **1,3,3 - Trimethyl - 2 - (N - methyl - N - (β - chloroethyl) - p - aminostyryl) - 3H - indole - 7,7,8,8 - tetracyanoquinodimethane complex**
$C_{12}H_4N_4^-$, $C_{22}H_{26}ClN_2^+$, $C_{12}H_4N_4$
For complete entry see 60.41

7.C **Morpholinium 7,7,8,8 - tetracyanoquinodimethane**
$2C_{12}H_4N_4^-$, $2C_4H_{10}NO^+$, $C_{12}H_4N_4$
For complete entry see 40.4

7.C **9 - Dicyanomethylene - 2,4,5,7 - tetranitro - fluorene**

$C_{16}H_4N_6O_8$

For complete entry see 28.6

UREA COMPOUNDS
(ALIPHATIC AND AROMATIC)

8.C **Urea - syn - 5 - nitro - 2 - furaldehyde oxime complex**
CH_4N_2O , $C_5H_4N_2O_4$
For complete entry see 60.2

8.C **α - D - Glucose - urea complex**
CH_4N_2O , $C_6H_{12}O_6$
For complete entry see 60.14

8.C **Estradiol - urea**
CH_4N_2O , $C_{18}H_{24}O_2$
For complete entry see 60.38

8.C **Urea - oxalic acid**
$2CH_4N_2O$. $C_2H_2O_4$
For complete entry see 60.3

8.1 **Hydroxyurea (neutron study)**
$CH_4N_2O_2$
W.E.Thiessen, H.A.Levy, B.D.Flaig
Amer. Cryst. Assoc., Abstr. Papers (Winter Meeting), 32, 1972

8.C **Pyridinium bromide - bis(thiourea) complex**
$2CH_4N_2S$, $C_5H_6N^+$, Br^-
For complete entry see 60.6

8.2 **Thiourea - lead(ii) formate complex**
$16CH_4N_2S$, $6CHO_2^-$, $3Pb^{2+}$
I.Goldberg, F.H.Herbstein *Acta Cryst. (B)*, **28**, 410, 1972
Residue 2 classified in 69

8.C **Guanidinium 5,5 - diethylbarbiturate dihydrate**
$CH_6N_3^+$, $C_8H_{11}N_2O_3^-$, $2H_2O$
For complete entry see 43.4

8.C **Diethylphosphorylguanidine - guanidinium chloride**
$0.5CH_6N_3^+$, $C_5H_{14}N_3O_3P$, $0.5Cl^-$
For complete entry see 8.9

8.3 **Nickel(ii) guanidinium sulfate hexahydrate**
$2CH_6N_3^+ . Ni^{2+} , 2O_4S^{2-} . 6H_2O$
C.N.Morimoto, E.C.Lingafelter
Amer. Cryst. Assoc., Abstr. Papers (Summer Meeting), 78, 1971

8.C **Guanidinium tetra - acetato - cerate monohydrate dimer**
$2CH_6N_3^+ . C_{16}H_{28}Ce_2O_{18}^{2-} . 2H_2O$
For complete entry see 81.43

8.4 **Thiocarbohydrazide dihydrochloride dihydrate**
$CH_8N_4S^{2+} . 2Cl^- . 2H_2O$
A.Braibanti, M.A.Pellinghelli, A.Tiripicchio, M.T.Camellini
Inorg. Chim. Acta, **5,** 523, 1971

8.5 **bis(Thiourea)iodine(i) iodide**
$C_2H_8IN_4S_2^+ . I^-$
G.H.-Y.Lin, H.Hope *Acta Cryst. (B),* **28,** 643, 1972

8.6 α,α' **- Dithio - bis(formamidinium) dichloride**
$C_2H_8N_4S_2^{2+} . 2Cl^-$
A.C.Villa, A.G.Manfredotti, M.Nardelli, M.E.V.Tani
Acta Cryst. (B), **28,** 356, 1972
Residue 1 also classified in 4

8.7 **N,N - Dimethyl - thiourea - S - trioxide**
$C_3H_8N_2O_3S$
W.Walter, J.Holst *J. Molec. Struct.,* **9,** 413, 1971

8.8 **2 - Guanido - ethane sulfinic acid dihydrate**
$C_3H_9N_3O_2S . 2H_2O$
J.Berthou, A.Laurent, A.Rimsky *C. R. Acad. Sci., Fr., C,* **274,** 157, 1972
Residue 1 also classified in 4

8.C **Potassium bis(3 - (n - propyl) - biuretato) cobaltate(iii) bis(1 - (n - propyl) - biuret)**
$2C_5H_{11}N_3O_2 . C_{10}H_{18}CoN_6O_4^- . K^+$
For complete entry see 79.10

8.9 **Diethylphosphorylguanidine - guanidinium chloride**
$C_5H_{14}N_3O_3P . 0.5CH_6N_3^+ . 0.5Cl^-$
D.L.Wampler, O.Kennard, J.C.Coppola, W.D.S.Motherwell, D.G.Watson
Amer. Cryst. Assoc., Abstr. Papers (Winter Meeting), 78, 1971
Residue 1 also classified in 64; residue 2 classified in 8

8.C **Morpholine biguanide hydrochloride**
$C_6H_{14}N_5O^+ . Cl^-$
For complete entry see 40.11

8.C **Dimethylglyoxal bis - guanylhydrazone dihydrochloride dihydrate**
$C_6H_{16}N_8^{2+} . 2Cl^- . 2H_2O$
For complete entry see 9.3

8.C **2 - Formylpyridine selenosemicarbazone**
$C_7H_8N_4Se$
For complete entry see 33.19

8.10 **Chlorguanide hydrochloride**
$C_{11}H_{16}ClN_5^+ , Cl^-$
L.A.Plastas. H.L.Ammon
Amer. Cryst. Assoc., Abstr. Papers (Winter Meeting), 31, 1972

8.11 **N,N′ - Dicyclohexylurea**
$C_{13}H_{24}N_2O$
V.M.Coiro, P.Giacomello, E.Giglio *Acta Cryst. (B),* **27,** 2112, 1971
Also classified in 21

8.12 **Stilbamidine di(ethanol - 2 - sulfonate) dihydrate**
$C_{16}H_{18}N_4^{2+} . 2C_2H_5O_4S^- . 2H_2O$
C.Courseille, B.Busetta, G.Comberton, M.Hospital
C. R. Acad. Sci., Fr., C, **272,** 1115, 1971
Residue 2 classified in 11

8.13 **Hydroxy - stilbamidine di - isoethionate dihydrate**
$C_{16}H_{18}N_4O^{2+} , 2C_2H_5O_4S^- . 2H_2O$
C.Courseille, B.Busetta, M.Hospital
C. R. Acad. Sci., Fr., C, **274,** 1921, 1972
Residue 1 also classified in 17; residue 2 classified in 11

NITROGEN-NITROGEN COMPOUNDS (ALIPHATIC AND AROMATIC)

9.1 **Azidoformamidinium chloride**

$CH_4N_5^+ , Cl^-$

H.Henke, H.Barnighausen *Acta Cryst. (B)*, **28**, 1100, 1972

9.C **Trimethylammonionitramidate**

$C_3H_9N_3O_2$

For complete entry see 3.6

9.2 **p - Benzenediazonium sulfonate**

$C_6H_4N_2O_3S$

C.Romming *Acta Chem. Scand.*, **26**, 523, 1972

Also classified in 11

9.C **Isonicotinic acid hydrazide**

$C_6H_7N_3O$

For complete entry see 33.17

9.3 **Dimethylglyoxal bis - guanylhydrazone dihydrochloride dihydrate**

$C_6H_{16}N_8^{2+} , 2Cl^- , 2H_2O$

J.W.Edmonds, W.C.Hamilton *Acta Cryst. (B)*, **28**, 1362, 1972

Residue 1 also classified in 8

9.4 **o - Bromobenzene - anti - diazocyanide**

$C_7H_4BrN_3$

I.Bo, B.Klewe, C.Romming *Acta Chem. Scand.*, **25**, 3261, 1971

Also classified in 7

9.C **3 - Azidotropone**

$C_7H_5N_3O$

For complete entry see 22.2

9.5 **o - Bromophenylazocarbamide (orthorhombic form)**

$C_7H_6BrN_3O$

G.D.Andreetti, L.Cavalca, A.G.Manfredotti, C.Pelizzi

J. Cryst. Mol. Struct., **1**, 225, 1971

9.6 **o - Bromophenylazocarbamide (monoclinic form)**

$C_7H_6BrN_3O$

G.D.Andreetti, L.Cavalca, A.G.Manfredotti, C.Pelizzi

J. Cryst. Mol. Struct., **1**, 225, 1971

9.C **2 - Formylpyridine selenosemicarbazone**
$C_7H_8N_4Se$
For complete entry see 33.19

9.7 **p - (N,N - Dimethylamino)phenyldiazonium chlorozincate**
$2C_8H_{10}N_3^+ . Cl_4Zn^{2-}$
Ya.M.Nesterova, M.A.Porai-Koshits *Zh. Strukt. Khim.*, **12,** 108, 1971
Residue 1 also classified in 16

9.8 **Azo - bis(isobutyronitrile)**
$C_8H_{12}N_4$
G.Argay, K.Sasvari *Acta Cryst. (B),* **27,** 1851, 1971
Also classified in 7

9.C **Trimethylammoniobenzamidate**
$C_{10}H_{14}N_2O$
For complete entry see 3.20

9.9 **cis - Azobenzene**
$C_{12}H_{10}N_2$
A.Mostad, C.Romming *Acta Chem. Scand.*, **25,** 3561, 1971

9.10 **Diazoaminobenzene (α form)**
$C_{12}H_{11}N_3$
V.F.Gladkova, Yu.D.Kondrashev *Kristallografija,* **16,** 929, 1971

9.11 **p,p' - Dibromo - benzalazine**
$C_{14}H_{10}Br_2N_2$
J.Marignan, J.L.Galigne, J.Falgueirettes *Acta Cryst. (B),* **28,** 93, 1972

9.C **2,2,5,5 - Tetramethyl - 1 - aza - cyclopentanone - 3 - azine - 3 - oxyl biradical**
$C_{16}H_{28}N_4O_2$
For complete entry see 12.13

9.12 **4 - Phenylazoazobenzene**
$C_{18}H_{14}N_4$
R.D.Gilardi, I.L.Karle *Acta Cryst. (B),* **28,** 1635, 1972

9.C **Ethyl p - azoxybenzoate**
$C_{18}H_{18}N_2O_5$
For complete entry see 13.20

9.13 **cis - 1,3 - Diphenyl - 2 - phenylazo - propene**
$C_{21}H_{18}N_2$
A.Foresti Serantoni, L.Riva di Sanseverino, G.Rosini
J. Chem. Soc. (B), 2372, 1971

9.14 **trans - 1,3 - Diphenyl - 2 - phenylazo - propene**
$C_{21}H_{18}N_2$
E.Foresti Serantoni, L.Riva di Sanseverino, G.Rosini
J. Chem. Soc. (B), 2372, 1971

9.C **1 - (2,5 - Dichlorophenylazo) - 2 - hydroxy - 3 - naphthoic acid 4 - chloro - 2,5 - dimethoxy - anilide**

$C_{25}H_{18}Cl_3N_3O_4$

For complete entry see 24.17

NITROGEN-OXYGEN COMPOUNDS (ALIPHATIC AND AROMATIC)

10.C **4 - Nitro - pyridine N - oxide**
$C_5H_4N_2O_3$
For complete entry see 33.4

10.C **syn - S - Methyl - O - (N - methylcarbamoyl)acetothiohydroximate**
$C_5H_{10}N_2O_2S$
For complete entry see 4.2

10.C **Picolinic acid N - oxide (at −100 ° C)**
$C_6H_5NO_3$
For complete entry see 33.14

10.C **2 - Hydroxymethylpyridine N - oxide**
$C_6H_7NO_2$
For complete entry see 33.16

10.C **syn - S - Cyanoethyl - O - (N - methylcarbamoyl)acetothiohydroximate**
$C_7H_{11}N_3O_2S$
For complete entry see 4.5

10.1 **Tetrathiazyl tetra - bis(trifluoromethyl)nitroxide**
$C_8F_{24}N_8O_4S_4$
R.A.Forder, G.M.Sheldrick *J. Fluorine Chem.*, **1**, 23, 1971

10.2 **Isonitroso - acetophenone potassium o - nitrophenolate complex**
$C_8H_7NO_2$, $C_6H_4NO_3^-$, K^+
M.A.Bush, M.R.Truter *J. Chem. Soc. (A)*, 745, 1971
Residue 2 classified in 6, 15

10.3 **Potassium bis(isonitroso - acetophenone) (at 98 ° K)**
$C_8H_7NO_2$, $C_8H_6NO_2^-$, K^+
M.A.Bush, H.Luth, M.R.Truter *J. Chem. Soc. (A)*, 740, 1971

10.C **anti - Ethyl - benzohydroximate**
$C_9H_{11}NO_2$
For complete entry see 13.12

10.C **(−) - N - Ethyl - N - methyl - aniline oxide (+) - 3 - bromocamphorsulfonate**
$C_9H_{14}NO^+$, $C_{10}H_{14}BrO_4S^-$
For complete entry see 16.7

10.C **N - oxyphenazine (at 20 ° C)**
$C_{12}H_8N_2O$
For complete entry see 36.9

10.C **N - Oxyphenazine (at –90 ° C)**
$C_{12}H_8N_2O$
For complete entry see 36.10

10.C **3 - (4 - Bromophenyl) - 6,7 - dihydro - 2 - hydroximinobenzofuran - 4(5)H - one (1 - 2 mixture of syn and anti isomers)**
$C_{14}H_{12}BrNO_3$
For complete entry see 38.17

10.C **Benzoyl - carvoxime**
$C_{17}H_{19}NO_2$
For complete entry see 21.13

SULPHUR AND SELENIUM COMPOUNDS

11.C **Stilbamidine di(ethanol - 2 - sulfonate) dihydrate**
$2C_2H_5O_4S^-, C_{16}H_{18}N_4^{2+}, 2H_2O$
For complete entry see 8.12

11.C **Hydroxy - stilbamidine di - isoethionate dihydrate**
$2C_2H_5O_4S^-, C_{16}H_{18}N_4O^{2+}, 2H_2O$
For complete entry see 8.13

11.C **Picrylsulphonic acid tetrahydrate**
2,4,6 - Trinitrobenzenesulphonic acid tetrahydrate
$C_6H_2N_3O_9S^-, H_5O_2^+, 2H_2O$
For complete entry see 15.2

11.C **2 - (2,3 - Dimethyl)butan - 1 - (2,3 - dimethyl)but - 2 - ene - methyl - sulfonium trinitrobenzene - sulfonate**
$C_6H_2N_3O_9S^-, C_{13}H_{27}S^+$
For complete entry see 11.8

11.1 **2,5 - Dibromobenzenesulfonic acid trihydrate**
$C_6H_3Br_2O_3S^-, H_7O_3^+$
J.-O.Lundgren *Acta Cryst. (B),* **28,** 475, 1972

11.2 **2,5 - Dichlorobenzenesulfonic acid trihydrate**
$C_6H_3Cl_2O_3S^-, H_7O_3^+$
J.-O.Lundgren, P.Lundin *Acta Cryst. (B),* **28,** 486, 1972

11.C **p - Benzenediazonium sulfonate**
$C_6H_4N_2O_3S$
For complete entry see 9.2

11.C **Sulfanilamide (α form)**
$C_6H_8N_2O_2S$
For complete entry see 16.3

11.C **5 - Sulfosalicylic acid trihydrate (neutron study)**
$C_7H_5O_6S^-, H_7O_3^+$
For complete entry see 13.2

11.3 **Oxonium p - toluenesulfonate**
4 - Methylbenzene - sulfonic acid monohydrate
$C_7H_7O_3S^-$, H_3O^+
S.K.Arora, M.Sundaralingam *Acta Cryst. (B),* **27,** 1293, 1971

11.C **1,5 - endo - Methylene - quinolizidinium p - toluene - sulfonate**
$C_7H_7O_3S^-$, $C_{10}H_{18}N^+$
For complete entry see 37.5

11.C **2,5 - Dithia - 1 - phenyl - 1 - thiophosphorus(v) - cyclopentane**
$C_8H_9PS_3$
For complete entry see 42.3

11.C **trans - 3 - (6 - Methyl - 2 - pyridylthio) - propenic acid**
$C_9H_9NO_2S$
For complete entry see 33.21

11.C **N′ - 2 - Thiazolyl - sulphanilamide (form i)**
Sulphathiazole
$C_9H_9N_3O_2S_2$
For complete entry see 41.9

11.C **N′ - 2 - Thiazolyl - sulphanilamide (form iii)**
Sulphathiazole
$C_9H_9N_3O_2S_2$
For complete entry see 41.10

11.4 **N - Toluene - p - sulfonyliminodimethylsulfur(iv)**
$C_9H_{13}NO_2S_2$
A.F.Cameron, N.J.Hair, D.G.Morris *J. Chem. Soc. (D),* 918, 1971

11.5 **3,3′ - Dithio - bis(2,4 - pentanedione)**
$C_{10}H_{14}O_4S_2$
L.F.Power, R.D.G.Jones *Inorg. Nucl. Chem. Letters,* **7,** 887, 1971

11.6 **3 - Phenyl - 2 - propene - 1,3 - dione - 1 - (dimethyl mercaptole)**
$C_{11}H_{12}OS_2$
I.P.Mellor, S.C.Nyburg *Acta Cryst. (B),* **27,** 1954, 1971

11.C **2 - Dimethylaminoethyl selenol - benzoate hydrochloride**
$C_{11}H_{16}NOSe^+$, Cl^-
For complete entry see 3.22

11.7 **Diphenyl disulfone**
$C_{12}H_{10}O_4S_2$
C.T.Kiers, A.Vos *Rec. Trav. Chim. Pays-Bas,* **91,** 126, 1972

11.C **4,4′ - Diamino - diphenyl - sulfone**
$C_{12}H_{12}N_2O_2S$
For complete entry see 16.10

11.C **Cyclohexyl - tosylate**
$C_{13}H_{18}O_3S$
For complete entry see 21.9

11.C **Cyclohexyl - tosylate (neutron study)**
$C_{13}H_{18}O_3S$
For complete entry see 21.10

11.8 **2 - (2,3 - Dimethyl)butan - 1 - (2,3 - dimethyl)but - 2 - ene - methyl - sulfonium trinitrobenzene - sulfonate**
$C_{13}H_{27}S^+ , C_6H_2N_3O_9S^-$
W.M.Barnes, M.Sundaralingam
Amer. Cryst. Assoc., Abstr. Papers (Summer Meeting), 103, 1971
Residue 2 classified in 15, 11

11.9 **bis(Bromophenylmethyl) - sulfoxide**
$C_{14}H_{12}Br_2OS$
B.B.Jarvis, S.D.Dutkey, H.L.Ammon *J. Amer. Chem. Soc.*, **94,** 2136, 1972

11.10 **Benzylidenimine tetrasulfide**
$C_{14}H_{12}N_2S_4$
J.C.Barrick, C.Calvo, F.P.Olsen *J. Chem. Soc. (D),* 1043, 1971

11.11 **Dibenzyl disulfide (space group no.15)**
$C_{14}H_{14}S_2$
R.Srinivasan, B.K.Vijayalakshmi *Acta Cryst. (B),* **28,** 2615, 1972

11.12 **p - Methoxybenzenesulfon - p - anisidide**
$C_{14}H_{15}NO_4S$
S.Pokrywiecki, W.L.Duax, C.M.Weeks, Y.Osawa, D.A.Norton
Amer. Cryst. Assoc., Abstr. Papers (Summer Meeting), 36, 1971
Also classified in 16, 17

11.13 **S - Propyl - S - phenyl - N - p - tolylsulfonyl sulfilimine**
$C_{16}H_{19}NO_2S_2$
A.Kalman, K.Sasvari *Cryst. Struct. Comm.*, **1,** 243, 1972

11.14 **N - Isopropyl - p - methoxybenzenesulfon - p - anisidide**
$C_{17}H_{21}NO_4S$
S.Pokrywiecki, W.L.Duax, C.M.Weeks, Y.Osawa, D.A.Norton
Amer. Cryst. Assoc., Abstr. Papers (Summer Meeting), 36, 1971
Also classified in 16, 17

11.15 **Dicinnamyl disulfide**
$C_{18}H_{18}S_2$
J.D.Lee, M.W.R.Bryant *Acta Cryst. (B),* **27,** 2325, 1971

11.C **1,4 - Naphthoquinone 2 - anilino - 3 - ethylmethylsulfonium ylide**
$C_{19}H_{17}NO_2S$
For complete entry see 25.7

11.16 **(2 - Phenyl - 4 - acetylphenoxy) - (2′,6′ - dimethylphenylimino) methanesulphenic acid**

$C_{23}H_{21}NO_3S$
K.Kato *Acta Cryst. (B),* **28,** 55. 1972
Also classified in 16

11.17 **Tetraphenyl orthothiocarbonate**

$C_{25}H_{20}S_4$
K.Kato *Acta Cryst. (B),* **28,** 606, 1972

11.C **N - Toluene - p - sulfonyliminotriphenylphosphorane**

$C_{25}H_{22}NO_2PS$
For complete entry see 64.34

11.18 **Diphenyl - bis(phenyl - di(trifluoromethyl)methoxy)sulfurane**

$C_{30}H_{20}F_{12}O_2S$
I.C.Paul, J.C.Martin, E.F.Perozzi *J. Amer. Chem. Soc.*, **93,** 6674, 1971

11.C **Glycerol - 1,2 - di(11 - bromoundecanoate) - 3 - p - toluenesulfonate**

$C_{32}H_{52}Br_2O_7S$
For complete entry see 1.14

11.C **2,3 - Dihydro - 1,3 - diphenyl - 2 - oxoindol - 3 - yl diphenyl(phenylcarbamoyl)methyl sulfide**

$C_{40}H_{30}N_2O_2S$
For complete entry see 35.39

CARBONIUM IONS, CARBANIONS, RADICALS

12.1 **Methyl - oxocarbonium tetrachloroaluminate**

$C_2H_3O^+$, $AlCl_4^-$

J.-M.Le Carpentier, R.Weiss *Acta Cryst. (B),* **28,** 1421, 1972

12.2 **Methyl - oxocarbonium hexachloroantimonate**

$C_2H_3O^+$, Cl_6Sb^-

J.-M.Le Carpentier, R.Weiss *Acta Cryst. (B),* **28,** 1421, 1972

12.3 **Dirubidium 1,1,2,2 - tetranitroethanedi - ide**

$C_2H_4O_8^{2-}$, $2Rb^+$

B.Klewe *Acta Chem. Scand.,* **26,** 1049, 1972

12.4 **Potassium N - methyl - α,α - dinitroacetamide**

$C_3H_4N_3O_5^-$, K^+

N.V.Grigor'eva, N.V.Margolis, I.V.Tselinskii, V.V.Mel'nikov, G.V.Makarenko *Zh. Strukt. Khim.,* **12,** 739, 1971

12.5 **Ethyl - oxocarbonium tetrachlorogallate**

$C_3H_5O^+$, Cl_4Ga^-

J.-M.Le Carpentier, R.Weiss *Acta Cryst. (B),* **28,** 1430, 1972

12.C **Potassium tricyanomethanide**

$C_4N_3^-$, K^+

For complete entry see 7.3

12.6 **Isopropyl - oxocarbonium hexachloroantimonate**

$C_4H_7O^+$, Cl_6Sb^-

J.-M.Le Carpentier, R.Weiss *Acta Cryst. (B),* **28,** 1430, 1972

12.C **Potassium dinitrotrichloro - cyclopentadienide**

$C_5Cl_3N_2O_4^-$, K^+

For complete entry see 20.1

12.C **Potassium 1,1,3 - tricyanopropan - 2 - one**

$C_6H_2N_3O^-$, K^+

For complete entry see 7.6

12.C **Ferrocinium tris(trichloroacetic acid)**

Dicyclopentadienyl iron tris(trichloroacetic acid)

$C_6H_3Cl_9O_6^-$, $C_{10}H_{10}Fe^+$

For complete entry see 73.3

12.7 **Rubidium 2,4,6 - trinitrophenyl - dinitromethane**

$C_7H_2N_5O_{10}^-$, Rb^+

N.V.Grigor'eva, N.V.Margolis, I.V.Tselinskii, V.V.Mel'nikov, G.V.Makarenko *Zh. Strukt. Khim.*, **12**, 938, 1971

Residue 1 also classified in 15

12.8 **Potassium p - chlorophenyldinitromethanide**

$C_7H_4ClN_2O_4^-$, K^+

B.Klewe, S.Ramsoy *Acta Chem. Scand.*, **26**, 1058, 1972

12.C **Potassium 2 - cyanomethyl - 1,1,3,3 - tetracyanopropene**

$C_9H_2N_5^-$, K^+

For complete entry see 7.11

12.C **2,2,5,5 - Tetramethyl - 3 - carbamidopyrroline - 1 - oxyl**

$C_9H_{15}N_2O_2$

For complete entry see 32.12

12.C **1,2,3 - tris(Dimethylamino)cyclopropenium perchlorate**

$C_9H_{18}N_3^+$, ClO_4^-

For complete entry see 20.4

12.9 **Dipotassium cis - hexacyanobutenedi - ide**

$C_{10}N_6^{2-}$, $2K^+$

E.Maverick, E.Goldish, J.Bernstein, K.N.Trueblood, S.Swaminathan, R.Hoffmann *J. Amer. Chem. Soc.*, **94**, 3364, 1972

Residue 1 also classified in 7

12.C **Quinolinium 2 - dicyanomethylene - 1,1,3,3 - tetracyanopropanedi - ide**

$C_{10}N_6^{2-}$, $2C_9H_8N^+$

For complete entry see 60.20

12.10 **Rubidium 7,7,8,8 - tetracyanoquinodimethane (at −160 ° C)**

$C_{12}H_4N_4^-$, Rb^+

A.Hoekstra, T.Spoelder, A.Vos *Acta Cryst. (B)*, **28**, 14, 1972

Residue 1 also classified in 7

12.11 **Perchlorodiphenylmethyl free radical**

$C_{13}Cl_{11}$

J.Silverman, L.J.Soltzberg, N.F.Yannoni, A.P.Krukonis *J. Phys. Chem.*, **75**, 1246, 1971

12.12 **4,4′ - bis(Dimethylamino)diphenylamine radical chlorate**

$C_{16}H_{21}N_3^+$, ClO_3^-

D.Hlavata *Acta Cryst. (B)*, **27**, 1483, 1971

Residue 1 also classified in 16

12.13 **2,2,5,5 - Tetramethyl - 1 - aza - cyclopentanone - 3 - azine - 3 - oxyl biradical**

$C_{16}H_{28}N_4O_2$

B.Chion, A.Capiomont, J.Lajzerowicz *Acta Cryst. (B)*, **28**, 618, 1972

Also classified in 32, 9

12.C **N - Methyl - N - benzoylmethylene - isoindolinium ylide**
$C_{17}H_{17}NO$
For complete entry see 35.23

12.14 **p - Methylbenzyl - dimethyl - benzoylmethylene - ammonium ylide monohydrate**
$C_{18}H_{21}NO \cdot H_2O$
N.A.Bailey, S.E.Hull, G.F.Kersting, J.Morrison
J. Chem. Soc. (D), 1429, 1971
Residue 1 also classified in 3

12.15 **Tri - (p - aminophenyl)carbonium perchlorate**
Pararosaniline perchlorate
$C_{19}H_{18}N_3^+ \cdot ClO_4^-$
L.L.Koh, K.Eriks *Acta Cryst. (B),* **27,** 1405, 1971
Residue 1 also classified in 16

12.C **Phenyl - di(tricarbonyliron - cyclobutadienyl)methyl fluoroborate**
$C_{21}H_{11}Fe_2O_6^+ \cdot BF_4^-$
For complete entry see 75.24

BENZOIC ACID DERIVATIVES

13.C **5 - Chlorosalicylic acid - theobromine complex**
$2C_7H_5ClO_3 . C_7H_8N_5O_2$
For complete entry see 60.17

13.1 **p - Nitrobenzoic acid (refinement of structure of Sakore and Paul, Acta Cryst.,21,715,1966)**
$C_7H_5NO_4$
S.S.Tavale, L.M.Pant *Acta Cryst. (B)*, **27**, 1479, 1971
Also classified in 15

13.C **Nicotinyl salicylate**
$C_7H_5O_3 , C_{10}H_{15}N_2$
For complete entry see 60.22

13.2 **5 - Sulfosalicylic acid trihydrate (neutron study)**
$C_7H_5O_6S^- . H_7O_3^+$
J.M.Williams, S.W.Petersen, H.A.Levy
Amer. Cryst. Assoc., Abstr. Papers (Winter Meeting), 51, 1972
Residue 1 also classified in 17, 11

13.3 **p - Aminobenzamide**
$C_7H_8N_2O$
M.Alleaume *Thesis, Bordeaux*, 57, 1967
Also classified in 16

13.4 **Terephthaloyl chloride**
$C_8H_4Cl_2O_2$
J.Leser, D.Rabinovich *Israel J. Chem.*, **9**, II, 1971

13.C **Phthalimide**
$C_8H_5NO_2$
For complete entry see 35.3

13.5 **Isophthalic acid**
$C_8H_6O_4$
R.Alcala, S.Martinez-Carrera *Acta Cryst. (B)*, **28**, 1671, 1972

13.6 **p - Toluic acid**
$C_8H_8O_2$
M.G.Takwale, L.M.Pant *Acta Cryst. (B)*, **27**, 1152, 1971

13.7 **N - Methylanthranilic acid**

$C_8H_9NO_2$

N.N.Dhaneshwar, L.M.Pant *Acta Cryst. (B),* **28,** 647, 1972

Also classified in 16

13.8 **1,2,3 - Benzenetricarboxylic acid dihydrate**

$C_9H_6O_6$, $2H_2O$

J.M.Fornies-Marquina, C.Courseille, B.Busetta, M.Hospital

Cryst. Struct. Comm., **1,** 47, 1972

13.9 **Ethyl 3,5 - dinitrobenzoate**

$C_9H_8N_2O_6$

D.L.Hughes, J.Trotter *J. Chem. Soc. (A),* 2358, 1971

Also classified in 15

13.10 **Toluene - α,2 - dicarboxylic acid**

Homophthalic acid

$C_9H_8O_4$

M.P.Gupta, M.Sahu *Acta Cryst. (B),* **27,** 2469, 1971

13.11 **2,3 - Dimethylbenzoic acid**

$C_9H_{10}O_2$

P.Smith, F.Florencio, S.Garcia-Blanco *Acta Cryst. (B),* **27,** 2255, 1971

13.12 **anti - Ethyl - benzohydroximate**

$C_9H_{11}NO_2$

I.K.Larsen *Acta Chem. Scand.,* **25,** 2409, 1971

Also classified in 10

13.13 **4 - Carbethoxyanilinium bis - (p - nitrophenyl)phosphate**

Benzocaine bis(p - nitrophenyl)phosphate

$C_9H_{12}NO_2^+$, $C_{12}H_8N_2O_8P^-$

J.Pletcher, M.Sax, C.S.Yoo *Acta Cryst. (B),* **28,** 378, 1972

Residue 1 also classified in 16; residue 2 classified in 46, 15

13.14 **Pyromellitic acid dihydrate**

Benzene - 1,2,4,5 - tetracarboxylic acid dihydrate

$C_{10}H_6O_8$, $2H_2O$

F.Takusagawa, K.Hirotsu, A.Shimada

Bull. Chem. Soc. Jap., **44,** 1274, 1971

13.15 **Dimethyl 3,6 - dichloro - 2,5 - dihydroxyterephthalate (yellow form)**

$C_{10}H_8Cl_2O_6$

S.R.Byrn, D.Y.Curtin, I.C.Paul *J. Amer. Chem. Soc.,* **94,** 890, 1972

Also classified in 17

13.16 **Dimethyl 3,6 - dichloro - 2,5 - dihydroxyterephthalate (white form)**

$C_{10}H_8Cl_2O_6$

S.R.Byrn, D.Y.Curtin, I.C.Paul *J. Amer. Chem. Soc.,* **94,** 890, 1972

Also classified in 17

13.C **2 - Dimethylaminoethyl selenol - benzoate hydrochloride**
$C_{11}H_{16}NOSe^{+}$, Cl^{-}
For complete entry see 3.22

13.C **Procaine hydrochloride**
2 - Diethylaminoethyl p - aminobenzoate hydrochloride
$C_{13}H_{21}N_{2}O_{2}^{+}$, Cl^{-}
For complete entry see 3.24

13.17 **2,2′ - Dibromodibenzoyl peroxide**
$C_{14}H_{8}Br_{2}O_{4}$
J.Z.Gougoutas, J.C.Clardy *Acta Cryst. (B),* **26,** 1999, 1970

13.18 **2,2′ - Di - iododibenzoyl peroxide**
$C_{14}H_{8}I_{2}O_{4}$
J.Z.Gougoutas, J.C.Clardy *Acta Cryst. (B),* **26,** 1999, 1970

13.19 **Dibenzoylacetylene**
$C_{16}H_{10}O_{2}$
R.J.Majeste, E.A.Meyers *Cryst. Struct. Comm.,* **1,** 231, 1972
Also classified in 5, 19

13.C **Benzoyl - carvoxime**
$C_{17}H_{19}NO_{2}$
For complete entry see 21.13

13.20 **Ethyl p - azoxybenzoate**
$C_{18}H_{18}N_{2}O_{5}$
W.R.Krigbaum, P.G.Barber *Acta Cryst. (B),* **27,** 1884, 1971
Also classified in 9

13.21 **(+) - trans - 2 - o - Tolylcyclohexanol 3 - nitro - 4 - bromobenzoate (absolute configuration)**
$C_{20}H_{20}BrNO_{4}$
A.Camerman, L.H.Jensen, T.G.Cochran, A.C.Huitric
J. Pharm. Sci., **59,** 1675, 1970
Also classified in 15, 21

13.22 **β - (p - Bromobenzoyloxy) - α - methyl - N,N - diethylcinnamide**
$C_{20}H_{22}BrNO_{3}$
B.Dammeier, W.Hoppe *Chem. Ber.,* **104,** 1674, 1971

13.C **Methyl α - (oxy - p - bromobenzoyl) - β - phenyl - β - (N - benzamido) - propionate (absolute configuration)**
$C_{24}H_{20}BrNO_{5}$
For complete entry see 59.19

13.23 **N - Methyl - N - benzyl - 2,4,6 - tri - t - butylbenzamide (monoclinic form)**
$C_{27}H_{39}NO$
A.E.Jungk, G.M.J.Schmidt *Chem. Ber.,* **104,** 3289, 1971

BENZOIC ACID SALTS (AMMONIUM, IA, IIA METALS)

No entries in this volume.

BENZENE NITRO COMPOUNDS

15.C **Potassium 4 - hydroxy - 5,7 - dinitrobenzfurazan monohydrate**

$C_6HN_4O_6^-$. K^+ . H_2O

For complete entry see 40.6

15.1 **Potassium picrate**

$C_6H_2N_3O_7^-$. K^+

G.J.Palenik *Acta Cryst. (B)*, **28**, 1633. 1972

Residue 1 also classified in 6

15.C **Serotonin picrate monohydrate**

$C_6H_2N_3O_7^-$. $C_{10}H_{13}N_2O^+$. H_2O

For complete entry see 35.11

15.C **tetrakis(Thiourea) lead(ii) picrate**

$2C_6H_2N_3O_7^-$. $C_4H_{16}N_8PbS_4^{2+}$

For complete entry see 69.8

15.2 **Picrylsulphonic acid tetrahydrate**

2,4,6 - Trinitrobenzenesulphonic acid tetrahydrate

$C_6H_2N_3O_9S^-$. $H_5O_2^+$. $2H_2O$

J.-O.Lundgren *Acta Cryst. (B)*, **28**, 1684, 1972

Residue 1 also classified in 11

15.C **2 - (2,3 - Dimethyl)butan - 1 - (2,3 - dimethyl)but - 2 - ene - methyl - sulfonium trinitrobenzene - sulfonate**

$C_6H_2N_3O_9S^-$. $C_{13}H_{27}S^+$

For complete entry see 11.8

15.C **4,6 - Dinitrobenzfurazan - 1 - oxide**

$C_6H_2N_4O_6$

For complete entry see 40.8

15.3 **1,3,5 - Trinitrobenzene (neutron study)**

$C_6H_3N_3O_6$

C.S.Choi, J.E.Abel *Acta Cryst. (B)*, **28**, 193, 1972

15.C **1,3,5 - Trinitrobenzene - 3 - formylbenzothiophene**

$C_6H_3N_3O_6$. C_9H_6OS

For complete entry see 60.8

15.C **1,3,5 - Trinitrobenzene - bis(N - t - butylsalicylidene - iminato) cobalt(ii) complex**

$C_6H_3N_3O_6 , C_{22}H_{28}CoN_2O_2$

For complete entry see 60.9

15.C **1,3,5 - Trinitrobenzene - bis(N - t - butylsalicylidene - iminato) copper(ii) complex**

$C_6H_3N_3O_6 , C_{22}H_{28}CuN_2O_2$

For complete entry see 60.10

15.C **1,3,5 - Trinitrobenzene - bis(N - t - butylsalicylidene - iminato) nickel(ii) complex**

$C_6H_3N_3O_6 , C_{22}H_{28}N_2NiO_2$

For complete entry see 60.11

15.C **Rubidium 2,4,6 - trinitrophenyl - dinitromethane**

$C_7H_2N_5O_{10}^- , Rb^+$

For complete entry see 12.7

15.C **p - Nitrobenzoic acid (refinement of structure of Sakore and Paul, Acta Cryst.,21,715,1966)**

$C_7H_5NO_4$

For complete entry see 13.1

15.4 **p - Nitrotoluene**

$C_7H_7NO_2$

J.V.Barve, L.M.Pant *Acta Cryst. (B)*, **27**, 1158, 1971

15.C **Isonitroso - acetophenone potassium o - nitrophenolate complex**

$C_8H_4NO_3^- , C_8H_7NO_2 , K^+$

For complete entry see 10.2

15.5 **2,4,6 - Trinitro - m - xylene**

$C_8H_7N_3O_6$

J.H.Bryden *Acta Cryst. (B)*, **28**, 1395, 1972

15.C **Ethyl 3,5 - dinitrobenzoate**

$C_9H_8N_2O_6$

For complete entry see 13.9

15.6 **N - Ethyl - N - p - nitrophenylcarbamoyl chloride**

$C_9H_9ClN_2O_3$

P.Ganis, G.Avitabile, S.Migdal, M.Goodman
J. Amer. Chem. Soc., **93**, 3328, 1971
Also classified in 16

15.C **1 - p - Nitrophenyl - 3 - methyl - 4 - bromo - pyrazole**

$C_{10}H_8BrN_3O_2$

For complete entry see 32.13

15.C **1,3 - Diaziridinyl - 2,4,6 - trinitrobenzene**

$C_{10}H_9N_5O_6$
For complete entry see 32.14

15.C **4 - Carbethoxyanilinium bis - (p - nitrophenyl)phosphate**

Benzocaine bis(p - nitrophenyl)phosphate

$C_{12}H_8N_2O_8P^-$, $C_9H_{12}NO_2^+$
For complete entry see 13.13

15.C **N,N′ - bis(4 - Ethoxyphenyl)acetamidinium bis - p - nitrophenylphosphate monohydrate**

Phenacaine bis - p - nitrophenylphosphate monohydrate

$C_{12}H_8N_2O_8P^-$, $C_{18}H_{23}N_2O_2^+$, H_2O
For complete entry see 16.17

15.C **1 - p - Nitrophenyl - 3 - methylperhydro - 2,9 - pyridoxazine (absolute configuration)**

$C_{15}H_{20}N_2O_3$
For complete entry see 40.21

15.C **anti - p - Bromobenzophenone oxime O - picryl ether**

$C_{19}H_{11}BrN_4O_7$
For complete entry see 17.12

15.C **syn - p - Bromobenzophenone oxime O - picryl ether**

$C_{19}H_{11}BrN_4O_7$
For complete entry see 17.13

15.C **(+) - trans - 2 - o - Tolylcyclohexanol 3 - nitro - 4 - bromobenzoate (absolute configuration)**

$C_{20}H_{20}BrNO_4$
For complete entry see 13.21

ANILINES

16.1 **2,4,6 - Trichloroaniline**

$C_6H_4Cl_3N$

V.G.Andrianov, A.F.Korotkevich, Yu.T.Struchkov

Zh. Strukt. Khim., **12,** 736, 1971

16.2 **p - Bromoanilinium hexafluorosilicate hydrate**

$2C_6H_7BrN^+ . F_6Si^{2-} . 2H_2O$

W.A.Denne, A.McL.Mathieson, M.F.Mackay

J. Cryst. Mol. Struct., **1,** 55, 1971

16.C **2 - Aminoresorcinol hydrochloride**

$C_6H_8NO_2^+ . Cl^-$

For complete entry see 17.4

16.3 **Sulfanilamide (α form)**

$C_6H_8N_2O_2S$

M.Alleaume *Thesis, Bordeaux,* 16, 1967

Also classified in 11

16.C **p - Aminobenzamide**

$C_7H_8N_2O$

For complete entry see 13.3

16.4 **p - Toluidinium bifluoride**

$C_7H_{10}N^+ . HF_2^-$

W.A.Denne, M.F.Mackay *J. Cryst. Mol. Struct.*, **1,** 311, 1971

16.5 **p - Toluidinium hexafluorosilicate hydrate**

$2C_7H_{10}N^+ . F_6Si^{2-} . 2H_2O$

W.A.Denne, A.McL.Mathieson, M.F.Mackay

J. Cryst. Mol. Struct., **1,** 55, 1971

16.C **N - Methylanthranilic acid**

$C_8H_9NO_2$

For complete entry see 13.7

16.C **p - (N,N - Dimethylamino)phenyldiazonium chlorozincate**

$2C_8H_{10}N_3^+ . Cl_4Zn^{2-}$

For complete entry see 9.7

16.C **N - Ethyl - N - p - nitrophenylcarbamoyl chloride**
$C_9H_9ClN_2O_3$
For complete entrv see 15.6

16.C **N′ - 2 - Thiazolyl - sulphanilamide (form i)**
Sulphathiazole
$C_9H_9N_3O_2S_2$
For complete entry see 41.9

16.C **N′ - 2 - Thiazolyl - sulphanilamide (form iii)**
Sulphathiazole
$C_9H_9N_3O_2S_2$
For complete entry see 41.10

16.6 **N - Phenylurethane**
$C_9H_{11}NO_2$
P.Ganis, G.Avitabile, S.Migdal, M.Goodman
J. Amer. Chem. Soc., **93**, 3328, 1971

16.C **4 - Carbethoxyanilinium bis - (p - nitrophenyl)phosphate**
Benzocaine bis(p - nitrophenyl)phosphate
$C_9H_{12}NO_2^+ . C_{12}H_8N_2O_8P^-$
For complete entry see 13.13

16.7 **(–) - N - Ethyl - N - methyl - aniline oxide (+) - 3 - bromocamphorsulfonate**
$C_9H_{14}NO^+ . C_{10}H_{14}BrO_4S^-$
R.L.Muntz, W.H.Pirkle, I.C.Paul *J. C. S. Perkin ii*, 483, 1972
Residue 1 also classified in 10; residue 2 classified in 52

16.8 **N,N,N′,N′ - Tetramethyl - p - diaminobenzene perchlorate (room temp.form)**
$C_{10}H_{17}N_2^+ . ClO_4^-$
J.L.de Boer, A.Vos *Acta Cryst. (B)*, **28**, 835, 1972

16.9 **N,N,N′,N′ - Tetramethyl - p - diaminobenzene perchlorate (low temp.form, at 110 ° K)**
$C_{10}H_{17}N_2^+ . ClO_4^-$
J.L.de Boer, A.Vos *Acta Cryst. (B)*, **28**, 839, 1972

16.C **Benzophenone - diphenylamine complex**
$C_{12}H_{11}N . C_{13}H_{10}O$
For complete entry see 60.26

16.10 **4,4′ - Diamino - diphenyl - sulfone**
$C_{12}H_{12}N_2O_2S$
M.Alleaume *Thesis, Bordeaux*, 49, 1967
Also classified in 11

16.11 **Neostigmine bromide**
$C_{12}H_{19}N_2O_2^+$, Br^-
P.Pauling, T.J.Petcher *J. Medicin. Chem., U. S. A.*, **14,** 1, 1971
Residue 1 also classified in 3

16.12 **N - (p - Chlorobenzylidene) - p - chloroaniline (metastable form)**
$C_{13}H_9Cl_2N$
J.Bernstein, G.M.J.Schmidt *J. C. S. Perkin ii,* 951, 1972

16.13 **N - (2,4 - Dichlorobenzylidene)aniline**
$C_{13}H_9Cl_2N$
J.Bernstein *J. C. S. Dalton,* 946, 1972

16.14 **N - Benzylidene - p - bromoaniline**
$C_{13}H_{10}BrN$
B.T.Blaylock, R.F.Bryan
Amer. Cryst. Assoc., Abstr. Papers (Winter Meeting), 21, 1972

16.C **Procaine hydrochloride**
2 - Diethylaminoethyl p - aminobenzoate hydrochloride
$C_{13}H_{21}N_2O_2^+$, Cl^-
For complete entry see 3.24

16.C **p - Methoxybenzenesulfon - p - anisidide**
$C_{14}H_{15}NO_4S$
For complete entry see 11.12

16.15 **Lidocaine hydrohexafluoroarsenate**
2 - Diethylamino - 2′,6′ - acetoxylidide hydrohexafluoroarsenate
$C_{14}H_{23}N_2O^+$, AsF_6^-
A.W.Hanson *Acta Cryst. (B),* **28,** 672, 1972
Residue 1 also classified in 3

16.16 **Di - p - tolylcarbodi - imide**
$C_{15}H_{14}N_2$
A.T.Vincent, P.J.Wheatley *J. C. S. Perkin ii,* 687, 1972

16.C **1 - Methyl - 3 - (N - methyl - 4′ - chloroanilino) - 5 - chloroindole**
$C_{16}H_{14}Cl_2N_2$
For complete entry see 35.21

16.C **N,N,N′,N′ - Tetramethylbenzidine - chloroanil complex**
$2C_{16}H_{20}N_2$. $C_6Cl_4O_2$
For complete entry see 60.33

16.C **4,4′ - bis(Dimethylamino)diphenylamine radical chlorate**
$C_{16}H_{21}N_3^+$, ClO_3^-
For complete entry see 12.12

16.C **N - (2 - Naphthylmethylene) - p - bromoaniline**
$C_{17}H_{12}BrN$
For complete entry see 24.11

16.C **N - Isopropyl - p - methoxybenzenesulfon - p - anisidide**
$C_{17}H_{21}NO_4S$
For complete entry see 11.14

16.C **1 - Cyano - 1 - (p - chlorophenyl) - 2 - (p - dimethylaminophenyl) - cyclopropane**
$C_{18}H_{17}ClN_2$
For complete entry see 20.10

16.17 **N,N′ - bis(4 - Ethoxyphenyl)acetamidinium bis - p - nitrophenylphosphate monohydrate**
Phenacaine bis - p - nitrophenylphosphate monohydrate
$C_{18}H_{23}N_2O_2^+$, $C_{12}H_8N_2O_8P^-$, H_2O
M.Sax, J.Pletcher, C.S.Yoo, J.M.Stewart
Acta Cryst. (B), **27,** 1635, 1971
Residue 1 also classified in 17; residue 2 classified in 46, 15

16.C **1,4 - Naphthoquinone 2 - anilino - 3 - ethylmethylsulfonium ylide**
$C_{19}H_{17}NO_2S$
For complete entry see 25.7

16.C **Tri - (p - aminophenyl)carbonium perchlorate**
Pararosaniline perchlorate
$C_{19}H_{18}N_3^+$, ClO_4^-
For complete entry see 12.15

16.C **4 - (4 - N,N - Diethylaminophenylimino) - 3 - methyl - 1 - phenyl - 2 - pyrazolin - 5 - one**
$C_{20}H_{22}N_4O$
For complete entry see 32.26

16.C **1,3,3 - Trimethyl - 2 - (N - methyl - N - (β - chloroethyl) - p - aminostyryl) - 3H - indole - 7,7,8,8 - tetracyanoquinodimethane complex**
$C_{22}H_{26}ClN_2^+$, $C_{12}H_4N_4^-$, $C_{12}H_4N_4$
For complete entry see 60.41

16.C **4 - (2,6 - Dimethyl - 4 - N,N - diethylaminophenylimino) - 3 - methyl - 1 - phenyl - 2 - pyrazolin - 5 - one**
$C_{22}H_{26}N_4O$
For complete entry see 32.29

16.C **(2 - Phenyl - 4 - acetylphenoxy) - (2′,6′ - dimethylphenylimino) methanesulphenic acid**
$C_{23}H_{21}NO_3S$
For complete entry see 11.16

16.C **1 - (2,5 - Dichlorophenylazo) - 2 - hydroxy - 3 - naphthoic acid 4 - chloro - 2,5 - dimethoxy - anilide**

$C_{25}H_{18}Cl_3N_3O_4$

For complete entry see 24.17

PHENOLS AND ETHERS

17.C **Triphenylarsine oxide - tetrachlorocatechol complex**
$C_6H_2Cl_4O_2 . C_{18}H_{15}AsO$
For complete entry see 60.35

17.C **17β - Hydroxy - 1,4 - androstadien - 3 - one p - bromophenol complex**
$C_6H_5BrO , C_{19}H_{26}O_2$
For complete entry see 60.39

17.1 **p - Chlorophenol (β form, at low temp.)**
C_6H_5ClO
M.Perrin, P.Michel *C. R. Acad. Sci., Fr., C,* **273,** 408, 1971

17.2 **Catechol**
$C_6H_6O_2$
H.Wunderlich, D.Mootz *Acta Cryst. (B),* **27,** 1684, 1971

17.C **Lumiflavin bromide - hydroquinone complex**
$1.5C_6H_6O_2 . C_{13}H_{13}N_2O_2^+ . Br^-$
For complete entry see 60.30

17.3 **Pyrogallol (at −150 ° C)**
$C_6H_6O_3$
P.Becker, H.Brusset, H.Gillier-Pandraud
C. R. Acad. Sci., Fr., C, **274,** 1043, 1972

17.C **Phloroglucinol - p - benzoquinone complex**
1,3,5 - Trihydroxybenzene - p - benzoquinone complex
$C_6H_6O_3 , 2C_6H_4O_2$
For complete entry see 60.12

17.4 **2 - Aminoresorcinol hydrochloride**
$C_6H_8NO_2^+ , Cl^-$
D.F.Grant, D.A.Price, J.P.G.Richards *Z. Kristallogr.*, **132,** 385, 1970
Residue 1 also classified in 16

17.C **5 - Chlorosalicylic acid - theobromine complex**
$2C_7H_5ClO_3 . C_7H_8N_5O_2$
For complete entry see 60.17

17.C **5 - Sulfosalicylic acid trihydrate (neutron study)**

$C_7H_5O_6S^-$. $H_7O_3^+$

For complete entry see 13.2

17.5 **o - Cresol (at –50 ° C)**

C_7H_8O

C.Bois *Acta Cryst. (B),* **28,** 25, 1972

17.C **Sodium phenoxyacetate hemihydrate**

$C_8H_7O_3^-$, Na^+ , $0.5H_2O$

For complete entry see 2.24

17.6 **2,3 - Dimethylphenol (at –150 ° C)**

$C_8H_{10}O$

H.Brusset, H.Gillier-Pandraud, A.Neuman

C. R. Acad. Sci., Fr., C, **274,** 948, 1972

17.7 **2,5 - Dimethylphenol (at –150 ° C)**

$C_8H_{10}O$

H.Brusset, H.Gillier-Pandraud, A.Neuman

C. R. Acad. Sci., Fr., C, **274,** 948, 1972

17.C **6 - Hydroxydopamine hydrochloride**

3,4,6 - Trihydroxyphenylethylamine hydrochloride

$C_8H_{12}NO_3^+$. Cl^-

For complete entry see 59.3

17.C **Adrenalone hydrochloride monohydrate**

$C_9H_{12}NO_3^+$. Cl^- . H_2O

For complete entry see 3.18

17.C **Dimethyl 3,6 - dichloro - 2,5 - dihydroxyterephthalate (yellow form)**

$C_{10}H_8Cl_2O_6$

For complete entry see 13.15

17.C **Dimethyl 3,6 - dichloro - 2,5 - dihydroxyterephthalate (white form)**

$C_{10}H_8Cl_2O_6$

For complete entry see 13.16

17.8 **4,4′ - Dihydroxydiphenyl**

$C_{12}H_{10}O_2$

N.A.Akhmed, M.S.Farag, A.Amin *Zh. Strukt. Khim.*, **12,** 738, 1971

17.C **DL - N - t - Butyl - 2 - (4 - hydroxy - 3 - hydroxymethylphenyl) - 2 - hydroxyethylamine**

Salbutamol

$C_{13}H_{21}NO_3$

For complete entry see 3.23

17.C **p - Methoxybenzenesulfon - p - anisidide**

$C_{14}H_{15}NO_4S$

For complete entry see 11.12

17.C **Hydroxy - stilbamidine di - isoethionate dihydrate**

$C_{16}H_{18}N_4O^{2+}$. $2C_2H_5O_4S^-$. $2H_2O$

For complete entry see 8.13

17.C **Athrotaxin**

$C_{17}H_{16}O_6$

For complete entry see 59.12

17.C **N - Isopropyl - p - methoxybenzenesulfon - p - anisidide**

$C_{17}H_{21}NO_4S$

For complete entry see 11.14

17.C **DL - N - (2 - (4 - Hydroxyphenyl) - 1 - methyl)ethyl - (2′ - (3,5 - dihydroxyphenyl) - 2′ - hydroxy)ethylamine hydrobromide**

$C_{17}H_{22}NO_4^+$. Br^-

For complete entry see 3.25

17.C **DL - N - (2 - (4 - Hydroxyphenyl) - 1 - methyl - ethyl) - 2 - (3,5 - dihydroxyphenyl) - 2 - hydroxy - ethylamine hydrobromide**

$C_{17}H_{22}NO_4^+$. Br^-

For complete entry see 3.26

17.9 **Dienestrol**

Cycladiene

$C_{18}H_{18}O_2$

J.M.Fornies-Marquina, C.Courseille, B.Busetta, M.Hospital

Acta Cryst. (B), **28,** 655, 1972

17.10 **Diethyl - stilbestrol**

$C_{18}H_{20}O_2$. C_2H_6O . H_2O

B.Busetta, F.Leroy, C.Courseille, M.Hospital

C. R. Acad. Sci., Fr., C, **272,** 1304, 1971

Residue 1 also classified in 51

17.11 **Diethyl - stilbestrol dimethylsulfoxide solvate**

$C_{18}H_{20}O_2$. C_2H_6OS

G.Comberton, F.Leroy *C. R. Acad. Sci., Fr., C,* **273,** 1160, 1971

Residue 1 also classified in 51

17.C **N,N′ - bis(4 - Ethoxyphenyl)acetamidinium bis - p - nitrophenylphosphate monohydrate**

Phenacaine bis - p - nitrophenylphosphate monohydrate

$C_{18}H_{23}N_2O_2^+$. $C_{12}H_8N_2O_8P^-$. H_2O

For complete entry see 16.17

17.12 **anti - p - Bromobenzophenone oxime O - picryl ether**
$C_{19}H_{11}BrN_4O_7$
J.D.McCullough Junior, I.C.Paul, D.Y.Curtin
J. Amer. Chem. Soc., **94,** 883, 1972
Also classified in 15

17.13 **syn - p - Bromobenzophenone oxime O - picryl ether**
$C_{19}H_{11}BrN_4O_7$
J.D.McCullough Junior, I.C.Paul, D.Y.Curtin
J. Amer. Chem. Soc., **94,** 883, 1972
Also classified in 15

17.14 **3,5 - Dibromo - p - hydroxy - triphenylmethane carbinol (red tautomeric form)**
$C_{19}H_{14}Br_2O_2$
C.Stora *Bull. Soc. Chim. Fr.*, **6,** 2153, 1971

17.C **bis(p - Methoxyphenyl) 2,5 - furane - dicarboxylate**
$C_{20}H_{16}O_7$
For complete entry see 38.24

17.C **16 - Bromo - myricanol nitromethane solvate (absolute configuration)**
$C_{21}H_{25}BrO_5 . CH_3NO_2$
For complete entry see 59.18

17.15 **1,1 - Di - (p - ethoxyphenyl) - 2,2 - dimethylpropane**
$C_{21}H_{28}O_2$
T.P.DeLacy, C.H.L.Kennard, G.Holan *J. Chem. Soc. (D)*, 930, 1971

17.C **p - (1,2,3,4 - Tetrahydro - 1 - naphthyl)phenoxy p - iodobenzoate (absolute configuration)**
$C_{23}H_{19}IO_2$
For complete entry see 24.16

17.C **1 - (2,5 - Dichlorophenylazo) - 2 - hydroxy - 3 - naphthoic acid 4 - chloro - 2,5 - dimethoxy - anilide**
$C_{25}H_{18}Cl_3N_3O_4$
For complete entry see 24.17

17.C **Ascochlorin p - bromobenzenesulfonate (absolute configuration)**
$C_{29}H_{32}BrClO_6S$
For complete entry see 50.10

BENZOQUINONES

18.1 **Tetrachloro - p - benzoquinone (at 110 ° K)**
Chloranil
$C_6Cl_4O_2$
K.J.van Weperen, G.J.Visser *Acta Cryst. (B),* **28,** 338, 1972

18.C **N,N,N′,N′ - Tetramethylbenzidine - chloroanil complex**
$C_6Cl_4O_2 . 2C_{16}H_{20}N_2$
For complete entry see 60.33

18.C **Thymine - p - benzoquinone complex**
$C_6H_4O_2 . C_5H_6N_2O_2$
For complete entry see 60.7

18.C **Phloroglucinol - p - benzoquinone complex**
1,3,5 - Trihydroxybenzene - p - benzoquinone complex
$2C_6H_4O_2 . C_6H_6O_3$
For complete entry see 60.12

18.2 **3,6 - Dichloro - 2,5 - bis(methylamino) - 1,4 - benzoquinone**
$C_8H_8Cl_2N_2O_2$
S.Kulpe *J. Prakt. Chem.*, **312,** 909, 1971

18.3 **2 - Methyl - 5,6 - dimethoxy - p - benzoquinone**
Fumagatin methyl ether
$C_9H_{10}O_4$
J.Silverman, I.Stam-Thole, C.H.Stam *Acta Cryst. (B),* **27,** 1846, 1971

18.4 **2,5 - Pentamethylene - imino - 1,4 - benzoquinone**
$C_{16}H_{22}N_2O_2$
S.Kulpe *Z. Chem.*, **11,** 466, 1972
Also classified in 33

BENZENE MISCELLANEOUS

19.C **Mesitylene - hexafluorobenzene complex (at −35 °C)**
$C_6F_6 . C_9H_{12}$
For complete entry see 60.21

19.1 **1,2,3 - Trichlorobenzene (neutron study)**
$C_6H_3Cl_3$
R.G.Hazell, M.S.Lehmann, G.S.Pawley *Acta Cryst. (B),* **28,** 1388, 1972

19.2 **p - Dichlorobenzene (monoclinic form, at −140 °C, new data)**
$C_6H_4Cl_2$
C.Panattoni, E.Frasson, S.Bezzi *Gazz. Chim. Ital.*, **93,** 813, 1963

19.3 **Monochlorobenzene (at 393 °K and 14.2kbars)**
C_6H_5Cl
D.Andre, R.Fourme, M.Renaud *Acta Cryst. (B),* **27,** 2371, 1971

19.4 **Benzene (form ii, at high pressure)**
C_6H_6
R.Fourme, D.Andre, M.Renaud *Acta Cryst. (B),* **27,** 1275, 1971

19.5 **Pentabromotoluene**
$C_7H_3Br_5$
W.R.Krigbaum, G.C.Wildman *Acta Cryst. (B),* **27,** 2353, 1971

19.6 **α - Bromoacetophenone**
C_8H_7BrO
M.P.Gupta, S.M.Prasad *Acta Cryst. (B),* **27,** 1649, 1971

19.7 **4 - Chloro - 3,5 - dimethyl - phenylcyanate**
C_9H_8ClNO
L.Kutschabsky, H.Schrauber *Z. Chem.*, **11,** 347, 1971

19.C **Mesitylene - hexafluorobenzene complex (at −35 °C)**
$C_9H_{12} . C_6F_6$
For complete entry see 60.21

19.C **p - Bromobenzylideneacetone**
$C_{10}H_9BrO$
For complete entry see 5.7

19.8 **1,4 - Dichloro - 2,3,5,6 - tetramethylbenzene**

$C_{10}H_{12}Cl_2$

J.-C.Messager, J.Blot *C. R. Acad. Sci., Fr.*, **272**, 684, 1971

19.9 **1,4 - Dichloro - 2,3,5,6 - tetramethylbenzene (at 122 ° K)**

$C_{10}H_{12}Cl_2$

J.-C.Messager, J.Blot *C. R. Acad. Sci., Fr.*, **272**, 684, 1971

19.C **N,N - Dimethyl - p - bromocinnamamide**

$C_{11}H_{12}BrNO$

For complete entry see 1.12

19.10 **Diphenyliodonium nitrate**

$C_{12}H_{10}I^+ . NO_3^-$

W.B.Wright, E.A.Meyers *Cryst. Struct. Comm.*, **1**, 95, 1972

19.C **Benzophenone - diphenylamine complex**

$C_{13}H_{10}O . C_{12}H_{11}N$

For complete entry see 60.26

19.11 **1,1 - bis(p - Chlorophenyl) - 2,2,2 - trichloroethane**

DDT

$C_{14}H_9Cl_5$

T.P.DeLacy, C.H.L.Kennard *J. Chem. Soc. (D)*, 1208, 1971

19.C **1,1 - Dimethyl - 2,5 - diphenyl - 1 - silacyclopentadiene diphenylacetylene complex**

$C_{14}H_{10} . C_{18}H_{18}Si$

For complete entry see 60.36

19.C **Dibenzoylacetylene**

$C_{16}H_{10}O_2$

For complete entry see 13.19

19.12 **p - Terphenyl (refinement of data of Dejace,Bull.Soc.Fr.Mineral. Cristallogr.,92,141,1969)**

$C_{18}H_{14}$

J.L.Baudour *Acta Cryst. (B)*, **28**, 1649, 1972

19.13 **Triphenylmethane**

$C_{19}H_{16}$

C.Riche, C.Pascard-Billy *C. R. Acad. Sci., Fr., C*, **274**, 846, 1972

19.14 **trans - 1 - Bromo - 2(p - bromophenyl) - 1,2 - diphenylethylene**

Broparestroe

$C_{22}H_{19}Br$

J.M.Fornies-Marquina, C.Courseille, B.Busetta, M.Hospital

Cryst. Struct. Comm., **1**, 261, 1972

19.C Pentaerythritol tetracinnamate

$C_{41}H_{36}O_8$

For complete entry see 5.9

MONOCYCLIC HYDROCARBONS (3, 4, 5-MEMBERED RINGS)

20.1 **Potassium dinitrotrichloro - cyclopentadienide**
$C_5Cl_3N_2O_4^-$. K^+
Y.Otaka, F.Marumo, Y.Saito *Acta Cryst. (B),* **28,** 1590, 1972
Residue 1 also classified in 12

20.2 **1,1 - Cyclobutanedicarboxylic acid**
$C_6H_8O_4$
L.Soltzberg, T.N.Margulis *J. Chem. Phys.*, **55,** 4907, 1971

20.C **1 - Aminocyclopentane carboxylic acid monohydrate**
$C_6H_{11}NO_2$. H_2O
For complete entry see 48.26

20.3 **1,1 - Cyclopentane - dicarboxylic acid**
$C_7H_{10}O_4$
T.N.Margulis
Amer. Cryst. Assoc., Abstr. Papers (Winter Meeting), 79, 1972

20.4 **1,2,3 - tris(Dimethylamino)cyclopropenium perchlorate**
$C_9H_{18}N_3^+$. ClO_4^-
A.T.Ku, M.Sundaralingam *J. Amer. Chem. Soc.*, **94,** 1688, 1972
Residue 1 also classified in 12

20.5 **1 - Chloro - 1 - phenylsulfonyl - 2,3 - dimethylcyclopropane**
$C_{11}H_{13}ClO_2S$
W.Saenger, C.H.Schwalbe *J. Org. Chem.*, **36,** 3401, 1971

20.6 **cis,trans,cis - 1,2,3,4 - Cyclobutanetetracarboxylic acid tetramethyl ester**
$C_{12}H_{16}O_8$
T.N.Margulis *J. Amer. Chem. Soc.*, **93,** 2193, 1971

20.7 **1,1′ - Dimethyl - bi(cyclopropyl) - 2,2′ - dicarboxylate**
$C_{12}H_{18}O_4$
C.Jongsma, H.van der Meer *Rec. Trav. Chim. Pays-Bas,* **90,** 33, 1971

20.8 **Diphenyl - cyclopropene - thione**
$C_{15}H_{10}S$
L.L.Reed, J.P.Schaefer *J. C. S. Chem. Comm.*, 528, 1972

20.C **N - Phenyl - 2 - (p - bromophenyl) - cyclopropane - 1,3 - dicarboximide**
$C_{17}H_{12}BrNO_2$
For complete entry see 32.21

20.C **N - Benzyl - 1,2 - dihydro - 2 - cyclopentadienylidene - pyridine**
$C_{17}H_{15}N$
For complete entry see 33.39

20.9 **3 - (4 - Bromocyclohexyl) - 4 - (3 - oxo - cyclopentyl) - hexane**
$C_{17}H_{29}BrO$
H.Wunderlich, D.Mootz *Acta Cryst. (B),* **27,** 2437, 1971
Also classified in 21

20.10 **1 - Cyano - 1 - (p - chlorophenyl) - 2 - (p - dimethylaminophenyl) - cyclopropane**
$C_{18}H_{17}ClN_2$
J.Meunier-Piret, M.van Meerssche *Bull. Soc. Chim. Belges,* **80,** 475, 1971
Also classified in 16

20.C **2 - (2′ - (β - Hydroxyethyl) - 2′ - methyl - 3′ - hydroxycyclopentyl) - 6 - methoxy - naphthalene**
$C_{20}H_{24}O_3$
For complete entry see 24.13

20.C **N - (2,6 - Dichlorophenyl) - 1,2 - dihydro - 1 - cyclopentadienylidene - isoquinoline**
$C_{21}H_{15}Cl_2N$
For complete entry see 35.27

20.11 **trans - 1,2 - Dichloro - 3,4 - bis(2,4,6 - trimethylbenzylidene)cyclobutane**
$C_{24}H_{26}Cl_2$
S.R.Byrn, E.Maverick, O.J.Muscio Junior, K.N.Trueblood, T.L.Jacobs
J. Amer. Chem. Soc., **93,** 6680, 1971

20.12 **5 - Bromo - 1,2,3,4 - tetraphenyl - cyclopentadiene**
$C_{29}H_{20}Br$
G.Evrard, P.Piret, M.van Meersche *Bull. Soc. Chim. Belges,* **80,** 159, 1971

20.13 **trans - 1,2 - Dichloro - 3,4 - bis(benzhydrylidene)cyclobutane**
$C_{30}H_{22}Cl_2$
S.R.Byrn, E.Maverick, O.J.Muscio Junior, K.N.Trueblood, T.L.Jacobs
J. Amer. Chem. Soc., **93,** 6680, 1971

20.14 **1,3 - Di - (p - bromobenzyl)cyclobutane - 2,4 - bis(2′ - spiro - (5′ - benzyl)cyclopentanone)**
$C_{38}H_{34}Br_2O_2$
D.A.Whiting *J. Chem. Soc. (C),* 3396, 1971
Also classified in 28

MONOCYCLIC HYDROCARBONS (6-MEMBERED RINGS)

21.C **Dodeca - sodium myo - inositol hexaphosphate octatriacontahydrate**
Sodium phytate octatriacontahydrate
$C_6H_6O_{24}P_6{}^{12-}$, $12Na^+$, $38H_2O$
For complete entry see 46.6

21.C **1,4 - Dideoxy - 1,4 - dinitro - neo - inositol - bis(tetrahydrothiophene - 1 - oxide) complex**
$C_6H_{10}N_2O_8$. $2C_4H_8OS$
For complete entry see 60.13

21.1 **epi-Inositol**
$C_6H_{12}O_6$
G.A.Jeffrey, H.S.Kim *Acta Cryst. (B),* **27,** 1812, 1971

21.C **Tetra - aquo - calcium myo - inositol bromide monohydrate**
$(C_6H_{20}CaO_{10}{}^{2+})_n$, $2nBr^-$, nH_2O
For complete entry see 67.6

21.C **1 - Aminocyclohexane carboxylic acid hydrochloride**
$C_7H_{14}NO_2{}^+$, Cl^-
For complete entry see 48.34

21.2 **Potassium hydrogen trans - cyclohexane - 1,4 - dicarboxylate**
$C_8H_{10}O_4{}^{2-}$, $2C_8H_{12}O_4$, $2K^+$
P.Luger, K.Plieth, G.Ruban *Acta Cryst. (B),* **28,** 699, 1972

21.3 **1,4 - trans - Cyclohexane dicarboxylic acid**
$C_8H_{12}O_4$
P.Luger, K.Plieth, G.Ruban *Acta Cryst. (B),* **28,** 706, 1972

21.4 **2,6 - Dichloro - 4 - t - butyl - cyclohexanone**
$C_{10}H_{16}Cl_2O$
A.Lectard, F.Metras, J.Petrissans, J.Gaultier
C. R. Acad. Sci., Fr., C, **272,** 2053, 1971

21.5 **trans - 2,4 - Dihydroxy - 2,4 - dimethylcyclohexane - trans - 1 - acetic acid γ - lactone**
$C_{10}H_{16}O_3$
R.M.Burnett, M.G.Rossmann *Acta Cryst. (B),* **27,** 1378, 1971

21.6 **cis - 2 - Chloro - 4 - t - butylcyclohexanone**
$C_{10}H_{17}ClO$
R.A.G.de Graaf, M.T.Giesen, E.W.M.Rutten, C.Romers
Acta Cryst. (B), **28,** 1576, 1972

21.7 **cis - 4 - t - Butylcyclohexane - 1 - carboxylic acid**
$C_{11}H_{20}O_2$
H.van Koningsveld *Acta Cryst. (B),* **28,** 1189, 1972

21.8 **3 - (1,1,5 - Trimethyl - 5 - cyclohexene - 6 - yl) - propenoic acid**
$C_{12}H_{18}O_2$
H.Schenk *Cryst. Struct. Comm.,* **1,** 143, 1972

21.9 **Cyclohexyl - tosylate**
$C_{13}H_{18}O_3S$
V.J.James, J.F.McConnell *Tetrahedron,* **27,** 5475, 1971
Also classified in 11

21.10 **Cyclohexyl - tosylate (neutron study)**
$C_{13}H_{18}O_3S$
V.J.James, J.F.McConnell *Tetrahedron,* **27,** 5475, 1971
Also classified in 11

21.C **N,N' - Dicyclohexylurea**
$C_{13}H_{24}N_2O$
For complete entry see 8.11

21.11 **3 - (p - Chlorophenyl) - 3,5,5 - trimethylcyclohexanone**
$C_{15}H_{19}ClO$
R.L.R.Towns, B.L.Shapiro *Cryst. Struct. Comm.,* **1,** 151, 1972

21.12 **cis - 1 - p - Bromophenyl - 4 - t - butylcyclohexane**
$C_{16}H_{23}Br$
G.Berti, B.Macchia, F.Macchia, S.Merlino, U.Muccini
Tetrahedron Letters, 3205, 1971

21.13 **Benzoyl - carvoxime**
$C_{17}H_{19}NO_2$
F.Baert, J.-P.Mornon, P.Herpin *C. R. Acad. Sci., Fr., C,* **273,** 231, 1971
Also classified in 10, 13

21.14 **cis - 4 - t - Butyl - cyclohexyl - p - toluenesulfonate (type B)**
$C_{17}H_{26}O_3S$
V.J.James, C.T.Grainger *Cryst. Struct. Comm.,* **1,** 111, 1972

21.C **3 - (4 - Bromocyclohexyl) - 4 - (3 - oxo - cyclopentyl) - hexane**
$C_{17}H_{29}BrO$
For complete entry see 20.9

21.15 **1,4 - Dicyclohexyl - cyclohexane (form 2)**
p - Tercyclohexane
$C_{18}H_{32}$
K.Sasvari *Cryst. Struct. Comm.*, **1,** 163, 1972

21.16 **1 - Bromo - 2 - benzoyl - 2 - phenylcyclohexane**
$C_{19}H_{19}BrO$
A.Ducruix, C.Pascard-Billy *Acta Cryst. (B),* **28,** 1848, 1972

21.C **Procyclidine hydrochloride**
α - Cyclohexyl - α - phenyl - β - (1 - pyrrolidine) - propanol hydrochloride
$C_{19}H_{30}NO^+$, Cl^-
For complete entry see 32.25

21.C **(+) - trans - 2 - o - Tolylcyclohexanol 3 - nitro - 4 - bromobenzoate (absolute configuration)**
$C_{20}H_{20}BrNO_4$
For complete entry see 13.21

21.17 **2,6 - Di - t - butyl - 4 - (p - bromophenyl)imino - 2,5 - cyclohexadiene - 1 - one**
$C_{20}H_{24}BrNO$
H.A.Fraterman, C.Romers *Rec. Trav. Chim. Pays-Bas,* **90,** 364, 1971

21.18 **anti - α - Acetoxy - α,2 - diphenyl(methylenecyclohexane)**
$C_{21}H_{22}O_2$
F.P.van Remoortere, J.J.Flynn *J. Amer. Chem. Soc.*, **93,** 5932, 1971

21.19 **syn - α - Acetoxy - α,2 - diphenyl(methylenecyclohexane)**
$C_{21}H_{22}O_2$
F.P.van Remoortere, J.J.Flynn *J. Amer. Chem. Soc.*, **93,** 5932, 1971

21.20 **cis,trans - 2,5 - Di - t - butylcyclohexyl toluene - p - sulfonate**
$C_{21}H_{34}O_3S$
D.H.Faber, C.Altona *J. Chem. Soc. (D),* 1210, 1971

21.21 **cis - 1,2 - bis(2 - Carboxy - 2 - propyl)cyclohexane**
$C_{24}H_{18}O_4$
R.P.Dodge, D.K.Wedegaertner, W.J.Darden Junior, E.B.Samuel, Q.Johnson *Tetrahedron Letters,* 1381, 1971

21.C **Ascochlorin p - bromobenzenesulfonate (absolute configuration)**
$C_{29}H_{33}BrClO_6S$
For complete entry see 50.10

21.C **Compound 8**
$C_{32}H_{41}N_2O_{12}^+$, Br^-
For complete entry see 35.37

21.22 **anti - 1,2,4,5 - Tetraphenyl - 3,6 - dicarbomethoxy - cyclohexa - 1,4 - diene**
$C_{34}H_{28}O_4$
M.J.Bennett, J.T.Purdham, S.Takada, S.Masamune
J. Amer. Chem. Soc., **93,** 4063, 1971

21.C **Datiscoside di - (p - iodobenzoate) (absolute configuration)**
$C_{52}H_{60}I_2O_{14}$
For complete entry see 59.29

MONOCYCLIC HYDROCARBONS (7, 8-MEMBERED RINGS)

22.1 **2 - Chlorotropone**

C_7H_5ClO

D.J.Watkin, T.A.Hamor *J. Chem. Soc. (B)*, 2167, 1971

22.2 **3 - Azidotropone**

$C_7H_5N_3O$

D.W.J.Cruickshank, G.Filippini, O.S.Mills

J. C. S. Chem. Comm., 101, 1972

Also classified in 9

22.C **1 - Aminocycloheptane carboxylic acid hydrobromide monohydrate**

$C_8H_{16}NO_2^+$, Br^-, H_2O

For complete entry see 48.37

22.C **1 - Aminocyclo - octane carboxylic acid hydrobromide**

$C_9H_{18}NO_2^+$, Br^-

For complete entry see 48.42

22.3 **3,5,7 - Tribromo - hinokitiol**

$C_{10}H_9Br_3O_2$

S.Ito, Y.Fukazawa, Y.Iitaka *Tetrahedron Letters*, 741, 1972

22.4 **5,7 - Dibromo - hinokitiol**

$C_{10}H_{10}Br_2O_2$

S.Ito, Y.Fukazawa, Y.Iitaka *Tetrahedron Letters*, 745, 1972

22.5 **3,7 - Dibromo - hinokitiol**

$C_{10}H_{10}Br_2O_2$

S.Ito, Y.Fukazawa, Y.Iitaka *Tetrahedron Letters*, 745, 1972

22.6 **4 - Isopropyl - tropolone**

β - Thujaplicin

$C_{10}H_{12}O_2$

J.E.Derry, T.A.Hamor *J. C. S. Perkin ii*, 694, 1972

22.7 **Perchloro - heptafulvalene**

$C_{14}Cl_{12}$

M.Ishimori, R.West, B.K.Teo, L.F.Dahl

J. Amer. Chem. Soc., **93**, 7101, 1971

22.8 **Tropolonyl p - chlorobenzoate**
$C_{14}H_9ClO_3$
J.P.Schaefer, L.L.Reed *J. Amer. Chem. Soc.*, **93**, 3902, 1971

22.C **Diaquo - calcium 2,4,6,8 - cyclo - octatetraene - 1,2 - dicarboxylate**
$C_{14}H_{10}CaO_6$
For complete entry see 67.14

22.9 **Heptafulvalene**
$C_{14}H_{12}$
R.Thomas, P.Coppens *Acta Cryst. (B)*, **28**, 1800, 1972

22.10 **Octamethyl - cyclo - octatetraene**
$C_{16}H_{24}$
J.Bordner, R.G.Parker, R.H.Stanford Junior
Acta Cryst. (B), **28**, 1069, 1972

22.C **2 - Troponyl - cyanomethylene - triphenylphosphonium betaine methanol solvate**
$C_{27}H_{20}NOP$, CH_4O
For complete entry see 64.35

22.C **2 - Troponyl - (ethoxycarbonyl - methylene - triphenylphosphonium)betaine**
$C_{29}H_{25}O_3P$
For complete entry see 64.38

MONOCYCLIC HYDROCARBONS
(9- AND HIGHER-MEMBERED RINGS)

23.1 **Cyclononane - mercury(ii) chloride complex**
$C_9H_{16}O$, Cl_2Hg
S.Dahl, P.Groth *Acta Chem. Scand.*, **25**, 1114, 1971

23.2 **cis,trans,cis - Cyclodeca - 2,4,8 - triene - 1,6 - dione**
$C_{10}H_{10}O_2$
O.Kennard, D.L.Wampler, J.C.Coppola, W.D.S.Motherwell, D.G.Watson, A.C.Larson *Acta Cryst. (B)*, **27**, 1116, 1971

23.3 **1,1,5,5 - Tetramethylcyclodecane - 8 - carboxylic acid**
$C_{15}H_{28}O_2$
J.D.Dunitz, H.Eser *Helv. Chim. Acta*, **50**, 1565, 1967

23.4 **bis(cis - Cyclodecene) silver nitrate**
$C_{20}H_{36}$, Ag^+ , NO_3^-
O.Ermer, H.Eser, J.D.Dunitz *Helv. Chim. Acta*, **54**, 2469, 1971
Residue 1 also classified in 75

23.5 **4,5.10,11 - bis(Tetramethylene) - 4,10 - cyclotridecadiene - 2,6,8,12 - tetrayne - 1 - one**
$C_{21}H_{16}O$
V.F.Duckworth, P.B.Hitchcock, R.Mason *J. Chem. Soc. (D)*, 963, 1971

23.6 **bis(1,1,4,4 - Tetramethyl - cis - cyclodec - 7 - ene) silver nitrate**
$C_{28}H_{52}$, Ag^+ , NO_3^-
O.Ermer, H.Eser, J.D.Dunitz *Helv. Chim. Acta*, **54**, 2469, 1971
Residue 1 also classified in 75

NAPHTHALENE COMPOUNDS

24.1 **Octafluoronaphthalene**
$C_{10}F_8$
G.S.Mandel, J.Donohue
Amer. Cryst. Assoc., Abstr. Papers (Winter Meeting), 42, 1972

24.2 **1,8 - Dinitroso - naphthalene**
$C_{10}H_6N_2O_2$
C.K.Prout, T.S.Cameron, R.M.A.Dunn, O.J.R.Hodder, D.Viterbo
Acta Cryst. (B), **27,** 1310, 1971
Also classified in 36

24.C **Lumiflavin - bis(naphthalene - 2,3 - diol) (yellow form)**
$2C_{10}H_8O_2, C_{13}H_{12}N_4O_2$
For complete entry see 60.27

24.C **10 - Propylisoalloxazine - bis(naphthalene - 2,3 - diol) complex**
$2C_{10}H_8O_2, C_{13}H_{12}N_4O_2$
For complete entry see 60.28

24.C **Lumiflavin bis(naphthalene - 2,3 - diol) trihydrate**
$2C_{10}H_8O_2, C_{13}H_{12}N_4O_2, 3H_2O$
For complete entry see 60.29

24.3 **racemic - trans - 1a,7b - Dihydro - oxireno(a)naphthalene - 3 - spiro - 2′ - oxiran - 2(3H) - one**
$C_{11}H_8O_3$
B.M.Gatehouse, D.J.Lloyd *J. C. S. Perkin ii,* 932, 1972
Also classified in 38

24.4 **4,5 - Dichloronaphthalic anhydride**
$C_{12}H_4Cl_2O_3$
L.N.Shok, G.A.Gol′der, L.A.Chetkina *Kristallografija,* **16,** 746, 1971
Also classified in 38

24.5 **4,5 - Dinitronaphthalic anhydride**
$C_{12}H_4N_2O_7$
J.Bordner, L.A.Jones *J. Cryst. Mol. Struct.,* **2,** 79, 1972
Also classified in 38

24.6 **5 - Bromo - N - hydroxy - naphthaloimide**

$C_{12}H_6BrNO_3$

L.N.Shok, G.A.Gol'der, L.A.Chetkina, M.A.Davydova

Kristallografija, **16,** 923, 1971

Also classified in 36

24.7 **Naphthalic anhydride**

$C_{12}H_6O_3$

L.N.Shok, L.A.Chetkina, M.G.Neigauz, G.A.Gol'der, E.M.Smelyanskaya, Yu.G.Fedorov *Kristallografija,* **16,** 500, 1971

Also classified in 38

24.8 **bis - 1,8 - Dimethylamino - naphthalene**

$C_{14}H_{18}N_2$

H.Einspahr, R.E.Marsh, J.D.Roberts, J.-B.Robert

Amer. Cryst. Assoc., Abstr. Papers (Winter Meeting), 90, 1972

24.C **tris(Hexafluoroacetylacetonato) copper(ii) 1 - dimethylammonium - 8 - dimethylamino - naphthalene**

$C_{14}H_{19}N_2^+$, $C_{15}H_3CuF_{18}O_6^-$

For complete entry see 77.7

24.C **tris(Hexafluoroacetylacetonato) magnesium 1 - dimethylammonium - 8 - dimethylamino - naphthalene**

$C_{14}H_{19}N_2^+$, $C_{15}H_3F_{18}MgO_6^-$

For complete entry see 67.17

24.9 **11,11,12,12 - Tetracyano - 1,4,naphthaquinodimethane**

$C_{16}H_6N_4$

F.Iwasaki *Acta Cryst. (B),* **27,** 1360, 1971

24.10 **Propanolol hydrochloride**

α - (N - Isopropyl - (3 - amino - 2 - hydroxy - n - propoxy))naphthalene hydrochloride

$C_{16}H_{22}NO_2^+$, Cl^-

M.Cotrait, J.Dangoumau *C. R. Acad. Sci., Fr., C,* **272,** 2057, 1971

24.11 **N - (2 - Naphthylmethylene) - p - bromoaniline**

$C_{17}H_{12}BrN$

B.T.Blaylock, R.F.Bryan

Amer. Cryst. Assoc., Abstr. Papers (Winter Meeting), 21, 1972

Also classified in 16

24.12 **N - (1 - α - Naphthylethyl) - N - (benzenesulfonyl)trichloromethane - sulfenamide**

$C_{19}H_{16}Cl_3NO_2S_2$

J.Kay, M.D.Glick, M.Rabano *J. Amer. Chem. Soc.,* **93,** 5224, 1971

24.13 **2 - (2′ - (β - Hydroxyethyl) - 2′ - methyl - 3′ - hydroxycyclopentyl) - 6 - methoxy - naphthalene**
$C_{20}H_{24}O_3$
E.G.Brain, F.Cassidy, A.W.Lake, P.J.Cox, G.A.Sim
J. C. S. Chem. Comm., 497, 1972
Also classified in 20

24.14 **1 - Acetyl - 5 - methyl - 4 - (2′ - (5′ - bromo - 6′ - methoxy)naphthyl) - 7 - oxa - bicyclo(3.2.1)octan - 6 - one**
$C_{21}H_{21}BrO_4$
E.G.Brain, F.Cassidy, A.W.Lake, P.J.Cox, G.A.Sim
J. C. S. Chem. Comm., 497, 1972
Also classified in 38

24.15 **1 - Acetyl - 5 - methyl - 4 - (2′ - (5′ - bromo - 6′ - methoxy)naphthyl) - 7 - oxa - bicyclo(3.2.1)octane**
$C_{21}H_{23}BrO_3$
E.G.Brain, F.Cassidy, A.W.Lake, P.J.Cox, G.A.Sim
J. C. S. Chem. Comm., 497, 1972
Also classified in 38

24.16 **p - (1,2,3,4 - Tetrahydro - 1 - naphthyl)phenoxy p - iodobenzoate (absolute configuration)**
$C_{23}H_{19}IO_2$
W.L.Bencze, B.Kiss, R.T.Puckett, N.Finch *Tetrahedron,* **26,** 5407, 1970
Also classified in 17

24.17 **1 - (2,5 - Dichlorophenylazo) - 2 - hydroxy - 3 - naphthoic acid 4 - chloro - 2,5 - dimethoxy - anilide**
$C_{25}H_{18}Cl_3N_3O_4$
D.Kobelt, E.F.Paulus, W.Kunstmann *Acta Cryst. (B),* **28,** 1319, 1972
Also classified in 9, 16, 17

24.18 **1,8 - bis(Phenylethynyl)naphthalene**
$C_{26}H_{16}$
A.E.Jungk, G.M.J.Schmidt *Chem. Ber.*, **104,** 3272, 1971

24.19 **1,4,5,8 - Tetraphenylnaphthalene**
$C_{34}H_{24}$
G.Evrard, P.Piret, M.van Meerssche *Acta Cryst. (B),* **28,** 497, 1972

NAPHTHOQUINONES

25.1 **3 - Iodonaphtho - 1,4 - quinone**
$C_{10}H_5IO_2$
J.Gaultier, C.Hauw, J.Housty, M.Schvoerer
C. R. Acad. Sci., Fr., C, **273,** 956, 1971

25.2 **3 - Iodo - 2 - hydroxynaphtho - 1,4 - quinone**
$C_{10}H_5IO_3$
C.Courseille, S.Geoffre, M.Schvoerer
C. R. Acad. Sci., Fr., C, **273,** 1633, 1971

25.3 **3 - Bromo - 4 - amino - 1,2 - naphthoquinone**
$C_{10}H_6BrNO_2$
D.Chasseau, G.Bravic *C. R. Acad. Sci., Fr., C,* **272,** 1215, 1971

25.4 **Juglone**
5 - Hydroxy - 1,4 - naphthoquinone
$C_{10}H_6O_3$
P.D.Cradwick, D.Hall *Acta Cryst. (B),* **27,** 1468, 1971

25.5 **Naphthazarin (form C,further refinement of the structure by Pascard - Billy,Acta Cryst.,15,519,1962)**
5,8 - Dihydroxy - 1,4 - naphthoquinone
$C_{10}H_6O_4$
P.D.Cradwick, D.Hall *Acta Cryst. (B),* **27,** 1990, 1971

25.6 **Naphthazarin (form A,further refinement of the structure by Pascard - Billy,Acta Cryst.,15,519,1962)**
5,8 - Dihydroxy - 1,4 - naphthoquinone
$C_{10}H_6O_4$
P.D.Cradwick, D.Hall *Acta Cryst. (B),* **27,** 1990, 1971

25.C **exo - 2,3 - Dichloro - 4a,8a - dicyano - 4a,5,8,8a - tetrahydro - 5,8 - methano - 1,4 - naphthoquinone**
$C_{13}H_6Cl_2N_2O_2$
For complete entry see 31.11

25.7 **1,4 - Naphthoquinone 2 - anilino - 3 - ethylmethylsulfonium ylide**
$C_{19}H_{17}NO_2S$
F.M.Lovell, D.B.Cosulich
Amer. Cryst. Assoc., Abstr. Papers (Summer Meeting), 104, 1971
Also classified in 16, 11

ANTHRACENE COMPOUNDS

26.C **Anthracene - 1,2,4,5 - tetracyanobenzene**
$C_{14}H_{10}$, $C_{10}H_2N_4$
For complete entry see 60.32

26.1 **9 - Methylanthracene**
$C_{15}H_{12}$
J.C.J.Bart, G.M.J.Schmidt *Israel J. Chem.*, **9**, 429, 1971

26.2 **9 - Methoxyanthracene**
$C_{15}H_{12}O$
J.C.J.Bart, G.M.J.Schmidt *Israel J. Chem.*, **9**, 429, 1971

26.3 **9,10 - bis(Chloromethyl)anthracene**
$C_{16}H_{12}Cl_2$
E.J.Gabe, J.P.Glusker *Acta Cryst. (B)*, **27**, 1925, 1971

26.4 **9 - Methoxycarbonylanthracene**
$C_{16}H_{12}O_2$
J.C.J.Bart, G.M.J.Schmidt *Israel J. Chem.*, **9**, 429, 1971

26.5 **Methyl 1β,4,4aβ,9,9aβ,10 - hexahydro - 9,10 - dioxo - anthr - 1 - yl - acetate**
$C_{17}H_{16}O_4$
G.D.Andreetti, G.Bocelli, L.Cavalca, P.Sgarabotto
Gazz. Chim. Ital., **101**, 735, 1971

26.6 **10 - Chloromethyl - 2,3,9 - trimethylanthracene**
$C_{18}H_{17}Cl$
A.Chomyn, J.P.Glusker, H.M.Berman, H.L.Carrell
Amer. Cryst. Assoc., Abstr. Papers (Winter Meeting), 23, 1972

26.7 **9 - t - Butyl - 9,10 - dihydroanthracene**
$C_{18}H_{20}$
T.Brennan, E.F.Putkey, M.Sundaralingam *J. Chem. Soc. (D)*, 1490, 1971

HYDROCARBONS (2 FUSED RINGS)

27.1 **cis - 6 - Chloro - trans - 7 - chloro - cis - bicyclo(3.2.0)heptan - 2 - one**
$C_7H_8Cl_2O$
F.P.Boer, P.P.North *J. C. S. Perkin ii,* 416, 1972

27.2 **(–) - Spiro(4.4)nonane - 1,6 - dione (at –160 ° C)**
$C_9H_{12}O_2$
C.Altona, R.A.G.de Graaf, C.H.Leeuwestein, C.Romers
J. Chem. Soc. (D), 1305, 1971

27.3 **Methyl 4 - Z - (chlorocyanomethylene) - 1,2,3,6 - tetrachlorobicyclo(3.1.0)hex - 2 - ene - 6 - syn - carboxylate**
$C_{10}H_4Cl_5NO_2$
R.Bau *Tetrahedron Letters,* 2081, 1972

27.4 **4 - Bromo - 3a - hydroxy - 7a - methyl - octahydroindene - 1,5 - dione**
$C_{10}H_{13}BrO_3$
H.van der Meer *Rec. Trav. Chim. Pays-Bas,* **90,** 529, 1971

27.5 **5,7 - Dibromo - 2,3 - benzotropone**
$C_{11}H_6Br_2O$
K.Ibata, T.Hata, H.Shimanouchi, Y.Sasada
J. C. S. Chem. Comm., 339, 1972

27.6 **5 - Chloro - 2,3 - benzotropone**
$C_{11}H_7ClO$
K.Ibata, T.Hata, H.Shimanouchi, Y.Sasada
J. C. S. Chem. Comm., 339, 1972

27.7 **3 - Iodo - 5 - oxa - tricyclo(5.3.1^{4,7})undeca - 6,10 - dione - 8 - carboxylic acid**
$C_{11}H_{11}IO_5$
C.A.Maier, J.A.Kapecki, I.C.Paul *J. Org. Chem.,* **36,** 1299, 1971
Also classified in 38

27.8 **Spirodienone II**
$C_{11}H_{14}O$
H.Koyama, T.Irie *J. C. S. Perkin ii,* 351, 1972

27.9 **N - exo - 6 - Bicyclo(3.1.0)hexyl p - bromosulfonamide**
$C_{12}H_{14}BrNO_2S$
M.F.Grostic, D.J.Duchamp, C.G.Chidester *J. Org. Chem.,* **36,** 2929, 1971

27.10 **2 - (p - Bromophenyl) - 1,3 - indanedione benzene solvate**
$2C_{15}H_9BrO_2, 0.5C_6H_6$
F.Bechtel, G.Bravic, J.Gaultier, C.Hauw
Cryst. Struct. Comm., **1**, 15, 1972

27.11 **4 - Chloro - 10,10,11,11 - tetracyano - bicyclo(9.2.0)undeca - 2,5,7 - triene**
$C_{15}H_9ClN_4$
J.Clardy, L.K.Read, M.J.Broadhurst, L.A.Paquette
J. Amer. Chem. Soc., **94**, 2904, 1972

27.12 **10,10,11,11 - Tetracyano - bicyclo(9.2.0)undeca - 2,5,7 - triene**
$C_{15}H_{10}N_4$
J.Clardy, L.K.Read, M.J.Broadhurst, L.A.Paquette
J. Amer. Chem. Soc., **94**, 2904, 1972

27.13 **3(a) - Dimethylamino - 2(a) - acetoxy - trans - decalin methiodide**
$C_{15}H_{28}NO_2^+, I^-$
E.Shefter, E.E.Smissman *J. Pharm. Sci.*, **60**, 1364, 1971
Residue 1 also classified in 3

27.14 **3 - Hydroxy - 9 - phenyl - 1,9 - dihydroazulen - 1 - one**
$C_{16}H_{12}O_2$
J.C.van de Grampel, A.J.Cuperus, A.Vos
Rec. Trav. Chim. Pays-Bas, **90**, 587, 1971

27.15 **8,8a - Dihydroxy - 7 - methoxy - 1,4 - dimethyl - decahydroazulene 8 - (p - bromobenzoate) (absolute configuration)**
$C_{20}H_{27}BrO_4$
S.Ito, H.Takeshita, M.Hirama, Y.Fukazawa *Tetrahedron Letters*, 9, 1972

27.16 **(±) - 1,2,3,4,4a,7,8,8a - Octahydro - 2α,4aβ,5,8aβ - tetramethylnaphthalene - 1β - ol p - bromobenzoate**
$C_{21}H_{27}BrO_2$
G.Saucy, R.E.Ireland, J.Bordner, R.E.Dickerson
J. Org. Chem., **36**, 1195, 1971

27.17 **Δ^{3a} - 4 - (m - Methoxybenzylsulfonyl) - 1,5 - diacetoxy - 7a - methylhexahydroindene**
$C_{22}H_{28}O_7S$
C.F.W.van de Ven, H.Schenk *Cryst. Struct. Comm.*, **1**, 147, 1972

27.18 **3,3,8,8 - Tetracarbomethoxy - bicyclo(8.4.0)tetradec - 5 - ene**
$C_{22}H_{32}O_8$
J.M.Jenks, S.H.Simonsen
Amer. Cryst. Assoc., Abstr. Papers (Winter Meeting), 91, 1972

27.19 Tetramethylverdene

6,7 - Dimethyl - 1,2 - diphenyl - 3 - (4′,5′ - dimethyl - 2′ - phenylethinylphenyl) - 4 - phenylethinyl - azulene

$C_{48}H_{36}$

N.Brodherr, P.Narayanan, K.Zechmeister, W.Hoppe

Ann. Chem., **750,** 53, 1971

HYDROCARBONS (3 FUSED RINGS)

28.1 **trans - 1,2 - Dibromoacenaphthene**
$C_{12}H_8Br_2$
M.-T.LeBihan, M.C.Perucaud *Acta Cryst. (B),* **28,** 629, 1972

28.2 **trans - 1,2 - Dichloroacenaphthene**
$C_{12}H_8Cl_2$
M.-T.LeBihan, M.C.Perucaud *Acta Cryst. (B),* **28,** 629, 1972

28.3 **9 - Fluorenone**
$C_{13}H_8O$
H.R.Luss, D.L.Smith *Acta Cryst. (B),* **28,** 884, 1972

28.4 **Phenanthrene (neutron study)**
$C_{14}H_{10}$
D.W.Jones, J.Yerkess *J. Cryst. Mol. Struct.,* **1,** 17, 1971

28.5 **Phenanthrene - antimony trichloride complex**
$C_{14}H_{10}, 2Cl_3Sb$
A.Demalde, A.Mangia, M.Nardelli, G.Pelizzi, M.E.V.Tani
Acta Cryst. (B), **28,** 147, 1972

28.6 **9 - Dicyanomethylene - 2,4,5,7 - tetranitro - fluorene**
$C_{16}H_4N_6O_8$
J.Silverman, N.F.Yannoni
Amer. Cryst. Assoc., Abstr. Papers (Winter Meeting), 90, 1972
Also classified in 7

28.7 **2 - (N,N - Diethylamino) - 2 - carbomethoxy - 2a,8a - dihydro - 2a - methoxycyclobuta(b)naphthalene - 3,8 - dione**
$C_{19}H_{21}NO_5$
J.A.Lerbscher, J.Trotter *J. Cryst. Mol. Struct.,* **1,** 355, 1971

28.8 **Levopimaric acid**
$C_{20}H_{30}O_2$
U.Weiss, W.B.Whalley, I.L.Karle *J. C. S. Chem. Comm.,* 16, 1972

28.9 **3 - Phenyl - 2 - cyclopentenone photodimer**
6,7 - Diphenyl - tricyclo(5.3.0.0^{2,6})decan - 3,10 - dione
$C_{22}H_{20}O_2$
A.V.Fratini, C.M.Shaw
Amer. Cryst. Assoc., Abstr. Papers (Winter Meeting), 92, 1972

28.C **3 - (10,11 - Dihydro - 5H - dibenzo(a,d)cyclohepten - 5 - ylidene) - 1 - ethyl - 2 - methylpyrrolidine hydrobromide**
Piroheptine hydrobromide
$C_{22}H_{26}N^{+}$, Br^{-}
For complete entry see 32.28

28.10 **8,9,13 - Trihydroxy - podocarpane 13 - p - bromobenzenesulfonate**
$C_{23}H_{33}BrO_{5}S$
S.G.Levine, I.Y.Chen, A.T.McPhail, P.Coggon
Tetrahedron Letters, 3459, 1971

28.11 **1,1′ - bis(Isopropoxycarbonyl) - 9,9′ - bisfluoroenylidene**
$C_{34}H_{28}O_{4}$
N.A.Bailey, S.E.Hull *J. Chem. Soc. (D),* 960, 1971

28.12 **anti - 1,2,4,5 - Tetraphenyl - 3,6 - dicarbomethoxy - tricyclo(3.1.0^{2,4})hexane**
$C_{34}H_{28}O_{4}$
M.J.Bennett, J.T.Purdham, S.Takada, S.Masamune
J. Amer. Chem. Soc., **93,** 4063, 1971

28.C **1,3 - Di - (p - bromobenzyl)cyclobutane - 2,4 - bis(2′ - spiro - (5′ - benzyl)cyclopentanone)**
$C_{38}H_{34}Br_{2}O_{2}$
For complete entry see 20.14

HYDROCARBONS (4 FUSED RINGS)

29.1 **3,3′ - Spiro - bis(bicyclo(3.1.0)hexane) - 2,2′ - dione (form A)**
$C_{11}H_{12}O_2$
F.H.Herbstein, H.Regev *J. Chem. Soc. (B),* 1696, 1971

29.2 **4,8 - Dihydro - dibenzo(cd,gh)pentalene**
$C_{14}H_{10}$
B.M.Trost, P.L.Kinson, C.A.Maier, I.C.Paul
J. Amer. Chem. Soc., **93,** 7275, 1971

29.3 **Pyrene**
$C_{16}H_{10}$
R.Allman *Z. Kristallogr.*, **132,** 129, 1970

29.4 **4b,9a - Dibromo - 9,20 - dihydroindeno(1,2 - a)indene**
$C_{16}H_{12}Br_2$
Y.Otaka, F.Marumo, Y.Saito *Acta Cryst. (B),* **27,** 2195, 1971

29.C **17β - Hydroxy - 8(9 - 10β)abeo - estra - 4 - en - 3,10 - dione**
$C_{18}H_{24}O_3$
For complete entry see 51.10

29.5 **9,10 - Dimethyl - 1,2 - benzanthracene (refinement of data of Sayre and Friedlander,Nature, 187,139,1960)**
$C_{20}H_{16}$
J.Iball *Nature,* **201,** 916, 1964

29.6 **5,8,10 - Trimethylbenzo(c)aceheptylene**
$C_{21}H_{18}$
H.J.Lindner *Tetrahedron Letters,* 3013, 1971

29.C **Achromycin hydrochloride (absolute configuration)**
$C_{22}H_{25}NO_8^+$, Cl^-
For complete entry see 50.6

29.7 **4b,5,6,6a,9,10,10a,10bα,11,12 - Decahydro - 2 - methoxy - 7(8H) - oxo - 4bβ,6aα,10aα - trimethylchrysene**
$C_{22}H_{30}O_2$
B.L.Trus, R.E.Marsh
Amer. Cryst. Assoc., Abstr. Papers (Winter Meeting), 74, 1972

29.8 **2,7 - Di - t - butylpyrene**

$C_{24}H_{26}$

A.C.Hazell, J.G.Lomborg *Acta Cryst. (B),* **28,** 1059, 1972

29.C **5,12a - Diacetyloxy - tetracycline**

$C_{26}H_{28}N_2O_{11}$

For complete entry see 50.9

29.C **Dehydro - ophiobolin - D diol mono - p - bromobenzenesulfonate**

$C_{30}H_{41}BrO_4S$

For complete entry see 59.23

HYDROCARBONS (5 OR MORE FUSED RINGS)

30.1 **4,8 - Dibromo - 3,7 - diketo - 3,4,4a,4b,7,8,8a,8b - octahydrodibenzo(a.g)biphenylene**

$C_{20}H_{14}Br_2O_2$
G.J.Kruger *J. Cryst. Mol. Struct.*, **1**, 271, 1971

30.2 **5,6 - Dihydrodibenz(a,h)anthracene**

$C_{22}H_{16}$
C.H.Wei, J.R.Einstein *Acta Cryst. (B)*, **28**, 1478, 1972

30.3 **5,6 - Dihydrodibenz(a,j)anthracene**

$C_{22}H_{16}$
C.H.Wei *Acta Cryst. (B)*, **28**, 1466, 1972

30.4′ **bis(2,2′ - Biphenylene)methane**

$C_{25}H_{16}$
H.Schenk *Acta Cryst. (B)*, **28**, 625, 1972

30.5 **(−) - 2 - Bromohexahelicene (absolute configuration)**

$C_{26}H_{15}Br$
D.A.Lightner, D.T.Hefelfinger, T.W.Powers, G.W.Frank, K.N.Trueblood
J. Amer. Chem. Soc., **94**, 3492, 1972

30.6 **2 - Methyl - hexahelicene**

$C_{27}H_{18}$
G.W.Frank, D.T.Hefelfinger, K.N.Trueblood, D.A.Lightner
Amer. Cryst. Assoc., Abstr. Papers (Winter Meeting), 89, 1972

30.7 **5,6,6a,6bα,7,8,12b,13,14,14a - Decahydro - 3 - ethoxy - 10 - methoxy - 6aβ,12bβ,14aα - trimethylpicene**

$C_{28}H_{36}O_2$
B.L.Trus, R.E.Marsh
Amer. Cryst. Assoc., Abstr. Papers (Winter Meeting), 74, 1972

30.8 **Pyreno(1′,2′.1,2) - pyrene**

$C_{30}H_{16}$
K.W.Muir, J.M.Robertson *Acta Cryst. (B)*, **28**, 879, 1972

30.C **Gymnemagenin**

$C_{30}H_{50}O_6$
For complete entry see 59.24

30.9 **Violanthrene - B**

$C_{34}H_{20}$

T.Maekawa *Sci. Papers Coll. gen. Educ., Univ. Tokyo,* **21,** 19, 1971

BRIDGED RING HYDROCARBONS

31.1 **exo - Tricyclo(3.2.1.0^{2,4})oct - 6 - ene - silver nitrate complex**

$C_8H_{10}Ag^+ , NO_3^-$

C.S.Gibbons, J.Trotter *J. Chem. Soc. (A),* 2058, 1971

Residue 1 also classified in 75

31.2 **(–) - 2 - exo - Aminonorbornane - 2 - carboxylic acid hydrobromide (absolute configuration)**

$C_8H_{14}NO_2^+ , Br^-$

P.A.Apgar, M.L.Ludwig *J. Amer. Chem. Soc.,* **94,** 964, 1972

Residue 1 also classified in 48

31.3 **Tricyclo(4.4.0.0^{2,8})dec - 3 - ene - 7,10 - dione**

$C_{10}H_{10}O_2$

C.S.Gibbons, J.Trotter *J. C. S. Perkin ii,* 737, 1972

31.4 **Bicyclo(2.2.2)octene - 2,3 - endo - dicarboxylic acid anhydride**

$C_{10}H_{12}O_3$

R.Destro, G.Filippini, C.M.Gramaccioli, M.Simonetta

Acta Cryst. (B), **27,** 2023, 1971

Also classified in 38

31.5 **7 - syn - 6 - endo - Dihydroxy - bicyclo(2.2.1)heptane - 2 - endo - carboxylic acid α - lactone**

$C_{10}H_{12}O_4$

R.M.Moriarty, H.Gopal, J.L.Flippen, J.Karle

Tetrahedron Letters, 351, 1972

Also classified in 38

31.6 **11,11 - Difluoro - 1,6 - methano(10)annulene**

$C_{11}H_8F_2$

C.M.Gramaccioli, M.Simonetta *Acta Cryst. (B),* **27,** 2231, 1971

31.7 **racemic - 8 - Carboxy - 2 - hydroxy - 2 - oxobicyclo(3.2.2)non - 6 - ene - 9,4 - carbolactone**

$C_{11}H_{10}O_6$

D.J.Pointer, J.B.Wilford, K.M.Chui *J. Chem. Soc. (B),* 895, 1971

31.8 **2,3,4 - Trichloro - pentacyclo($6.4.0.0^{2,4}.0^{3,10}.0^{5,9}$) dodeca - 6,11 - diene**
$C_{12}H_9Cl_3$
W.L.Mock, C.M.Sprecher, R.F.Stewart, M.G.Northolt
J. Amer. Chem. Soc., **94**, 2015, 1972

31.9 **trans - 9,10 - Pentacyclo($4.4.0.0^{2,5}.0^{3,8}.0^{4,7}$) decane - dicarboxylic acid**
$C_{12}H_{12}O_4$
J.P.Schaefer, K.K.Walthers *Tetrahedron*, **27**, 5281, 1971

31.10 **anti - Tricyclo($4.2.2.0^{2,5}$)deca - 3,9 - diene - 7,8 - endo - dicarboxylic anhydride**
$C_{12}H_{14}O_3$
G.Filippini, M.Simonetta
Atti Accad. Nazion. Lincei. R.C., Cl. Sci. Fis. Mat. Nat., **49**, 389, 1970

31.11 **exo - 2,3 - Dichloro - 4a,8a - dicyano - 4a,5,8,8a - tetrahydro - 5,8 - methano - 1,4 - naphthoquinone**
$C_{13}H_6Cl_2N_2O_2$
D.J.Pointer, J.B.Wilford, O.J.R.Hodder *J. Chem. Soc. (B)*, 2009, 1971
Also classified in 25

31.12 **Decahydro(1,2,4.5,6,8)dimetheno - s - indacene - 3,7 - dione**
Binor - S - dione
$C_{14}H_{12}O_2$
F.P.Boer, M.A.Neuman, R.J.Roth, T.J.Katz
J. Amer. Chem. Soc., **93**, 4436, 1971

31.13 **Norbornadiene dimer silver nitrate complex**
$C_{14}H_{16}Ag^+$, NO_3^-
G.D.Smith, C.N.Caughlan
Amer. Cryst. Assoc., Abstr. Papers (Summer Meeting), 80, 1971
Residue 1 also classified in 75

31.14 **(4.4.4)Propellatriene**
$C_{14}H_{18}$
O.Ermer, R.Gerdil, J.D.Dunitz *Helv. Chim. Acta*, **54**, 2476, 1971

31.15 **(±) - 2 - exo - Norbornanol p - toluenesulfonate**
(±) - Bicyclo(2.2.1)heptane - 2 - exo - ol p - toluenesulfonate
$C_{14}H_{18}O_3S$
C.Altona, M.Sundaralingam *Acta Cryst. (B)*, **28**, 1806, 1972

31.16 **(4.4.4)Propellane**
$C_{14}H_{24}$
O.Ermer, R.Gerdil, J.D.Dunitz *Helv. Chim. Acta*, **54**, 2476, 1971

31.17 **1,1,2,2,9,9,10,10 - Octafluoro - (2.2)paracyclophane**
$C_{16}H_8F_8$
H.Hope, J.Bernstein, K.N.Trueblood *Acta Cryst. (B)*, **28**, 1733, 1972

31.18 **(2.2)Paracyclophane**

$C_{16}H_{16}$

H.Hope, J.Bernstein, K.N.Trueblood *Acta Cryst. (B)*, **28**, 1733, 1972

31.C **exo - 7,11 - Methano - 6a,7,7a,10a,11,11a - hexahydrobenz(c)indeno(5,6 - e)thiazine - 5(H) - 6 - 'oxide**

$C_{16}H_{17}NOS$

For complete entry see 41.27

31.19 **Hexacyclo ($10,3,1,0^{2,10},0^{3,7},0^{6,15},0^{9,14}$) hexadecane**

Ethanocongressane

$C_{16}H_{22}$

S.T.Rao, M.Sundaralingam *Acta Cryst. (B)*, **28**, 694, 1972

31.20 **1,6.8,13 - Propane - 1,3 - diylidene(14)annulene**

$C_{17}H_{14}$

A.Gavezzotti, A.Mugnoli, M.Raimondi, M.Simonetta

J. C. S. Perkin ii, 425, 1972

31.21 **7 - Methoxycarbonyl - anti - 1,6.8,13 - bismethano(14)annulene**

$C_{18}H_{12}O_2$

C.M.Gramaccioli, A.Mimum, A.Mugnoli, M.Simonetta

J. Chem. Soc. (D), 796, 1971

31.22 **1,6.8,13 - Butane - diylidene(14)annulene**

$C_{18}H_{16}$

C.M.Gramacciolo, A.Mugnoli, T.Pilati, M.Raimondi, M.Simonetta

J. Chem. Soc. (D), 973, 1971

31.23 **endo - 2,3 - Dichloro - 4a,8a - dicyano - 4a,5,8,8a - tetrahydro - 5,7 - dimethyl - (2,2 - dimethylethano) - 1,4, - naphthoquinone**

$C_{18}H_{16}Cl_2N_2O_2$

D.J.Pointer, J.B.Wilford, O.J.R.Hodder *J. Chem. Soc. (B)*, 2009, 1971

31.C **(3.3)Paracyclophane - tetracyanoethylene complex**

$C_{18}H_{20}$, C_6N_4

For complete entry see 60.37

31.24 **Bicyclo(5.3.1)undeca - 7 - en - 11 - one - 1 - carboxylic acid p - chloroanilide**

$C_{18}H_{20}ClNO_2$

A.F.Cameron, G.Jamieson *J. Chem. Soc. (B)*, 1581, 1971

31.25 **Cedrone dimethylformamide solvate**

$C_{18}H_{20}O_6$, $2C_3H_7NO$

J.A.Beisler, J.V.Silverton *Acta Cryst. (B)*, **28**, 298, 1972

31.26 **2 - Bromo - 11 - ethyl - 5,9 - dimethoxy - tetracyclo($5.4.1.1^{4,12}.1^{8,11}$)tetradecan - 3 - one**

$C_{18}H_{27}BrO_3$

R.F.Dunphy, H.Lynton *Canad. J. Chem.*, **49**, 2497, 1971

31.27 **Triptycene**

$C_{20}H_{14}$

R.G.Hazell, G.S.Pawley, C.E.L.Petersen

J. Cryst. Mol. Struct., **1**, 319, 1971

31.C **16 - Bromo - myricanol nitromethane solvate (absolute configuration)**

$C_{21}H_{25}BrO_5$, CH_3NO_2

For complete entry see 59.18

31.28 **5 - t - Butoxy - 14 - carbomethoxy - 12 - hydroxy - 4 - methyl - tetracyclo (9.2.2.0^{1,9}.0^{4,8})pentadecane**

$C_{21}H_{34}O_4$

J.W.Scott, W.Vetter, W.E.Oberhansli, A.Furst

Tetrahedron Letters, 1719, 1972

31.29 **11 - (p - Bromophenyl) - 1,2,5,5,9 - pentamethyl - 4 - oxatris - cyclo(7,2,1,0^{3,8})dodeca - 2,7 - dien - 12 - one**

Adduct of 1 - tocoquinone and p - bromostyrene

$C_{22}H_{25}BrO_2$

O.Lefebvre-Soubeyran *Acta Cryst. (B),* **27,** 1218, 1971

31.C **1,3 - Diadamantyl - aziridinone**

$C_{22}H_{33}NO$

For complete entry see 32.30

31.30 **5,6,6a,7,7a,11b - Hexahydro - 2,3,5,5,7,7,9,10 - octamethyl - 6,11a,11c - metheno - 1H - benzo(c) - fluorene - 1,4,8,11(4aH) - tetrone**

Plastoquinone - 1 photoproduct

$C_{26}H_{30}O_4$

W.H.Watson, J.E.Whinnery, D.Creed, H.Werbin, E.T.Strom

J. C. S. Chem. Comm., 743, 1972

31.31 **Tetramethyl 3,3′ - bi - tricyclo(3.2.2.0^{2,4})nona - 6,8 - diene 6,6,7,7′ - tetracarboxylate**

$C_{26}H_{30}O_8$

J.Bordner, G.H.Wahl Junior *J. Org. Chem.*, **36,** 3630, 1971

31.32 **1,4,12,12 - Tetramethyl - 13 - hydroxy - tetracyclo(9.4.1^{4,8}.0.0^{2,8})pentadec - 6 - ene - 5,14 - dione p - bromobenzoate**

$C_{27}H_{31}BrO_4$

M.Laing, P.Sommerville, D.Hanouskova, K.H.Pegel, L.P.L.Piacenza, L.Phillips, E.S.Waight *J. C. S. Chem. Comm.*, 196, 1972

HETERO-NITROGEN
(3, 4, 5-MEMBERED MONOCYCLIC)

32.1 **Imidazole**
$C_3H_4N_2$
Yu.A.Omel'chenko, Yu.D.Kondrashev *Kristallografija*, **16**, 115, 1971

32.2 **Imidazolium dihydrogen phosphate**
$C_3H_5N_2^+ , H_2O_4P^-$
R.H.Blessing, E.L.McGandy *J. Amer. Chem. Soc.*, **94**, 4034, 1972

32.3 **3 - Methyl - 3 - pyrazolin - 5 - one**
$C_4H_6N_2O$
W.H.DeCamp, J.M.Stewart *Acta Cryst. (B)*, **27**, 1227, 1971

32.4 **1 - Acetyl - 4 - bromopyrazole**
$C_5H_5BrN_2O$
J.Lapasset, A.Escande *C. R. Acad. Sci., Fr., C*, **273**, 728, 1971

32.5 **1,3 - Dimethyl - 4 - imino - 5 - oxo - 2 - thione - imidazolidine**
$C_5H_7N_3OS$
T.Kinoshita, S.Sato, C.Tamura *Tetrahedron Letters*, 3695, 1971

32.6 **1,3 - Dimethylimidazole - 2(3H) - thione (space group Bm2$_1$b)**
$C_5H_8N_2S$
G.B.Ansell *J. C. S. Perkin ii*, 841, 1972

32.7 **1,3 - Dimethylimidazole - 2(3H) - thione (space group Bmmb)**
$C_5H_8N_2S$
G.B.Ansell *J. C. S. Perkin ii*, 841, 1972

32.8 **Histamine tetrachlorocobaltate(ii)**
$C_5H_{11}N_3^{2+} , Cl_4Co^{2-}$
J.J.Bonnet, Y.Jeannin *Acta Cryst. (B)*, **28**, 1079, 1972
Residue 1 also classified in 3

32.C **β - (3 - Pyrazolyl) - L - alanine**
$C_6H_9N_3O_2$
For complete entry see 48.21

32.C **N - Methyl pyro - L - glutamic acid amide**
5 - (N - Methylcarboxamido) - pyrrolid - 2 - one
$C_6H_{10}N_2O_2$
For complete entry see 48.23

32.C **L - Histidine hydrochloride monohydrate**
$C_6H_{10}N_3O_2^+$, Cl^-, H_2O
For complete entry see 48.24

32.C **L - Histidinium dihydrogen orthophosphate orthophosphoric acid**
$C_6H_{10}N_3O_2^+$, $H_2O_4P^-$, H_3O_4P
For complete entry see 48.25

32.9 **5,5 - bis(Trifluoromethyl) - 1 - (4 - methyl - pyrazol - 1 - en - 3 - yl) - 1,2,3 - triazol - 2 - ene**
$C_8H_9F_6N_5$
A.Gieren, K.Burger, J.Fehn *Angew. Chem.*, **84,** 212, 1972

32.10 **(E) - 2,3 - Dichloro - 1 - (pyrrolidin - 2 - ylidene) - Δ^2 - pyrrolinium bromide dihydrate**
$C_8H_{11}Cl_2N_2^+$, Br^-, $2H_2O$
F.L.Suddath Junior, A.C.Baillie, J.A.Bertrand, J.R.Dyer
J. Chem. Soc. (D), 1153, 1971

32.C **L - N - Acetylhistidine monohydrate**
$C_8H_{11}N_3O_3$, H_2O
For complete entry see 48.36

32.11 **1 - (2 - Chloroethyl) - 3 - (4 - carbamoylpyrazol - 3 - yl) - Δ^2 - 1,2,3 - triazolium chloride**
$C_8H_{12}ClN_6O^+$, Cl^-
W.H.Watson, J.E.Whinnery, K.C.Go
Amer. Cryst. Assoc., Abstr. Papers (Winter Meeting), 22, 1972

32.12 **2,2,5,5 - Tetramethyl - 3 - carbamidopyrroline - 1 - oxyl**
$C_9H_{15}N_2O_2$
J.W.Turley, F.P.Boer *Acta Cryst. (B),* **28,** 1641, 1972
Also classified in 12

32.13 **1 - p - Nitrophenyl - 3 - methyl - 4 - bromo - pyrazole**
$C_{10}H_8BrN_3O_2$
J.Lapasset, J.Falgueirettes *Acta Cryst. (B),* **28,** 791, 1972
Also classified in 15

32.14 **1,3 - Diaziridinyl - 2,4,6 - trinitrobenzene**
$C_{10}H_9N_5O_6$
C.Dickinson, J.R.Holden, E.G.Boonstra, J.M.Stewart
Amer. Cryst. Assoc., Abstr. Papers (Summer Meeting), 74, 1971
Also classified in 15

32.C **6 - Histaminopurine dihydrate**
$C_{10}H_{11}N_7$, $2H_2O$
For complete entry see 44.20

32.15 **Antipyrine**
1 - Phenyl - 2,3 - dimethyl - 5 - pyrazolone
$C_{11}H_{12}N_2O$
M.Vijayan *Curr. Sci.*, **40,** 489, 1971

32.16 **4,5 - Di - t - butylimidazole (at −160 ° C)**
$C_{11}H_{20}N_2$
G.J.Visser, A.Vos *Acta Cryst. (B),* **27,** 1802, 1971

32.17 **Ethyl 4 - acetyl - 3 - ethyl - 5 - methylpyrrole - 2 - carboxylate**
$C_{12}H_{17}NO_3$
R.Bonnettt, M.B.Hursthouse, S.Neidle *J. C. S. Perkin ii,* 902, 1972

32.C **Aminopyrine - barbital complex**
$C_{13}H_{17}N_3O$, $C_8H_{12}N_2O_3$
For complete entry see 60.31

32.18 **1 - Phenyl - 2,3,5,5 - tetramethyl - 2 - pyrazolinium perchlorate**
$C_{13}H_{19}N_2^+$, ClO_4^-
J.-L.Aubagnac, J.Elguero, B.Rerat, C.Rerat, Y.Uesu
C. R. Acad. Sci., Fr., C, **274,** 1192, 1972

32.19 **1 - Benzyl - 1,3,3 - trimethylazetidinium iodide**
$C_{13}H_{20}N^+$, I^-
R.L.Towns, L.M.Trefonas *J. Amer. Chem. Soc.*, **93,** 1761, 1971

32.20 **Diphenylhydantoin**
$C_{15}H_{12}N_2O_2$
A.Camerman, N.Camerman *Acta Cryst. (B),* **27,** 2205, 1971

32.C **2,2,5,5 - Tetramethyl - 1 - aza - cyclopentanone - 3 - azine - 3 - oxyl biradical**
$C_{16}H_{28}N_4O_2$
For complete entry see 12.13

32.21 **N - Phenyl - 2 - (p - bromophenyl) - cyclopropane - 1,3 - dicarboximide**
$C_{17}H_{12}BrNO_2$
J.P.Declerq, P.Piret, M.van Meerssche *Acta Cryst. (B),* **28,** 328, 1972
Also classified in 20

32.C **Tosyl - L - prolyl - L - hydroxyproline monohydrate (refinement)**
$C_{17}H_{22}N_2O_6S$, H_2O
For complete entry see 48.52

32.22 **1 - (α - Benzoyloxy - benzylideneamino) - 4,5 - dimethyl - 1,2,3 - triazole**
$C_{18}H_{16}N_4O_2$
H.Bauer, A.J.Boulton, W.Fedeli, A.R.Katritzky, A.Majid-Hamid, F.Mazza, A.Vaciago *J. C. S. Perkin ii,* 662, 1972

32.23 **trans - 3 - Phenyl - 3 - methoxycarbonyl - 4 - p - methoxyphenyl - 1 - pyrazoline**

$C_{18}H_{18}N_2O_3$

M.-P.Rousseaux, J.Meunier-Piret, J.-P.Putzeys, G.Germain, M.van Meerssche *Acta Cryst. (B)*, **28**, 1720, 1972

32.24 **5 - Hydroxy - 3 - phenyl - 1 - (3 - methyl - 1 - isoquinolyl)pyrazole**

$C_{19}H_{15}N_3O$

G.S.D.King, H.Reimlinger *Chem. Ber.*, **104**, 2694, 1971

Also classified in 35

32.25 **Procyclidine hydrochloride**

α - Cyclohexyl - α - phenyl - β - (1 - pyrrolidine) - propanol hydrochloride

$C_{19}H_{30}NO^+$, Cl^-

N.Camerman, A.Camerman *Molec. Pharm.*, **7**, 406, 1971

Residue 1 also classified in 21

32.26 **4 - (4 - N,N - Diethylaminophenylimino) - 3 - methyl - 1 - phenyl - 2 - pyrazolin - 5 - one**

$C_{20}H_{22}N_4O$

D.L.Smith, E.K.Barrett *Acta Cryst. (B)*, **27**, 2043, 1971

Also classified in 16

32.27 **1,3 - Diphenyl - 4 - (p - chlorobenzylidene) - 5 - pyrazolone**

$C_{22}H_{15}ClN_2O$

B.Bovio, S.Locchi *Cryst. Struct. Comm.*, **1**, 253, 1972

32.28 **3 - (10,11 - Dihydro - 5H - dibenzo(a,d)cyclohepten - 5 - ylidene) - 1 - ethyl - 2 - methylpyrrolidine hydrobromide**

Piroheptine hydrobromide

$C_{22}H_{26}N^+$, Br^-

Y.Tokuma, H.Nojima, Y.Morimoto *Bull. Chem. Soc. Jap.*, **44**, 2665, 1971

Residue 1 also classified in 28

32.29 **4 - (2,6 - Dimethyl - 4 - N,N - diethylaminophenylimino) - 3 - methyl - 1 - phenyl - 2 - pyrazolin - 5 - one**

$C_{22}H_{26}N_4O$

D.L.Smith, E.K.Barrett *Acta Cryst. (B)*, **27**, 2043, 1971

Also classified in 16

32.30 **1,3 - Diadamantyl - aziridinone**

$C_{22}H_{33}NO$

A.H.-J.Wang, I.C.Paul, E.R.Talaty, A.E.Dupuy Junior

J. C. S. Chem. Comm., 43, 1972

Also classified in 31

HETERO-NITROGEN
(6-MEMBERED MONOCYCLIC)

33.1 **2 - Amino - 3 - chloropyrazine**
$C_4H_4ClN_3$
J.C.Morrow, B.P.Huddle *Acta Cryst. (B),* **28,** 1748, 1972

33.2 **Piperazine thiomolybdate**
$C_4H_{12}N_2^{2+}$, MoS_4^{2-}
P.A.Koz'min, Z.V.Popova *Zh. Strukt. Khim.,* **12,** 99, 1971

33.3 **2,3,5,6 - Tetrachloro - 4 - hydroxypyridine**
C_5HCl_4NO
F.P.Boer, J.W.Turley, F.P.van Remoortere
J. C. S. Chem. Comm., 573, 1972

33.C **2 - Pyridone - 6 - chloro - 2 - hydroxypyridine complex**
C_5H_4ClNO , C_5H_5NO
For complete entry see 60.5

33.4 **4 - Nitro - pyridine N - oxide**
$C_5H_4N_2O_3$
F.K.Ross, P.Coppens
Amer. Cryst. Assoc., Abstr. Papers (Winter Meeting), 27, 1972
Also classified in 10

33.5 **4 - Nitropyridine - N - oxide trans - dichlorodiaquo copper(ii) complex**
$2C_5H_4N_2O_3$, $H_4Cl_2CuO_2$
R.J.Williams, D.T.Cromer, W.H.Watson
Acta Cryst. (B), **27,** 1619, 1971

33.C **Pyridine - N - oxide trichloroacetic acid complex**
C_5H_5NO , $C_2HCl_3O_2$
For complete entry see 60.4

33.C **2 - Pyridone - 6 - chloro - 2 - hydroxypyridine complex**
C_5H_5NO , C_5H_4ClNO
For complete entry see 60.5

33.6 **Pyrazine - 2 - carboxamide (β form)**
$C_5H_5N_3O$
G.Ro, H.Sorum *Acta Cryst. (B),* **28,** 991, 1972

33.7 **Pyrazine - 2 - carboxamide (δ form)**
$C_5H_5N_3O$
G.Ro, H.Sorum *Acta Cryst. (B),* **28,** 1677, 1972

33.C **Pyridinium bromide - bis(thiourea) complex**
$C_5H_6N^+ , 2CH_4N_2S , Br^-$
For complete entry see 60.6

33.8 **Pyridinium di - μ - hydroperoxo - tetraperoxo - dioxo - dimolybdate(vi)**
$2C_5H_6N^+ , H_2Mo_2O_{14}^{2-}$
J.-M.Le Carpentier, A.Mitschler, R.Weiss
Acta Cryst. (B), **28,** 1288, 1972

33.9 **Pyridinium μ - oxo - tetraperoxo - dioxo - diaquo - dimolybdate(vi)**
$2C_5H_6N^+ , H_4Mo_2O_{13}^{2-}$
J.-M.Le Carpentier, A.Mitschler, R.Weiss
Acta Cryst. (B), **28,** 1288, 1972

33.10 **Piperidinium p - bromobenzoate**
$C_5H_{12}N^+ , C_7H_4BrO_2^-$
S.Kashino, Y.Sumida, M.Haisa *Acta Cryst. (B),* **28,** 1374, 1972

33.11 **Piperidinium p - chlorobenzoate**
$C_5H_{12}N^+ , C_7H_4ClO_2^-$
S.Kashino, Y.Sumida, M.Haisa *Acta Cryst. (B),* **28,** 1374, 1972

33.12 **Piperidinium hexathiotetra - arsenate**
$2C_5H_{12}N^+ , As_4S_6^{2-}$
E.J.Porter, G.M.Sheldrick *J. Chem. Soc. (A),* 3130, 1971

33.13 **4 - Cyanopyridine**
$C_6H_4N_2$
M.Laing, N.Sparrow, P.Sommerville *Acta Cryst. (B),* **27,** 1986, 1971
Also classified in 7

33.14 **Picolinic acid N - oxide (at $-100\,^\circ$ C)**
$C_6H_5NO_3$
M.Laing, C.Nicholson *J. S. Afr. Chem. Inst.,* **24,** 186, 1971
Also classified in 10

33.15 **2 - Thioformamido - pyridine**
$C_6H_6N_2S$
T.C.Downie, W.Harrison, E.S.Raper, M.A.Hepworth
Acta Cryst. (B), **28,** 283, 1972

33.16 **2 - Hydroxymethylpyridine N - oxide**
$C_6H_7NO_2$
R.Desiderato, J.C.Terry *J. Heterocycl. Chem.,* **8,** 617, 1971
Also classified in 10

33.17 **Isonicotinic acid hydrazide**
$C_6H_7N_3O$
L.H.Jensen *J. Amer. Chem. Soc.*, **76,** 4663, 1954
Also classified in 9

33.C **4 - Methylpyridinium triphenylphosphine - tribromozincate**
$C_6H_8N^+$, $C_{18}H_{15}Br_3PZn^-$
For complete entry see 86.10

33.C **2(S) - Carboxy - 4(R),5(S) - dihydroxypiperidine hydrochloride (absolute configuration)**
$C_6H_{12}NO_4^+$, Cl^-
For complete entry see 59.2

33.18 **3,5 - Dichloro - 4 - hydroxy - 2,6 - dimethylpyridine**
$C_7H_7Cl_2NO$
F.P.Boer, J.W.Turley, F.P.van Remoortere
J. C. S. Chem. Comm., 573, 1972

33.19 **2 - Formylpyridine selenosemicarbazone**
$C_7H_8N_4Se$
A.Conde, A.Lopez-Castro, R.Marquez
Cryst. Struct. Comm., **1,** 155, 1972
Also classified in 8, 9

33.20 **4 - Ethylpyridinium tetrabromoferrate(iii)**
$C_7H_{10}N^+$, Br_4Fe^-
M.L.Nackert, R.A.Jacobson *Acta Cryst. (B)*, **27,** 1658, 1971

33.21 **trans - 3 - (6 - Methyl - 2 - pyridylthio) - propenic acid**
$C_9H_9NO_2S$
P.Groth, K.Davidkov, A.Aasen *Acta Chem. Scand.*, **26,** 1141, 1972
Also classified in 11

33.22 **Pyridoxol 5′ - methylphosphonate**
$C_9H_{14}NO_5P$
F.E.Cole, B.Lachmann, W.Korytnyk
Amer. Cryst. Assoc., Abstr. Papers (Winter Meeting), 91, 1972
Also classified in 64

33.23 **Hexamethyl - melamine**
2,4,6 - tris(Dimethylamino) - 1,3,5 - triazine
$C_9H_{18}N_6$
G.J.Bullen, D.J.Corney, F.S.Stephens *J. C. S. Perkin ii,* 642, 1972

33.24 **2,2′ - Bipyridinium tetrabromocobaltate(ii)**
$C_{10}H_{10}N_2^{2+}$, Br_4Co^{2-}
S.Koda, S.Ooi, H.Kuroya *Bull. Chem. Soc. Jap.*, **44,** 1597, 1971

33.25 **Bipyridinium pentachloromanganese(iii)**
$C_{10}H_{10}N_2^{2+}$, Cl_5Mn^{2-}
I.Bernal, N.Elliott, R.Lalancette *J. Chem. Soc. (D)*, 803, 1971

33.C **3 - Phenyl - 7 - bromoisoxazolo(4,5 - d)pyridazin - 4(5H) - one**
$C_{11}H_6BrN_3O_2$
For complete entry see 40.16

33.C **Cyclo - glycyl - L - tyrosyl**
L - 3 - (4 - Hydroxybenzyl) - 2,5 - piperazinedione
$C_{11}H_{12}N_2O_3$
For complete entry see 48.45

33.26 **N - (2,6 - Dichlorobenzyl) - 1,2 - dihydropyridin - 2 - one**
$C_{12}H_9Cl_2NO$
G.L.Wheeler, H.L.Ammon
Amer. Cryst. Assoc., Abstr. Papers (Winter Meeting), 29, 1972

33.27 **2,2′ - bis(6 - Methyl - 3 - pyridinol)**
$C_{12}H_{12}N_2O_2$
L.H.Vogt Junior, J.G.Wirth *J. Amer. Chem. Soc.*, **93**, 5402, 1971

33.28 **N,N′ - Dimethyl - 4,4′ - bipyridylium dibromide**
Paraquat dibromide
$C_{12}H_{14}N_2^{2+}$, $2Br^-$
J.H.Russell, S.C.Wallwork *Acta Cryst. (B)*, **28**, 1527, 1972

33.29 **N,N′ - Dimethyl - 4,4′ - bipyridylium dichloride**
Paraquat dichloride
$C_{12}H_{14}N_2^{2+}$, $2Cl^-$
J.H.Russell, S.C.Wallwork *Acta Cryst. (B)*, **28**, 1527, 1972

33.30 **N,N′ - Dimethyl - 4,4′ - bipyridylium di - iodide**
Paraquat di - iodide
$C_{12}H_{14}N_2^{2+}$, $2I^-$
J.H.Russell, S.C.Wallwork *Acta Cryst. (B)*, **28**, 1527, 1972

33.C **Cyclo - L - seryl - L - tyrosyl monohydrate**
L - 3 - (4 - Hydroxybenzyl) - L - 6 - hydroxymethyl - 2,5 - piperazinedione monohydrate
$C_{12}H_{14}N_2O_4$, H_2O
For complete entry see 48.48

33.31 **1 - (3′,4′ - Dichlorobenzyloxy) - 4,6 - diamino - 1,2 - dihydrotriazine hydrochloride**
$C_{12}H_{16}Cl_2N_5O^+$, Cl^-
L.A.Plastas, H.L.Ammon
Amer. Cryst. Assoc., Abstr. Papers (Winter Meeting), 31, 1972

33.C **β - Thalidomide**
N - (β - Glutarimido) - phthalimide
$C_{13}H_{10}N_2O_4$
For complete entry see 35.15

33.C **D - α - Thalidomide**
D - N - (α - Glutarimido) - phthalimide
$C_{13}H_{10}N_2O_4$
For complete entry see 35.16

33.32 **3,6 - Diphenyl - s - tetrazine**
$C_{14}H_{10}N_4$
N.A.Ahmed, A.I.Kitaigorodskij *Acta Cryst. (B),* **28,** 739, 1972

33.33 **1,2 - bis - (2 - Methylpyridinium)ethane tetracyanonickelate(ii) trihydrate**
$C_{14}H_{18}N_2^{2+}$, $C_4N_4Ni^{2-}$, $3H_2O$
L.D.C.Bok, S.S.Basson, J.G.Leipoldt, G.F.S.Wessels
Acta Cryst. (B), **27,** 1233, 1971

33.34 **2 - ((2) - Dimethylaminoethyl - 2 - thienylamino)pyridine hydrochloride**
$C_{14}H_{20}N_3S^+$, Cl^-
G.R.Clark, G.J.Palenik *J. Amer. Chem. Soc.,* **94,** 4005, 1972
Residue 1 also classified in 39, 3

33.35 **α - Promedol alcohol**
1,2a,5e - Trimethyl - 4e - phenylpiperidin - 4a - ol
$C_{14}H_{21}NO$
W.H.De Camp, F.R.Ahmed
Amer. Cryst. Assoc., Abstr. Papers (Winter Meeting), 24, 1972

33.36 **(±) - β - Promedol alcohol (monoclinic form)**
(±) - β - 1,2,5 - Trimethyl - 4 - phenylpiperidin - 4 - ol
$C_{14}H_{21}NO$
W.H.De Camp, F.R.Ahmed *Acta Cryst. (B),* **28,** 1796, 1972

33.37 **(±) - γ - Promedol alcohol**
(±) - γ - 1,2,5 - Trimethyl - 4 - phenylpiperidin - 4 - ol
$C_{14}H_{21}NO$
W.H.De Camp, F.R.Ahmed *Acta Cryst. (B),* **28,** 1791, 1972

33.38 **DL - Bromopheniramine hydrogen maleate**
1 - (p - Bromophenyl) - 1 - (2′ - pyridyl) - 3 - N,N - dimethylpropylamine hydrogen maleate
$C_{16}H_{20}BrN_2^+$, $C_4H_3O_4^-$
M.N.G.James, G.J.B.Williams *J. Medicin. Chem., U. S. A.,* **14,** 670, 1971
Residue 2 classified in 2

33.C **2,5 - Pentamethylene - imino - 1,4 - benzoquinone**
$C_{16}H_{22}N_2O_2$
For complete entry see 18.4

33.39 **N - Benzyl - 1,2 - dihydro - 2 - cyclopentadienylidene - pyridine**
$C_{17}H_{15}N$
G.L.Wheeler, H.L.Ammon
Amer. Cryst. Assoc., Abstr. Papers (Winter Meeting), 29, 1972
Also classified in 20

33.40 **2,6 - bis(Bromomethyl) - 1,4 - diphenyl - piperazine**
$C_{18}H_{20}Br_2N_2$
B.Morosin, J.Howatson *J. C. S. Perkin ii,* 1087, 1972

33.41 **2,5 - Distyryl - pyrazine**
$C_{20}H_{16}N_2$
Y.Sasada, H.Shimanouchi, H.Nakanishi, M.Hasegawa
Bull. Chem. Soc. Jap., **44,** 1262, 1971

33.C **Methixene hydrochloride monohydrate**
9 - (N - Methyl - 3 - piperidylmethyl) - thioxanthene hydrochloride monohydrate
$C_{20}H_{24}NS^+, Cl^-, H_2O$
For complete entry see 39.40

33.42 **1 - Benzyl - 1 - ethyl - 4 - phenylpiperidinium chloride**
$C_{20}H_{26}N^+, Cl^-$
R.P.Duke, R.A.Y.Jones, A.R.Katritzky, J.R.Carruthers, W.Fedeli, F.Mazza, A.Vaciago *J. C. S. Chem. Comm.*, 455, 1972

33.43 **4′ - Fluoro - 4 - (1 - (4 - hydroxy - 4 - (p - fluorophenyl)piperidino))butyrophenone**
$C_{21}H_{23}F_2NO_2$
M.H.J.Koch, G.Germain *Acta Cryst. (B),* **28,** 121, 1972

33.44 **4′ - Fluoro - 4 - (1 - (4 - hydroxy - 4 - (p - fluorophenyl)piperidino)) butyrophenone hydrochloride**
$C_{21}H_{24}F_2NO_2^+, Cl^-$
M.H.J.Koch, G.Germain *Acta Cryst. (B),* **28,** 121, 1972

33.45 **1 - Benzyl - 1 - isopropyl - 4 - phenylpiperidinium chloride**
$C_{21}H_{28}N^+, Cl^-$
R.P.Duke, R.A.Y.Jones, A.R.Katritzky, J.R.Carruthers, W.Fedeli, F.Mazza, A.Vaciago *J. C. S. Chem. Comm.*, 455, 1972

33.46 **N,N′ - Dibenzyl - 4,4′ - bipyridylium di - iodide**
$C_{24}H_{22}N_2^{2+}, 2I^-$
J.H.Russell, S.C.Wallwork *Acta Cryst. (B),* **27,** 2473, 1971

33.47 **Ethyl 2 - benzamido - 5 - benzoyl - 4 - dimethylamino - 6 - thioxonicotinate**
$C_{24}H_{23}N_3O_4S$
R.W.Carney, J.Wojtkunski, B.Fechtig, R.T.Puckett, B.Biffar, G.DeStevens
J. Org. Chem., **36,** 2602, 1971

HETERO-NITROGEN
(7- AND HIGHER-MEMBERED MONOCYCLIC)

34.1 **N - Methyl - thiocapryl - lactam**

$C_9H_{17}NS$
J.L.Flippen, R.D.Gilardi
Amer. Cryst. Assoc., Abstr. Papers (Winter Meeting), 93, 1972

34.2 **N - Methyl - thiolauryl - lactam**

$C_{13}H_{25}NS$
J.L.Flippen, R.D.Gilardi
Amer. Cryst. Assoc., Abstr. Papers (Winter Meeting), 93, 1972

34.3 **11 - Chloro - 8,12b - dihydro - 2,8 - dimethyl - 12b - phenyl - 4H(1,3)oxazino(3,2 - d)(1,4)benzodiazepine - 4,7(6H) - dione**

$C_{20}H_{17}ClN_2O_3$
J.Szmuszkovicz, C.G.Chidester, D.J.Duchamp, F.A.MacKellar, G.Slomp
Tetrahedron Letters, 3665, 1971
Also classified in 40

HETERO-NITROGEN (2 FUSED RINGS)

35.1 **Lumazine sesquihydrate**
Pteridine - 2,4 - dione sesquihydrate
$C_6H_4N_4O_2$, $1.5H_2O$
R.Norrestam, B.Stensland, E.Soderberg *Acta Cryst. (B),* **28,** 659, 1972

35.C **trans - 3,4 - Methylene - L - proline monohydrate**
$C_6H_9NO_2$, H_2O
For complete entry see 48.20

35.C **cis - 3,4 - Methylene - L - proline hydrochloride monohydrate**
$C_6H_{10}NO_2^+$, Cl^-, H_2O
For complete entry see 48.22

35.2 **Spinacine dihydrate**
4,5,6,7 - Tetrahydro - 1H - imidazo(4,5 - c)pyridin - 6 - carboxylic acid dihydrate
$C_7H_9N_3O_2$, $2H_2O$
G.D.Andreetti, L.Cavalca, P.Sgarabotto *Gazz. Chim. Ital.*, **101,** 625, 1971

35.3 **Phthalimide**
$C_8H_5NO_2$
E.Matzat *Acta Cryst. (B),* **28,** 415, 1972
Also classified in 13

35.4 **2,6 - Dimethyl - 4,8 - dichloro - 2H,6H - pyridazino(4,5 - d)pyridazin - 1,5 - dione**
$C_8H_6Cl_2N_4O_2$
C.Sabelli, P.F.Zanazzi *Acta Cryst. (B),* **28,** 1173, 1972

35.5 **1,8 - Naphthyridine**
$C_8H_6N_2$
A.Clearfield, M.J.Sims, P.Singh *Acta Cryst. (B),* **28,** 350, 1972

35.6 **1 - endo - Carboxypyrrolizidine hydrobromide**
$C_8H_{14}NO_2^+$, Br^-
E.Soderberg *Acta Chem. Scand.*, **25,** 615, 1971

35.7 **4 - Azoniaspiro(3.5)nonane perchlorate**
$C_8H_{16}N^+$, $C_{11}O_4^-$
H.M.Zacharis, L.M.Trefonas *J. Amer. Chem. Soc.*, **93,** 2935, 1971

35.C **Quinolinium 2 - dicyanomethylene - 1,1,3,3 - tetracyanopropanedi - ide**
$2C_9H_8N^+$, $C_{10}N_6^{2-}$
For complete entry see 60.20

35.8 **N - Acetyl - 5,6 - dihydrofuro(2,3 - b)pyrid - 2 - one**
$C_9H_9NO_3$
R.H.Good, G.Jones, J.R.Phipps, G.Ferguson, W.C.Marsh
Tetrahedron Letters, 609, 1972
Also classified in 38

35.9 **N,N′ - Dimethyl - benzimidazolium tetracyanoquinodimethane**
$C_9H_{11}N_2^+$, $C_{12}H_4N_4^-$
D.Chasseau, J.Gaultier, C.Hauw *C. R. Acad. Sci., Fr., C*, **274**, 1434, 1972
Residue 2 classified in 7

35.10 **1,4,5,8 - Tetramethoxy - pyridazino(4,5 - d)pyridazine**
$C_{10}H_{12}N_4O_4$
L.Fanfani, P.F.Zanazzi, C.Sabelli *Acta Cryst. (B)*, **28**, 1178, 1972

35.11 **Serotonin picrate monohydrate**
$C_{10}H_{13}N_2O^+$, $C_6H_2N_3O_7^-$, H_2O
U.Thewalt, C.E.Bugg *Acta Cryst. (B)*, **28**, 82, 1972
Residue 2 classified in 6, 15

35.C **Cyclo - L - prolyl - L - leucyl**
$C_{11}H_{18}N_2O_2$
For complete entry see 48.47

35.12 **N - Cyano - N - methyl - trans - decahydroquinolinium fluoroborate**
$C_{11}H_{19}N_2^+$, BF_4^-
S.Abidi, G.Fodor, C.S.Huber, I.Miura, K.Nakanishi
Tetrahedron Letters, 355, 1972

35.13 **N′,N′ - (2 - Chlorobenzylidene) - histamine**
$C_{12}H_{12}ClN_3$
A.C.Villa, A.G.Manfredotti, M.Nardelli, G.Pelizzi
J. Cryst. Mol. Struct., **1**, 123, 1971

35.14 **2 - Amino - 7 - methyl - cyclonona - 3,5,8 - triene carboxylic acid lactam**
$C_{12}H_{13}NO$
L.A.Paquette, M.J.Broadhurst, C.Lee, J.Clardy
J. Amer. Chem. Soc., **94**, 630, 1972

35.C **N - n - Propylquinolinium bis(7,7,8,8 - tetracyanoquinodimethane)**
$C_{12}H_{14}N^+$, $C_{12}H_4N_4^-$, $C_{12}H_4N_4$
For complete entry see 60.25

35.15 **β - Thalidomide**
N - (β - Glutarimido) - phthalimide
$C_{13}H_{10}N_2O_4$
F.M.Lovell
Amer. Cryst. Assoc., Abstr. Papers (Winter Meeting), 30, 1971
Also classified in 33

35.16 **D - α - Thalidomide**
D - N - (α - Glutarimido) - phthalimide
$C_{13}H_{10}N_2O_4$
F.M.Lovell
Amer. Cryst. Assoc., Abstr. Papers (Winter Meeting), 30, 1971
Also classified in 33

35.C **2 - Oxo - indoline - 3 - spiro - 2′ - (3′ - chloro - 5′,5′ - dimethyl - 4′ - oxothiolane - 1′ - oxide)**
$C_{13}H_{12}ClNO_3S$
For complete entry see 39.16

35.17 **1 - (p - Iodobenzenesulfonyl) - 1 - azaspiro(2.5)octane**
$C_{13}H_{16}INO_2S$
H.M.Zacharis, L.M.Trefonas *J. Heterocycl. Chem.*, **7,** 1301, 1970

35.18 **3 - Phenyl - 2,4 - (1H,3H) - quinazoline - dione**
$C_{14}H_{10}N_2O_2$
Y.Kitano, M.Kashiwagi, Y.Kinoshita *Acta Cryst. (B),* **28,** 1223, 1972

35.19 **N - (2,6 - Dichlorobenzyl) - 1,2 - dihydro - isoquinolin - 1 - one**
$C_{16}H_{11}Cl_2NO$
G.L.Wheeler, H.L.Ammon
Amer. Cryst. Assoc., Abstr. Papers (Winter Meeting), 29, 1972

35.20 **Diazepam**
7 - Chloro - 1,3 - dihydro - 1 - methyl - 5 - phenyl - 2H - 1,4 - benzodiazepin - 2 - one
$C_{16}H_{13}ClN_2O$
A.Camerman, N.Camerman *J. Amer. Chem. Soc.*, **94,** 268, 1972

35.21 **1 - Methyl - 3 - (N - methyl - 4′ - chloroanilino) - 5 - chloroindole**
$C_{16}H_{14}Cl_2N_2$
M.Zocchi, G.Tieghi, A.Albinati *Cryst. Struct. Comm.*, **1,** 139, 1972
Also classified in 16

35.22 **2,3 - Di - t - butylquinoxaline (at −160 ° C)**
$C_{16}H_{22}N_2$
G.J.Visser, A.Vos *Acta Cryst. (B),* **27,** 1793, 1971

35.23 **N - Methyl - N - benzoylmethylene - isoindolinium ylide**
$C_{17}H_{17}NO$
N.A.Bailey, S.E.Hull, G.F.Kersting, J.Morrison
J. Chem. Soc. (D), 1429, 1971
Also classified in 12

35.C **5 - Hydroxy - 3 - phenyl - 1 - (3 - methyl - 1 - isoquinolyl)pyrazole**
$C_{19}H_{15}N_3O$
For complete entry see 32.24

35.24 **Indomethacin**
1 - (p - Chlorobenzoyl) - 5 - methoxy - 2 - methylindole - 3 - acetic acid
$C_{19}H_{16}ClNO_4$
T.J.Kistenmacher, R.E.Marsh *J. Amer. Chem. Soc.*, **94,** 1340, 1972

35.25 **Anhydro - 1 - phenyl - 3 - phenylimino - s - triazolo(4,3 - a)pyridinium hydroxide**
$C_{19}H_{16}N_4$
R.J.Grout, T.J.King, M.W.Partridge *J. Chem. Soc. (D),* 898, 1971

35.26 **Dimethyl 2,5 - diphenyl - 1,3a,4,6a - tetra - azapentalene - 3,6 - dicarboxylate**
$C_{20}H_{16}N_4O_4$
M.Brufani, G.Casini, W.Fedeli, F.Mazza, A.Vaciago
Gazz. Chim. Ital., **101,** 322, 1971

35.27 **N - (2,6 - Dichlorophenyl) - 1,2 - dihydro - 1 - cyclopentadienylidene - isoquinoline**
$C_{21}H_{15}Cl_2N$
G.L.Wheeler, H.L.Ammon
Amer. Cryst. Assoc., Abstr. Papers (Winter Meeting), 29, 1972
Also classified in 20

35.28 **Compound II**
$C_{21}H_{19}ClN_2O_4$
C.G.Chidester, D.J.Duchamp, J.Szmuszkovicz
Amer. Cryst. Assoc., Abstr. Papers (Winter Meeting), 32, 1972

35.29 **Compound III**
$C_{21}H_{19}ClN_2O_4$
C.G.Chidester, D.J.Duchamp, J.Szmuszkovicz
Amer. Cryst. Assoc., Abstr. Papers (Winter Meeting), 32, 1972

35.30 **2 - Cyano - 2 - (3 - cyano - 4 - diethylamino - 1H - quinolylidene - 2) - N,N - diethylacetamide**
$C_{21}H_{25}N_5O$
J.A.Lerbscher, J.Trotter *J. Cryst. Mol. Struct.*, **2,** 67, 1972

35.31 **1,1′ - Diethyl - 2,2′ - cyanine bromide**
$C_{22}H_{23}N_2^+$, Br^-
H.Yoshioka, K.Nakatsu *Chem. Phys. Letters,* **11,** 255, 1971

35.32 **1,1′ - Diethyl - 4,4′ - cyanine bromide**
$C_{22}H_{23}N_2^+$, Br^-
H.Yoshioka, K.Nakatsu *Chem. Phys. Letters,* **11,** 255, 1971

35.C **1,3,3 - Trimethyl - 2 - (N - methyl - N - (β - chloroethyl) - p - aminostyryl) - 3H - indole - 7,7,8,8 - tetracyanoquinodimethane complex**
$C_{22}H_{26}ClN_2^+$, $C_{12}H_4N_4^-$, $C_{12}H_4N_4$
For complete entry see 60.41

35.33 **N,N′ - Diethyl - pseudoisocyanin chloride monohydrate**
$C_{23}H_{23}N_2^+$, Cl^-, H_2O
B.Dammeier, W.Hoppe *Acta Cryst. (B),* **27,** 2364, 1971

35.34 **Dimethyl 2,3 - dihydro - 2 - (3 - indolyl) - 1 - methyl - benz(b)azepine - 3,4 - dicarboxylate**
$C_{24}H_{24}N_2O_4$
R.M.Acheson, J.N.Bridson *J. Chem. Soc. (D),* 1225, 1971

35.C **Desalipactamycate tosylate (absolute configuration)**
$C_{25}H_{31}N_3O_8S$
For complete entry see 50.7

35.C **3,5 - Epidithio - 2,5 - diphenyl - 2,4 - pentadienylidene - 3 - aminoquinoline**
$C_{26}H_{18}N_2S_2$
For complete entry see 41.34

35.35 **Dimethyl 3 - cyclohexyl - 7 - phenyl - 3 - azabicyclo(6.4.0)dodeca - 1(8),4,6,9,11 - pentaene - 2 - one 5,6 - dicarboxylate**
$C_{27}H_{27}NO_5$
A.Padwa, P.Sackman, E.Shefter, E.Vega *J. C. S. Chem. Comm.*, 680, 1972

35.36 **3,3′ - Diphenyl - 1,1′ - bi(isoindolylidene)**
$C_{28}H_{18}N_2$
E.Carstensen-Oeser *Chem. Ber.*, **104,** 3108, 1971

35.C **O,O - Dimethylipecoside sesquihydrate**
$C_{28}H_{38}NO_{12}$, $1.5H_2O$
For complete entry see 52.6

35.37 **Compound 8**
$C_{32}H_{41}N_2O_{12}^+$, Br^-
J.Z.Gougoutas, W.Saenger *J. Org. Chem.*, **36,** 3632, 1971
Residue 1 also classified in 21

35.38 **1 - (4 - Bromo - 2,6 - dimethylphenyl) - 2 - (4 - bromo - 2,6 - dimethylphenyl)imino - 1,2 - dihydro - 3,4,5,5,6,7 - hexacarbomethoxy - 5H - 1 - pyridine acetone solvate**
$C_{36}H_{34}Br_2N_2O_{12}$, C_3H_6O
Y.Suzuki, Y.Iitaka *Bull. Chem. Soc. Jap.*, **44,** 56, 1971

35.39 **2,3 - Dihydro - 1,3 - diphenyl - 2 - oxoindol - 3 - yl diphenyl(phenylcarbamoyl)methyl sulfide**

$C_{40}H_{30}N_2O_2S$

Y.Kai, N.Yasuoka, N.Kasai, T.Minami, K.Yamataka, Y.Ohshiro, T.Agawa *J. Chem. Soc. (D),* 1532, 1971

Also classified in 11

HETERO-NITROGEN (MORE THAN 2 FUSED RINGS)

36.C **cis - anti - Uracil photodimer**
$C_8H_8N_4O_4$
For complete entry see 44.17

36.1 **8 - Hydroxyquinoline - N - oxide**
$C_9H_7NO_2$
R.Desiderato, J.C.Terry, G.R.Freeman, H.A.Levy
Acta Cryst. (B), **27,** 2443, 1971

36.C **1,8 - Dinitroso - naphthalene**
$C_{10}H_6N_2O_2$
For complete entry see 24.2

36.C **cis - syn - 6 - Methyluracil photodimer monohydrate**
$C_{10}H_{12}N_4O_4 . H_2O$
For complete entry see 44.21

36.2 **5 - Keto - 1,5 - dihydrobenz(cd)indole**
$C_{11}H_7NO$
M.B.Laing, R.A.Sparks, K.N.Trueblood
Acta Cryst. (B), **28,** 1920, 1972

36.3 **bis(10 - Methyl - isoalloxazine) sesqui(silver(i) formate) hydrate**
$2C_{11}H_8N_4O_2 . 1.5CHO_2^- . 1.5Ag^+ . xH_2O$
G.D.Sproul, C.J.Fritchie Junior
Amer. Cryst. Assoc., Abstr. Papers (Winter Meeting), 33, 1972

36.C **5 - Bromo - N - hydroxy - naphthaloimide**
$C_{12}H_6BrNO_3$
For complete entry see 24.6

36.4 **9,10 - Diazaphenanthrene**
$C_{12}H_8N_2$
H.van der Meer *Acta Cryst. (B),* **28,** 367, 1972

36.5 **Phenazine**
$C_{12}H_8N_2$
A.M.Glazer *Phil. Trans. R. Soc.,* **266,** 593+, 1970

36.C **7,7,8,8 - Tetracyanoquinodimethane - phenazine complex**
$C_{12}H_8N_2 . C_{12}H_4N_4$
For complete entry see 60.23

36.6 **Phenazine - N - oxyphenazine (92 - 8 mixture)**
$C_{12}H_8N_2 . C_{12}H_8N_2O$
A.M.Glazer *Phil. Trans. R. Soc.*, **266,** 593+, 1970

36.7 **Phenazine - N - oxyphenazine (48 - 52 mixture)**
$C_{12}H_8N_2 . C_{12}H_8N_2O$
A.M.Glazer *Phil. Trans. R. Soc.*, **266,** 593+, 1970

36.8 **Phenazine - N - oxyphenazine (19 - 81 mixture)**
$C_{12}H_8N_2 . C_{12}H_8N_2O$
A.M.Glazer *Phil. Trans. R. Soc.*, **266,** 593+, 1970

36.9 **N - oxyphenazine (at 20 ° C)**
$C_{12}H_8N_2O$
A.M.Glazer *Phil. Trans. R. Soc.*, **266,** 593+, 1970
Also classified in 10

36.10 **N - Oxyphenazine (at –90 ° C)**
$C_{12}H_8N_2O$
A.M.Glazer *Phil. Trans. R. Soc.*, **266,** 593+, 1970
Also classified in 10

36.11 **Dipyrido(1,2 - a.2′,1′ - c)pyrazinium dibromide monohydrate**
$C_{12}H_{10}N_2^{2+} . 2Br^- . H_2O$
J.E.Derry, T.A.Hamor *Acta Cryst. (B)*, **28,** 1244, 1972

36.C **Methyl orotate trans - syn - dimer dihydrate**
$C_{12}H_{12}N_4O_8 . 2H_2O$
For complete entry see 44.23

36.C **Lumiflavin - bis(naphthalene - 2,3 - diol) (yellow form)**
$C_{13}H_{12}N_4O_2 . 2C_{10}H_8O_2$
For complete entry see 60.27

36.C **10 - Propylisoalloxazine - bis(naphthalene - 2,3 - diol) complex**
$C_{13}H_{12}N_4O_2 . 2C_{10}H_8O_2$
For complete entry see 60.28

36.C **Lumiflavin bis(naphthalene - 2,3 - diol) trihydrate**
$C_{13}H_{12}N_4O_2 . 2C_{10}H_8O_2 . 3H_2O$
For complete entry see 60.29

36.C **Lumiflavin bromide - hydroquinone complex**
$C_{13}H_{13}N_2O_2^+ . 1.5C_6H_6O_2 . Br^-$
For complete entry see 60.30

36.12 **4,5 - Iminophenanthrene (tetragonal form)**
4H - Benzo(def)carbazole
$C_{14}H_9N$
V.Ern, L.J.Guggenberger, G.J.Sloan *J. Chem. Phys.*, **54,** 5371, 1971

36.13 **3 - Methyl - lumiflavin**
3,7,8,10 - Tetramethyl - isoalloxazine
$C_{14}H_{14}N_4O_2$
R.Norrestam, B.Stensland *Acta Cryst. (B),* **28,** 440, 1972

36.14 **trans - 6,8 - Dibromo - 4a,9 - dimethyl - 1,2,3,4,4a,9a - hexahydrocarbazole**
$C_{14}H_{17}Br_2N$
O.L.Chapman, G.L.Eian, A.Bloom, J.Clardy
J. Amer. Chem. Soc., **93,** 2918, 1971

36.15 **9 - Bromo - 1,3,7,8,10 - pentamethyl - 1,5 - dihydroisoalloxazine**
$C_{15}H_{17}BrN_4O_2$
R.Norrestam, M.von Glehn *Acta Cryst. (B),* **28,** 434, 1972

36.C **Thymine phototrimer monohydrate**
$C_{15}H_{20}N_6O_7 . H_2O$
For complete entry see 44.26

36.16 **1 - Chlorobenz(a)phenazine - 7N - oxide**
$C_{16}H_9ClN_2O$
B.Bovio, S.Locchi *J. Cryst. Mol. Struct.*, **1,** 325, 1971

36.C **Latumcidin selenate (absolute configuration)**
$C_{16}H_{12}NO^+ . HO_3Se^-$
For complete entry see 50.4

36.17 **5 - Acetyl - 9 - bromo - 1,3,7,8 - tetramethyl - 2,5 - dihydroalloxazine**
$C_{16}H_{17}BrN_4O_3$
M.Leijonmarck, P.-E.Werner *Acta Chem. Scand.*, **25,** 2273, 1971

36.18 **10 - Chloro - 3,4 - dihydro - 4,4,7,8 - tetramethyl - 2 - methoxy - pyrimido(5,4 - b)quinoline**
$C_{16}H_{18}ClN_3O$
E.Shefter, E.M.Levine, P.Sackman, T.J.Bardos
Cryst. Struct. Comm., **1,** 283, 1972

36.19 **5 - Ethyl - 3,7,8,10 - tetramethylisoalloxazinium perchlorate**
$C_{16}H_{19}N_4O_2^+ . ClO_4^-$
R.Norrestam, O.Tillberg *Acta Cryst. (B),* **28,** 1704, 1972

36.20 **8 - Chloro - 1 - methyl - 6 - phenyl - 4H - s - triazolo(4,3 - a)(1,4)benzodiazepine hemihydrobromide ethanol solvate**
$C_{17}H_{13}ClN_4 . C_{17}H_{14}ClN_4^+ . Br^- . 2C_2H_6O$
J.B.Hester Junior, D.J.Duchamp, C.G.Chidester
Tetrahedron Letters, 1609, 1971

36.C **1 - Carboxy - 2,3 - dihydro - 8,9 - dimethoxy - 4 - oxo - 5 - methyl - 1H - 3A,5 - diaza - 10β - azonia - acephenenthrene hydroxide inner salt tetrahydrate**

$C_{17}H_{19}N_3O_5 . 4H_2O$

For complete entry see 59.13

36.C **8 - Azaestradiol**

$C_{17}H_{23}NO_2$

For complete entry see 51.1

36.21 **2 - (2 - (6 - Chloro - 2 - methoxy - 9 - acridinylamino)ethylamino)ethanol**

$C_{18}H_{20}ClN_3O_2$

J.P.Glusker, J.A.Minkin, W.Orehowsky Junior

Acta Cryst. (B), **28,** 1, 1972

36.22 **1,2,3,4,5,6 - Hexamethyl - 7 - p - bromophenyl - 7,8,9 - triazatricyclo(4.3.0.0^{2,5})nona - 3,8 - diene**

$C_{18}H_{22}BrN_3$

L.A.Paquette, R.J.Haluska, M.R.Short, L.K.Read, J.Clardy

J. Amer. Chem. Soc., **94,** 529, 1972

36.23 **4a - Allyl - 3,5,7,8,10 - pentamethyl - 4a,5 - dihydroisoalloxazine**

$C_{18}H_{22}N_4O_2$

R.Norrestam *Acta Cryst. (B),* **28,** 1713, 1972

36.C **12 - Keto - 17 - deoxo - 8 - azaestrone methyl ether hydrobromide**

$C_{18}H_{24}NO_2^+ . Br^-$

For complete entry see 51.7

36.24 **5,5 - Diethyl - 3,7,8,10 - tetramethyl - 1,5 - dihydroisoalloxazine trihydrate**

$C_{18}H_{24}N_4O_2 . 3H_2O$

P.-E.Werner, B.Linnros, M.Leijonmarck

Acta Chem. Scand., **25,** 1297, 1971

36.25 **Diethyl 2 - nitrophenazine - 1 - bromomalonate**

$C_{19}H_{16}BrN_3O_6$

B.Bovio, S.Locchi *R. C. Ist. Lombardo Sci. A,* **104,** 924, 1970

36.26 **2 - (3 - (6 - Chloro - 2 - methoxy - 9 - acridinylamino)propylamino)ethanol**

$C_{19}H_{22}ClN_3O_2$

H.L.Carrell *Acta Cryst. (B),* **28,** 1754, 1972

36.27 **2,3,4,4a,9,9a - Hexahydro - 2 - methyl - 9 - cis - phenyl - 1H - indeno(2,1 - c)pyridine hydrobromide**

$C_{19}H_{22}N^+ . Br^-$

F.H.Allen, D.Rogers, P.G.H.Troughton *Acta Cryst. (B),* **27,** 1325, 1971

36.28 **(4aRS,4bSR,13bRS) - 12 - Bromo - 1 - (p - bromophenyl) - 1,4a,4b,5,6,13b - hexahydro - 4H - dipyridazino - (1,6 - a.4,3 - c)quinoline**

$C_{20}H_{18}Br_2N_4$

I.K.Larsen *Acta Cryst. (B),* **28,** 1136, 1972

36.29 **racemic - (4aRS,4bSR,13bRS) - 12 - Bromo - 1 - (p - bromophenyl) - 1,4a,4b,5,6,13b - hexahydro - 4H - dipyridazino - (1,6 - a.4,3 - c)quinoline**
$C_{20}H_{18}Br_2N_4$
I.K.Larsen *Acta Cryst. (B),* **28,** 1136, 1972

36.30 **racemic - (4aRS,4bRS,13bRS) - 12 - Bromo - 1 - (p - bromophenyl) - 1,4a,4b,5,6,13b - hexahydro - 4H - dipyridazino - (1,6 - a.4,3 - c)quinoline**
$C_{20}H_{18}Br_2N_4$
I.K.Larsen *Acta Cryst. (B),* **28,** 1136, 1972

36.C **Compound I**
$C_{21}H_{19}ClN_2O_4$
For complete entry see 40.23

36.31 **Ethidium bromide monohydrate**
2,7 - Diamino - 9 - phenyl - 10 - ethyl - phenanthridinium bromide monohydrate
$C_{21}H_{20}N_3^+$, Br^- , H_2O
E.Subramanian, J.Trotter, C.E.Bugg *J. Cryst. Mol. Struct.,* **1,** 3, 1971

36.32 **2 - Methoxy - 6 - chloro - 9 - (3′ - (ethyl - 2″ - hydroxyethyl)aminopropylamino) acridine**
$C_{21}H_{26}ClN_3O_2$
H.M.Berman, J.P.Glusker *Acta Cryst. (B),* **28,** 590, 1972

36.33 **cis - 5 - Acetyl - 5,5a,6,7,8,10,11,11a - octahydro - 9H - cyclo - oct(b)indol - 9 - one p - bromophenylhydrazone**
$C_{22}H_{24}BrN_3O$
D.J.Duchamp, C.G.Chidester *Acta Cryst. (B),* **28,** 1092, 1972

36.34 **12 - (p - Bromobenzoxycarbonyl) - 8 - benzyl - 1 - hydroxy - 11 - methyl - 6,9,12 - triaza - tricyclo(7.3.0.0^{2,6})dodeca - 7,10 - dione**
$C_{25}H_{26}BrN_3O_5$
S.Cerrini, W.Fedeli, F.Mazza *J. Chem. Soc. (D),* 1607, 1971

36.C **Surugatoxin heptahydrate (absolute configuration)**
$C_{25}H_{26}BrN_5O_{13}$, $7H_2O$
For complete entry see 59.21

36.35 **1 - Methyl - 2′ - (p - tolylsulphonamido) - 2 - (p - tolylsulphonylimino)indoline - 3 - spirocyclopentane**
$C_{27}H_{29}N_3O_4S_2$
I.J.Tickle, C.K.Prout *J. Chem. Soc. (C),* 3401, 1971

36.36 **4,4 - Dichloro - 2a - aza - A - homo - coprostane - 3 - one acetone solvate**
$C_{27}H_{45}Cl_2NO$, $0.5C_3H_6O$
H.Altenburg, D.Mootz, B.Berking *Acta Cryst. (B),* **28,** 567, 1972
Residue 1 also classified in 51

36.37 **bis(10 - Methyl - 9 - acridine)monoaza - monomethinecyanine perchlorate**

$C_{28}H_{22}N_3^+ , ClO_4^-$

J.Preuss, A.Gieren, K.Zechmeister, E.Daltrozzo, W.Hoppe, V.Zanker

Chem. Ber., **105**, 203, 1972

36.C **Isozygosporin A p - bromobenzoate isopropanol solvate (absolute configuration)**

$C_{37}H_{40}BrNO_7 , C_3H_8O$

For complete entry see 50.12

HETERO-NITROGEN
(BRIDGED RING SYSTEMS)

37.C **Bromoform - hexamethylenetetramine complex (at −35 ° C)**
$C_6H_{12}N_4, 2CHBr_3$
For complete entry see 60.1

37.1 **Hexamethylenetetramine - magnesium dichromate hexahydrate**
$2C_6H_{12}N_4, H_{12}MgO_6^{2+}, Cr_2O_7^{2-}$
F.Dahan *C. R. Acad. Sci., Fr., C,* **273,** 805, 1971

37.2 **1,3 - Dinitro - 6 - hydroxymethylene - 3,4 - methano - nortropane**
$C_9H_{13}N_3O_5$
H.Schenk, P.Benci *Acta Cryst. (B),* **28,** 538, 1972

37.3 **1,3 - Dinitro - 6 - hydroxymethylene - N - hydroxy - 3,4 - methano - nortropane**
$C_9H_{13}N_3O_6$
H.Schenk, P.Benci *Acta Cryst. (B),* **28,** 538, 1972

37.4 **1,4 - bis(2′ - Chloroethyl) - 1,4 - diazabicyclo(2.2.1)heptane diperchlorate**
$C_9H_{18}Cl_2N_2^{2+}, 2ClO_4^-$
D.J.Abraham, R.D.Rosenstein, G.R.Pettit
J. Medicin. Chem., U. S. A., **14,** 1141, 1971

37.C **1,4 - Dimethyl - 1,4 - diazabicyclo(2.2.2)octane o - benzoquinonedi - imine(tetracyano) iron(iii)**
$C_{10}H_6FeN_6^{2-}, C_8H_{18}N_2^{2+}$
For complete entry see 83.34

37.5 **1,5 - endo - Methylene - quinolizidinium p - toluene - sulfonate**
$C_{10}H_{18}N^+, C_7H_7O_3S^-$
C.S.Huber *Acta Cryst. (B),* **25,** 1140, 1969
Residue 2 classified in 11

37.6 **1 - Azabicyclo(3.3.3)undecane hydrochloride**
$C_{10}H_{20}N^+, Cl^-$
N.J.Leonard, J.C.Coll, A.H.-J.Wang, R.J.Missavage, I.C.Paul
J. Amer. Chem. Soc., **93,** 4628, 1971

37.7 **2,9 - Diacetyl - 9 - aza - bicyclo(4.2.1)nona - 2 - ene**
$C_{12}H_{17}NO_2$
C.P.Huber
Amer. Cryst. Assoc., Abstr. Papers (Summer Meeting), 32, 1971

37.8 **12 - Methyl - 11,13 - dioxo - 12 - aza - pentacyclo(4.4.3.0^{1,6}.0^{2,10}.0^{5,7}) trideca - 3,8 - diene**
$C_{13}H_{11}NO_2$
K.J.Hwang, J.Donohue, C.Tsai *Acta Cryst. (B),* **28,** 1727, 1972

37.9 **exo - 2 - Methoxy - 3 - aza - 4 - keto - 7,8 - benzobicyclo(4.2.1)nonene**
$C_{13}H_{15}NO_2$
P.H.Watts Junior, H.L.Ammon, P.H.Mazzochi, H.J.Tamburin, W.J.Kopecky
Amer. Cryst. Assoc., Abstr. Papers (Winter Meeting), 30, 1972

37.10 **7 - Methoxy - 5,6 - dimethyl - 5 - aza - benzo(2,3)bicyclo(2.2.2)octa - 2,5 - diene iodide**
$C_{14}H_{18}NO^+$, I^-
C.K.Bradsher, F.H.Day, A.T.McPhail, P.-S.Wong
Tetrahedron Letters, 4205, 1971

37.11 **N - Methylazepine trans - dimer dihydrobromide dihydrate**
$C_{14}H_{20}N_2^{2+}$, $2Br^-$, $2H_2O$
S.Gottlicher, G.Habermehl *Chem. Ber.,* **104,** 524, 1971

37.12 **9 - Benzoyl - 3α - bromo - 9 - azabicyclo(3.3.1)nonan - 2 - one**
$C_{15}H_{16}BrNO_2$
P.D.Cradwick, G.A.Sim *J. Chem. Soc. (B),* 2218, 1971

37.13 **9 - (p - Iodophenyl) - 9 - azatetracyclo(5.3.1.0^{2,6}.0^{8,10})undec - 4 - ene**
$C_{16}H_{16}IN$
J.N.Brown, R.L.R.Towns, L.M.Trefonas
J. Heterocycl. Chem., **7,** 1321, 1970

37.14 **1,3,4,7,8,9,10,11 - Octahydro - 2H,5H - 4a,11.10a,5 - bis(iminomethano) dibenzo(a,e)cyclo - octene - 13,16 - dione**
5,6,7,8 - Tetrahydro - 2 - quinolone photodimer
$C_{18}H_{22}N_2O_2$
J.N.Brown, R.L.R.Towns, L.M.Trefonas
J. Amer. Chem. Soc., **93,** 7012, 1971

37.15 **6 - Bromoquinaldine dimer**
$C_{20}H_{20}Br_2N_2$
I.W.Elliott, J.Bordner *Tetrahedron Letters,* 4481, 1971

37.16 **4,8,9,10,11,15,19,20 - Octahydro - 2,4,6,13,15,17 - hexamethyl - 9,5′10,14, - dimetheneodipyrimido - (4,5 - j′5′,4′ - n)(1,3,8) - tetra - aza - cyclohexadecine - 1,3,7,12,16,18 - (2H,6H,13H,17H) - hexone trihydrate**
$C_{24}H_{28}N_8O_6$, $3H_2O$
J.L.Flippen, R.D.Gilardi, I.L.Karle *Acta Cryst. (B),* **28,** 360, 1972
Residue 1 also classified in 44

HETERO-OXYGEN

38.1 **Tetrahydrofuran - mercury(ii) bromide**
C_4H_8O , Br_2Hg
M.Frey, H.Leligny, M.Ledesert
Bull. Soc. Fr. Mineral. Cristallogr., **94,** 467, 1971

38.C **Urea - syn - 5 - nitro - 2 - furaldehyde oxime complex**
$C_5H_4N_2O_4$. CH_4N_2O
For complete entry see 60.2

38.2 **Tetracyanoethylene oxide**
C_6N_4O
D.A.Matthews, J.Swanson, M.H.Mueller, G.D.Stucky
J. Amer. Chem. Soc., **93,** 5945, 1971
Also classified in 7

38.3 **Tetracyanoethylene oxide (neutron study)**
C_6N_4O
D.A.Matthews, J.Swanson, M.H.Mueller, G.D.Stucky
J. Amer. Chem. Soc., **93,** 5945, 1971
Also classified in 7

38.4 **Tetracyanoethylene oxide (discussion of results of Matthews et al., J.Amer.Chem.Soc.,93,5945,1971)**
C_6N_4O
D.A.Matthews, G.D.Stucky *J. Amer. Chem. Soc.*, **93,** 5954, 1971
Also classified in 7

38.C **1,6.2,3 - Dianhydro - β - D - gulopyranose**
$C_6H_8O_4$
For complete entry see 45.3

38.5 **cis - 1,4,5,8 - Tetraoxa - decalin**
Hexahydro - p - dioxino(2,3 - b) - p - dioxin
$C_6H_{10}O_4$
B.Fuchs, I.Goldberg, U.Shmueli *J. C. S. Perkin ii,* 357, 1972

38.6 **7 - Oxa - bicyclo(2.2.1)hept - 5 - ene - 2,3 - exo - dicarboxylic anhydride**
$C_8H_6O_4$
S.Baggio, A.Barriola, P.K.de Perazzo *J. C. S. Perkin ii,* 934, 1972

38.7 **Dilactophorbic acid chloride (absolute configuration)**
(5R,8R) 8 - Carboxy - 2,6 dioxo - 1,7 - dioxo(4,4) - spirononane
$C_8H_7ClO_5$
E.Rosenqvist *Acta Chem. Scand.*, **25,** 3111, 1971
Also classified in 1

38.8 **Dimethyl 3,4 - furandicarboxylate**
$C_8H_8O_5$
Y.Okada, K.Nakatsu, A.Shimada *Bull. Chem. Soc. Jap.*, **44,** 928, 1971

38.9 **cis - 9,10 - bis(Bromomethyl) - 1,4,5,8 - tetraoxa - decalin**
4a,8a - bis(Bromomethyl)hexahydro - p - dioxino(2,3 - b) - p - dioxin
$C_8H_{12}Br_2O_4$
B.Fuchs, I.Goldberg, U.Shmueli *J. C. S. Perkin ii,* 357, 1972

38.C **N - Acetyl - 5,6 - dihydrofuro(2,3 - b)pyrid - 2 - one**
$C_9H_9NO_3$
For complete entry see 35.8

38.10 **9,9 - Dichloro - trans,trans - bicyclo(6.1.0)nonane - 4,5 - epoxide**
$C_9H_{12}Cl_2O$
J.Deyrup, M.Betkouski, W.Szabo, M.Mathew, G.J.Palenik
J. Amer. Chem. Soc., **94,** 2147, 1972

38.11 **3 - Chloromethyl - 1,5,7 - trimethyl - 2,4,6,8 - tetraoxa - tricyclo(3.3.1$0^{3,7}$)nonane**
$C_9H_{13}ClO_4$
R.P.Dodge *Cryst. Struct. Comm.*, **1,** 173, 1972

38.C **Bicyclo(2.2.2)octene - 2,3 - endo - dicarboxylic acid anhydride**
$C_{10}H_{12}O_3$
For complete entry see 31.4

38.C **7 - syn - 6 - endo - Dihydroxy - bicyclo(2.2.1)heptane - 2 - endo - carboxylic acid α - lactone**
$C_{10}H_{12}O_4$
For complete entry see 31.5

38.12 **1α,5β - Dimethyl - 2 - oxo - 3 - oxabicyclo(3.3.0)octane - 6β - carboxylic acid**
($\pm$) - trans - π - Camphanic acid
$C_{10}H_{14}O_4$
D.W.Hudson, O.S.Mills *J. C. S. Chem. Comm.*, 647, 1972

38.C **racemic - trans - 1a,7b - Dihydro - oxireno(a)naphthalene - 3 - spiro - 2′ - oxiran - 2(3H) - one**
$C_{11}H_8O_3$
For complete entry see 24.3

38.C **3 - Iodo - 5 - oxa - tricyclo(5.3.1^{4,7})undeca - 6,10 - dione - 8 - carboxylic acid**
$C_{11}H_{11}IO_5$
For complete entry see 27.7

38.C **4,5 - Dichloronaphthalic anhydride**
$C_{12}H_4Cl_2O_3$
For complete entry see 24.4

38.13 **2,3,7,8 - Tetrachlorodibenzo - p - dioxin**
$C_{12}H_4Cl_4O_2$
F.P.Boer, F.P.van Remoortere, P.P.North, M.A.Neuman
Acta Cryst. (B), **28**, 1023, 1972

38.C **4,5 - Dinitronaphthalic anhydride**
$C_{12}H_4N_2O_7$
For complete entry see 24.5

38.14 **2,7 - Dichlorodibenzo - p - dioxin**
$C_{12}H_6Cl_2O_2$
F.P.Boer, P.P.North *Acta Cryst. (B)*, **28**, 1613, 1972

38.C **Naphthalic anhydride**
$C_{12}H_6O_3$
For complete entry see 24.7

38.15 **Dibenzofuran**
$C_{12}H_8O$
O.Dideberg, L.Dupont, J.M.Andre *Acta Cryst. (B)*, **28**, 1002, 1972

38.C **Dehydro - L - ascorbic acid dimer**
$C_{12}H_{12}O_{12}$
For complete entry see 45.17

38.16 **1,5,9,13 - Tetraoxacyclohexadecane**
$C_{12}H_{24}O_4$
P.Groth *Acta Chem. Scand.*, **25**, 725, 1971

38.C **Citrinin**
$C_{13}H_{14}O_5$
For complete entry see 59.6

38.17 **3 - (4 - Bromophenyl) - 6,7 - dihydro - 2 - hydroximinobenzofuran - 4(5)H - one (1 - 2 mixture of syn and anti isomers)**
$C_{14}H_{12}BrNO_3$
G.B.Ansell, D.W.Moore, A.T.Nielsen *J. Chem. Soc. (B)*, 2376, 1971
Also classified in 10

38.18 **3 - Benzoyl - 1,7.4,5 - dianhydro - 1 - hydroxymethyl - cyclohexane - 1,3,4,5 - tetrol**
$C_{14}H_{14}O_3$
C.Riche *C. R. Acad. Sci., Fr., C*, **274**, 150, 1972

38.C **Spectinomycin dihydrobromide pentahydrate (absolute configuration)**
$C_{14}H_{28}N_2O_8^{2+}$, $2Br^-$, $5H_2O$
For complete entry see 50.2

38.19 **5,6,7,7 - Tetrachloro - 16,17 - syn - dioxahexacyclo($9.2.2.1.1^{4,8}.0^{4,14}.0^{8,15}$) - heptadeca - 5,12 - diene**
$C_{15}H_{12}Cl_4O_2$
M.A.Battiste, L.A.Kapicak, M.Mathew, G.J.Palenik
J. Chem. Soc. (D), 1536, 1971

38.C **Tetrahydropentalenolactone bromohydrin acetone solvate (absolute configuration)**
$C_{15}H_{21}BrO_5$, $0.5C_3H_6O$
For complete entry see 50.3

38.C **Johnstonol (absolute configuration)**
$C_{15}H_{21}Br_2ClO_3$
For complete entry see 59.8

38.C **4 - Ethyl - 1 - hydroxy - 4,8,8,10,10 - pentamethyl - 7,9 - dioxo - 2,3 - dioxabicyclo(4.4.0)dec - 5 - ene (monoclinic form)**
$C_{15}H_{22}O_5$
For complete entry see 59.9

38.20 **5,7 - Dibromo - 2 - (3′,5′ - dibromo - 2′ - hydroxyphenyl) - 2 - methoxymethoxy - 3(2H) - benzofuranone**
$C_{16}H_{10}Br_4O_5$
J.I.Leenhouts *Rec. Trav. Chim. Pays-Bas*, **90,** 385, 1971

38.C **Latumcidin selenate (absolute configuration)**
$C_{16}H_{12}NO^+$, HO_4Se^-
For complete entry see 50.4

38.21 **2,3 - Dihydroxy - 3 - hydroxymethyl - 2,4 - dimethyl - 5 - methylene - adipic acid (1,3a.6,3)dilactone p - bromobenzoate (absolute configuration)**
$C_{16}H_{14}BrO_6$
C.G.Gordon-Gray, R.B.Wells, N.Hallak, M.B.Hursthouse, S.Neidle, T.P.Toube *Tetrahedron Letters*, 707, 1972

38.C **Brefeldin A**
$C_{16}H_{24}O_4$
For complete entry see 59.10

38.C **Aflatoxin b$_1$ (orthorhombic form)**
$C_{17}H_{12}O_6$
For complete entry see 59.11

38.C **Athrotaxin**
$C_{17}H_{16}O_6$
For complete entry see 59.12

38.22 **3 - (1 - Phenylpropyl) - 4 - hydroxy - coumarin**
Marcoumar
$C_{18}H_{16}O_3$
G.Bravic, J.Gaultier, C.Hauw *C. R. Acad. Sci., Fr., C,* **272,** 1112, 1971

38.C **4 - p - Hydroxyphenyl - 2,2,4 - trimethylchroman - n - heptanol**
Dianin's compound - n - heptanol
$6C_{18}H_{20}O_2$, $C_7H_{16}O$
For complete entry see 61.2

38.23 **3,3′ - Methylene - bis(6 - bromo - 4 - hydroxycoumarin)**
$C_{19}H_{10}Br_2O_6$
N.W.Alcock, E.Hough *Acta Cryst. (B),* **28,** 1957, 1972

38.C **3′,5,5′,6 - Tetramethoxyflavone**
$C_{19}H_{18}O_6$
For complete entry see 59.14

38.C **Viridin**
$C_{20}H_{16}O_6$
For complete entry see 59.15

38.24 **bis(p - Methoxyphenyl) 2,5 - furane - dicarboxylate**
$C_{20}H_{16}O_7$
W.C.Bryson, S.H.Simonsen
Amer. Cryst. Assoc., Abstr. Papers (Winter Meeting), 28, 1972
Also classified in 17

38.C **Carpanone carbon tetrachloride solvate**
$C_{20}H_{18}O_6$, CCl_4
For complete entry see 59.16

38.C **Kromycin**
$C_{20}H_{30}O_5$
For complete entry see 50.5

38.C **1 - Acetyl - 5 - methyl - 4 - (2′ - (5′ - bromo - 6′ - methoxy)naphthyl) - 7 - oxa - bicyclo(3.2.1)octan - 6 - one**
$C_{21}H_{21}BrO_4$
For complete entry see 24.14

38.C **1 - Acetyl - 5 - methyl - 4 - (2′ - (5′ - bromo - 6′ - methoxy)naphthyl) - 7 - oxa - bicyclo(3.2.1)octane**
$C_{21}H_{23}BrO_3$
For complete entry see 24.15

38.C **O(3) - Acetyl - 16α,17 - dibromo - gibberellic acid**
$C_{21}H_{24}Br_2O_7$
For complete entry see 59.17

38.25 **Pyrethrosin 3 - o - chlorophenylisoxazoline derivative**
$C_{24}H_{26}ClNO_6$
E.J.Gabe, S.Neidle, D.Rogers, C.E.Nordman *J. Chem. Soc. (D),* 559, 1971
Also classified in 40, 53

38.C **Sterigmatocystin p - bromobenzoate**
$C_{25}H_{15}BrO_7$
For complete entry see 59.20

38.C **2,3.17,18 - Dibenzo - 1,4,7,10,13,16,19,22,25,28 - decaoxacyclotriaconta - 2,17 - diene - potassium iodide**
Dibenzo - 30 - crown - 10 potassium iodide
$C_{28}H_{40}KO_{10}^+$, I^-
For complete entry see 67.25

38.26 **2,3.17,18 - Dibenzo - 1,4,7,10,13,16,19,22,25,28 - decaoxacyclotriaconta - 2,17 - diene**
Dibenzo - 30 - crown - 10
$C_{28}H_{40}O_{10}$
M.A.Bush, M.R.Truter *J. C. S. Perkin ii,* 345, 1972

38.C **Phebalin**
$C_{30}H_{28}O_6$
For complete entry see 59.22

38.C **Phragmalin iodoacetate**
$C_{31}H_{37}IO_{12}$
For complete entry see 59.25

38.C **15 - Acetyl - 12 - benzoyl - 3,6 - di(iodoacetyl) - 4,17,17 - trimethyl - 3,6,9.11,12,15 - hexahydroxy - 16 - oxa - tetracyclo(12.2.0.1^{7,11}.0^{4,13})heptadeca - 7 - ene - 5 - one (absolute configuration)**
$C_{33}H_{38}I_2O_{12}$
For complete entry see 59.26

38.C **Monensin monohydrate**
$C_{36}H_{62}O_{11}$, H_2O
For complete entry see 50.11

38.C **Isorubratoxin B bis(p - bromophenylhydrazide) methyl ester**
$C_{39}H_{44}Br_2N_4O_{10}$
For complete entry see 59.27

38.C **Di - (p - bromobenzoyl)pederin ethanol solvate (absolute configuration)**
$C_{39}H_{51}Br_2NO_{11}$, C_2H_6O
For complete entry see 59.28

38.C **De - valino - boromycin rubidium salt methanol solvate**
$C_{40}H_{64}BO_{14}^-$, Rb^+, $2CH_4O$
For complete entry see 50.13

38.C **Grisorixin thallium salt monohydrate (absolute configuration)**

$C_{40}H_{67}O_{10}Tl$, H_2O

For complete entry see 50.14

38.C **Antibiotic X - 206 silver salt (absolute configuration)**

$C_{45}H_{77}AgO_{13}$

For complete entry see 50.15

38.C **Potassium dianemycin**

$C_{47}H_{76}KO_{14}$

For complete entry see 50.16

38.C **N - Iodoacetyl - amphotericin B tritetrahydrofurane solvate monohydrate**

$C_{49}H_{74}INO_{18}$, $3C_4H_8O$, H_2O

For complete entry see 50.17

38.C **Stretovaricin C triacetate cyclic p - bromophenyl - boronate methylene chloride solvate**

$C_{52}H_{59}BBrNO_{17}$, CH_2Cl_2

For complete entry see 50.18

HETERO-SULPHUR AND HETERO-SELENIUM

39.1 **1 - Thiacyclobutane - 3 - carboxylic acid 1 - oxide (high melting form)**
$C_4H_6O_3S$
S.Abrahamsson, G.Rehnberg *Acta Chem. Scand.*, **26**, 494, 1972

39.C **1,4 - Dideoxy - 1,4 - dinitro - neo - inositol - bis(tetrahydrothiophene - 1 - oxide) complex**
$2C_4H_8OS$, $C_6H_{10}N_2O_8$
For complete entry see 60.13

39.2 **6a - Selenaselenophthene**
$C_5H_4Se_3$
A.Hordvik, K.Julshamn *Acta Chem. Scand.*, **25**, 2507, 1971

39.3 **1,3,5,7,9 - Pentathio - cyclodecane (monoclinic form)**
$C_5H_{10}S_5$
G.Valle, A.Piazzesi, A.Del Pra *Cryst. Struct. Comm.*, **1**, 289, 1972

39.4 **1,3,5,7,9 - Pentathio - cyclodecane (orthorhombic form)**
$C_5H_{10}S_5$
G.Valle, A.Piazzesi, V.Busetti *Cryst. Struct. Comm.*, **1**, 293, 1972

39.5 **2.2′ - Bi - 1,3 - dithiole**
$C_6H_4S_4$
W.F.Cooper, N.C.Kenny, J.W.Edmonds, A.Nagel, F.Wudl, P.Coppens
J. Chem. Soc. (D), 889, 1971

39.6 **4,5 - Dihydrothiepin 1,1 - dioxide**
$C_6H_8O_2S$
H.L.Ammon, M.R.Smith, E.Kelso *Acta Cryst. (B),* **28,** 246, 1972

39.7 **1,2,3,4 - Tetrathiadecalin**
$C_6H_{10}S_4$
F.Feher, A.Klaeren, K.-H.Linke *Acta Cryst. (B),* **28,** 534, 1972

39.8 **2 - Thiabicyclo(6.1.0)non - 6 - en - 3 - one**
$C_8H_{10}OS$
A.Padwa, A.Battisti *J. Amer. Chem. Soc.*, **94,** 521, 1972

39.9 **DL - 6 - Thioctic acid**
α - DL - Lipoic acid
$C_8H_{14}O_2S_2$
R.M.Stroud, C.H.Carlisle *Acta Cryst. (B)*, **28**, 304, 1972
Also classified in 1

39.C **1,3,5 - Trinitrobenzene - 3 - formylbenzothiophene**
C_9H_6OS , $C_6H_3N_3O_6$
For complete entry see 60.8

39.10 **4 - Phenyl - 1,2 - dithiolium chloride hydrate**
$C_9H_7S_2^+$, Cl^- , $0.79H_2O$
F.Grundtvig, A.Hordvik *Acta Chem. Scand.*, **25**, 1567, 1971

39.11 **trans - 2 - Chloro - 3 - morpholino - 2,4,4 - trimethylthietane - 1,1 - dioxide**
$C_{10}H_{18}ClNO_3S$
G.D.Andreetti, L.Cavalca, P.Sgarabotto *Gazz. Chim. Ital.*, **101**, 440, 1971
Also classified in 40

39.12 **Ethyl 2 - amino - 4,5,6,7 - tetrahydrobenzo(b)thiophene - 3 - carboxylate**
$C_{11}H_{15}NO_2S$
S.H.Simonsen
Amer. Cryst. Assoc., Abstr. Papers (Winter Meeting), 29, 1972

39.13 **2 - Methyl - 4 - phenyl - thiothiophthene**
$C_{12}H_{10}S_3$
A.Hordvik, K.Julshamn *Acta Chem. Scand.*, **25**, 1835, 1971

39.14 **3 - Amino - 2 - methylthio - 5 - phenyl - 6a - thiathiophthene**
$C_{12}H_{11}NS_4$
A.J.Barnett, R.J.S.Beer, B.V.Karaoghlanian, E.C.Llaguno, I.C.Paul
J. C. S. Chem. Comm., 836, 1972

39.15 **7,8,15,16 - Tetrathia - dispiro(5.2.5.2)hexadecane**
$C_{12}H_{20}S_4$
A.Zalkin, H.Ruben, D.H.Templeton
Amer. Cryst. Assoc., Abstr. Papers (Summer Meeting), 73, 1971

39.16 **2 - Oxo - indoline - 3 - spiro - 2′ - (3′ - chloro - 5′,5′ - dimethyl - 4′ - oxothiolane - 1′ - oxide)**
$C_{13}H_{12}ClNO_3S$
J.Bergman, S.Abrahamsson, B.Dahlen *Tetrahedron*, **27**, 6143, 1971
Also classified in 35

39.17 **12 - Chloro - 13 - dioxo - 13 - thia - tricyclo(5.4.3.0)tetradeca - 9 - ene**
$C_{13}H_{19}ClO_2S$
L.A.Paquette, R.E.Wingard Junior, J.C.Philips, G.L.Thompson, L.K.Read, J.Clardy *J. Amer. Chem. Soc.*, **93**, 4508, 1971

39.18 **(1)Benzothieno(2,3 - b)(1)benzothiophene disulfone**
$C_{14}H_8O_4S_2$
I.Goldberg, U.Shmueli *Acta Cryst. (B)*, **27**, 2173, 1971

39.19 **(1)Benzothieno(2,3 - b)(1)benzothiophene**
$C_{14}H_8S_2$
I.Goldberg, U.Shmueli *Acta Cryst. (B)*, **27**, 2164, 1971

39.20 **2,7 - Dimethyl - thianthrene**
$C_{14}H_{12}S_2$
C.H.Wei *Acta Cryst. (B)*, **27**, 1523, 1971

39.21 **Dimethyl 1,3 - dimethyl - thieno(3,4 - d)thiepin dicarboxylate**
$C_{14}H_{14}O_4S_2$
S.C.Jain, H.M.Sobell
Amer. Cryst. Assoc., Abstr. Papers (Summer Meeting), 103, 1971

39.22 **4,4′ - Dicarboxy - 2,2′,5,5′ - tetramethyl - 3,3′ - biselenienyl**
$C_{14}H_{14}O_4Se_2$
B.Aurivillius *Chem. Scr.*, **1**, 25, 1971

39.23 **7 - (5 - t - Butyl - 1,2 - dithiole - 3 - ylidene) - 4,5,6,7 - tetrahydro - 1,2 - benzodithiole - 3 - thione**
$C_{14}H_{16}S_5$
J.Sletten *Acta Chem. Scand.*, **26**, 873, 1972

39.C **2 - ((2) - Dimethylaminoethyl - 2 - thienylamino)pyridine hydrochloride**
$C_{14}H_{20}N_3S^+$, Cl^-
For complete entry see 33.34

39.24 **3,5 - bis(Pivaloylmethylene) - 1,2,4 - trithiolane**
$C_{14}H_{20}O_2S_3$
I.P.Mellor, S.C.Nyburg *Acta Cryst. (B)*, **27**, 1959, 1971

39.25 **cis - 2,4 - Diphenylthietane - trans - 1 - monoxide**
$C_{15}H_{14}OS$
G.L.Hardgrove Junior, J.S.Bratholdt, M.M.Lien
Amer. Cryst. Assoc., Abstr. Papers (Summer Meeting), 104, 1971

39.26 **2 - (5 - Phenyl - 1,2 - dithiole - 3 - ylidene)cyclohexanone**
$C_{15}H_{14}OS_2$
R.Pinel, Y.Mollier, E.C.Llaguno, I.C.Paul *J. Chem. Soc. (D)*, 1352, 1971

39.27 **2,5 - Diphenyl - 1,4 - dithiin 1 - oxide**
$C_{16}H_{12}OS_2$
G.Bandoli, C.Panattoni, D.A.Clemente, E.Tondello, A.Dondoni, A.Mangini *J. Chem. Soc. (B)*, 1407, 1971

39.28 **2 - t - Butyl - 4,5 - (1 - (1,3 - dithiolane - 2 - ylidene)tetramethylene) - 1,6,6a - thiathiophthene**
$C_{16}H_{20}S_5$
J.Sletten *Acta Chem. Scand.*, **25**, 3577, 1971

39.29 **1,3 - Dimethyl - 5,7 - bis(2′ - isopropanol)thieno(3,4 - d)thiepin**
$C_{16}H_{23}O_2S_2$
S.C.Jain, H.M.Sobell
Amer. Cryst. Assoc., Abstr. Papers (Summer Meeting), 103, 1971

39.30 **2,5 - Diphenyl - thiothiophthene**
$C_{17}H_{12}S_3$
A.Hordvik *Acta Chem. Scand.*, **25**, 1583, 1971

39.31 **2,4 - Di(benzoyl - methylene) - 1,4 - dithiacyclobutane**
Acetophenone desaurin
$C_{18}H_{12}O_2S_2$
T.R.Lynch, I.P.Mellor, S.C.Nyburg *Acta Cryst. (B)*, **27**, 1948, 1971

39.32 **2,5 - Diphenyl - 3 - methyl - 6a - thiathiophthene**
$C_{18}H_{13}S_3$
A.Hordvik, O.Sjolset, L.J.Saethre *Acta Chem. Scand.*, **26**, 1297, 1972

39.33 **2,11,20 - Trithia(3.3.3)(1,3,5)cyclophane**
$C_{18}H_{18}S_3$
A.W.Hanson, E.W.Macaulay *Acta Cryst. (B)*, **28**, 1255, 1972

39.C **4 - p - Hydroxyphenyl - 2,2,4 - trimethylthiochroman 2,5,5 - trimethylhex - 3 - yn - 2 - ol**
$6C_{18}H_{20}OS \cdot C_9H_{16}O$
For complete entry see 61.1

39.34 **syn - 2,11 - Dithia - 9,18, - dimethyl(3.3)metacyclophane**
$C_{18}H_{20}S_2$
B.R.Davis, I.Bernal *J. Chem. Soc. (B)*, 2307, 1971

39.35 **9,10 - Dihydro - 4 - (3 - dimethylamino - propylidene) - 4H - benzo(4,5)cyclohepta(1,2 - b)thiophene hydrochloride**
$C_{18}H_{22}NS^+ , Cl^-$
J.M.Bastian, H.P.Weber *Helv. Chim. Acta*, **54**, 293, 1971
Residue 1 also classified in 3

39.36 **2 - (p - Dimethylaminophenyl) - 4 - phenyl - 6,6a - dithiafurophthene**
$C_{19}H_{17}NOS_2$
A.Hordvik, L.J.Saethre *Acta Chem. Scand.*, **26**, 849, 1972
Also classified in 42

39.37 **N,S - trans - N - (p - Bromophenylcarbamoyl)thiamine anhydride**
$C_{19}H_{20}BrN_5O_2S$
H.Nakai, H.Koyama *J. Chem. Soc. (B)*, 1525, 1971
Also classified in 44

39.38 **N,S - cis - N - (p - Bromophenylcarbamoyl)thiamine anhydride**
$C_{19}H_{20}BrN_5O_2S$
H.Nakai, H.Koyama *J. C. S. Perkin ii,* 248, 1972
Also classified in 44

39.39 **2 - (Morpholine - carboxamide) - 4 - methyl - 5 - methylthio - 2H - thiopyran - 3 - carboxylic acid p - bromobenzyl ester**
$C_{20}H_{22}BrNO_4S_2$
R.Kalish, A.E.Smith, E.J.Smutny *Tetrahedron Letters,* 2241, 1971
Also classified in 40

39.40 **Methixene hydrochloride monohydrate**
9 - (N - Methyl - 3 - piperidylmethyl) - thioxanthene hydrochloride monohydrate
$C_{20}H_{24}NS^+, Cl^-, H_2O$
S.S.C.Chu
Amer. Cryst. Assoc., Abstr. Papers (Winter Meeting), 24, 1972
Residue 1 also classified in 33

39.C **9β - 3 - Methoxy - 17 - acetoxy - 7 - thiaestra - 1,3,5(10),8(14) - tetraene**
$C_{20}H_{24}O_3S$
For complete entry see 51.19

39.41 **7,15,17,19 - Tetraethoxy - 2,3,4,5,10,11,12,13 - octathiatricyclo(12.2.2.2^{6,9})eicosa - 6,8,14,16,17,19 - hexaene**
$C_{20}H_{24}O_4S_8$
J.S.Ricci Junior, I.Bernal *J. Chem. Soc. (B),* 1928, 1971

39.42 **2,6 - bis(5 - t - Butyl - 1,2 - dithiole - 3 - ylidene) - cyclohexanethione**
$C_{20}H_{24}S_5$
R.Kristensen, J.Sletten *Acta Chem. Scand.,* **25,** 2366, 1971

39.43 **2 - (p - Bromobenzoyl) - 5 - isopropyl - 7 - methyl - 8H - azuleno(1,8 - bc)thiophene**
$C_{22}H_{19}BrOS$
H.L.Ammon, L.L.Replogle, P.H.Watts Junior, K.Katsumoto, J.M.Stewart
J. Amer. Chem. Soc., **93,** 2196, 1971

39.44 **2,3,4 - Triphenyl - thiothiophthene**
$C_{23}H_{16}S_3$
A.Hordvik *Acta Chem. Scand.,* **25,** 1822, 1971

39.45 **1H,4H - Naphtho(1,8)diselenepine dimer**
$C_{24}H_{20}Se_4$
S.Aleby *Acta Cryst. (B),* **28,** 1509, 1972

39.C **3,5 - Epidithio - 2,5 - diphenyl - 2,4 - pentadienylidene - 3 - aminoquinoline**
$C_{26}H_{18}N_2S_2$
For complete entry see 41.34

39.46 **5 - Phenyl - 5 - (p - bromophenyl) - 2 - (benzoyl - (p - bromophenyl) - methylene) dithiolan - 4 - one**

$C_{29}H_{18}Br_2O_2S_2$

K.Dichmann, D.Bichan, S.C.Nyburg, P.Yates

Tetrahedron Letters, 3649, 1971

39.47 **Tetraphenylthieno(3,4 - c)thiophene**

$C_{30}H_{20}S_2$

M.D.Glick, R.E.Cook *Acta Cryst. (B),* **28,** 1336, 1972

HETERO-(NITROGEN AND OXYGEN)

40.1 **2 - Oxazolidinone**
$C_3H_5NO_2$
J.W.Turley *Acta Cryst. (B),* **28,** 140, 1972

40.2 **7 - Amino - furazano(3,4 - d)pyrimidine**
$C_4H_3N_5O$
E.Shefter, B.E.Evans, E.C.Taylor *J. Amer. Chem. Soc.,* **93,** 7281, 1971

40.3 **Muscimol**
$C_4H_6N_2O_2$
L.Brehm, H.Hjeds, P.Krogsgaard-Larsen
Acta Chem. Scand., **26,** 1298, 1972

40.4 **Morpholinium 7,7,8,8 - tetracyanoquinodimethane**
$2C_4H_{10}NO^+$, $2C_{12}H_4N_4^-$, $C_{12}H_4N_4$
T.Sundaresan, S.C.Wallwork *Acta Cryst. (B),* **28,** 491, 1972
Residue 2 classified in 7

40.5 **4,5 - Dimethyl - 6 - chloro - 1,2,3 - oxathiazin - 2,2 - dioxide**
$C_5H_6ClNO_3S$
D.Kobelt, E.F.Paulus, K.Clauss *Tetrahedron Letters,* 3627, 1971
Also classified in 41

40.C **Triethyl phosphate - benzotrifurazan complex**
$C_6N_6O_3$, $C_6H_{15}O_4P$
For complete entry see 60.15

40.C **Triethyl phosphate - benzotrifurazan complex (at −120 ° C)**
$C_6N_6O_3$, $C_6H_{15}O_4P$
For complete entry see 60.16

40.6 **Potassium 4 - hydroxy - 5,7 - dinitrobenzfurazan monohydrate**
$C_6HN_4O_6^-$, K^+ , H_2O
M.Mathew, G.J.Palenik *Acta Cryst. (B),* **27,** 1388, 1971
Residue 1 also classified in 6, 15

40.7 **5,6 - Dichlorobenzfurazan - 1 - oxide**
$C_6H_2Cl_2N_2O_2$
D.Britton, J.Konnert, J.Hamer, L.M.Trefonas
Acta Cryst. (B), **28,** 1123, 1972

40.8 **4,6 - Dinitrobenzfurazan - 1 - oxide**
$C_6H_2N_4O_6$
C.K.Prout, O.J.R.Hodder, D.Viterbo *Acta Cryst. (B),* **28,** 1523, 1972
Also classified in 15

40.9 **5 - Bromobenzfurazan - 1 - oxide**
$C_6H_3BrN_2O_2$
D.Britton, G.L.Hardgrove, R.Hegstrom, G.V.Nelson
Acta Cryst. (B), **28,** 1121, 1972

40.10 **5 - Iodobenzfurazan - 1 - oxide**
$C_6H_3IN_2O_2$
R.C.Gehrz, D.Britton *Acta Cryst. (B),* **28,** 1126, 1972

40.11 **Morpholine biguanide hydrochloride**
$C_6H_{14}N_5O^+$, Cl^-
R.Handa, N.N.Saha *J. Cryst. Mol. Struct.,* **1,** 235, 1971
Residue 1 also classified in 8

40.12 **5 - Methylbenzfurazan - 1 - oxide**
$C_7H_6N_2O_2$
D.Britton, W.E.Noland *Acta Cryst. (B),* **28,** 1116, 1972

40.13 **2 - Cyano - 5 - bromobenz(f)(1,3)oxazepine**
$C_{10}H_5BrN_2O$
O.Simonsen *Acta Chem. Scand.,* **25,** 2666, 1971

40.14 **cis - 2 - Isopropyl - 3 - (p - nitrophenyl) - oxaziridine**
$C_{10}H_{12}N_2O_3$
J.F.Cannon, J.Daly, J.V.Silverton, D.R.Boyd, D.M.Jerina
J. C. S. Perkin ii, 1137, 1972

40.15 **Decahydropyrazino(2,3 - b)pyrazino - 1,6′.4,2′.5,3′.8,5 - dioxane**
$C_{10}H_{14}N_4O_2$
R.D.Gilardi *Acta Cryst. (B),* **28,** 742, 1972

40.C **trans - 2 - Chloro - 3 - morpholino - 2,4,4 - trimethylthietane - 1,1 - dioxide**
$C_{10}H_{18}ClNO_3S$
For complete entry see 39.11

40.16 **3 - Phenyl - 7 - bromoisoxazolo(4,5 - d)pyridazin - 4(5H) - one**
$C_{11}H_6BrN_3O_2$
B.Bovio, S.Locchi *J. Cryst. Mol. Struct.,* **2,** 89, 1972
Also classified in 33

40.17 **3 - Methyl - 4 - benzylidene - isoxazoline - 5 - one**
$C_{11}H_9NO_2$
J.Meunier-Piret, P.Piret, G.Germain, J.-P.Putzeys, M.van Meerssche
Acta Cryst. (B), **28,** 1308, 1972

40.18 **1,7,10,16 - Tetraoxa - 4,13 - diaza - cyclo - octadiene**
$C_{12}H_{26}N_2O_4$
M.Herceg, R.Weiss *Bull. Soc. Chim. Fr.*, 549, 1972

40.19 **bis - (p - Chlorophenyl)furazan N - oxide**
$C_{14}H_8Cl_2N_2O_2$
A.Battaglia, A.Dondoni, C.Panattoni, G.Bandoli, D.A.Clemente
Tetrahedron Letters, 2907, 1971

40.20 **2 - Phenyl - 7 - bromo - benz(d)(1,3)oxazepine**
$C_{15}H_{10}BrNO$
B.Jensen *Acta Cryst. (B),* **28,** 771, 1972

40.21 **1 - p - Nitrophenyl - 3 - methylperhydro - 2,9 - pyridoxazine (absolute configuration)**
$C_{15}H_{20}N_2O_3$
C.S.Huber *Acta Cryst. (B),* **28,** 37, 1972
Also classified in 15

40.22 **2 - p - Bromophenyl - 3,4 - dimethyl - 5 - phenyl - oxazolidine (absolute configuration)**
$C_{17}H_{18}BrNO$
L.Neelakantan, J.A.Molin-Case *J. Org. Chem.*, **36,** 2261, 1971

40.C **11 - Chloro - 8,12b - dihydro - 2,8 - dimethyl - 12b - phenyl - 4H(1,3)oxazino(3,2 - d)(1,4)benzodiazepine - 4,7(6H) - dione**
$C_{20}H_{17}ClN_2O_3$
For complete entry see 34.3

40.C **2 - (Morpholine - carboxamide) - 4 - methyl - 5 - methylthio - 2H - thiopyran - 3 - carboxylic acid p - bromobenzyl ester**
$C_{20}H_{22}BrNO_4S_2$
For complete entry see 39.39

40.23 **Compound I**
$C_{21}H_{19}ClN_2O_4$
C.G.Chidester, D.J.Duchamp, J.Szmuszkovicz
Amer. Cryst. Assoc., Abstr. Papers (Winter Meeting), 32, 1972
Also classified in 36

40.C **Pyrethrosin 3 - o - chlorophenylisoxazoline derivative**
$C_{24}H_{26}ClNO_6$
For complete entry see 38.25

40.24 **Photoproduct from 3 - O,23 - O,N - triacetylisojervine - 11β - ol nitrite (absolute configuration)**
$C_{33}H_{47}IN_2O_6$
H.Suginome, T.Tsuneno, N.Sato, T.Masamune, H.Shimanouchi, Y.Tsuchida, Y.Sasada *Tetrahedron Letters,* 661, 1972

40.25 **2,4,5,7 - Tetraphenyl - 6 - (4 - bromophenyl) - 1,3 - oxazepine**

$C_{35}H_{24}BrNO$

B.Jensen *Acta Cryst. (B)*, **28,** 774, 1972

40.C **Actinomycin C_1 - deoxyguanosine complex dodecahydrate (form ii)**

$C_{62}H_{86}N_{12}O_{16}$, $2C_{10}H_{13}N_5O_4$, $12H_2O$

For complete entry see 60.43

40.C **Actinomycin C_1 - deoxyguanosine complex dodecahydrate (form ii, further refinement)**

$C_{62}H_{86}N_{12}O_{16}$, $2C_{10}H_{13}N_5O_4$, $12H_2O$

For complete entry see 60.44

HETERO-(NITROGEN AND SULPHUR)

41.1 **5 - Amino - 2 - thiol - 1,3,4 - thiadiazole**
$C_2H_3N_3S_2$
T.C.Downie, W.Harrison, E.S.Raper, M.A.Hepworth
Acta Cryst. (B), **28,** 1584, 1972

41.2 **3,5 - Diamino - 1,2,4 - dithiazolium iodide**
Thiouret hydroiodide
$C_2H_4N_3S_2^+$, I^-
P.F.Rodesiler, E.L.Amma *Acta Cryst. (B),* **27,** 1687, 1971

41.3 **7 - Amino - 1,2,5 - thiadiazolo(3,4 - d)pyrimidine**
$C_4H_3N_5S$
E.Shefter, B.E.Evans, E.C.Taylor *J. Amer. Chem. Soc.,* **93,** 7281, 1971

41.4 **2,5 - Dimethyl - thiadiazole**
$C_4H_6N_2S$
Z.P.Povet'eva, Z.V.Zvonkova *Kristallografija,* **16,** 1032, 1971

41.C **4,5 - Dimethyl - 6 - chloro - 1,2,3 - oxathiazin - 2,2 - dioxide**
$C_5H_6ClNO_3S$
For complete entry see 40.5

41.5 **2,3,4 - Trimethyl - thiazolium bromide**
$C_6H_{10}NS^+$, Br^-
G.Pepe, M.Pierrot, M.Chanon *Tetrahedron Letters,* 2651, 1972

41.6 **5 - Chloro - 2,1 - benzisothiazole**
C_7H_4ClNS
M.Davis, M.F.Mackay, W.A.Denne *J. C. S. Perkin ii,* 565, 1972

41.7 **2 - Mercapto - benzothiazole**
$C_7H_5NS_2$
J.P.Chesick, J.Donohue *Acta Cryst. (B),* **27,** 1441, 1971

41.C **cis - syn - 1 - Thiauracil photodimer**
$C_8H_6N_2O_4S_2$
For complete entry see 44.16

41.8 **2 - Amino - 5 - phenyl - thiazolin - 4 - one (form ii)**
$C_9H_9N_2OS$
J.-P.Mornon, R.Bally *C. R. Acad. Sci., Fr., C,* **272,** 761, 1971

41.9 **N′ - 2 - Thiazolyl - sulphanilamide (form i)**
Sulphathiazole
$C_9H_9N_3O_2S_2$
G.J.Kruger, G.Gafner *Acta Cryst. (B)*, **28**, 272, 1972
Also classified in 16, 11

41.10 **N′ - 2 - Thiazolyl - sulphanilamide (form iii)**
Sulphathiazole
$C_9H_9N_3O_2S_2$
G.J.Kruger, G.Gafner *Acta Cryst. (B)*, **28**, 272, 1972
Also classified in 16, 11

41.11 **7 - Bromo - 3 - carboxy - 2,5 - dimethyl - 1 - oxo - dihydrothiazolo(3,2 - a)pyridinium - 8 - hydroxylate**
$C_{10}H_{10}BrNO_4S$
N.Thorup *Acta Chem. Scand.*, **25**, 1353, 1971

41.12 **1,6 - Dimethyl - 3,4 - trimethylene - 6a - thia - azophthene**
$C_{10}H_{14}N_2S$
A.Hordvik, K.Julshamn *Acta Chem. Scand.*, **26**, 343, 1972

41.C **Luciferin (absolute configuration)**
2 - (6 - Hydroxy - 2 - benzo - thiazolyl) - 2 - thiazoline - 4 - carboxylic acid
$C_{11}H_8N_2O_3S_2$
For complete entry see 59.4

41.C **D - (–) - Luciferin (absolute configuration)**
2 - (6 - Hydroxy - 2 - benzo - thiazolyl) - 2 - thiazoline - 4 - carboxylic acid
$C_{11}H_8N_2O_3S_2$
For complete entry see 59.5

41.13 **5 - Methoxycarbonyl - methylene - 2 - piperidino - delta^2 - thiazolin - 4 - one**
$C_{11}H_{14}N_2O_3S$
A.F.Cameron, N.J.Hair *J. Chem. Soc. (B)*, 1733, 1971

41.14 **Acenaphtho(1,2 - c) - 1,2,5 - thiadiazole**
$C_{12}H_6N_2S$
J.P.Schaefer, S.K.Arora *J. Chem. Soc. (D)*, 1623, 1971

41.15 **4,4′ - Diacetoxy - 5,5′ - dimethyl - 2,2′ - bithiazolyl**
$C_{12}H_{12}N_2O_4S_2$
K.J.Palmer, R.Y.Wong, K.S.Lee *Acta Cryst. (B)*, **27**, 1817, 1971

41.16 **Dimethyl 1,3 - dimethyl - 8 - thia - 2 - azabicyclo(3.2.1)oct - 3 - ene - 4,7 - dicarboxylate**
$C_{12}H_{17}NO_4S$
J.L.Flippen *Acta Cryst. (B)*, **28**, 1519, 1972

41.17 **Thiamine chloride monohydrate**

$C_{12}H_{17}N_4OS^+$, Cl^-, H_2O
J.Pletcher, M.Sax
Amer. Cryst. Assoc., Abstr. Papers (Summer Meeting), 85, 1971
Residue 1 also classified in 44

41.18 **Thiamine pyrophosphate hydrochloride**

$C_{12}H_{19}N_4O_7P_2S^+$, Cl^-
J.Pletcher, M.Sax *J. Amer. Chem. Soc.*, **94,** 3998, 1972
Residue 1 also classified in 44, 46

41.19 **N - Methyl - phenothiazine**

$C_{13}H_{11}NS$
N.I.Wakayama *Bull. Chem. Soc. Jap.*, **44,** 2847, 1971

41.C **7,7,8,8 - Tetracyanoquinodimethane - N - methylphenothiazine complex**

$C_{13}H_{11}NS$, $C_{12}H_4N_4$
For complete entry see 60.24

41.20 **7 - Chloro - 1,3,5 - trimethyl - 5H - pyrimido(5,4 - b)(1,4)benzothiazine - 2,4(1H,3H) - dione**

$C_{13}H_{12}ClN_3O_2S$
J.P.Schaefer, L.L.Reed *J. Amer. Chem. Soc.*, **94,** 908, 1972

41.21 **3 - (2′ - Diethylammoniumethoxy) - 1,2 - benzisothiazole tetrachlorocobaltate(ii)**

$2C_{13}H_{19}N_2OS^+$, Cl_4Co^{2-}
A.C.Bonamartini, M.Nardelli, C.Palmieri
Acta Cryst. (B), **28,** 1207, 1972

41.22 **3 - (2 - Diethylammoniumethoxy) - 1,2 - benzisothiazole tetrachlorocuprate**

$2C_{13}H_{19}N_2OS^+$, Cl_4Cu^{2-}
A.C.Bonamartini, M.Nardelli, C.Palmieri, C.Pelizzi
Acta Cryst. (B), **27,** 1775, 1971
Residue 1 also classified in 3

41.23 **2,5 - Diphenyl - 3,4 - diaza - 1,6,6a - trithiapentalene**

$C_{15}H_{10}N_2S_3$
A.Hordvik, L.Milje *J. C. S. Chem. Comm.*, 182, 1972

41.24 **Methyl 3 - phenyl - benzo(2,3)thiazolium - 2 - carboxylate chloride**

$C_{15}H_{12}NO_2S^+$, Cl^-
D.B.Cosulich, F.M.Lovell
Amer. Cryst. Assoc., Abstr. Papers (Winter Meeting), 26, 1972

41.25 **2 - Phenylamino - 5 - phenylthiazolin - 4 - one**

$C_{15}H_{12}N_2OS$
R.Bally, J.-P.Mornon *C. R. Acad. Sci., Fr., C,* **274,** 609, 1972

41.26 **2,5 - Dianilino - 3,4 - diaza - 1,6,6a - trithiapentalene**

$C_{15}H_{12}N_4S_3$
A.Hordvik, P.Oftedal *J. C. S. Chem. Comm.*, 543, 1972

41.27 **exo - 7,11 - Methano - 6a,7,7a,10a,11,11a - hexahydrobenz(c)indeno(5,6 - e)thiazine - 5(H) - 6 - oxide**

$C_{16}H_{17}NOS$
D.J.Pointer, J.B.Wilford, K.M.Chui *J. C. S. Perkin ii,* 1134, 1972
Also classified in 31

41.28 **3 - Chloro - 10 - (3′ - dimethylaminopropyl) - phenothiazine hydrochloride (high temp. form)**

$C_{17}H_{20}ClN_2S^+$, Cl^-
M.-R.Dorignac-Calas, P.Marsau *C. R. Acad. Sci., Fr., C,* **274,** 1806, 1972
Residue 1 also classified in 3

41.29 **bis(2,3 - Dimethylbenzothiazoline)**

$C_{18}H_{20}N_2S_2$
E.M.Srenger *C. R. Acad. Sci., Fr., C,* **272,** 1208, 1971

41.30 **3 - Methoxy - 10 - (3′ - dimethylaminopropyl) - phenothiazine hydrogen maleate**

$C_{18}H_{23}N_2OS^+$, $C_4H_3O_4^-$
P.Marsau, J.Gauthier *C. R. Acad. Sci., Fr., C,* **274,** 1915, 1972
Residue 1 also classified in 3

41.31 **10 - (2 - Diethylamino - propyl) - phenothiazine hydrochloride**
Isothazine hydrochloride

$C_{19}H_{25}N_2S^+$, Cl^-
P.Marsau, M. R.Calas *Acta Cryst. (B),* **27,** 2058, 1971

41.32 **5 - Benzoylimino - 3 - phenyl - 2 - (4 - bromophenyl) - 2,5 - dihydro - 1,2,4 - thiadiazole**

$C_{21}H_{14}BrN_3OS$
A.Kutoglu, H.Jepsen *Chem. Ber.*, **105,** 125, 1972

41.C **3,3 - Diethylthiacarbocyanine - tetracyanoquinodimethane complex**

$C_{21}H_{21}N_2S_2^+$, $C_{12}H_4N_4^-$, $C_{12}H_4N_4$
For complete entry see 60.40

41.33 **5 - Phenyl - 4 - methyl - 1,3,4 - thiadiazolyl - 2 - (1′ - phenyl - 2′ - benzyl - ethylene - 2′ - thiolate)**

$C_{24}H_{20}N_2S_2$
R.M.Moriarty, R.Mukherjee, J.L.Flippen, J.Karle
J. Chem. Soc. (D), 1436, 1971

41.34 **3,5 - Epidithio - 2,5 - diphenyl - 2,4 - pentadienylidene - 3 - aminoquinoline**

$C_{26}H_{18}N_2S_2$
F.Leung, S.C.Nyburg *Canad. J. Chem.*, **49,** 167, 1971
Also classified in 39, 35

HETERO-MIXED MISCELLANEOUS

42.1 **6a - Selenathiophthene**

$C_5H_4S_2Se$

A.Hordvik, K.Julshamn *Acta Chem. Scand.*, **25**, 1895, 1971

42.C **5,5 - Dimethyl - 2 - chloro - 2 - oxo - 1,3,2 - dioxaphosphorinane (at –40 ° C)**

$C_5H_{10}ClO_3P$

For complete entry see 64.5

42.C **Triethanolamine borate**

$C_6H_{12}BNO_3$

For complete entry see 62.10

42.2 **2 - Hydroxymethyl - 6 - methoxy - 1,4 - oxathian - S - oxide (absolute configuration)**

$C_6H_{12}O_4S$

D.J.Watkins, T.A.Hamor *J. Chem. Soc. (B),* 1692, 1971

42.C **1,1,4,4 - Tetramethyl - 1,4 - diaza - 2,5 - diboracyclohexane**

$C_6H_{16}B_2N_2$

For complete entry see 62.11

42.3 **2,5 - Dithia - 1 - phenyl - 1 - thiophosphorus(v) - cyclopentane**

$C_8H_9PS_3$

J.D.Lee, G.W.Goodacre *Acta Cryst. (B),* **27,** 1841, 1971

Also classified in 11, 64

42.C **Bicyclic phosphorane from 2 - amino - 2 - methylpropan - 1 - ol**

$C_8H_{19}N_2O_2P$

For complete entry see 64.11

42.4 **2 - (p - Methylphenyl) - 2,4 - dioxo - 1,2,3 - oxathiazoline**

$C_9H_9NO_3S$

D.Kobelt, E.F.Paulus, G.Lohaus *Tetrahedron Letters,* 4243, 1971

42.5 **4 - p - Bromophenyl - buta - 1,3 - diene - 1,4 - sultone**

$C_{10}H_7BrO_3S$

W.E.Barnett, M.G.Newton, J.A.McCormack

J. C. S. Chem. Comm., 264, 1972

42.6 **3,9 - Dimethoxy(1,2)benzoxathiolo(2,3 - b)(1,2)benzoxathiole - 6 - S^{iv}**

$C_{15}H_{12}O_4S$
R.D.Gilardi, I.L.Karle *Acta Cryst. (B)*, **27**, 1073, 1971

42.C **trans - Methyl meso - hydrobenzoin phosphite**

$C_{15}H_{15}O_3P$
For complete entry see 64.20

42.C **2 - (p - Dimethylaminophenyl) - 4 - phenyl - 6,6a - dithiafurophthene**

$C_{19}H_{17}NOS_2$
For complete entry see 39.36

42.C **10,10'(5H,5'H) - Spirobiphenophosphazinium chloride**

$C_{24}H_{18}N_2P^+ , Cl^-$
For complete entry see 64.32

BARBITURATES

43.1 **5 - Ethylbarbituric acid (monoclinic form)**
$C_6H_8N_2O_3$
B.M.Gatehouse, B.M.Craven *Acta Cryst. (B)*, **27**, 1337, 1971

43.2 **5 - Hydroxy - 5 - ethylbarbituric acid**
$C_6H_8N_2O_4$
B.M.Gatehouse, B.M.Craven *Acta Cryst. (B)*, **27**, 1337, 1971

43.3 **Sodium 5,5 - diethylbarbiturate**
Sodium barbital
$C_8H_{11}N_2O_3^-$, Na^+
B.Berking, B.M.Craven *Acta Cryst. (B)*, **27**, 1107, 1971

43.4 **Guanidinium 5,5 - diethylbarbiturate dihydrate**
$C_8H_{11}N_2O_3^-$, $CH_6N_3^+$, $2H_2O$
R.J.McClure, B.M.Craven
Amer. Cryst. Assoc., Abstr. Papers (Winter Meeting), 71, 1972
Residue 2 classified in 8

43.5 **Calcium barbital trihydrate**
$2C_8H_{11}N_2O_3^-$, Ca^{2+}, $3H_2O$
B.Berking *Acta Cryst. (B)*, **28**, 98, 1972

43.6 **5,5 - Diethylbarbituric acid (form iv)**
Barbital
$C_8H_{12}N_2O_3$
B.M.Craven, E.A.Vizzini *Acta Cryst. (B)*, **27**, 1917, 1971

43.C **9 - Ethyladenine - barbital complex**
$C_8H_{12}N_2O_3$, $C_7H_9N_5$
For complete entry see 60.18

43.C **Aminopyrine - barbital complex**
$C_8H_{12}N_2O_3$, $C_{13}H_{17}N_3O$
For complete entry see 60.31

43.C **Barbital - caffeine complex**
$2C_8H_{12}N_2O_3$, $C_8H_{10}N_4O_2$
For complete entry see 60.19

43.7 **Sodium 1 - methyl - 5,5 - diethylbarbiturate**
Sodium metharbital
$C_9H_{13}N_2O_3^-$, Na^+
B.Berking *Acta Cryst. (B),* **28,** 1539, 1972

43.8 **5 - Ethyl - 5 - butyl - barbituric acid**
$C_{10}H_{16}N_2O_3$
J.-P.Bideau *C. R. Acad. Sci., Fr., C,* **272,** 757, 1971

43.9 **5 - Ethyl - 5 - (3,5 - dimethyl - n - butyl) - barbituric acid**
γ - Methylamobarbital
$C_{12}H_{20}N_2O_3$
G.L.Gartland, B.M.Craven *Acta Cryst. (B),* **27,** 1909, 1971

PYRIMIDINES AND PURINES

44.1 **5 - Fluoropyrimidine - 2 - one (monoclinic form)**
$C_4H_3FN_2O$
S.Furberg, C.S.Petersen *Acta Chem. Scand.*, **26,** 760, 1972

44.2 **bis(Uracil) mercury(ii) chloride complex**
$2C_4H_4N_2O_2$, Cl_2Hg
J.A.Carrabine, M.Sundaralingam *Biochemistry,* **10,** 292, 1971

44.3 **bis(Dihydrouracil) mercury(ii) chloride complex**
$2C_4H_6N_2O_2$, Cl_2Hg
J.A.Carrabine, M.Sundaralingam *Biochemistry,* **10,** 292, 1971

44.4 **Ammonium orotate monohydrate**
Ammonium uracil - 6 - carboxylate monohydrate
$C_5H_3N_2O_4^-$, H_4N^+, H_2O
J.Solbakk *Acta Chem. Scand.*, **25,** 3006, 1971

44.5 **Potassium hydrogen bis(5 - bromo - 3 - hydroxy - 6 - methyluracil)**
$C_5H_4BrN_2O_3^-$, $C_5H_5BrN_2O_3$, K^+
M.R.Truter, B.L.Vickery *J. Chem. Soc. (A),* 2077, 1971

44.6 **Rubidium hydrogen bis(5 - bromo - 3 - hydroxy - 6 - methyluracil)**
$C_5H_4BrN_2O_3^-$, $C_5H_5BrN_2O_3$, Rb^+
M.R.Truter, B.L.Vickery *J. Chem. Soc. (A),* 2077, 1971

44.7 **3 - Hydroxyxanthine dihydrate**
$C_5H_4N_4O_2$, $2H_2O$
W.E.Thiessen, H.A.Levy, B.D.Flaig
Amer. Cryst. Assoc., Abstr. Papers (Winter Meeting), 32, 1972

44.8 **Guanine monohydrate**
$C_5H_5N_5O$, H_2O
U.Thewalt, C.E.Bugg, R.E.Marsh *Acta Cryst. (B),* **27,** 2358, 1971

44.C **Thymine - p - benzoquinone complex**
$C_5H_6N_2O_2$, $C_6H_4O_2$
For complete entry see 60.7

44.9 **5 - Diazo - 6 - methoxy - 6 - hydrouracil**

$C_5H_6N_4O_3$

D.J.Abraham, T.G.Cochran, R.D.Rosenstein

J. Amer. Chem. Soc., **93,** 6279, 1971

44.10 **Adenine - 1 - oxide sulfate**

$C_5H_7N_5O^{2+}$, O_4S^{2-}

P.Prusiner, M.Sundaralingam

Amer. Cryst. Assoc., Abstr. Papers (Summer Meeting), 88, 1971

44.11 **5,6 - Dihydro - 1 - methyl - 4 - thiouracil**

$C_5H_8N_2OS$

M.J.E.Hewlins *J. C. S. Perkin ii,* 275, 1972

44.12 **1 - Methylcytosine hydrochloride**

$C_5H_8N_3O^+$, Cl^-

B.L.Trus, R.E.Marsh *Acta Cryst. (B),* **28,** 1834, 1972

44.13 **5,6 - Dimethyl - uracil - 1,3 - disulfofluoride**

$C_6H_6F_2N_2O_6S_2$

D.Kobelt, E.F.Paulus, K.Clauss *Tetrahedron Letters,* 3627, 1971

44.14 **1 - Ethyl - 5 - bromouracil (form i)**

$C_6H_7BrN_2O_2$

H.Mizuno, N.Nakanishi, T.Fujiwara, K.Tomita, T.Tsukihara, T.Ashida, M.Kakudo *Biochem. Biophys. Res. Comm.*, **41,** 1161, 1970

44.15 **1 - Ethyl - 5 - bromouracil (form ii)**

$C_6H_7BrN_2O_2$

H.Mizuno, N.Nakanishi, T.Fujiwara, K.Tomita, T.Tsukihara, T.Ashida, M.Kakudo *Biochem. Biophys. Res. Comm.*, **41,** 1161, 1970

44.C **5 - Chlorosalicylic acid - theobromine complex**

$C_7H_8N_5O_2$, $2C_7H_5ClO_3$

For complete entry see 60.17

44.C **9 - Ethyladenine - barbital complex**

$C_7H_9N_5$, $C_8H_{12}N_2O_3$

For complete entry see 60.18

44.16 **cis - syn - 1 - Thiauracil photodimer**

$C_8H_6N_2O_4S_2$

J.B.Bremner, R.N.Warrener, E.Adman, L.H.Jensen

J. Amer. Chem. Soc., **93,** 4574, 1971

Also classified in 41

44.17 **cis - anti - Uracil photodimer**

$C_8H_8N_4O_4$

J.Konnert, I.L.Karle *J. Cryst. Mol. Struct.*, **1,** 107, 1971

Also classified in 36

44.C **Barbital - caffeine complex**
$C_8H_{10}N_4O_2 , 2C_8H_{12}N_2O_3$
For complete entry see 60.19

44.18 **2′ - Deoxy - 5 - diazo - 6 - hydro - 0[6],5′ - cyclouridine hemihydrate**
$C_9H_{10}N_4O_5 , 0.5H_2O$
D.J.Abraham, T.G.Cochran, R.D.Rosenstein
J. Amer. Chem. Soc., **93,** 6279, 1971
Residue 1 also classified in 45

44.C **2′ - Chloro - 2′ - deoxyuridine**
$C_9H_{11}ClN_2O_5$
For complete entry see 47.1

44.C **Disodium uridine - 3′ - phosphate tetrahydrate**
$C_9H_{11}N_2O_9P^{2-} , 2Na^+ , 4H_2O$
For complete entry see 47.2

44.19 **N[6] - (Δ[2] - Isopentenyl) - 2 - methylthioadenine**
$C_9H_{11}N_5S$
H.Y.Lin, R.K.McMullan, M.Sundaralingam
Amer. Cryst. Assoc., Abstr. Papers (Summer Meeting), 90, 1971

44.C **2,4 - Dithiouridine monohydrate**
$C_9H_{12}N_2O_4S_2 , H_2O$
For complete entry see 47.3

44.C **2 - Thio - isouridine**
2 - Thio - (3 - β - D - ribofuranosyl) - uracil
$C_9H_{12}N_2O_5S$
For complete entry see 47.4

44.C **1 - β - D - Arabinofuranosyl - 4 - thiouracil monohydrate**
$C_9H_{12}N_2O_5S , H_2O$
For complete entry see 47.5

44.C **Arabinosyl - thymine**
$C_9H_{13}N_2O_6$
For complete entry see 47.6

44.C **Dihydrouridine hemihydrate**
$C_9H_{14}N_2O_6 , 0.5H_2O$
For complete entry see 47.7

44.C **Dihydrouridine hemihydrate**
$C_9H_{14}N_2O_6 , 0.5H_2O$
For complete entry see 47.8

44.C **Deoxycytidine 5′ - phosphate monohydrate (diffractometer data)**
$C_9H_{14}N_3O_7P , H_2O$
For complete entry see 47.9

44.C **Deoxycytidine 5′ - phosphate monohydrate (photographic data)**
$C_9H_{14}N_3O_7P$, H_2O
For complete entry see 47.10

44.C **5 - Trifluoromethyl - 2′ - deoxyuridine**
$C_{10}H_{11}F_3N_2O_5$
For complete entry see 47.11

44.20 **6 - Histaminopurine dihydrate**
$C_{10}H_{11}N_7$, $2H_2O$
U.Thewalt, C.E.Bugg *Acta Cryst. (B)*, **28,** 1767, 1972
Residue 1 also classified in 32

44.C **Riboflavin - 5′ - bromo - 5′ - deoxyadenosine complex trihydrate**
$C_{10}H_{12}BrN_5O_3$, $C_{17}H_{20}N_4O_6$, $3H_2O$
For complete entry see 60.34

44.21 **cis - syn - 6 - Methyluracil photodimer monohydrate**
$C_{10}H_{12}N_4O_4$, H_2O
J.W.Gibson, I.L.Karle *J. Cryst. Mol. Struct.*, **1,** 115, 1971
Residue 1 also classified in 36

44.C **Adenosine - 3′,5′ - cyclic(5′ - phosphonate) monohydrate**
$C_{10}H_{12}N_5O_5P$, H_2O
For complete entry see 47.12

44.C **Rubidium adenosine - 5′ - diphosphate trihydrate**
$C_{10}H_{12}N_5O_{10}P_2^{3-}$, $3Rb^+$, $3H_2O$
For complete entry see 47.13

44.C **Actinomycin C_1 - deoxyguanosine complex dodecahydrate (form ii)**
$2C_{10}H_{13}N_5O_4$, $C_{62}H_{86}N_{12}O_{16}$, $12H_2O$
For complete entry see 60.43

44.C **Actinomycin C_1 - deoxyguanosine complex dodecahydrate (form ii, further refinement)**
$2C_{10}H_{13}N_5O_4$, $C_{62}H_{86}N_{12}O_{16}$, $12H_2O$
For complete entry see 60.44

44.C **6 - Thioguanosine monohydrate**
$C_{10}H_{13}N_5O_4S$, H_2O
For complete entry see 47.14

44.C **Adenosine - 3′ - phosphonate ethanol solvate**
$C_{10}H_{14}N_5O_6P$, C_2H_6O
For complete entry see 47.15

44.22 **N^6 - (Δ^2 - Isopentenyl) - 2 - methylthioadenine**
$C_{11}H_{15}N_5S$
R.K.McMullan, M.Sundaralingam *J. Amer. Chem. Soc.*, **93,** 7050, 1971

44.C **3′ - Deoxy - 3′ - (dihydroxyphosphinylmethyl)adenosine ethanol solvate**
$C_{11}H_{16}N_5O_6P$, C_2H_6O
For complete entry see 47.16

44.C **5′ - Methylammonium - 5′ - deoxyadenosine iodide monohydrate**
$C_{11}H_{17}N_6O_3^+$, I^- , H_2O
For complete entry see 47.17

44.23 **Methyl orotate trans - syn - dimer dihydrate**
$C_{12}H_{12}N_4O_8$, $2H_2O$
G.I.Birnbaum *Acta Cryst. (B),* **28,** 1248, 1972
Residue 1 also classified in 36

44.C **3′ - O - Acetyl - 4 - thiothymidine**
$C_{12}H_{16}N_2O_5S$
For complete entry see 47.18

44.C **Thiamine chloride monohydrate**
$C_{12}H_{17}N_4OS^+$, Cl^- , H_2O
For complete entry see 41.17

44.C **N[2] - Dimethylguanosine**
$C_{12}H_{17}N_5O_5$
For complete entry see 47.19

44.C **Thiamine pyrophosphate hydrochloride**
$C_{12}H_{19}N_4O_7P_2S^+$, Cl^-
For complete entry see 41.18

44.24 **6 - (2 - (N - p - Bromophenyl - 1 - imino - prop - 2 - en - 3 - ol)) - purine**
$C_{14}H_{10}BrN_5O$
A.R.Kalyanaraman, R.Srinivasan
Indian J. Pure Appl. Phys., **9,** 215, 1971

44.25 **6 - Chloro - 9 - (3,4 - di - O - acetyl - 2 - deoxy - β - D - ribopyranosyl)purine**
$C_{14}H_{15}ClN_4O_5$
D.J.Abraham, R.D.Rosenstein, T.G.Cochran, E.E.Leutzinger,
L.B.Townsend *Tetrahedron Letters,* 2353, 1971
Also classified in 45

44.26 **Thymine phototrimer monohydrate**
$C_{15}H_{20}N_6O_7$, H_2O
J.L.Flippen, I.L.Karle *J. Amer. Chem. Soc.*, **93,** 2762, 1971
Residue 1 also classified in 36

44.C **N,S - trans - N - (p - Bromophenylcarbamoyl)thiamine anhydride**
$C_{19}H_{20}BrN_5O_2S$
For complete entry see 39.37

44.C **N,S - cis - N - (p - Bromophenylcarbamoyl)thiamine anhydride**

$C_{19}H_{20}BrN_5O_2S$

For complete entry see 39.38

44.C **Uridine - 3′,5′ - adenosine phosphate**

$C_{19}H_{23}N_7O_{12}P$

For complete entry see 47.20

44.C **4,8,9,10,11,15,19,20 - Octahydro - 2,4,6,13,15,17 - hexamethyl - 9,5′10,14, - dimetheneodipyrimido - (4,5 - j′5′,4′ - n)(1,3,8) - tetra - aza - cyclohexadecine - 1,3,7,12,16,18 - (2H,6H,13H,17H) - hexone trihydrate**

$C_{24}H_{28}N_8O_6 , 3H_2O$

For complete entry see 37.16

CARBOHYDRATES

45.1 α - **Xylose**
$C_5H_{10}O_5$
A.Hordvik *Acta Chem. Scand.*, **25**, 215, 1971

45.2 **Barium 2 - sulfuryl - L - ascorbate dihydrate**
$2C_6H_7O_9S^-, Ba^{2+}, 2H_2O$
B.W.McClelland, J.R.Einstein
Amer. Cryst. Assoc., Abstr. Papers (Winter Meeting), 25, 1972

45.3 **1,6.2,3 - Dianhydro -** β **- D - gulopyranose**
$C_6H_8O_4$
B.Berking, N.C.Seeman *Acta Cryst. (B),* **27**, 1752, 1971
Also classified in 38

45.4 **Methyl 1,5 - dithio -** α **- D - ribopyranoside hemihydrate**
$C_6H_{12}O_3S_2, 0.5H_2O$
R.L.Girling, G.A.Jeffrey *Carbohyd. Res.*, **18**, 339, 1971

45.5 **Methyl 5 - thio -** α **- D - ribopyranoside**
$C_6H_{12}O_4S$
R.L.Girling
Amer. Cryst. Assoc., Abstr. Papers (Summer Meeting), 86, 1971

45.6 **Methyl 1 - thio -** α **- D - ribopyranoside**
$C_6H_{12}O_4S$
R.L.Girling, G.A.Jeffrey *Carbohyd. Res.*, **18**, 339, 1971

45.7 α - **Rhamnose monohydrate**
6 - Deoxymannose monohydrate
$C_6H_{12}O_5, H_2O$
R.C.G.Killean, J.L.Lawrence, V.C.Sharma
Acta Cryst. (B), **27**, 1707, 1971

45.C α **- D - Glucose - urea complex**
$C_6H_{12}O_6, CH_4N_2O$
For complete entry see 60.14

45.8 **1,3 - O - D - Mannitol borate monohydrate**
1,3 - (Hydroxyborylene) - D - mannitol monohydrate
$C_6H_{13}BO_8, H_2O$
J.C.Wallace
Amer. Cryst. Assoc., Abstr. Papers (Winter Meeting), 21, 1972
Residue 1 also classified in 62

45.9 **Allitol**
$C_6H_{14}O_6$
N.Azarnia, G.A.Jeffrey, M.S.Shen *Acta Cryst. (B)*, **28**, 1007, 1972

45.10 **D - Glucitol (form A)**
$C_6H_{14}O_6$
Y.J.Park, G.A.Jeffrey, W.C.Hamilton *Acta Cryst. (B)*, **27**, 2393, 1971

45.11 **D - Glucitol (form A,neutron study)**
$C_6H_{14}O_6$
Y.J.Park, G.A.Jeffrey, W.C.Hamilton *Acta Cryst. (B)*, **27**, 2393, 1971

45.12 **D - Iditol**
$C_6H_{14}O_6$
N.Azarnia, G.A.Jeffrey, M.S.Shen *Acta Cryst. (B)*, **28**, 1007, 1972

45.C **Tri - aquo - calcium α - galactose bromide**
$(C_6H_{18}CaO_9^{2+})_n, 2nBr^-$
For complete entry see 67.5

45.13 **Methyl 6 - deoxy - 6 - methylsulfinyl - α - D - glucopyranoside (absolute configuration)**
$C_8H_{16}O_6S$
B.Lindberg, P.Kierkegaard *Acta Chem. Scand.*, **25**, 1139, 1971

45.C **2′ - Deoxy - 5 - diazo - 6 - hydro - 0^6,5′ - cyclouridine hemihydrate**
$C_9H_{10}N_4O_5, 0.5H_2O$
For complete entry see 44.18

45.C **2′ - Chloro - 2′ - deoxyuridine**
$C_9H_{11}ClN_2O_5$
For complete entry see 47.1

45.C **Disodium uridine - 3′ - phosphate tetrahydrate**
$C_9H_{11}N_2O_9P^{2-}, 2Na^+, 4H_2O$
For complete entry see 47.2

45.C **2,4 - Dithiouridine monohydrate**
$C_9H_{12}N_2O_4S_2, H_2O$
For complete entry see 47.3

45.C **2 - Thio - isouridine**
2 - Thio - (3 - β - D - ribofuranosyl) - uracil
$C_9H_{12}N_2O_5S$
For complete entry see 47.4

45.C **1 - β - D - Arabinofuranosyl - 4 - thiouracil monohydrate**
$C_9H_{12}N_2O_5S$, H_2O
For complete entry see 47.5

45.C **Arabinosyl - thymine**
$C_9H_{13}N_2O_6$
For complete entry see 47.6

45.C **Dihydrouridine hemihydrate**
$C_9H_{14}N_2O_6$, $0.5H_2O$
For complete entry see 47.7

45.C **Dihydrouridine hemihydrate**
$C_9H_{14}N_2O_6$, $0.5H_2O$
For complete entry see 47.8

45.C **Deoxycytidine 5′ - phosphate monohydrate (diffractometer data)**
$C_9H_{14}N_3O_7P$, H_2O
For complete entry see 47.9

45.C **Deoxycytidine 5′ - phosphate monohydrate (photographic data)**
$C_9H_{14}N_3O_7P$, H_2O
For complete entry see 47.10

45.C **5 - Trifluoromethyl - 2′ - deoxyuridine**
$C_{10}H_{11}F_3N_2O_5$
For complete entry see 47.11

45.C **Riboflavin - 5′ - bromo - 5′ - deoxyadenosine complex trihydrate**
$C_{10}H_{12}BrN_5O_3$, $C_{17}H_{20}N_4O_6$, $3H_2O$
For complete entry see 60.34

45.C **Adenosine - 3′,5′ - cyclic(5′ - phosphonate) monohydrate**
$C_{10}H_{12}N_5O_5P$, H_2O
For complete entry see 47.12

45.C **Rubidium adenosine - 5′ - diphosphate trihydrate**
$C_{10}H_{12}N_5O_{10}P_2^{3-}$, $3Rb^+$, $3H_2O$
For complete entry see 47.13

45.C **Actinomycin C₁ - deoxyguanosine complex dodecahydrate (form ii)**
$2C_{10}H_{13}N_5O_4$, $C_{62}H_{86}N_{12}O_{16}$, $12H_2O$
For complete entry see 60.43

45.C **Actinomycin C_1 - deoxyguanosine complex dodecahydrate (form ii, further refinement)**
$2C_{10}H_{13}N_5O_4$, $C_{62}H_{86}N_{12}O_{16}$, $12H_2O$
For complete entry see 60.44

45.C **6 - Thioguanosine monohydrate**
$C_{10}H_{13}N_5O_4S$, H_2O
For complete entry see 47.14

45.C **Adenosine - 3′ - phosphonate ethanol solvate**
$C_{10}H_{14}N_5O_6P$, C_2H_6O
For complete entry see 47.15

45.14 **Ethyl 2 - S - ethyl - 1,2 - dithio - α - D - mannofuranoside**
$C_{10}H_{20}O_4S_2$
A.Ducruix, C.Pascard-Billy *Acta Cryst. (B),* **28,** 1195, 1972

45.15 **Tri - O - acetyl - β - D - arabinopyranosyl bromide**
$C_{11}H_{15}BrO_7$
J.D.Mokren, P.W.R.Corfield
Amer. Cryst. Assoc., Abstr. Papers (Summer Meeting), 86, 1971

45.C **3′ - Deoxy - 3′ - (dihydroxyphosphinylmethyl)adenosine ethanol solvate**
$C_{11}H_{16}N_5O_6P$, C_2H_6O
For complete entry see 47.16

45.16 **3,6 - (Acetyl - epimino) - 3,6 - dideoxy - 1,2 - O - isopropylidene - β - L - idofuranose**
$C_{11}H_{17}NO_5$
J.S.Brimacombe, J.Iball, J.N.Low *J. C. S. Perkin ii,* 937, 1972

45.C **5′ - Methylammonium - 5′ - deoxyadenosine iodide monohydrate**
$C_{11}H_{17}N_6O_3^+$, I^-, H_2O
For complete entry see 47.17

45.17 **Dehydro - L - ascorbic acid dimer**
$C_{12}H_{12}O_{12}$
J.Hvoslef *Acta Cryst. (B),* **28,** 916, 1972
Also classified in 38

45.C **3′ - O - Acetyl - 4 - thiothymidine**
$C_{12}H_{16}N_2O_5S$
For complete entry see 47.18

45.18 **3,6 - Anhydro - α - D - glucosyl - 1,4.3,6 - dianhydro - β - D - fructoside**
$C_{12}H_{16}O_8$
N.W.Isaacs, C.H.L.Kennard *J. C. S. Perkin ii,* 582, 1972

45.C **N^2 - Dimethylguanosine**
$C_{12}H_{17}N_5O_5$
For complete entry see 47.19

45.19 **Sucrose**
$C_{12}H_{22}O_{11}$
J.C.Hanson, L.C.Sieker, L.H.Jensen
Amer. Cryst. Assoc., Abstr. Papers (Winter Meeting), 26, 1971

45.20 **α - Lactose monohydrate**
$C_{12}H_{22}O_{11}$, H_2O
C.A.Beevers, H.N.Hansen *Acta Cryst. (B),* **27,** 1323, 1971

45.21 **α,α - Trehalose dihydrate**
$C_{12}H_{22}O_{11}$, $2H_2O$
G.M.Brown, D.C.Rohrer, B.Berking, C.A.Beevers, R.O.Gould, R.Simpson
Amer. Cryst. Assoc., Abstr. Papers (Winter Meeting), 94, 1972

45.C **Tetra - aquo - calcium lactose bromide trihydrate**
$(C_{12}H_{34}CaO_{15}^{2+})_n$, $2nBr^-$, $3nH_2O$
For complete entry see 67.13

45.C **6 - Chloro - 9 - (3,4 - di - O - acetyl - 2 - deoxy - β - D - ribopyranosyl)purine**
$C_{14}H_{15}ClN_4O_5$
For complete entry see 44.25

45.22 **3 - O - Acetyl - 1,2.4,5 - di - O - isopropylidene - α - D - glucoseptanose**
$C_{14}H_{22}O_7$
E.T.Pallister, N.C.Stephenson, J.D.Stevens
J. C. S. Chem. Comm., 98, 1972

45.C **Riboflavin - 5′ - bromo - 5′ - deoxyadenosine complex trihydrate**
$C_{17}H_{20}N_4O_6$, $C_{10}H_{12}BrN_5O_3$, $3H_2O$
For complete entry see 60.34

45.23 **N - (p - Bromobenzyl) - nogalonamide (absolute configuration)**
$C_{17}H_{26}BrNO_5$
P.F.Wiley, D.J.Duchamp, V.Hsiung, C.G.Chidester
J. Org. Chem., **36,** 2670, 1971

45.24 **1 - Kestose**
$C_{18}H_{32}O_{16}$
G.A.Jeffrey, Y.J.Park *Acta Cryst. (B),* **28,** 257, 1972

45.25 **Planteose dihydrate**
α - D - Galactopyranosyl - (1,6) - β - D - fructofuranosyl - (2,1) - α - D - glucopyranoside dihydrate
$C_{18}H_{32}O_{16}$, $2H_2O$
D.C.Rohrer *Acta Cryst. (B),* **28,** 425, 1972

45.C **Uridine - 3′,5′ - adenosine phosphate**
$C_{19}H_{23}N_7O_{12}P$
For complete entry see 47.20

45.C **O,O - Dimethylipecoside sesquihydrate**

$C_{28}H_{38}NO_{12}$, $1.5H_2O$

For complete entry see 52.6

45.C **Paeniflorin bromo - ethanolysis product (absolute configuration)**

$C_{33}H_{39}BrO_{15}$

For complete entry see 51.54

PHOSPHATES

46.1 **Methyl diammonium phosphate dihydrate**

$CH_3O_4P^{2-}$, $2H_4N^+$, $2H_2O$

F.Garbassi, L.Giarda, G.Fagherazzi *Acta Cryst. (B),* **28,** 1665, 1972

46.2 **Dipotassium ethylphosphate tetrahydrate**

$C_2H_5O_4P^{2-}$, $2K^+$, $4H_2O$

W.S.McDonald, D.W.J.Cruickshank *Acta Cryst. (B),* **27,** 1315, 1971

46.3 **Disodium DL - glycerol 3 - phosphate hexahydrate**

$C_3H_7O_6P^{2-}$, $2Na^+$, $6H_2O$

R.H.Fenn, G.E.Marshall *J. Chem. Soc. (D),* 984, 1971

46.C **Hexa - aquo - disodium DL - α - glycerophosphate**

$C_3H_{19}Na_2O_{12}P$

For complete entry see 67.3

46.C **O - Phospho - L - threonine**

$C_4H_{10}NO_6P$

For complete entry see 48.12

46.4 **O - (- L - α - Glyceryl - phosphoryl) - ethanolamine monohydrate (absolute configuration)**

$C_5H_{14}NO_6P$, H_2O

G.T.DeTitta, B.M.Craven *Nature New Biology,* **233,** 118, 1971

Residue 1 also classified in 3

46.5 **Catechol cyclic phosphate**

$C_6H_5O_4P$

F.P.Boer *Acta Cryst. (B),* **28,** 1201, 1972

46.6 **Dodeca - sodium myo - inositol hexaphosphate octatriacontahydrate**

Sodium phytate octatriacontahydrate

$C_6H_6O_{24}P_6^{12-}$, $12Na^+$, $38H_2O$

G.E.Blank, J.Pletcher, M.Sax

Biochem. Biophys. Res. Comm., **44,** 319, 1971

Residue 1 also classified in 21

46.C **Triethyl phosphate - benzotrifurazan complex**

$C_6H_{15}O_4P$, $C_6N_6O_3$

For complete entry see 60.15

46.C **Triethyl phosphate - benzotrifurazan complex (at −120 ° C)**
$C_6H_{15}O_4P$, $C_6N_6O_3$
For complete entry see 60.16

46.C **Disodium uridine - 3′ - phosphate tetrahydrate**
$C_9H_{11}N_2O_9P^{2-}$, $2Na^+$, $4H_2O$
For complete entry see 47.2

46.C **Deoxycytidine 5′ - phosphate monohydrate (diffractometer data)**
$C_9H_{14}N_3O_7P$, H_2O
For complete entry see 47.9

46.C **Deoxycytidine 5′ - phosphate monohydrate (photographic data)**
$C_9H_{14}N_3O_7P$, H_2O
For complete entry see 47.10

46.C **Rubidium adenosine - 5′ - diphosphate trihydrate**
$C_{10}H_{12}N_5O_{10}P_2^{3-}$, $3Rb^+$, $3H_2O$
For complete entry see 47.13

46.C **4 - Carbethoxyanilinium bis - (p - nitrophenyl)phosphate**
Benzocaine bis(p - nitrophenyl)phosphate
$C_{12}H_8N_2O_8P^-$, $C_9H_{12}NO_2^+$
For complete entry see 13.13

46.C **N,N′ - bis(4 - Ethoxyphenyl)acetamidinium bis - p - nitrophenylphosphate monohydrate**
Phenacaine bis - p - nitrophenylphosphate monohydrate
$C_{12}H_8N_2O_8P^-$, $C_{18}H_{23}N_2O_2^+$, H_2O
For complete entry see 16.17

46.C **Thiamine pyrophosphate hydrochloride**
$C_{12}H_{19}N_4O_7P_2S^+$, Cl^-
For complete entry see 41.18

46.7 **Triethylammonium tris(o - phenylenedioxy) phosphate**
$C_{18}H_{12}O_6P^-$, $C_6H_{16}N^+$
H.R.Allcock, E.C.Bissell *J. C. S. Chem. Comm.*, 676, 1972
Residue 2 classified in 3

46.C **Uridine - 3′,5′ - adenosine phosphate**
$C_{19}H_{23}N_7O_{12}P$
For complete entry see 47.20

NUCLEOSIDES AND NUCLEOTIDES

47.1 **2′ - Chloro - 2′ - deoxyuridine**

$C_9H_{11}ClN_2O_5$

D.Suck, W.Saenger, J.Hobbs *Biochim. Biophys. Acta,* **259,** 157, 1972

Also classified in 44, 45

47.2 **Disodium uridine - 3′ - phosphate tetrahydrate**

$C_9H_{11}N_2O_9P^{2-}$, $2Na^+$, $4H_2O$

M.A.Viswamitra, B.S.Reddy, M.N.G.James, G.J.B.Williams

Acta Cryst. (B), **28,** 1108, 1972

Residue 1 also classified in 44, 45, 46

47.3 **2,4 - Dithiouridine monohydrate**

$C_9H_{12}N_2O_4S_2$, H_2O

W.Saenger, D.Suck *Acta Cryst. (B),* **27,** 1178, 1971

Residue 1 also classified in 44, 45

47.4 **2 - Thio - isouridine**

2 - Thio - (3 - β - D - ribofuranosyl) - uracil

$C_9H_{12}N_2O_5S$

L.M.Jacob, B.C.Pal, J.R.Einstein

Amer. Cryst. Assoc., Abstr. Papers (Winter Meeting), 69, 1972

Also classified in 44, 45

47.5 **1 - β - D - Arabinofuranosyl - 4 - thiouracil monohydrate**

$C_9H_{12}N_2O_5S$, H_2O

W.Saenger *J. Amer. Chem. Soc.,* **94,** 621, 1972

Residue 1 also classified in 44, 45

47.6 **Arabinosyl - thymine**

$C_9H_{13}N_2O_6$

P.Tougard *C. R. Acad. Sci., Fr., C,* **273,** 878, 1971

Also classified in 44, 45

47.7 **Dihydrouridine hemihydrate**

$C_9H_{14}N_2O_6$, $0.5H_2O$

D.Suck, W.Saenger, K.Zechmeister *Acta Cryst. (B),* **28,** 596, 1972

Residue 1 also classified in 44, 45

47.8 **Dihydrouridine hemihydrate**

$C_9H_{14}N_2O_6$, $0.5H_2O$

M.Sundaralingam, S.T.Rao, J.Abola *Science,* **172,** 725, 1971

Residue 1 also classified in 44, 45

47.9 **Deoxycytidine 5′ - phosphate monohydrate (diffractometer data)**

$C_9H_{14}N_3O_7P$, H_2O

M.A.Viswamitra, B.Swaminatha Reddy, G.H.-Y.Lin, M.Sundaralingam

J. Amer. Chem. Soc., **93,** 4565, 1971

Residue 1 also classified in 44, 45, 46

47.10 **Deoxycytidine 5′ - phosphate monohydrate (photographic data)**

$C_9H_{14}N_3O_7P$, H_2O

M.A.Viswamitra, B.Swaminatha Reddy, G.H.-Y.Lin, M.Sundaralingam

J. Amer. Chem. Soc., **93,** 4565, 1971

Residue 1 also classified in 44, 45, 46

47.11 **5 - Trifluoromethyl - 2′ - deoxyuridine**

$C_{10}H_{11}F_3N_2O_5$

A.H.Tench

Amer. Cryst. Assoc., Abstr. Papers (Winter Meeting), 70, 1972

Also classified in 44, 45

47.C **Riboflavin - 5′ - bromo - 5′ - deoxyadenosine complex trihydrate**

$C_{10}H_{12}BrN_5O_3$, $C_{17}H_{20}N_4O_6$, $3H_2O$

For complete entry see 60.34

47.12 **Adenosine - 3′,5′ - cyclic(5′ - phosphonate) monohydrate**

$C_{10}H_{12}N_5O_5P$, H_2O

J.Abola, S.Hecht, M.Sundaralingam

Amer. Cryst. Assoc., Abstr. Papers (Summer Meeting), 89, 1971

Residue 1 also classified in 44, 45, 64

47.13 **Rubidium adenosine - 5′ - diphosphate trihydrate**

$C_{10}H_{12}N_5O_{10}P_2^{3-}$, $3Rb^+$, $3H_2O$

F.A.Muller, A.B.DeLuke

Amer. Cryst. Assoc., Abstr. Papers (Winter Meeting), 26, 1971

Residue 1 also classified in 44, 45, 46

47.C **Actinomycin C$_1$ - deoxyguanosine complex dodecahydrate (form ii)**

$2C_{10}H_{13}N_5O_4$, $C_{62}H_{86}N_{12}O_{16}$, $12H_2O$

For complete entry see 60.43

47.C **Actinomycin C$_1$ - deoxyguanosine complex dodecahydrate (form ii, further refinement)**

$2C_{10}H_{13}N_5O_4$, $C_{62}H_{86}N_{12}O_{16}$, $12H_2O$

For complete entry see 60.44

47.14 **6 - Thioguanosine monohydrate**
$C_{10}H_{13}N_5O_4S$, H_2O
C.E.Bugg, U.Thewalt
Amer. Cryst. Assoc., Abstr. Papers (Winter Meeting), 69, 1972
Residue 1 also classified in 44, 45

47.15 **Adenosine - 3′ - phosphonate ethanol solvate**
$C_{10}H_{14}N_5O_6P$, C_2H_6O
J.Abola, S.Hecht, M.Sundaralingam
Amer. Cryst. Assoc., Abstr. Papers (Summer Meeting), 89, 1971
Residue 1 also classified in 44, 45, 64

47.16 **3′ - Deoxy - 3′ - (dihydroxyphosphinylmethyl)adenosine ethanol solvate**
$C_{11}H_{16}N_5O_6P$, C_2H_6O
S.M.Hecht, M.Sundaralingam *J. Amer. Chem. Soc.*, **94,** 4314, 1972
Residue 1 also classified in 44, 45, 64

47.17 **5′ - Methylammonium - 5′ - deoxyadenosine iodide monohydrate**
$C_{11}H_{17}N_6O_3^+$, I^- , H_2O
W.Saenger *J. Amer. Chem. Soc.*, **93,** 3035, 1971
Residue 1 also classified in 44, 45

47.18 **3′ - O - Acetyl - 4 - thiothymidine**
$C_{12}H_{16}N_2O_5S$
W.Saenger, D.Suck *Acta Cryst. (B),* **27,** 2105, 1971
Also classified in 44, 45

47.19 **N[2] - Dimethylguanosine**
$C_{12}H_{17}N_5O_5$
T.F.Brennan, C.Weeks, E.Shefter, S.T.Rao, M.Sundaralingam
Amer. Cryst. Assoc., Abstr. Papers (Summer Meeting), 89, 1971
Also classified in 44, 45

47.20 **Uridine - 3′,5′ - adenosine phosphate**
$C_{19}H_{23}N_7O_{12}P$
J.L.Sussman, N.C.Seeman, H.M.Berman, S.H.Kim
Amer. Cryst. Assoc., Abstr. Papers (Summer Meeting), 87, 1971
Also classified in 44, 45, 46

AMINO-ACIDS AND PEPTIDES

48.1 **Glycine (α form,neutron study)**
$C_2H_5NO_2$
P.-G.Jonsson, A.Kvick *Acta Cryst. (B),* **28,** 1827, 1972

48.2 **Triglycine sulfate (ferroelectric form,neutron study)**
Glycine diglycinium sulfate
$C_2H_5NO_2, 2C_2H_6NO_2^+, O_4S^{2-}$
M.I.Kay, R.Kleinberg
Amer. Cryst. Assoc., Abstr. Papers (Winter Meeting), 77, 1972

48.3 **Triglycine sulfate (ferroelectric form, neutron study)**
Glycine diglycinium sulfate
$C_2H_5NO_2, 2C_2H_6NO_2^+, O_4S^{2-}$
V.M.Padmanabhan, V.S.Yadav *Curr. Sci.,* **40,** 60, 1971

48.4 **Diglycine sodium iodide monohydrate**
$2C_2H_5NO_2, Na^+, I^-, H_2O$
J.Verbist, J.-P.Putzeys, P.Piret, M.van Meerssche
Acta Cryst. (B), **27,** 1190, 1971

48.5 **L - Alanine (neutron study)**
$C_3H_7NO_2$
M.S.Lehmann, T.F.Koetzle, W.C.Hamilton
J. Amer. Chem. Soc., **94,** 2657, 1972

48.6 **L - Cysteine (orthorhombic form)**
$C_3H_7NO_2S$
K.A.Kerr
Amer. Cryst. Assoc., Abstr. Papers (Winter Meeting), 67, 1972

48.7 **L - (–) - Serine**
$C_3H_7NO_3$
E.Benedetti, C.Pedone, A.Sirigu *Cryst. Struct. Comm.,* **1,** 35, 1972

48.8 **L - Serine monohydrate (neutron study)**
$C_3H_7NO_3, H_2O$
M.N.Frey, M.S.Lehmann, T.F.Koetzle, W.C.Hamilton
Amer. Cryst. Assoc., Abstr. Papers (Winter Meeting), 85, 1972

48.9 **Glycylglycine - lithium bromide**
$C_4H_8N_2O_3$, Li^+, Br^-
R.Meulemans, P.Piret, M.van Meerssche
Bull. Soc. Chim. Belges, **80,** 73, 1971

48.10 **L - Asparagine monohydrate (neutron study)**
$C_4H_8N_2O_3$, H_2O
W.C.Hamilton, J.F.Verbist, T.F.Koetzle, M.S.Lehmann
Amer. Cryst. Assoc., Abstr. Papers (Winter Meeting), 86, 1972

48.11 **Glycylglycine - calcium chloride**
$2C_4H_8N_2O_3$, Ca^{2+}, $2Cl^-$
R.Meulemans, P.Piret, M.van Meerssche
Bull. Soc. Chim. Belges, **80,** 73, 1971

48.12 **O - Phospho - L - threonine**
$C_4H_{10}NO_6P$
C.J.Slone, F.E.Cole
Amer. Cryst. Assoc., Abstr. Papers (Summer Meeting), 56, 1971
Also classified in 46

48.13 **L - α,γ - Diaminobutyric acid monohydrochloride**
$C_4H_{11}N_2O_2^+$, Cl^-
P.S.Naganathan, K.Venkatesan *Acta Cryst. (B),* **27,** 2159, 1971
Residue 1 also classified in 1, 3

48.14 **L - α,γ - Diaminobutyric acid monohydrochloride**
$C_4H_{11}N_2O_2^+$, Cl^-
H.Hinazumi, T.Mitsui *Acta Cryst. (B),* **27,** 2152, 1971
Residue 1 also classified in 1, 3

48.15 **Pyroglutamic acid**
5 - Oxo - proline
$C_5H_7NO_3$
V.Pattabhi, K.Venkatesan *Cryst. Struct. Comm.,* **1,** 87, 1972

48.16 **L - Glutamine hydrochloride**
$C_5H_{11}N_2O_3^+$, Cl^-
N.Shamala, K.Venkatesan *Cryst. Struct. Comm.,* **1,** 227, 1972

48.17 **Glycyl - L - alanine hydrochloride**
$C_5H_{11}N_2O_3^+$, Cl^-
P.S.Naganathan, K.Venkatesan *Acta Cryst. (B),* **28,** 552, 1972

48.18 **L - Penicillamine hydrochloride monohydrate**
$C_5H_{12}NO_2S^+$, Cl^-, H_2O
S.N.Rao, R.Parthasarathy, F.E.Cole
Amer. Cryst. Assoc., Abstr. Papers (Summer Meeting), 71, 1971

48.19 **DL - Ornithine hydrobromide**
$C_5H_{13}N_2O_2^+$, Br^-
A.R.Kalyanaraman, R.Srinivasan *Acta Cryst. (B),* **27,** 1420, 1971

48.20 **trans - 3,4 - Methylene - L - proline monohydrate**
$C_6H_9NO_2$, H_2O
Y.Fujimoto, F.Irreverre, J.Karle, I.L.Karle, B.Witkop
J. Amer. Chem. Soc., **93,** 3471, 1971
Residue 1 also classified in 35

48.21 **β - (3 - Pyrazolyl) - L - alanine**
$C_6H_9N_3O_2$
N.C.Seeman, E.L.McGandy, R.D.Rosenstein
J. Amer. Chem. Soc., **94,** 1717, 1972
Also classified in 32

48.22 **cis - 3,4 - Methylene - L - proline hydrochloride monohydrate**
$C_6H_{10}NO_2^+$, Cl^-, H_2O
Y.Fujimoto, F.Irreverre, J.Karle, I.L.Karle, B.Witkop
J. Amer. Chem. Soc., **93,** 3471, 1971
Residue 1 also classified in 35

48.23 **N - Methyl pyro - L - glutamic acid amide**
5 - (N - Methylcarboxamido) - pyrrolid - 2 - one
$C_6H_{10}N_2O_2$
A.Aubry, M.Marraud, J.Protas, J.Neel
C. R. Acad. Sci., Fr., C, **274,** 1378, 1972
Also classified in 32

48.24 **L - Histidine hydrochloride monohydrate**
$C_6H_{10}N_3O_2^+$, Cl^-, H_2O
K.Oda, H.Koyama *Acta Cryst. (B),* **28,** 639, 1972
Residue 1 also classified in 32

48.25 **L - Histidinium dihydrogen orthophosphate orthophosphoric acid**
$C_6H_{10}N_3O_2^+$, $H_2O_4P^-$, H_3O_4P
R.H.Blessing, E.L.McGandy
Amer. Cryst. Assoc., Abstr. Papers (Winter Meeting), 68, 1972
Residue 1 also classified in 32

48.26 **1 - Aminocyclopentane carboxylic acid monohydrate**
$C_6H_{11}NO_2$, H_2O
M.Mallikarjunan, K.K.Chacko, R.Zand
J. Cryst. Mol. Struct., **2,** 53, 1972
Residue 1 also classified in 20

48.27 **Glycylglycylglycine lithium bromide**
$C_6H_{11}N_3O_4$, Li^+, Br^-
R.Meulemans, P.Piret, M.van Meerssche
Acta Cryst. (B), **27,** 1187, 1971

48.C **2(S) - Carboxy - 4(R),5(S) - dihydroxypiperidine hydrochloride (absolute configuration)**
$C_6H_{12}NO_4^+$, Cl^-
For complete entry see 59.2

48.28 **L - Isoleucine**
$C_6H_{13}NO_2$
K.Torii, Y.Iitaka *Acta Cryst. (B),* **27,** 2237, 1971

48.29 **Di - L - leucine hydrochloride**
$C_6H_{13}NO_2$, $C_6H_{14}NO_2^+$, Cl^-
L.Golic, W.C.Hamilton *Acta Cryst. (B),* **28,** 1265, 1972

48.30 **DL - Lanthionine dihydrochloride**
$C_6H_{14}N_2O_4S^{2+}$, $2Cl^-$
R.E.Rosenfield Junior, R.Parthasarathy
Amer. Cryst. Assoc., Abstr. Papers (Summer Meeting), 55, 1971

48.31 **L - Cystine dihydrobromide dihydrate**
$C_6H_{14}N_2O_4S_2^{2+}$, $2Br^-$, $2H_2O$
R.E.Rosenfield Junior, R.Parthasarathy
Amer. Cryst. Assoc., Abstr. Papers (Summer Meeting), 55, 1971

48.32 **L - Citrulline hydrochloride**
$C_6H_{14}N_3O_3^+$, Cl^-
P.S.Naganathan, K.Venkatesan *Acta Cryst. (B),* **27,** 1079, 1971

48.33 **L - Citrulline hydrochloride**
$C_6H_{14}N_3O_3^+$, Cl^-
T.Ashida, K.Funakoshi, T.Tsukihara, T.Ueki, M.Kakudo
Acta Cryst. (B), **28,** 1367, 1972

48.34 **1 - Aminocyclohexane carboxylic acid hydrochloride**
$C_7H_{14}NO_2^+$, Cl^-
K.K.Chacko, R.Srinivasan, R.Zand *J. Cryst. Mol. Struct.,* **1,** 261, 1971
Residue 1 also classified in 21

48.35 **L - homoCitrulline hydrochloride**
$C_7H_{16}N_3O_3^+$, Cl^-
T.Ashida, K.Funakoshi, T.Tsukihara, T.Ueki, M.Kakudo
Acta Cryst. (B), **28,** 1367, 1972

48.36 **L - N - Acetylhistidine monohydrate**
$C_8H_{11}N_3O_3$, H_2O
T.J.Kistenmacher, R.E.Marsh *Science,* **172,** 945, 1971
Residue 1 also classified in 32

48.C **(–) - 2 - exo - Aminonorbornane - 2 - carboxylic acid hydrobromide (absolute configuration)**
$C_8H_{14}NO_2^+$, Br^-
For complete entry see 31.2

48.37 **1 - Aminocycloheptane carboxylic acid hydrobromide monohydrate**
$C_8H_{16}NO_2^+$, Br^- , H_2O
K.K.Chacko, R.Srinivasan, R.Zand *J. Cryst. Mol. Struct.*, **1,** 213, 1971
Residue 1 also classified in 22

48.38 **Hippuric acid**
N - Benzoylglycine
$C_9H_9NO_3$
W.Harrison, S.Rettig, J.Trotter *J. C. S. Perkin ii,* 1036, 1972

48.39 **L - Tyrosine**
$C_9H_{11}NO_3$
A.Mostad, H.M.Nissen, C.Romming *Tetrahedron Letters,* 2131, 1971

48.40 **L - 3,4 - Dihydroxyphenylalanine (absolute configuration)**
$C_9H_{11}NO_4$
A.Mostad, T.Ottersen, C.Romming *Acta Chem. Scand.*, **25,** 3549, 1971

48.41 **L - Phenylalanine hydrochloride (refinement)**
$C_9H_{12}NO_2^+$, Cl^-
G.V.Gurskaya *Kristallografija,* **9,** 839, 1964

48.42 **1 - Aminocyclo - octane carboxylic acid hydrobromide**
$C_9H_{18}NO_2^+$, Br^-
T.Srikrishnan, R.Srinivasan, R.Zand *J. Cryst. Mol. Struct.*, **1,** 199, 1971
Residue 1 also classified in 22

48.43 **N - Acetyl - methionyl - dimethylamide**
$C_9H_{18}N_2O_2S$
A.Aubry, M.Marraud, J.Protas, J.Neel
C. R. Acad. Sci., Fr., C, **273,** 959, 1971

48.C **tris(Sarcosine) calcium chloride (room temp.form)**
$(C_9H_{24}CaN_3O_6^{2+})_n$, $2Cl^-$
For complete entry see 67.9

48.C **Ethylenediamine tetra - acetic acid**
$C_{10}H_{16}N_2O_8$
For complete entry see 1.10

48.44 **D - Penicillamine disulfide dihydrochloride**
$C_{10}H_{22}N_2O_4S_2^{2+}$, $2Cl^-$
R.E.Rosenfield Junior, R.Parthasarathy
Amer. Cryst. Assoc., Abstr. Papers (Summer Meeting), 55, 1971

48.45 **Cyclo - glycyl - L - tyrosyl**
L - 3 - (4 - Hydroxybenzyl) - 2,5 - piperazinedione
$C_{11}H_{12}N_2O_3$
L.E.Webb, C.-F.Lin *J. Amer. Chem. Soc.*, **93,** 3818, 1971
Also classified in 33

48.46 **Glycyl - L - phenylalanine hydrochloride monohydrate (absolute configuration)**
$C_{11}H_{15}N_2O_3^+$, Cl^-, H_2O
M.Cotrait *C. R. Acad. Sci., Fr., C,* **273,** 1239, 1971

48.47 **Cyclo - L - prolyl - L - leucyl**
$C_{11}H_{18}N_2O_2$
I.L.Karle *J. Amer. Chem. Soc.,* **94,** 81, 1972
Also classified in 35

48.48 **Cyclo - L - seryl - L - tyrosyl monohydrate**
L - 3 - (4 - Hydroxybenzyl) - L - 6 - hydroxymethyl - 2,5 - piperazinedione monohydrate
$C_{12}H_{14}N_2O_4$, H_2O
L.E.Webb, C.-F.Lin
Amer. Cryst. Assoc., Abstr. Papers (Summer Meeting), 56, 1971
Residue 1 also classified in 33

48.49 **N - Acetyl - L - tyrosine ethyl ester monohydrate**
$C_{13}H_{17}NO_4$, H_2O
A.F.Pieret, F.Durant, G.Germain, M.Koch
Cryst. Struct. Comm., **1,** 75, 1972

48.50 **L - 3,5,3′ - Tri - iodothyronine**
$C_{15}H_{13}I_3NO_4$
N.Camerman, A.Camerman
Amer. Cryst. Assoc., Abstr. Papers (Winter Meeting), 74, 1972

48.51 **Cyclo - (L - prolyl - L - prolyl - L - hydroxyprolyl)**
$C_{15}H_{21}N_3O_4$
G.Kartha, G.Ambady
Amer. Cryst. Assoc., Abstr. Papers (Winter Meeting), 67, 1972

48.52 **Tosyl - L - prolyl - L - hydroxyproline monohydrate (refinement)**
$C_{17}H_{22}N_2O_6S$, H_2O
M.N.Sabesan, K.Venkatesan *Acta Cryst. (B),* **27,** 1879, 1971
Residue 1 also classified in 32

48.53 **t - Butyloxycarbonyl - L - prolyl - L - prolyl - glycine monohydrate**
$C_{17}H_{27}N_3O_6$, H_2O
J.M.Hudson, B.Shaw, J.M.Schurr, L.H.Jensen
Amer. Cryst. Assoc., Abstr. Papers (Winter Meeting), 66, 1972

48.54 **Methyl D - phenylalanyl - L - phenylalanate hydrobromide**
$C_{19}H_{23}N_2O_3^+$, Br^-
V.Pattabhi, K.Venkatesan *Indian J. Pure Appl. Phys.,* **8,** 795, 1970

48.55 **N - (Bromoacetyl) - L - phenylalanyl - L - phenylalanine ethyl ester**
$C_{22}H_{25}BrN_2O_4$
C.H.Wei, D.G.Doherty, J.R.Einstein *Acta Cryst. (B),* **28,** 907, 1972

48.56 **N - (Chloroacetyl) - L - phenylalanyl - L - phenylalanine ethyl ester**
$C_{22}H_{25}ClN_2O_4$
C.H.Wei, D.G.Doherty, J.R.Einstein *Acta Cryst. (B)*, **28**, 907, 1972

48.C **Viomycin dihydrobromide hydrochloride trihydrate**
$C_{25}H_{46}N_{13}O_{10}^{3+}$, $2Br^-$, Cl^-, $3H_2O$
For complete entry see 50.8

48.57 **o - Bromocarbobenzoxy - glycyl - L - prolyl - L - leucyl - glycyl - L - proline ethylacetate solvate monohydrate**
$C_{28}H_{38}BrN_5O_8$, $0.75C_4H_8O_2$, H_2O
T.Ueki, S.Bando, T.Ashida, M.Kakudo *Acta Cryst. (B)*, **27**, 2219, 1971

48.C **Actinomycin C_1 - deoxyguanosine complex dodecahydrate (form ii)**
$C_{62}H_{86}N_{12}O_{16}$, $2C_{10}H_{13}N_5O_4$, $12H_2O$
For complete entry see 60.43

48.C **Actinomycin C_1 - deoxyguanosine complex dodecahydrate (form ii, further refinement)**
$C_{62}H_{86}N_{12}O_{16}$, $2C_{10}H_{13}N_5O_4$, $12H_2O$
For complete entry see 60.44

PORPHYRINS AND CORRINS

49.1 **Porphine**

$C_{20}H_{14}N_4$

B.M.L.Chen, A.Tulinsky *J. Amer. Chem. Soc.*, **94**, 4144, 1972

49.2 **Nickel(ii) 1,8,8,13,13 - pentamethyl - 5 - cyano - trans - corrin**

$C_{25}H_{30}N_5Ni^+$, Cl^- , xCH_4O

J.D.Dunitz, E.F.Meyer Junior *Helv. Chim. Acta,* **54**, 77, 1971

49.3 **Nickel(ii) 15 - cyano - 2,2,7,7,12,12 - hexamethyl - 1 - methylene - AD - seco - corrin perchlorate methyl acetate solvate**

$C_{27}H_{34}N_5Ni^+$, ClO_4^- , $0.8C_3H_6O_2$

M.Currie, J.D.Dunitz *Makromol. Chem.*, **54**, 98, 1971

49.4 **Palladium(ii) 15 - cyano - 2,2,7,7,12,12 - hexamethyl - 1 - methylene - AD - seco - corrin perchlorate acetone solvate**

$C_{27}H_{34}N_5Pd^+$, ClO_4^- , $0.6C_3H_6O$

M.Currie, J.D.Dunitz *Helv. Chim. Acta,* **54**, 98, 1971

49.5 **Platinum(ii) 15 - cyano - 2,2,7,7,12,12 - hexamethyl - 1 - methylene - AD - seco - corrin perchlorate**

$C_{27}H_{34}N_5Pt^+$, ClO_4^-

M.Currie, J.D.Dunitz *Helv. Chim. Acta,* **54**, 98, 1971

49.6 **Nickel(ii) 5 - cyano - 16 - ethoxy - 1,8,8,13,13 - pentamethyl - 14,methylene - CD - seco - corrin perchlorate**

$C_{27}H_{36}N_5NiO^+$, ClO_4^-

M.Dobler, J.D.Dunitz *Helv. Chim. Acta,* **54**, 90, 1971

49.7 **Dichloro - phthalocyaninato - tin(iv)**

$C_{32}H_{16}Cl_2N_8Sn$

D.Rogers, R.S.Osborn *J. Chem. Soc. (D),* 840, 1971

49.8 **Magnesium phthalocyanine di - pyridine monohydrate**

$C_{32}H_{16}MgN_8$, $2C_5H_5N$, H_2O

M.S.Fischer, D.H.Templeton, A.Zalkin, M.Calvin
J. Amer. Chem. Soc., **93**, 2622, 1971

49.9 $\alpha,\beta,\gamma,\delta$ **- Tetra - n - propylporphine**

$C_{32}H_{38}N_4$

P.W.Codding, A.Tulinsky *J. Amer. Chem. Soc.*, **94**, 4151, 1972

49.10 **Nickel(ii) deoxophylloerythrin methyl ester 1,2 - dichloroethane solvate**
$C_{34}H_{36}N_4NiO_2$, $0.5C_2H_4Cl_2$
R.C.Pettersen *J. Amer. Chem. Soc.*, **93**, 5629, 1971

49.11 **Tetrabenzo - monoaza - porphine**
$C_{35}H_{21}N_5$
I.M.Das, B.Chaudhuri *Acta Cryst. (B)*, **28**, 579, 1972

49.12 **Methyl pheophorbide a**
$C_{36}H_{38}N_4O_5$
M.S.Fischer, D.H.Templeton, A.Zalkin, M.Calvin
J. Amer. Chem. Soc., **94**, 3613, 1972

49.13 **n - Hexylamine - zinc phthalocyanine n - hexylamine solvate**
$C_{38}H_{31}N_9Zn$, $0.5C_6H_{15}N$
T.Kobayashi, T.Ashida, N.Uyeda, E.Suito, M.Kakudo
Bull. Chem. Soc. Jap., **44**, 2095, 1971

49.14 **Methyl - triethylsiloxy - silicon phthalocyanine**
$C_{39}H_{34}N_8OSi_2$
K.Knox, A.D'Addario
Amer. Cryst. Assoc., Abstr. Papers (Summer Meeting), 61, 1967
Also classified in 63

49.15 **Octaethyl - porphinato(monopyridine) zinc(ii)**
$C_{41}H_{49}N_5Zn$
D.L.Cullen, E.F.Meyer Junior
Amer. Cryst. Assoc., Abstr. Papers (Summer Meeting), 82, 1971

49.16 **5 - Benzoyloxy - octaethyl - porphyrin**
$C_{43}H_{50}N_4O_2$
M.B.Hursthouse, S.Neidle *J. C. S. Chem. Comm.*, 449, 1972

49.17 **(Tetraphenylporphine) - silver - tetraphenylporphine solid solution**
$0.54C_{44}H_{28}AgN_4$, $0.46C_{44}H_{30}N_4$
M.L.Schneider *J. C. S. Dalton*, 1093, 1972

49.18 **bis(4 - Methylpyridine) - phthalocyanine - iron(ii) 4 - methylpyridine solvate**
$C_{44}H_{30}FeN_{10}$, $2C_6H_7N$
T.Kobayashi, F.Kurokawa, T.Ashida, N.Uyeda, E.Suito
J. Chem. Soc. (D), 1631, 1971

49.19 **Dicarbonyl - meso - tetraphenylporphinato - ruthenium(ii)**
$C_{46}H_{28}N_4O_2Ru$
D.Cullen, E.Meyer Junior, T.S.Srivastava, M.Tsutsui
J. C. S. Chem. Comm., 584, 1972

49.20 **Cobyric acid undecahydrate**
$C_{46}H_{66}CoN_{11}O_9$, $11H_2O$
K.Venkatesan, D.Dale, D.C.Hodgkin, C.E.Nockolds, F.H.Moore,
B.H.O'Connor *Proc. R. Soc., A*, **323**, 455, 1971

49.21 **bis(Imidazole) - $\alpha,\beta,\gamma,\delta$ - tetraphenylporphinato - iron(iii) chloride methanol solvate**

$C_{50}H_{36}FeN_8^+$, Cl^- , $4CH_4O$
D.M.Collins, R.Countryman, J.L.Hoard
J. Amer. Chem. Soc., **94,** 2066, 1972

49.22 **bis(Piperidine) - $\alpha,\beta,\gamma,\delta$ - tetraphenylporphinato - iron(ii)**

$C_{54}H_{50}FeN_6$
L.J.Radonovich, A.Bloom, J.L.Hoard *J. Amer. Chem. Soc.*, **94,** 2073, 1972

49.23 **neoVitamin B_{12} hydrate**

$C_{63}H_{88}CoN_{14}O_{14}P$, $20H_2O$
H.Stoeckli-Evans, E.Edmond, D.C.Hodgkin *J. C. S. Perkin ii,* 605, 1972

49.24 μ **- Oxo - bis($\alpha,\beta,\gamma,\delta$ - tetraphenylporphinato - iron(iii))**

$C_{88}H_{56}Fe_2N_8O$
A.U.Hoffman, D.M.Collins, V.W.Day, E.B.Fleischer, T.S.Srivastava, J.L.Hoard *J. Amer. Chem. Soc.*, **94,** 3620, 1972

ANTIBIOTICS

50.1 **Chloroemphenicol**

$C_{12}H_{12}Cl_2N_2O_4$
M.Sundaralingam, H.Y.Lin, S.K.Arora
Amer. Cryst. Assoc., Abstr. Papers (Summer Meeting), 71, 1971

50.2 **Spectinomycin dihydrobromide pentahydrate (absolute configuration)**

$C_{14}H_{28}N_2O_8^{2+}$, $2Br^-$, $5H_2O$
T.G.Cochran, D.J.Abraham, L.L.Martin *J. C. S. Chem. Comm.*, 494, 1972
Residue 1 also classified in 38

50.3 **Tetrahydropentalenolactone bromohydrin acetone solvate (absolute configuration)**

$C_{15}H_{21}BrO_5$, $0.5C_3H_6O$
D.J.Duchamp, C.G.Chidester *Acta Cryst. (B),* **28,** 173, 1972
Residue 1 also classified in 38

50.4 **Latumcidin selenate (absolute configuration)**

$C_{16}H_{12}NO^+$, HO_4Se^-
Y.Kono, S.Takeuchi, H.Yonehara, F.Marumo, Y.Saito
Acta Cryst. (B), **27,** 2341, 1971
Residue 1 also classified in 36, 38

50.5 **Kromycin**

$C_{20}H_{30}O_5$
C.Tsai, J.J.Stezowiski, R.E.Hughes *J. Amer. Chem. Soc.*, **93,** 7286, 1971
Also classified in 38

50.6 **Achromycin hydrochloride (absolute configuration)**

$C_{22}H_{25}NO_8^+$, Cl^-
K.Kamiya, M.Asai, Y.Wada, M.Nishikawa *Experientia,* **27,** 363, 1971
Residue 1 also classified in 29

50.7 **Desalipactamycate tosylate (absolute configuration)**

$C_{25}H_{31}N_3O_8S$
D.J.Duchamp
Amer. Cryst. Assoc., Abstr. Papers (Winter Meeting), 23, 1972
Also classified in 35

50.8 **Viomycin dihydrobromide hydrochloride trihydrate**

$C_{25}H_{46}N_{13}O_{10}^{3+}$, $2Br^-$, Cl^- , $3H_2O$
B.W.Bycroft *J. C. S. Chem. Comm.*, 660, 1972
Residue 1 also classified in 48

50.9 **5,12a - Diacetyloxy - tetracycline**

$C_{26}H_{28}N_2O_{11}$
R.B.von Dreele, R.E.Hughes *J. Amer. Chem. Soc.*, **93,** 7290, 1971
Also classified in 29

50.10 **Ascochlorin p - bromobenzenesulfonate (absolute configuration)**

$C_{29}H_{33}BrClO_6S$
Y.Nawata, Y.Iitaka *Bull. Chem. Soc. Jap.*, **44,** 2652, 1971
Also classified in 21, 17

50.11 **Monensin monohydrate**

$C_{36}H_{62}O_{11}$, H_2O
W.K.Lutz, F.K.Winkler, J.D.Dunitz *Helv. Chim. Acta,* **54,** 1103, 1971
Residue 1 also classified in 38

50.12 **Isozygosporin A p - bromobenzoate isopropanol solvate (absolute configuration)**

$C_{37}H_{40}BrNO_7$, C_3H_8O
Y.Tsukuda, H.Koyama *J. C. S. Perkin ii,* 739, 1972
Residue 1 also classified in 36

50.13 **De - valino - boromycin rubidium salt methanol solvate**

$C_{40}H_{64}BO_{14}^-$, Rb^+ , $2CH_4O$
J.D.Dunitz, D.M.Hawley, D.Miklos, D.N.J.White, Yu.Berlin, R.Marusic, V.Prelog *Helv. Chim. Acta,* **54,** 1709, 1971
Residue 1 also classified in 62, 38

50.14 **Grisorixin thallium salt monohydrate (absolute configuration)**

$C_{40}H_{67}O_{10}Tl$, H_2O
M.Alleaume, D.Hickel *J. C. S. Chem. Comm.*, 175, 1972
Residue 1 also classified in 38, 68

50.15 **Antibiotic X - 206 silver salt (absolute configuration)**

$C_{45}H_{77}AgO_{13}$
J.F.Blount, J.W.Westley *J. Chem. Soc. (D),* 927, 1971
Also classified in 38

50.16 **Potassium dianemycin**

$C_{47}H_{76}KO_{14}$
E.W.Czerinski, L.K.Steinrauf
Amer. Cryst. Assoc., Abstr. Papers (Summer Meeting), 70, 1971
Also classified in 38

50.17 **N - Iodoacetyl - amphotericin B tritetrahydrofurane solvate monohydrate**

$C_{49}H_{74}INO_{18}, 3C_4H_8O, H_2O$
P.Ganis, G.Avitabile, W.Mechlinski, C.P.Schaffner
J. Amer. Chem. Soc., **93,** 4560, 1971
Residue 1 also classified in 38

50.18 **Stretovaricin C triacetate cyclic p - bromophenyl - boronate methylene chloride solvate**

$C_{52}H_{59}BBrNO_{17}, CH_2Cl_2$
A.H.-J.Wang, I.C.Paul, K.L.Rinehart Junior, F.J.Antosz
J. Amer. Chem. Soc., **93,** 6275, 1971
Residue 1 also classified in 38, 62

50.19 **Valinomycin**

$C_{54}H_{90}N_6O_{18}$
W.L.Duax, H.Hauptman
Amer. Cryst. Assoc., Abstr. Papers (Winter Meeting), 84, 1972

50.C **Actinomycin C_1 - deoxyguanosine complex dodecahydrate (form ii)**

$C_{62}H_{86}N_{12}O_{16}, 2C_{10}H_{13}N_5O_4, 12H_2O$
For complete entry see 60.43

50.C **Actinomycin C_1 - deoxyguanosine complex dodecahydrate (form ii, further refinement)**

$C_{62}H_{86}N_{12}O_{16}, 2C_{10}H_{13}N_5O_4, 12H_2O$
For complete entry see 60.44

STEROIDS

51.1 **8 - Azaestradiol**

$C_{17}H_{23}NO_2$

J.N.Brown, L.M.Trefonas *J. Amer. Chem. Soc.*, **94,** 4311, 1972

Also classified in 36

51.C **Diethyl - stilbestrol**

$C_{18}H_{20}O_2$, C_2H_6O, H_2O

For complete entry see 17.10

51.C **Diethyl - stilbestrol dimethylsulfoxide solvate**

$C_{18}H_{20}O_2$, C_2H_6OS

For complete entry see 17.11

51.2 **2,4 - Dibromoestradiol (form A)**

$C_{18}H_{22}Br_2O_2$

V.Cody, F.DeJarnette, W.Duax, D.A.Norton

Acta Cryst. (B), **27,** 2458, 1971

51.3 **2,4 - Dibromoestradiol (form B)**

$C_{18}H_{22}Br_2O_2$

V.Cody, F.DeJarnette, W.Duax, D.A.Norton

Acta Cryst. (B), **27,** 2458, 1971

51.4 **Estrone (form i)**

3 - Hydroxyestra - 1,3,5(10) - trien - 17 - one

$C_{18}H_{22}O_2$

B.Busetta, C.Courseille, M.Hospital

C. R. Acad. Sci., Fr., C, **272,** 1211, 1971

51.5 **Estrone (form iii)**

3 - Hydroxyestra - 1,3,5(10) - trien - 17 - one

$C_{18}H_{22}O_2$

B.Busetta, C.Courseille, F.Leroy, M.Hospital

C. R. Acad. Sci., Fr., C, **274,** 153, 1972

51.6 **Estrone (form ii)**

3 - Hydroxyestra - 1,3,5(10) - trien - 17 - one

$C_{18}H_{22}O_2$

T.D.J.Debaerdemaeker *Cryst. Struct. Comm.*, **1,** 39, 1972

51.7 **12 - Keto - 17 - deoxo - 8 - azaestrone methyl ether hydrobromide**
$C_{18}H_{24}NO_2^+$, Br^-
J.N.Brown, R.L.R.Towns, L.M.Trefonas
J. Heterocycl. Chem., **8**, 273, 1971
Residue 1 also classified in 36

51.8 **Estradiol hemihydrate**
$C_{18}H_{24}O_2$, $0.5H_2O$
B.Busetta, M.Hospital *Acta Cryst. (B)*, **28**, 560, 1972

51.C **Estradiol - urea**
$C_{18}H_{24}O_2$, CH_4N_2O
For complete entry see 60.38

51.9 **Estradiol propanol solvate**
$C_{18}H_{24}O_2$, C_3H_8O
B.Busetta, C.Courseille, S.Geoffre, M.Hospital
Acta Cryst. (B), **28**, 1349, 1972

51.10 **17β - Hydroxy - 8(9 - 10β)abeo - estra - 4 - en - 3,10 - dione**
$C_{18}H_{24}O_3$
M.O.Chaney, N.D.Jones *Cryst. Struct. Comm.*, **1**, 197, 1972
Also classified in 29

51.11 **Estro - p - quinol methyl ether**
$C_{19}H_{24}O_3$
P.Narayanan, K.Zechmeister, W.Hoppe *Z. Kristallogr.*, **132**, 411, 1970

51.12 **Androst - 4 - en - 3,17 - dione**
$C_{19}H_{26}O_2$
B.Busetta, G.Comberton, C.Courseille, M.Hospital
Cryst. Struct. Comm., **1**, 129, 1972

51.C **17β - Hydroxy - 1,4 - androstadien - 3 - one p - bromophenol complex**
$C_{19}H_{26}O_2$, C_6H_5BrO
For complete entry see 60.39

51.13 **10,13 - bis(Demethyl) - androstan - 3 - one 17 - bromoacetate**
$C_{19}H_{27}BrO_3$
W.C.Marsh, G.Ferguson, D.F.Rendle
Amer. Cryst. Assoc., Abstr. Papers (Winter Meeting), 72, 1972

51.14 **16β,17β - Dibromoandrostane**
$C_{19}H_{30}Br_2$
N.Mandel, J.Donohue *Acta Cryst. (B)*, **28**, 308, 1972

51.15 **5α - Androstan - 3β - ol - 17 - one**
epi Androsterone
$C_{19}H_{30}O_2$
C.M.Weeks, A.Cooper, D.A.Norton, H.Hauptman, J.Fisher
Acta Cryst. (B), **27**, 1562, 1971

51.16 **17β - Hydroxy - androstan - 3 - one monohydrate**
$C_{19}H_{30}O_2, H_2O$
B.Busetta, C.Courseille, J.M.Fornies-Marquina, M.Hospital
Cryst. Struct. Comm., **1**, 43, 1972

51.17 **5β - Androstane - 3α,17β - diol**
$C_{19}H_{32}O_2$
C.M.Weeks, A.Cooper, D.A.Norton, H.Hauptman, J.Fisher
Acta Cryst. (B), **27**, 1562, 1971

51.18 **3α,17β - Dihydroxy - 5α - androstane**
$C_{19}H_{32}O_2$
C.Precigoux, B.Busetta, C.Courseille, M.Hospital
Cryst. Struct. Comm., **1**, 265, 1972

51.19 **9β - 3 - Methoxy - 17 - acetoxy - 7 - thiaestra - 1,3,5(10),8(14) - tetraene**
$C_{20}H_{24}O_3S$
C.F.W.van de Ven, H.Schenk *Cryst. Struct. Comm.*, **1**, 121, 1972
Also classified in 39

51.20 **Estra - 5(10) - ene - 17 - ol - 3 - one iodoacetate**
$C_{20}H_{27}IO_3$
R.R.Sobti, J.Bordner, S.G.Levine *J. Amer. Chem. Soc.*, **93**, 5588, 1971

51.21 **19 - nor - retroTestosterone acetate**
$C_{20}H_{28}O_3$
B.Busetta, C.Courseille, M.Hospital *Cryst. Struct. Comm.*, **1**, 235, 1972

51.22 **13 - Demethyl - androstan - 3 - one 17 - bromoacetate**
$C_{20}H_{29}BrO_3$
W.C.Marsh, G.Ferguson, D.F.Rendle
Amer. Cryst. Assoc., Abstr. Papers (Winter Meeting), 72, 1972

51.23 **2α - Bromo - 17α - methyl - 5α,14β - androstan - 3α - ol**
$C_{20}H_{33}BrO$
A.Chiaroni, C.Pascard-Billy *Acta Cryst. (B)*, **28**, 1085, 1972

51.24 **Aldosterone 18 - acetal - 20 - hemiacetal (monoclinic form)**
$C_{21}H_{28}O_5$
W.L.Duax, H.Hauptman, C.M.Weeks, D.A.Norton
J. Chem. Soc. (D), 1055, 1971

51.25 **Cortisone**
$C_{21}H_{28}O_5$
J.P.Declerq, G.Germain, M.van Meerssche
Cryst. Struct. Comm., **1**, 13, 1972

51.26 **9α - Fluorocortisol**
$C_{21}H_{29}FO_5$
C.M.Weeks, W.L.Duax
Amer. Cryst. Assoc., Abstr. Papers (Winter Meeting), 73, 1972

51.27 **6α - Fluorocortisol**

$C_{21}H_{29}FO_5$
W.L.Duax, C.M.Weeks, H.Hauptman *Acta Cryst. (B)*, **28**, 1857, 1972

51.28 **9α - Fluoro - Δ^4 - pregnene - 11β,17α,21 - triol - 3,20 - dione**

$C_{21}H_{29}FO_5$
L.Dupont, O.Dideberg, H.Campsteyn
Cryst. Struct. Comm., **1**, 177, 1972

51.29 **Progesterone**

$C_{21}H_{30}O_2$
H.Campsteyn, L.Dupont, O.Dideberg
Cryst. Struct. Comm., **1**, 219, 1972

51.30 **17α - Hydroxyprogesterone**

$C_{21}H_{30}O_3$
J.P.Declerq, G.Germain, M.van Meerssche
Cryst. Struct. Comm., **1**, 9, 1972

51.31 **6α - Methyl - 11β,17α,21β - trihydroxy - 1,4 - pregnadiene - 3,20 - dione**

$C_{22}H_{30}O_5$
J.P.Declerq, G.Germain, M.van Meerssche
Cryst. Struct. Comm., **1**, 5, 1972

51.32 **13 - Demethyl - 4,4 - dimethyl - androst - 5 - ene 17 - iodoacetate**

$C_{22}H_{32}IO_3$
W.C.Marsh, G.Ferguson, D.F.Rendle
Amer. Cryst. Assoc., Abstr. Papers (Winter Meeting), 72, 1972

51.33 **17β - Trimethylsiloxy - 4 - androsten - 3 - one**
Silandrone

$C_{22}H_{36}O_2Si$
C.M.Weeks, H.Hauptman, D.A.Norton
Cryst. Struct. Comm., **1**, 79, 1972
Also classified in 63

51.34 **Cortisone acetate (form ii)**

$C_{23}H_{30}O_6$
J.P.Declerq, G.Germain, M.van Meerssche
An. R. Soc. Esp. Fis. Quim., A, **1**, 59, 1972

51.35 **2β - Hydroxytestosterone 2 - acetate 17 - chloroacetate methanol solvate**

$C_{23}H_{31}ClO_5$, CH_4O
W.L.Duax, C.Eger, S.Pokrywiecki, Y.Osawa
J. Medicin. Chem., U. S. A., **14**, 295, 1971

51.36 **2β - Hydroxytestosterone diacetate p - bromophenol solvate**

$C_{23}H_{32}O_5$, C_6H_5BrO
W.L.Duax, C.Eger, S.Pokrywiecki, Y.Osawa
J. Medicin. Chem., U. S. A., **14**, 295, 1971

51.37 **Strophanthidin hemihydrate**
$C_{23}H_{32}O_6$, $0.5H_2O$
J.L.Flippen, R.D.Gilardi
Amer. Cryst. Assoc., Abstr. Papers (Summer Meeting), 33, 1971

51.38 **N - Cyano - N - methyl - 18 - amino - 20 - bromo - C - nor - D - homo - pregnane**
$C_{23}H_{37}BrN_2$
J.Guilhem *Acta Cryst. (B),* **28,** 291, 1972

51.39 **7α - Acetylthio - 3 - oxo - 17α - 4 - pregnene - 21,17β - carbolactone**
$C_{24}H_{32}O_4S$
O.Dideberg, L.Dupont *Cryst. Struct. Comm.,* **1,** 99, 1972

51.C **Deoxycholic acid - acetic acid complex**
$C_{24}H_{40}O_4$, $C_2H_4O_2$
For complete entry see 60.42

51.40 **16α - Hydroxy - 3 - methoxy - D - norestra - 1,3,5(10) - triene p - bromobenzenesulfonate**
$C_{25}H_{29}BrO_4S$
P.Coggon, A.T.McPhail, S.G.Levine, R.Misra
J. Chem. Soc. (D), 1133, 1971

51.41 **16β - Hydroxy - 3 - methoxy - D - norestra - 1,3,5(10) - triene p - bromobenzenesulfonate**
$C_{25}H_{29}BrO_4S$
P.Coggon, A.T.McPhail, S.G.Levine, R.Misra
J. Chem. Soc. (D), 1133, 1971

51.42 **2β - Methyl - 17β - p - bromobenzenesulfonyloxy - 4 - estren - 3 - one**
$C_{25}H_{31}BrO_4S$
V.Cody, K.Yasuda, W.L.Duax, D.A.Norton
Amer. Cryst. Assoc., Abstr. Papers (Summer Meeting), 37, 1971

51.43 **3β - p - Bromobenzoyloxy - 13β - androst - 5 - en - 17 - one**
$C_{26}H_{31}BrO_3$
J.C.Portheine, C.Romers, E.W.M.Rutten *Acta Cryst. (B),* **28,** 849, 1972

51.44 **3β - p - Bromobenzoyloxy - 13β - androst - 5 - en - 17 - one (at -180 ° C)**
$C_{26}H_{31}BrO_3$
J.C.Portheine, C.Romers, E.W.M.Rutten *Acta Cryst. (B),* **28,** 849, 1972

51.45 **3β - Bromoacetoxy - 16α - ethyl - 16^2 - cyano - 16^2,21 - cyclo - 5α - pregna - 17,21 - diene**
$C_{26}H_{34}BrNO_2$
D.R.Pollard, F.R.Ahmed *Acta Cryst. (B),* **27,** 1976, 1971

51.46 **17α - Hydroxy - 7,7 - dimethyl - 8α,14β - estr - 4 - en - 3 - one p - bromobenzoate**
$C_{27}H_{33}BrO_3$
D.Lednicer, D.E.Emmert, C.G.Chidester, D.J.Duchamp
J. Org. Chem., **36,** 3260, 1971

51.47 **17β - Bromobenzoyloxy - 2,2 - dimethyl - 5α - androstan - 3 - one**
$C_{27}H_{37}BrO_3$
V.Cody, W.L.Duax, F.E.DeJarnette, D.A.Norton, Y.Osawa
Amer. Cryst. Assoc., Abstr. Papers (Winter Meeting), 72, 1972

51.C **4,4 - Dichloro - 2a - aza - A - homo - coprostane - 3 - one acetone solvate**
$C_{27}H_{45}Cl_2NO$, $0.5C_3H_6O$
For complete entry see 36.36

51.48 **22,26 - Epimino - 5α - cholestan - 3β,26α,20 - triol hydroiodide methanol solvate**
$C_{27}H_{48}NO_3^+$, I^- , CH_4O
E.Hohne, I.Seidel, G.Adam, D.Voigt, K.Schreiber
J. Prakt. Chem., **313,** 51, 1971

51.49 **Withanolide E dihydrate**
$C_{28}H_{38}O_7$, $2H_2O$
D.Lavie, I.Kirson, E.Glotter, D.Rabinovich, Z.Shakked
J. C. S. Chem. Comm., 877, 1972

51.50 **2β - Bromo - 2α - methyl - 5α - cholestan - 3 - one**
$C_{28}H_{55}BrO$
V.Cody, W.L.Duax, F.E.DeJarnette, D.A.Norton, Y.Osawa
Amer. Cryst. Assoc., Abstr. Papers (Winter Meeting), 72, 1972

51.51 **2α - Bromo - 2β - methyl - 5α - cholestan - 3 - one**
$C_{28}H_{55}BrO$
V.Cody, W.L.Duax, F.E.DeJarnette, D.A.Norton, Y.Osawa
Amer. Cryst. Assoc., Abstr. Papers (Winter Meeting), 72, 1972

51.52 **Desoxycholic acid p - bromoanilide**
$C_{30}H_{44}BrNO_3$
J.P.Schaefer, L.L.Reed *Acta Cryst. (B),* **28,** 1743, 1972

51.53 **17 - Hydroxyprogesterone 10 - chloro - 9 - keto - deconate**
$C_{31}H_{45}ClO_5$
W.H.Watson, J.E.Whinnery, K.C.Go
Amer. Cryst. Assoc., Abstr. Papers (Winter Meeting), 22, 1972

51.C **Abieslactone**
$C_{31}H_{48}O_3$
For complete entry see 56.2

51.54 **Paeniflorin bromo - ethanolysis product (absolute configuration)**
$C_{33}H_{39}BrO_{15}$
M.Kaneda, Y.Iitaka *Acta Cryst. (B),* **28,** 1411, 1972
Also classified in 45

51.55 **26 - Hydroxycholest - 4 - en - 3 - one p - bromobenzoate**
$C_{34}H_{47}BrO_3$
D.J.Duchamp, C.G.Chidester, J.A.F.Wickramasinghe, E.Caspi, B.Yagen
J. Amer. Chem. Soc., **93,** 6283, 1971

51.56 **(20S) - 20 - Chloro - 3β,16β - di - (p - bromobenzoyloxy) - 5α - pregnane**
$C_{35}H_{41}Br_2ClO_4$
H.-H.Worch, E.Hohne, G.Adam, K.Schreiber
J. Prakt. Chem., **312,** 1043, 1970

51.57 **23 - Demethylgorgosterol p - iodobenzoate (absolute configuration)**
22,23 - Methylene - 24 - methylcholest - 5 - ene - 3β - ol p - iodobenzoate
$C_{36}H_{51}FO_2$
E.L.Enwall, D.van der Helm, I.N.Hsu, T.Pattabhiraman, F.J.Schmitz, R.L.Spraggins, A.J.Weinheimer *J. C. S. Chem. Comm.*, 215, 1972

51.58 **9 - Oxo - 9,11 - secogorgost - 5 - ene - 3β - 11 - diol 11 - acetate p - iodobenzoate (absolute configuration)**
22,23 - Methylene - 23,24 - dimethyl - 9 - oxo - 9,11 - secocholest - 5 - ene - 3β,11 - diol 11 - acetate p - iodobenzoate
$C_{39}H_{55}IO_5$
E.L.Enwall, D.van der Helm, I.N.Hsu, T.Pattabhiraman, F.J.Schmitz, R.L.Spraggins, A.J.Weinheimer *J. C. S. Chem. Comm.*, 215, 1972

51.C **Datiscoside di - (p - iodobenzoate) (absolute configuration)**
$C_{52}H_{60}I_2O_{14}$
For complete entry see 59.29

MONOTERPENES

52.C **(–) - N - Ethyl - N - methyl - aniline oxide (+) - 3 - bromocamphorsulfonate**
$C_{10}H_{14}BrO_4S^-$, $C_9H_{14}NO^+$
For complete entry see 16.7

52.1 **trans - 2,8 - Dihydroxy - 1(7) - p - menthene**
$C_{10}H_{18}O_2$
W.E.Scott, G.F.Richards *J. Org. Chem.*, **36**, 63, 1971

52.2 **3 - Dimethylaminomethyl - 2(10) - pinene hydrobromide**
$C_{13}H_{24}N^+$, Br^-
L.Kutschabsky, G.Reck *J. Prakt. Chem.*, **312**, 896, 1970

52.3 **2 - Bromo - 6 - dimethylaminomethyl - fenchane hydrobromide**
$C_{13}H_{25}BrN^+$, Br^-
L.Kutschabsky, G.Reck *J. Prakt. Chem.*, **312**, 896, 1970

52.4 **(–) - Menthyl methyl phenylphosphonate**
$C_{17}H_{26}O_3P$
W.B.Farnham, K.Mislow, N.Mandel, J.Donohue
J. C. S. Chem. Comm., 120, 1972
Also classified in 64

52.5 **Menthyl S - methyl - phenylphosphonothioate (absolute configuration)**
$C_{17}H_{27}O_2PS$
J.Donohue, N.Mandel, W.B.Farnham, R.K.Murray Junior, K.Mislow, H.P.Benschop *J. Amer. Chem. Soc.*, **93**, 3792, 1971
Also classified in 64

52.6 **O,O - Dimethylipecoside sesquihydrate**
$C_{28}H_{38}NO_{12}$, $1.5H_2O$
O.Kennard, P.J.Roberts, N.W.Isaacs, F.H.Allen, W.D.S.Motherwell, K.H.Gibson, A.R.Battersby *J. Chem. Soc. (D)*, 899, 1971
Residue 1 also classified in 45, 35

SESQUITERPENES

53.1 **3 - Bromoanhydro - dehydro - dihydro - pulchellin**

$C_{15}H_{19}BrO_3$

K.Aota, C.N.Caughlan, M.T.Emerson, W.Herz, S.Inayama, M.-ul-Haque

J. Org. Chem., **35**, 1448, 1970

53.2 **Pacifenol (absolute configuration)**

$C_{15}H_{21}Br_2ClO_2$

J.J.Sims, W.Fenical, R.M.Wing, P.Radlick

J. Amer. Chem. Soc., **93**, 3774, 1971

53.3 **Pseudoclovene - B dibromide (absolute configuration)**

$C_{15}H_{24}Br_2$

R.I.Crane, C.Eck, W.Parker, A.B.Penrose, T.F.W.McKillop, D.M.Hawley, J.M.Robertson *J. C. S. Chem. Comm.*, 385, 1972

53.4 **Bromo - gaillardin (absolute configuration)**

$C_{17}H_{21}BrO_5$

T.A.Dullforce, G.A.Sim, D.N.J.White *J. Chem. Soc. (B)*, 1399, 1971

53.5 **Centaurepensin (absolute configuration)**

$C_{19}H_{24}Cl_2O_7$

J.Harley-Mason, A.T.Hewson, O.Kennard, R.C.Pettersen

J. C. S. Chem. Comm., 460, 1972

53,6 **Melampodin (absolute configuration)**

$C_{21}H_{24}O_9$

S.Neidle, D.Rogers *J. C. S. Chem. Comm.*, 140, 1972

53.7 **Illudol derivative (absolute configuration)**

$C_{21}H_{27}BrN_2O_2$

P.D.Cradwick, G.A.Sim *J. Chem. Soc. (D)*, 431, 1971

53.8 **Isocollybolide**

$C_{22}H_{20}O_7$

C.Pascard-Billy *Acta Cryst. (B)*, **28**, 331, 1972

53.9 **Deacetyl - dihydro - gaillardin p - bromobenzoate (absolute configuration)**

$C_{22}H_{25}BrO_5$

T.A.Dullforce, G.A.Sim, D.N.J.White *J. Chem. Soc. (B)*, 1399, 1971

53.C **Pyrethrosin 3 - o - chlorophenylisoxazoline derivative**

$C_{24}H_{26}ClNO_6$

For complete entry see 38.25

53.10 **Liatrin diol mono - o - bromobenzoate**

$C_{27}H_{31}BrO_7$

S.M.Kupchan, V.H.Davies, T.Fujita, M.R.Cox, R.F.Bryan

J. Amer. Chem. Soc., **93,** 4916, 1971

53.C **Maytoline methiodide methanol solvate**

$C_{30}H_{40}NO_{13}^{+}$, I^{-} , $0.5CH_4O$

For complete entry see 58.36

53.11 **Cedryl chromate (absolute configuration)**

$C_{30}H_{50}CrO_4$

V.Amirthalingam, D.F.Grant, A.Senol *Acta Cryst. (B)*, **28,** 1340, 1972

DITERPENES

54.1 **racemic(3S,2′R,4′S) - 4 - (4′ - p - Bromobenzoyloxy - 2′ - hydroxy - 2′,6′,6′ - trimethyl - cyclohexylidene)but - 3 - en - 2 - one**
Allenic ketone - II
$C_{20}H_{23}BrO_4$
T.E.DeVille, J.Hora, M.B.Hursthouse, T.P.Toube, B.C.L.Weedon
J. Chem. Soc. (D), 1231, 1970

54.2 **(3R,2′R,4′S) - 4 - (4′ - p - Bromobenzoyloxy - 2′ - hydroxy - 2′,6′,6′ - trimethyl - cyclohexylidene)but - 3 - en - 2 - one (absolute configuration)**
Allenic ketone - III
$C_{20}H_{23}BrO_4$
T.E.DeVille, M.B.Hursthouse, S.E.Russell, B.C.L.Weedon
J. Chem. Soc. (D), 1311, 1969

54.3 **2 - Keto - 3 - bromo - tetrahydro - isobulbin - A (absolute configuration)**
$C_{20}H_{25}BrO_7$
K.Kamiya, Y.Wada, T.Komori, M.Arita, T.Kawasaki
Tetrahedron Letters, 1869, 1972

54.4 **Jatrophone dihydrobromide (absolute configuration)**
$C_{20}H_{26}Br_2O_3$
R.C.Haltiwanger, R.F.Bryan *J. Chem. Soc. (B),* 1598, 1971

54.5 **11 - cis - Retinal**
$C_{20}H_{28}O$
R.D.Gilardi, I.L.Karle, J.Karle
Amer. Cryst. Assoc., Abstr. Papers (Summer Meeting), 72, 1971

54.6 **all - trans - Retinal**
$C_{20}H_{28}O$
R.D.Gilardi, I.L.Karle, J.Karle
Amer. Cryst. Assoc., Abstr. Papers (Summer Meeting), 72, 1971

54.7 **all - trans - Retinal**
$C_{20}H_{28}O$
T.Hamanaka, T.Mitsui, T.Ashida, M.Kakudo
Acta Cryst. (B), **28,** 214, 1972

54.8 **Phorbol**

$C_{20}H_{28}O_6$

F.Brandl, M.Rohrl, K.Zechmeister, W.Hoppe

Acta Cryst. (B), **27,** 1718, 1971

54.9 **(–) - Kaur - 15 - en - 19 - al**

$C_{20}H_{30}O$

I.L.Karle *Acta Cryst. (B),* **28,** 585, 1972

54.10 **Glutierolide**

$C_{21}H_{31}ClO_5$

W.B.T.Cruse, M.N.G.James, A.A.Al-Shamma, J.K.Beal, R.W.Doskotch

J. Chem. Soc. (D), 1278, 1971

54.11 **7β - Hydroxykaurenolide p - bromobenzenesulfonate (absolute configuration)**

$C_{26}H_{31}BrO_5S$

J.R.Hanson, G.M.McLaughlin, G.A.Sim *J. C. S. Perkin ii,* 1124, 1972

54.12 **7 - Hydroxy - lathyrol - 3,5,7 - triacetate (absolute configuration)**

$C_{26}H_{36}O_8$

P.Narayanan, M.Rohrl, K.Zechmeister, D.W.Engel, W.Hoppe, E.Hecker, W.Adolf *Tetrahedron Letters,* 1325, 1971

54.13 **Portulal p - bromophenylsulfonylhydrazone (absolute configuration)**

$C_{26}H_{37}BrN_2O_5S$

S.Yamazaki, S.Tamura, F.Marumo, Y.Saito

Acta Cryst. (B), **27,** 2097, 1971

54.14 **Beyeran - 3 - α - ol p - bromobenzenesulfonate (absolute configuration)**

$C_{26}H_{37}BrO_3S$

J.R.Hanson, G.M.McLaughlin, G.A.Sim *J. C. S. Perkin ii,* 1124, 1972

54.15 **p - Bromobenzoyl - crepital**

$C_{31}H_{31}BrO_{10}$

K.Sakata, K.Kawazu, T.Mitsui, N.Masaki

Tetrahedron Letters, 1141, 1971

54.16 **neoPhorbol - 13,20 - diacetate - 3 - p - bromobenzoate (absolute configuration)**

$C_{31}H_{35}BrO_9$

F.Brandl, M.Rohrl, K.Zechmeister, W.Hoppe

Acta Cryst. (B), **27,** 1718, 1971

54.17 **12 - Hydroxydaphnetoxin tri(bromoacetate)**

$C_{33}H_{33}Br_3O_{12}$

J.Coetzer, M.J.Pieterse *Acta Cryst. (B),* **28,** 620, 1972

54.18 **Clerodendrin A p - bromobenzoate chlorohydrin ethanol solvate (absolute configuration)**

$C_{35}H_{46}BrClO_{13}$, C_2H_6O

N.Kato, S.Shibayama, K.Munakata, C.Katayama

J. Chem. Soc. (D), 1632, 1971

54.19 **Fusicoccin A p - iodobenzenesulfonate (absolute configuration)**

$C_{42}H_{59}IO_{14}S$

M.Brufani, S.Cerrini, W.Fedeli, A.Vaciago *J. Chem. Soc. (B),* 2021, 1971

Also classified in 55

54.C **Staphisine methiodide ethanol solvate**

$C_{44}H_{63}N_2O_2^+$, I^-, C_2H_6O

For complete entry see 58.44

SESTERTERPENES

55.C **Fusicoccin A p - iodobenzenesulfonate (absolute configuration)**
$C_{42}H_{59}IO_{14}S$
For complete entry see 54.19

TRITERPENES

56.1 **Baccharis oxide**

$C_{30}H_{50}O$

F.Mo, T.Anthonsen, T.Bruun *Acta Chem. Scand.*, **26**, 1287, 1972

56.2 **Abieslactone**

$C_{31}H_{48}O_3$

J.P.Kutney, N.D.Westcott, F.H.Allen, N.W.Isaacs, O.Kennard, W.D.S.Motherwell *Tetrahedron Letters*, 3463, 1971

Also classified in 51

56.3 **Katonic acid keto acetate**

$C_{32}H_{48}O_3$

W.E.Thiessen, H.A.Levy, W.G.Dauben, G.H.Beasley, D.A.Cox *J. Amer. Chem. Soc.*, **93**, 4312, 1971

56.4 **3β - Acetoxy - 20 - hydroxylupane**

$C_{32}H_{54}O_3$

W.H.Watson, H.-Y.Ting, X.A.Dominguez *Acta Cryst. (B)*, **28**, 8, 1972

56.5 **6 - O - p - Bromobenzoyl - zeorin**

$C_{37}H_{55}BrO_3$

T.Nakanishi, H.Yamauchi, T.Fujiwara, K.Tomita *Tetrahedron Letters*, 1157, 1971

56.6 **p - Bromophenacyl retigerate A (absolute configuration)**

$C_{38}H_{53}BrO_5$

R.Takahashi, Y.Iitaka *Acta Cryst. (B)*, **28**, 764, 1972

TETRATERPENES

57.1 **9,10 - trans - β - Ionylidene - γ - crotonic acid**
$C_{17}H_{24}O_2$
B.Koch *Acta Cryst. (B)*, **28**, 1151, 1972

ALKALOIDS

58.C **Nicotinyl salicylate**
$C_{10}H_{15}N_2$, $C_7H_5O_3$
For complete entry see 60.22

58.1 **Bromoanhydrotetrodoic lactone hydrobromide**
$C_{11}H_{15}BrN_3O_7^+$, Br^-
A.Furusaki, Y.Tomiie, I.Nitta *Bull. Chem. Soc. Jap.*, **43,** 3325, 1970

58.2 **Tetrodotoxin hydrobromide**
$C_{11}H_{18}N_3O_8^+$, Br^-
A.Furusaki, Y.Tomiie, I.Nitta *Bull. Chem. Soc. Jap.*, **43,** 3332, 1970

58.3 (–) - **Anhalonine hydrobromide (absolute configuration)**
$C_{12}H_{16}NO_3^+$, Br^-
A.Brossi, J.F.Blount, J.O'Brien, S.Teitel
J. Amer. Chem. Soc., **93,** 6248, 1971

58.4 **Tecomanine methoperchlorate (absolute configuration)**
$C_{12}H_{20}NO^+$, ClO_4^-
G.Jones, G.Ferguson, W.C.Marsh *J. Chem. Soc. (D),* 994, 1971

58.5 **Alkaloid C methiodide (from tecomanine,absolute configuration)**
Oxygenated skytanthine methiodide
$C_{12}H_{24}NO^+$, I^-
G.Jones, G.Ferguson, W.C.Marsh *J. Chem. Soc. (D),* 994, 1971

58.6 (+) - O - **Methyl - anhalonidine hydrobromide (absolute configuration)**
$C_{13}H_{20}NO_3^+$, Br^-
A.Brossi, J.F.Blount, J.O'Brien, S.Teitel
J. Amer. Chem. Soc., **93,** 6248, 1971

58.7 **Alchorneine bromomethylate (absolute configuration)**
$C_{13}H_{22}N_3O^+$, Br^-
M.Cesario, J.Guilhem *Acta Cryst. (B),* **28,** 151, 1972

58.8 **Coccinellin hemihydrochloride**
$C_{13}H_{23}NO$, $C_{13}H_{24}NO^+$, Cl^-
R.Karlsson, D.Losman *J. C. S. Chem. Comm.*, 626, 1972

58.9 **Porantherine hydrobromide (absolute configuration)**

$C_{15}H_{24}N^+$, Br^-

W.A.Denne, S.R.Johns, J.A.Lamberton, A.McL.Mathieson

Tetrahedron Letters, 3107, 1971

58.10 **Poranthericine hydrobromide (absolute configuration)**

$C_{15}H_{28}NO^+$, Br^-

W.A.Denne, S.R.Johns, J.A.Lamberton, A.McL.Mathieson, H.Suares

Tetrahedron Letters, 1767, 1972

58.11 **Porantheridine hydrobromide (absolute configuration)**

$C_{15}H_{28}NO^+$, Br^-

W.A.Denne, S.R.Johns, J.A.Lamberton, A.McL.Mathieson, H.Suares

Tetrahedron Letters, 1767, 1972

58.12 **(+) - Isopilosine (absolute configuration)**

$C_{16}H_{18}N_2O_3$

W.E.Oberhansli *Cryst. Struct. Comm.*, **1,** 203, 1972

58.13 **Fulvine**

$C_{16}H_{23}NO_5$

S.Wodak, J.L.Sussman, C.Levinthal

Amer. Cryst. Assoc., Abstr. Papers (Summer Meeting), 34, 1971

58.14 **Lycopecurine hydrobromide (α form, absolute configuration)**

$C_{16}H_{26}NO^+$, Br^-

W.A.Ayer, N.Masaki *Canad. J. Chem.*, **49,** 524, 1971

58.15 **Heliotrine**

$C_{16}H_{27}NO_5$

S.Wodak

Amer. Cryst. Assoc., Abstr. Papers (Winter Meeting), 25, 1972

58.16 **Lycopodine Hofmann base hydrobromide**

$C_{16}H_{28}NO^+$, Br^-

N.Chin-You, D.B.MacLean, A.Prakash, C.Calvo

Canad. J. Chem., **49,** 3240, 1971

58.17 **(–) - O - Methyl - laurepukin monohydrate**

1,2 - Methylenedioxy - 11 - methoxy - 6a,β - aporphine - 6β - oxide monohydrate

$C_{19}H_{19}NO_4$, H_2O

W.E.Oberhansli *Helv. Chim. Acta,* **54,** 1389, 1971

58.18 **Alkaloid F hydrobromide from argemone grandiflora sweet**

$C_{20}H_{18}NO_5^+$, Br^-

K.A.Kerr

Amer. Cryst. Assoc., Abstr. Papers (Winter Meeting), 79, 1971

58.19 **Sceletium alkaloid A_4**
$C_{20}H_{24}N_2O_2$
P.W.Jeffs, P.A.Luhan, A.T.McPhail, N.H.Martin
J. Chem. Soc. (D), 1466, 1971

58.20 **Delnudine hydrochloride**
$C_{20}H_{26}NO_3^+ , Cl^-$
K.B.Birnbaum *Acta Cryst. (B),* **27,** 1169, 1971

58.21 **Thalphenine iodide dihydrate (absolute configuration)**
$C_{21}H_{22}NO_4^+ , I^- , 2H_2O$
M.Shamma, J.L.Moniot, S.Y.Yao, J.A.Stanko
J. C. S. Chem. Comm., 408, 1972

58.22 **N - Methyl - rhoeagenine iodide (absolute configuration)**
$C_{21}H_{22}NO_6^+ , I^-$
C.S.Huber *Acta Cryst. (B),* **28,** 982, 1972

58.23 **Sewarine N^4 - methiodide (absolute configuration)**
$C_{21}H_{25}N_2O_3^+ , I^-$
J.M.Karle, P.W.Le Quesne *J. C. S. Chem. Comm.,* 416, 1972

58.24 **Clivorine monohydrate (at −160 ° C)**
$C_{21}H_{27}NO_7 , H_2O$
K.B.Birnbaum, A.Klasek, P.Sedmera, G.Snatzke, L.F.Johnson, F.Santavy
Tetrahedron Letters, 3421, 1971

58.25 **Yohimbine hydrochloride**
$C_{21}H_{27}N_2O_3^+ , Cl^-$
G.Ambady, G.Kartha
Amer. Cryst. Assoc., Abstr. Papers (Winter Meeting), 75, 1972

58.26 **Narciclasine tetra - acetate**
$C_{22}H_{21}NO_{11}$
A.Immirzi, C.Fuganti *J. C. S. Chem. Comm.,* 240, 1972

58.27 **Delphinine aromatic product acid oxalate**
$C_{22}H_{30}NO_5^+ , C_2HO_4^-$
K.B.Birnbaum *Acta Cryst. (B),* **28,** 1551, 1972

58.28 **Daphnilactone B benzene solvate**
$C_{22}H_{31}NO_2 , 0.5C_6H_6$
K.Sasaki, Y.Hirata *Tetrahedron Letters,* 1891, 1972

58.29 **Daphnilactone A**
$C_{23}H_{35}NO_2$
K.Sasaki, Y.Hirata *Tetrahedron Letters,* 1275, 1972

58.30 **Acetyl colchicum cornigerum alkaloid 2 methiodide**
$C_{24}H_{32}NO_6^+ , I^-$
A.R.Battersby, R.Ramage, A.F.Cameron, C.Hannaway, F.Santavy
J. Chem. Soc. (C), 3514, 1971

58.31 **Methyl N - bromoacetyl - homosecodaphniphyllate (absolute configuration)**
$C_{25}H_{38}BrNO_3$
K.Sasaki, Y.Hirata *J. Chem. Soc. (B)*, 1565, 1971

58.32 **4 - Demethyl - hasubanonine p - bromobenzenesulfonate (absolute configuration)**
$C_{26}H_{28}BrNO_7S$
D.N.J.White, A.T.McPhail, G.A.Sim *J. C. S. Perkin ii*, 1280, 1972

58.33 **Obscurinervine hydrobromide**
$C_{26}H_{31}N_2O_5^+ , Br^-$
J.Kahrl, T.Gebreyesus, C.Djerassi *Tetrahedron Letters*, 2527, 1971

58.34 **Vertaline hydrobromide (absolute configuration)**
$C_{26}H_{32}NO_5^+ , Br^-$
J.A.Hamilton, L.K.Steinrauf *J. Amer. Chem. Soc.*, **93**, 2939, 1971

58.35 **O,O,O,O - Tetra - acetyl - azomethine - anopteryl alcohol methiodide (absolute configuration)**
$C_{29}H_{39}NO_9^+ , I^-$
W.A.Denne, S.R.Johns, J.A.Lamberton, A.McL.Mathieson, H.Suares
Tetrahedron Letters, 2727, 1972

58.36 **Maytoline methiodide methanol solvate**
$C_{30}H_{40}NO_{13}^+ , I^- , 0.5CH_4O$
R.F.Bryan, R.M.Smith *J. Chem. Soc. (B)*, 2159, 1971
Residue 1 also classified in 53

58.37 **Zygacine acetonide hydroiodide acetone solvate**
$C_{32}H_{50}NO_8^+ , I^- , 2C_3H_6O$
R.Restivo, R.F.Bryan
Amer. Cryst. Assoc., Abstr. Papers (Winter Meeting), 75, 1972

58.38 **Vakognavine hydriodide**
$C_{34}H_{38}NO_{10}^+ , I^-$
S.W.Pelletier, K.N.Iyer, L.H.Wright, M.G.Newton, N.Singh
J. Amer. Chem. Soc., **93**, 5942, 1971

58.39 **Bromoacetyl - neoevonine monohydrate (absolute configuration)**
$C_{36}H_{42}BrNO_{17} , H_2O$
Y.Sasaki, Y.Hirata *J. C. S. Perkin ii*, 1268, 1972

58.40 **3 - Bromopropyl - maytansine (absolute configuration)**

$C_{37}H_{51}BrClN_3O_{10}$

S.M.Kupchan, Y.Komoda, W.A.Court, G.J.Thomas, R.M.Smith, A.Karim, C.J.Gilmore, R.C.Haltiwanger, R.F.Bryan

J. Amer. Chem. Soc., **94,** 1354, 1972

58.41 **Dihydro - O - methyl - cancentrine - methine hydrobromide (absolute configuration)**

$C_{38}H_{41}N_2O_7^+$, Br^-

G.R.Clark, G.J.Palenik *J. C. S. Perkin ii,* 1219, 1972

58.42 **C - Curarine di - iodide**

$C_{40}H_{44}N_4O^{2+}$, $2I^-$

N.D.Jones, W.Nowacki *J. C. S. Chem. Comm.*, 805, 1972

58.43 **Serpentinine dihydrobromide dihydrate (absolute configuration)**

$C_{42}H_{46}N_4O_5^{2+}$, $2Br^-$, $2H_2O$

H.Irie, K.Ishizuka, S.Kawashima, N.Masaki, K.Osaki, T.Shungu, S.Uyeo, H.Kaneko, S.Naruto *J. C. S. Chem. Comm.*, 871, 1972

58.44 **Staphisine methiodide ethanol solvate**

$C_{44}H_{63}N_2O_2^+$, I^-, C_2H_6O

S.W.Pelletier, A.H.Kapadi, L.H.Wright, S.W.Page, M.G.Newton

J. Amer. Chem. Soc., **94,** 1754, 1972

Residue 1 also classified in 54

MISCELLANEOUS NATURAL PRODUCTS

59.1 **Choline chloride**

$C_5H_{14}NO^+ , Cl^-$

J.Hjortas, H.Sorum *Acta Cryst. (B),* **27**, 1320, 1971

Residue 1 also classified in 3

59.2 **2(S) - Carboxy - 4(R),5(S) - dihydroxypiperidine hydrochloride (absolute configuration)**

$C_6H_{12}NO_4^+ , Cl^-$

G.Evrard, F.Durant, M.Marlier *Cryst. Struct. Comm.*, **1,** 215, 1972

Residue 1 also classified in 33, 48

59.3 **6 - Hydroxydopamine hydrochloride**

3,4,6 - Trihydroxyphenylethylamine hydrochloride

$C_8H_{12}NO_3^+ , Cl^-$

M.Kolderup, A.Mostad, C.Romming *Acta Chem. Scand.*, **26,** 483, 1972

Residue 1 also classified in 17, 3

59.4 **Luciferin (absolute configuration)**

2 - (6 - Hydroxy - 2 - benzo - thiazolyl) - 2 - thiazoline - 4 - carboxylic acid

$C_{11}H_8N_2O_3S_2$

R.H.Stanford, D.Dennis

Amer. Cryst. Assoc., Abstr. Papers (Winter Meeting), 73, 1971

Also classified in 41

59.5 **D - (–) - Luciferin (absolute configuration)**

2 - (6 - Hydroxy - 2 - benzo - thiazolyl) - 2 - thiazoline - 4 - carboxylic acid

$C_{11}H_8N_2O_3S_2$

G.E.Blank, J.Pletcher, M.Sax

Biochem. Biophys. Res. Comm., **42,** 583, 1971

Also classified in 41

59.6 **Citrinin**

$C_{13}H_{14}O_5$

O.R.Rodig, M.Shiro, Q.Fernando *J. Chem. Soc. (D),* 1553, 1971

Also classified in 38

59.7 **Dibromoleucodrin**

$C_{15}H_{14}Br_2O_8$

R.D.Diamand, D.Rogers *Proc. Chem. Soc.*, 63, 1964

59.8 **Johnstonol (absolute configuration)**
$C_{15}H_{21}Br_2ClO_3$
J.J.Sims, W.Fenical, R.M.Wing, P.Radlick *Tetrahedron Letters,* 195, 1972
Also classified in 38

59.9 **4 - Ethyl - 1 - hydroxy - 4,8,8,10,10 - pentamethyl - 7,9 - dioxo - 2,3 - dioxabicyclo(4.4.0)dec - 5 - ene (monoclinic form)**
$C_{15}H_{22}O_5$
M.Sterns *J. Cryst. Mol. Struct.,* **1,** 373, 1971
Also classified in 38

59.10 **Brefeldin A**
$C_{16}H_{24}O_4$
H.P.Weber, D.Hauser, H.P.Sigg *Helv. Chim. Acta,* **54,** 2763, 1971
Also classified in 38

59.11 **Aflatoxin b$_1$ (orthorhombic form)**
$C_{17}H_{12}O_6$
T.C.Van Soest *Acta Cryst. (B),* **26,** 1947, 1970
Also classified in 38

59.12 **Athrotaxin**
$C_{17}H_{16}O_6$
P.Daniels, H.Erdtman, K.Nishimura, T.Norin, P.Kierkegaard, A.-M.Pilotti
J. C. S. Chem. Comm., 246, 1972
Also classified in 38, 17

59.13 **1 - Carboxy - 2,3 - dihydro - 8,9 - dimethoxy - 4 - oxo - 5 - methyl - 1H - 3A,5 - diaza - 10β - azonia - acephenenthrene hydroxide inner salt tetrahydrate**
$C_{17}H_{19}N_3O_5$, $4H_2O$
I.L.Karle, J.Karle *Acta Cryst. (B),* **27,** 1891, 1971
Residue 1 also classified in 36

59.C **Riboflavin - 5′ - bromo - 5′ - deoxyadenosine complex trihydrate**
$C_{17}H_{20}N_4O_6$, $C_{10}H_{12}BrN_5O_3$, $3H_2O$
For complete entry see 60.34

59.14 **3′,5,5′,6 - Tetramethoxyflavone**
$C_{19}H_{18}O_6$
H.-Y.Ting, W.H.Watson, X.A.Dominguez
Acta Cryst. (B), **28,** 1046, 1972
Also classified in 38

59.15 **Viridin**
$C_{20}H_{16}O_6$
S.Neidle, D.Rogers, M.B.Hursthouse *J. C. S. Perkin ii,* 760, 1972
Also classified in 38

59.16 **Carpanone carbon tetrachloride solvate**
$C_{20}H_{18}O_6$, CCl_4
O.L.Chapman, M.R.Engel, J.P.Springer, J.C.Clardy
J. Amer. Chem. Soc., **93**, 6696, 1971
Residue 1 also classified in 38

59.17 **O(3) - Acetyl - 16α,17 - dibromo - gibberellic acid**
$C_{21}H_{24}Br_2O_7$
E.Hohne, G.Schneider, K.Schreiber *J. Prakt. Chem.*, **312**, 816, 1970
Also classified in 38

59.18 **16 - Bromo - myricanol nitromethane solvate (absolute configuration)**
$C_{21}H_{25}BrO_5$, CH_3NO_2
M.J.Begley, R.V.M.Campbell, L.Crombie, B.Tuck, D.A.Whiting
J. Chem. Soc. (C), 3634, 1971
Residue 1 also classified in 31, 17

59.19 **Methyl α - (oxy - p - bromobenzoyl) - β - phenyl - β - (N - benzamido) - propionate (absolute configuration)**
$C_{24}H_{20}BrNO_5$
M.C.Wani, H.L.Taylor, M.E.Wall, P.Coggon, A.T.McPhail
J. Amer. Chem. Soc., **93**, 2325, 1971
Also classified in 13

59.20 **Sterigmatocystin p - bromobenzoate**
$C_{25}H_{15}BrO_7$
N.Tanaka, Y.Katsube, Y.Hatsuda, T.Hamasaki, M.Ishida
Bull. Chem. Soc. Jap., **43**, 3635, 1970
Also classified in 38

59.21 **Surugatoxin heptahydrate (absolute configuration)**
$C_{25}H_{26}BrN_5O_{13}$, $7H_2O$
T.Kosuge, H.Zenda, A.Ochiai, N.Masaki, M.Noguchi, S.Kimura, H.Narita
Tetrahedron Letters, 2545, 1972
Residue 1 also classified in 36

59.22 **Phebalin**
$C_{30}H_{28}O_6$
K.Brown, R.C.Cambie, D.Hall *Chem. and Industry*, 1020, 1971
Also classified in 38

59.23 **Dehydro - ophiobolin - D diol mono - p - bromobenzenesulfonate**
$C_{30}H_{41}BrO_4S$
S.Nozoe, A.Itai, Y.Iitaka *J. Chem. Soc. (D)*, 872, 1971
Also classified in 29

59.24 **Gymnemagenin**
$C_{30}H_{50}O_6$
R.Hoge, C.E.Nordman
Amer. Cryst. Assoc., Abstr. Papers (Summer Meeting), 35, 1971
Also classified in 30

59.25 **Phragmalin iodoacetate**
$C_{31}H_{37}IO_{12}$
J.Coetzer, W.J.Baxter, G.Gafner *Acta Cryst. (B),* **27**, 1434, 1971
Also classified in 38

59.26 **15 - Acetyl - 12 - benzoyl - 3,6 - di(iodoacetyl) - 4,17,17 - trimethyl - 3,6,9.11,12,15 - hexahydroxy - 16 - oxa - tetracyclo(12.2.0.1^{7,11}.0^{4,13})heptadeca - 7 - ene - 5 - one (absolute configuration)**
$C_{33}H_{38}I_2O_{12}$
M.C.Wani, H.L.Taylor, M.E.Wall, P.Coggon, A.T.McPhail
J. Amer. Chem. Soc., **93**, 2325, 1971
Also classified in 38

59.27 **Isorubratoxin B bis(p - bromophenylhydrazide) methyl ester**
$C_{39}H_{44}Br_2N_4O_{10}$
G.Buchi, K.M.Snader, J.D.White, J.Z.Gougoutas, S.Singh
J. Amer. Chem. Soc., **92**, 6638, 1970
Also classified in 38

59.28 **Di - (p - bromobenzoyl)pederin ethanol solvate (absolute configuration)**
$C_{39}H_{51}Br_2NO_{11}$, C_2H_6O
A.B.Corradi, A.Mangia, M.Nardelli, G.Pelizzi
Gazz. Chim. Ital., **101**, 591, 1971
Residue 1 also classified in 38

59.29 **Datiscoside di - (p - iodobenzoate) (absolute configuration)**
$C_{52}H_{60}I_2O_{14}$
S.M.Kupchan, C.W.Sigel, L.J.Guttman, R.J.Restivo, R.F.Bryan
J. Amer. Chem. Soc., **94**, 1353, 1972
Also classified in 51, 21

MOLECULAR COMPLEXES

60.1 **Bromoform - hexamethylenetetramine complex (at −35 ° C)**
$2CHBr_3, C_6H_{12}N_4$
T.Dahl, O.Hassel *Acta Chem. Scand.*, **25,** 2168, 1971
Residue 1 also classified in 5; residue 2 classified in 60, 37

60.2 **Urea - syn - 5 - nitro - 2 - furaldehyde oxime complex**
$CH_4N_2O, C_5H_4N_2O_4$
M.Mathew, G.J.Palenik *J. C. S. Perkin ii,* 1033, 1972
Residue 1 also classified in 8; residue 2 classified in 60, 38

60.C α **- D - Glucose - urea complex**
$CH_4N_2O, C_6H_{12}O_6$
For complete entry see 60.14

60.C **Estradiol - urea**
$CH_4N_2O, C_{18}H_{24}O_2$
For complete entry see 60.38

60.3 **Urea - oxalic acid**
$2CH_4N_2O, C_2H_2O_4$
S.Harkema, J.W.Bates, A.M.Weyenberg, D.Feil
Acta Cryst. (B), **28,** 1646, 1972
Residue 1 also classified in 8; residue 2 classified in 60, 1

60.C **Pyridinium bromide - bis(thiourea) complex**
$2CH_4N_2S, C_5H_6N^+, Br^-$
For complete entry see 60.6

60.C **Pyridine - N - oxide trichloroacetic acid complex**
$C_2HCl_3O_2, C_5H_5NO$
For complete entry see 60.4

60.C **Urea - oxalic acid**
$C_2H_2O_4, 2CH_4N_2O$
For complete entry see 60.3

60.C **Deoxycholic acid - acetic acid complex**
$C_2H_4O_2, C_{24}H_{40}O_4$
For complete entry see 60.42

60.C **1,4 - Dideoxy - 1,4 - dinitro - neo - inositol - bis(tetrahydrothiophene - 1 - oxide) complex**
$2C_4H_8OS$, $C_6H_{10}N_2O_8$
For complete entry see 60.13

60.C **2 - Pyridone - 6 - chloro - 2 - hydroxypyridine complex**
C_5H_4ClNO , C_5H_5NO
For complete entry see 60.5

60.C **Urea - syn - 5 - nitro - 2 - furaldehyde oxime complex**
$C_5H_4N_2O_4$, CH_4N_2O
For complete entry see 60.2

60.4 **Pyridine - N - oxide trichloroacetic acid complex**
C_5H_5NO , $C_2HCl_3O_2$
L.Golic, D.Hadzi, F.Lazarini *J. Chem. Soc. (D),* 860, 1971
Residue 1 also classified in 33; residue 2 classified in 60, 1

60.5 **2 - Pyridone - 6 - chloro - 2 - hydroxypyridine complex**
C_5H_5NO , C_5H_4ClNO
J.Almlof, A.Kvick, I.Olovsson *Acta Cryst. (B),* **27,** 1201, 1971
Residue 1 also classified in 33; residue 2 classified in 60, 33

60.6 **Pyridinium bromide - bis(thiourea) complex**
$C_5H_6N^+$, $2CH_4N_2S$, Br^-
M.R.Truter, B.L.Vickery *Acta Cryst. (B),* **28,** 387, 1972
Residue 1 also classified in 33; residue 2 classified in 60, 8

60.7 **Thymine - p - benzoquinone complex**
$C_5H_6N_2O_2$, $C_6H_4O_2$
T.Sakurai, M.Okunuki *Acta Cryst. (B),* **27,** 1445, 1971
Residue 1 also classified in 44; residue 2 classified in 60, 18

60.C **N,N,N′,N′ - Tetramethylbenzidine - chloroanil complex**
$C_6Cl_4O_2$, $2C_{16}H_{20}N_2$
For complete entry see 60.33

60.C **Mesitylene - hexafluorobenzene complex (at −35 ° C)**
C_6F_6 , C_9H_{12}
For complete entry see 60.21

60.C **(3.3)Paracyclophane - tetracyanoethylene complex**
C_6N_4 , $C_{18}H_{20}$
For complete entry see 60.37

60.C **Triethyl phosphate - benzotrifurazan complex**
$C_6N_6O_3$, $C_6H_{15}O_4P$
For complete entry see 60.15

60.C **Triethyl phosphate - benzotrifurazan complex (at −120 ° C)**
$C_6N_6O_3$, $C_6H_{15}O_4P$
For complete entry see 60.16

60.C **Triphenylarsine oxide - tetrachlorocatechol complex**
$C_6H_2Cl_4O_2$, $C_{18}H_{15}AsO$
For complete entry see 60.35

60.8 **1,3,5 - Trinitrobenzene - 3 - formylbenzothiophene**
$C_6H_3N_3O_6$, C_9H_6OS
R.Pascard, C.Pascard-Billy *Acta Cryst. (B)*, **28**, 1926, 1972
Residue 1 also classified in 15; residue 2 classified in 60, 39

60.9 **1,3,5 - Trinitrobenzene - bis(N - t - butylsalicylidene - iminato) cobalt(ii) complex**
$C_6H_3N_3O_6$, $C_{22}H_{28}CoN_2O_2$
E.E.Castellano, O.J.R.Hodder, C.K.Prout, P.J.Sadler
J. Chem. Soc. (A), 2620, 1971
Residue 1 also classified in 15; residue 2 classified in 60, 78

60.10 **1,3,5 - Trinitrobenzene - bis(N - t - butylsalicylidene - iminato) copper(ii) complex**
$C_6H_3N_3O_6$, $C_{22}H_{28}CuN_2O_2$
E.E.Castellano, O.J.R.Hodder, C.K.Prout, P.J.Sadler
J. Chem. Soc. (A), 2620, 1971
Residue 1 also classified in 15; residue 2 classified in 60, 78

60.11 **1,3,5 - Trinitrobenzene - bis(N - t - butylsalicylidene - iminato) nickel(ii) complex**
$C_6H_3N_3O_6$, $C_{22}H_{28}N_2NiO_2$
E.E.Castellano, O.J.R.Hodder, C.K.Prout, P.J.Sadler
J. Chem. Soc. (A), 2620, 1971
Residue 1 also classified in 15; residue 2 classified in 60, 78

60.C **Thymine - p - benzoquinone complex**
$C_6H_4O_2$, $C_5H_6N_2O_2$
For complete entry see 60.7

60.C **Phloroglucinol - p - benzoquinone complex**
1,3,5 - Trihydroxybenzene - p - benzoquinone complex
$2C_6H_4O_2$, $C_6H_6O_3$
For complete entry see 60.12

60.C **17β - Hydroxy - 1,4 - androstadien - 3 - one p - bromophenol complex**
C_6H_5BrO, $C_{19}H_{26}O_2$
For complete entry see 60.39

60.C **Lumiflavin bromide - hydroquinone complex**
$1.5C_6H_6O_2$, $C_{13}H_{13}N_2O_2^+$, Br^-
For complete entry see 60.30

60.12 **Phloroglucinol - p - benzoquinone complex**
1,3,5 - Trihydroxybenzene - p - benzoquinone complex
$C_6H_6O_3, 2C_6H_4O_2$
T.Sakurai, H.Tagawa *Acta Cryst. (B),* **27,** 1453, 1971
Residue 1 also classified in 17; residue 2 classified in 60, 18

60.13 **1,4 - Dideoxy - 1,4 - dinitro - neo - inositol - bis(tetrahydrothiophene - 1 - oxide) complex**
$C_6H_{10}N_2O_8, 2C_4H_8OS$
R.Dodge, Q.Johnson, W.Selig *Cryst. Struct. Comm.,* **1,** 181, 1972
Residue 1 also classified in 21; residue 2 classified in 60, 39

60.C **Bromoform - hexamethylenetetramine complex (at −35 ° C)**
$C_6H_{12}N_4, 2CHBr_3$
For complete entry see 60.1

60.14 **α - D - Glucose - urea complex**
$C_6H_{12}O_6, CH_4N_2O$
R.L.Snyder, R.D.Rosenstein *Acta Cryst. (B),* **27,** 1969, 1971
Residue 1 also classified in 45; residue 2 classified in 60, 8

60.15 **Triethyl phosphate - benzotrifurazan complex**
$C_6H_{15}O_4P, C_6N_6O_3$
T.S.Cameron, C.K.Prout *Acta Cryst. (B),* **28,** 447, 1972
Residue 1 also classified in 46; residue 2 classified in 60, 40

60.16 **Triethyl phosphate - benzotrifurazan complex (at −120 ° C)**
$C_6H_{15}O_4P, C_6N_6O_3$
T.S.Cameron, C.K.Prout *Acta Cryst. (B),* **28,** 447, 1972
Residue 1 also classified in 46; residue 2 classified in 60, 40

60.17 **5 - Chlorosalicylic acid - theobromine complex**
$2C_7H_5ClO_3, C_7H_8N_5O_2$
E.Shefter, T.F.Brennan, P.Sackman *Chem. Pharm. Bull.,* **19,** 746, 1971
Residue 1 also classified in 13, 17; residue 2 classified in 60, 44

60.C **Nicotinyl salicylate**
$C_7H_5O_3, C_{10}H_{15}N_2$
For complete entry see 60.22

60.C **5 - Chlorosalicylic acid - theobromine complex**
$C_7H_8N_5O_2, 2C_7H_5ClO_3$
For complete entry see 60.17

60.18 **9 - Ethyladenine - barbital complex**
$C_7H_9N_5, C_8H_{12}N_2O_3$
D.Voet *Amer. Cryst. Assoc., Abstr. Papers (Winter Meeting),* 74, 1971
Residue 1 also classified in 44; residue 2 classified in 60, 43

60.C **Barbital - caffeine complex**
$C_8H_{10}N_4O_2$, $2C_8H_{12}N_2O_3$
For complete entry see 60.19

60.C **9 - Ethyladenine - barbital complex**
$C_8H_{12}N_2O_3$, $C_7H_9N_5$
For complete entry see 60.18

60.C **Aminopyrine - barbital complex**
$C_8H_{12}N_2O_3$, $C_{13}H_{17}N_3O$
For complete entry see 60.31

60.19 **Barbital - caffeine complex**
$2C_8H_{12}N_2O_3$, $C_8H_{10}N_4O_2$
B.M.Craven, G.L.Gartland *J. Pharm. Sci.*, **59,** 1666, 1970
Residue 1 also classified in 43; residue 2 classified in 60, 44

60.C **1,3,5 - Trinitrobenzene - 3 - formylbenzothiophene**
C_9H_6OS , $C_6H_3N_3O_6$
For complete entry see 60.8

60.20 **Quinolinium 2 - dicyanomethylene - 1,1,3,3 - tetracyanopropanedi - ide**
$2C_9H_8N^+$, $C_{10}N_6^{2-}$
S.Sakanoue, N.Yasuoka, N.Kasai, M.Kakudo
Bull. Chem. Soc. Jap., **44,** 1, 1971
Residue 1 also classified in 35; residue 2 classified in 60, 12, 7

60.21 **Mesitylene - hexafluorobenzene complex (at –35 ° C)**
C_9H_{12} , C_6F_6
T.Dahl *Acta Chem. Scand.*, **25,** 1031, 1971
Residue 1 also classified in 19; residue 2 classified in 60, 19

60.C **Quinolinium 2 - dicyanomethylene - 1,1,3,3 - tetracyanopropanedi - ide**
$C_{10}N_6^{2-}$, $2C_9H_8N^+$
For complete entry see 60.20

60.C **Anthracene - 1,2,4,5 - tetracyanobenzene**
$C_{10}H_2N_4$, $C_{14}H_{10}$
For complete entry see 60.32

60.C **Lumiflavin - bis(naphthalene - 2,3 - diol) (yellow form)**
$2C_{10}H_8O_2$, $C_{13}H_{12}N_4O_2$
For complete entry see 60.27

60.C **10 - Propylisoalloxazine - bis(naphthalene - 2,3 - diol) complex**
$2C_{10}H_8O_2$, $C_{13}H_{12}N_4O_2$
For complete entry see 60.28

60.C **Lumiflavin bis(naphthalene - 2,3 - diol) trihydrate**
$2C_{10}H_8O_2$, $C_{13}H_{12}N_4O_2$, $3H_2O$
For complete entry see 60.29

60.C **Riboflavin - 5′ - bromo - 5′ - deoxyadenosine complex trihydrate**
$C_{10}H_{12}BrN_5O_3, C_{17}H_{20}N_4O_6, 3H_2O$
For complete entry see 60.34

60.C **Actinomycin C_1 - deoxyguanosine complex dodecahydrate (form ii)**
$2C_{10}H_{13}N_5O_4, C_{62}H_{86}N_{12}O_{16}, 12H_2O$
For complete entry see 60.43

60.C **Actinomycin C_1 - deoxyguanosine complex dodecahydrate (form ii, further refinement)**
$2C_{10}H_{13}N_5O_4, C_{62}H_{86}N_{12}O_{16}, 12H_2O$
For complete entry see 60.44

60.22 **Nicotinyl salicylate**
$C_{10}H_{15}N_2, C_7H_5O_3$
H.S.Kim, G.A.Jeffrey *Acta Cryst. (B),* **27,** 1123, 1971
Residue 1 also classified in 58; residue 2 classified in 60, 13

60.23 **7,7,8,8 - Tetracyanoquinodimethane - phenazine complex**
$C_{12}H_4N_4, C_{12}H_8N_2$
I.Goldberg, U.Shmueli *Nature Phys. Sci.,* **234,** 36, 1971
Residue 1 also classified in 7; residue 2 classified in 60, 36

60.24 **7,7,8,8 - Tetracyanoquinodimethane - N - methylphenothiazine complex**
$C_{12}H_4N_4, C_{13}H_{11}NS$
H.Kobayashi, Y.Saito *Bull. Chem. Soc. Jap.,* **44,** 1444, 1971
Residue 1 also classified in 7; residue 2 classified in 60, 41

60.C **N - n - Propylquinolinium bis(7,7,8,8 - tetracyanoquinodimethane)**
$C_{12}H_4N_4^-, C_{12}H_{14}N^+, C_{12}H_4N_4$
For complete entry see 60.25

60.C **3,3 - Diethylthiacarbocyanine - tetracyanoquinodimethane complex**
$C_{12}H_4N_4^-, C_{21}H_{21}N_2S_2^+, C_{12}H_4N_4$
For complete entry see 60.40

60.C **1,3,3 - Trimethyl - 2 - (N - methyl - N - (β - chloroethyl) - p - aminostyryl) - 3H - indole - 7,7,8,8 - tetracyanoquinodimethane complex**
$C_{12}H_4N_4^-, C_{22}H_{26}ClN_2^+, C_{12}H_4N_4$
For complete entry see 60.41

60.C **7,7,8,8 - Tetracyanoquinodimethane - phenazine complex**
$C_{12}H_8N_2, C_{12}H_4N_4$
For complete entry see 60.23

60.C **Benzophenone - diphenylamine complex**
$C_{12}H_{11}N, C_{13}H_{10}O$
For complete entry see 60.26

60.25 **N - n - Propylquinolinium bis(7,7,8,8 - tetracyanoquinodimethane)**
$C_{12}H_{14}N^+$, $C_{12}H_4N_4^-$, $C_{12}H_4N_4$
T.Sundaresan, S.C.Wallwork *Acta Cryst. (B),* **28,** 1163, 1972
Residue 1 also classified in 35; residue 2 classified in 60, 7

60.26 **Benzophenone - diphenylamine complex**
$C_{13}H_{10}O$, $C_{12}H_{11}N$
C.Brassy, J.-P.Mornon *C. R. Acad. Sci., Fr., C,* **274,** 1728, 1972
Residue 1 also classified in 19; residue 2 classified in 60, 16

60.C **7,7,8,8 - Tetracyanoquinodimethane - N - methylphenothiazine complex**
$C_{13}H_{11}NS$, $C_{12}H_4N_4$
For complete entry see 60.24

60.27 **Lumiflavin - bis(naphthalene - 2,3 - diol) (yellow form)**
$C_{13}H_{12}N_4O_2$, $2C_{10}H_8O_2$
B.L.Trus, J.L.Wells, R.M.Johnstone, C.J.Fritchie Junior, R.E.Marsh
J. Chem. Soc. (D), 751, 1971
Residue 1 also classified in 36; residue 2 classified in 60, 24

60.28 **10 - Propylisoalloxazine - bis(naphthalene - 2,3 - diol) complex**
$C_{13}H_{12}N_4O_2$, $2C_{10}H_8O_2$
M.-C.Kuo, J.B.R.Dunn, C.J.Fritchie Junior
J. C. S. Chem. Comm., 205, 1972
Residue 1 also classified in 36; residue 2 classified in 60, 24

60.29 **Lumiflavin bis(naphthalene - 2,3 - diol) trihydrate**
$C_{13}H_{12}N_4O_2$, $2C_{10}H_8O_2$, $3H_2O$
C.J.Fritchie Junior, R.M.Johnstone
Amer. Cryst. Assoc., Abstr. Papers (Winter Meeting), 33, 1972
Residue 1 also classified in 36; residue 2 classified in 60, 24

60.30 **Lumiflavin bromide - hydroquinone complex**
$C_{13}H_{13}N_2O_2^+$, $1.5C_6H_6O_2$, Br^-
O.Tillberg, R.Norrestam *Acta Cryst. (B),* **28,** 890, 1972
Residue 1 also classified in 36; residue 2 classified in 60, 17

60.31 **Aminopyrine - barbital complex**
$C_{13}H_{17}N_3O$, $C_8H_{12}N_2O_3$
S.Kiryu *J. Pharm. Sci.,* **60,** 699, 1971
Residue 1 also classified in 32; residue 2 classified in 60, 43

60.32 **Anthracene - 1,2,4,5 - tetracyanobenzene**
$C_{14}H_{10}$, $C_{10}H_2N_4$
H.Tsuchiya, F.Marumo, Y.Saito *Acta Cryst. (B),* **28,** 1935, 1972
Residue 1 also classified in 26; residue 2 classified in 60, 7

60.C **1,1 - Dimethyl - 2,5 - diphenyl - 1 - silacyclopentadiene diphenylacetylene complex**
$C_{14}H_{10}$, $C_{18}H_{18}Si$
For complete entry see 60.36

60.33 **N,N,N′,N′ - Tetramethylbenzidine - chloroanil complex**
$2C_{16}H_{20}N_2 , C_6Cl_4O_2$
K.Yakushi, I.Ikemoto, H.Kuroda *Acta Cryst. (B),* **27,** 1710, 1971
Residue 1 also classified in 16; residue 2 classified in 60, 18

60.34 **Riboflavin - 5′ - bromo - 5′ - deoxyadenosine complex trihydrate**
$C_{17}H_{20}N_4O_6 , C_{10}H_{12}BrN_5O_3 , 3H_2O$
D.Voet, A.Rich *Proc. Nation. Acad. Sci. U. S. A.*, **68,** 1151, 1971
Residue 1 also classified in 59, 45; residue 2 classified in 64, 47, 44, 45

60.35 **Triphenylarsine oxide - tetrachlorocatechol complex**
$C_{18}H_{15}AsO , C_6H_2Cl_4O_2$
F.F.Farris, W.R.Robinson *J. Organometal. Chem.*, **31,** 375, 1971
Residue 1 also classified in 65; residue 2 classified in 60, 17

60.36 **1,1 - Dimethyl - 2,5 - diphenyl - 1 - silacyclopentadiene diphenylacetylene complex**
$C_{18}H_{18}Si , C_{14}H_{10}$
J.Clardy, T.J.Barton *J. C. S. Chem. Comm.*, 690, 1972
Residue 1 also classified in 63; residue 2 classified in 60, 5, 19

60.37 **(3.3)Paracyclophane - tetracyanoethylene complex**
$C_{18}H_{20} , C_6N_4$
J.Bernstein, K.N.Trueblood *Acta Cryst. (B),* **27,** 2078, 1971
Residue 1 also classified in 31; residue 2 classified in 60, 7

60.38 **Estradiol - urea**
$C_{18}H_{24}O_2 , CH_4N_2O$
W.L.Duax *Acta Cryst. (B),* **28,** 1864, 1972
Residue 1 also classified in 51; residue 2 classified in 60, 8

60.39 **17β - Hydroxy - 1,4 - androstadien - 3 - one p - bromophenol complex**
$C_{19}H_{26}O_2 , C_6H_5BrO$
W.L.Duax, D.A.Norton, S.Pokrywiecki, C.Eger *Steroids,* **18,** 525, 1971
Residue 1 also classified in 51; residue 2 classified in 60, 17

60.40 **3,3 - Diethylthiacarbocyanine - tetracyanoquinodimethane complex**
$C_{21}H_{21}N_2S_2^+ , C_{12}H_4N_4^- , C_{12}H_4N_4$
D.N.Fedutin, I.F.Shchegolev, V.B.Stryukov, E.B.Yagubskii, A.V.Zvarykina, L.O.Atovmyan, V.F.Kaminskii, R.P.Shibaeva
Phys. Status Solidi, **48,** 87, 1971
Residue 1 also classified in 41; residue 2 classified in 60, 7

60.41 **1,3,3 - Trimethyl - 2 - (N - methyl - N - (β - chloroethyl) - p - aminostyryl) - 3H - indole - 7,7,8,8 - tetracyanoquinodimethane complex**
$C_{22}H_{26}ClN_2^+ , C_{12}H_4N_4^- , C_{12}H_4N_4$
R.P.Shibaeva, L.O.Atovmyan, L.P.Rozenberg
Tetrahedron Letters, 3303, 1971
Residue 1 also classified in 35, 16; residue 2 classified in 60, 7

60.C **1,3,5 - Trinitrobenzene - bis(N - t - butylsalicylidene - iminato) cobalt(ii) complex**
$C_{22}H_{28}CoN_2O_2, C_6H_3N_3O_6$
For complete entry see 60.9

60.C **1,3,5 - Trinitrobenzene - bis(N - t - butylsalicylidene - iminato) copper(ii) complex**
$C_{22}H_{28}CuN_2O_2, C_6H_3N_3O_6$
For complete entry see 60.10

60.C **1,3,5 - Trinitrobenzene - bis(N - t - butylsalicylidene - iminato) nickel(ii) complex**
$C_{22}H_{28}N_2NiO_2, C_6H_3N_3O_6$
For complete entry see 60.11

60.42 **Deoxycholic acid - acetic acid complex**
$C_{24}H_{40}O_4, C_2H_4O_2$
B.M.Craven, G.T.DeTitta *J. C. S. Chem. Comm.*, 530, 1972
Residue 1 also classified in 51; residue 2 classified in 60, 1

60.43 **Actinomycin C$_1$ - deoxyguanosine complex dodecahydrate (form ii)**
$C_{62}H_{86}N_{12}O_{16}, 2C_{10}H_{13}N_5O_4, 12H_2O$
H.M.Sobell, T.D.Sakore, S.C.Jain, C.E.Nordman
Amer. Cryst. Assoc., Abstr. Papers (Winter Meeting), 73, 1971
Residue 1 also classified in 50, 40, 48; residue 2 classified in 64, 47, 44, 45

60.44 **Actinomycin C$_1$ - deoxyguanosine complex dodecahydrate (form ii, further refinement)**
$C_{62}H_{86}N_{12}O_{16}, 2C_{10}H_{13}N_5O_4, 12H_2O$
H.M.Sobell, S.C.Jain
Amer. Cryst. Assoc., Abstr. Papers (Summer Meeting), 88, 1971
Residue 1 also classified in 50, 40, 48; residue 2 classified in 64, 47, 44, 45

CLATHRATES

61.1 **4 - p - Hydroxyphenyl - 2,2,4 - trimethylthiochroman 2,5,5 - trimethylhex - 3 - yn - 2 - ol**
$6C_{18}H_{20}OS$, $C_9H_{16}O$
D.D.MacNicol, F.B.Wilson *J. Chem. Soc. (D),* 786, 1971
Residue 1 also classified in 39; residue 2 classified in 5

61.2 **4 - p - Hydroxyphenyl - 2,2,4 - trimethylchroman - n - heptanol**
Dianin's compound - n - heptanol
$6C_{18}H_{20}O_2$, $C_7H_{16}O$
J.L.Flippen, J.Karle *J. Phys. Chem.*, **75,** 3566, 1971
Residue 1 also classified in 38; residue 2 classified in 5

BORON COMPOUNDS

62.C **Ethylenediamine - bis(borane)**

$C_2H_{14}B_2N_2$

For complete entry see 3.5

62.1 **1 - Methyl - 1 - galla - 2,4 - dicarba - closo - heptaborane(7)**

$C_3H_8B_4Ga$

R.N.Grimes, W.J.Rademaker, M.L.Denniston, R.F.Bryan, P.T.Greene

J. Amer. Chem. Soc., **94,** 1865, 1972

Also classified in 68

62.2 **Trimethylamino - boron tribromide**

$C_3H_9BBr_3N$

P.H.Clippard, J.C.Hanson, R.C.Taylor

J. Cryst. Mol. Struct., **1,** 363, 1971

Also classified in 3

62.3 **Trimethylamino - boron trichloride**

$C_3H_9BCl_3N$

P.H.Clippard, J.C.Hanson, R.C.Taylor

J. Cryst. Mol. Struct., **1,** 363, 1971

Also classified in 3

62.4 **Trimethylamino - boron tri - iodide**

$C_3H_9BI_3N$

P.H.Clippard, J.C.Hanson, R.C.Taylor

J. Cryst. Mol. Struct., **1,** 363, 1971

Also classified in 3

62.C **1 - Bromo - μ - trimethylsilyl - pentaborane(9)**

$C_3H_{16}B_5BrSi$

For complete entry see 63.1

62.5 **Methoxy - 6 - dimethylsulfido - dodecahydrononaborane**

$C_3H_{21}B_9OS$

V.Subrtova *Collect. Czechosl. Chem. Communic.*, **36,** 4034, 1971

62.6 **nido - Carborane compound (at −170 ° C)**

$C_4H_{15}B_7$

J.C.Huffman, W.E.Streib *J. C. S. Chem. Comm.*, 665, 1972

62.7 **3 - Ethyl - 3 - alumina - 1,2 - dicarba - closo - dodecaborane(12)**

$C_4H_{16}AlB_9$

M.R.Churchill, A.H.Reis Junior *J. C. S. Dalton,* 1317, 1972

Also classified in 68

62.8 **7,8 - μ - Dimethylalumina - 1,2 - dicarba - nido - undecaborane(13) (at −100 ° C)**

$C_4H_{18}AlB_9$

M.R.Churchill, A.H.Reis Junior *J. C. S. Dalton,* 1314, 1972

Also classified in 68

62.9 **Tetraethylammonium 2,2′ - commo - bis(nonahydrodicarba - 2 - cobalta - closo - decaborate) (tetragonal form)**

$C_4H_{18}B_{14}Co^-$, $C_8H_{20}N^+$

D.Saint Clair, A.Zalkin, D.H.Templeton *Inorg. Chem.,* **11,** 377, 1972

Residue 2 classified in 3

62.C **bis(π - 7,8 - Dicarbaundecaboran(13) - 10 - ylthio)methyliumato(4 -) cobalt**

$C_5H_{21}B_{18}CoS_2$

For complete entry see 71.2

62.10 **Triethanolamine borate**

$C_6H_{12}BNO_3$

Z.Taira, K.Osaki *Inorg. Nucl. Chem. Letters,* **7,** 509, 1971

Also classified in 42

62.C **1,3 - O - D - Mannitol borate monohydrate**

1,3 - (Hydroxyborylene) - D - mannitol monohydrate

$C_6H_{13}BO_8$, H_2O

For complete entry see 45.8

62.11 **1,1,4,4 - Tetramethyl - 1,4 - diaza - 2,5 - diboracyclohexane**

$C_6H_{16}B_2N_2$

T.H.Hseu, V.Schomaker

Amer. Cryst. Assoc., Abstr. Papers (Summer Meeting), 73, 1971

Also classified in 42

62.12 **B - tris(Dimethylamino)borazine**

$C_6H_{21}B_3N_6$

H.Hess, B.Reiser *Z. Anorg. Allg. Chem.,* **381,** 91, 1971

62.C **Cesium - 3,3′ - commo - bis(nonahydro - 1,2 - dimethyl - 1,2 - dicarba - 3 - chroma - closo - dodecaborate) monohydrate**

$C_8H_{30}B_{18}Cr^-$, Cs^+, H_2O

For complete entry see 71.5

62.13 **Compound B**

$C_9H_{26}B_{17}CoN^-$, $C_8H_{20}N^+$

M.R.Churchill, K.Gold *J. C. S. Chem. Comm.,* 901, 1972

Residue 2 classified in 3

62.14 **Benzoylacetonato boron difluoride (at −180 ° C)**

$C_{10}H_9BF_2O_2$
A.W.Hanson, E.W.Macaulay *Acta Cryst. (B)*, **28**, 1961, 1972

62.C **bis(bis(Diethylether) - μ - (dodecahydro - nido - decaborato) cadmium)**

$C_{16}H_{64}B_{20}Cd_2O_4$
For complete entry see 84.9

62.C **(Fluoro(6,6′,6″ - phosphinidyne - tris(α - picolinaldehyde oximato))borato) iron tetrafluoroborate methylene dichloride solvate**

$C_{18}H_{12}BFFeN_6O_3P^+$, BF_4^- , CH_2Cl_2
For complete entry see 83.86

62.C **bis(3,5 - Dimethylpyrazolylborato)dicarbonyl - trihapto - cycloheptatrienyl molybdenum**

$C_{19}H_{23}BMoN_4O_2$
For complete entry see 75.21

62.C **Compound X**

$C_{24}H_{20}B^+$, $C_9H_{17}N_6Pt^-$
For complete entry see 71.7

62.C **Bromo - (bis - (2 - diethylaminoethyl) - (2 - diphenylphosphinoethyl)amine) nickel(ii) tetraphenylborate**

$C_{24}H_{20}B^-$, $C_{26}H_{42}BrN_2NiP^+$
For complete entry see 83.126

62.C **(tris(o - Diphenylphosphinophenyl)phosphine) - chlorocobalt(ii) tetraphenylborate**

$C_{24}H_{20}B^-$, $C_{54}H_{42}ClCoP_4^+$
For complete entry see 86.56

62.C **Nitrosyl - bis(1,2 - bis(diphenylphosphino)ethane) ruthenium(0) tetraphenylborate acetone solvate**

$C_{24}H_{20}B^-$, $C_{58}H_{48}NOP_4Ru^+$, C_3H_6O
For complete entry see 86.63

62.C **Triphenylphosphine tris(o - diphenylphosphinophenyl)phosphine iridium(i) tetraphenylborate**

$C_{24}H_{20}B^-$, $C_{72}H_{57}IrP_5^+$
For complete entry see 86.64

62.15 **2,2,4,4 - Tetraiodo - 1,1,3,3 - tetraphenyl - cyclodiborataphosphoniane**

$C_{24}H_{20}B_2I_4P_2$
G.J.Bullen, P.R.Mallinson *J. C. S. Dalton*, 1143, 1972
Also classified in 64

62.16 **Diphenylmethylene - aminodimesitylborane**

$C_{31}H_{32}BN$
G.J.Bullen, K.Wade *J. Chem. Soc. (D)*, 1122, 1971

62.C **De - valino - boromycin rubidium salt methanol solvate**

$C_{40}H_{64}BO_{14}^{-}$, Rb^{+}, $2CH_4O$

For complete entry see 50.13

62.C **Stretovaricin C triacetate cyclic p - bromophenyl - boronate methylene chloride solvate**

$C_{52}H_{59}BBrNO_{17}$, CH_2Cl_2

For complete entry see 50.18

SILICON COMPOUNDS

63.1 **1 - Bromo - μ - trimethylsilyl - pentaborane(9)**
$C_3H_{16}B_5BrSi$
J.C.Calabrese, L.F.Dahl *J. Amer. Chem. Soc.*, **93**, 6042, 1971
Also classified in 62

63.2 **2,2,3,3,7,7 - Hexafluoro - 2,3,7 - trisilanorborn - 5 - ene**
$C_4H_4F_6Si_3$
C.S.Liu, S.C.Nyburg, J.T.Szymanski, J.C.Thompson
J. C. S. Dalton, 1129, 1972

63.3 **Dimethylsilicon dicyanide**
$C_4H_6N_2Si$
J.Konnert, D.Britton, Y.M.Chow *Acta Cryst. (B),* **28**, 180, 1972

63.4 **2,2′ - Bipyridyl - tetrafluoro - silicon(iv)**
$C_{10}H_8F_4N_2Si$
A.D.Adley, P.H.Bird, A.R.Fraser, M.Onyszchuk
Inorg. Chem., **11**, 1402, 1972

63.C **Tricarbonyl((trimethylsilyl) - π - cyclopentadienyl) rhenium**
$C_{11}H_{13}O_3ReSi$
For complete entry see 73.9

63.C **tris(Di(dimethylsilyl)amino) - nitrosyl - chromium(ii)**
$C_{12}H_{36}CrN_4OSi_3$
For complete entry see 83.61

63.5 **μ_3 - Trimethylsilylimido - μ_3 - carbonyl - tris(tricarbonyl iron)**
$C_{13}H_9Fe_3NO_{10}Si$
B.L.Barnett, C.Kruger *Angew. Chem.*, **83**, 969, 1971
Also classified in 83

63.6 **bis(Diethylsilicon) - di - μ - hydrido - bis(tetracarbonyl - tungsten)**
$C_{16}H_{22}O_8Si_2W_2$
M.J.Bennett, K.A.Simpson *J. Amer. Chem. Soc.*, **93**, 7156, 1971

63.C **bis(Tetrahydrofuran) - bis(di(dimethylsilyl)amino) chromium(ii)**
$C_{16}H_{40}CrN_2O_2Si_4$
For complete entry see 83.85

63.7 **Cyclic silicon nitrogen compound**

$C_{17}H_{49}N_3Si_7$
D.Mootz, J.Fayos, A.Zinnius *Angew. Chem.*, **84**, 27, 1972

63.C **1,1 - Dimethyl - 2,5 - diphenyl - 1 - silacyclopentadiene diphenylacetylene complex**

$C_{18}H_{18}Si$, $C_{14}H_{10}$
For complete entry see 60.36

63.8 **9 - Methyl - 9 - (p - bromophenyl) - 9,10 - dihydro - 9 - sila - 10 - oxa - phenanthrene**

$C_{19}H_{15}BrOSi$
A.I.Gusev, V.E.Shklover, E.A.Chernyshev, T.L.Krasnova, Yu.T.Struchkov
Zh. Strukt. Khim., **12**, 282, 1971

63.9 **3 - Bromo - 2,2 - diphenyl - 2 - sila - Δ^3 - 1 - tetralone**

$C_{21}H_{15}BrOSi$
J.-P.Vidal, J.-L.Galigne, J.Falgueirettes
C. R. Acad. Sci., Fr., C, **272**, 1852, 1971

63.C **17β - Trimethylsiloxy - 4 - androsten - 3 - one**

Silandrone
$C_{22}H_{36}O_2Si$
For complete entry see 51.33

63.10 **Tetraphenyl silane**

$C_{24}H_{20}Si$
C.Glidewell, G.M.Sheldrick *J. Chem. Soc. (A)*, 317, 1971

63.11 **Tetraphenyl silane**

$C_{24}H_{20}Si$
P.C.Chieh
Amer. Cryst. Assoc., Abstr. Papers (Summer Meeting), 84, 1971

63.C **bis - μ - (Trimethylsilylmethylidyne) tetrakis(trimethylsilylmethyl) diniobium(v)**

$C_{24}H_{62}Nb_2Si_6$
For complete entry see 71.52

63.C **hexakis(Trimethylsilyl) dimolybdenum**

$C_{24}H_{66}Mo_2Si_6$
For complete entry see 71.53

63.C **Chloro - bis(π - cyclopentadienyl)(triphenylsilyl) zirconium(iv)**

$C_{28}H_{25}ClSiZr$
For complete entry see 73.30

63.12 **(+) - α - (1 - Naphthyl - phenyl - methyl - silyl)benzyl p - bromobenzoate (absolute configuration)**

$C_{31}H_{25}BrO_2Si$
S.C.Nyburg, A.G.Brook, J.D.Pascoe, J.T.Szymanski
Acta Cryst. (B), **28**, 1785, 1972

63.13 **(+) - Phenyl - triphenylsilyl - carbinol (–) - p - bromobenzoate (absolute configuration)**

$C_{32}H_{25}BrO_2Si$
K.T.Black, H.Hope *J. Amer. Chem. Soc.*, **93**, 3053, 1971

63.C **Methyl - triethylsiloxy - silicon phthalocyanine**

$C_{39}H_{34}N_8OSi_2$
For complete entry see 49.14

63.C **Chloro - bis(triphenylphosphine) platinum trimethylsilylmethilide**

$C_{40}H_{41}ClPtSi$
For complete entry see 71.81

PHOSPHORUS COMPOUNDS

64.1 **Trisodium phosphonoformate hexahydrate**

$CO_5P^{3-}, 3Na^+, 6H_2O$

R.R.Naqvi, P.J.Wheatley, E.Foresti-Serantoni

J. Chem. Soc. (A), 2751, 1971

64.2 **Trisodium phosphonoformate hexahydrate (independent data set)**

$CO_5P^{3-}, 3Na^+, 6H_2O$

R.R.Naqvi, P.J.Wheatley, E.Foresti-Serantoni

J. Chem. Soc. (A), 2751, 1971

64.3 **Calcium cyclo - tetraphosphonate decahydrate**

$C_4H_8O_{12}P_4^{4-}, 2Ca^{2+}, 10H_2O$

E.Philippot, J.C.Jumas, G.Brun, M.Maurin

Cryst. Struct. Comm., **1,** 103, 1972

64.4 **bis(Bromomethyl)acetoxyphosphine oxide**

$C_5H_9Br_2O_3P$

J.C.Clardy, G.K.McEwen, J.A.Mosbo, J.G.Verkade

J. Amer. Chem. Soc., **93,** 6937, 1971

64.5 **5,5 - Dimethyl - 2 - chloro - 2 - oxo - 1,3,2 - dioxaphosphorinane (at −40 ° C)**

$C_5H_{10}ClO_3P$

L.Silver, R.Rudman *Acta Cryst. (B),* **28,** 574, 1972

Also classified in 42

64.6 **Tetramethylformamidinium - phosphonate**

$C_5H_{13}N_2O_3P$

J.J.Daly *J. C. S. Dalton,* 1048, 1972

64.C **Diethylphosphorylguanidine - guanidinium chloride**

$C_5H_{14}N_3O_3P, 0.5CH_6N_3^+, 0.5Cl^-$

For complete entry see 8.9

64.7 **Tri(ethynyl)phosphine**

C_6H_3P

J.Kroon, J.B.Hulscher, A.F.Peerdeman *J. Molec. Struct.,* **7,** 217, 1971

64.8 **2,2,4 - Trichloro - 4,6,6 - tris(dimethylamino)cyclotriphosphazatriene**

$C_6H_{18}Cl_3N_6P_3$

F.R.Ahmed, D.R.Pollard *Acta Cryst. (B),* **28,** 513, 1972

64.9 **tris(1,2 - Dimethylhydrazino)diphosphine**
$C_6H_{18}N_6P_2$
W.van Doorne, G.W.Hunt, R.W.Perry, A.W.Cordes
Inorg. Chem., **10,** 2591, 1971

64.C **2,5 - Dithia - 1 - phenyl - 1 - thiophosphorus(v) - cyclopentane**
$C_8H_9PS_3$
For complete entry see 42.3

64.10 **1,3 - Di(t - butyl) - 2,4 - dichlorodiazadiphosphetidine**
$C_8H_{18}Cl_2N_2P_2$
K.W.Muir, J.F.Nixon *J. Chem. Soc. (D),* 1405, 1971

64.11 **Bicyclic phosphorane from 2 - amino - 2 - methylpropan - 1 - ol**
$C_8H_{19}N_2O_2P$
M.G.Newton, J.E.Collier
Amer. Cryst. Assoc., Abstr. Papers (Summer Meeting), 105, 1971
Also classified in 42

64.C **Lead diethyldithiophosphate**
$C_8H_{20}O_4P_2PbS_4$
For complete entry see 69.14

64.12 **Octamethoxy - cyclotetraphosphazene**
$C_8H_{24}N_4O_8P_4$
G.B.Ansell, G.J.Bullen *J. Chem. Soc. (A),* 2498, 1971

64.C **Pyridoxol 5′ - methylphosphonate**
$C_9H_{14}NO_5P$
For complete entry see 33.22

64.C **Adenosine - 3′,5′ - cyclic(5′ - phosphonate) monohydrate**
$C_{10}H_{12}N_5O_5P$, H_2O
For complete entry see 47.12

64.C **Adenosine - 3′ - phosphonate ethanol solvate**
$C_{10}H_{14}N_5O_6P$, C_2H_6O
For complete entry see 47.15

64.13 **trans - 1 - Methyl - 4 - t - butyl - 4 - phosphorinanol**
$C_{10}H_{21}OP$
A.T.McPhail, P.A.Luhan, S.I.Featherman, L.D.Quin
J. Amer. Chem. Soc., **94,** 2126, 1972

64.14 **1 - Phenyl - 4 - phosphorinanone**
$C_{11}H_{13}OP$
A.T.McPhail, J.J.Breen, L.D.Quin *J. Amer. Chem. Soc.*, **93,** 2574, 1971

64.C **3′ - Deoxy - 3′ - (dihydroxyphosphinylmethyl)adenosine ethanol solvate**
$C_{11}H_{16}N_5O_6P$, C_2H_6O
For complete entry see 47.16

64.15 **Bi - (2,2,4,4 - tetrachloro - 6 - phenyl - cyclotriphosphazatrien - 6 - yl)**
$C_{12}H_{10}Cl_8N_6P_6$
H.Zoer, A.J.Wagner *Cryst. Struct. Comm.*, **1,** 17, 1972

64.16 **1 - Iodomethyl - 4 - methyl - 1 - phenyl - 1 - phosphacyclopentane iodide**
$C_{12}H_{17}IP^+, I^-$
A.Fitzgerald, C.N.Caughlan, G.D.Smith
Amer. Cryst. Assoc., Abstr. Papers (Summer Meeting), 106, 1971

64.C **Lead(ii) O,O′ - di - isopropyl - phosphorodithioate**
$C_{12}H_{28}O_4P_2PbS_4$
For complete entry see 69.24

64.17 **Metaphosphoric acid diethylamide trimer (orthorhombic form)**
$C_{12}H_{30}N_3O_6P_3$
V.D.Cherepenskii-Malov, A.I.Gusev, I.A.Nuretdinov, Yu.T.Struchkov
Zh. Strukt. Khim., **12,** 126, 1971

64.18 **8 - Phenyl - 8 - methyl - 8 - phosphonium - bicyclo(3.2.1)octan - 3 - one iodide**
$C_{14}H_{18}OP^+, I^-$
Z.Zurr, U,Shmueli *Israel J. Chem.*, **9,** V, 1971

64.19 **2,2,3,3,4 - Pentamethyl - 1 - phenylphosphetane - 1 - oxide**
$C_{14}H_{21}OP$
A.Fitzgerald, C.N.Caughlan, G.D.Smith
Amer. Cryst. Assoc., Abstr. Papers (Winter Meeting), 28, 1972

64.20 **trans - Methyl meso - hydrobenzoin phosphite**
$C_{15}H_{15}O_3P$
B.S.Campbell, M.G.Newton
Amer. Cryst. Assoc., Abstr. Papers (Summer Meeting), 105, 1971
Also classified in 42

64.21 **6 - Chloro - 5,6,7,12 - tetrahydro - 2,5,7,10 - tetramethyl - dibenzo(d,g)(1,3,2)diaza phosphocine 6 - oxide**
$C_{17}H_{20}ClN_2OP$
T.S.Cameron *J. C. S. Perkin ii,* 591, 1972

64.C **(–) - Menthyl methyl phenylphosphonate**
$C_{17}H_{26}O_3P$
For complete entry see 52.4

64.C **Menthyl S - methyl - phenylphosphonothioate (absolute configuration)**
$C_{17}H_{27}O_2PS$
For complete entry see 52.5

64.22 **p - Bromophenyl - diphenylphosphine - oxide**
$C_{18}H_{14}BrOP$
W.Dreissig, K.Plieth *Acta Cryst. (B),* **27,** 1140, 1971

64.23 **p - Chlorophenyl - diphenylphosphine - oxide**
$C_{18}H_{14}ClOP$
W.Dreissig, K.Plieth *Acta Cryst. (B)*, **27**, 1146, 1971

64.24 **Methyl - triphenyl - phosphonium bis(α,α,α',α' - tetracyanoquinodimethanide)**
$C_{19}H_{18}P^{+}$, $C_{12}H_{4}N_{4}^{-}$, $C_{12}H_{4}N_{4}$
A.T.McPhail, G.M.Semeniuk, D.B.Chesnut *J. Chem. Soc. (A)*, 2174, 1971
Residue 2 classified in 7

64.25 **2,2 - Diphenyl - 2 - (bis(trifluoromethyl)methoxy) - 3 - methyl - 4,4 - bis(trifluoromethyl) - 2,2 - dihydro - 1,2 - oxaphosphetane**
$C_{20}H_{15}F_{12}O_{2}P$
M.-ul-Haque, C.N.Caughlan, F.Ramirez, J.F.Pilot, C.P.Smith
J. Amer. Chem. Soc., **93**, 5229, 1971

64.26 **Bicyclic phosphorane from (–) - ephedrine**
$C_{20}H_{27}N_{2}O_{2}P$
M.G.Newton, J.E.Collier
Amer. Cryst. Assoc., Abstr. Papers (Summer Meeting), 105, 1971

64.27 **Octa(dimethylamino) - cyclotetraphosphazene tetracarbonyl - tungsten**
$C_{20}H_{48}N_{12}O_{4}P_{4}W$
H.P.Calhoun, N.L.Paddock, J.Trotter, J.N.Wingfield
J. C. S. Chem. Comm., 875, 1972
Also classified in 83

64.28 **Tri - o - tolylphosphine oxide**
$C_{21}H_{21}OP$
R.A.Shaw, M.Woods, T.S.Cameron, B.Dahlen
Chem. and Industry, 151, 1971

64.29 **Tri - o - tolylphosphine**
$C_{21}H_{21}P$
R.A.Shaw, M.Woods, T.S.Cameron, B.Dahlen
Chem. and Industry, 151, 1971

64.30 **Tri - o - tolylphosphine selenide**
$C_{21}H_{21}PSe$
R.A.Shaw, M.Woods, T.S.Cameron, B.Dahlen
Chem. and Industry, 151, 1971

64.31 **1,2,5 - Triphenylphosphole**
$C_{22}H_{17}P$
W.P.Ozbirn, R.A.Jacobson, J.C.Clardy *J. Chem. Soc. (D)*, 1062, 1971

64.32 **10,10'(5H,5'H) - Spirobiphenophosphazinium chloride**
$C_{24}H_{18}N_{2}P^{+}$, Cl^{-}
R.N.Jenkins, L.D.Freedman, J.Bordner *J. Chem. Soc. (D)*, 1213, 1971
Residue 1 also classified in **42**

64.C **2,2,4,4 - Tetraiodo - 1,1,3,3 - tetraphenyl - cyclodiborataphosphoniane**
$C_{24}H_{20}B_2I_4P_2$
For complete entry see 62.15

64.33 **cis - 4,6 - Dimethyl - 2 - oxo - 2 - triphenylmethyl - 1,3,2 - dioxaphosphorinan**
$C_{24}H_{25}O_3P$
M.G.B.Drew, J.Rodgers *Acta Cryst. (B),* **28,** 924, 1972

64.34 **N - Toluene - p - sulfonyliminotriphenylphosphorane**
$C_{25}H_{22}NO_2PS$
A.F.Cameron, N.J.Hair, D.G.Morris *J. Chem. Soc. (D),* 918, 1971
Also classified in 11

64.35 **2 - Troponyl - cyanomethylene - triphenylphosphonium betaine methanol solvate**
$C_{27}H_{20}NOP$, CH_4O
I.Kawamoto, T.Hata, Y.Kishida, C.Tamura
Tetrahedron Letters, 2417, 1971
Residue 1 also classified in 22

64.36 **1,1 - bis(Dimethylamino) - 2,4,6 - triphenylphosphorin**
$C_{27}H_{29}N_2P$
U.Thewalt, C.E.Bugg *Acta Cryst. (B),* **28,** 871, 1972

64.37 **2,cis - 4,trans - 6,trans - 8 - tetrakis(Methylamino) - 2,4,6,8 - tetraphenyl - cyclotetraphosphazene**
$C_{28}H_{36}N_8P_4$
G.J.Bullen, P.R.Mallinson *J. C. S. Dalton,* 1412, 1972

64.38 **2 - Troponyl - (ethoxycarbonyl - methylene - triphenylphosphonium)betaine**
$C_{29}H_{25}O_3P$
I.Kawamoto, T.Hata, Y.Kishida, C.Tamura
Tetrahedron Letters, 1611, 1972
Also classified in 22

64.39 **1,1,5,8 - Tetraphenyl - 3 - p - bromophenyl - 1 - phospha - 2,4,9 - trioxabicyclo(4.3.0)nona - 5,7 - diene**
$C_{35}H_{26}BrO_3P$
D.D.Swank, C.N.Caughlan, F.Ramirez, J.F.Pilot
J. Amer. Chem. Soc., **93,** 5236, 1971

64.40 **tris(2,2′ - Dioxybiphenyl)cyclotriphosphazene**
$C_{36}H_{24}N_3O_6P_3$
H.R.Allcock, M.T.Stein, J.A.Stanko *J. Amer. Chem. Soc.,* **93,** 3173, 1971

64.41 **bis(Triphenylphosphine)iminium molybdenum nickel carbonyl**
$2C_{36}H_{30}NP_2^+$, $C_{16}Mo_2Ni_3O_{16}^{2-}$
J.K.Ruff, R.P.White Junior, L.F.Dahl *J. Amer. Chem. Soc.,* **93,** 2159, 1971

64.42 **bis(Triphenylphosphine)iminium tungsten nickel carbonyl**
$2C_{36}H_{30}NP_2^+$, $C_{16}Ni_3O_{16}W_2^{2-}$
J.K.Ruff, R.P.White Junior, L.F.Dahl *J. Amer. Chem. Soc.*, **93**, 2159, 1971

64.43 **bis(Triphenylphosphoranylidene)methane**
Hexaphenylcarbodiphosphorane
$C_{37}H_{30}P_2$
A.T.Vincent, P.J.Wheatley *J. C. S. Dalton*, 617, 1972

64.44 **Triphenylphosphonium 1 - (2,3,4 - triphenyl - 5 - acetyl - cyclopentadienylide)**
$C_{43}H_{33}OP$
G.Ferguson, F.C.March, D.F.Rendle
Amer. Cryst. Assoc., Abstr. Papers (Winter Meeting), 82, 1972

64.45 **Ethylene - 1,1 - bis(triphenylphosphonium) - 2,2 - bis(phenylamide)**
$C_{50}H_{40}N_2P_2$
F.K.Ross, W.C.Hamilton, F.Ramirez *Acta Cryst. (B)*, **27**, 2331, 1971

ARSENIC COMPOUNDS

65.1 **Trimethylarsenic dibromide**
$C_3H_9AsBr_3$
M.B.Hursthouse, I.A.Steer *J. Organometal. Chem.*, **27**, C11, 1971

65.2 **Trimethylarsenic dichloride**
$C_3H_9AsCl_2$
M.B.Hursthouse, I.A.Steer *J. Organometal. Chem.*, **27**, C11, 1971

65.3 **n - Propylarsonic acid**
$C_3H_9AsO_3$
M.R.Smith, R.A.Zingaro, E.A.Meyers
J. Organometal. Chem., **27**, 341, 1971

65.4 **Dithiothreitol arsenite**
1 - Arsa - 2,7 - dithia - 4 - hydroxy - 8 - oxabicyclo(3.2.1)octane
$C_4H_7AsO_2S_2$
W.B.T.Cruse, M.N.G.James *Acta Cryst. (B)*, **28**, 1325, 1972

65.5 **Triethylammonium phenylthioarsenate**
$C_6H_6AsO_2S^-$, $C_6H_{16}N^+$
L.G.McCrae, R.W.Perry, C.K.Fair, A.Hunt, A.W.Cordes
Inorg. Chem., **11**, 618, 1972
Residue 2 classified in 3

65.6 **Trimethylarsenic iron tetracarbonyl**
$C_7H_9AsFeO_4$
J.-J.Legendre, C.Girard, M.Huber *Bull. Soc. Chim. Fr.*, **6**, 1998, 1971

65.7 **Arsenic trichloride dipyridyl**
$C_{10}H_8AsCl_3N_2$
J.U.Cameron, R.C.G.Killean *Cryst. Struct. Comm.*, **1**, 31, 1972

65.8 **Diphenyldiarsenic trisulfide**
$C_{12}H_{10}As_2S_3$
A.W.Cordes, P.D.Gwinup, M.C.Malmstrom *Inorg. Chem.*, **11**, 836, 1972

65.C **Triphenylarsine oxide - tetrachlorocatechol complex**
$C_{18}H_{15}AsO$. $C_6H_2Cl_4O_2$
For complete entry see 60.35

65.9 **Methyl - triphenyl - arsonium bis(α,α,α',α' - tetracyanoquinodimethanide)**
$C_{19}H_{18}As^{+}$. $C_{12}H_{4}N_{4}^{-}$, $C_{12}H_{4}N_{4}$
A.T.McPhail, G.M.Semeniuk, D.B.Chesnut *J. Chem. Soc. (A)*, 2174, 1971
Residue 2 classified in 7

65.10 **Tetraphenylarsonium tri - iodide (at −160 ° C)**
$C_{24}H_{20}As^{+}$. I_{3}^{-}
J.Runsink, S.Swan-Walstra, T.Migchelsen
Acta Cryst. (B), **28**, 1331, 1972

65.C **Tetraphenylarsonium tetrakis(1 - isopropyltetrazol - 5 - ato) aurate(iii)**
$C_{24}H_{20}As^{+}$. $C_{16}H_{28}AuN_{16}^{-}$
For complete entry see 71.34

65.11 **Tetraphenylarsonium di - μ - azido - bis(diazido - palladate(ii))**
$2C_{24}H_{20}As^{+}$. $N_{18}Pd_{2}^{2-}$
W.P.Fehlhammer, L.F.Dahl *J. Amer. Chem. Soc.*, **94**, 3377, 1972

65.12 **Triphenylarsonium 1 - (3,4 - dibenzoyl - cyclopentadienylide)**
$C_{37}H_{27}AsO_{2}$
G.Ferguson, F.C.March, D.F.Rendle
Amer. Cryst. Assoc., Abstr. Papers (Winter Meeting), 82, 1972

65.13 **Triphenylarsonium 2 - acetyl - 3,4,5 - triphenylcyclopentadienylide**
$C_{43}H_{33}AsO$
G.Ferguson, D.F.Rendle, D.Lloyd, M.I.C.Singer
J. Chem. Soc. (D), 1647, 1971

65.14 **1 - (2,3,4 - Triphenyl - 5 - acetyl - cyclopenta - 1,3 - dienyl) - triphenyl - arsonium perchlorate**
$C_{43}H_{34}AsO^{+}$. ClO_{4}^{-}
G.Ferguson, F.C.March, D.F.Rendle
Amer. Cryst. Assoc., Abstr. Papers (Winter Meeting), 82, 1972

ANTIMONY AND BISMUTH COMPOUNDS

66.1 **Di - μ - methoxy - bis(tetrachloroantimony)**
$C_2H_6Cl_8O_2Sb_2$
H.Preiss *Z. Anorg. Allg. Chem.*, **380,** 65, 1971

66.2 **Succinyl chloride bis(pentachloro - antimony)**
$C_4H_4Cl_{12}O_2Sb_2$
J.-M.Le Carpentier, R.Weiss *Acta Cryst. (B),* **28,** 1442, 1972

66.3 **μ - Oxo - bis(trimethyl antimony) perchlorate**
$C_6H_{18}OSb_2^{2+}$, $2ClO_4^-$
F.C.March, G.Ferguson, D.R.Ridley, R.G.Goel
Amer. Cryst. Assoc., Abstr. Papers (Winter Meeting), 53, 1972

66.4 **μ - Oxo - bis(trimethyl antimony) azide**
$C_6H_{18}OSb_2^{2+}$, $2N_3^-$
F.C.March, G.Ferguson, D.R.Ridley, R.G.Goel
Amer. Cryst. Assoc., Abstr. Papers (Winter Meeting), 53, 1972

66.5 **Trimethylantimony iron tetracarbonyl**
$C_7H_9FeO_5Sb$
J.-J.Legendre, C.Girard, M.Huber *Bull. Soc. Chim. Fr.*, **6,** 1998, 1971

66.6 **D - Potassium antimony tartrate trihydrate**
$C_8H_4O_{12}Sb_2^{2-}$, $2K^+$, $3H_2O$
M.Hsiang-Ch'i *Ko Hsueh Tung Pao,* **17,** 502, 1966

66.7 **bis(1 - Oxopyridine - 2 - thiolato)phenyl - bismuth**
$C_{16}H_{13}BiN_2O_2S_2$
J.D.Curry, R.J.Jandacek *J. C. S. Dalton,* 1120, 1972

66.8 **Trichloro - tris(3 - sulfanilamido - 6 - methoxy - pyridazine) bismuth(iii)**
$C_{33}H_{36}BiCl_3N_{12}O_9S_3$
M.B.Ferrari, L.C.Capacchi, L.Cavalca, G.F.Gasparri
Acta Cryst. (B), **28,** 1169, 1972

66.9 **μ - Oxo - bis(triphenyl - bismuth) diperchlorate**
$C_{36}H_{30}Bi_2O^{2+}$, $2ClO_4^-$
G.Ferguson, R.G.Goel, F.C.March, D.R.Ridley, H.S.Prasad
J. Chem. Soc. (D), 1547, 1971

66.10 **μ - Oxo - bis(triphenyl - azido - antimony)**

$C_{36}H_{30}N_6OSb_2$

G.Ferguson, R.G.Goel, F.C.March, D.R.Ridley, H.S.Prasad

J. Chem. Soc. (D), 1547, 1971

GROUPS IA AND IIA COMPOUNDS

67.1 **Calcium hydrazinecarboxylate**
$(C_2H_6CaN_4O_4)_n$
A.Braibanti, A.M.M.Lanfredi, M.A.Pellinghelli, A.Tiripicchio
Acta Cryst. (B), **27,** 2448, 1971

67.2 **Aquo calcium hydrazine carboxylate**
$(C_2H_8CaN_4O_5)_n$
A.Braibanti, A.M.M.Lanfredi, M.A.Pellinghelli, A.Tiripicchio
Acta Cryst. (B), **27,** 2261, 1971

67.3 **Hexa - aquo - disodium DL - α - glycerophosphate**
$C_3H_{19}Na_2O_{12}P$
T.Taga, M.Senma, K.Osaki *J. C. S. Chem. Comm.*, 465, 1972
Also classified in 46

67.C **Tetra(trimethylene)dichromium tetra(lithium etherate)**
$4C_4H_{10}LiO^+ , C_{12}H_{14}Cr_2^{4-}$
For complete entry see 71.12

67.4 **Diaquo - calcium nitrilotriacetate**
$(C_6H_{11}CaNO_8)_n$
S.H.Whitlow *Acta Cryst. (B),* **28,** 1914, 1972

67.C **Potassium 2,2′ - (diethoxy)diethylether bis(cyclo - octatetraenyl) cerium(iii)**
$C_6H_{14}KO_3^+ , C_{16}H_{16}Ce^-$
For complete entry see 75.13

67.5 **Tri - aquo - calcium α - galactose bromide**
$(C_6H_{18}CaO_9^{2+})_n , 2nBr^-$
C.E.Bugg, W.J.Cook *J. C. S. Chem. Comm.*, 727, 1972
Residue 1 also classified in 45

67.6 **Tetra - aquo - calcium myo - inositol bromide monohydrate**
$(C_6H_{20}CaO_{10}^{2+})_n , 2nBr^- , nH_2O$
C.E.Bugg, W.J.Cook *J. C. S. Chem. Comm.*, 727, 1972
Residue 1 also classified in 21

67.7 **Tetra - aquo strontium dipicolinate**
$(C_7H_{11}NO_8Sr)_n$
K.J.Palmer, R.Y.Wong, J.C.Lewis *Acta Cryst. (B),* **28,** 223, 1972

67.8 **Aquo - barium methacrylate (neutron study)**
$(C_8H_{12}BaO_5)_n$
N.W.Isaacs, J.J.van der Zee, K.G.Shields, J.V.Tillack, D.H.Wheeler, F.H.Moore, C.H.L.Kennard *Cryst. Struct. Comm.*, **1**, 193, 1972

67.9 **tris(Sarcosine) calcium chloride (room temp.form)**
$(C_9H_{24}CaN_3O_6^{2+})_n$, $2nCl^-$
T.Ashida, S.Bando, M.Kakudo *Acta Cryst. (B)*, **28**, 1560, 1972
Residue 1 also classified in 48

67.10 **bis(Cyclopentadienyl) beryllium (at −120 ° C)**
$C_{10}H_{10}Be$
C.Wong, T.Lee, K.Chao, S.Lee *Acta Cryst. (B)*, **28**, 1662, 1972

67.11 **bis(Acetylacetonato) beryllium**
$C_{10}H_{14}BeO_4$
J.M.Stewart, B.Morosin
Amer. Cryst. Assoc., Abstr. Papers (Winter Meeting), 84, 1972

67.12 **Potassium 1,7,10,16 - tetraoxa - 4,13 - diaza - cyclo - octadecane thiocyanate**
$C_{12}H_{26}KN_2O_4^+$, CNS^-
D.Moras, B.Metz, M.Herceg, R.Weiss *Bull. Soc. Chim. Fr.*, 551, 1972

67.13 **Tetra - aquo - calcium lactose bromide trihydrate**
$(C_{12}H_{34}CaO_{15}^{2+})_n$, $2nBr^-$, $3nH_2O$
C.E.Bugg, W.J.Cook *J. C. S. Chem. Comm.*, 727, 1972
Residue 1 also classified in 45

67.14 **Diaquo - calcium 2,4,6,8 - cyclo - octatetraene - 1,2 - dicarboxylate**
$C_{14}H_{10}CaO_6$
D.A.Wright, K.Seff, D.P.Schoemaker *J. Cryst. Mol. Struct.*, **2**, 41, 1972
Also classified in 22

67.15 **Aquo - (2,3 - benzo - 1,4,7,10,13 - pentaoxacyclopentadec - 2 - ene)sodium iodide**
Aquo(benzo - 15 - crown - 5)sodium iodide
$C_{14}H_{24}NaO_6^+$, I^-
M.A.Bush, M.R.Truter *J. C. S. Perkin ii*, 341, 1972

67.16 **bis(μ - Propynyl) - di(methyl - trimethylamino - beryllium)**
$C_{14}H_{30}Be_2N_2$
B.Morosin, J.Howatson *J. Organometal. Chem.*, **29**, 7, 1971

67.17 **tris(Hexafluoroacetylacetonato) magnesium 1 - dimethylammonium - 8 - dimethylamino - naphthalene**
$C_{15}H_3F_{18}MgO_6^-$, $C_{14}H_{19}N_2^+$
M.R.Truter, B.L.Vickery *J. C. S. Dalton*, 395, 1972
Residue 2 classified in 24

67.18 **Dimethyl - bis(quinuclidine) beryllium**
$C_{16}H_{32}BeN_2$
C.D.Whitt, J.L.Atwood *J. Organometal. Chem.*, **32,** 17, 1971

67.19 **1,4,7,10,13,16 - Hexaoxa - cyclo - octadecane - potassium p - toluenesulfonate complex**
$C_{19}H_{31}KO_9S$
P.Groth *Acta Chem. Scand.*, **25,** 3189, 1971

67.20 **Aquo - (4,7,13,16,21,24 - hexaoxa - 1,10 - diazabicyclo(8.8.8) hexacosane) isothiocyanatobarium(+) thiocyanate**
$C_{19}H_{38}BaN_3O_7S^+$, CNS^-
B.Metz, D.Moras, R.Weiss *J. Amer. Chem. Soc.*, **93,** 1806, 1971

67.21 **2,3,11,12 - Dibenzo - 1,4,7,10,13,16 - hexaoxacyclo - octadeca - 2,11 - diene - sodium bromide complex dihydrate**
Dibenzo - 18 - crown - 6 sodium bromide complex dihydrate
$C_{20}H_{26}BrNaO_7$, $C_{20}H_{28}NaO_8^+$, Br^-, $2H_2O$
M.A.Bush, M.R.Truter *J. Chem. Soc. (B),* 1440, 1971

67.22 **Diaquo(4,7,10,16,19,24,27 - heptaoxa - 1,13 - diazabicyclo(11.8.8)nonacosane) barium(2+) dithiocyanate**
$C_{20}H_{44}BaN_2O_9^{2+}$, $2CNS^-$
B.Metz, D.Moras, R.Weiss *J. Amer. Chem. Soc.*, **93,** 1806, 1971

67.23 **cis,syn,cis - Dicyclohexyl - 18 - crown - 6 barium dithiocyanate complex monohydrate**
2,5,8,15,18,21 - Hexaoxatricyclo - (20,4.0,0^{9,14}) - hexacosane barium dithiocyanate complex monohydrate
$C_{22}H_{38}BaN_2O_7S_2$
N.K.Dalley, D.E.Smith, R.M.Izatt, J.J.Christensen
J. C. S. Chem. Comm., 90, 1972

67.24 **bis(Potassium thiocyanate) dibenzo - 24 - crown - 8 complex**
$C_{26}H_{32}K_2N_2O_8S_2$
D.E.Fenton, M.Mercer, N.S.Poonia, M.R.Truter
J. C. S. Chem. Comm., 66, 1972

67.25 **2,3.17,18 - Dibenzo - 1,4,7,10,13,16,19,22,25,28 - decaoxacyclotriaconta - 2,17 - diene - potassium iodide**
Dibenzo - 30 - crown - 10 potassium iodide
$C_{28}H_{40}KO_{10}^+$, I^-
M.A.Bush, M.R.Truter *J. C. S. Perkin ii,* 345, 1972
Residue 1 also classified in 38

67.26 **Diethyl - hexachloro - hexa(tetrahydrofuran) - tetramagnesium**
$C_{28}H_{58}Cl_6Mg_4O_6$
J.Toney, G.D.Stucky *J. Organometal. Chem.*, **28,** 5, 1971

67.27 **tris - p,p' - Diaminodiphenylmethane sodium chloride complex**
$(C_{39}H_{42}N_6Na^+)_n$, nCl^-
J.A.J.Jarvis, P.G.Owston *J. Chem. Soc. (D),* 1403, 1971

67.28 **tris - p,p' - Diaminodiphenylmethane sodium chloride complex**
$(C_{39}H_{42}N_6Na^+)_n$, nCl^-
J.W.Swardstrom, L.A.Duvall, D.P.Miller
Amer. Cryst. Assoc., Abstr. Papers (Summer Meeting), 107, 1971

GROUP III COMPOUNDS

68.1 **Trichloro - aluminium propionyl chloride**
$C_3H_5AlCl_4O$
J.-M.Le Carpentier, R.Weiss *Acta Cryst. (B),* **28,** 1437, 1972

68.C **1 - Methyl - 1 - galla - 2,4 - dicarba - closo - heptaborane(7)**
$C_3H_8B_4Ga$
For complete entry see 62.1

68.C **3 - Ethyl - 3 - alumina - 1,2 - dicarba - closo - dodecaborane(12)**
$C_4H_{16}AlB_9$
For complete entry see 62.7

68.C **7,8 - μ - Dimethylalumina - 1,2 - dicarba - nido - undecaborane(13) (at −100 ° C)**
$C_4H_{18}AlB_9$
For complete entry see 62.8

68.2 **Hexamethyl - dialuminium (at −170 ° C)**
$C_6H_{18}Al_2$
J.C.Huffman, W.E.Streib *J. Chem. Soc. (D),* 911, 1971

68.3 **Aziridinyl - gallane trimer**
$C_6H_{18}Ga_3N_3$
W.Harrison, A.Storr, J.Trotter *J. Chem. Soc. (D),* 1101, 1971

68.4 **(Hydrogen ethylenediaminetetra - acetato) aquogallate(iii)**
$C_{10}H_{15}GaN_2O_9$
C.H.L.Kennard *Inorg. Chim. Acta,* **1,** 347, 1967

68.5 **Trimethyl(quinuclidine) aluminium**
$C_{10}H_{22}AlN$
C.D.Whitt, L.M.Parker, J.L.Atwood
J. Organometal. Chem., **32,** 291, 1971

68.6 **bis(Methylamino) tetra(dimethylamino) tetra(aluminium chloride)**
$C_{10}H_{30}Al_4Cl_4N_6$
U.Thewalt, I.Kawada *Chem. Ber.,* **103,** 2754, 1970

68.7 **tris(Tetraethylammonium) tris - (1,2 - dicyanoethylene - 1,2 - dithiolato) indate(iii)**

$C_{12}InN_6S_6^{3-}$, $3C_8H_{20}N^+$
F.W.B.Einstein, R.D.G.Jones *J. Chem. Soc. (A)*, 2762, 1971
Residue 2 classified in 3

68.8 **Trichloro(1,10 - phenanthroline) thallium(iii)**

$C_{12}H_8Cl_3N_2Tl$
W.J.Baxter, G.Gafner *Inorg. Chem.*, **11**, 176, 1972

68.C **Quasiracemic mixture of cobalt tris(acetylacetonate) and aluminium tris(acetylacetonate) (absolute configuration)**

$C_{15}H_{21}CoO_6$, $C_{15}H_{21}AlO_6$
For complete entry see 77.9

68.9 **Dimethyl((Z) - N - phenylbenzimidato)(trimethylamine N - oxide) aluminum**

$C_{18}H_{25}AlN_2O_2$
Y.Kai, N.Yasuoka, N.Kasai, M.Kakudo, H.Yasuda, H.Tani
J. Chem. Soc. (D), 940, 1971

68.10 **bis(2 - Methyl - 8 - quinolinolato) chlorogallium(iii)**

$C_{20}H_{16}ClGaN_2O_2$
M.Shiro, Q.Fernando *Anal. Chem.*, **43**, 1222, 1971

68.11 **cis - Dichloro - bis(2,2′ - bipyridyl) gallium(iii) tetrachlorogallate(iii)**

$C_{20}H_{16}Cl_2GaN_4^+$, Cl_4Ga^-
R.Restivo, G.J.Palenik *J. C. S. Dalton*, 341, 1972

68.C **Di - μ - (tricarbonyl - π - cyclopentadienyl - tungsten - OO′)bis(dimethyl - aluminium)**

$C_{20}H_{22}Al_2O_6W_2$
For complete entry see 73.24

68.12 **Dimethylaluminium N - phenylbenzimidate dimer**

$C_{30}H_{32}Al_2N_2O_2$
Y.Kai, N.Yasuoka, N.Kasai, M.Kakudo
J. Organometal. Chem., **32**, 165, 1971

68.C **Aluminium tris(μ - carbonyl - cyclopentadienyl(dicarbonyl)tungsten) - tris(tetrahydrofuran)**

$C_{36}H_{39}AlO_{12}W_3$
For complete entry see 73.31

68.13 **Lithium tetrakis(di - t - butylmethyleneamino)aluminate**

$C_{36}H_{72}AlLiN_4$
H.M.M.Shearer, R.Snaith, J.D.Sowerby, K.Wade
J. Chem. Soc. (D), 1275, 1971

68.C **Grisorixin thallium salt monohydrate (absolute configuration)**

$C_{40}H_{67}O_{10}Tl$, H_2O
For complete entry see 50.14

68.14 Diphenyl - aluminium nitride tetramer

$C_{48}H_{40}Al_4N_4$

T.R.R.McDonald, W.S.McDonald *Acta Cryst. (B)*, **28,** 1619, 1972

GERMANIUM, TIN, LEAD COMPOUNDS

69.C **Thiourea - lead(ii) formate complex**
$6CHO_2^-, 16CH_4N_2S, 3Pb^{2+}$
For complete entry see 8.2

69.1 **Methyltin trinitrate**
$CH_3N_3O_9Sn$
G.S.Brownlee, A.Walker, S.C.Nyburg, J.T.Szymanski
J. Chem. Soc. (D), 1073, 1971

69.2 **Dimethyltin bis(fluorosulfate)**
$(C_2H_6F_2O_6S_2Sn)_n$
F.H.Allen, J.A.Lerbscher, J.Trotter *J. Chem. Soc. (A),* 2507, 1971

69.C **Tri(cyclopentadienyl - molybdenum)tetrasulfide trimethyl - dichloro - tin**
$C_3H_9Cl_2Sn^-, C_{15}H_{15}Mo_3S_4^+$
For complete entry see 73.14

69.3 **Germanium tetrachloride - trimethylamine**
$C_3H_9Cl_4GeN$
M.S.Bilton, M.Webster *J. C. S. Dalton,* 722, 1972

69.4 **tris(Methylimino - dichloro - germanium)**
$C_3H_9Cl_6Ge_3N_3$
M.Ziegler, J.Weiss *Z. Naturforsch., B,* **26,** 735, 1971

69.5 **Trimethyl aquo tin nitrate**
$C_3H_{11}NO_4Sn$
R.E.Drew, F.W.B.Einstein *Acta Cryst. (B),* **28,** 345, 1972

69.6 **Dimethylgermanium dicyanide**
$C_4H_6GeN_2$
J.Konnert, D.Britton, Y.M.Chow *Acta Cryst. (B),* **28,** 180, 1972

69.7 **Dimethyltin dicyanide**
$C_4H_6N_2Sn$
J.Konnert, D.Britton, Y.M.Chow *Acta Cryst. (B),* **28,** 180, 1972

69.8 **tetrakis(Thiourea) lead(ii) picrate**
$C_4H_{16}N_8PbS_4^{2+}, 2C_6H_2N_3O_7^-$
F.H.Herbstein, M.Kaftory *Acta Cryst. (B),* **28,** 405, 1972
Residue 2 classified in 6, 15

69.9 **Trimethyltin(iv) dicyanamide**
$(C_5H_9N_3Sn)_n$
Y.M.Chow *Inorg. Chem.*, **10**, 1938, 1971

69.10 **Dimethyltin(iv) bis(dicyanamide)**
$(C_6H_6N_6Sn)_n$
Y.M.Chow *Inorg. Chem.*, **10**, 1938, 1971

69.11 **Benzene - tin(ii) tetrachloroaluminate complex**
$C_6H_6Sn^{2+}$, $2AlCl_4^-$
E.L.Amma, H.Luth, P.F.Rodesiler, M.S.Weininger, A.G.Gash
Amer. Cryst. Assoc., Abstr. Papers (Summer Meeting), 39, 1971

69.12 **Trimethyltin prop - 1 - ynyl - sulfinate**
$(C_6H_{12}O_2SSn)_n$
D.Ginderow, M.Huber *C. R. Acad. Sci., Fr., C,* **274**, 1919, 1972

69.13 **hexakis(Thiourea) lead(ii) perchlorate**
$C_6H_{24}N_{12}PbS_6^{2+}$, $2ClO_4^-$
I.Goldberg. F.H.Herbstein *Acta Cryst. (B),* **28**, 400, 1972

69.14 **Lead diethyldithiophosphate**
$C_8H_{20}O_4P_2PbS_4$
T.Ito *Acta Cryst. (B),* **28**, 1034, 1972
Also classified in 64

69.15 **Tetraethylammonium tris(ethylxanthato) lead(ii)**
$C_9H_{15}O_3PbS_6^-$, $C_8H_{20}N^+$
W.G.Mumme, G.Winter *Inorg. Nucl. Chem. Letters,* **7**, 505, 1971
Residue 2 classified in 3

69.16 **2,2′ - Bipyridyl - tetrafluoro - germanium(iv) nitromethane solvate**
$C_{10}H_8F_4GeNi$, CH_3NO_2
A.D.Adley, P.H.Bird, A.R.Fraser, M.Onyszchuk
Inorg. Chem., **11**, 1402, 1972

69.17 **2,2′ - Bipyridyl - tetrafluoro - tin(iv) nitromethane solvate**
$C_{10}H_8F_4N_2Sn$, CH_3NO_2
A.D.Adley, P.H.Bird, A.R.Fraser, M.Onyszchuk
Inorg. Chem., **11**, 1402, 1972

69.18 **Distannous ethylenediaminetetra - acetate dihydrate**
$(C_{10}H_{12}N_2O_8Sn_2)_n$, $2nH_2O$
F.P.van Remoortere, J.J.Flynn, F.P.Boer, P.P.North
Inorg. Chem., **10**, 1511, 1971

69.C **π - Butadiene - π - cyclopentadienyl - (dichloro - methyl - germyl) iron (orthorhombic form)**
$C_{10}H_{14}Cl_2FeGe$
For complete entry see 72.5

69.C **π - Butadiene - π - cyclopentadienyl - (dichloro - methyl - germyl) iron (triclinic form)**

$C_{10}H_{14}Cl_2FeGe$

For complete entry see 72.6

69.19 **Aquo tin(iv) ethylenediaminetetra - acetate**

$C_{10}H_{14}N_2O_9Sn$

F.P.van Remoortere, J.J.Flynn, F.P.Boer *Inorg. Chem.*, **10,** 2313, 1971

69.20 **Lead(ii) diethyldithiocarbamate**

$C_{10}H_{20}N_2PbS_4$

H.Iwasaki, H.Hagihara *Acta Cryst. (B),* **28,** 507, 1972

69.21 **Dichloro(diphenyl)tin**

$C_{12}H_{10}Cl_2Sn$

P.T.Greene, R.F.Bryan *J. Chem. Soc. (A),* 2549, 1971

69.22 **Di - μ - chloro - bis(benzene - tin) tetrachloroaluminate**

$C_{12}H_{12}Cl_2Sn_2^{2+}$, $2AlCl_4^-$

M.S.Weininger, P.F.Rodesiler, A.G.Gash, E.L.Amma

J. Amer. Chem. Soc., **94,** 2135, 1972

69.23 **Di - μ - dimethyl - stannylene - bis(tetracarbonyl iron)**

$C_{12}H_{12}Fe_2O_8Sn_2$

C.J.Gilmore, P.Woodward *J. C. S. Dalton,* 1387, 1972

69.24 **Lead(ii) O,O′ - di - isopropyl - phosphorodithioate**

$C_{12}H_{28}O_4P_2PbS_4$

S.L.Lawton, G.T.Kokotailo *Inorg. Chem.*, **11,** 363, 1972

Also classified in 64

69.25 **μ - (Dimethylstannado) - μ - hydrido - dodecacarbonyl - trirhenium**

$C_{14}H_7O_{12}Re_3Sn$

B.T.Huie, C.M.Knobler

Amer. Cryst. Assoc., Abstr. Papers (Winter Meeting), 48, 1972

69.26 **Tri - μ - (dimethylgermanio) - tris(tricarbonyl - ruthenium)**

$C_{15}H_{15}Ge_3O_9Ru_3$

J.Howard, P.Woodward *J. Chem. Soc. (A),* 3648, 1971

69.C **μ - (Tetracarbonyliron) - μ - (di - μ - carbonyl - di(cyclopentadienylcobalt)) - di(dichloro - germanium)**

$C_{16}H_{10}Cl_4Co_2FeGe_2O_6$

For complete entry see 73.17

69.27 **Di - μ - chloro - bis(p - xylene - tin) tetrachloroaluminate**

$C_{16}H_{20}Cl_2Sn_2^{2+}$, $2AlCl_4^-$

M.S.Weininger, P.F.Rodesiler, A.G.Gash, E.L.Amma

J. Amer. Chem. Soc., **94,** 2135, 1972

69.28 **Lead 8 - mercaptoquinolinolate**
$C_{18}H_{12}N_2PbS_2$
V.M.Agre, E.A.Shugam *Zh. Strukt. Khim.*, **12**, 102, 1971

69.29 **Triphenyltin chloride**
$C_{18}H_{15}ClSn$
N.G.Bokii, G.N.Zakharova, Yu.T.Struchkov
Zh. Strukt. Khim., **11**, 895, 1970

69.C **Dichloro - bis(2,3,5,6 - tetrahapto - norbornadiene - dicarbonyl - cobalt) tin(iv)**
$C_{18}H_{16}Cl_2Co_2O_4Sn$
For complete entry see 75.20

69.30 **Tetramethylammonium tri(acetato) - diphenyl - plumbate(iv)**
$C_{18}H_{19}O_6Pb^-$, $C_4H_{12}N^+$
N.W.Alcock *J. C. S. Dalton,* 1189, 1972
Residue 2 classified in 3

69.31 **tetrakis(N,N - Diethyldithiocarbamato) tin(iv)**
$C_{20}H_{40}N_4S_8Sn$
C.S.Harreld, E.O.Schlemper *Acta Cryst. (B),* **27,** 1964, 1971

69.32 **Tetraphenyl germane (data set 1)**
$C_{24}H_{20}Ge$
P.C.Chieh *J. Chem. Soc. (A),* 3243, 1971

69.33 **Tetraphenyl - germane (data set 2)**
$C_{24}H_{20}Ge$
P.C.Chieh *J. Chem. Soc. (A),* 3243, 1971

69.34 **Tetraphenyltin**
$C_{24}H_{20}Sn$
N.A.Akhmed, G.G.Aleksandrov *Zh. Strukt. Khim.*, **11,** 891, 1970

69.35 **Triphenyl - (7 - cyclohepta - 1,3,5 - trienyl) tin**
$C_{25}H_{22}Sn$
J.E.Weidenborner, R.B.Larrabee, A.L.Bednowitz
J. Amer. Chem. Soc., **94,** 4140, 1972

69.36 **Tetrabenzyl - tin (at –40 ° C)**
$C_{28}H_{28}Sn$
G.R.Davies, J.A.J.Jarvis, B.T.Kilbourn *J. Chem. Soc. (D),* 1511, 1971

69.C **Diphenyl - bis(2,3,5,6 - tetrahapto - norbornadiene - dicarbonyl - cobalt) tin(iv)**
$C_{30}H_{26}Co_2O_4Sn$
For complete entry see 75.29

69.C **cis - (Hydroxydiphenylgermyl) - phenyl - bis(triethylphosphine) platinum(ii)**

$C_{30}H_{46}GeOP_2Pt$
For complete entry see 71.68

69.37 **Triphenyl tin N - benzoyl - N - phenylhydroxamate**

$C_{31}H_{25}NO_2Sn$
T.J.King, P.G.Harrison *J. C. S. Chem. Comm.*, 815, 1972

69.38 **Hexaphenyl - diplumbane**

$C_{36}H_{30}Pb_2$
H.Preut, H.J.Haupt, F.Huber *Z. Anorg. Allg. Chem.*, **388,** 165, 1972

69.C **(1,2 - bis(Diphenylphosphino)hexafluorocyclopentene) - carbonyl - iron - trimethyltin**

$C_{37}H_{34}F_6FeP_2Sn$
For complete entry see 73.32

69.C **Di((benzenesulfinato) - μ - hydroxo - phenylstannio) - tetracarbonyl - di - π - cyclopentadienyl - di - iron**

$C_{38}H_{32}Fe_2O_{10}S_2Sn_2$
For complete entry see 73.33

TELLURIUM COMPOUNDS

70.1 **β - Chloroethyl - tellurium - trichloride**
$(C_2H_4Cl_4Te)_n$
D.Kobelt, E.F.Paulus *Angew. Chem.*, **83,** 81, 1971

70.2 **Di - iodo(dimethyl) tellurium (α form)**
$C_2H_6I_2Te$
L.Y.Y.Chan, F.W.B.Einstein *J. C. S. Dalton,* 316, 1972

70.3 **tetrakis(Selenourea) tellurium(ii) dichloride**
$C_4H_{16}N_8Se_4Te^{2+}$, $2Cl^-$
S.Hauge, M.Tysseland *Acta Chem. Scand.*, **25,** 3072, 1971

70.4 **α - Tellurophene - carboxylic acid**
$C_5H_4O_2Te$
L.Fanfani, A.Nunzi, P.F.Zanazzi, A.R.Zanzari
Cryst. Struct. Comm., **1,** 273, 1972

70.5 **bis(2 - Chloropropyl) tellurium dichloride**
$C_6H_{12}Cl_4Te$
D.Kobelt, E.F.Paulus *J. Organometal. Chem.*, **27,** C63, 1971

70.6 **trans - Tellurium - di(selenocyanate) - di(ethylenethiourea)**
$C_8H_{12}N_6S_2Se_2Te$
K.Ase, K.Boyum, O.Fcss, K.Maroy *Acta Chem. Scand.*, **25,** 2457, 1971

70.7 **trans - Tellurium - di(thiocyanate) - di(ethylenethiourea)**
$C_8H_{12}N_6S_4Te$
K.Ase, K.Boyum, O.Foss, K.Maroy *Acta Chem. Scand.*, **25,** 2457, 1971

70.8 **trans - Diselenocyanato - bis(trimethylenethiourea) tellurium(ii)**
$C_{10}H_{16}N_6S_2Se_2Te$
K.Ase, O.Foss, I.Roti *Acta Chem. Scand.*, **25,** 3808, 1971

70.9 **bis(N,N - Diethyldithiocarbamato) tellurium(ii)**
$C_{10}H_{20}N_2S_4Te$
C.Fabiani, R.Spagna, A.Vaciago, L.Zambonelli
Acta Cryst. (B), **27,** 1499, 1971

70.10 **Phenoxtellurine di - iodide**

$C_{12}H_8I_2OTe$
J.D.McCullough
Amer. Cryst. Assoc., Abstr. Papers (Winter Meeting), 63, 1972

70.11 **Tellurium di - benzenthiosulfonate**

$C_{12}H_{10}O_4S_4Te$
K.Ase *Acta Chem. Scand.*, **25**, 838, 1971

70.12 **trans - Diselenocyanate - bis(tetramethylthiourea) tellurium(ii)**

$C_{12}H_{24}N_6S_2Se_2Te$
K.Ase, O.Foss, I.Roti *Acta Chem. Scand.*, **25**, 3808, 1971

70.13 **trans - Dithiourea - bis(tetramethylthiourea) tellurium(ii) bromide**

$C_{12}H_{32}N_8S_4Te^{2+}$, $2Br^-$
O.P.Anderson *Acta Chem. Scand.*, **25**, 3593, 1971

70.14 **trans - Dithiourea - bis(tetramethylthiourea) tellurium(ii) chloride**

$C_{12}H_{32}N_8S_4Te^{2+}$, $2Cl^-$
O.P.Anderson *Acta Chem. Scand.*, **25**, 3593, 1971

70.15 **trans - Tellurium - di(benzenethiosulfonate) - di(ethylenethiourea)**

$C_{18}H_{22}N_4O_4S_6Te$
K.Ase, K.Maartmann-Moe, J.O.Solheim
Acta Chem. Scand., **25**, 2467, 1971

70.16 **tetrakis(D ethyldithiocarbamato) tellurium(iv)**

$C_{20}H_{40}N_4S_8Te$
S.Esperas, S.Husebye, S.E.Svaeren *Acta Chem. Scand.*, **25**, 3539, 1971

70.17 **tris(Diethyldithiocarbamato) - phenyl - tellurium(iv)**

$C_{21}H_{35}N_3S_6Te$
S.Esperas, S.Husebye, S.E.Svaeren *Acta Chem. Scand.*, **25**, 3539, 1971

TRANSITION METAL-C COMPOUNDS

71.1 **1,1 - Dichloro - 1 - pallada - 2,5 - di(methylamino) - 3,4 - diazacyclopentadiene**

$C_4H_{10}Cl_2N_4Pd$

W.M.Butler, J.H.Enemark *Inorg. Chem.*, **10,** 2416, 1971

71.2 **bis(π - 7,8 - Dicarbaundecaboran(13) - 10 - ylthio)methyliumato(4 -) cobalt**

$C_5H_{21}B_{18}CoS_2$

M.R.Churchill, K.Gold *Inorg. Chem.*, **10,** 1928, 1971

Also classified in 62

71.3 **Tetramethyl - bis(N - methyl - N - nitrosohydroxylaminato) tungsten(vi)**

$C_6H_{18}N_4O_4W$

S.R.Fletcher, A.Shortland, A.C.Skapski, G.Wilkinson

J. C. S. Chem. Comm., 922, 1972

Also classified in 84

71.4 **Cyclopentadienyl(trimethyl) platinum(iv)**

$C_8H_{14}Pt$

G.W.Adamson, J.C.J.Bart, J.J.Daly *J. Chem. Soc. (A),* 2616, 1971

Also classified in 73

71.5 **Cesium - 3,3′ - commo - bis(nonahydro - 1,2 - dimethyl - 1,2 - dicarba - 3 - chroma - closo - dodecaborate) monohydrate**

$C_8H_{30}B_{18}Cr^-$, Cs^+, H_2O

D.Saint Clair, A.Zalkin, D.H.Templeton *Inorg. Chem.*, **10,** 2587, 1971

Residue 1 also classified in 62

71.6 **π - Cyclopentadienyl - ethylene - tetrafluoroethylene - rhodium**

$C_9H_9F_4Rh$

L.J.Guggenberger, R.Cramer *J. Amer. Chem. Soc.*, **94,** 3779, 1972

Also classified in 73

71.7 **Compound X**

$C_9H_{17}N_6Pt^-$, $C_{24}H_{20}B^+$

W.M.Butler, J.H.Enemark, A.L.Balch

Amer. Cryst. Assoc., Abstr. Papers (Winter Meeting), 47, 1972

Residue 2 classified in 62

71.8 **Pentacarbonyl - (dimethylamino(ethoxy)carbene) - chromium(0)**

$C_{10}H_{11}CrNO_6$
G.Huttner, B.Krieg *Chem. Ber.*, **105**, 67, 1972

71.9 **Dicarbonyl - 3 - (π - (2 - cyclohexadienyl)) - σ - propenoyl iron**

$C_{11}H_7FeO_3$
P.J.van Vuuren, R.J.Fletterick, J.Meinwald, R.E.Hughes
J. Amer. Chem. Soc., **93**, 4394, 1971
Also classified in 75

71.10 **bis(Cyclopentadienyl) - methyl - nitrosyl - molybdenum**

$C_{11}H_{13}MoNO$
F.A.Cotton, G.A.Rusholme *J. Amer. Chem. Soc.*, **94**, 402, 1972
Also classified in 73

71.11 **Tetracyano - tetra(N - methylcyano) molybdenum**

$C_{12}H_{12}MoN_8$
F.H.Cano, D.W.J.Cruickshank *J. Chem. Soc. (D)*, 1617, 1971

71.12 **Tetra(trimethylene)dichromium tetra(lithium etherate)**

$C_{12}H_{14}Cr_2^{4-}$, $4C_4H_{10}LiO^+$
J.Krausse, G.Schodl *J. Organometal. Chem.*, **27**, 59, 1971
Residue 2 classified in 67

71.13 **bis - (Di - μ - allyl - (di - μ - chloro - diplatinum))**

$C_{12}H_{20}Cl_4Pt_4$
G.Raper, W.S.McDonald *J. C. S. Dalton*, 265, 1972
Also classified in 72

71.14 **Pentacarbonyl(methyl(phenylthio)carbene) chromium**

$C_{13}H_8CrO_5S$
R.J.Hoare, O.S.Mills *J. C. S. Dalton*, 653, 1972

71.15 **Nitrogeno molybdenum chelate compound**

$C_{13}H_{14}MoN_2O_5^+$, F_6P^-
C.K.Prout, T.S.Cameron, A.R.Gent *Acta Cryst. (B)*, **28**, 32, 1972
Residue 1 also classified in 73, 83

71.16 **Dichloro(pyridinium propylide)pyridine - platinum(ii)**

$C_{13}H_{16}Cl_2N_2Pt$
M.Keeton, R.Mason, D.R.Russell *J. Organometal. Chem.*, **33**, 259, 1971
Also classified in 83

71.C **1,6 - Dichloro - 2,3 - trimethylene - 4,5 - bis(pyridine) platinum(iv)**

$C_{13}H_{16}Cl_2N_2Pt$
For complete entry see 83.63

71.17 **Tetrachloro(pyridinium propylide)pyridine - platinum(iv) chloroform solvate**

$C_{13}H_{16}Cl_4N_2Pt$, $CHCl_3$
M.Keeton, R.Mason, D.R.Russell *J. Organometal. Chem.*, **33**, 259, 1971
Residue 1 also classified in 83

71.18 **Ferrocenylmethyl(dimethyl)ammonium tetrachlorozincate monohydrate**
$2C_{13}H_{18}FeN^{+} . Cl_4Zn^{2-} . H_2O$
C.S.Gibbons, J.Trotter *J. Chem. Soc. (A),* 2659, 1971
Residue 1 also classified in 73

71.19 **cis - Dichloro(diethyl(phenyl)phosphine) (ethylisocyanide) platinum(ii)**
$C_{13}H_{20}Cl_2NPPt$
B.Jovanovic, L.Manojlovic-Muir *J. C. S. Dalton,* 1176, 1972
Also classified in 86

71.20 **bis(Methylaminomethyl)methylamino - iron(ii) tetrakis(methylisocyanide) hexafluorophosphate**
$C_{13}H_{23}FeN_7^{2+} , 2F_6P^-$
J.Miller, A.L.Balch, J.H.Enemark *J. Amer. Chem. Soc.,* **93,** 4613, 1971
Residue 1 also classified in 72

71.21 **cis - Dichloro - bis(phenylisocyanide) platinum(ii)**
$C_{14}H_{10}Cl_2N_2Pt$
B.Jovanovic, L.Manojlovic-Muir, K.W.Muir *J. C. S. Dalton,* 1178, 1972

71.22 **2 - Mercuri - 4 - methylphenol 2′ - nitroso - 4′ - methylphenolate**
$C_{14}H_{13}HgNO_3$
Y.Kobayashi, Y.Iitaka, Y.Kido *Bull. Chem. Soc. Jap.,* **43,** 3070, 1970
Also classified in 83, 84

71.23 **Chloro - (2 - methoxycyclo - octa - 1,5 - dienyl)pyridine - platinum**
$C_{14}H_{18}ClNOPt$
G.Bombieri, E.Forsellini, R.Graziani *J. C. S. Dalton,* 525, 1972
Also classified in 75

71.24 **Chloro - aquo - bis(trimethylarsine)tetrakis(trifluoromethyl) rhodiacyclopentadiene**
$C_{14}H_{20}As_2ClF_{12}ORh$
J.T.Mague *J. Amer. Chem. Soc.,* **93,** 350, 1971
Also classified in 86

71.25 **N - (2 - cis - Dichloro - platinum - hex - 5 - en - yl) - α - methylbenzylamine (form I,absolute configuration)**
$C_{14}H_{20}Cl_2NPt$
C.Pedone, E.Benedetti *J. Organometal. Chem.,* **31,** 403, 1971
Also classified in 72

71.26 **N - (2 - cis - Dichloro - platinum - hex - 5 - en - yl) - α - methylbenzylamine (form II,absolute configuration)**
$C_{14}H_{20}Cl_2NPt$
C.Pedone, E.Benedetti *J. Organometal. Chem.,* **31,** 403, 1971
Also classified in 72

71.27 **(1 - Methyl - 3 - ethyl(σ - 1,3 - h^2,π - 1,2,3 - h^3) - allyl)nonacarbonyl - triruthenium**
$C_{15}H_{10}O_9Ru$
M.Evans, M.Hursthouse, E.W.Randall, E.Rosenberg, L.Milone, M.Valle
J. C. S. Chem. Comm., 545, 1972

71.28 **Allene - trimer hexacarbonyl di - iron (form ii)**
$C_{15}H_{12}Fe_2O_6$
N.Yasuda, Y.Kai, N.Yasuoka, N.Kasai, M.Kakudo
J. C. S. Chem. Comm., 157, 1972
Also classified in 72

71.29 **Allene - trimer hexacarbonyl di - iron (form iii)**
$C_{15}H_{12}Fe_2O_6$
N.Yasuda, Y.Kai, N.Yasuoka, N.Kasai, M.Kakudo
J. C. S. Chem. Comm., 157, 1972
Also classified in 72

71.30 **Pentacarbonyl - (cyclohexylamino - (1 - methoxyvinyl) - carbene) chromium(0)**
$C_{15}H_{17}CrNO_6$
G.Huttner, S.Lange *Chem. Ber.*, **103,** 3149, 1970

71.31 **Tricarbonyl - (tetracyanobicyclo(4.2.1)nonanediyl) iron**
$C_{16}H_8FeN_4O_3$
J.Weaver, P.Woodward *J. Chem. Soc. (A),* 3521, 1971
Also classified in 75

71.32 **5 - bis(N - t - Butylcyano) - 2,2,4,4 - tetra(trifluoromethyl) - 5 - nickela - 1,3 - oxazolidine**
$C_{16}H_{19}F_{12}N_3NiO$
R.Countryman, B.R.Penfold *J. Chem. Soc. (D),* 1598, 1971
Also classified in 84

71.33 **bis(Acetylacetonato - μ - allyl - platinum)**
$C_{16}H_{24}O_4Pt_2$
G.Raper, W.S.McDonald *J. C. S. Dalton,* 265, 1972
Also classified in 72, 77

71.34 **Tetraphenylarsonium tetrakis(1 - isopropyltetrazol - 5 - ato) aurate(iii)**
$C_{16}H_{28}AuN_{16}^-$, $C_{24}H_{20}As^+$
W.P.Fehlhammer, L.F.Dahl *J. Amer. Chem. Soc.*, **94,** 3370, 1972
Residue 2 classified in 65

71.35 **Tetramethylammonium carbido(hexadeca - carbonyl)hexa - iron**
$C_{17}Fe_6O_{16}^{2-}$, $2C_4H_{12}N^+$
M.R.Churchill, J.Wormald, J.Knight, M.J.Mays
J. Amer. Chem. Soc., **93,** 3073, 1971
Residue 2 classified in 3

71.36 **o - (Tetracarbonyl manganese) - benzylideneaniline**
$C_{17}H_{10}MnNO_4$
M.I.Bruce, B.L.Goodall, M.Z.Iqbal, F.G.A.Stone, R.J.Doedens, R.G.Little
J. Chem. Soc. (D), 1595, 1971
Also classified in 83

71.37 **1 - (2 - (Dimethylarsino) - 3,3,4,4 - tetrafluorocyclobutenyl) - dimethylarsino - tri - iron nonacarbonyl**
$C_{17}H_{12}As_2F_4Fe_3O_9$
F.W.B.Einstein, A.-M.Pilotti, R.Restivo *Inorg. Chem.*, **10,** 1947, 1971
Also classified in 86

71.38 **bis(Cyclopentadienyl) - (1 - phenyl - 2 - carboxylato) titanium**
$C_{17}H_{14}O_2Ti$
G.G.Aleksandrov, Yu.T.Struchkov *Zh. Strukt. Khim.*, **12,** 667, 1971
Also classified in 73, 81

71.39 **(Methinyl tricobalt enneacarbonyl) acetylene dicobalt hexacarbonyl complex**
$C_{18}HCo_5O_{15}$
R.J.Dellaca, B.R.Penfold, B.H.Robinson, J.L.Spencer
Inorg. Chem., **9,** 2197, 1970
Also classified in 72

71.40 **N,N′ - Ethylene - bis(salicylideneiminato) - ethyl cobalt(iii)**
$C_{18}H_{19}CoN_2O_2$
M.Calligaris, D.Minichelli, G.Nardin, L.Randaccio
J. Chem. Soc. (A), 2720, 1971
Also classified in 78

71.41 **N - Methyl - N - phenyl - N′ - methyl - N′ - (o - phenyl - chloropalladium) biacetyl osazone**
$C_{18}H_{21}ClN_4Pd$
G.Bombieri, L.Caglioti, L.Cattalini, E.Forsellini, F.Gasparrini, R.Graziani, P.A.Vigato *J. Chem. Soc. (D),* 1415, 1971
Also classified in 83

71.42 **bis(Methinyl - tricobalt - enneacarbonyl)**
$C_{20}Co_6O_{18}$
M.D.Brice, B.R.Penfold *Inorg. Chem.*, **11,** 1381, 1972

71.43 **Dibromo - (2 - bromo - 1 - (o - diphenylphosphinophenyl)ethyl) gold**
$C_{20}H_{17}AuBr_3P$
M.A.Bennett, K.Hoskins, W.R.Kneen, R.S.Nyholm, P.B.Hitchcock, R.Mason, G.B.Robertson, A.D.C.Towl
J. Amer. Chem. Soc., **93,** 4591, 1971
Also classified in 86

71.44 **Niobocene dimer**

$C_{20}H_{20}Nb_2$

L.J.Guggenberger, F.N.Tebbe *J. Amer. Chem. Soc.*, **93,** 5924, 1971

Also classified in 73

71.45 **Nonacarbonyl(cyclododecatrienetriyl)hydrido - triruthenium**

$C_{21}H_{16}O_9Ru_3$

A.Cox, P.Woodward *J. Chem. Soc. (A),* 3599, 1971

Also classified in 75

71.46 **Dibromo - (2 - bromo - 1 - (o - diphenylphosphinobenzyl)ethyl) gold**

$C_{21}H_{19}AuBr_3P$

M.A.Bennett, K.Hoskins, W.R.Kneen, R.S.Nyholm, P.B.Hitchcock, R.Mason, G.B.Robertson, A.D.C.Towl

J. Amer. Chem. Soc., **93,** 4591, 1971

Also classified in 86

71.47 **Di - π - cyclopentadienyl - diphenyl - titanium**

$C_{22}H_{20}Ti$

V.Kocman, J.C.Rucklidge, R.J.O'Brien, W.Santo

J. Chem. Soc. (D), 1340, 1971

Also classified in 73

71.48 **bis(t - Butyl isocyanide)(azobenzene) nickel(0)**

$C_{22}H_{28}N_4Ni$

R.S.Dickson, J.A.Ibers *J. Amer. Chem. Soc.*, **94,** 2988, 1972

Also classified in 83

71.49 **Tribromo - (o - styryldimethylarsine) - 2 - ethoxy - 2 - (o - dimethylarsinophenyl) - ethyl platinum**

$C_{22}H_{31}As_2Br_3OPt$

M.A.Bennett, K.Hoskins, W.R.Kneen, R.S.Nyholm, R.Mason, P.B.Hitchcock, G.B.Robertson, A.D.C.Towl

J. Amer. Chem. Soc., **93,** 4592, 1971

Also classified in 86

71.50 **1 - (Triphenylphosphonium) - 2 - (bromo - tetracarbonyl - manganese(ii)) - acetylene**

$C_{24}H_{15}BrMnO_4P$

S.Z.Goldberg, E.M.Duesler, K.N.Raymond *Inorg. Chem.*, **11,** 1397, 1972

71.51 **Acetylacetonato - ((1 - pentamethylcyclopenta - 2,4 - dienyl) - (p - tolyl)methyl) - palladium**

$C_{24}H_{32}O_2Pd$

C.Calvo, T.Hosokawa, H.Reinheimer, P.M.Maitlis

J. Amer. Chem. Soc., **94,** 3237, 1972

Also classified in 73, 77

71.52 **bis - μ - (Trimethylsilylmethylidyne) tetrakis(trimethylsilylmethyl) diniobium(v)**

$C_{24}H_{62}Nb_2Si_6$
F.Huq, W.Mowat, A.C.Skapski, G.Wilkinson
J. Chem. Soc. (D), 1477, 1971
Also classified in 63

71.53 **hexakis(Trimethylsilyl) dimolybdenum**

$C_{24}H_{66}Mo_2Si_6$
F.Huq, W.Mowat, A.Shortland, A.C.Skapski, G.Wilkinson
J. Chem. Soc. (D), 1079, 1971
Also classified in 63

71.54 **Compound iv**

$C_{25}H_{16}Fe_3O_9$
A.H.-J.Wang, I.C.Paul, G.N.Schrauzer *J. C. S. Chem. Comm.*, 736, 1972
Also classified in 72, 75

71.55 **Compound iv benzene solvate**

$C_{25}H_{16}Fe_3O_9$, $0.5C_6H_6$
A.H.-J.Wang, I.C.Paul, G.N.Schrauzer *J. C. S. Chem. Comm.*, 736, 1972
Residue 1 also classified in 72, 75

71.56 **Tricarbonyl(3 - (o - (diphenyl - phosphino)phenyl) - 2 - buten - 2 - olato) manganese**

$C_{25}H_{20}MnO_4P$
M.A.Bennett, G.B.Robertson, R.Watt, P.O.Whimp
J. Chem. Soc. (D), 752, 1971
Also classified in 86, 84

71.57 **Malononitrilo - (1,2 - propane - N,N′ - bis(salicylideneiminato)) cobalt(iii) pyridine**

$C_{25}H_{22}CoN_5O_2$
N.A.Bailey, B.M.Higson, E.D.McKenzie
Inorg. Nucl. Chem. Letters, **7,** 591, 1971
Also classified in 78, 83

71.58 **Carboxymethyl - palladium (triphenylphosphine)pyridine**

$C_{25}H_{22}NO_2PPd$
S.Baba, T.Ogura, S.Kawaguchi, H.Tokunan, Y.Kai, N.Kasai
J. C. S. Chem. Comm., 910, 1972
Also classified in 81, 86, 83

71.59 **Tetramethylcyclobutadiene - (trifluoromethyl) - bis(dimethylphenylphosphine) platinum(ii) hexafluoroantimonate**

$C_{25}H_{34}F_3P_2Pt^+$, F_6Sb^-
D.B.Crump, N.C.Payne
Amer. Cryst. Assoc., Abstr. Papers (Winter Meeting), 46, 1972
Residue 1 also classified in 72, 86

71.60 **Di - μ - acetato - bis(2 - methallyl - norbornyl) dipalladium(ii)**
$C_{26}H_{40}O_4Pd_2$
M.Zocchi, G.Tieghi, A.Albinati *J. Organometal. Chem.*, **33,** C47, 1971
Also classified in 72, 81

71.61 **μ - (3,6 - Diphenylpyridazine) - hexacarbonyl di - iron maleic anhydride**
$C_{26}H_{142}FeN_2O_9$
H.A.Patel, A.J.Carty, M.Mathew, G.J.Palenik
J. C. S. Chem. Comm., 810, 1972
Also classified in 83

71.62 **Tetrabenzyl - hafnium (at —40 ° C)**
$C_{28}H_{28}Hf$
G.R.Davies, J.A.J.Jarvis, B.T.Kilbourn *J. Chem. Soc. (D),* 1511, 1971

71.63 **Tetrabenzyl - titanium**
$C_{28}H_{28}Ti$
I.W.Bassi, G.Allegra, R.Scordamaglia, G.Chioccola
J. Amer. Chem. Soc., **93,** 3787, 1971

71.64 **Tetrabenzyl - titanium (at —40 ° C)**
$C_{28}H_{28}Ti$
G.R.Davies, J.A.J.Jarvis, B.T.Kilbourn *J. Chem. Soc. (D),* 1511, 1971

71.65 **(Isoprene dimer) nickel tri(cyclohexyl)phosphine**
$C_{28}H_{49}NiP$
B.Barnett, B.Bussemeier, P.Heimbach, P.W.Jolly, C.Kruger, I.Tkatchenko, G.Wilke *Tetrahedron Letters,* 1457, 1972
Also classified in 72, 86

71.66 **Molybdenum dithiocarbamate compound**
$C_{28}H_{56}Mo_2N_4S_8$
L.Ricard, J.Estienne, R.Weiss *J. C. S. Chem. Comm.*, 906, 1972
Also classified in 80, 85

71.67 **bis(2 - (1,3 - Diphenyl)imidazolyl) mercury(ii) perchlorate**
$C_{30}H_{24}HgN_4^{2+}$, $2ClO_4^-$
P.Luger, G.Ruban *Acta Cryst. (B),* **27,** 2276, 1971

71.68 **cis - (Hydroxydiphenylgermyl) - phenyl - bis(triethylphosphine) platinum(ii)**
$C_{30}H_{46}GeOP_2Pt$
R.J.D.Gee, H.M.Powell *J. Chem. Soc. (A),* 1956, 1971
Also classified in 69, 86

71.69 **Iodo - hexakis(t - butyl isocyanide) molybdenum(ii) iodide**
$C_{30}H_{54}IMoN_6^+$, I^-
D.F.Lewis, S.J.Lippard *Inorg. Chem.*, **11,** 621, 1972

71.70 **1 - (Cyclopentadienyl - triphenylphosphine - ruthenium) - 1,2,3,4 - tetra(trifluoromethyl) - buta - 1,3 - diene**
$C_{31}H_{21}F_{12}PRu$
T.Blackmore, M.I.Bruce, F.G.A.Stone, R.E.Davis, A.Garza
J. Chem. Soc. (D), 852, 1971
Also classified in 72, 86

71.71 **Compound Z**
$C_{37}H_{24}O_7Os_3P_2$
C.W.Bradford, R.S.Nyholm, G.J.Gainsford, J.M.Guss, P.R.Ireland, R.Mason *J. C. S. Dalton,* 87, 1972
Also classified in 86

71.72 **bis(Triphenylphosphine) - methyl - platinum iodosulfone**
$C_{37}H_{33}IO_2P_2PtS$
M.R.Snow, J.McDonald, F.Basolo, J.A.Ibers
J. Amer. Chem. Soc., **94,** 2526, 1972
Also classified in 86

71.73 **Compound Y**
$C_{38}H_{24}O_8Os_3P_2$
C.W.Bradford, R.S.Nyholm, G.J.Gainsford, J.M.Guss, P.R.Ireland, R.Mason *J. C. S. Chem. Comm.,* 87, 1972
Also classified in 86

71.74 **trans - bis(Methyldiphenylphosphine) - (σ - pentafluorophenyl) - (σ - pentachlorophenyl) - nickel(ii)**
$C_{38}H_{26}Cl_5F_5NiP_2$
M.R.Churchill, M.V.Veidis *J. Chem. Soc. (A),* 3463, 1971
Also classified in 86

71.75 **trans - bis(Methyldiphenylphosphino) - bis(σ - pentafluorophenyl) nickel(ii)**
$C_{38}H_{26}F_{10}NiP_2$
M.R.Churchill, M.V.Veidis *J. C. S. Dalton,* 670, 1972
Also classified in 86

71.76 **3,3 - Di(trifluoromethyl) - 1 - nickela - 2 - oxa - cyclopropane bis(triphenylphosphine)**
$C_{39}H_{30}F_6NiOP_2$
R.Countryman, B.R.Penfold *J. Chem. Soc. (D),* 1598, 1971
Also classified in 84, 86

71.77 **cis - 1,2 - bis(Triphenylphosphine - gold) - 1,2 - bis(trifluoromethyl)ethylene**
$C_{40}H_{30}Au_2F_6P_2$
C.J.Gilmore, P.Woodward *J. Chem. Soc. (D),* 1233, 1971
Also classified in 86

71.78 **Cyano(cyanoacetylido)bis(triphenylphosphine) platinum(ii)**
$C_{40}H_{30}N_2P_2Pt$
W.H.Baddley, C.Panattoni, G.Bandoli, D.A.Clemente, U.Belluco
J. Amer. Chem. Soc., **93,** 5590, 1971
Also classified in 86

71.79 **Diphenyl - bis(2,9 - dimethyl - 1,10 - phenanthroline) mercury(ii)**
$C_{40}H_{34}HgN_4$
A.J.Canty, B.M.Gatehouse *Acta Cryst. (B)*, **28,** 1872, 1972
Also classified in 83

71.80 **Anhydrodi(carboxymethyl) palladium - bis(triphenylphosphine)**
$C_{40}H_{34}O_3P_2Pd$
S.Baba, T.Ogura, S.Kawaguchi, H.Tokunan, Y.Kai, N.Kasai
J. C. S. Chem. Comm., 910, 1972
Also classified in 86

71.81 **Chloro - bis(triphenylphosphine) platinum trimethylsilylmethilide**
$C_{40}H_{41}ClPtSi$
M.R.Collier, C.Eaborn, B.Jovanovic, M.F.Lapperts, L.Manojlovic-Muir, K.W.Muir, M.M.Truelock *J. C. S. Chem. Comm.*, 613, 1972
Also classified in 86, 63

71.82 **(4 - Methyl - 2 - cupriobenzyl)dimethylamine tetramer**
$C_{40}H_{56}Cu_4N_4$
J.M.Guss, R.Mason, I.Sotofte, G.van Koten, J.G.Noltes
J. C. S. Chem. Comm., 446, 1972
Also classified in 83

71.83 **Di - μ - chloro - bis(chloro - di(4 - methylpyridino) - (3 - (3,5 - dimethyl - 5 - hydroxymethyl - tetrahydrofuran)methyl) rhodium)**
$C_{40}H_{58}Cl_4N_4O_4Rh_2$
J.A.Evans, D.R.Russell, A.Bright, B.L.Shaw *J. Chem. Soc. (D)*, 841, 1971
Also classified in 83

71.84 **bis(Trifluoromethyl)diazomethane bis(triphenylphosphine) platinum(0) methylene chloride solvate**
$C_{42}H_{30}F_{12}N_2P_2Pt$, $0.4CH_2Cl_2$
J.Clemens, R.E.Davis, M.Green, J.D.Oliver, F.G.A.Stone
J. Chem. Soc. (A), 1095, 1971
Residue 1 also classified in 83, 86

71.85 **Chloro - carbonyl - bis(triphenylphosphine) - (5 - fluoro - 2 - diazo - phenyl) iridium tetrafluoroborate acetone solvate**
$C_{43}H_{34}ClFIrN_2OP_2^+$, BF_4^- , C_3H_6O
F.W.B.Einstein, A.B.Gilchrist, G.W.Rayner-Canham, D.Sutton
J. Amer. Chem. Soc., **94,** 645, 1972
Residue 1 also classified in 86, 83

71.86 **Di - μ - hexafluorobut - 2 - enylene - bis(cis - triphenylphosphine - nitrosyl - iridium(i))**
$C_{44}H_{30}F_{12}Ir_2N_2O_2P_2$
J.Clemens, M.Green, M.-C.Kuo, C.J.Fritchie Junior, J.T.Mague, F.G.A.Stone *J. C. S. Chem. Comm.*, 53, 1972

71.87 **Chloro - rhodium norbornadiene hexafluorobut - 2 - yne adduct tetramer**
$C_{44}H_{32}Cl_4F_{24}Rh_4$
J.A.Evans, R.D.W.Kemmitt, B.Y.Kimura, D.R.Russell
J. C. S. Chem. Comm., 509, 1972
Also classified in 75

71.88 **Diphenyl - bis(2,4,7,9 - tetramethyl - 1,10 - phenanthroline) mercury(ii)**
$C_{44}H_{42}HgN_4$
A.J.Canty, B.M.Gatehouse *Acta Cryst. (B),* **28,** 1872, 1972
Also classified in 83

71.89 **Compound X**
$C_{45}H_{30}O_9Os_3P_2$
C.W.Bradford, R.S.Nyholm, G.J.Gainsford, J.M.Guss, P.R.Ireland, R.Mason *J. C. S. Chem. Comm.*, 87, 1972
Also classified in 86

71.90 **1,1 - bis(Triphenylphosphine) - 2,3 - diphenyl - 1 - platinacyclobut - 2 - ene - 4 - one**
$C_{48}H_{40}OP_2Pt$
W.Wong, S.J.Singer, W.D.Pitts, S.F.Watkins, W.H.Baddley
J. C. S. Chem. Comm., 672, 1972
Also classified in 86

71.91 **tetrakis(Pentafluorophenyl) - μ - bis(diphenylarsino)methane - dimercury(ii)**
$C_{49}H_{22}As_2F_{20}Hg_2$
A.J.Canty, B.M.Gatehouse *J. C. S. Dalton,* 511, 1972
Also classified in 86

71.92 **1,2,3,4 - Tetrahydro - 1,4 - dioxo - naphthaleno(2,3 - c)(1′ - bis(triphenylphosphine)chloro - rhoda) cyclopentadiene monohydrate ethanol solvate**
$C_{60}H_{44}ClO_2P_2Rh \cdot H_2O \cdot 0.5C_2H_6O$
E.Muller, E.Langer, H.Jakle, H.Muhm, W.Hoppe, R.Graziani, A.Gieren, F.Brandl *Z. Naturforsch., B,* **20,** 305, 1971
Residue 1 also classified in 86

71.93 **5,5,7,8,9,10 - Hexaphenyl - 6 - (triphenylphosphino) - 5,6 - dihydro - 5 - phospha - 6 - rhodabenzocyclo - octene toluene solvate**
$C_{64}H_{51}P_2Rh \cdot C_7H_8$
J.S.Ricci, J.A.Ibers *J. Organometal. Chem.*, **27,** 261, 1971
Residue 1 also classified in 86

71.94 Tetracopper - bis(triphenylphosphine - iridium - tetra(phenylacetylide))
$C_{100}H_{70}Cu_4Ir_2P_2$
O.M.A.Salah, M.I.Bruce, M.R.Churchill, S.A.Bezman
J. C. S. Chem. Comm., 858, 1972
Also classified in 72, 86

METAL PI-COMPLEXES (OPEN-CHAIN)

72.1 **Potassium trichloroethylene platinate monohydrate**
Zeise's salt
$C_2H_4Cl_3Pt^-, K^+, H_2O$
G.B.Bokii, G.A.Kukina *Zh. Strukt. Khim.*, **6**, 706, 1965

72.2 **Potassium trichloroethylene platinate monohydrate**
Zeise's salt
$C_2H_4Cl_3Pt^-, K^+, H_2O$
J.A.J.Jarvis, B.T.Kilbourn, P.G.Owston *Acta Cryst. (B)*, **27**, 366, 1971

72.3 **Trichloro(π - trans - but - 2 - en - 1,4 - diammonium) platinum(ii) chloride hemihydrate**
$C_4H_{12}Cl_3N_2Pt^+, Cl^-, 0.5H_2O$
R.Spagna, L.Zambonelli *J. Chem. Soc. (A)*, 2544, 1971

72.4 **π - Allyl - di(thiourea) nickel(ii) chloride**
$C_5H_{13}N_4NiS_2^+, Cl^-$
A.Sirigu *Inorg. Chem.*, **9**, 2245, 1970
Residue 1 also classified in 79

72.5 **π - Butadiene - π - cyclopentadienyl - (dichloro - methyl - germyl) iron (orthorhombic form)**
$C_{10}H_{14}Cl_2FeGe$
V.G.Andrianov, V.P.Martynov, Yu.T.Struchkov
Zh. Strukt. Khim., **12**, 866, 1971
Also classified in 69, 73

72.6 **π - Butadiene - π - cyclopentadienyl - (dichloro - methyl - germyl) iron (triclinic form)**
$C_{10}H_{14}Cl_2FeGe$
V.G.Andrianov, V.P.Martynov, Yu.T.Struchkov
Zh. Strukt. Khim., **12**, 866, 1971
Also classified in 69, 73

72.7 **Cinnamaldehyde iron tricarbonyl**
$C_{12}H_8FeO_4$
A.de Cian, R.Weiss
Nat. Proprietes Liason Coordin. Coll. Internation. Paris, 261, 1970

72.8 **(+) - cis - Dichloro - (1 - butene) - (α - methylbenzylamine) platinum(ii) (absolute configuration)**
$C_{12}H_{19}Cl_2NPt$
C.Pedone, E.Benedetti *J. Organometal. Chem.*, **29,** 443, 1971
Also classified in 83

72.C **bis - (Di - μ - allyl - (di - μ - chloro - diplatinum))**
$C_{12}H_{20}Cl_4Pt_4$
For complete entry see 71.13

72.9 **Tetra - allyl - dimolybdenum**
$C_{12}H_{20}Mo_2$
F.A.Cotton, J.R.Pipal *J. Amer. Chem. Soc.*, **93,** 5441, 1971

72.10 **μ - Chloro - chloro - di - π - allyl(cyclohexanone oxime) dipalladium**
$C_{12}H_{21}Cl_2NOPd_2$
Y.Kitano, T.Kajimoto, M.Kashiwagi, Y.Kinoshita
J. Organometal. Chem., **33,** 123, 1971
Also classified in 83

72.11 **trans - Dichloro(benzylamino)(3 - methyl - 1 - pentene) platinum(ii)**
$C_{13}H_{21}Cl_2NPt$
S.Merlino, R.Lazzaroni, G.Montagnoli
J. Organometal. Chem., **30,** C93, 1971
Also classified in 83

72.C **bis(Methylaminomethyl)methylamino - iron(ii) tetrakis(methylisocyanide) hexafluorophosphate**
$C_{13}H_{23}FeN_7^{2+}$, $2F_6P^-$
For complete entry see 71.20

72.C **N - (2 - cis - Dichloro - platinum - hex - 5 - en - yl) - α - methylbenzylamine (form I,absolute configuration)**
$C_{14}H_{20}Cl_2NPt$
For complete entry see 71.25

72.C **N - (2 - cis - Dichloro - platinum - hex - 5 - en - yl) - α - methylbenzylamine (form II,absolute configuration)**
$C_{14}H_{20}Cl_2NPt$
For complete entry see 71.26

72.C **Allene - trimer hexacarbonyl di - iron (form ii)**
$C_{15}H_{12}Fe_2O_6$
For complete entry see 71.28

72.C **Allene - trimer hexacarbonyl di - iron (form iii)**
$C_{15}H_{12}Fe_2O_6$
For complete entry see 71.29

72.12 **π - Allyl - dihydrobis(3,5 - dimethyl - 1 - pyrazolyl)borato - dicarbonylmolybdenum**
$C_{15}H_{21}BMoN_4O_2$
C.A.Kosky, P.Ganis, G.Avitabile *Acta Cryst. (B)*, **27**, 1859, 1971
Also classified in 83

72.13 **bis(Cyclopentadienyl) - 2,2′ - bi - π - allyl - bis(nickel)**
$C_{16}H_{18}Ni_2$
A.E.Smith *Inorg. Chem.*, **11**, 165, 1972
Also classified in 73

72.C **bis(Acetylacetonato - μ - allyl - platinum)**
$C_{16}H_{24}O_4Pt_2$
For complete entry see 71.33

72.14 **(2 - Neopentyl - π - allyl) palladium chloride dimer**
$C_{16}H_{30}Cl_2Pd_2$
M.X.Minasjanc, Yu.T.Struchkov *Arm. Chem. Zh.*, **24**, 569, 1971

72.15 **Tricarbonyl - 1 - syn - (1′,2′ - dihydro - 2′ - oxo - 1′ - oxa - azulen - 3′ - yl) - pentahapto - pentadienylmanganese**
$C_{17}H_{11}MnO_5$
M.J.Barrow, O.S.Mills, F.Haque, P.L.Pauson
J. Chem. Soc. (D), 1239, 1971

72.C **(Methinyl tricobalt enneacarbonyl) acetylene dicobalt hexacarbonyl complex**
$C_{18}HCo_5O_{15}$
For complete entry see 71.39

72.16 **Dichloro(2,7 - dimethyl - octa - 2,6 - diene - 1,8 - diyl) ruthenium(iv) dimer**
$C_{20}H_{32}Cl_4Ru_2$
A.Colombo, G.Allegra *Acta Cryst. (B)*, **27**, 1653, 1971

72.17 **Hexacarbonyl - (4,4 - diphenyl - 2 - azabuta - 2,3 - diene) di - iron**
$C_{21}H_{13}Fe_2NO_6$
K.Ogawa, A.Torii, H.Kobayashi-Tamura, T.Watanabe, T.Yoshida, S.Otsuka *J. Chem. Soc. (D)*, 991, 1971

72.18 **Bromo(tri - o - vinylphenyl)phosphine - rhodium(i)**
$C_{24}H_{21}BrPRh$
C.Nave, M.R.Truter *J. Chem. Soc. (D)*, 1253, 1971
Also classified in 86

72.C **Compound iv**
$C_{25}H_{16}Fe_3O_9$
For complete entry see 71.54

72.C **Compound iv benzene solvate**
$C_{25}H_{16}Fe_3O_9 , 0.5C_6H_6$
For complete entry see 71.55

72.C **Tetramethylcyclobutadiene - (trifluoromethyl) - bis(dimethylphenylphosphine) platinum(ii) hexafluoroantimonate**
$C_{25}H_{34}F_3P_2Pt^+$, F_6Sb^-
For complete entry see 71.59

72.C **Di - μ - acetato - bis(2 - methallyl - norbornyl) dipalladium(ii)**
$C_{26}H_{40}O_4Pd_2$
For complete entry see 71.60

72.19 **(Allene tetramer)(1,3 - diphenylpropan - 1,3 - dionato) rhodium**
$C_{27}H_{27}O_2Rh$
G.Pantini, P.Racanelli, A.Immirzi, L.Porri
J. Organometal. Chem., **33,** C17, 1971
Also classified in 77

72.C **(Isoprene dimer) nickel tri(cyclohexyl)phosphine**
$C_{28}H_{49}NiP$
For complete entry see 71.65

72.C **1 - (Cyclopentadienyl - triphenylphosphine - ruthenium) - 1,2,3,4 - tetra(trifluoromethyl) - buta - 1,3 - diene**
$C_{31}H_{21}F_{12}PRu$
For complete entry see 71.70

72.20 **(Tetraphenylbutatriene) tetracarbonyl - iron**
$C_{32}H_{20}FeO_4$
D.Bright, O.S.Mills *J. Chem. Soc. (A),* 1979, 1971

72.21 **Octacarbonyl - μ - (1,2,3,4 - tetraphenylbut - 2 - ene - 1,1,4,4 - tetrayl) - triangulo - triosmium**
$C_{36}H_{20}O_8Os_3$
G.Ferraris, G.Gervasio *J. C. S. Dalton,* 1057, 1972

72.22 **1 - Chloro - 1,2,2 - trifluoroethylene bis(triphenylphosphine) platinum(0) complex**
$C_{38}H_{30}ClF_3P_2Pt$
J.N.Francis, A.McAdam, J.A.Ibers *J. Organometal. Chem.*, **29,** 131, 1971
Also classified in 86

72.23 **1,1 - Dichloro - 2,2 - difluoro - ethylene bis(triphenylphosphine) platinum(0) complex**
$C_{38}H_{30}Cl_2F_2P_2Pt$
J.N.Francis, A.McAdam, J.A.Ibers *J. Organometal. Chem.*, **29,** 131, 1971
Also classified in 86

72.24 **Tetrachloroethylene bis(triphenylphosphine) platinum(0) complex**
$C_{38}H_{30}Cl_4P_2Pt$
J.N.Francis, A.McAdam, J.A.Ibers *J. Organometal. Chem.*, **29,** 131, 1971
Also classified in 86

72.25 **bis(Triphenylphosphine)(ethylene) nickel**
$C_{38}H_{34}NiP_2$
P.-T.Cheng, C.D.Cook, C.H.Koo, S.C.Nyburg, M.T.Shiomi
Acta Cryst. (B), **27,** 1904, 1971
Also classified in 86

72.26 **bis(Triphenylphosphine) platinum ethylene complex**
$C_{38}H_{34}P_2Pt$
P.-T.Cheng, C.D.Cook, S.C.Nyburg, K.Y.Wan
Inorg. Chem., **10,** 2210, 1971
Also classified in 86

72.27 **bis(Triphenylphosphine)allene - platinum**
$C_{39}H_{34}P_2Pt$
M.Kadonaga, N.Yasucka, N.Kasai *J. Chem. Soc. (D),* 1597, 1971
Also classified in 86

72.28 **μ - Allyl - μ - iodo - bis(triphenylphosphine palladium) benzene solvate**
$C_{39}H_{35}IP_2Pd_2$, C_6H_6
Y.Kobayashi, Y.Iitaka, H.Yamazaki *Acta Cryst. (B),* **28,** 899, 1972
Residue 1 also classified in 86

72.29 **bis(Diphenyl - trifluoromethylethynyl - phosphine) - decacarbonyl - tetracobalt**
$C_{40}H_{20}Co_4F_6O_{10}P_2$
N.K.Hota, H.A.Patel, A.J.Carty, M.Mathew, G.J.Palenik
J. Organometal. Chem., **32,** C55, 1971
Also classified in 86

72.30 **1,1 - Dichloro - 2,2 - dicyanoethylene bis(triphenylphosphine) platinum(0) complex**
$C_{40}H_{30}Cl_2N_2P_2Pt$
A.McAdam, J.N.Francis, J.A.Ibers *J. Organometal. Chem.,* **29,** 149, 1971
Also classified in 86

72.31 **bis(Triphenylphosphine) - bis - π - allyl - ruthenium toluene solvate**
$C_{42}H_{40}P_2Ru$, C_7H_8
A.E.Smith
Amer. Cryst. Assoc., Abstr. Papers (Summer Meeting), 83, 1971
Residue 1 also classified in 86

72.32 **tris(Diphenylacetylene) tungsten - carbonyl**
$C_{43}H_{30}OW$
R.M.Laine, R.E.Moriarty, R.Bau *J. Amer. Chem. Soc.,* **94,** 1402, 1972

72.C **(Cyano(dicyanomethyl)ketiminato)carbonyl (tetracyanoethylene) bis(triphenylphosphine) iridium benzene solvate**
$C_{49}H_{31}IrN_8OP_2$, $0.5C_6H_6$
For complete entry see 86.45

72.C **Tetracopper - bis(triphenylphosphine - iridium - tetra(phenylacetylide))**
$C_{100}H_{70}Cu_4Ir_2P_2$
For complete entry see 71.94

METAL PI-COMPLEXES (CYCLOPENTADIENE)

73.C **Cyclopentadienyl(trimethyl) platinum(iv)**
$C_8H_{14}Pt$
For complete entry see 71.4

73.C **π - Cyclopentadienyl - ethylene - tetrafluoroethylene - rhodium**
$C_9H_9F_4Rh$
For complete entry see 71.6

73.1 **Decachloro - ruthenocene**
$C_{10}Cl_{10}Ru$
G.M.Brown, F.L.Hedberg, H.Rosenberg *J. C. S. Chem. Comm.*, 5, 1972

73.2 **Difluoro - bis(π - cyclopentadienyl) - zirconium(iv)**
$C_{10}H_{10}F_2Zr$
M.A.Bush, G.A.Sim *J. Chem. Soc. (A)*, 2225, 1971

73.3 **Ferrocinium tris(trichloroacetic acid)**
Dicyclopentadienyl iron tris(trichloroacetic acid)
$C_{10}H_{10}Fe^+$, $C_6H_3Cl_9O_6^-$
A.W.Schlueter, H.B.Gray
Amer. Cryst. Assoc., Abstr. Papers (Summer Meeting), 41, 1971
Residue 2 classified in 12

73.4 **Iodo - bis(cyclopentadienyl) ruthenium tri - iodide**
$C_{10}H_{10}IRu^+$, I_3^-
A.W.Schlueter, H.B.Gray
Amer. Cryst. Assoc., Abstr. Papers (Summer Meeting), 41, 1971

73.5 **Di - iodo - bis(π - cyclopentadienyl) zirconium(iv)**
$C_{10}H_{10}I_2Zr$
M.A.Bush, G.A.Sim *J. Chem. Soc. (A)*, 2225, 1971

73.6 **Cyclopentadienyl - nitro - manganese bis - μ - nitrosyl cyclopentadienyl - nitrosyl - manganese**
$C_{10}H_{10}Mn_2N_4O_5$
J.L.Calderon, F.A.Cotton, B.G.DeBoer, N.Martinez
J. Chem. Soc. (D), 1476, 1971

73.7 **bis(π - Cyclopentadienyl)tetrasulfido - tungsten(iv)**
$C_{10}H_{10}S_4W$
B.R.Davis, I.Bernal, H.Kopf *Angew. Chem.*, **83**, 1018, 1972

73.8 **Dicyclopentadienyl titanium pentasulfide**
$C_{10}H_{10}S_5Ti$
E.F.Epstein, I.Bernal, H.Kopf *J. Organometal. Chem.*, **26**, 229, 1971

73.C **π - Butadiene - π - cyclopentadienyl - (dichloro - methyl - germyl) iron (orthorhombic form)**
$C_{10}H_{14}Cl_2FeGe$
For complete entry see 72.5

73.C **π - Butadiene - π - cyclopentadienyl - (dichloro - methyl - germyl) iron (triclinic form)**
$C_{10}H_{14}Cl_2FeGe$
For complete entry see 72.6

73.C **bis(Cyclopentadienyl) - methyl - nitrosyl - molybdenum**
$C_{11}H_{13}MoNO$
For complete entry see 71.10

73.9 **Tricarbonyl((trimethylsilyl) - π - cyclopentadienyl) rhenium**
$C_{11}H_{13}O_3ReSi$
W.Harrison, J.Trotter *J. C. S. Dalton*, 678, 1972
Also classified in 63

73.10 **trans - β - Ferrocenyl - acrylonitrile**
$C_{13}H_{11}FeN$
T.E.Borovyak, V.E.Shklover, A.I.Gusev, S.P.Gubin, A.A.Koridze, Yu.T.Struchkov *Zh. Strukt. Khim.*, **11**, 1087, 1970

73.11 **(1,1' - Trimethylene - dicyclopentadienyl) dichloro - titanium**
$C_{13}H_{14}Cl_2Ti$
B.R.Davis, I.Bernal *J. Organometal. Chem.*, **30**, 75, 1971

73.C **Nitrogeno molybdenum chelate compound**
$C_{13}H_{14}MoN_2O_5^+$, F_6P^-
For complete entry see 71.15

73.C **Ferrocenylmethyl(dimethyl)ammonium tetrachlorozincate monohydrate**
$2C_{13}H_{18}FeN^+$, Cl_4Zn^{2-} , H_2O
For complete entry see 71.18

73.12 **2,2 - Dicyanovinyl - ferrocene**
$C_{14}H_9FeN_2$
A.P.Krukonis, J.Silverman, N.F.Yannoni *Acta Cryst. (B)*, **28**, 987, 1972

73.13 **bis(Cyclopentadienyl)pentacarbonyl - dirhenium**
$C_{15}H_{10}O_5Re_2$
A.S.Foust, J.K.Hoyano, W.A.G.Graham
J. Organometal. Chem., **32**, C65, 1971

73.14 **Tri(cyclopentadienyl - molybdenum)tetrasulfide trimethyl - dichloro - tin**
$C_{15}H_{15}Mo_3S_4^+$, $C_3H_9Cl_2Sn^-$
P.J.Vergamini, H.Vahrenkamp, L.F.Dahl
J. Amer. Chem. Soc., **93,** 6327, 1971
Residue 2 classified in 69

73.15 **Di - iodo(carbonyl)(ferrocene - 1,1′ - bis(dimethylarsine)) nickel(ii)**
$C_{15}H_{20}As_2FeI_2NiO$
C.G.Pierpont, R.Eisenberg *Inorg. Chem.*, **11,** 828, 1972
Also classified in 86

73.16 **Di - μ - carbonyl - nonacarbonyl - (π - cyclopentadienyl - rhodio)tri - iron**
$C_{16}H_5Fe_3O_{11}Rh$
M.R.Churchill, M.V.Veidis *J. Chem. Soc. (A),* 2995, 1971

73.17 **μ - (Tetracarbonyliron) - μ - (di - μ - carbonyl - di(cyclopentadienylcobalt)) - di(dichloro - germanium)**
$C_{16}H_{10}Cl_4Co_2FeGe_2O_6$
M.Elder, W.L.Hutcheon *J. C. S. Dalton,* 175, 1972
Also classified in 69

73.C **bis(Cyclopentadienyl) - 2,2′ - bi - π - allyl - bis(nickel)**
$C_{16}H_{18}Ni_2$
For complete entry see 72.13

73.C **bis(Cyclopentadienyl) - (1 - phenyl - 2 - carboxylato) titanium**
$C_{17}H_{14}O_2Ti$
For complete entry see 71.38

73.18 **Tri - μ - carbonyl - pentacarbonyl - bis(π - cyclopentadienyl rhodio) di - iron**
$C_{18}H_{10}Fe_2O_8Rh_2$
M.R.Churchill, M.V.Veidis *J. Chem. Soc. (A),* 2170, 1971

73.19 **Di - μ - thio - n - butyl(bis - π - cyclopentadienyl) molybdenum - iron - dichloride**
$C_{18}H_{28}Cl_2FeMoS_2$
T.S.Cameron, C.K.Prout *Acta Cryst. (B),* **28,** 453, 1972
Also classified in 85

73.20 **1′ - Acetyl - 1 - benzoyl - ferrocene**
$C_{19}H_{16}FeO_2$
G.Calvarin, D.Weigel *Acta Cryst. (B),* **27,** 1253, 1971

73.21 **π - Indenyl rhodium π - duroquinone**
$C_{19}H_{19}O_2Rh$
G.G.Aleksandrov, Yu.T.Struchkov *Zh. Strukt. Khim.*, **12,** 120, 1971
Also classified in 74

73.22 **Dicarbonyl - π - cyclopentadienyl - iodo(tri(n - butyl)phosphine) molybdenum(ii)**
$C_{19}H_{32}IMoO_2P$
R.H.Fenn, J.H.Cross *J. Chem. Soc. (A)*, 3312, 1971
Also classified in 86

73.C **Niobocene dimer**
$C_{20}H_{20}Nb_2$
For complete entry see 71.44

73.C **μ - (5 - Cyclopentadienylcyclopentadiene) - bis(π - cyclopentadienyl - platinum)**
$C_{20}H_{20}Pt_2$
For complete entry see 75.23

73.23 **Tetra(cyclopentadienyl) titanium**
$C_{20}H_{20}Ti$
J.L.Calderon, F.A.Cotton, B.G.DeBoer, J.Takats
J. Amer. Chem. Soc., **93,** 3592, 1971

73.24 **Di - μ - (tricarbonyl - π - cyclopentadienyl - tungsten - OO′)bis(dimethyl - aluminium)**
$C_{20}H_{22}Al_2O_6W_2$
G.J.Gainsford, R.R.Schrieke, J.D.Smith *J. C. S. Chem. Comm.*, 650, 1972
Also classified in 68

73.C **Cyclopentadienyl(tricarbonyl) molybdenum - μ₄ - sulfido - (cyclopentadienyl(dicarbonyl)molybdenum - bis(tricarbonyl rhenium) - μ₃ - sulfide)**
$C_{21}H_{10}Mo_2O_{11}Re_2S_2$
For complete entry see 85.24

73.25 **Chloro - tri(cyclopentadienyl - dicarbonyl - iron) antimony tetrachloroferrate(ii) methylene chloride solvate**
$2C_{21}H_{15}ClFe_3O_6Sb^+$, Cl_4Fe^{2-}, CH_2Cl_2
T.Toan, L.F.Dahl *J. Amer. Chem. Soc.*, **93,** 2654, 1971

73.C **Di - π - cyclopentadienyl - diphenyl - titanium**
$C_{22}H_{20}Ti$
For complete entry see 71.47

73.26 **2,1′ - Trimethylene - 1 - (α - phenyl - α - hydroxypropyl) - ferrocene (form i)**
$C_{22}H_{24}FeO$
C.Lecomte, Y.Dusausoy, C.Moise, J.Protas, J.Tirouflet
C. R. Acad. Sci., Fr., C, **273,** 952, 1971

73.27 **tetrakis(Cyclopentadienyl - iron - carbonyl) hexafluorophosphate**
$C_{24}H_{20}Fe_2O_4^+$, F_6P^-
Trinh-Toan, W.P.Fehlhammer, L.F.Dahl
J. Amer. Chem. Soc., **94,** 3389, 1972

73.28 **tetrakis(Cyclopentadienyl - iron - carbonyl)**
$C_{24}H_{20}Fe_4O_4$
M.A.Neuman, Trinh-Toan, L.F.Dahl *J. Amer. Chem. Soc.*, **94**, 3383, 1972

73.29 **bis(bis(π - Cyclopentadienyl) niobium - bis(μ - methanethiolato)) nickel tetrafluoroborate dihydrate**
$C_{24}H_{32}Nb_2NiS_4^{2+}$. $2BF_4^-$. $2H_2O$
W.E.Douglas, M.L.H.Greene, C.K.Prout, G.V.Rees
J. Chem. Soc. (D), 896, 1971
Residue 1 also classified in 85

73.C **Acetylacetonato - ((1 - pentamethylcyclopenta - 2,4 - dienyl) - (p - tolyl)methyl) - palladium**
$C_{24}H_{32}O_2Pd$
For complete entry see 71.51

73.C **π - Cyclopentadienyl - π - triphenylcyclopropenyl nickel**
$C_{26}H_{20}Ni$
For complete entry see 75.27

73.30 **Chloro - bis(π - cyclopentadienyl)(triphenylsilyl) zirconium(iv)**
$C_{28}H_{25}ClSiZr$
K.M.Muir *J. Chem. Soc. (A)*, 2663, 1971
Also classified in 63

73.31 **Aluminium tris(μ - carbonyl - cyclopentadienyl(dicarbonyl)tungsten) - tris(tetrahydrofuran)**
$C_{36}H_{39}AlO_{12}W_3$
R.B.Petersen, J.J.Stezowski, C.Wan, J.M.Burlitch, R.E.Hughes
J. Amer. Chem. Soc., **93**, 3532, 1971
Also classified in 68

73.32 **(1,2 - bis(Diphenylphosphino)hexafluorocyclopentene) - carbonyl - iron - trimethyltin**
$C_{37}H_{34}F_6FeP_2Sn$
F.W.B.Einstein, R.Restivo *Inorg. Chim. Acta*, **5**, 501, 1971
Also classified in 69, 86

73.33 **Di((benzenesulfinato) - μ - hydroxo - phenylstannio) - tetracarbonyl - di - π - cyclopentadienyl - di - iron**
$C_{38}H_{32}Fe_2O_{10}S_2Sn_2$
R.Restivo, R.F.Bryan *J. Chem. Soc. (A)*, 3364, 1971
Also classified in 69

METAL PI-COMPLEXES (ARENE)

74.1 **Benzene - uranium(iii) tris(tetrachloroaluminium)**
$C_6H_6U^{3+}$, $3AlCl_4^-$
M.Cesari, U.Pedretti, A.Zazzetta, G.Lugli, W.Marconi
Inorg. Chim. Acta, **5,** 439, 1971

74.2 **DL - o - Hydroxy - acetyl - benchrotrene**
DL - (o - Hydroxy - acetyl - benzene chromium tricarbonyl)
$C_{11}H_8CrO_5$
Y.Dusausoy, J.Protas, J.Besancon *C. R. Acad. Sci., Fr., C,* **272,** 282, 1971

74.3 **Acenaphthylene - silver perchlorate**
$(C_{12}H_8AgClO_4)_n$
P.F.Rodesiler, E.L.Amma *Inorg. Chem.,* **11,** 388, 1972

74.4 **L - o - Methoxy - acetyl - benchrotrene**
L - (o - Methoxy - acetyl - benzene chromium tricarbonyl)
$C_{12}H_{10}CrO_5$
Y.Dusausoy, J.Protas, J.Besancon *C. R. Acad. Sci., Fr., C,* **272,** 282, 1971

74.5 **Acetylacetonato - rhodium duroquinone complex (discussion)**
$C_{15}H_{19}O_4Rh$
G.G.Aleksandrov, Yu.T.Struchkov, V.S.Khandkarova, S.P.Gubin
J. Organometal. Chem., **25,** 243, 1970
Also classified in 77

74.6 **2 - Methyl - (1′ - hydroxy - 1′ - phenyl - propyl) - benzene chromium tricarbonyl (m.p. 89 ° C)**
2 - Methyl - (1′ - hydroxy - 1′ - phenyl - propyl)benchrotrene
$C_{19}H_{18}CrO_4$
Y.Dusausoy, J.Besancon, J.Protas *C. R. Acad. Sci., Fr., C,* **274,** 774, 1972

74.7 **2 - Methyl - (1′ - hydroxy - 1′ - phenyl - propyl) - benzene chromium tricarbonyl (m.p. 163 ° C)**
2 - Methyl - (1′ - hydroxy - 1′ - phenyl - propyl)benchrotrene
$C_{19}H_{18}CrO_4$
Y.Dusausoy, J.Besancon, J.Protas *C. R. Acad. Sci., Fr., C,* **274,** 774, 1972

74.C **π - Indenyl rhodium π - duroquinone**
$C_{19}H_{19}O_2Rh$
For complete entry see 73.21

74.8 **Acenaphthene - silver perchlorate**
$C_{24}H_{20}Ag_2Cl_2O_8$
P.F.Rodesiler, E.L.Amma *Inorg. Chem.*, **11**, 388, 1972

74.9 **bis(Cyclohexylbenzene) silver(i) perchlorate**
$C_{24}H_{32}Ag^+ . ClO_4^-$
E.A.H.Griffith, E.L.Amma *J. Amer. Chem. Soc.*, **93**, 3167, 1971

74.10 **bis(Hexamethyl - benzene) ruthenium(0)**
$C_{24}H_{36}Ru$
G.Huttner, S.Lange, E.O.Fischer *Angew. Chem.*, **83**, 579, 1971

74.11 **bis(Benzene) copper(ii) trifluoroacetate complex**
$C_{32}H_{24}Cu_2F_{12}O_8$
E.L.Amma, H.Luth, P.F.Rodesiler, M.S.Weininger, A.G.Gash
Amer. Cryst. Assoc., Abstr. Papers (Summer Meeting), 39, 1971
Also classified in 81

METAL PI-COMPLEXES (MISCELLANEOUS RING SYSTEMS)

75.1 **Tricarbonyl - (N - methylpyrrole) chromium(0)**
$C_8H_7CrNO_3$
G.Huttner, O.S.Mills *Chem. Ber.*, **105**, 301, 1972

75.C **exo - Tricyclo(3.2.1.0^{2,4})oct - 6 - ene - silver nitrate complex**
$C_8H_{10}Ag^+ , NO_3^-$
For complete entry see 31.1

75.2 **Azulene - triruthenium heptacarbonyl**
$C_{10}H_8O_7Ru_3$
M.R.Churchill, F.R.Scholer, J.Wormald
J. Organometal. Chem., **28**, C21, 1971

75.C **Dicarbonyl - 3 - (π - (2 - cyclohexadienyl)) - σ - propenoyl iron**
$C_{11}H_7FeO_3$
For complete entry see 71.9

75.3 **Tricarbonyl ruthenium - 1,2,3,6 - tetrahapto - 5 - cyano - cyclo - octadiene**
$C_{12}H_{11}NO_3Ru$
F.A.Cotton, M.D.LaPrade, B.F.G.Johnson, J.Lewis
J. Amer. Chem. Soc., **93**, 4626, 1971

75.4 **Tricarbonyl - (1 - isopropoxycarbonyl - 1,2 - diazepine) iron**
$C_{12}H_{12}FeN_2O_5$
R.Allmann *Angew. Chem.*, **82**, 982, 1970

75.5 **bis(Methoxyborinato) cobalt**
$C_{12}H_{16}B_2CoO_2$
G.Huttner, B.Krieg *Angew. Chem.*, **84**, 29, 1972

75.6 **Cycloheptatrienyl di - iron hexacarbonyl**
$C_{13}H_8Fe_2O_6$
F.A.Cotton, B.G.DeBoer, T.J.Marks *J. Amer. Chem. Soc.*, **93**, 5069, 1971

75.7 **bis(1,3 - Cyclohexadiene)monocarbonyl - iron**
$C_{13}H_{16}FeO$
C.Kruger, Y.-H.Tsay *J. Organometal. Chem.*, **33**, 59, 1971

75.8 **(5,7 - Dimethyl - 4H - cyclohepta(c)thiophene) chromium tricarbonyl**
$C_{14}H_{12}CrO_3S$
Y.Dusausoy, R.Guilard, J.Protas *C. R. Acad. Sci., Fr., C,* **273,** 228, 1971

75.9 **Tricarbonyl(bicyclo(4.4.1)undeca - 1,3,5 - triene) chromium**
$C_{14}H_{14}CrO_3$
M.J.Barrow, O.S.Mills *J. Chem. Soc. (A),* 1982, 1971

75.C **Norbornadiene dimer silver nitrate complex**
$C_{14}H_{16}Ag^+ . NO_3^-$
For complete entry see 31.13

75.C **Chloro - (2 - methoxycyclo - octa - 1,5 - dienyl)pyridine - platinum**
$C_{14}H_{18}ClNOPt$
For complete entry see 71.23

75.10 **5,7,8 - Trimethyl - 8H - cycloheptatrieno - (b) - thiophene chromium tricarbonyl**
$C_{15}H_{14}CrO_3S$
Y.Dusausoy, R.Guilard, J.Protas, J.Tirouflet
C. R. Acad. Sci., Fr., C, **272,** 2134, 1971

75.11 **Tricarbonyl(hexaethylborazine) chromium(0)**
$C_{15}H_{30}B_3CrN_3O_3$
G.Huttner, B.Krieg *Angew. Chem.,* **83,** 541, 1971

75.C **Tricarbonyl - (tetracyanobicyclo(4.2.1)nonanediyl) iron**
$C_{16}H_8FeN_4O_3$
For complete entry see 71.31

75.12 **(Bicyclo(6.2.0)deca - 2,4,6 - triene)hexacarbonyl - di - iron**
$C_{16}H_{12}Fe_2O_6$
F.A.Cotton, B.A.Frenz
Amer. Cryst. Assoc., Abstr. Papers (Winter Meeting), 42, 1972

75.13 **Potassium 2,2′ - (diethoxy)diethylether bis(cyclo - octatetraenyl) cerium(iii)**
$C_{16}H_{16}Ce^- . C_6H_{14}KO_3^+$
K.O.Hodgson, K.N.Raymond
Amer. Cryst. Assoc., Abstr. Papers (Winter Meeting), 50, 1972
Residue 2 classified in 67

75.14 **bis(Cyclo - octatetraenyl) thorium(iv)**
$C_{16}H_{16}Th$
A.Avdeef, K.N.Raymond, K.O.Hodgson, A.Zalkin
Inorg. Chem., **11,** 1083, 1972

75.15 **bis(Cyclo - octatetraenyl) uranium(iv)**
$C_{16}H_{16}U$
A.Avdeef, K.N.Raymond, K.O.Hodgson, A.Zalkin
Inorg. Chem., **11,** 1083, 1972

75.16 **Tetracarbonyl - (hexamethyl - bicyclo(2.2.0)hexa - 2,5 - diene)**
$C_{16}H_{18}CrO_4$
G.Huttner, O.S.Mills *J. Organometal. Chem.*, **29**, 275, 1971

75.17 **12 - Oxa(4.4.3)propella - 2,4,7,9 - tetracene bis(iron tricarbonyl) (unsymmetrical form)**
12 - Oxa - tricyclo(4.4.3.0)trideca - 2,4,7.9 - tetraene bis(tricarbonyl iron)
$C_{18}H_{12}Fe_2O_7$
G.I.Birnbaum *J. Amer. Chem. Soc.*, **94**, 2455, 1972

75.18 **12 - Oxa(4.4.3)propella - 2,4,7,9 - tetraene bis(iron tricarbonyl) (symmetrical form)**
12 - Oxa - tricyclo(4.4.3.0)trideca - 2,4,7.9 - tetraene bis(tricarbonyl iron)
$C_{18}H_{12}Fe_2O_7$
K.B.Birnbaum *Acta Cryst. (B)*, **28**, 161, 1972

75.19 **Indene silver(i) perchlorate dimer**
$C_{18}H_{16}Ag_2ClO_4^+ . ClO_4^-$
P.F.Rodesiler, E.A.H.Griffith, B.L.Amma
J. Amer. Chem. Soc., **94**, 761, 1972

75.20 **Dichloro - bis(2,3,5,6 - tetrahapto - norbornadiene - dicarbonyl - cobalt) tin(iv)**
$C_{18}H_{16}Cl_2Co_2O_4Sn$
F.P.Boer, J.J.Flynn *J. Amer. Chem. Soc.*, **93**, 6495, 1971
Also classified in 69

75.21 **bis(3,5 - Dimethylpyrazolylborato)dicarbonyl - trihapto - cycloheptatrienyl molybdenum**
$C_{19}H_{23}BMoN_4O_2$
F.A.Cotton, J.L.Calderon, M.Jeremic, A.Shaver
J. C. S. Chem. Comm., 777, 1972
Also classified in 83, 62

75.22 **Hexacarbonyl - trans - 6a,12a - dihydro - octalene - dichromium(0)**
$C_{20}H_{14}Cr_2O_6$
K.Stockel, F.Sondheimer, T.A.Clarke, M.Guss, R.Mason
J. Amer. Chem. Soc., **93**, 2571, 1971

75.23 **μ - (5 - Cyclopentadienylcyclopentadiene) - bis(π - cyclopentadienyl - platinum)**
$C_{20}H_{20}Pt_2$
K.K.Cheung, R.J.Cross, K.P.Forrest, R.Wardle, M.Mercer
J. Chem. Soc. (D), 875, 1971
Also classified in 73

75.C **bis(cis - Cyclodecene) silver nitrate**
$C_{20}H_{36} , Ag^+ , NO_3^-$
For complete entry see 23.4

75.24 **Phenyl - di(tricarbonyliron - cyclobutadienyl)methyl fluoroborate**
$C_{21}H_{11}Fe_2O_6^+$, BF_4^-
R.E.Davis, H.D.Simpson, N.Grice, R.Pettit
J. Amer. Chem. Soc., **93,** 6688, 1971
Residue 1 also classified in 12

75.C **Nonacarbonyl(cyclododecatrienetriyl)hydrido - triruthenium**
$C_{21}H_{16}O_9Ru_3$
For complete entry see 71.45

75.25 **bis(1,3,5,7 - Tetramethylcyclo - octatetraenyl) uranium(iv)**
$C_{24}H_{32}U$
K.O.Hodgson, D.Dempf, K.N.Raymond *J. Chem. Soc. (D),* 1592, 1971

75.C **Compound iv**
$C_{25}H_{16}Fe_3O_9$
For complete entry see 71.54

75.C **Compound iv benzene solvate**
$C_{25}H_{16}Fe_3O_9$, $0.5C_6H_6$
For complete entry see 71.55

75.26 **(1,5 - Cyclo - octadiene) - bis(dimethyl - phenyl - phosphine) - methyl iridium(i)**
$C_{25}H_{37}IrP_2$
M.R.Churchill, S.A.Bezman *J. Organometal. Chem.*, **31,** C43, 1971
Also classified in 86

75.27 **π - Cyclopentadienyl - π - triphenylcyclopropenyl nickel**
$C_{26}H_{20}Ni$
R.M.Tuggle, D.L.Weaver *Inorg. Chem.*, **10,** 1504, 1971
Also classified in 73

75.28 **Tri - indenyl - uranium chloride**
$C_{27}H_{21}ClU$
J.H.Burns, P.G.Lauberau *Inorg. Chem.*, **10,** 2789, 1971

75.C **bis(1,1,4,4 - Tetramethyl - cis - cyclodec - 7 - ene) silver nitrate**
$C_{28}H_{52}$, Ag^+, NO_3^-
For complete entry see 23.6

75.29 **Diphenyl - bis(2,3,5,6 - tetrahapto - norbornadiene - dicarbonyl - cobalt) tin(iv)**
$C_{30}H_{26}Co_2O_4Sn$
F.P.Boer, J.J.Flynn *J. Amer. Chem. Soc.*, **93,** 6495, 1971
Also classified in 69

75.30 **π - (Triphenylcyclopropenyl)chloro - (dipyridine) nickel(0) pyridine solvate**
$C_{31}H_{25}ClN_2Ni$, C_5H_5N
R.M.Tuggle, D.L.Weaver *Inorg. Chem.*, **10,** 2599, 1971
Residue 1 also classified in 83

75.31 **Di - μ - chloro - di(π - cyclo - octatetraene - bis(tetrahydrofuran) cerium)**
$C_{32}H_{48}Ce_2Cl_2O_4$
K.O.Hodgson, K.N.Raymond *Inorg. Chem.*, **11**, 171, 1972
Also classified in 84

75.32 **(1,5 - Cyclo - octadiene) - (1,2 - bis(diphenylphosphino)ethane) - methyl iridium(i)**
$C_{35}H_{39}IrP_2$
M.R.Churchill, S.A.Bezman *J. Organometal. Chem.*, **31**, C43, 1971
Also classified in 86

75.33 **bis(Triphenylphosphine) - (1,2 - methylcyclopropene) platinum**
$C_{41}H_{38}P_2Pt$
J.P.Visser, A.J.Schipperijn, J.Lukas, D.Bright, J.J.de Boer
J. Chem. Soc. (D), 1266, 1971
Also classified in 86

75.34 **Cyclohexyne - bis(triphenylphosphine) - platinum(0)**
$C_{42}H_{38}P_2Pt$
G.B.Robertson, P.O.Whimp *J. Organometal. Chem.*, **32**, C69, 1971
Also classified in 86

75.35 **Cycloheptyne - platinum - bis(triphenylphosphine)**
$C_{43}H_{40}P_2Pt$
M.A.Bennett, G.B.Robertson, P.O.Whimp, T.Yoshida
J. Amer. Chem. Soc., **93**, 3797, 1971
Also classified in 86

75.C **Chloro - rhodium norbornadiene hexafluorobut - 2 - yne adduct tetramer**
$C_{44}H_{32}Cl_4F_{24}Rh_4$
For complete entry see 71.87

METAL COMPLEXES (ETHYLENEDIAMINE)

76.1 **Dithiocyanato - ethylenediamine - copper(ii)**
$C_4H_8CuN_4S_2$
J.Garaj, M.Dunaj-Jurco, O.Lindgren
Collect. Czechosl. Chem. Communic., **36,** 3863, 1971

76.2 **cis - Dichloro(meso - 2,3 - diaminobutane) palladium(ii)**
$C_4H_{12}Cl_2N_2Pd$
T.Ito, F.Marumo, Y.Saito *Acta Cryst. (B),* **27,** 1695, 1971

76.3 **$(+)_{589}$ - Dichloro - bis(ethylenediamine) cobalt(iii) chloride monohydrate (absolute configuration)**
$C_4H_{16}Cl_2CoN_4^+$, Cl^-, H_2O
K.Matsumoto, S.Ooi, H.Kuroya *Bull. Chem. Soc. Jap.*, **43,** 3801, 1970

76.4 **bis(Ethylenediamine) copper(ii) selenosulfate**
$C_4H_{16}CuN_4^{2+}$, O_3SSe^{2-}
N.V.Podberezskava, S.V.Borisov, V.V.Bakakin
Zh. Strukt. Khim., **12,** 840, 1971

76.5 **bis(Ethylendiamine) copper(ii) thiosulfate**
$C_4H_{16}CuN_4^{2+}$, $O_3S_2^{2-}$
N.V.Podberezskava, S.V.Borisov, V.V.Bakakin
Zh. Strukt. Khim., **12,** 840, 1971

76.6 **bis(Ethylenediamine) copper(ii) thiocyanate perchlorate**
$C_4H_{16}CuN_4^{2+}$, CNS^-, ClO_4^-
M.Cannas, G.Carta, G.Marongiu *J. Chem. Soc. (D),* 1462, 1971

76.7 **bis(Ethylenediamine)nitro - nickel iodide**
$C_4H_{16}N_5NiO_2^+$, I^-
L.Kh.Minacheva, A.S.Antsyshkina, M.A.Porai-Koshits, A.E.Shvelashvili
Zh. Strukt. Khim., **11,** 936, 1970

76.8 **Aquo - bis(ethylenediamine) copper(ii) di - (catena - di - μ - cyanocuprate(i))**
$nC_4H_{18}CuN_4O^{2+}$, $(C_4Cu_2N_4^{2-})_n$
R.J.Williams, A.C.Larson, D.T.Cromer *Acta Cryst. (B),* **28,** 858, 1972

76.9 **Di - μ - oxo - bis(penta - ammine - ruthenium) - bis(ethylendiamine) ruthenium hexachloride**

$C_4H_{46}N_{14}O_2Ru_3^{6+}$, $6Cl^-$

P.M.Smith. T.Fealey, J.E.Earley. J.V.Silverton

Inorg. Chem., **10**, 1943, 1971

76.10 **(+)$_{589}$ - Dicyano - bis(ethylenediamine) cobalt(iii) chloride monohydrate (absolute configuration)**

$C_6H_{16}CoN_6^+$, Cl^- , H_2O

K.Matsumoto, S.Ooi, H.Kuroya *Bull. Chem. Soc. Jap.*, **44**, 2721, 1971

76.11 **Chloro - ethylenediamine - (α - amino - (2 - aminoethylimino) - acetamide) cobalt(iii) chloride monohydrate**

$C_6H_{20}ClCoN_6^{2+}$, $2Cl^-$, H_2O

D.A.Buckingham, B.M.Foxman, A.M.Sargeson, A.Zanella

J. Amer. Chem. Soc., **94**, 1007, 1972

Residue 1 also classified in 83

76.12 **bis(1,3 - Diaminopropane) copper(ii) thiocyanate**

$C_6H_{20}CuN_4^{2+}$, $2CNS^-$

G.D.Andreetti, L.Cavalca. P.Sgarabotto *Gazz. Chim. Ital.*, **101**, 483, 1971

Residue 1 also classified in 83

76.13 **tris(Ethylenediamine) cobalt(iii) hexachlorocadmate(ii) dichloride dihydrate**

$2C_6H_{24}CoN_6^{3+}$, $CdCl_6^{4-}$, $2Cl^-$, $2H_2O$

J.T.Veal, D.J.Hodgson *Inorg. Chem.*, **11**, 597, 1972

76.14 **tris(Ethylenediamine) cobalt(iii) monohydrogen - phosphate nonahydrate**

$2C_6H_{24}CoN_6^{3+}$, $3HO_4P^{2-}$, $9H_2O$

E.N.Duesler, K.N.Raymond *Inorg. Chem.*, **10**, 1486, 1971

76.15 **(N,N′ - Di(2 - aminoethyl)malondiamidato) nickel(ii) trihydrate**

$C_7H_{14}N_4NiO_2$, $3H_2O$

R.M.Lewis, G.H.Nancollas, P.Coppens *Inorg. Chem.*, **11**, 1371, 1972

Residue 1 also classified in 83

76.16 **Potassium nitro(ethylenediaminetriacetato) cobaltate(iii) sesquihydrate**

$C_8H_{11}CoN_3O_8$, K^+ , $1.5H_2O$

J.D.Bell, G.L.Blackmer

Amer. Cryst. Assoc., Abstr. Papers (Winter Meeting), 44, 1972

Residue 1 also classified in 81, 82

76.C **Sodium(+)$_{546}$ ethylenediamine - bis(malonato) cobalt(iii) dihydrate (absolute configuration)**

$C_8H_{12}CoN_2O_8^-$, Na^+ , $2H_2O$

For complete entry see 81.24

76.17 **β_2 - Triethylenetetra - amine(glycinato) cobalt(iii) dichloride monohydrate**
$C_8H_{22}CoN_5O_2^{2+}$, $2Cl^-$, H_2O
R.J.Dellaca, V.Janson, W.T.Robinson, D.A.Buckingham, L.G.Marzilli, I.E.Mazwell, K.R.Turnbull, A.M.Sargeson *J. C. S. Chem. Comm.*, 57, 1972
Residue 1 also classified in 82

76.18 **β_1 - Triethylenetetra - amine(glycinato) cobalt(iii) di - iodide hemihydrate**
$2C_8H_{22}CoN_5O_2^{2+}$, $4I^-$, H_2O
R.J.Dellaca, V.Janson, W.T.Robinson, D.A.Buckingham, L.G.Marzilli, I.E.Mazwell, K.R.Turnbull, A.M.Sargeson *J. C. S. Chem. Comm.*, 57, 1972
Residue 1 also classified in 82

76.19 **$(+)_{546}$ - trans - Dinitro(1,10 - diamino - 4,7 - diazadecane) cobalt(iii) bromide (absolute configuration)**
$C_8H_{22}CoN_6O_4^+$, Br^-
N.C.Payne *Inorg. Chem.*, **11,** 1376, 1972
Residue 1 also classified in 83

76.20 **$(-)_{589}$ - trans - Dinitro - (L - 3,8 - dimethyltriethylenetetramine) - cobalt(iii) perchlorate**
$C_8H_{22}CoN_6O_4^+$, ClO_4^-
M.Ito, F.Marumo, Y.Saito *Acta Cryst. (B),* **28,** 463, 1972

76.21 **$(-)_{589}$ - cis - Dinitro - (L - 3,8 - dimethyltriethylenetetramine) - cobalt(iii) perchlorate**
$C_8H_{22}CoN_6O_4^+$, ClO_4^-
M.Ito, F.Marumo, Y.Saito *Acta Cryst. (B),* **28,** 457, 1972

76.22 **Chloro - (tetraethylenepentamine) cobalt(iii) tetrachlorozincate**
$C_8H_{23}ClCoN_5^{2+}$, Cl_4Zn^{2-}
A.Mangia, M.Nardelli, C.Pelizzi, G.Pelizzi
Gazz. Chim. Ital., **101,** 91, 1971

76.23 **(4 - (2 - Aminoethyl) - 1,4,7,10 - tetra - azadecane) azidocobalt(iii) nitrate monohydrate**
$C_8H_{23}CoN_8^{2+}$, $2NO_3^-$, H_2O
I.E.Maxwell *Inorg. Chem.*, **10,** 1782, 1971

76.24 **Chloro - (ethylenediamine)(dipropylenetriamine) cobalt(iii) iodide monohydrate (α form)**
$C_8H_{25}ClCoN_5^{2+}$, $2I^-$, H_2O
D.A.House, P.R.Ireland, I.E.Maxwell, W.T.Robinson
Inorg. Chim. Acta, **5,** 397, 1971
Residue 1 also classified in 83

76.25 **bis(Diethylenetriamine) cobalt(iii) bromide**
$C_8H_{26}CoN_6^{3+}$, $3Br^-$
M.Kobayashi, F.Marumo, Y.Saito *Acta Cryst. (B),* **28,** 470, 1972

76.26 **bis(Diethylenetriamine) zinc(ii) nitrate**
$C_8H_{26}N_6Zn^{2+}$, $2NO_3^-$
M.Zocchi, A.Albinati, G.Tieghi *Cryst. Struct. Comm.*, **1,** 135, 1972

76.27 **Chloro - bis(ethylenediamine) nickel(ii) perchlorate dimer**
$C_8H_{32}Cl_2N_8Ni_2^{2+}$, $2ClO_4^-$
L.K.Minacheva, A.S.Antsyshkina, M.A.Porai-Koshits
Zh. Strukt. Khim., **12,** 845, 1971

76.28 **μ - Amido - μ - superoxo - bis(bis(ethylenediamine) cobalt(iii)) tetranitrate monohydrate**
$C_8H_{34}Co_2N_9O_2^{4+}$, $4NO_3^-$, H_2O
U.Thewalt, R.E.Marsh *Inorg. Chem.*, **11,** 351, 1972

76.29 **μ - Amido - μ - sulfato - bis(bis(ethylendiamine) cobalt(iii)) tribromide**
$C_8H_{34}Co_2N_9O_4S^{3+}$, $3Br^-$
U.Thewalt *Acta Cryst. (B)*, **27,** 1744, 1971

76.30 **μ - Amido - μ - hydroxo - bis(bis(ethylenediamine) cobalt(iii)) tetranitrate monohydrate**
$C_8H_{35}Co_2N_9O^{4+}$, $4NO_3^-$, H_2O
U.Thewalt, R.E.Marsh *Inorg. Chem.*, **10,** 1789, 1971

76.31 **cis - (–) - tris((–) - Propylene - 1,2 - diamine) cobalt(iii) bromide (re - interpretation)**
$C_9H_{30}CoN_6^{3+}$, $3Br^-$
P.F.Crossing, M.R.Snow *J. C. S. Dalton*, 295, 1972

76.32 **$(-)_{589}$ tris(1,2 - Diaminopropane) cobalt(iii) $(+)_{589}$ tris(malonato) chromium(iii) trihydrate (absolute configuration)**
$C_9H_{30}CoN_6^{3+}$, $C_9H_6CrO_{12}^{3-}$, $3H_2O$
K.R.Butler, M.R.Snow *J. Chem. Soc. (D)*, 550, 1971
Residue 2 classified in 81

76.33 **Sodium samarium ethylenediaminetetra - acetate octahydrate (neutron study)**
$C_{10}H_{12}N_2O_8Sm^-$, Na^+ , $8H_2O$
T.F.Koetzle, W.C.Hamilton
Amer. Cryst. Assoc., Abstr. Papers (Winter Meeting), 86, 1972
Residue 1 also classified in 81, 82

76.34 **Trisodium (ethylenediaminetetra - acetato) dioxovanadate(v) tetrahydrate**
$C_{10}H_{12}N_2O_{10}V^{3-}$, $3Na^+$, $4H_2O$
W.R.Scheidt, R.Countryman, J.L.Hoard
J. Amer. Chem. Soc., **93,** 3878, 1971
Residue 1 also classified in 81, 82

76.35 **Lithium aquo manganese(ii) ethylenediaminetetra - acetate tetrahydrate**
$C_{10}H_{14}MnN_2O_9^{2-} . 2Li^+ . 4H_2O$
T.N.Polynova, N.N.Anan'eva, M.A.Porai-Koshits, L.I.Martynenko. N.I.Pechurova *Zh. Strukt. Khim.*, **12,** 335, 1971
Residue 1 also classified in 81, 82

76.36 **Ammonium (dihydrogenethylenediaminetetra - acetato) dioxovanadate(v) trihydrate**
$C_{10}H_{14}N_2O_{10}V^- . H_4N^+ . 3H_2O$
W.R.Scheidt, D.M.Collins, J.L.Hoard *J. Amer. Chem. Soc.*, **93,** 3873, 1971
Residue 1 also classified in 81, 82

76.37 **(Hydrogen ethylenediaminetetra - acetato)aquoferrate(iii)**
$C_{10}H_{15}FeN_2O_9$
C.H.L.Kennard *Inorg. Chim. Acta,* **1,** 347, 1967
Also classified in 81, 82

76.38 **N - (1 - Methyl - 3 - oxo - butylidene) - N' - (1 - methyl - 2 - isonitroso - 3 - oxobutylidene)ethylenediamine copper(ii)**
$C_{12}H_{17}CuN_3O_3$
M.B.Cingi, A.C.Villa, A.G.Manfredotti, C.Guastini, M.Nardelli
Acta Cryst. (B), **28,** 1075, 1972
Also classified in 83

76.C **N,N' - Ethylene - bis(acetylacetoniminato)nitrosyl - cobalt**
$C_{12}H_{18}CoN_3O_3$
For complete entry see 77.5

76.39 **tris(Di - μ - hydroxo - bis(ethylenediamine) cobalt(iii)) cobalt(iii) tris(dithionate) octahydrate**
$C_{12}H_{54}Co_4N_{12}O_6^{6+} . 3O_6S_2^{2-} . 8H_2O$
U.Thewalt *Chem. Ber.*, **104,** 2657, 1971

76.40 **$(-)_{589}$ - tris(+trans - 1,2 - Diaminocyclopentane) cobalt(iii) chloride tetrahydrate**
$C_{15}H_{36}CoN_6^{3+} . 3Cl^- . 4H_2O$
M.Ito, F.Marumo, Y.Saito *Acta Cryst. (B),* **27,** 2187, 1971
Residue 1 also classified in 83

76.C **N,N' - bis(Salicylal) - ethylenedi - iminato - nickel**
$C_{16}H_{14}N_2NiO_2$
For complete entry see 78.3

76.41 **bis - (1,1,1,5,5,5 - Hexafluoropentane - 2,4 - dionato)bis - (N,N - dimethylethylenediamine) copper(ii)**
$C_{18}H_{23}CuF_{15}N_4O_4$
M.A.Bush, D.E.Fenton *J. Chem. Soc. (A),* 2446, 1971
Also classified in 77

76.42 **bis(Salicylidene) - triethylenetetramine nickel(ii) hexahydrate**
$C_{20}H_{24}N_4NiO_2 , 6H_2O$
P.D.Cradwick, M.E.Cradwick, G.G.Dodson, D.Hall, T.N.Waters
Acta Cryst. (B), **28**, 45, 1972
Residue 1 also classified in 78

76.C **N,N′ - Ethylene - bis(benzoylacetoniminato)nitrosyl - cobalt**
$C_{22}H_{22}CoN_3O_3$
For complete entry see 77.16

76.C **N,N′ - Ethylene bis(salicylaldehydeiminato) cobalt(ii) (oxygen - inactive form)**
$C_{32}H_{28}Co_2N_4O_4$
For complete entry see 78.14

76.43 **(N,N′ - Dimethyl - N,N′ - bis(2 - diphenylphosphinoethyl)ethylenediamine) - bromo - cobalt(ii) hexafluorophosphate**
$C_{32}H_{38}BrCoN_2P_2^+ , F_6P^-$
A.Bianchi, C.A.Ghilardi, C.Mealli, L.Sacconi
J. C. S. Chem. Comm., 651, 1972
Residue 1 also classified in 86

76.44 **(N,N′ - Dimethyl - N,N′ - bis(2 - diphenylphosphinoethyl)ethylenediamine) - bromo - nickel(ii) bromide butanol solvate**
$C_{32}H_{38}BrN_2NiP_2^+ , Br^- , 0.5C_4H_{10}O$
A.Bianchi, C.A.Ghilardi, C.Mealli, L.Sacconi
J. C. S. Chem. Comm., 651, 1972
Residue 1 also classified in 86

METAL COMPLEXES (ACETYLACETONE)

77.1 **Di - μ - chloro - bis(dichloro - acetylacetonato - titanium(iv))**
$C_{10}H_{14}Cl_6O_4Ti$
N.Serpone, P.H.Bird, D.G.Bickley, D.W.Thompson
J. C. S. Chem. Comm., 217, 1972

77.2 **Palladium(ii) diacetylacetonate**
$C_{10}H_{14}O_4Pd$
A.N.Knyazeva, E.A.Shugam, L.M.Shkol'nikova
Zh. Strukt. Khim., **11**, 938, 1970

77.3 **Dibromo - bis(acetylacetone) nickel(ii)**
$C_{10}H_{16}Br_2NiO_4$
S.Koda, S.Ooi, H.Kuroya, I.Isobe, Y.Nakamura, S.Kawaguchi
J. Chem. Soc. (D), 1321, 1971

77.4 **bis(Acetylacetone) - diaquo - nickel(ii) perchlorate**
$C_{10}H_{20}NiO_6^{2+}$, $2ClO_4^-$
K.Anzenhofer, T.G.Hewitt *Z. Kristallogr.*, **134**, 54, 1971

77.5 **N,N' - Ethylene - bis(acetylacetoniminato)nitrosyl - cobalt**
$C_{12}H_{18}CoN_3O_3$
R.Wiest, R.Weiss *J. Organometal. Chem.*, **30**, C33, 1971
Also classified in 76

77.C **bis(Thioisobutyrylacetonato) nickel(ii)**
$C_{14}H_{22}NiO_2S_2$
For complete entry see 85.20

77.6 **bis(Acetylacetonato) nickel(ii) di - ethanol complex**
$C_{14}H_{32}NiO_6$
C.E.Pfluger, T.S.Burke
Amer. Cryst. Assoc., Abstr. Papers (Summer Meeting), 77, 1971
Also classified in 84

77.7 **tris(Hexafluoroacetylacetonato) copper(ii) 1 - dimethylammonium - 8 - dimethylamino - naphthalene**
$C_{15}H_3CuF_{18}O_6^-$, $C_{14}H_{19}N_2^+$
M.R.Truter, B.L.Vickery *J. C. S. Dalton*, 395, 1972
Residue 2 classified in 24

77.C **Acetylacetonato - rhodium duroquinone complex (discussion)**
$C_{15}H_{19}O_4Rh$
For complete entry see 74.5

77.8 **Chloro - tris(acetylacetonato) zirconium(iv)**
$C_{15}H_{21}ClO_6Zr$
R.B.VonDreele, J.J.Stezowski, R.C.Fay
J. Amer. Chem. Soc., **93,** 2887, 1971

77.9 **Quasiracemic mixture of cobalt tris(acetylacetonate) and aluminium tris(acetylacetonate) (absolute configuration)**
$C_{15}H_{21}CoO_6$, $C_{15}H_{21}AlO_6$
R.B.von Dreele, R.C.Fray *J. Amer. Chem. Soc.*, **93,** 4936, 1971
Residue 1 also classified in 68

77.10 **Vanadium(iii) tris(acetylacetonate)**
$C_{15}H_{21}O_6V$
F.Sanz-Ruiz, S.Martinez-Carrera, S.Garcia-Blanco
J. Prakt. Chem., **66,** 309, 1970

77.11 **Diaquo - tris(acetylacetonato) holmium(iii) dihydrate**
$C_{15}H_{25}HoO_8$. $2H_2O$
L.A.Aslanov, E.F.Korytnyi, M.A.Porai-Koshits
Zh. Strukt. Khim., **12,** 661, 1971

77.12 **Diaquo - neodymium(iii) tris(acetylacetonate)**
$C_{15}H_{25}NdO_8$
L.A.Aslanov, M.A.Porai-Koshits. M.O.Dekaprilevich
Zh. Strukt. Khim., **12,** 470, 1971

77.13 **(4 - Methylpyridine - N - oxide) - bis(hexafluoroacetylacetonato) copper(ii)**
$(C_{16}H_9CuF_{12}NO_5)_n$
G.H.Schreiber, G.J.Palenik
Amer. Cryst. Assoc., Abstr. Papers (Summer Meeting), 49, 1968
Also classified in 84

77.C **bis(Acetylacetonato - μ - allyl - platinum)**
$C_{16}H_{24}O_4Pt_2$
For complete entry see 71.33

77.C **bis - (1,1,1,5,5,5 - Hexafluoropentane - 2,4 - dionato)bis - (N,N - dimethylethylenediamine) copper(ii)**
$C_{18}H_{23}CuF_{15}N_4O_4$
For complete entry see 76.41

77.14 **(N,N′ - Ethylene - bis(salicylideneiminato))acetylacetonato - cobalt(iii) hydrate**
$C_{21}H_{21}CoN_2O_4$, $0.7H_2O$
M.Calligaris, G.Manzini, G.Nardin, L.Randaccio
J. C. S. Dalton, 543, 1972
Residue 1 also classified in 78

77.15 **4 - Phenylpyridine vanadyl - bis(acetylacetonate)**
$C_{21}H_{23}NO_5V$
M.R.Caira. J.M.Haigh, L.R.Nassimbeni
Inorg. Nucl. Chem. Letters, **8,** 109, 1972
Also classified in 83

77.16 **N,N' - Ethylene - bis(benzoylacetoniminato)nitrosyl - cobalt**
$C_{22}H_{22}CoN_3O_3$
R.Wiest, R.Weiss *J. Organometal. Chem.,* **30,** C33, 1971
Also classified in 76

77.17 **bis(2,2,6,6 - Tetramethylheptane - 5 - thiolo - 3 - onato) nickel(ii)**
$C_{22}H_{38}NiO_2S_2$
J.Coetzer, J.C.A.Boeyens *J. Cryst. Mol. Struct.,* **1,** 277, 1971
Also classified in 85

77.C **Acetylacetonato - ((1 - pentamethylcyclopenta - 2,4 - dienyl) - (p - tolyl)methyl) - palladium**
$C_{24}H_{32}O_2Pd$
For complete entry see 71.51

77.C **Benzoylacetonato - (N,N' - ethylene - bis(salicylideneiminato)) cobalt(iii) sesquihydrate**
$C_{26}H_{23}CoN_2O_4$, $1.5H_2O$
For complete entry see 78.13

77.18 **Allylamine bis(acetylacetonato) manganese(ii) dimer**
$C_{26}H_{42}Mn_2N_2O_8$
S.Koda, S.Ooi, H.Kuroya. Y.Nishikawa. Y.Nakamura. S.Kawaguchi
Inorg. Nucl. Chem. Letters, **8,** 89. 1972
Also classified in 83

77.C **(Allene tetramer)(1,3 - diphenylpropan - 1,3 - dionato) rhodium**
$C_{27}H_{27}O_2Rh$
For complete entry see 72.19

77.19 **tris(1,1,1,2,2,3,3 - Heptafluoro - 7,7 - dimethyl - 4,6 - octanedionato)aquo - lutetium(iii)**
$C_{30}H_{32}F_{21}LuO_7$
J.C.A.Boeyens, J.P.R.de Villiers *J. Cryst. Mol. Struct.,* **1,** 297, 1971

77.20 **bis(Acetylacetonato)cyclohexylamine - cobalt(ii) dimer**
bis(2,4 - Pentanedionato)cyclohexylamine - cobalt(ii) dimer
$C_{32}H_{54}Co_2N_2O_8$
J.A.Bertrand, A.R.Kalyanaraman *Inorg. Chim. Acta,* **5,** 167, 1971
Also classified in 83

77.21 **tris(2,2,6,6 - Tetramethylheptane - 3,5 - dionato) erbium(iii)**
$C_{33}H_{57}ErO_6$
J.P.R.de Villiers, J.C.A.Boeyens *Acta Cryst. (B),* **27,** 2335, 1971

77.22 **tris(2,2,6,6 - Tetramethylheptane - 3,5 - dionato)aquo - dysprosium(iii)**
$C_{33}H_{59}DyO_7$
C.S.Erasmus, J.C.A.Boeyens *J. Cryst. Mol. Struct.*, **1**, 83, 1971

77.23 **tris(2,2,6,6 - Tetramethylheptane - 3,5 - dionato) - dipyridine - europium(iii)**
$C_{43}H_{67}EuN_2O_6$
R.E.Cramer, K.Seff *J. C. S. Chem. Comm.*, 400, 1972
Also classified in 83

77.24 **(2,2,6,6 - Tetramethylheptane - 3,5 - dionato) holmium(iii) bis(4 - picoline)**
$C_{45}H_{71}HoN_2O_6$
W.DeW.Horrocks Junior, J.P.Sipe III, J.R.Luber
J. Amer. Chem. Soc., **93**, 5258, 1971
Also classified in 83

METAL COMPLEXES
(SALICYLIC DERIVATIVES)

78.1 **N - Salicylidene - α - aminoisobutyrato - aquo - copper(ii)**
$C_{11}H_{13}CuNO_4$
H.Fujimaki, I.Oonishi, F.Muto, A.Nakahara, Y.Komiyama
Bull. Chem. Soc. Jap., **44**, 28, 1971

78.2 **Dichloro - N,N′ - ethylene - bis(salicylideneiminato) titanium(iv) tetrahydrofuran solvate**
$C_{16}H_{14}Cl_2N_2O_2Ti$, C_4H_8O
G.Gilli, D.W.J.Cruickshank, R.L.Beddoes, O.S.Mills
Acta Cryst. (B), **28**, 1889, 1972

78.3 **N,N′ - bis(Salicylal) - ethylenedi - iminato - nickel**
$C_{16}H_{14}N_2NiO_2$
L.M.Shkol'nikova, E.M.Yumal', E.A.Shugam, V.A.Voblikova
Zh. Strukt. Khim., **11**, 886, 1970
Also classified in 76

78.4 **trans - bis(Chloro(N - methylsalicylaldimino) copper(ii))**
$C_{16}H_{16}Cl_2Cu_2N_2O_2$
E.Sinn, W.T.Robinson *J. C. S. Chem. Comm.*, 359, 1972

78.5 **bis(o - Hydroxyacetophenone - iminato) copper(ii)**
$C_{16}H_{16}CuN_2O_2$
G.Marongiu, E.C.Lingafelter *Acta Cryst. (B)*, **27**, 1195, 1971
Also classified in 83, 84

78.6 **Compound IIa**
$C_{18}H_{18}Cl_2Cu_2N_2O_2$
E.Sinn, W.T.Robinson *J. C. S. Chem. Comm.*, 359, 1972

78.C **N,N′ - Ethylene - bis(salicylideneiminato) - ethyl cobalt(iii)**
$C_{18}H_{19}CoN_2O_2$
For complete entry see 71.40

78.7 **trans - bis(Chloro - (N - ethylsalicylaldimino) copper(ii))**
$C_{18}H_{20}Cl_2Cu_2N_2O_2$
E.Sinn, W.T.Robinson *J. C. S. Chem. Comm.*, 359, 1972

78.8 **bis(2 - Aminoethylsalicylideneiminato) chromium(iii) iodide**
$C_{18}H_{22}CrN_4O_2^+ , I^-$
A.P.Gardner, B.M.Gatehouse, J.C.B.White
Acta Cryst. (B), **27**, 1505, 1971

78.9 **bis(Salicylidene - γ - iminopropyl)amine - nickel(ii)**
$C_{20}H_{23}N_3NiO_2$
M.Seleborg, S.L.Holt, B.Post *Inorg. Chem.*, **10**, 1501, 1971

78.10 **Chloro - N - (2 - hydroxypropyl)salicylaldimino copper(ii) dimer**
$C_{20}H_{24}Cl_2Cu_2N_2O_4$
J.A.Bertrand, J.A.Kelley, J.L.Breece *Inorg. Chim. Acta*, **4**, 247, 1970

78.C **bis(Salicylidene) - triethylenetetramine nickel(ii) hexahydrate**
$C_{20}H_{24}N_4NiO_2 , 6H_2O$
For complete entry see 76.42

78.C **(N,N' - Ethylene - bis(salicylideneiminato))acetylacetonato - cobalt(iii) hydrate**
$C_{21}H_{21}CoN_2O_4 , 0.7H_2O$
For complete entry see 77.14

78.11 **Uranyl N,N' - o - phenylene - bis(salicylideneiminate) ethanol complex**
$C_{22}H_{20}N_2O_5U$
G.Bandoli, D.A.Clemente, U.Croatto, M.Vidali, P.A.Vigato
J. Chem. Soc. (D), 1330, 1971

78.C **1,3,5 - Trinitrobenzene - bis(N - t - butylsalicylidene - iminato) cobalt(ii) complex**
$C_{22}H_{28}CoN_2O_2 , C_6H_3N_3O_6$
For complete entry see 60.9

78.C **1,3,5 - Trinitrobenzene - bis(N - t - butylsalicylidene - iminato) copper(ii) complex**
$C_{22}H_{28}CuN_2O_2 , C_6H_3N_3O_6$
For complete entry see 60.10

78.C **1,3,5 - Trinitrobenzene - bis(N - t - butylsalicylidene - iminato) nickel(ii) complex**
$C_{22}H_{28}N_2NiO_2 , C_6H_3N_3O_6$
For complete entry see 60.11

78.12 **bis(N - β - Dimethylaminoethyl - salicylaldiminato) copper(ii)**
$C_{22}H_{30}CuN_4O_2$
P.C.Chieh, G.J.Palenik *Inorg. Chem.*, **11**, 816, 1972

78.C **Malononitrilo - (1,2 - propane - N,N' - bis(salicylideneiminato)) cobalt(iii) pyridine**
$C_{25}H_{22}CoN_5O_2$
For complete entry see 71.57

78.13 **Benzoylacetonato - (N,N′ - ethylene - bis(salicylideneiminato)) cobalt(iii) sesquihydrate**

$C_{26}H_{23}CoN_2O_4$, $1.5H_2O$
N.A.Bailey, B.M.Higson, E.D.McKenzie *J. C. S. Dalton,* 503, 1972
Residue 1 also classified in 77

78.14 **N,N′ - Ethylene bis(salicylaldehydeiminato) cobalt(ii) (oxygen - inactive form)**

$C_{32}H_{28}Co_2N_4O_4$
R.DeIasi, S.L.Holt, B.Pcst *Inorg. Chem.*, **10,** 1498, 1971
Also classified in 76, 84

78.15 **μ - Peroxo - bis - (3,3′ - di - imino - di - n - propylamine - bis - salicylaldehyde cobalt(iii)) toluene solvate**

$C_{40}H_{46}Co_2N_6O_6$, C_7H_8
L.A.Lindblom, W.P.Schaefer, R.E.Marsh
Acta Cryst. (B), **27,** 1461, 1971

78.16 **μ - Oxo - bis(bis(N - n - propylsalicylideneiminato) iron(iii))**

$C_{40}H_{48}Fe_2N_4O_5$
J.E.Davies, B.M.Gatehouse *Cryst. Struct. Comm.*, **1,** 115, 1972

78.17 **tetrakis(Aquo - (N - 2 - pyridylsalicylaldiminato) copper(ii)) tetranitrate**

$C_{48}H_{44}Cu_4N_8O_8^{4+}$. $4NO_3^-$
J.Drummond, J.S.Wood *J. C. S. Dalton,* 365, 1972

METAL COMPLEXES (THIOUREA)

79.1 **Dichloro - bis(thiourea) cobalt(ii)**

$C_2H_8Cl_2CoN_4S_2$

P.Domiano, A.Tiripicchio *Cryst. Struct. Comm.*, **1**, 107, 1972

79.2 **bis(Thiosemicarbazide) copper(ii) nitrate**

$C_2H_{10}CuN_6S_2^{2+}$, $2NO_3^-$

A.C.Villa, A.G.Manfreddoti, C.Guastini

Cryst. Struct. Comm., **1**, 207, 1972

79.3 **bis(Thiosemicarbazide) copper(ii) sulfate**

$C_2H_{10}CuN_6S_2^{2+}$, O_4S^{2-}

A.C.Villa, A.G.Manfredotti, C.Guastini

Cryst. Struct. Comm., **1**, 125, 1972

79.4 **bis(Thiocarbohydrazide) - dichlorido - cadmium**

$C_2H_{12}CdCl_2N_8S_2$

F.Bigoli, A.Braibanti, A.M.M.Lanfredi, A.Tiripicchio, M.T.Camellini

Inorg. Chim. Acta, **5**, 392, 1971

Also classified in 83, 86

79.C **π - Allyl - di(thiourea) nickel(ii) chloride**

$C_5H_{13}N_4NiS_2^+$, Cl^-

For complete entry see 72.4

79.5 **Ethylenebis(biguanidine) nickel(ii) chloride monohydrate**

$C_6H_{16}N_{10}Ni^{2+}$, $2Cl^-$, H_2O

D.L.Ward, C.N.Caughlan, G.D.Smith *Acta Cryst. (B),* **27**, 1541, 1971

Residue 1 also classified in 83

79.6 **hexakis(Thiourea) dicopper(i) tetrafluoroborate**

$C_6H_{24}Cu_2N_{12}S_6^{2+}$, $2BF_4^-$

I.F. Taylor Junior, P.Boldrini, E.L. Amma

Amer. Chem. Soc., Abstr. Papers, 208, 1969

79.7 **hexakis(Thiourea) dicopper(i) perchlorate**

$C_6H_{24}Cu_2N_{12}S_6^{2+}$, $2ClO_4^-$

F.Hanic, E.Durcanska *Inorg. Chim. Acta,* **3**, 293, 1969

79.8 **Titanium hexa(urea) perchlorate**

$C_6H_{24}N_{12}O_6Ti^{3+}$, $3ClO_4^-$

B.N.Figgis, L.G.B.Wadley, J.Graham *Acta Cryst. (B),* **28,** 187, 1972

Residue 1 also classified in 83

79.9 **nonakis(Thiourea) tetracopper(i) nitrate**

$(C_9H_{36}Cu_4N_{18}S_9^{4+})_n$, $4nNO_3^-$

R.G.Vranka, E.L.Amma *J. Amer. Chem. Soc.,* **88,** 4270, 1966

79.10 **Potassium bis(3 - (n - propyl) - biuretato) cobaltate(iii) bis(1 - (n - propyl) - biuret)**

$C_{10}H_{18}CoN_6O_4^-$, $2C_5H_{11}N_3O_2$, K^+

J.J.Bour, P.T.Beurskens, J.J.Steggarda *J. C. S. Chem. Comm.,* 221, 1972

Residue 1 also classified in 83; residue 2 classified in 8

79.11 **Dichloro - bis(N,N′ - diethylthiourea) zinc(ii)**

$C_{10}H_{24}Cl_2N_4S_2Zn$

M.Bonamico, G.Dessy, V.Fares, L.Scaramuzza

J. Chem. Soc. (A), 3195, 1971

79.12 **tetrakis(N,N′ - Diallylthiourea) nickel(ii) iodide**

$C_{28}H_{48}N_8NiS_4^{2+}$, $2I^-$

A.Chiesi, A.Mangia, M.Nardelli, G.Pelizzi

J. Cryst. Mol. Struct., **1,** 285, 1971

METAL COMPLEXES (THIOCARBAMATE OR XANTHATE)

80.1 **bis(Dimethyldithiocarbamato)nitrosyl - cobalt**

$C_6H_{12}CoN_3OS_4$

J.H.Enemark, R.D.Feltham *J. C. S. Dalton,* 718, 1972

80.2 **Chromium(iii) tris(O - ethylxanthate)**

$C_9H_{15}CrO_3S_6$

S.Merlino, F.Sartori *Acta Cryst. (B),* **28,** 972, 1972

80.3 **Iodo - bis(N,N - diethyldithiocarbamato) iron(iii)**

$C_{10}H_{20}FeIN_2S_4$

P.C.Healy, A.H.White, B.F.Hoskins *J. C. S. Dalton,* 1369, 1972

80.4 **bis(N,N - Diethyldithiocarbamato) palladium(ii)**

$C_{10}H_{20}N_2PdS_4$

P.T.Beurskens, J.A.Cras, T.W.Hummelink, J.H.Noordik
J. Cryst. Mol. Struct., **1,** 253, 1971

80.5 **Nitrido - bis(N,N - diethyldithiocarbamato) rhenium(v)**

$C_{10}H_{20}N_3ReS_4$

S.R.Fletcher, A.C.Skapski *J. C. S. Dalton,* 1079, 1972

80.6 **Bromo - dimethoxy - bis(diethyldithiocarbamato) niobium(ii)**

$C_{12}H_{26}BrNbO_2S_4$

J.W.Moncrief, D.C.Pantaleo, N.E.Smith
Inorg. Nucl. Chem. Letters, **7,** 255, 1971
Also classified in 84

80.7 **Chloro - dimethoxy - bis(diethyldithiocarbamato) niobium(v)**

$C_{12}H_{26}ClNbO_2S_4$

J.W.Moncrief, D.C.Pantaleo, N.E.Smith
Inorg. Nucl. Chem. Letters, **7,** 255, 1971
Also classified in 84

80.8 **bis(N,N - Diethyldithiocarbamato)(cis - 1,2 - bis(trifluoromethyl) - ethylene - 1,2 - dithiolato) - iron**

$C_{14}H_{20}F_6FeN_2S_6$

D.L.Johnston, W.L.Rohrbaugh, W.DeW.Horrocks Junior
Inorg. Chem., **10,** 1474, 1971
Also classified in 85

80.9 **tris(1 - Pyrrolidine - carbodithioato) iron(iii)**
tris(N - Tetramethylene - dithiocarbamato) iron(iii)
$C_{15}H_{24}FeN_3S_6$
P.C.Healy, A.H.White *J. C. S. Dalton,* 1163, 1972

80.10 **tris(t - Butyl thioxanthato) iron(iii)**
$C_{15}H_{27}FeS_9$
D.F.Lewis, S.J.Lippard, J.A.Zubieta *Inorg. Chem.*, **11,** 823, 1972

80.11 **bis(N,N - Di - n - butyldithiocarbamato) gold(iii) dibromoargentate(i)**
$C_{18}H_{36}AuN_2S_4^+$, $AgBr_2^-$
J.A.Cras, J.H.Noordik, P.T.Beurskens, A.M.Verhoeven
J. Cryst. Mol. Struct., **1,** 155, 1971

80.12 **bis(N,N - Di - n - butyldithiocarbamato) gold(iii) bis(1,2 - dicyanoethene - 1,2 - dithiolato) aurate(iii)**
$C_{18}H_{36}AuN_2S_4^+$, $C_8AuN_4S_4^-$
J.H.Noordik, P.T.Beurskens *J. Cryst. Mol. Struct.*, **1,** 339, 1971
Residue 2 classified in 85

80.13 **μ - Oxo - bis(oxo - bis(N,N - diethyldithiocarbamato) rhenium(v))**
$C_{20}H_{40}N_4O_3Re_2S_8$
D.G.Tisley, R.A.Walton, D.L.Wills
Inorg. Nucl. Chem. Letters, **7,** 523, 1971

80.14 **μ - Oxo - bis(oxo - bis(N,N - diethyldithiocarbamato) rhenium(v))**
$C_{20}H_{40}N_4O_3Re_2S_8$
S.R.Fletcher, A.C.Skapski *J. C. S. Dalton,* 1073, 1972

80.15 **tetrakis(N,N - Diethyldithiocarbamato) titanium(iv)**
$C_{20}H_{40}N_4S_8Ti$
M.Colapietro, A.Vaciago, D.C.Bradley, M.B.Hursthouse, I.F.Rendall
J. C. S. Dalton, 1052, 1972

80.16 **Triphenylphosphine - (N,N - diethyldithiocarbamato) gold(i)**
$C_{23}H_{25}AuNPS_2$
J.G.Wijnhoven, W.P.J.H.Bosman, P.T.Beurskens
J. Cryst. Mol. Struct., **2,** 7, 1972
Also classified in 86

80.17 **tris(N - Methyl - N - phenyl - dithiocarbamato) iron(iii)**
$C_{24}H_{24}FeN_3S_6$
P.C.Healy, A.H.White *J. C. S. Dalton,* 1163, 1972

80.18 **tris(N,N - Di - n - butyldiselenocarbamato) nickel(iv) bromide**
$C_{27}H_{54}N_3NiSe_6^+$, Br^-
P.T.Beurskens, J.A.Cras *J. Cryst. Mol. Struct.*, **1,** 63, 1971
Residue 1 also classified in 85

80.C **Molybdenum dithiocarbamate compound**

$C_{28}H_{56}Mo_2N_4S_8$
For complete entry see 71.66

80.19 **Zinc isopropylxanthate tetramer**

$C_{32}H_{56}O_8S_{16}Zn_4$
T.Ito *Acta Cryst. (B)*, **28**, 1697, 1972

80.20 **Silver(i) dipropylmonothiocarbamate hexamer**

$C_{42}H_{84}Ag_6N_6O_6S_6$
P.Jennische, R.Hesse *Acta Chem. Scand.*, **25**, 423, 1971

METAL COMPLEXES (CARBOXYLIC ACID)

81.1 **catena - μ - Oxalato - ammine - copper(ii)**
$(C_2H_3CuNO_4)_n$
L.Cavalca, A.C.Villa, A.G.Manfredotti, A.A.G.Tomlinson
J. C. S. Dalton, 391, 1972

81.2 **bis(Hydrazinecarboxylato) zinc**
$C_2H_6N_4O_4Zn$
F.Bigoli, A.Braibanti, A.Tiripicchio, M.T.Camellini
Acta Cryst. (B), **27,** 2453, 1971
Also classified in 83

81.3 **Ammonium diperoxo(dioxalato) niobate monohydrate**
$C_4NbO_{12}^{3-}$, $3H_4N^+$, H_2O
G.Mathern, R.Weiss *Acta Cryst. (B),* **27,** 1572, 1971

81.4 **Yttrium dihydronium oxalate dihydrate (neutron study,at 25 ° C)**
$C_4O_8Y^-$, $H_5O_2^+$, $2H_2O$
C.K.Johnson, G.D.Brunton
Amer. Cryst. Assoc., Abstr. Papers (Winter Meeting), 77, 1972

81.5 **Yttrium dihydronium oxalate dihydrate (52 percent deuterated,neutron study,at 25 ° C)**
$C_4O_8Y^-$, $H_5O_2^+$, $2H_2O$
C.K.Johnson, G.D.Brunton
Amer. Cryst. Assoc., Abstr. Papers (Winter Meeting), 77, 1972

81.6 **Yttrium dihydronium oxalate dihydrate (52 percent deuterated,neutron study,at 60 ° C)**
$C_4O_8Y^-$, $H_5O_2^+$, $2H_2O$
C.K.Johnson, G.D.Brunton
Amer. Cryst. Assoc., Abstr. Papers (Winter Meeting), 77, 1972

81.7 **Yttrium dihydronium oxalate dihydrate (52 percent deuterated,neutron study, at −155 ° C)**
$C_4O_8Y^-$, $H_5O_2^+$, $2H_2O$
C.K.Johnson, G.D.Brunton
Amer. Cryst. Assoc., Abstr. Papers (Winter Meeting), 77, 1972

81.8 **Yttrium dihydronium oxalate dihydrate (91 percent deuterated,neutron study,at 25 ° C)**
$C_4O_8Y^-$, $H_5O_2^+$, $2H_2O$
C.K.Johnson, G.D.Brunton
Amer. Cryst. Assoc., Abstr. Papers (Winter Meeting), 77, 1972

81.9 **Triammonium bis(oxalato) dioxovanadate(v) dihydrate**
$C_4O_{10}V^{3-}$, $3H_4N^+$, $2H_2O$
W.R.Scheidt, C.Tsai, J.L.Hoard *J. Amer. Chem. Soc.*, **93**, 3867, 1971

81.10 **Triammonium bis(oxalato) dioxovanadate(v) dihydrate**
$C_4O_{10}V^{3-}$, $3H_4N^+$, $2H_2O$
L.O.Atovmyan, Y.A.Sokolova *Zh. Strukt. Khim.*, **12**, 934, 1971

81.11 **Aquo - oxo - hydroxo - bis(oxalato)niobic acid tetrahydrate**
$C_4H_3NbO_{11}^{2-}$, $2H^+$, $4H_2O$
N.Galesic, B.Matkovic, M.Herceg, M.Sljukic
J. less-Common Metals, Netherl., **25**, 234, 1971

81.12 **aquo - Copper(ii) maleate**
$C_4H_4CuO_5$
C.K.Prout, J.R.Carruthers, F.J.C.Rossotti *J. Chem. Soc. (A)*, 3342, 1971

81.13 **Copper(ii) d - tartrate trihydrate**
$(C_4H_4CuO_6)_n$, $3nH_2O$
C.K.Prout, J.R.Carruthers, F.J.C.Rossotti *J. Chem. Soc. (A)*, 3336, 1971

81.14 **Copper(ii) meso - tartrate trihydrate**
$(C_4H_4CuO_6)_n$, $3nH_2O$
C.K.Prout, J.R.Carruthers, F.J.C.Rossotti *J. Chem. Soc. (A)*, 3336, 1971

81.15 **Diaquo - bis(amidoxalato) cobalt(ii) dihydrate**
$C_4H_8CoN_2O_8$, $2H_2O$
M.A.Pellinghelli, A.Tiripicchio, M.T.Camellini
Acta Cryst. (B), **28**, 998, 1972

81.16 **bis(Amidoxalato) - diaquo - zinc**
$C_4H_8N_2O_8Zn$
A.Braibanti, M.A.Pellinghelli, A.Tiripicchio, M.T.Camellini
Acta Cryst. (B), **27**, 1240, 1971

81.17 **Diaquo - cadmium diacetate**
$(C_4H_{10}CdO_6)_n$
W.Harrison, J.Trotter *J. C. S. Dalton*, 956, 1972

81.18 **Triaquo(iminodiacetato) neodymium(iii) chloride**
$(C_4H_{11}NNdO_7^+)_n$, nCl^-
A.Oskarsson *Acta Chem. Scand.*, **25**, 1206, 1971
Residue 1 also classified in 82

81.C **tris(Hydroxyacetato) erbium(iii) dihydrate**
Erbium(iii) glycolate dihydrate
$C_4H_{14}ErO_{10}^-$, $C_8H_{12}ErO_{12}^+$
For complete entry see 81.25

81.19 **Ammonium oxo - trioxalato - niobate monohydrate**
$C_6NbO_{13}^{3-}$, $3H_4N^+$, H_2O
G.Mathern, R.Weiss *Acta Cryst. (B),* **27,** 1610, 1971

81.20 **Potassium tris(oxalato) rhodate(iii) hydrate**
$C_6O_{12}Rh^{3-}$, $3K^+$, $4.5H_2O$
B.C.Dalzell, K.Eriks *J. Amer. Chem. Soc.,* **93,** 4298, 1971

81.21 **tris(Hydroxyacetato) europium(iii) (monoclinic form)**
tris(Glycolato) europium(iii)
$(C_6H_9EuO_9)_n$
I.Grenthe *Acta Chem. Scand.,* **25,** 3347, 1971

81.22 **Diaquo - ethylenedithiodiacetato - nickel(ii)**
$C_6H_{12}NiO_6S_2$
J.Podlahova, J.Loub, C.Novak *Acta Cryst. (B),* **28,** 1623, 1972
Also classified in 85

81.23 **aquo - Copper(ii) phthalate**
$C_8H_6CuO_5$
C.K.Prout, J.R.Carruthers, F.J.C.Rossotti *J. Chem. Soc. (A),* 3350, 1971

81.C **Potassium nitro(ethylenediaminetriacetato) cobaltate(iii) sesquihydrate**
$C_8H_{11}CoN_3O_8$, K^+ , $1.5H_2O$
For complete entry see 76.16

81.24 **Sodium(+)$_{546}$ ethylenediamine - bis(malonato) cobalt(iii) dihydrate (absolute configuration)**
$C_8H_{12}CoN_2O_8^-$, Na^+ , $2H_2O$
K.R.Butler, M.R.Snow *J. Chem. Soc. (D),* 550, 1971
Residue 1 also classified in 76

81.25 **tris(Hydroxyacetato) erbium(iii) dihydrate**
Erbium(iii) glycolate dihydrate
$C_8H_{12}ErO_{12}^+$, $C_4H_{14}ErO_{10}^-$
I.Grenthe *Acta Chem. Scand.,* **25,** 3721, 1971
Residue 2 classified in 81

81.26 **Tetra - μ - acetato - di(aquo - chromium)**
$C_8H_{16}Cr_2O_{10}$
F.A.Cotton, B.G.DeBoer, M.D.LaPrade, J.R.Pipal, D.A.Ucko
Acta Cryst. (B), **27,** 1664, 1971

81.27 **Copper diaquo - acetate (neutron study)**
$C_8H_{16}Cu_2O_{10}$
R.Chidambaram, G.M.Brown *Cryst. Struct. Comm.,* **1,** 269, 1972

81.28 **Tetra - μ - acetato - di(aquo - rhodium(ii))**

$C_8H_{16}O_{10}Rh_2$
F.A.Cotton, B.G.DeBoer, M.D.LaPrade, J.R.Pipal, D.A.Ucko
Acta Cryst. (B), **27**, 1664, 1971

81.C **$(-)_{589}$ tris(1,2 - Diaminopropane) cobalt(iii) $(+)_{589}$ tris(malonato) chromium(iii) trihydrate (absolute configuration)**

$C_9H_6CrO_{12}^{3-}$, $C_9H_{30}CoN_6^{3+}$, $3H_2O$
For complete entry see 76.32

81.29 **bis(μ - Acetato)hexacarbonyl - diosmium**

$C_{10}H_6O_{10}Os_2$
J.G.Bullitt, F.A.Cotton *Inorg. Chim. Acta*, **5**, 406, 1971

81.C **Sodium samarium ethylenediaminetetra - acetate octahydrate (neutron study)**

$C_{10}H_{12}N_2O_8Sm^-$, Na^+, $8H_2O$
For complete entry see 76.33

81.C **Trisodium (ethylenediaminetetra - acetato) dioxovanadate(v) tetrahydrate**

$C_{10}H_{12}N_2O_{10}V^{3-}$, $3Na^+$, $4H_2O$
For complete entry see 76.34

81.30 **bis(Imidazole) - bis(acetato) - cobalt**

$C_{10}H_{14}CoN_4O_4$
A.Gadet *C. R. Acad. Sci., Fr., C*, **272**, 1299, 1971
Also classified in 83

81.C **Lithium aquo manganese(ii) ethylenediaminetetra - acetate tetrahydrate**

$C_{10}H_{14}MnN_2O_9^{2-}$, $2Li^+$, $4H_2O$
For complete entry see 76.35

81.C **Ammonium (dihydrogenethylenediaminetetra - acetato) dioxovanadate(v) trihydrate**

$C_{10}H_{14}N_2O_{10}V^-$, H_4N^+, $3H_2O$
For complete entry see 76.36

81.C **(Hydrogen ethylenediaminetetra - acetato)aquoferrate(iii)**

$C_{10}H_{15}FeN_2O_9$
For complete entry see 76.37

81.C **$(-)_{546}$ - Sodium - bis(trans - N - methyl - (S) - alaninato) - oxalato - cobaltate(iii) dihydrate**

$C_{10}H_{16}CoN_2O_8^-$, Na^+, $2H_2O$
For complete entry see 82.10

81.31 **Potassium $(-)_{546}$ - trimethylenediamine - tetra - acetato - cobaltate(iii) dihydrate (absolute configuration)**

$C_{11}H_{14}CoN_2O_8^-$, K^+, $2H_2O$
R.Nagao, F.Marumo, Y.Saito *Acta Cryst. (B)*, **28**, 1852, 1972
Residue 1 also classified in 82

81.32 **Diaquo bis(picolinato) copper(ii)**
$C_{12}H_{12}CuN_2O_6$
A.Takenaka, H.Utsumi, T.Yamamoto, A.Furusaki, I.Nitta
J. Chem. Soc. Jap., Pure Chem. Sect., **91,** 928, 1970
Also classified in 83

81.33 **trans - Diaquo - bis(picolinato) nickel(ii) dihydrate**
$C_{12}H_{12}N_2NiO_6$, $2H_2O$
A.Takenaka, H.Utsumi, N.Ishihara, A.Furusaki, I.Nitta
J. Chem. Soc. Jap., Pure Chem. Sect., **91,** 921, 1970
Residue 1 also classified in 83

81.34 **trans - Diaquo - bis(picolinato) nickel(ii) dihydrate**
$C_{12}H_{12}N_2NiO_6$, $2H_2O$
H.Loiseleur *Acta Cryst. (B),* **28,** 816, 1972
Residue 1 also classified in 83

81.35 **trans - Diaquo - bis(picolinato) zinc dihydrate**
$C_{12}H_{12}N_2O_6Zn$, $2H_2O$
A.Takenaka, H.Utsumi, N.Ishihara, A.Furusaki, I.Nitta
J. Chem. Soc. Jap., Pure Chem. Sect., **91,** 921, 1970
Residue 1 also classified in 83

81.36 **bis(Pyridine - 2,3 - dicarboxylato) silver(ii) dihydrate**
$C_{14}H_8AgN_2O_8$, $2H_2O$
M.G.B.Drew, R.W.Matthews, R.A.Walton *J. Chem. Soc. (A),* 2959, 1971
Residue 1 also classified in 83, 84

81.37 **Nickel(ii) bis(hydrogen pyridine - 2,6 - dicarboxylate) trihydrate**
$C_{14}H_8N_2NiO_8$, $3H_2O$
H.Gaw, W.R.Robinson, R.A.Walton
Inorg. Nucl. Chem. Letters, **7,** 695, 1971
Residue 1 also classified in 83

81.38 **Diaquo - zinc 2 - pyridyl - acetate**
$C_{14}H_{16}N_2O_6Zn$
R.Faure, H.Loiseleur *Acta Cryst. (B),* **28,** 811, 1972
Also classified in 83

81.39 **tris(μ - Acetato - μ - acetoximato palladium(ii)) benzene solvate**
$C_{15}H_{27}N_3O_9Pd_3$, $0.5C_6H_6$
A.Mawby, G.E.Pringle *J. Inorg. Nucl. Chem.*, **33,** 1989, 1971
Residue 1 also classified in 83

81.40 **Copper(ii) dichloroacetate bis(α - picoline)**
$C_{16}H_{16}Cl_4CuN_2O_4$
G.Davey, F.S.Stephens *J. Chem. Soc. (A),* 2577, 1971
Also classified in 83

81.41 **bis(Methoxyacetato)bis(pyridine) copper(ii) tetrahydrate**
$C_{16}H_{20}CuO_6 , 4H_2O$
C.K.Prout, M.J.Barrow, F.J.C.Rossotti *J. Chem. Soc. (A)*, 3326, 1971
Residue 1 also classified in 83

81.42 **bis(Phenoxyacetato) triaquocopper(ii)**
$C_{16}H_{20}CuO_9$
C.V.Goebel, R.J.Doedens *Inorg. Chem.*, **10**, 2607, 1971

81.43 **Guanidinium tetra - acetato - cerate monohydrate dimer**
$C_{16}H_{28}Ce_2O_{18}^{2-} , 2CH_6N_3^+ , 2H_2O$
G.G.Sadikov, G.A.Kukina, M.A.Porai-Koshits
Zh. Strukt. Khim., **12**, 859, 1971
Residue 2 classified in 8

81.44 **Copper(ii) butyrate dimer**
$C_{16}H_{28}Cu_2O_8$
M.J.Bird, T.R.Lomer *Acta Cryst. (B)*, **28**, 242, 1972

81.C **bis(Cyclopentadienyl) - (1 - phenyl - 2 - carboxylato) titanium**
$C_{17}H_{14}O_2Ti$
For complete entry see 71.38

81.45 **Diaquo - erbium(iii) isonicotinate**
$C_{18}H_{16}ErN_3O_8$
L.A.Aslanov, I.K.Abdul'minev, R.A.Chupakhina, M.A.Porai-Koshits
Zh. Strukt. Khim., **12**, 936, 1971

81.46 **Diaquo - tris(isonicotinato) lanthanum(iii)**
$(C_{18}H_{16}LaN_3O_8)_n$
J.Kay, J.W.Moore, M.D.Glick
Amer. Cryst. Assoc., Abstr. Papers (Summer Meeting), 82, 1971

81.47 **Diaquo - tri(nicotinic acid) holmium(iii) hexa(isothiocyanato) chromate(iii) dihydrate**
$(C_{18}H_{19}HoN_3O_8^{3+})_n , nC_6CrN_6S_6^{3-} , 2nH_2O$
J.Kay, J.W.Moore, M.D.Glick
Amer. Cryst. Assoc., Abstr. Papers (Summer Meeting), 82, 1971

81.48 **Diacetato - copper(ii) - bis(p - toluidine) trihydrate**
$C_{18}H_{24}CuN_2O_4 , 3H_2O$
R.C.Komson, A.T.McPhail, F.E.Mabbs, J.K.Porter
J. Chem. Soc. (A), 3447, 1971
Residue 1 also classified in 83

81.49 **bis(Methoxyacetato)tetrakis(imidazole) copper(ii)**
$C_{18}H_{26}CuN_8O_6$
C.K.Prout, G.B.Allison, F.J.C.Rossotti *J. Chem. Soc. (A)*, 3331, 1971
Also classified in 83

81.50 **tris(Dimethoxyethane) - tri - μ - trifluoroacetato - μ_3 - chloro - μ_3 - sulfato - tricobalt(ii)**

$C_{18}H_{30}ClCo_3F_9O_{16}$

J.Estienne, R.Weiss *J. C. S. Chem. Comm.*, 862, 1972

Also classified in 84

81.51 **Trisodium tris(pyridine - 2,6 - dicarboxylato) neodymium(iii) hydrate**

$C_{21}H_9N_3NdO_{12}^{3-}$, $3Na^+$, $15H_2O$

J.Albertsson *Acta Chem. Scand.*, **26**, 1023, 1972

Residue 1 also classified in 83

81.52 **Trisodium tris(pyridine - 2,6 - dicarboxylato) ytterbium(iii) sodium perchlorate decahydrate**

$C_{21}H_9N_3O_{12}Yb^{3-}$, ClO_4^-, $4Na^+$, $10H_2O$

J.Albertsson *Acta Chem. Scand.*, **26**, 1005, 1972

Residue 1 also classified in 83

81.53 **Trisodium tris(pyridine - 2,6 - dicarboxylato) ytterbium(iii) hydrate**

$C_{21}H_9N_3O_{12}Yb^{3-}$, $3Na^+$, $13H_2O$

J.Albertsson *Acta Chem. Scand.*, **26**, 985, 1972

Residue 1 also classified in 83

81.C **Carboxymethyl - palladium (triphenylphosphine)pyridine**

$C_{25}H_{22}NO_2PPd$

For complete entry see 71.58

81.54 **bis(Phenoxyacetato)aquo - bis(pyridine) copper(ii)**

$C_{26}H_{26}CuN_2O_7$

C.K.Prout, M.J.Barrow, F.J.C.Rossotti *J. Chem. Soc. (A)*, 3326, 1971

Also classified in 83

81.C **Di - μ - acetato - bis(2 - methallyl - norbornyl) dipalladium(ii)**

$C_{26}H_{40}O_4Pd_2$

For complete entry see 71.60

81.C **bis(Benzene) copper(ii) trifluoroacetate complex**

$C_{32}H_{24}Cu_2F_{12}O_8$

For complete entry see 74.11

81.55 **bis(3 - Hydroxy - 4 - phenyl - 2,2,3 - trimethylcyclohexane - carboxylato) disilver(i) dihydrate**

$C_{32}H_{42}Ag_2O_6$, $2H_2O$

P.Coggon, A.T.McPhail *J. C. S. Chem. Comm.*, 91, 1972

81.56 **Diaquo - lanthanum(iii) nicotinate dimer**

$C_{36}H_{32}La_2N_6O_{16}$

J.W.Moore, M.D.Glick, W.A.Baker Junior

J. Amer. Chem. Soc., **94**, 1858, 1972

81.57 **Diaquo - praseodymium(iii) nicotinate dimer**
$C_{36}H_{32}N_6O_{16}Pr_2$
I.K.Abdul'minev, L.A.Aslanov, M.A.Porai-Koshits
Zh. Strukt. Khim., **12,** 935, 1971

81.58 **Diaquo - samarium(iii) nicotinate dimer**
$C_{36}H_{32}N_6O_{16}Sm_2$
J.W.Moore, M.D.Glick, W.A.Baker Junior
J. Amer. Chem. Soc., **94,** 1858, 1972

81.59 **Di - μ - hippurato - di(diaquo - hippurato - copper(ii)) tetrahydrate**
$C_{36}H_{40}Cu_2N_4O_{16}, 4H_2O$
J.N.Brown, H.R.Eichelberger, E.Schaeffer, M.L.Good, L.M.Trefonas
J. Amer. Chem. Soc., **93,** 6290, 1971
Residue 1 also classified in 82

81.60 **Compound Q**
$C_{48}H_{30}Cu_4F_{18}N_4O_{14}$
R.G.Little, D.B.W.Yawney, R.J.Doedens *J. C. S. Chem. Comm.*, 228, 1972
Also classified in 83

81.61 **Di - μ - acetato - bis(dimethylglyoximato)bis(triphenylphosphine) dirhodium(ii) monohydrate**
$C_{48}H_{48}N_4O_8P_2Rh_2, H_2O$
J.Halpern, E.Kimura, J.Molin-Case, C.S.Wong
J. Chem. Soc. (D), 1207, 1971
Residue 1 also classified in 86, 83

81.62 **Hexa(acetato) - tris(triphenylphosphine) - oxotriruthenium**
$C_{66}H_{63}O_{13}P_3Ru_3$
F.A.Cotton, J.G.Norman, A.Spencer, G.Wilkinson
J. Chem. Soc. (D), 967, 1971
Also classified in 86

METAL COMPLEXES (AMINO-ACID)

82.1 **Glycinato - silver(i)**
$(C_2H_4AgNO_2)_n$
C.B.Acland, H.C.Freeman *J. Chem. Soc. (D)*, 1016, 1971

82.2 **Glycinato - silver(i) hemihydrate**
$(C_2H_4AgNO_2)_{2n}$, nH_2O
C.B.Acland, H.C.Freeman *J. Chem. Soc. (D)*, 1016, 1971

82.3 **bis(Thioglycinato) nickel(ii)**
$C_4H_8N_2NiO_2S_2$
J.R.Ruble, K.Seff *Acta Cryst. (B)*, **28**, 1272, 1972
Also classified in 85

82.4 **Glycine silver(i) nitrate**
$C_4H_{10}Ag_2N_2O_4^{2+}$, $2NO_3^-$
J.K.M.Rao, M.A.Viswamitra *Acta Cryst. (B)*, **28**, 1484, 1972

82.5 **Glycine silver(i) nitrate (at −135 ° C)**
$C_4H_{10}Ag_2N_2O_4^{2+}$, $2NO_3^-$
J.K.M.Rao, M.A.Viswamitra *Acta Cryst. (B)*, **28**, 1484, 1972

82.C **Triaquo(iminodiacetato) neodymium(iii) chloride**
$(C_4H_{11}NNdO_7^+)_n$, nCl^-
For complete entry see 81.18

82.6 **Pyruvylideneglycinato - aquo - copper(ii) dihydrate**
$C_5H_7CuNO_5$, $2H_2O$
A.Torri, H.Tamura-Kobayashi, K.Ogawa, T.Watanabe
Z. Kristallogr., **133**, 179, 1971

82.7 **tris(Glycinato) chromium(iii) monohydrate**
$C_6H_{12}CrN_3O_6$, H_2O
R.F.Bryan, P.T.Greene, P.F.Stokely, E.W.Wilson Junior
Inorg. Chem., **10**, 1468, 1971

82.C **Potassium nitro(ethylenediaminetriacetato) cobaltate(iii) sesquihydrate**
$C_8H_{11}CoN_3O_8$, K^+, $1.5H_2O$
For complete entry see 76.16

82.8 **Glycylglycine - silver(i) nitrate**
$C_8H_{16}Ag_2N_4O_6^{2+}$, $2NO_3^-$
C.B.Acland, H.C.Freeman *J. Chem. Soc. (D)*, 1016, 1971

82.9 **Di - μ - hydroxo - bis(bis(glycinato) chromium(iii))**
$C_8H_{18}Cr_2N_4O_{10}$
D.L.Lewis, J.T.Veal, W.E.Hatfield, D.J.Hodgson
Amer. Cryst. Assoc., Abstr. Papers (Winter Meeting), 79, 1972

82.C **β_2 - Triethylenetetra - amine(glycinato) cobalt(iii) dichloride monohydrate**
$C_8H_{22}CoN_5O_2^{2+}$, $2Cl^-$, H_2O
For complete entry see 76.17

82.C **β_1 - Triethylenetetra - amine(glycinato) cobalt(iii) di - iodide hemihydrate**
$2C_8H_{22}CoN_5O_2^{2+}$, $4I^-$, H_2O
For complete entry see 76.18

82.C **Sodium samarium ethylenediaminetetra - acetate octahydrate (neutron study)**
$C_{10}H_{12}N_2O_8Sm^-$, Na^+ , $8H_2O$
For complete entry see 76.33

82.C **Trisodium (ethylenediaminetetra - acetato) dioxovanadate(v) tetrahydrate**
$C_{10}H_{12}N_2O_{10}V^{3-}$, $3Na^+$, $4H_2O$
For complete entry see 76.34

82.C **Lithium aquo manganese(ii) ethylenediaminetetra - acetate tetrahydrate**
$C_{10}H_{14}MnN_2O_9^{2-}$, $2Li^+$, $4H_2O$
For complete entry see 76.35

82.C **Ammonium (dihydrogenethylenediaminetetra - acetato) dioxovanadate(v) trihydrate**
$C_{10}H_{14}N_2O_{10}V^-$, H_4N^+ , $3H_2O$
For complete entry see 76.36

82.C **(Hydrogen ethylenediaminetetra - acetato)aquoferrate(iii)**
$C_{10}H_{15}FeN_2O_9$
For complete entry see 76.37

82.10 **$(-)_{546}$ - Sodium - bis(trans - N - methyl - (S) - alaninato) - oxalato - cobaltate(iii) dihydrate**
$C_{10}H_{16}CoN_2O_8^-$, Na^+ , $2H_2O$
G.W.Svetich, A.A.Voge, J.G.Brushmiller, E.A.Berends
J. C. S. Chem. Comm., 701, 1972
Residue 1 also classified in 81

82.11 **1,5 - Diazacyclo - octane - N,N′ - diacetato(aquo) nickel(ii) dihydrate**
$C_{10}H_{18}N_2NiO_5$, $2H_2O$
D.O.Nielson, M.L.Larsen, R.D.Willett, J.I.Legg
J. Amer. Chem. Soc., **93**, 5079, 1971

82.C **Potassium $(-)_{546}$ - trimethylenediamine - tetra - acetato - cobaltate(iii) dihydrate (absolute configuration)**

$C_{11}H_{14}CoN_2O_8^-$, K^+, $2H_2O$

For complete entry see 81.31

82.12 **bis(L - Histidinato) zinc(ii) dihydrate (refinement of data of Kretsinger et al.,Acta Cryst.,16,651,1963)**

$C_{12}H_{16}N_6O_4Zn$, $2H_2O$

T.J.Kistenmacher *Acta Cryst. (B),* **28,** 1302, 1972

82.13 **Di - μ - sulfido - bis(oxo - (L - histidinato) molybdenum(v)) trihydrate**

$C_{12}H_{20}Mo_2N_6O_6S_2$, $3H_2O$

B.Spivack, A.P.Gaughan, Z.Dori *J. Amer. Chem. Soc.,* **93,** 5265, 1971

82.14 **bis(L - Phenylalaninato) copper(ii)**

$C_{18}H_{20}CuN_2O_4$

D.van der Helm, M.B.Lawson, E.L.Enwall

Acta Cryst. (B), **27,** 2411, 1971

82.15 **Copper(ii) glycyl - L - leucyl - L - tyrosinate dimer octahydrate diethylether solvate**

$C_{34}H_{46}Cu_2N_6O_{10}$, $8H_2O$, $C_4H_{10}O$

W.A.Franks, D.van der Helm *Acta Cryst. (B),* **27,** 1299, 1971

82.C **Di - μ - hippurato - di(diaquo - hippurato - copper(ii)) tetrahydrate**

$C_{36}H_{40}Cu_2N_4O_{16}$, $4H_2O$

For complete entry see 81.59

82.C **bis(bis(Tricyclohexylphosphine) nickel)dinitrogen**

$C_{72}H_{132}N_2Ni_2P_4$

For complete entry see 86.66

METAL COMPLEXES (NITROGEN LIGAND)

83.1 **catena - Di - μ - chloro - semicarbazide - copper(ii) (monoclinic form)**
$(CH_5Cl_2CuN_3O)_n$
A.C.Villa, A.G.Manfredotti, M.Nardelli, G.Pelizzi
J. Cryst. Mol. Struct., **1**, 245, 1971
Also classified in 84

83.2 **catena - Di - μ - chloro - semicarbazide - copper(ii) (orthorhombic form)**
$(CH_5Cl_2CuN_3O)_n$
A.C.Villa, A.G.Manfredotti, M.Nardelli, G.Pelizzi
J. Cryst. Mol. Struct., **1**, 245, 1971
Also classified in 84

83.3 **cis - Tetrachloro - di(formonitrile) - titanium(iv)**
$C_2H_2Cl_4N_2Ti$
G.Constant, J.C.Daran, Y.Jeannin *Acta Cryst. (B)*, **27**, 2388, 1971

83.4 **Acetonitrile copper(i) bromide**
$(C_2H_3BrCuN)_n$
M.Massaux, M.J.Bernard, M.-T.Le Bihan
Acta Cryst. (B), **27**, 2419, 1971

83.5 **Di - μ - chloro(dicyanodiamide) cadmium(ii)**
$(C_2H_4CdCl_2N_4)_n$
A.C.Villa, L.Coghi, A.Mangia, M.Nardelli, G.Pelizzi
J. Cryst. Mol. Struct., **1**, 291, 1971

83.C **bis(Hydrazinecarboxylato) zinc**
$C_2H_6N_4O_4Zn$
For complete entry see 81.2

83.C **bis(Thiocarbohydrazide) - dichlorido - cadmium**
$C_2H_{12}CdCl_2N_8S_2$
For complete entry see 79.4

83.6 **(N - Methyl - N,N - bis(methyleneamino - N' - (dithionitrite))amino) nickel(ii)**
$C_3H_7N_5NiS_4$
U.Thewalt *Z. Anorg. Allg. Chem.*, **374**, 259, 1970
Also classified in 85

83.7 **catena - Di - μ - chloro - chloro - (1,3 - diaminopropane) cadmium(ii)**
$(C_3H_{10}CdCl_2N_2)_n$
G.D.Andreetti, L.Cavalca, M.A.Pellinghelli, P.Sgarabotto
Gazz. Chim. Ital., **101**, 488, 1971

83.8 **Tetrachloro(trichloroacetonitrile)pentachloroethyl nitrido - tungsten(vi)**
$C_4Cl_{12}N_2W$
M.G.B.Drew, K.C.Moss, N.Rolfe
Inorg. Nucl. Chem. Letters, **7**, 1219, 1971

83.9 **Diethylamino - titanium trichloride**
$(C_4H_{10}Cl_3NTi)_n$
J.Fayos, D.Mootz *Z. Anorg. Allg. Chem.*, **380**, 196, 1971

83.10 **bis(Dihydroxo - boroxalene - diamide - dioximato) nickel(ii) tetrahydrate**
$C_4H_{12}B_2N_8NiO_8 , 4H_2O$
W.Fedder, H.G.von Schnering, F.Umland
Z. Anorg. Allg. Chem., **382**, 123, 1971

83.11 **Pyridazine copper(i) cyanide**
$(C_5H_4CuN_3)_n$
D.T.Cromer, A.C.Larson *Acta Cryst. (B)*, **28**, 1052, 1972

83.12 **1 - (2 - Aminoethyl)biguanide - isothiacyanato - copper(ii) thiocyanate**
$C_5H_{12}CuN_7S^+ , CNS^-$
G.D.Andreetti, L.Coghi, M.Nardelli, P.Sgarabotto
J. Cryst. Mol. Struct., **1**, 147, 1971

83.13 **Chloro - pentakis(methylamine) cobalt(iii) nitrate**
$C_5H_{25}ClCoN_5^{2+} , 2NO_3^-$
B.M.Foxman *J. C. S. Chem. Comm.*, 515, 1972

83.14 **4 - Cyanopyridine copper(i) cyanide**
$C_6H_4CuN_3$
D.T.Cromer, A.C.Larson *Acta Cryst. (B)*, **28**, 1052, 1972

83.15 **bis(Imidazole) silver(i) nitrate**
$C_6H_8AgN_4^+ , NO_3^-$
C.B.Acland, H.C.Freeman *J. Chem. Soc. (D)*, 1016, 1971

83.16 **bis(Imidazole) silver(i) nitrate**
$C_6H_8AgN_4^+ , NO_3^-$
C.-J.Antti, B.K.S.Lundberg *Acta Chem. Scand.*, **25**, 1758, 1971

83.17 **Chloro - (triethanolamine) zinc(ii)**
$C_6H_{14}ClNO_3Zn$
H.Follner *Acta Cryst. (B)*, **28**, 157, 1972
Also classified in 84

83.C **Ethylenebis(biguanidine) nickel(ii) chloride monohydrate**
$C_6H_{16}N_{10}Ni^{2+}$, $2Cl^-$, H_2O
For complete entry see 79.5

83.18 **Triethanolamine - aquo - zinc(ii) - μ - chloro - trichlorozinc(ii)**
$C_6H_{17}Cl_4NO_4Zn_2$
H.Follner *Z. Anorg. Allg. Chem.*, **387**, 43, 1972
Also classified in 84

83.19 **bis(Trimethylamine) titanium tribromide**
$C_6H_{18}Br_3N_2Ti$
P.T.Greene, B.J.Russ, J.S.Wood *J. Chem. Soc. (A)*, 3636, 1971

83.20 **bis(Trimethylamine) chromium trichloride**
$C_6H_{18}Cl_3CrN_2$
P.T.Greene, B.J.Russ, J.S.Wood *J. Chem. Soc. (A)*, 3636, 1971

83.C **tris(2 - Aminoethanolato) cobalt(iii) - tris(2 - aminoethanol) nickel(ii) di - iodide**
$C_6H_{18}N_3NiO_3^-$, $C_6H_{21}CoN_3O_3^{3+}$, $2I^-$
For complete entry see 83.21

83.C **Chloro - ethylenediamine - (α - amino - (2 - aminoethylimino) - acetamide) cobalt(iii) chloride monohydrate**
$C_6H_{20}ClCoN_6^{2+}$, $2Cl^-$, H_2O
For complete entry see 76.11

83.C **bis(1,3 - Diaminopropane) copper(ii) thiocyanate**
$C_6H_{20}CuN_4^{2+}$, $2CNS^-$
For complete entry see 76.12

83.21 **tris(2 - Aminoethanolato) cobalt(iii) - tris(2 - aminoethanol) nickel(ii) di - iodide**
$C_6H_{21}CoN_3O_3^{3+}$, $C_6H_{18}N_3NiO_3^-$, $2I^-$
J.A.Bertrand, W.J.Howard, A.R.Kalyanaraman
J. Chem. Soc. (D), 437, 1971
Residue 1 also classified in 84; residue 2 classified in 83, 84

83.C **tris(1,3 - Diaminopropane) nickel(ii) diaquo - bis(1,3 - diaminopropane) nickel(ii) tetrachloride monohydrate**
$C_6H_{24}N_4NiO_2^{2+}$, $C_9H_{30}N_6Ni^{2+}$, $4Cl^-$, H_2O
For complete entry see 83.33

83.C **Titanium hexa(urea) perchlorate**
$C_6H_{24}N_{12}O_6Ti^{3+}$, $3ClO_4^-$
For complete entry see 79.8

83.22 **2 - (2 - Aminoethyl)pyridine - dibromo - copper(ii)**
$(C_7H_{10}Br_2CuN_2)_n$
P.Singh, V.C.Copeland, W.E.Hatfield, D.J.Hodgson
Amer. Cryst. Assoc., Abstr. Papers (Winter Meeting), 49, 1972

83.23 **2 - (2 - Aminoethyl)pyridine - dichloro - copper(ii)**
$(C_7H_{10}Cl_2CuN_2)_n$
P.Singh, V.C.Copeland, W.E.Hatfield, D.J.Hodgson
Amer. Cryst. Assoc., Abstr. Papers (Winter Meeting), 49, 1972

83.C **(N,N′ - Di(2 - aminoethyl)malondiamidato) nickel(ii) trihydrate**
$C_7H_{14}N_4NiO_2$, $3H_2O$
For complete entry see 76.15

83.24 **cis - (Carbonato - bis(trimethylenediamine) cobalt(iii)) perchlorate**
$C_7H_{20}CoN_4O_3^+$, ClO_4^-
R.J.Geue, M.R.Snow *J. Chem. Soc. (A),* 2981, 1971

83.25 **Dichloro - bis(pyrazine) cobalt(ii)**
$(C_8H_8Cl_2CoN_4)_n$
P.W.Carreck, M.Goldstein, E.M.McPartlin, W.D.Unsworth
J. Chem. Soc. (D), 1634, 1971

83.26 **bis(Diaminomaleonitrile)dichloro - palladium**
$C_8H_8Cl_2N_8Pd$
M.G.Miles, M.B.Hursthouse, A.G.Robinson
J. Inorg. Nucl. Chem., **33,** 2015, 1971

83.27 **Diaquo - copper(ii) diviolurate dihydrate**
$C_8H_8CuN_6O_{10}$, $2H_2O$
M.Hamelin *Acta Cryst. (B),* **28,** 228, 1972
Residue 1 also classified in 84

83.28 **bis(Cytosine) dichlorocopper(ii)**
$C_8H_{10}Cl_2CuN_6O_2$
M.Sundaralingam, J.A.Carrabine *J. Molec. Biol.*, **61,** 287, 1971

83.29 **trans - Diammine - bis(N - methylimidazole) platinum(ii) chloride dihydrate**
$C_8H_{18}N_6Pt^{2+}$, $2Cl^-$, $2H_2O$
J.W.Carmichael, N.Chan, A.W.Cordes, C.K.Fair, D.A.Johnson
Inorg. Chem., **11,** 1117, 1972

83.30 **Dichloro - tetrakis(acetaldoxime) nickel(ii)**
$C_8H_{20}Cl_2N_4NiO_4$
M.E.Stone, B.E.Robertson, E.Stanley *J. Chem. Soc. (A),* 3632, 1971

83.C **$(+)_{546}$ - trans - Dinitro(1,10 - diamino - 4,7 - diazadecane) cobalt(iii) bromide (absolute configuration)**
$C_8H_{22}CoN_6O_4^+$, Br^-
For complete entry see 76.19

83.31 **$(+)_{470}$ - cis - (1,9 - Diamino - 4 - methyl - 3,7 - diazanonane) - bis(nitrito) cobalt(iii) bromide (absolute configuration)**
$C_8H_{22}CoN_6O_4^+$, Br^-
P.W.R.Corfield, J.C.Dabrowiak, E.S.Gore
Amer. Cryst. Assoc., Abstr. Papers (Summer Meeting), 76, 1971

83.C **Chloro - (ethylenediamine)(dipropylenetriamine) cobalt(iii) iodide monohydrate (α form)**
$C_8H_{25}ClCoN_5^{2+}$, $2I^-$, H_2O
For complete entry see 76.24

83.32 **Bromo - tris(3 - aminopropyl)amine - cobalt(ii) bromide ethanol solvate**
$C_9H_{24}BrCoN_4^+$, Br^- , $0.5C_2H_6O$
J.L.Shafer, K.N.Raymond *Inorg. Chem.*, **10,** 1799, 1971

83.33 **tris(1,3 - Diaminopropane) nickel(ii) diaquo - bis(1,3 - diaminopropane) nickel(ii) tetrachloride monohydrate**
$C_9H_{30}N_6Ni^{2+}$, $C_6H_{24}N_4NiO_2^{2+}$, $4Cl^-$, H_2O
G.D.Andreetti, L.Cavalca, P.Sgarabotto *Gazz. Chim. Ital.*, **101,** 494, 1971
Residue 2 classified in 83

83.34 **1,4 - Dimethyl - 1,4 - diazabicyclo(2.2.2)octane o - benzoquinonedi - imine(tetracyano) iron(iii)**
$C_{10}H_6FeN_6^{2-}$, $C_8H_{18}N_2^{2+}$
G.G.Christoph, V.Goedken
Amer. Cryst. Assoc., Abstr. Papers (Summer Meeting), 76, 1971
Residue 1 also classified in 37

83.35 **Silver(i) dipyridine nitrate monohydrate**
$C_{10}H_{10}AgN_2^+$, NO_3^- , H_2O
S.Menchetti, G.Rossi, V.Tazzoli
R. C. Ist. Lombardo Sci. A, **104,** 309, 1970

83.36 **bis(H - Pyrrole - 2 - aldimine) copper(ii)**
$C_{10}H_{10}CuN_4$
R.Tewari, R.C.Srivastava *Acta Cryst. (B),* **27,** 1644, 1971

83.37 **Dinitrato - bis(pyridine) copper(ii) pyridine solvate**
$C_{10}H_{10}CuN_4O_6$, $0.5C_5H_5N$
A.F.Cameron, K.P.Forrest, D.W.Taylor, R.H.Nuttall
J. Chem. Soc. (A), 2492, 1971
Residue 1 also classified in 84

83.38 **Dinitrato - bis(pyridine) zinc(ii)**
$C_{10}H_{10}N_4O_6Zn$
A.F.Cameron, D.W.Taylor, R.H.Nuttall *J. Chem. Soc. (A),* 3402, 1971

83.39 **Guanine hydrochloride copper(ii) chloride monohydrate dimer**
$C_{10}H_{12}Cl_6Cu_2N_{10}O_2$, $2H_2O$
J.P.Declerq, M.Debbaudt, M.van Meerssche
Bull. Soc. Chim. Belges, **80,** 527, 1971

83.40 **Guanine hydrochloride copper(ii) chloride monohydrate dimer**
$C_{10}H_{12}Cl_6Cu_2N_{10}O_2$, $2H_2O$
M.Sundaralingam, J.A.Carrabine *J. Molec. Biol.*, **61,** 287, 1971

83.C **bis(Imidazole) - bis(acetato) - cobalt**
$C_{10}H_{14}CoN_4O_4$
For complete entry see 81.30

83.41 **bis - Aquo - bis - nitrato - bis - pyridine - nickel(ii)**
$C_{10}H_{14}N_2NiO_8$
A.F.Cameron, D.W.Taylor, R.H.Nuttall *J. C. S. Dalton,* 422, 1972

83.C **Potassium bis(3 - (n - propyl) - biuretato) cobaltate(iii) bis(1 - (n - propyl) - biuret)**
$C_{10}H_{18}CoN_6O_4^-$, $2C_5H_{11}N_3O_2$, K^+
For complete entry see 79.10

83.42 **bis(2 - Amino - 2 - methyl - 3 - butanone oximato) nickel(ii) chloride monohydrate (neutron study)**
$C_{10}H_{23}N_4NiO_2^+$, Cl^-, H_2O
E.O.Schlemper, W.C.Hamilton, S.J.La Placa
J. Chem. Phys., **54,** 3990, 1971

83.43 **μ - 2,3 - (2,3 - Diazabicyclo(2.2.1)heptane)diyl - bis(tricarbonyl iron)**
$C_{11}H_8Fe_2N_2O_6$
R.G.Little, R.J.Doedens *Inorg. Chem.*, **11,** 1392, 1972

83.44 **(4 - Methyliminopentane - 2,3 - dione - 3 - oximato)(4 - iminopentane - 2,3 - dione - 3 - oximato) nickel(ii)**
$C_{11}H_{16}N_4NiO_4$
M.J.Lacey, C.G.Macdonald, J.F.McConnell, J.S.Shannon
J. Chem. Soc. (D), 1206, 1971
Also classified in 84

83.45 **11,13 - Dimethyl - 1,4,7,10 - tetra - azacyclotrideca - 10,12 - dienato nickel(ii) perchlorate**
$C_{11}H_{21}N_4Ni^+$, ClO_4^-
M.F.Richardson, R.E.Sievers *J. Amer. Chem. Soc.*, **94,** 4134, 1972

83.46 **Oxo - diperoxo - hexamethylphosphoramido - pyridino - molybdate(vi)**
$C_{11}H_{23}MoN_4O_6P$
J.-M.Le Carpentier, R.Schlupp, R.Weiss
Acta Cryst. (B), **28,** 1278, 1972
Also classified in 84

83.47 **Potassium triperoxo - (o - phenanthroline) niobate trihydrate**
$C_{12}H_8N_2NbO_6^-$, K^+, $3H_2O$
G.Mathern, R.Weiss *Acta Cryst. (B),* **27,** 1582, 1971

83.48 **Potassium triperoxo - (o - phenanthroline) niobate trihydrate perhydrate**
$C_{12}H_8N_2NbO_6^-$, K^+, $3H_2O$, H_2O_2
G.Mathern, R.Weiss *Acta Cryst. (B),* **27,** 1582, 1971

83.C **Diaquo bis(picolinato) copper(ii)**
$C_{12}H_{12}CuN_2O_6$
For complete entry see 81.32

83.C **trans - Diaquo - bis(picolinato) nickel(ii) dihydrate**
$C_{12}H_{12}N_2NiO_6$, $2H_2O$
For complete entry see 81.33

83.C **trans - Diaquo - bis(picolinato) nickel(ii) dihydrate**
$C_{12}H_{12}N_2NiO_6$, $2H_2O$
For complete entry see 81.34

83.C **trans - Diaquo - bis(picolinato) zinc dihydrate**
$C_{12}H_{12}N_2O_6Zn$, $2H_2O$
For complete entry see 81.35

83.49 **bis(γ - Picoline) zinc(ii) dibromide**
$C_{12}H_{14}Br_2N_2Zn$
L.Fanfani, A.Nunzi, P.F.Zanazzi *Acta Cryst. (B),* **28,** 323, 1972

83.50 **Mono - (N,N - diethylnicotinamide) cadmium dithiocyanate**
$(C_{12}H_{14}CdN_4OS_2)_n$
F.Bigoli, A.Braibanti, M.A.Pellinghelli, A.Tiripicchio
Acta Cryst. (B), **28,** 962, 1972
Also classified in 84, 85

83.51 **Dichloro - bis(4 - methylpyridine) zinc(ii)**
$C_{12}H_{14}Cl_2N_2Zn$
H.Lynton, M.C.Sears *Canad. J. Chem.,* **49,** 3418, 1971

83.52 **Dinitrato - bis(α - picoline) copper(ii) (form ii)**
$C_{12}H_{14}CuN_4O_6$
A.F.Cameron, D.W.Taylor, R.H.Nuttall *J. C. S. Dalton,* 58, 1972

83.53 **Dinitrato - bis(α - picoline) copper(ii) (form i)**
$C_{12}H_{14}CuN_4O_6$
A.F.Cameron, D.W.Taylor, R.H.Nuttall *J. C. S. Dalton,* 58, 1972

83.54 **Diaquo - bis(9 - methylhypoxanthine) copper(ii) chloride trihydrate**
$C_{12}H_{16}CuN_8O_4^{2+}$, $2Cl^-$, $3H_2O$
E.Sletten *J. Chem. Soc. (D),* 558, 1971

83.55 **bis(N - Nicotinato)tetra - aquo - zinc(ii)**
$C_{12}H_{16}N_2O_8Zn$
M.B.Cingi, P.Domiano, C.Guastini, A.Musatti, M.Nardelli
Gazz. Chim. Ital., **101,** 455, 1971

83.C **N - (1 - Methyl - 3 - oxo - butylidene) - N′ - (1 - methyl - 2 - isonitroso - 3 - oxobutylidene)ethylenediamine copper(ii)**
$C_{12}H_{17}CuN_3O_3$
For complete entry see 76.38

83.56 **Cobalt(ii) 1,8 - bis(fluoroboro) - 2,7,9,14,15,20 - hexaoxa - 3,6,10,13,16,19 - hexa - aza - 4,5,11,12,17,18 - hexamethylbicyclo(6,6.6)eicosa - 3,5,10,12,16,18 - hexane**

$C_{12}H_{18}B_2CoF_2N_6O_6$
G.A.Zakrzewski, C.A.Ghilardi, E.C.Lingafelter
J. Amer. Chem. Soc., **93,** 4411, 1971

83.57 **Cobalt(iii) 1,8 - bis(fluoroboro) - 2,7,9,14,15,20 - hexa - oxa - 3,6,10,13,16,19 - hexa - aza - 4,5,11,12,17,18 - hexamethylbicyclo(6.6.6)eicosa - 3,5,10,12,16,18 - hexaene tetrafluoroborate**

$C_{12}H_{18}B_2CoF_2N_6O_6^+$, BF_4^-
G.A.Zakrzewski, C.A.Ghilardi, E.C.Lingafelter
J. Amer. Chem. Soc., **93,** 4411, 1971

83.C **(+) - cis - Dichloro - (1 - butene) - (α - methylbenzylamine) platinum(ii) (absolute configuration)**

$C_{12}H_{19}Cl_2NPt$
For complete entry see 72.8

83.58 **Nitrato - bis(pentamethylenetetrazole) silver(i)**

$C_{12}H_{20}AgN_9O_3$
R.L.Bodner, A.I.Popov *Inorg. Chem.*, **11,** 1410, 1972

83.C **μ - Chloro - chloro - di - π - allyl(cyclohexanone oxime) dipalladium**

$C_{12}H_{21}Cl_2NOPd_2$
For complete entry see 72.10

83.59 **Bromo - (2 - diethylaminoethanolato) copper(ii) dimer**

$C_{12}H_{28}Br_2Cu_2N_2O_2$
A.Pajunen, M.Lehtonen *Suomen Kemistil. (B),* **44,** 200, 1971
Also classified in 84

83.60 **Di(tri(ethanol)amino) nickel(ii) dinitrate**

$C_{12}H_{30}N_2NiO_6^{2+}$, $2NO_3^-$
K.Nielsen, R.G.Hazell, S.E.Rasmussen *Acta Chem. Scand.*, **26,** 889, 1972
Residue 1 also classified in 84

83.61 **tris(Di(dimethylsilyl)amino) - nitrosyl - chromium(ii)**

$C_{12}H_{36}CrN_4OSi_3$
D.C.Bradley, M.B.Hursthouse, C.W.Newing, A.J.Welch
J. C. S. Chem. Comm., 567, 1972
Also classified in 63

83.C **μ_3 - Trimethylsilylimido - μ_3 - carbonyl - tris(tricarbonyl iron)**

$C_{13}H_9Fe_3NO_{10}Si$
For complete entry see 63.5

83.62 **Dibromo - (N - 2 - methylthiophenyl - 2′ - pyridylmethyleneimine) zinc**
$C_{13}H_{12}Br_2N_2SZn$
A.Mangia, M.Nardelli, C.Palmieri, G.Pelizzi
J. Cryst. Mol. Struct., **2**, 99, 1972
Also classified in 85

83.C **Dichloro - (N - 2 - methylthiophenyl - 2′ - pyridylmethyleneimine) copper(ii)**
$C_{13}H_{12}Cl_2CuN_2S$
For complete entry see 85.14

83.C **Nitrogeno molybdenum chelate compound**
$C_{13}H_{14}MoN_2O_5^+$, F_6P^-
For complete entry see 71.15

83.C **Dichloro(pyridinium propylide)pyridine - platinum(ii)**
$C_{13}H_{16}Cl_2N_2Pt$
For complete entry see 71.16

83.63 **1,6 - Dichloro - 2,3 - trimethylene - 4,5 - bis(pyridine) platinum(iv)**
$C_{13}H_{16}Cl_2N_2Pt$
R.D.Gillard, M.Keeton, R.Mason, M.F.Pilbrow, D.R.Russell
J. Organometal. Chem., **33**, 247, 1971
Also classified in 71

83.C **Tetrachloro(pyridinium propylide)pyridine - platinum(iv) chloroform solvate**
$C_{13}H_{16}Cl_4N_2Pt$, $CHCl_3$
For complete entry see 71.17

83.C **3 - Formyl - 5 - methyl - salicylaldehyde - di - thiosemicarbazone(ethoxy) dinickel(ii) dimethylformamide solvate**
$C_{13}H_{16}N_6Ni_2O_2S_2$, $2C_3H_7NO$
For complete entry see 85.15

83.C **trans - Dichloro(benzylamino)(3 - methyl - 1 - pentene) platinum(ii)**
$C_{13}H_{21}Cl_2NPt$
For complete entry see 72.11

83.C **bis(Pyridine - 2,3 - dicarboxylato) silver(ii) dihydrate**
$C_{14}H_8AgN_2O_8$, $2H_2O$
For complete entry see 81.36

83.C **Nickel(ii) bis(hydrogen pyridine - 2,6 - dicarboxylate) trihydrate**
$C_{14}H_8N_2NiO_8$, $3H_2O$
For complete entry see 81.37

83.64 **bis(Pyridine - 2 - (N - cyanocarboxamidato))aquo - copper(ii)**
$C_{14}H_{10}CuN_6O_3$
A.C.Bonamartini, A.Montenero, M.Nardelli, C.Palmieri, C.Pelizzi
J. Cryst. Mol. Struct., **1**, 389, 1971

83.65 **Aquo - glyoxal - bis(2 - hydroxyanil)dioxo - uranium**
$C_{14}H_{12}N_2O_5U$
G.Bandoli, L.Cattalini, D.A.Clemente, M.Vidali, P.A.Vigato
J. C. S. Chem. Comm., 344, 1972
Also classified in 84

83.66 **Dinitro - (2,9 - dimethyl - 1,10 - phenanthroline) palladium(ii)**
$C_{14}H_{12}N_4O_4Pd$
J.Fridrichsons, A.McL.Mathieson, L.F.Power
J. Cryst. Mol. Struct., **1**, 333, 1971

83.C **2 - Mercuri - 4 - methylphenol 2′ - nitroso - 4′ - methylphenolate**
$C_{14}H_{13}HgNO_3$
For complete entry see 71.22

83.67 **bis(1 - Methyl - 3 - o - chlorophenyl - triazene 1 - oxide) cobalt(ii)**
$C_{14}H_{14}Cl_2CoN_6O_2$
G.L.Dwivedi, R.C.Srivastava *Acta Cryst. (B),* **27,** 2316, 1971
Also classified in 84

83.68 **bis(Pyridine - 2 - acetamide) copper(ii) perchlorate**
$C_{14}H_{16}CuN_4O_2^{2+}$, $2ClO_4^-$
M.Sekizaki, F.Marumo, K.Yamasaki, Y.Saito
Bull. Chem. Soc. Jap., **44,** 1731, 1971
Residue 1 also classified in 84

83.C **Diaquo - zinc 2 - pyridyl - acetate**
$C_{14}H_{16}N_2O_6Zn$
For complete entry see 81.38

83.69 **Dibromo - bis(2,3 - dimethylpyridine) copper(ii)**
$C_{14}H_{18}Br_2CuN_2$
W.Stahlin, H.R.Oswald *Acta Cryst. (B),* **27,** 1368, 1971

83.70 **Dichloro - bis(2,3 - dimethylpyridine) copper(ii)**
$C_{14}H_{18}Cl_2CuN_2$
W.Stahlin, H.R.Oswald *Acta Cryst. (B),* **27,** 1368, 1971

83.71 **Di - μ - methoxo - bis(chloro - 2 - methylpyridine - copper(ii))**
$C_{14}H_{20}Cl_2Cu_2N_2O_2$
M.Sterns *J. Cryst. Mol. Struct.*, **1,** 383, 1971
Also classified in 84

83.72 **bis(Dimethylglyoximato)di - imidazole - iron(ii) di - methanol solvate**
$C_{14}H_{22}FeN_8O_4$, $2CH_4O$
K.Bowman, A.P.Gaughan, Z.Dori *J. Amer. Chem. Soc.*, **94,** 727, 1972

83.73 **meso - Chloro - aquo - (6,16 - dimethyl - 1,5,9,13 - tetra - azacyclohexadeca - 6,16 - diene) nickel(ii) chloride**
$C_{14}H_{30}ClN_4NiO^+$, Cl^-
J.F.Myers, C.H.L.Kennard *J. C. S. Chem. Comm.*, 77, 1972

83.74 **Dichloro(N - (methoxy(6 - methyl - 2 - pyridyl)methyl)benzothiazolin - 2 - ylideneimine) cobalt(ii)**

$C_{15}H_{15}Cl_2CoN_3OS$

A.Mangia, M.Nardelli, C.Pelizzi, G.Pelizzi *J. C. S. Dalton,* 996, 1972

83.75 **Chloro(methyl - di((6 - methyl - 2 - pyridyl)methyl)amine) palladium(ii) chloride**

$C_{15}H_{19}ClN_3Pd^+ , Cl^-$

M.G.B.Drew, M.J.Riedl, J.Rodgers *J. C. S. Dalton,* 234, 1972

83.76 **Trichloro - (N,N - bis(6 - methyl - 2 - pyridylmethyl)methylamine) titanium(iii)**

$C_{15}H_{19}Cl_3N_3Ti$

R.K.Collins, M.G.B.Drew, J.Rodgers *J. C. S. Dalton,* 899, 1972

83.C **π - Allyl - dihydrobis(3,5 - dimethyl - 1 - pyrazolyl)borato - dicarbonylmolybdenum**

$C_{15}H_{21}BMoN_4O_2$

For complete entry see 72.12

83.77 **Chloro - penta - aquo - terpyridyl - praseodymium(iii) chloride trihydrate**

$C_{15}H_{21}ClN_3O_5Pr^{2+} , 2Cl^- , 3H_2O$

L.J.Radonovich, M.D.Glick *Inorg. Chem.,* **10,** 1463, 1971

83.C **tris(μ - Acetato - μ - acetoximato palladium(ii)) benzene solvate**

$C_{15}H_{27}N_3O_9Pd_3 , 0.5C_6H_6$

For complete entry see 81.39

83.C **$(-)_{589}$ - tris(+trans - 1,2 - Diaminocyclopentane) cobalt(iii) chloride tetrahydrate**

$C_{15}H_{36}CoN_6^{3+} , 3Cl^- , 4H_2O$

For complete entry see 76.40

83.78 **$(+)_{546}$ - tris(2,4 - Diaminopentane) cobalt(iii) chloride monohydrate**

$C_{15}H_{42}CoN_6^{3+} , 3Cl^- , H_2O$

A.Kobayashi, F.Marumo, Y.Saito, J.Fujita, F.Mizukami

Inorg. Nucl. Chem. Letters, **7,** 777, 1971

83.C **Copper(ii) dichloroacetate bis(α - picoline)**

$C_{16}H_{16}Cl_4CuN_2O_4$

For complete entry see 81.40

83.C **bis(o - Hydroxyacetophenone - iminato) copper(ii)**

$C_{16}H_{16}CuN_2O_2$

For complete entry see 78.5

83.79 **Potassium tetrakis(succinimidato) copper(ii)**

$C_{16}H_{16}CuN_4O_8^{2-} , 2K^+$

L.Srinivasan, M.R.Taylor

Amer. Cryst. Assoc., Abstr. Papers (Winter Meeting), 48, 1972

83.80 **Lithium tetrakis(succinimidato) copper(ii)**
$C_{16}H_{16}CuN_4O_8^{2-}$, $2Li^+$
L.Srinivasan, M.R.Taylor
Amer. Cryst. Assoc., Abstr. Papers (Winter Meeting), 48, 1972

83.C **bis(Methoxyacetato)bis(pyridine) copper(ii) tetrahydrate**
$C_{16}H_{20}CuO_6$, $4H_2O$
For complete entry see 81.41

83.81 **Dibromo - tetrakis(5 - methylpyrazole) manganese(ii)**
$C_{16}H_{24}Br_2MnN_8$
J.Reedijk, B.A.Stork-Blaisse, G.C.Verschoor *norg. Chem.*, **10**, 2594, 1971

83.82 **Nitrato - tetrakis(2 - methylimidazole) cobalt(ii) nitrate ethanol solvate (form B)**
$C_{16}H_{24}CoN_9O_3^+$, NO_3^- , $0.5C_2H_6O$
F.Akhtar, F.Huq, A.C.Skapski *J. C. S. Dalton,* 1353, 1972

83.83 **(5,7,7,12,12,14 - Hexamethyl - 1,4,8,11 - tetra - azacyclotetradeca - 4,8,10,14 - tetraene)nickel(ii) perchlorate**
$C_{16}H_{28}N_4Ni^{2+}$, $2ClO_4^-$
I.E.Maxwell, M.F.Bailey *J. C. S. Dalton,* 935, 1972

83.84 **racemic - (5,7,7,12,14,14 - Hexamethyl - 1,4,8,11 - tetra - azacyclotetradeca - 4,11 - diene) nickel(ii) perchlorate (form Aα)**
$C_{16}H_{32}N_4Ni^{2+}$, $2ClO_4^-$
M.F.Bailey, I.E.Maxwell *J. C. S. Dalton,* 938, 1972

83.85 **bis(Tetrahydrofuran) - bis(di(dimethylsilyl)amino) chromium(ii)**
$C_{16}H_{40}CrN_2O_2Si_4$
D.C.Bradley, M.B.Hursthouse, C.W.Newing, A.J.Welch
J. C. S. Chem. Comm., 567, 1972
Also classified in 84, 63

83.C **o - (Tetracarbonyl manganese) - benzylideneaniline**
$C_{17}H_{10}MnNO_4$
For complete entry see 71.36

83.86 **(Fluoro(6,6′,6″ - phosphinidyne - tris(α - picolinaldehyde oximato))borato) iron tetrafluoroborate methylene dichloride solvate**
$C_{18}H_{12}BFFeN_6O_3P^+$, BF_4^- , CH_2Cl_2
M.R.Churchill, A.H.Reis Junior *J. Chem. Soc. (D),* 1307, 1971
Residue 1 also classified in 62

83.C **Dioxo - bis(8 - hydroxyquinolinato) molybdenum(vi)**
$C_{18}H_{12}MoN_2O_4$
For complete entry see 84.10

83.87 **(2,2′ - Bipyridine) - pyridine - tricarbonyl - molybdenum(iii)**
$C_{18}H_{13}MoN_3O_3$
A.Griffiths *J. Cryst. Mol. Struct.*, **1**, 75, 1971

83.88 **μ_3 - Hydroxo - tri - μ - (pyridine - 2 - carboxaldehyde oximato) - μ_3 - sulphato - tricopper(ii) hydrate**
$C_{18}H_{16}Cu_3N_6O_5S$. $16.3H_2O$
R.Beckett, B.F.Hoskins *J. C. S. Dalton,* 291, 1972
Residue 1 also classified in 84

83.89 **bis(O - Ethyl - N - phenyl - thiocarbamato) palladium(ii)**
$C_{18}H_{20}N_2O_2PdS_2$
L.Gastaldi, P.Porta *Gazz. Chim. Ital.,* **101,** 641, 1971
Also classified in 85

83.C **N - Methyl - N - phenyl - N′ - methyl - N′ - (o - phenyl - chloropalladium) biacetyl osazone**
$C_{18}H_{21}ClN_4Pd$
For complete entry see 71.41

83.90 **tris(2 - Picoline) copper(i) perchlorate**
$C_{18}H_{21}CuN_3^+$, ClO_4^-
A.H.Lewin, R.J.Michl, P.Ganis, U.Lepore
J. C. S. Chem. Comm., 661, 1972

83.91 **hexakis(Imidazole) cadmium(ii) nitrate**
$C_{18}H_{24}CdN_{12}^{2+}$, $2NO_3^-$
A.D.Mighell, A.Santoro *Acta Cryst. (B),* **27,** 2089, 1971

83.92 **hexakis(Imidazole) cadmium(ii) hydroxide nitrate tetrahydrate**
$C_{18}H_{24}CdN_{12}^{2+}$, HO^- , NO_3^- , $4H_2O$
A.D.Mighell, A.Santoro *Acta Cryst. (B),* **27,** 2089, 1971

83.93 **hexakis(Imidazole) cadmium(ii) carbonate pentahydrate**
$C_{18}H_{24}CdN_{12}^{2+}$, CO_3^{2-} . $5H_2O$
B.-M.Antti, B.K.S.Lundberg, N.Ingri *J. C. S. Chem. Comm.,* 712, 1972

83.94 **hexakis(Imidazole) cobalt(ii) nitrate**
$C_{18}H_{24}CoN_{12}^{2+}$, $2NO_3^-$
E.Prince, A.D.Mighell, C.W.Reimann, A.Santoro
Cryst. Struct. Comm., **1,** 247, 1972

83.95 **hexakis(Imidazole) cobalt(ii) nitrate (neutron study)**
$C_{18}H_{24}CoN_{12}^{2+}$, $2NO_3^-$
E.Prince, A.D.Mighell, C.W.Reimann, A.Santoro
Cryst. Struct. Comm., **1,** 247, 1972

83.96 **Hexa - imidazole cobalt(ii) carbonate pentahydrate**
$C_{18}H_{24}CoN_{12}^{2+}$, CO_3^{2-} . $5H_2O$
R.Strandberg, B.K.S.Lundberg *Acta Chem. Scand.,* **25,** 1767, 1971

83.97 **Hexa(imidazole) cobalt(ii) acetate monohydrate**
$C_{18}H_{24}CoN_{12}^{2+}$, $2C_2H_3O_2^-$, H_2O
A.Gadet *C. R. Acad. Sci., Fr., C,* **274,** 263, 1972

83.C **Diacetato - copper(ii) - bis(p - toluidine) trihydrate**
$C_{18}H_{24}CuN_2O_4$, $3H_2O$
For complete entry see 81.48

83.98 **bis(N - t - Butylpyrrole - 2 - carboxaldimino) cobalt(ii)**
$C_{18}H_{26}CoN_4$
C.H.Wei *Inorg. Chem.*, **11,** 1100, 1972

83.C **bis(Methoxyacetato)tetrakis(imidazole) copper(ii)**
$C_{18}H_{26}CuN_8O_6$
For complete entry see 81.49

83.99 **bis(Ethane - 1,2 - bis(2′ - imidazoline)) - di - isothiocyanato nickel(ii)**
$C_{18}H_{28}N_{10}NiS_2$
S.H.Simonsen, J.Haley
Amer. Cryst. Assoc., Abstr. Papers (Winter Meeting), 66, 1970

83.100 **Di - μ - hydroxo - bis(2 - (2 - ethylaminoethyl)pyridine - copper(ii)) perchlorate**
$C_{18}H_{32}Cu_2N_4O_2^{2+}$, $2ClO_4^-$
D.L.Lewis, J.T.Veal, W.E.Hatfield, D.J.Hodgson
Amer. Cryst. Assoc., Abstr. Papers (Winter Meeting), 79, 1972

83.101 **((1RS,3SR,8RS,10SR) - 3,5,7,7,10,12,14,14 - Octamethyl - 1,4,8,11 - tetra - azacyclotetradeca - 4,11 - diene) nickel(ii) perchlorate**
$C_{18}H_{36}N_4Ni^{2+}$, $2ClO_4^-$
D.A.Swann, T.N.Waters, N.F.Curtis *J. C. S. Dalton,* 1115, 1972

83.C **bis(3,5 - Dimethylpyrazolylborato)dicarbonyl - trihapto - cycloheptatrienyl molybdenum**
$C_{19}H_{23}BMoN_4O_2$
For complete entry see 75.21

83.102 **Nitrato - bis(2,2′ - dipyridyl) cobalt(iii) hydroxide nitrate tetrahydrate**
$C_{20}H_{16}CoN_5O_3^{2+}$, HO^- , NO_3^- , $4H_2O$
C.W.Reimann, M.Zocchi, A.D.Mighell, A.Santoro
Acta Cryst. (B), **27,** 2211, 1971

83.103 **bis(2,2′ - Bipyridine) copper(ii) perchlorate**
$C_{20}H_{16}CuN_4^{2+}$, $2ClO_4^-$
H.Nakai *Bull. Chem. Soc. Jap.*, **44,** 2412, 1971

83.104 **bis(2 - Methyl - 8 - quinolinolato) oxo vanadium(iv)**
$C_{20}H_{16}N_2O_3V$
M.Shiro, Q.Fernando *Anal. Chem.*, **43,** 1222, 1971
Also classified in 84

83.105 **bis(2,2′ - Bipyridyl) palladium(ii) nitrate monohydrate**
$C_{20}H_{16}N_4Pd^{2+}$, $2NO_3^-$, H_2O
M.Hinamoto, S.Ooi, H.Kuroya *J. C. S. Chem. Comm.*, 356, 1972

83.106 **bis(2,2′ - Bipyridyl) palladium(ii) nitrate monohydrate**
$C_{20}H_{16}N_4Pd^{2+}$, $2NO_3^-$, H_2O
A.J.Carty, P.C.Chieh *J. C. S. Chem. Comm.*, 158, 1972

83.107 **Ammine - bis(2,2′ - bipyridyl) copper(ii) tetrafluoroborate**
$C_{20}H_{19}CuN_5^{2+}$, $2BF_4^-$
F.S.Stephens *J. C. S. Dalton,* 1350, 1972

83.108 **Dichloro - tetra - μ - adenine - dicopper(ii) chloride hexahydrate**
$C_{20}H_{20}Cl_2Cu_2N_{20}^{2+}$, $2Cl^-$, $6H_2O$
P.de Meester, A.C.Skapski *J. Chem. Soc. (A),* 2167, 1971

83.109 **trans - Dichloro - cis - dipyridine tungsten(iii) di - μ - chloro cis - dichloro - trans - dipyridine - tungsten(iii) acetone solvate**
$C_{20}H_{20}Cl_6N_4W_2$, xC_3H_6O
R.B.Jackson, W.E.Streib *Inorg. Chem.*, **10,** 1760, 1971

83.110 **trans - Dioxo - tetrapyridine - rhenium(v) chloride dihydrate**
$C_{20}H_{20}NO_2Re^+$, Cl^- , $2H_2O$
C.Calvo, N.Krishnamachari, C.J.L.Lock
J. Cryst. Mol. Struct., **1,** 161, 1971

83.111 **Nitrato - bis(2 - methyl - 8 - aminoquinoline) nickel(ii) nitrate**
$C_{20}H_{20}N_5NiO_3^+$, NO_3^-
L.F.Power, A.M.Tait *Inorg. Nucl. Chem. Letters,* **7,** 337, 1971

83.112 **Mercury bis(N,N - dimethyl - N′ - phenacetylhydrazine)**
$C_{20}H_{26}HgN_4O_2$
W.Oppolzer, H.P.Weber *Tetrahedron Letters,* 1711, 1972

83.113 **bis(O,O′ - Diethyl - dithiophosphato) - 1,10 - phenanthroline - nickel(ii)**
$C_{20}H_{28}N_2NiO_4P_2S_4$
D.C.Craig, E.T.Pallister, N.C.Stephenson
Acta Cryst. (B), **27,** 1163, 1971
Also classified in 85

83.114 **bis(Dimethyl - phenyl - phosphine) - bis(5 - methyltetrazolato) palladium**
$C_{20}H_{30}N_8P_2Pd$
G.B.Ansell
Amer. Cryst. Assoc., Abstr. Papers (Summer Meeting), 83, 1971
Also classified in 86

83.C **Di - μ - iodo - bis - ((o - dimethylaminophenyl)dimethylarsine) - dicopper(i)**
$C_{20}H_{32}As_2Cu_2I_2N_2$
For complete entry see 86.12

83.C **Octa(dimethylamino) - cyclotetraphosphazene tetracarbonyl - tungsten**
$C_{20}H_{48}N_{12}O_4P_4W$
For complete entry see 64.27

83.C **Trisodium tris(pyridine - 2,6 - dicarboxylato) neodymium(iii) hydrate**
$C_{21}H_9N_3NdO_{12}^{3-} . 3Na^+ . 15H_2O$
For complete entry see 81.51

83.C **Trisodium tris(pyridine - 2,6 - dicarboxylato) ytterbium(iii) sodium perchlorate decahydrate**
$C_{21}H_9N_3O_{12}Yb^{3-} . ClO_4^- . 4Na^+ . 10H_2O$
For complete entry see 81.52

83.C **Trisodium tris(pyridine - 2,6 - dicarboxylato) ytterbium(iii) hydrate**
$C_{21}H_9N_3O_{12}Yb^{3-} . 3Na^+ . 13H_2O$
For complete entry see 81.53

83.C **2,6 - Di(N - (benzene - 2 - thiolato)acetylimino) pyridine zinc**
$C_{21}H_{17}N_3S_2Zn$
For complete entry see 85.25

83.C **Di - isothiocyanato - bis(2,2′ - bipyridylamine) copper(ii) isothiocyanato - bis(2,2′ - bipyridylamine) copper(ii) perchlorate**
$C_{21}H_{18}CuN_7S^+ , C_{22}H_{18}CuN_8S_2 . ClO_4^-$
For complete entry see 83.117

83.C **4 - Phenylpyridine vanadyl - bis(acetylacetonate)**
$C_{21}H_{23}NO_5V$
For complete entry see 77.15

83.115 **bis(Dipyridyl) iron(ii) dithiocyanate (form ii, at 295 ° K)**
$C_{22}H_{16}FeN_6S_2$
E.Konig, K.J.Watson *Chem. Phys. Letters,* **6,** 457, 1970

83.116 **bis(Bipyridyl) iron(ii) dithiocyanate (form ii, at 100 ° K)**
$C_{22}H_{16}FeN_6S_2$
E.Konig, K.J.Watson *Chem. Phys. Letters,* **6,** 457, 1970

83.117 **Di - isothiocyanato - bis(2,2′ - bipyridylamine) copper(ii) isothiocyanato - bis(2,2′ - bipyridylamine) copper(ii) perchlorate**
$C_{22}H_{18}CuN_8S_2 , C_{21}H_{18}CuN_7S^+ . ClO_4^-$
J.E.Johnson, R.A.Jacobson
Amer. Cryst. Assoc., Abstr. Papers (Summer Meeting), 79, 1971
Residue 2 classified in 83

83.C **bis(t - Butyl isocyanide)(azobenzene) nickel(0)**
$C_{22}H_{28}N_4Ni$
For complete entry see 71.48

83.118 **bis(5,5 - Diethylbarbiturato) - bis(imidazole) nickel(ii) (orange form)**
$C_{22}H_{30}N_8NiO_6$
B.C.Wang
Amer. Cryst. Assoc., Abstr. Papers (Winter Meeting), 43, 1972
Also classified in 84

83.119 **bis(5,5 - Diethylbarbiturato) - bis(imidazole) nickel (green form)**
$C_{22}H_{30}N_8NiO_6$
B.C.Wang
Amer. Cryst. Assoc., Abstr. Papers (Winter Meeting), 43, 1972
Also classified in 84

83.120 **Chloro - pyridino - cis - 1,12 - bis(dimethylglyoximato)dodecane - cobalt(iii) methylene chloride solvate**
$C_{23}H_{37}ClCoN_5O_4 , CH_2Cl_2$
M.W.Bartlett, J.D.Dunitz *Helv. Chim. Acta,* **54,** 2753, 1971

83.121 **cis,cis - 1,3,5 - tris(Pyridine - 2 - aldimino)cyclohexane - nickel(ii) perchlorate**
$C_{24}H_{24}N_6Ni^{2+} , 2ClO_4^-$
E.B.Fleischer, A.E.Gebala, D.R.Swift *J. Chem. Soc. (D),* 1280, 1971

83.122 **Cobalt(ii) chloride - dodeca(dimethylamino)cyclohexaphosphazene complex chloroform solvate**
$2C_{24}H_{72}ClCoN_{18}P_6^+ , Cl_6Co_2^{2-} , 2CHCl_3$
W.Harrison, N.L.Paddock, J.Trotter, J.N.Wingfield
J. C. S. Chem. Comm., 23, 1972

83.123 **bis(Pentacarbonyl - manganese) - 2,2′.6′,2″ - terpyridyl cadmium**
$C_{25}H_{11}CdMn_2N_3O_{10}$
W.Clegg, P.J.Wheatley *J. C. S. Chem. Comm.,* 760, 1972

83.C **Malononitrilo - (1,2 - propane - N,N′ - bis(salicylideneiminato)) cobalt(iii) pyridine**
$C_{25}H_{22}CoN_5O_2$
For complete entry see 71.57

83.C **Carboxymethyl - palladium (triphenylphosphine)pyridine**
$C_{25}H_{22}NO_2PPd$
For complete entry see 71.58

83.124 **cis - Dithiocyanato - bis(1,10 - phenanthroline) mercury(ii)**
$C_{26}H_{16}HgN_6S_2$
A.L.Beauchamp, B.Saperas, R.Rivest *Canad. J. Chem.,* **49,** 3579, 1971

83.C **Zinc(ii) dithizonate**
$C_{26}H_{22}N_8S_2Zn$
For complete entry see 85.29

83.C **Zinc(ii) dithizonate**
$C_{26}H_{22}N_8S_2Zn$
For complete entry see 85.30

83.C **bis(Phenoxyacetato)aquo - bis(pyridine) copper(ii)**
$C_{26}H_{26}CuN_2O_7$
For complete entry see 81.54

83.125 **bis(Isothiocyanato)tetrakis(4 - methylpyridine) nickel(ii)**
$C_{26}H_{28}N_6NiS_2$
G.D.Andreetti, G.Bocelli, P.Sgarabotto
Cryst. Struct. Comm., **1,** 51, 1972

83.126 **Bromo - (bis - (2 - diethylaminoethyl) - (2 - diphenylphosphinoethyl)amine) nickel(ii) tetraphenylborate**
$C_{26}H_{42}BrN_2NiP^+ , C_{24}H_{20}B^-$
I.Bertini, P.Dapporto, G.Fallani, L.Sacconi *Inorg. Chem.*, **10,** 1703, 1971
Residue 1 also classified in 86; residue 2 classified in 62

83.C **Allylamine bis(acetylacetonato) manganese(ii) dimer**
$C_{26}H_{42}Mn_2N_2O_8$
For complete entry see 77.18

83.C **μ - (3,6 - Diphenylpyridazine) - hexacarbonyl di - iron maleic anhydride**
$C_{26}H_{142}FeN_2O_9$
For complete entry see 71.61

83.127 **bis(3 - Phenyl - 5 - pyridyl(2) - pyrazolato) nickel(ii)**
$C_{28}H_{20}N_6Ni$
J.Sieler, H.Hennig *Z. Anorg. Allg. Chem.*, **381,** 219, 1971

83.128 **8 - Hydroxyquinolinato(carbonyl)(triphenylphosphine) rhodium(i)**
$C_{28}H_{21}NO_2PRh$
L.G.Kuzmina, Y.S.Varshavskii, N.G.Bokii, Yu.T.Struchkov,
T.G.Cherkasova *Zh. Strukt. Khim.*, **12,** 653, 1971
Also classified in 84, 86

83.129 **bis(Isothiocyanato)tetrakis(4 - vinylpyridine) cobalt(ii)**
$C_{30}H_{28}CoN_6S_2$
G.D.Andreetti, P.Sgarabotto *Cryst. Struct. Comm.*, **1,** 55, 1972

83.130 **Di(bis(trimethylsilyl)amino) cobalt(ii) - triphenylphosphine**
$C_{30}H_{51}CoN_2PSi_4$
D.C.Bradley, M.B.Hursthouse, R.J.Smallwood, A.J.Welch
J. C. S. Chem. Comm., 872, 1972
Also classified in 86

83.C **π - (Triphenylcyclopropenyl)chloro - (dipyridine) nickel(0) pyridine solvate**
$C_{31}H_{25}ClN_2Ni , C_5H_5N$
For complete entry see 75.30

83.131 **Compound Ni_2L_2**
$C_{32}H_{32}N_{12}Ni_2$
N.A.Bailey, T.A.James, J.A.McCleverty, E.D.McKenzie, R.D.Moore,
J.M.Worthington *J. C. S. Chem. Comm.*, 681, 1972

83.C **bis(Acetylacetonato)cyclohexylamine - cobalt(ii) dimer**
bis(2,4 - Pentanedionato)cyclohexylamine - cobalt(ii) dimer
$C_{32}H_{54}Co_2N_2O_8$
For complete entry see 77.20

83.C **bis(O - Methyl phenylthiocarbamato)(triphenylphosphine) palladium(ii)**
$C_{34}H_{33}N_2O_2PPdS_2$
For complete entry see 85.33

83.132 **tetrakis(5 - Bromo - 8 - quinolinolato) tungsten(iv) benzene solvate**
$C_{36}H_{20}Br_4N_4O_4W$, C_6H_6
W.D.Bonds Junior, R.D.Archer, W.C.Hamilton
Inorg. Chem., **10,** 1764, 1971
Residue 1 also classified in 84

83.133 **tris(1,10 - Phenanthroline) nickel(ii) pentacarbonylmanganate (at −123 ° C)**
$C_{36}H_{24}N_6Ni^{2+}$, $2C_5MnO_5^-$
B.A.Frenz, J.A.Ibers *Inorg. Chem.*, **11,** 1109, 1972

83.C **3,6 - Diphenylpyridazino - di - iron - triphenylphosphine pentacarbonyl**
$C_{39}H_{27}Fe_2N_2O_5P$
For complete entry see 86.36

83.C **tris(2,2,6,6 - Tetramethylheptane - 3,5 - dionato) lutetium(iii) β - picoline complex**
$C_{39}H_{64}LuNO_6$
For complete entry see 84.16

83.134 **Di - μ - oxo - tetrakis(2,2′ - bipyridine) dimanganese(iii,iv) perchlorate trihydrate**
$C_{40}H_{32}Mn_2N_8O_2^{3+}$, $3ClO_4^-$, $3H_2O$
P.M.Plaksin, R.C.Stoufer, M.Mathew, G.J.Palenik
J. Amer. Chem. Soc., **94,** 212, 1972

83.C **Diphenyl - bis(2,9 - dimethyl - 1,10 - phenanthroline) mercury(ii)**
$C_{40}H_{34}HgN_4$
For complete entry see 71.79

83.135 **Di(iodo - bis(2,2′ - bipyridylamine) copper(ii)) iodide perchlorate**
$C_{40}H_{36}Cu_2I_2N_{12}^{2+}$, I^- , ClO_4^-
J.E.Johnson, R.A.Jacobson
Amer. Cryst. Assoc., Abstr. Papers (Summer Meeting), 79, 1971

83.C **(4 - Methyl - 2 - cupriobenzyl)dimethylamine tetramer**
$C_{40}H_{56}Cu_4N_4$
For complete entry see 71.82

83.C **Di - μ - chloro - bis(chloro - di(4 - methylpyridino) - (3 - (3,5 - dimethyl - 5 - hydroxymethyl - tetrahydrofuran)methyl) rhodium)**
$C_{40}H_{58}Cl_4N_4O_4Rh_2$
For complete entry see 71.83

83.C **bis(Trifluoromethyl)diazomethane bis(triphenylphosphine) platinum(0) methylene chloride solvate**
$C_{42}H_{30}F_{12}N_2P_2Pt$, $0.4CH_2Cl_2$
For complete entry see 71.84

83.C **Di - iodo - tris(2 - diphenylphosphinoethyl)amine - cobalt(ii)**
$C_{42}H_{42}CoI_2NP_3$
For complete entry see 86.44

83.136 **bis(Trimethylsilyl)amino - nickel(i) - bis(triphenylphosphine)**
$C_{42}H_{48}NNiP_2Si_2$
D.C.Bradley, M.B.Hursthouse, R.J.Smallwood, A.J.Welch
J. C. S. Chem. Comm., 872, 1972
Also classified in 86

83.C **Chloro - carbonyl - bis(triphenylphosphine) - (5 - fluoro - 2 - diazo - phenyl) iridium tetrafluoroborate acetone solvate**
$C_{43}H_{34}ClFIrN_2OP_2^+$, BF_4^- , C_3H_6O
For complete entry see 71.85

83.C **tris(2,2,6,6 - Tetramethylheptane - 3,5 - dionato) - dipyridine - europium(iii)**
$C_{43}H_{67}EuN_2O_6$
For complete entry see 77.23

83.C **Diphenyl - bis(2,4,7,9 - tetramethyl - 1,10 - phenanthroline) mercury(ii)**
$C_{44}H_{42}HgN_4$
For complete entry see 71.88

83.137 **Ferroverdin sodium salt carbon tetrachloride methanol solvate**
$C_{45}H_{30}FeN_3O_{12}^-$, Na^+ , $3CCl_4$, $3CH_4O$
S.Candeloro, D.Grdenic, N.Taylor, B.Thompson, M.Viswamitra, D.C.Hodgkin *Nature,* **224,** 589, 1969
Residue 1 also classified in 84

83.138 **Ferroverdin sodium salt acetone solvate**
$C_{45}H_{30}FeN_3O_{12}^-$, Na^+ , $2C_3H_6O$
S.Candeloro, D.Grdenic, N.Taylor, B.Thompson, M.Viswamitra, D.C.Hodgkin *Nature,* **224,** 589, 1969
Residue 1 also classified in 84

83.C **(2,2,6,6 - Tetramethylheptane - 3,5 - dionato) holmium(iii) bis(4 - picoline)**
$C_{45}H_{71}HoN_2O_6$
For complete entry see 77.24

83.C **Compound Q**
$C_{48}H_{30}Cu_4F_{18}N_4O_{14}$
For complete entry see 81.60

83.C **Di - μ - acetato - bis(dimethylglyoximato)bis(triphenylphosphine) dirhodium(ii) monohydrate**

$C_{48}H_{48}N_4O_8P_2Rh_2$, H_2O

For complete entry see 81.61

83.C **(Cyano(dicyanomethyl)ketiminato)carbonyl (tetracyanoethylene) bis(triphenylphosphine) iridium benzene solvate**

$C_{49}H_{31}IrN_8OP_2$, $0.5C_6H_6$

For complete entry see 86.45

83.139 **Carbonyl - bis(triphenylphosphine) - (1,4 - di(p - fluorophenyl)tetrazene) iridium tetrafluoroborate benzene solvate**

$C_{49}H_{38}F_2IrN_4OP_2^+$, BF_4^- , C_6H_6

F.W.B.Einstein, A.B.Gilchrist, G.W.Rayner-Canham, D.Sutton

J. Amer. Chem. Soc., **93,** 1826, 1971

Residue 1 also classified in 86

83.C **bis(Triphenylphosphine)tetrakis (dimethylglyoximato) dirhodium monohydrate propanol solvate**

$C_{52}H_{58}N_8O_8P_2Rh_2$, C_3H_8O , H_2O

For complete entry see 86.54

83.C **Di - 2 - (5 - perfluoromethyltetrazolato) - μ - 1,2 - bis(diphenylphosphino) ethane - bis(1,2 - (diphenylphosphino) - ethane) dicopper(i)**

$C_{82}H_{72}Cu_2F_6N_8P_6$

For complete entry see 86.69

METAL COMPLEXES (OXYGEN LIGAND)

84.C **catena - Di - μ - chloro - semicarbazide - copper(ii) (monoclinic form)**
$(CH_5Cl_2CuN_3O)_n$
For complete entry see 83.1

84.C **catena - Di - μ - chloro - semicarbazide - copper(ii) (orthorhombic form)**
$(CH_5Cl_2CuN_3O)_n$
For complete entry see 83.2

84.1 **Cobalt(ii) monoglycerolate**
$(C_3H_6CoO_3)_n$
P.G.Slade, E.W.Radoslovich, M.Raupach
Acta Cryst. (B), **27**, 2432, 1971

84.2 **1,2 - Dimethoxyethane - dichloro - thio - tungsten μ - oxo tetrachloro - thio - tungsten**
$C_4H_{10}Cl_6O_3S_2W_2$
D.Britnell, M.G.B.Drew, G.W.A.Fowles, D.A.Rice
J. C. S. Chem. Comm., 462, 1972

84.C **Chloro - (triethanolamine) zinc(ii)**
$C_6H_{14}ClNO_3Zn$
For complete entry see 83.17

84.C **Triethanolamine - aquo - zinc(ii) - μ - chloro - trichlorozinc(ii)**
$C_6H_{17}Cl_4NO_4Zn_2$
For complete entry see 83.18

84.C **tris(2 - Aminoethanolato) cobalt(iii) - tris(2 - aminoethanol) nickel(ii) di - iodide**
$C_6H_{18}N_3NiO_3^-$, $C_6H_{21}CoN_3O_3^{3+}$, $2I^-$
For complete entry see 83.21

84.C **Tetramethyl - bis(N - methyl - N - nitrosohydroxylaminato) tungsten(vi)**
$C_6H_{18}N_4O_4W$
For complete entry see 71.3

84.3 **Oxo - diperoxo - hexamethylphosphoramido - aquo - molybdate(vi)**
$C_6H_{20}MoN_3O_7P$
J.-M.Le Carpentier, R.Schlupp, R.Weiss
Acta Cryst. (B), **28**, 1278, 1972

84.C **tris(2 - Aminoethanolato) cobalt(iii) - tris(2 - aminoethanol) nickel(ii) di - iodide**

$C_6H_{21}CoN_3O_3^{3+}$, $C_6H_{18}N_3NiO_3^-$, $2I^-$

For complete entry see 83.21

84.4 **Zinc hexa(urea) nitrate**

$C_6H_{24}N_{12}O_6Zn^{2+}$, $2NO_3^-$

W.van de Giesen, C.H.Stam *Cryst. Struct. Comm.*, **1**, 257, 1972

84.C **Diaquo - copper(ii) diviolurate dihydrate**

$C_8H_8CuN_6O_{10}$, $2H_2O$

For complete entry see 83.27

84.5 **Mercury(ii) bromide bis(dioxan)**

$C_8H_{16}Br_2HgO_4$

M.Frey, J.-C.Monier *Acta Cryst. (B)*, **27**, 2487, 1971

84.6 **Tetra(dimethylsulfoxide) lanthanum(iii) nitrate**

$C_8H_{24}LaN_3O_{13}S_4$

K.K.Bhandary, H.Manohar *Indian J. Chem.*, **9**, 275, 1971

84.7 **tetrakis(Acetamide) - bis(aquo) nickel(ii) dichloride**

$C_8H_{24}N_4NiO_6^{2+}$, $2Cl^-$

M.E.Stone, B.E.Robertson, E.Stanley *J. Chem. Soc. (A)*, 3632, 1971

84.C **Dinitrato - bis(pyridine) copper(ii) pyridine solvate**

$C_{10}H_{10}CuN_4O_6$, $0.5C_5H_5N$

For complete entry see 83.37

84.C **(4 - Methyliminopentane - 2,3 - dione - 3 - oximato)(4 - iminopentane - 2,3 - dione - 3 - oximato) nickel(ii)**

$C_{11}H_{16}N_4NiO_4$

For complete entry see 83.44

84.C **Oxo - diperoxo - hexamethylphosphoramido - pyridino - molybdate(vi)**

$C_{11}H_{23}MoN_4O_6P$

For complete entry see 83.46

84.C **Mono - (N,N - diethylnicotinamide) cadmium dithiocyanate**

$(C_{12}H_{14}CdN_4OS_2)_n$

For complete entry see 83.50

84.C **Bromo - dimethoxy - bis(diethyldithiocarbamato) niobium(ii)**

$C_{12}H_{26}BrNbO_2S_4$

For complete entry see 80.6

84.C **Chloro - dimethoxy - bis(diethyldithiocarbamato) niobium(v)**

$C_{12}H_{26}ClNbO_2S_4$

For complete entry see 80.7

84.C **Bromo - (2 - diethylaminoethanolato) copper(ii) dimer**
$C_{12}H_{28}Br_2Cu_2N_2O_2$
For complete entry see 83.59

84.C **Di(tri(ethanol)amino) nickel(ii) dinitrate**
$C_{12}H_{30}N_2NiO_6^{2+}$, $2NO_3^-$
For complete entry see 83.60

84.C **3 - Formyl - 5 - methyl - salicylaldehyde - di - thiosemicarbazone(ethoxy) dinickel(ii) dimethylformamide solvate**
$C_{13}H_{16}N_6Ni_2O_2S_2$, $2C_3H_7NO$
For complete entry see 85.15

84.C **bis(Pyridine - 2,3 - dicarboxylato) silver(ii) dihydrate**
$C_{14}H_8AgN_2O_8$, $2H_2O$
For complete entry see 81.36

84.C **Aquo - glyoxal - bis(2 - hydroxyanil)dioxo - uranium**
$C_{14}H_{12}N_2O_5U$
For complete entry see 83.65

84.C **2 - Mercuri - 4 - methylphenol 2′ - nitroso - 4′ - methylphenolate**
$C_{14}H_{13}HgNO_3$
For complete entry see 71.22

84.C **bis(1 - Methyl - 3 - o - chlorophenyl - triazene 1 - oxide) cobalt(ii)**
$C_{14}H_{14}Cl_2CoN_6O_2$
For complete entry see 83.67

84.C **bis(Pyridine - 2 - acetamide) copper(ii) perchlorate**
$C_{14}H_{16}CuN_4O_2^{2+}$, $2ClO_4^-$
For complete entry see 83.68

84.C **Di - μ - methoxo - bis(chloro - 2 - methylpyridine - copper(ii))**
$C_{14}H_{20}Cl_2Cu_2N_2O_2$
For complete entry see 83.71

84.8 **Di - μ - (pyridine oxide) - bis(dichloro - dimethylsulfoxide - copper(ii))**
$C_{14}H_{22}Cl_4Cu_2N_2O_4S_2$
R.J.Williams, W.H.Watson, A.C.Larson
Amer. Cryst. Assoc., Abstr. Papers (Winter Meeting), 49, 1972

84.C **bis(Acetylacetonato) nickel(ii) di - ethanol complex**
$C_{14}H_{32}NiO_6$
For complete entry see 77.6

84.C **(4 - Methylpyridine - N - oxide) - bis(hexafluoroacetylacetonato) copper(ii)**
$(C_{16}H_9CuF_{12}NO_5)_n$
For complete entry see 77.13

84.C **bis(o - Hydroxyacetophenone - iminato) copper(ii)**
$C_{16}H_{16}CuN_2O_2$
For complete entry see 78.5

84.C **5 - bis(N - t - Butylcyano) - 2,2,4,4 - tetra(trifluoromethyl) - 5 - nickela - 1,3 - oxazolidine**
$C_{16}H_{19}F_{12}N_3NiO$
For complete entry see 71.32

84.C **bis(Tetrahydrofuran) - bis(di(dimethylsilyl)amino) chromium(ii)**
$C_{16}H_{40}CrN_2O_2Si_4$
For complete entry see 83.85

84.9 **bis(bis(Diethylether) - μ - (dodecahydro - nido - decaborato) cadmium)**
$C_{16}H_{64}B_{20}Cd_2O_4$
N.N.Greenwood, J.A.McGinnety, J.D.Owen *J. C. S. Dalton,* 989, 1972
Also classified in 62

84.10 **Dioxo - bis(8 - hydroxyquinolinato) molybdenum(vi)**
$C_{18}H_{12}MoN_2O_4$
L.O.Atovmyan, Y.A.Sokolova *Zh. Strukt. Khim.*, **12,** 851, 1971
Also classified in 83

84.C **μ_3 - Hydroxo - tri - μ - (pyridine - 2 - carboxaldehyde oximato) - μ_3 - sulphato - tricopper(ii) hydrate**
$C_{18}H_{16}Cu_3N_6O_5S$, $16.3H_2O$
For complete entry see 83.88

84.C **tris(Dimethoxyethane) - tri - μ - trifluoroacetato - μ_3 - chloro - μ_3 - sulfato - tricobalt(ii)**
$C_{18}H_{30}ClCo_3F_9O_{16}$
For complete entry see 81.50

84.C **bis(2 - Methyl - 8 - quinolinolato) oxo vanadium(iv)**
$C_{20}H_{16}N_2O_3V$
For complete entry see 83.104

84.11 **Dibromo - bis(pyridine - N - oxide) copper(ii) dimer**
$C_{20}H_{20}Br_4Cu_2N_4O_4$
A.D.Mighell, C.W.Reimann, A.Santoro *Acta Cryst. (B),* **28,** 126, 1972

84.C **bis(Dithioacetato)dioxo(triphenylphosphine oxide) uranium(vi)**
$C_{22}H_{21}O_3PS_4U$
For complete entry see 85.26

84.12 **Dichloro - bis(antipyrine) zinc**
$C_{22}H_{24}Cl_2N_4O_2Zn$
M.Biagini Cingi, C.Guastini, A.Musatti, M.Nardelli
Acta Cryst. (B), **28,** 667, 1972

84.C **bis(5,5 - Diethylbarbiturato) - bis(imidazole) nickel(ii) (orange form)**
$C_{22}H_{30}N_8NiO_6$
For complete entry see 83.118

84.C **bis(5,5 - Diethylbarbiturato) - bis(imidazole) nickel (green form)**
$C_{22}H_{30}N_8NiO_6$
For complete entry see 83.119

84.C **Tricarbonyl(3 - (o - (diphenyl - phosphino)phenyl) - 2 - buten - 2 - olato) manganese**
$C_{25}H_{20}MnO_4P$
For complete entry see 71.56

84.13 **catena - μ - bis(1,2 - Diphenylphosphinyl)ethane - dichloro copper(ii)**
$(C_{26}H_{24}Cl_2CuO_2P_2)_n$
M.Mathew, G.J.Palenik *Inorg. Chim. Acta,* **5,** 573, 1971

84.C **8 - Hydroxyquinolinato(carbonyl)(triphenylphosphine) rhodium(i)**
$C_{28}H_{21}NO_2PRh$
For complete entry see 83.128

84.C **bis(N,N - Diethyl diselenocarbamate)dioxo(triphenylarsine oxide) uranium(vi)**
$C_{28}H_{35}AsN_2O_3Se_4U$
For complete entry see 85.32

84.C **N,N′ - Ethylene bis(salicylaldehydeiminato) cobalt(ii) (oxygen - inactive form)**
$C_{32}H_{28}Co_2N_4O_4$
For complete entry see 78.14

84.C **Di - μ - chloro - di(π - cyclo - octatetraene - bis(tetrahydrofuran) cerium)**
$C_{32}H_{48}Ce_2Cl_2O_4$
For complete entry see 75.31

84.C **tetrakis(5 - Bromo - 8 - quinolinolato) tungsten(iv) benzene solvate**
$C_{36}H_{20}Br_4N_4O_4W \, , C_6H_6$
For complete entry see 83.132

84.14 **Dichloro - bis(triphenylphosphine oxide) copper(ii)**
$C_{36}H_{35}Cl_2CuO_2P_2$
J.A.Bertrand, A.R.Kalyanaraman *Inorg. Chim. Acta,* **5,** 341, 1971

84.15 **Titanium ethoxide hydrolysis product**
$C_{38}H_{95}O_{24}Ti_7$
D.L.Ward, C.N.Caughlan
Amer. Cryst. Assoc., Abstr. Papers (Summer Meeting), 84, 1971

84.C **3,3 - Di(trifluoromethyl) - 1 - nickela - 2 - oxa - cyclopropane bis(triphenylphosphine)**

$C_{39}H_{30}F_6NiOP_2$

For complete entry see 71.76

84.16 **tris(2,2,6,6 - Tetramethylheptane - 3,5 - dionato) lutetium(iii) β - picoline complex**

$C_{39}H_{64}LuNO_6$

S.J.Wasson, D.E.Sands, W.F.Wagner

Amer. Cryst. Assoc., Abstr. Papers (Summer Meeting), 109, 1971

Also classified in 83

84.17 **octakis(Pyridine - N - oxide) lanthanum(iii) perchlorate (form i)**

$C_{40}H_{40}LaN_8O_8^{3+}$, $3ClO_4^-$

A.R.Al-Karaghouli, J.S.Wood *J. C. S. Chem. Comm.*, 516, 1972

84.C **Ferroverdin sodium salt carbon tetrachloride methanol solvate**

$C_{45}H_{30}FeN_3O_{12}^-$, Na^+, $3CCl_4$, $3CH_4O$

For complete entry see 83.137

84.C **Ferroverdin sodium salt acetone solvate**

$C_{45}H_{30}FeN_3O_{12}^-$, Na^+, $2C_3H_6O$

For complete entry see 83.138

84.18 **sym - trans - Di - μ - phenoxy - hexaphenoxy - diphenol - dititanium(iv)**

$C_{60}H_{52}O_{10}Ti_2$

G.W.Svetich, A.A.Voge *Acta Cryst. (B),* **28,** 1760, 1972

METAL COMPLEXES
(SULPHUR OR SELENIUM LIGAND)

85.1 **Dimethylsulfoxide - pentammine - ruthenium(ii) hexafluorophosphate**
$C_2H_{21}N_5ORuS^{2+}$, $2F_6P^-$
F.C.March, G.Ferguson *Canad. J. Chem.*, **49**, 3590, 1971

85.C **(N - Methyl - N,N - bis(methyleneamino - N′ - (dithionitrite))amino) nickel(ii)**
$C_3H_7N_5NiS_4$
For complete entry see 83.6

85.2 **Nickel bis(ethylene - 1,2 - dithiolene)**
$C_4H_4NiS_4$
K.W.Browall, L.V.Interrante, J.S.Kasper
J. Amer. Chem. Soc., **93**, 6289, 1971

85.3 **3 - Methylrhodanine - copper(i) iodide**
$(C_4H_5CuINOS_2)_n$
F.G.Moers, W.P.J.H.Bosman, P.T.Beurskens
J. Cryst. Mol. Struct., **2**, 23, 1972

85.C **bis(Thioglycinato) nickel(ii)**
$C_4H_8N_2NiO_2S_2$
For complete entry see 82.3

85.4 **Di - iodo - N,N,N′,N′ - tetramethylthiuramdisulfide - mercury(ii)**
$C_6H_{12}HgI_2N_2S_4$
P.T.Beurskens, J.A.Cras, J.H.Noordik, A.M.Spruijt
J. Cryst. Mol. Struct., **1**, 93, 1971

85.C **Diaquo - ethylenedithiodiacetato - nickel(ii)**
$C_6H_{12}NiO_6S_2$
For complete entry see 81.22

85.5 **tris(O,O′ - Dimethyldithiophosphato) cobalt(iii)**
$C_6H_{18}CoO_6P_3S_6$
J.F.McConnell, A.Schwartz *Acta Cryst. (B)*, **28**, 1546, 1972

85.C **bis(N,N - Di - n - butyldithiocarbamato) gold(iii) bis(1,2 - dicyanoethene - 1,2 - dithiolato) aurate(iii)**
$C_8AuN_4S_4^-$, $C_{18}H_{36}AuN_2S_4^+$
For complete entry see 80.12

85.6 **bis(Ethylene - 1,2 - dithiolene) palladium dimer**
$C_8H_8Pd_2S_8$
J.S.Kasper, K.W.Browall, T.Bursh, L.V.Interrante
Amer. Cryst. Assoc., Abstr. Papers (Winter Meeting), 46, 1972

85.7 **bis(Ethylene - 1,2 - dithiolene) platinum dimer**
$C_8H_8Pt_2S_8$
J.S.Kasper, K.W.Browall, T.Bursh, L.V.Interrante
Amer. Cryst. Assoc., Abstr. Papers (Winter Meeting), 46, 1972

85.8 **tetrakis(Dithioacetato) vanadium(iv)**
$C_8H_{12}S_8V$
L.Fanfani, A.Nunzi, P.F.Zanazzi, A.R.Zanari
Acta Cryst. (B), **28,** 1298, 1972

85.9 **bis(Imidotetramethyldithiodiphosphino - S,S) iron(ii)**
$C_8H_{14}FeN_2P_4S_4$
M.R.Churchill, J.Wormald *Inorg. Chem.*, **10,** 1778, 1971

85.10 **tetrakis(Thioacetamide) nickel(ii) bromide**
$C_8H_{20}N_4NiS_4^{2+}$, $2Br^-$
W.A.Spofford III, P.Boldrini, E.L.Amma *Inorg. Chim. Acta,* **5,** 70, 1971

85.11 **Tetracarbonyl(dithiobenzoato) rhenium(i)**
$C_{11}H_5O_4ReS_2$
G.Thiele, G.Liehr *Chem. Ber.*, **104,** 1877, 1971

85.12 **tris - (cis - 1,2 - Di(trifluoromethyl)ethylene - 1,2 - diselenato) molybdenum**
$C_{12}F_{18}MoSe_6$
C.G.Pierpont, R.Eisenberg *J. Chem. Soc. (A),* 2285, 1971

85.C **Mono - (N,N - diethylnicotinamide) cadmium dithiocyanate**
$(C_{12}H_{14}CdN_4OS_2)_n$
For complete entry see 83.50

85.13 **Dehydrodithizone dichloro - mercury(ii)**
$C_{13}H_{10}Cl_2HgN_4S$
W.J.Kozarek, Q.Fernando *J. C. S. Chem. Comm.*, 604, 1972

85.C **Dibromo - (N - 2 - methylthiophenyl - 2′ - pyridylmethyleneimine) zinc**
$C_{13}H_{12}Br_2N_2SZn$
For complete entry see 83.62

85.14 **Dichloro - (N - 2 - methylthiophenyl - 2′ - pyridylmethyleneimine) copper(ii)**
$C_{13}H_{12}Cl_2CuN_2S$
A.Mangia, M.Nardelli, G.Pelizzi, C.Pelizzi
J. Cryst. Mol. Struct., **1,** 139, 1971
Also classified in 83

85.15 **3 - Formyl - 5 - methyl - salicylaldehyde - di - thiosemicarbazone(ethoxy) dinickel(ii) dimethylformamide solvate**

$C_{13}H_{16}N_6Ni_2O_2S_2$, $2C_3H_7NO$

B.F.Hoskins, R.Robsin, H.Schaap *Inorg. Nucl. Chem. Letters,* **8,** 21, 1972

Residue 1 also classified in 83, 84

85.16 **bis(Dithiotropolonato) nickel(ii)**

$C_{14}H_{10}NiS_4$

G.P.Khare, A.J.Schultz, R.Eisenberg *J. Amer. Chem. Soc.,* **93,** 3597, 1971

85.17 **Nickel(ii) bis(trithioperoxybenzoate)**

$C_{14}H_{10}NiS_6$

M.Bonamico, G.Dessy, V.Fares, L.Scaramuzza

J. Chem. Soc. (A), 3191, 1971

85.18 **bis(Dithiobenzoato) zinc(ii)**

$C_{14}H_{10}S_4Zn$

M.Bonamico, G.Dessy, V.Fares, L.Scaramuzza

Cryst. Struct. Comm., **1,** 63, 1972

85.19 **Zinc(ii) bis(trithioperoxybenzoate)**

$C_{14}H_{10}S_6Zn$

M.Bonamico, G.Dessy, V.Fares, L.Scaramuzza

J. Chem. Soc. (A), 3191, 1971

85.C **bis(N,N - Diethyldithiocarbamato)(cis - 1,2 - bis(trifluoromethyl) - ethylene - 1,2 - dithiolato) - iron**

$C_{14}H_{20}F_6FeN_2S_6$

For complete entry see 80.8

85.20 **bis(Thioisobutyrylacetonato) nickel(ii)**

$C_{14}H_{22}NiO_2S_2$

J.Sieler, P.Thomas, E.Uhlemann, E.Hohne

Z. Anorg. Allg. Chem., **380,** 160, 1971

Also classified in 77

85.21 **tetrakis(Tricarbonyl - μ_3 - methanethiolato - rhenium)**

$C_{16}H_{12}O_{12}Re_4S_4$

W.Harrison, W.C.Marsh, J.Trotter *J. C. S. Dalton,* 1009, 1972

85.C **bis(Dimethyl - o - thiolophenylarsine) palladium(ii) pyridine solvate**

$C_{16}H_{20}As_2PdS_2$, C_5H_5N

For complete entry see 86.5

85.22 **tetrakis(Ethyl thioxanthato) - bis - μ - (ethylthio) - dicobalt(iii)**

$C_{16}H_{30}Co_2S_{14}$

D.F.Lewis, S.J.Lippard, J.A.Zubieta *J. Amer. Chem. Soc.,* **94,** 1563, 1972

85.23 **bis(cis - 1 - Mercapto - 2 - p - bromobenzoyl - ethylene) nickel(ii)**

$C_{18}H_{12}Br_2NiO_2S_2$

L.Kutschabsky, L.Beyer *Z. Chem.,* **11,** 30, 1971

85.C **bis(O - Ethyl - N - phenyl - thiocarbamato) palladium(ii)**
$C_{18}H_{20}N_2O_2PdS_2$
For complete entry see 83.89

85.C **Di - μ - thio - n - butyl(bis - π - cyclopentadienyl) molybdenum - iron - dichloride**
$C_{18}H_{28}Cl_2FeMoS_2$
For complete entry see 73.19

85.C **bis(O,O′ - Diethyl - dithiophosphato) - 1,10 - phenanthroline - nickel(ii)**
$C_{20}H_{28}N_2NiO_4P_2S_4$
For complete entry see 83.113

85.24 **Cyclopentadienyl(tricarbonyl) molybdenum - μ_4 - sulfido - (cyclopentadienyl(dicarbonyl)molybdenum - bis(tricarbonyl rhenium) - μ_3 - sulfide)**
$C_{21}H_{10}Mo_2O_{11}Re_2S_2$
P.J.Vergamini, H.Vahrenkamp, L.F.Dahl
J. Amer. Chem. Soc., **93,** 6326, 1971
Also classified in 73

85.25 **2,6 - Di(N - (benzene - 2 - thiolato)acetylimino) pyridine zinc**
$C_{21}H_{17}N_3S_2Zn$
L.F.Lindoy, D.H.Busch, V.Goedken *J. C. S. Chem. Comm.*, 683, 1972
Also classified in 83

85.26 **bis(Dithioacetato)dioxo(triphenylphosphine oxide) uranium(vi)**
$C_{22}H_{21}O_3PS_4U$
G.Bombieri, U.Croatto, E.Forsellini, B.Zarli, R.Graziani
J. C. S. Dalton, 560, 1972
Also classified in 84

85.C **bis(2,2,6,6 - Tetramethylheptane - 5 - thiolo - 3 - onato) nickel(ii)**
$C_{22}H_{38}NiO_2S_2$
For complete entry see 77.17

85.27 **cis - Dichloro - bis(4,4′ - dichlorodiphenyl sulfide) platinum(ii)**
$C_{24}H_{16}Cl_6PtS_2$
W.A.Spofford III, E.L.Amma, C.V.Senoff *Inorg. Chem.*, **10,** 2309, 1971

85.C **bis(bis(π - Cyclopentadienyl) niobium - bis(μ - methanethiolato)) nickel tetrafluoroborate dihydrate**
$C_{24}H_{32}Nb_2NiS_4^{2+} . 2BF_4^- . 2H_2O$
For complete entry see 73.29

85.28 **Copper(i) O,O′ - di - isopropylphosphorodithioate tetramer**
$C_{24}H_{56}Cu_4O_8P_4S_8$
S.L.Lawton, W.J.Rohrbaugh, G.T.Kokotailo *Inorg. Chem.*, **11,** 612, 1972

85.29 **Zinc(ii) dithizonate**
$C_{26}H_{22}N_8S_2Zn$
A.Mawby, H.M.N.H.Irving *Anal. Chim. Acta,* **55,** 269, 1971
Also classified in 83

85.30 **Zinc(ii) dithizonate**
$C_{26}H_{22}N_8S_2Zn$
K.S.Math, H.Freiser *Talanta,* **18,** 435, 1971
Also classified in 83

85.C **tris(N,N - Di - n - butyldiselenocarbamato) nickel(iv) bromide**
$C_{27}H_{54}N_3NiSe_6^+$, Br^-
For complete entry see 80.18

85.31 **tetrakis(Dithiobenzoato) vanadium(iv)**
$C_{28}H_{20}S_8V$
M.Bonamico, G.Dessy, V.Fares, L.Scaramuzza
Cryst. Struct. Comm., **1,** 91, 1972

85.32 **bis(N,N - Diethyl diselenocarbamate)dioxo(triphenylarsine oxide) uranium(vi)**
$C_{28}H_{35}AsN_2O_3Se_4U$
B.Zarli, R.Graziani, E.Forsellini, U.Croatto, G.Bombieri
J. Chem. Soc. (D), 1501, 1971
Also classified in 84

85.C **Molybdenum dithiocarbamate compound**
$C_{28}H_{56}Mo_2N_4S_8$
For complete entry see 71.66

85.33 **bis(O - Methyl phenylthiocarbamato)(triphenylphosphine) palladium(ii)**
$C_{34}H_{33}N_2O_2PPdS_2$
C.Furlani, T.Tarantelli, L.Gastaldi, P.Porta *J. Chem. Soc. (A),* 3778, 1971
Also classified in 83, 86

85.34 **Platinum bis(triphenylphosphine) dithiofluoroformate bifluoride**
$C_{37}H_{30}FP_2PtS_2^+$, HF_2^-
J.A.Evans, M.J.Hacker, R.D.W.Kemmitt, D.R.Russell, J.Stocks
J. C. S. Chem. Comm., 72, 1972
Residue 1 also classified in 86

85.35 **Platinum(ii) dithiocumate dimer**
Platinum(ii) di(p - isopropyl - dithiobenzoate) dimer
$C_{40}H_{44}Pt_2S_8$
J.P.Fackler Junior *J. Amer. Chem. Soc.,* **94,** 1009, 1972

85.36 **bis(Diphenylphosphinodithioato) (triphenylphosphine) palladium**
$C_{42}H_{35}P_3PdS_4$
J.M.C.Alison, T.A.Stephensen, R.O.Gould *J. Chem. Soc. (A),* 3690, 1971
Also classified in 86

85.37 **(bis(o - Dimethylarsinophenyl) - (o - methylthiophenyl)arsine) - bromo - nickel(ii) perchlorate chlorobenzene solvate**

$C_{43}H_{35}As_3BrNiS^+$, ClO_4^- , C_6H_5Cl
M.Mathew, G.J.Palenik, G.Dyer, D.W.Meek
J. C. S. Chem. Comm., 379, 1972
Residue 1 also classified in 86

85.38 **bis(Pyridine - 2 - thiolato) - bis(triphenylphosphine) ruthenium(ii)**

$C_{46}H_{38}N_2P_2RuS_2$
S.R.Fletcher, A.C.Skapski *J. C. S. Dalton*, 635, 1972
Also classified in 86

85.39 **Dicarbonyl - bis(triphenylphosphine) - tris(toluene - 3,4 - dithiolato)di - iridium(iii)**

$C_{59}H_{48}Ir_2O_2P_2S_6$
G.P.Khare, R.Eisenberg *Inorg. Chem.*, **11**, 1385, 1972
Also classified in 86

85.40 **Di - μ - pentafluorobenzenethiolato - trans - bis((pentafluorobenzenthiolato) - (triphenylphosphine) palladium(ii)) ethanol solvate**

$C_{60}H_{30}F_{20}P_2Pd_2S_4$, $2C_2H_6O$
R.H.Fenn, G.R.Segrott *J. C. S. Dalton*, 330, 1972
Residue 1 also classified in 86

METAL COMPLEXES (P, AS, SB LIGAND)

86.C **bis(Thiocarbohydrazide) - dichlorido - cadmium**
$C_2H_{12}CdCl_2N_8S_2$
For complete entry see 79.4

86.1 **(1,2 - bis(Dimethylarsino)trifluoroethane) tetracarbonyl molybdenum**
$C_{10}H_{13}As_2F_3MoO_4$
I.W.Nowell, J.Trotter *J. Chem. Soc. (A),* 2922, 1971

86.C **cis - Dichloro(diethyl(phenyl)phosphine) (ethylisocyanide) platinum(ii)**
$C_{13}H_{20}Cl_2NPPt$
For complete entry see 71.19

86.C **Chloro - aquo - bis(trimethylarsine)tetrakis(trifluoromethyl) rhodiacyclopentadiene**
$C_{14}H_{20}As_2ClF_{12}ORh$
For complete entry see 71.24

86.C **Di - iodo(carbonyl)(ferrocene - 1,1′ - bis(dimethylarsine)) nickel(ii)**
$C_{15}H_{20}As_2FeI_2NiO$
For complete entry see 73.15

86.2 **Di(pentafluorophenyl)diarsine - iron - tetracarbonyl**
Decafluoro - arsenobenzene - iron - tetracarbonyl
$C_{16}As_2F_{10}FeO_4$
P.S.Elmes, P.Leverett, B.O.West *J. Chem. Soc. (D),* 747, 1971

86.3 **1,2 - bis(Dimethylarsino) - 3,3,4,4 - tetrafluorocyclobutene bis(iodo - tetracarbonyl - manganese)**
$C_{16}H_{12}As_2F_4I_2Mn_2O_8$
J.P.Crow, W.R.Cullen, F.L.Hou, L.Y.Y.Chan, F.W.B.Einstein
J. Chem. Soc. (D), 1229, 1971

86.4 **1,2 - bis(Dimethylarsino) - 3,3,4,4 - tetrafluorocyclobutene dimanganese octacarbonyl**
$C_{16}H_{12}As_2F_4Mn_2O_8$
J.P.Crow, W.R.Cullen, F.L.Hou, L.Y.Y.Chan, F.W.B.Einstein
J. Chem. Soc. (D), 1229, 1971

86.5 **bis(Dimethyl - o - thiolophenylarsine) palladium(ii) pyridine solvate**
$C_{16}H_{20}As_2PdS_2$, C_5H_5N
J.P.Beale, N.C.Stephenson *Acta Cryst. (B),* **28,** 557, 1972
Residue 1 also classified in 85

86.6 **Di - μ - diethylphosphido - bis(tetracarbonyl - molybdenum)**
$C_{16}H_{20}MoO_8P_2$
L.R.Nassimbeni *Inorg. Nucl. Chem. Letters,* **7,** 909, 1971

86.7 **Trichloro(2 - diphenylphosphine oxide)ethyl - dimethylammonium copper(ii)**
$C_{16}H_{21}Cl_3CuNOP$
H.D.Caughman, M.G.Newton, R.C.Taylor
Amer. Cryst. Assoc., Abstr. Papers (Summer Meeting), 80, 1971

86.8 **cis - Dichloro - bis(dimethylphenylphosphine) palladium(ii)**
$C_{16}H_{22}Cl_2P_2Pd$
L.L.Martin, R.A.Jacobson *Inorg. Chem.,* **10,** 1795, 1971

86.C **1 - (2 - (Dimethylarsino) - 3,3,4,4 - tetrafluorocyclobutenyl) - dimethylarsino - tri - iron nonacarbonyl**
$C_{17}H_{12}As_2F_4Fe_3O_9$
For complete entry see 71.37

86.9 **Ethylidene - heptacarbonyl - μ - (1,2 - bis(dimethylarsino) tetrafluorocyclobutene) - triangulo - tricobalt**
$C_{17}H_{15}As_2Co_3F_4O_7$
F.W.B.Einstein, R.D.G.Jones *Inorg. Chem.,* **11,** 395, 1972

86.10 **4 - Methylpyridinium triphenylphosphine - tribromozincate**
$C_{18}H_{15}Br_3PZn^-$, $C_6H_8N^+$
R.E.DeSimone, G.D.Stucky *Inorg. Chem.,* **10,** 1808, 1971
Residue 2 classified in 33

86.C **Dicarbonyl - π - cyclopentadienyl - iodo(tri(n - butyl)phosphine) molybdenum(ii)**
$C_{19}H_{32}IMoO_2P$
For complete entry see 73.22

86.11 **Dicarbonyl - nitrosyl - triphenylphosphine - cobalt(0)**
$C_{20}H_{15}CoNO_3P$
D.L.Ward, C.N.Caughlan, G.E.Voecks, P.W.Jennings
Acta Cryst. (B), **28,** 1949, 1972

86.C **Dibromo - (2 - bromo - 1 - (o - diphenylphosphinophenyl)ethyl) gold**
$C_{20}H_{17}AuBr_3P$
For complete entry see 71.43

86.C **bis(Dimethyl - phenyl - phosphine) - bis(5 - methyltetrazolato) palladium**
$C_{20}H_{30}N_8P_2Pd$
For complete entry see 83.114

86.12 **Di - μ - iodo - bis - ((o - dimethylaminophenyl)dimethylarsine) - dicopper(i)**
$C_{20}H_{32}As_2Cu_2I_2N_2$
R.Graziani, G.Bombieri, E.Forsellini *J. Chem. Soc. (A),* 2331, 1971
Also classified in 83

86.13 **Nitrosyl - bis(o - phenylene - bis(dimethylarsine)) cobalt diperchlorate**
$C_{20}H_{32}As_4CoNO^{2+}$, $2ClO_4^-$
J.H.Enemark, R.D.Feltham
Amer. Cryst. Assoc., Abstr. Papers (Winter Meeting), 44, 1972

86.14 **Dichloro - hydrido - bis(t - butyl - di - n - propyl - phosphine) rhodium(iii)**
$C_{20}H_{47}Cl_2P_2Rh$
C.Masters, B.L.Shaw *J. Chem. Soc. (A),* 3679, 1971

86.C **Dibromo - (2 - bromo - 1 - (o - diphenylphosphinobenzyl)ethyl) gold**
$C_{21}H_{19}AuBr_3P$
For complete entry see 71.46

86.15 **Chloro(tetracarbonyl) manganese triphenylphosphine complex**
$C_{22}H_{15}ClMnO_4P$
H.Vahrenkamp *Chem. Ber.,* **104,** 449, 1971

86.16 **Tetracarbonyl - (1 - (dimethylarsino) - 2 - (diphenylphosphino)tetrafluoro - cyclobutene) iron**
$C_{22}H_{16}AsF_4FeO_4P$
F.W.B.Einstein, R.D.G.Jones *J. C. S. Dalton,* 442, 1972

86.17 **cis - Dichloro - bis(benzyl - Δ^3 - phospholen) nickel(ii)**
$C_{22}H_{26}Cl_2NiP_2$
A.T.McPhail, R.C.Komson, J.F.Engel, L.D.Quin
J. C. S. Dalton, 874, 1972

86.C **Tribromo - (o - styryldimethylarsine) - 2 - ethoxy - 2 - (o - dimethylarsinophenyl) - ethyl platinum**
$C_{22}H_{31}As_2Br_3OPt$
For complete entry see 71.49

86.C **Triphenylphosphine - (N,N - diethyldithiocarbamato) gold(i)**
$C_{23}H_{25}AuNPS_2$
For complete entry see 80.16

86.C **Bromo(tri - o - vinylphenyl)phosphine - rhodium(i)**
$C_{24}H_{21}BrPRh$
For complete entry see 72.18

86.18 **Di - μ - (1,2 - bis(dimethylarsino)tetrafluorocyclobutene) - octacarbonyl - tetracobalt**
$C_{24}H_{24}As_4Co_4F_8O_8$
F.W.B.Einstein, R.D.G.Jones *J. Chem. Soc. (A),* 3359, 1971

86.19 **cis - mer - Oxodichlorotris(dimethylphenylphosphine) molybdenum(iv)**
$C_{24}H_{33}Cl_2MoOP_3$
L.Manojlovic-Muir *J. Chem. Soc. (A),* 2796, 1971

86.C **Tricarbonyl(3 - (o - (diphenyl - phosphino)phenyl) - 2 - buten - 2 - olato) manganese**
$C_{25}H_{20}MnO_4P$
For complete entry see 71.56

86.C **Carboxymethyl - palladium (triphenylphosphine)pyridine**
$C_{25}H_{22}NO_2PPd$
For complete entry see 71.58

86.C **Tetramethylcyclobutadiene - (trifluoromethyl) - bis(dimethylphenylphosphine) platinum(ii) hexafluoroantimonate**
$C_{25}H_{34}F_3P_2Pt^+$, F_6Sb^-
For complete entry see 71.59

86.C **(1,5 - Cyclo - octadiene) - bis(dimethyl - phenyl - phosphine) - methyl iridium(i)**
$C_{25}H_{37}IrP_2$
For complete entry see 75.26

86.20 **Dichloro - nitrosyl - bis(methyl - diphenyl - phosphine) cobalt**
$C_{26}H_{26}Cl_2CoNOP$
C.S.Pratt, J.A.Ibers, P.Farnham, C.Reed, J.P.Collman
Amer. Cryst. Assoc., Abstr. Papers (Winter Meeting), 43, 1972

86.21 **Triphenylphosphine - cis - bis(difluoro - diethylamino - phosphine) - chloro - rhodium(i)**
$C_{26}H_{35}ClF_4N_2P_3Rh$
M.A.Bennett, G.B.Robertson, T.W.Turney, P.O.Whimp
J. Chem. Soc. (D), 762, 1971

86.C **Bromo - (bis - (2 - diethylaminoethyl) - (2 - diphenylphosphinoethyl)amine) nickel(ii) tetraphenylborate**
$C_{26}H_{42}BrN_2NiP^+$, $C_{24}H_{20}B^-$
For complete entry see 83.126

86.22 **μ - (Carbonyl - triphenylphosphine - platinio)octacarbonyl - di - iron**
$C_{27}H_{15}Fe_2O_9PPt$
R.Mason, J.Zubieta, A.T.T.Hsieh, J.Knight, M.J.Mays
J. C. S. Chem. Comm., 200, 1972

86.C **8 - Hydroxyquinolinato(carbonyl)(triphenylphosphine) rhodium(i)**
$C_{28}H_{21}NO_2PRh$
For complete entry see 83.128

86.C **(Isoprene dimer) nickel tri(cyclohexyl)phosphine**
$C_{28}H_{49}NiP$
For complete entry see 71.65

86.23 **Dibromo - tricarbonyl(1,2 - bis(diphenylphosphino)ethane) molybdenum(ii) acetone solvate**

$C_{29}H_{24}Br_2MoO_3P_2 , C_3H_6O$
M.G.B.Drew *J. C. S. Dalton,* 1329, 1972

86.24 **bis(Diphenylethylphosphine)chlorocarbonyl - iridium(i) oxygen complex**

$C_{29}H_{30}ClIrO_3P_2$
M.S.Weininger, I.F.Taylor Junior, E.L.Amma
J. Chem. Soc. (D), 1172, 1971

86.25 **(1,2 - bis(Diphenylphosphino)ethane) - tetracarbonyl - chromium (at 2 ° C)**

$C_{30}H_{24}CrO_4P_2$
M.J.Bennett, F.A.Cotton, M.D.LaPrade
Acta Cryst. (B), **27,** 1899, 1971

86.26 **cis,mer - Dichloro - tris(diethyl - phenyl - phosphine)oxo - molybdenum(iv)**

$C_{30}H_{45}Cl_2MoOP_3$
L.Manojlovic-Muir, K.W.Muir *J. C. S. Dalton,* 686, 1972

86.27 **pentakis(2,8,9 - Trioxa - 1 - phospha - adamantane) nickel(ii) perchlorate**

$C_{30}H_{45}NiO_{15}P_5^{2+} , 2ClO_4^-$
E.F.Riedel, R.A.Jacobson *Inorg. Chim. Acta,* **4,** 407, 1970

86.C **cis - (Hydroxydiphenylgermyl) - phenyl - bis(triethylphosphine) platinum(ii)**

$C_{30}H_{46}GeOP_2Pt$
For complete entry see 71.68

86.C **Di(bis(trimethylsilyl)amino) cobalt(ii) - triphenylphosphine**

$C_{30}H_{51}CoN_2PSi_4$
For complete entry see 83.130

86.C **1 - (Cyclopentadienyl - triphenylphosphine - ruthenium) - 1,2,3,4 - tetra(trifluoromethyl) - buta - 1,3 - diene**

$C_{31}H_{21}F_{12}PRu$
For complete entry see 71.70

86.C **(N,N′ - Dimethyl - N,N′ - bis(2 - diphenylphosphinoethyl)ethylenediamine) - bromo - cobalt(ii) hexafluorophosphate**

$C_{32}H_{38}BrCoN_2P_2^+ , F_6P^-$
For complete entry see 76.43

86.C **(N,N′ - Dimethyl - N,N′ - bis(2 - diphenylphosphinoethyl)ethylenediamine) - bromo - nickel(ii) bromide butanol solvate**

$C_{32}H_{38}BrN_2NiP_2^+ , Br^- , 0.5C_4H_{10}O$
For complete entry see 76.44

86.28 **Di - μ - carbonyl - heptacarbonyl - tris((dimethyl)phenylphosphine) - tri - iron**

$C_{33}H_{33}Fe_3O_9P_3$
G.Raper, W.S.McDonald *J. Chem. Soc. (A),* 3430, 1971

86.C **bis(O - Methyl phenylthiocarbamato)(triphenylphosphine) palladium(ii)**
$C_{34}H_{33}N_2O_2PPdS_2$
For complete entry see 85.33

86.C **(1,5 - Cyclo - octadiene) - (1,2 - bis(diphenylphosphino)ethane) - methyl iridium(i)**
$C_{35}H_{39}IrP_2$
For complete entry see 75.32

86.29 **Chloro - dinitrosyl - bis(triphenylphosphine) ruthenium hexafluorophosphate benzene solvate**
$C_{36}H_{30}ClN_2O_2P_2Ru^+$, F_6P^- , C_6H_6
C.G.Pierpont, R.Eisenberg *Inorg. Chem.*, **11,** 1088, 1972

86.30 **Azido - nitrosyl - bis(triphenylphosphine) nickel**
$C_{36}H_{30}N_4NiOP_2$
J.H.Enemark *Inorg. Chem.*, **10,** 1952, 1971

86.31 **bis(Triphenylphosphine) - dioxo - platinum chloroform solvate**
$C_{36}H_{30}O_2P_2Pt$, $2CHCl_3$
P.-T.Cheng, C.D.Cook, S.C.Nyburg, K.Y.Wan
Canad. J. Chem., **49,** 3772, 1971

86.32 **(Phenyl - di(3 - (diphenylphosphino)propyl) - phosphine) - chloro - nitrosyl - rhodium hexafluorophosphate**
$C_{36}H_{37}ClNOP_3Rh^+$, F_6P^-
R.M.Kirchner, J.A.Ibers
Amer. Cryst. Assoc., Abstr. Papers (Winter Meeting), 45, 1972

86.33 **(Phenyl - di(3 - (diphenylphosphino)propyl) - phosphine) - chloro - rhodium**
$C_{36}H_{37}ClP_3Rh$
R.M.Kirchner, J.A.Ibers
Amer. Cryst. Assoc., Abstr. Papers (Winter Meeting), 45, 1972

86.34 **Nitrato - bis(tricyclohexylphosphine) copper(i)**
$C_{36}H_{66}CuNO_3P_2$
W.A.Anderson, A.J.Carty, G.J.Palenik, G.Schreiber
Canad. J. Chem., **49,** 761, 1971

86.C **Compound Z**
$C_{37}H_{24}O_7Os_3P_2$
For complete entry see 71.71

86.C **Platinum bis(triphenylphosphine) dithiofluoroformate bifluoride**
$C_{37}H_{30}FP_2PtS_2^+$, HF_2^-
For complete entry see 85.34

86.C **bis(Triphenylphosphine) - methyl - platinum iodosulfone**
$C_{37}H_{33}IO_2P_2PtS$
For complete entry see 71.72

86.C **(1,2 - bis(Diphenylphosphino)hexafluorocyclopentene) - carbonyl - iron - trimethyltin**

$C_{37}H_{34}F_6FeP_2Sn$
For complete entry see 73.32

86.C **Compound Y**

$C_{38}H_{24}O_8Os_3P_2$
For complete entry see 71.73

86.C **trans - bis(Methyldiphenylphosphine) - (σ - pentafluorophenyl) - (σ - pentachlorophenyl) - nickel(ii)**

$C_{38}H_{26}Cl_5F_5NiP_2$
For complete entry see 71.74

86.C **trans - bis(Methyldiphenylphosphino) - bis(σ - pentafluorophenyl) nickel(ii)**

$C_{38}H_{26}F_{10}NiP_2$
For complete entry see 71.75

86.C **1 - Chloro - 1,2,2 - trifluoroethylene bis(triphenylphosphine) platinum(0) complex**

$C_{38}H_{30}ClF_3P_2Pt$
For complete entry see 72.22

86.C **1,1 - Dichloro - 2,2 - difluoro - ethylene bis(triphenylphosphine) platinum(0) complex**

$C_{38}H_{30}Cl_2F_2P_2Pt$
For complete entry see 72.23

86.C **Tetrachloroethylene bis(triphenylphosphine) platinum(0) complex**

$C_{38}H_{30}Cl_4P_2Pt$
For complete entry see 72.24

86.35 **Dicarbonyl - nitrosyl - bis(triphenylphosphine) osmium perchlorate dichloromethane solvate**

$C_{38}H_{30}NO_3OsP_2^+$, ClO_4^-, CH_2Cl_2
G.R.Clark, K.R.Grundy, W.R.Roper, J.M.Waters, K.R.Whittle
J. C. S. Chem. Comm., 119, 1972

86.C **bis(Triphenylphosphine)(ethylene) nickel**

$C_{38}H_{34}NiP_2$
For complete entry see 72.25

86.C **bis(Triphenylphosphine) platinum ethylene complex**

$C_{38}H_{34}P_2Pt$
For complete entry see 72.26

86.36 **3,6 - Diphenylpyridazino - di - iron - triphenylphosphine pentacarbonyl**

$C_{39}H_{27}Fe_2N_2O_5P$
L.G.Kuz'mina, N.G.Bokii, Yu.T.Struchkov, A.V.Arutyunyan, L.V.Rybin, M.I.Rybinskaya *Zh. Strukt. Khim.*, **12,** 875, 1971
Also classified in 83

86.C **3,3 - Di(trifluoromethyl) - 1 - nickela - 2 - oxa - cyclopropane bis(triphenylphosphine)**
$C_{39}H_{30}F_6NiOP_2$
For complete entry see 71.76

86.37 **Dibromo - tris - (5 - methyl - 5H - dibenzophosphole) - platinum(ii)**
$C_{39}H_{33}Br_2P_3Pt$
H.M.Powell, K.M.Chui *J. Chem. Soc. (D)*, 1037, 1971

86.C **bis(Triphenylphosphine)allene - platinum**
$C_{39}H_{34}P_2Pt$
For complete entry see 72.27

86.C **μ - Allyl - μ - iodo - bis(triphenylphosphine palladium) benzene solvate**
$C_{39}H_{35}IP_2Pd_2$, C_6H_6
For complete entry see 72.28

86.C **bis(Diphenyl - trifluoromethylethynyl - phosphine) - decacarbonyl - tetracobalt**
$C_{40}H_{20}Co_4F_6O_{10}P_2$
For complete entry see 72.29

86.C **cis - 1,2 - bis(Triphenylphosphine - gold) - 1,2 - bis(trifluoromethyl)ethylene**
$C_{40}H_{30}Au_2F_6P_2$
For complete entry see 71.77

86.C **1,1 - Dichloro - 2,2 - dicyanoethylene bis(triphenylphosphine) platinum(0) complex**
$C_{40}H_{30}Cl_2N_2P_2Pt$
For complete entry see 72.30

86.38 **trans - bis(Triphenylphosphite) tetracarbonyl chromium(0)**
$C_{40}H_{30}CrO_{10}P_2$
H.S.Preston, J.M.Stewart, H.J.Plastas, S.O.Grim
Inorg. Chem., **11,** 161, 1972

86.C **Cyano(cyanoacetylido)bis(triphenylphosphine) platinum(ii)**
$C_{40}H_{30}N_2P_2Pt$
For complete entry see 71.78

86.C **Anhydrodi(carboxymethyl) palladium - bis(triphenylphosphine)**
$C_{40}H_{34}O_3P_2Pd$
For complete entry see 71.80

86.C **Chloro - bis(triphenylphosphine) platinum trimethylsilylmethilide**
$C_{40}H_{41}ClPtSi$
For complete entry see 71.81

86.39 **cis - Dihydro - tetrakis(diethylphenylphosphonite) iron(ii) (data set D)**
$C_{40}H_{62}FeO_8P_4$
L.J.Guggenberger, D.D.Titus, M.T.Flood, R.E.Marsh, A.A.Orio, H.B.Gray *J. Amer. Chem. Soc.*, **94,** 1135, 1972

86.40 **cis - Dihydro - tetrakis(diethylphenylphosphonite) iron(iii) (data set C)**
$C_{40}H_{62}FeO_8P_4$
L.J.Guggenberger, D.D.Titus, M.T.Flood, R.E.Marsh, A.A.Orio, H.B.Gray *J. Amer. Chem. Soc.*, **94,** 1135, 1972

86.C **bis(Triphenylphosphine) - (1,2 - methylcyclopropene) platinum**
$C_{41}H_{38}P_2Pt$
For complete entry see 75.33

86.C **bis(Trifluoromethyl)diazomethane bis(triphenylphosphine) platinum(0) methylene chloride solvate**
$C_{42}H_{30}F_{12}N_2P_2Pt$, $0.4CH_2Cl_2$
For complete entry see 71.84

86.41 **Pentacarbonyl chromium(0) - (phenyl - bis(2 - diphenylphosphinoethyl)phosphine) - bromo - manganese(i) tricarbonyl**
$C_{42}H_{33}BrCrMnO_8P_3$
M.L.Schneider, N.J.Coville, I.S.Butler *J. C. S. Chem. Comm.*, 799, 1972

86.C **bis(Diphenylphosphinodithioato) (triphenylphosphine) palladium**
$C_{42}H_{35}P_3PdS_4$
For complete entry see 85.36

86.C **Cyclohexyne - bis(triphenylphosphine) - platinum(0)**
$C_{42}H_{38}P_2Pt$
For complete entry see 75.34

86.42 **Dibromo - tris(5 - ethyl - 5H - dibenzophosphole) platinum(ii)**
$C_{42}H_{39}Br_2P_2Pt$
H.M.Powell, K.M.Chui *J. Chem. Soc. (D)*, 1037, 1971

86.43 **Dibromo - tris(5 - ethyl - 5H - dibenzophosphole) palladium(ii)**
$C_{42}H_{39}Br_2P_3Pd$
H.M.Powell, K.M.Chui *J. Chem. Soc. (D)*, 1037, 1971

86.C **bis(Triphenylphosphine) - bis - π - allyl - ruthenium toluene solvate**
$C_{42}H_{40}P_2Ru$, C_7H_8
For complete entry see 72.31

86.44 **Di - iodo - tris(2 - diphenylphosphinoethyl)amine - cobalt(ii)**
$C_{42}H_{42}CoI_2NP_3$
C.Mealli, P.L.Orioli, L.Sacconi *J. Chem. Soc. (A)*, 2691, 1971
Also classified in 83

86.C **bis(Trimethylsilyl)amino - nickel(i) - bis(triphenylphosphine)**

$C_{42}H_{48}NNiP_2Si_2$
For complete entry see 83.136

86.C **Chloro - carbonyl - bis(triphenylphosphine) - (5 - fluoro - 2 - diazo - phenyl) iridium tetrafluoroborate acetone solvate**

$C_{43}H_{34}ClFIrN_2OP_2^+$, BF_4^-, C_3H_6O
For complete entry see 71.85

86.C **(bis(o - Dimethylarsinophenyl) - (o - methylthiophenyl)arsine) - bromo - nickel(ii) perchlorate chlorobenzene solvate**

$C_{43}H_{35}As_3BrNiS^+$, ClO_4^-, C_6H_5Cl
For complete entry see 85.37

86.C **Cycloheptyne - platinum - bis(triphenylphosphine)**

$C_{43}H_{40}P_2Pt$
For complete entry see 75.35

86.C **Compound X**

$C_{45}H_{30}O_9Os_3P_2$
For complete entry see 71.89

86.C **bis(Pyridine - 2 - thiolato) - bis(triphenylphosphine) ruthenium(ii)**

$C_{46}H_{38}N_2P_2RuS_2$
For complete entry see 85.38

86.C **1,1 - bis(Triphenylphosphine) - 2,3 - diphenyl - 1 - platinacyclobut - 2 - ene - 4 - one**

$C_{48}H_{40}OP_2Pt$
For complete entry see 71.90

86.C **Di - μ - acetato - bis(dimethylglyoximato)bis(triphenylphosphine) dirhodium(ii) monohydrate**

$C_{48}H_{48}N_4O_8P_2Rh_2$, H_2O
For complete entry see 81.61

86.C **tetrakis(Pentafluorophenyl) - μ - bis(diphenylarsino)methane - dimercury(ii)**

$C_{49}H_{22}As_2F_{20}Hg_2$
For complete entry see 71.91

86.45 **(Cyano(dicyanomethyl)ketiminato)carbonyl (tetracyanoethylene) bis(triphenylphosphine) iridium benzene solvate**

$C_{49}H_{31}IrN_8OP_2$, $0.5C_6H_6$
J.S.Ricci, J.A.Ibers *J. Amer. Chem. Soc.*, **93**, 2391, 1971
Residue 1 also classified in 72, 83

86.C **Carbonyl - bis(triphenylphosphine) - (1,4 - di(p - fluorophenyl)tetrazene) iridium tetrafluoroborate benzene solvate**

$C_{49}H_{38}F_2IrN_4OP_2^+$, BF_4^-, C_6H_6
For complete entry see 83.139

86.46 **bis(bis(Dicyclohexylphosphino)methane) nickel**

$C_{50}H_{92}NiP_4$

C.Kruger, Y.-H.Tsay *Acta Cryst. (B)*, **28**, 1941, 1972

86.47 **bis(bis(Diphenylarsino)methane)dibromo - dicarbonyl - molybdenum(ii)**

$C_{52}H_{44}As_4Br_2MoO_2$

M.G.B.Drew *J. C. S. Dalton*, 626, 1972

86.48 **Dioxygen - bis(cis - 1,2 - bis(diphenylphosphino) - ethylene)cobalt tetrafluoroborate benzene solvate**

$C_{52}H_{44}CoO_2P_4^+$, BF_4^-, $2C_6H_6$

N.W.Terry III, E.L.Amma, L.Vaska *J. Amer. Chem. Soc.*, **94**, 653, 1972

86.49 **Chloro - bis(bis(1,2 - diphenylphosphino)ethane) cobalt(ii) trichlorostannate(ii)**

$C_{52}H_{48}ClCoP_4^+$, Cl_3Sn^-

J.K.Stalick, D.W.Meek, B.Y.K.Ho, J.J.Zuckerman

J. C. S. Chem. Comm., 630, 1972

86.50 **Chloro - bis(bis(1,2 - diphenylphosphino)ethane) cobalt(ii) trichlorostannate(ii) chlorobenzene solvate**

$C_{52}H_{48}ClCoP_4^+$, Cl_3Sn^-, C_6H_5Cl

J.K.Stalick, D.W.Meek, B.Y.K.Ho, J.J.Zuckerman

J. C. S. Chem. Comm., 630, 1972

86.51 **Disulfur - bis(bis(diphenylphosphino)ethane) iridium(i) chloride acetonitrile solvate**

$C_{52}H_{48}IrP_4S_2^+$, Cl^-, C_2H_3N

W.D.Bonds Junior, J.A.Ibers *J. Amer. Chem. Soc.*, **94**, 3413, 1972

86.52 **trans - bis(Dinitrogen)bis(1,2 - bis(diphenylphospino)ethane) molybdenum(0)**

$C_{52}H_{48}MoN_4P_4$

T.Uchida, Y.Uchida, M.Hidai, T.Kodama

Bull. Chem. Soc. Jap., **44**, 2883, 1971

86.53 **Tetra(methyl - diphenyl - phosphine) iridium(i) tetrafluoroborate cyclohexane solvate**

$C_{52}H_{52}IrP_4^+$, BF_4^-, C_6H_{12}

G.R.Clark, C.A.Reed, W.R.Roper, B.W.Skelton, T.N.Waters

J. Chem. Soc. (D), 758, 1971

86.54 **bis(Triphenylphosphine)tetrakis (dimethylglyoximato) dirhodium monohydrate propanol solvate**

$C_{52}H_{58}N_8O_8P_2Rh_2$, C_3H_8O, H_2O

K.G.Caulton, F.A.Cotton *J. Amer. Chem. Soc.*, **93**, 1914, 1971

Residue 1 also classified in 83

86.55 **Dibromo - tricarbonyl - bis(bis(diphenylarsino)methane) tungsten(ii)**
$C_{53}H_{44}As_4Br_2O_3W$
M.G.B.Drew, A.W.Johans, A.P.Wolters, I.B.Tomkins
J. Chem. Soc. (D), 819, 1971

86.56 **(tris(o - Diphenylphosphinophenyl)phosphine) - chlorocobalt(ii) tetraphenylborate**
$C_{54}H_{42}ClCoP_4^+$, $C_{24}H_{20}B^-$
T.L.Blundell, H.M.Powell *Acta Cryst. (B)*, **27**, 2304, 1971
Residue 2 classified in 62

86.57 **Di - μ - chloro - tris(triphenylphosphine) dicopper(i)**
$C_{54}H_{45}Cl_2Cu_2P_3$
V.G.Albano, P.L.Bellon, G.Ciani, M.Manassero *J. C. S. Dalton*, 171, 1972

86.58 **tris(Triphenylphosphine)nitrosyl - iridium**
$C_{54}H_{45}IrNOP_3$
V.G.Albano, P.Bellon, M.Sansoni *J. Chem. Soc. (A)*, 2420, 1971

86.59 **Hydrido - nitrosyl - tris(triphenylphosphine) iridium(i) perchlorate**
$C_{54}H_{46}IrNOP_3^+$, ClO_4^-
D.M.P.Mingos, J.A.Ibers *Inorg. Chem.*, **10**, 1479, 1971

86.60 **Hydrido - nitrosyl - tris(triphenylphosphine) ruthenium**
$C_{54}H_{46}NOP_3Ru$
C.G.Pierpont, R.Eisenberg *Inorg. Chem.*, **11**, 1094, 1972

86.61 **Di - μ - thiocyanato - tetrakis(methyldiphenylphosphine) dicopper(i)**
$C_{54}H_{52}Cu_2N_2P_4S_2$
A.P.Gaughan, R.F.Ziolo, Z.Dori *Inorg. Chim. Acta*, **4**, 640, 1970

86.62 **Di(1,2 - bis(diphenylphosphino)hexafluorocyclopentene) rhodium(i) cis - dicarbonyl - dichloro - rhodate(i)**
$C_{58}H_{40}F_{12}P_4Rh^+$, $C_2Cl_2O_2Rh^-$
F.W.B.Einstein, C.R.S.M.Hampton *Canad. J. Chem.*, **49**, 1901, 1971

86.63 **Nitrosyl - bis(1,2 - bis(diphenylphosphino)ethane) ruthenium(0) tetraphenylborate acetone solvate**
$C_{58}H_{48}NOP_4Ru^+$, $C_{24}H_{20}B^-$, C_3H_6O
C.G.Pierpont, A.Pucci, R.Eisenberg *J. Amer. Chem. Soc.*, **93**, 3050, 1971
Residue 2 classified in 62

86.C **Dicarbonyl - bis(triphenylphosphine) - tris(toluene - 3,4 - dithiolato)di - iridium(iii)**
$C_{59}H_{48}Ir_2O_2P_2S_6$
For complete entry see 85.39

86.C **Di - μ - pentafluorobenzenethiolato - trans - bis((pentafluorobenzenthiolato) - (triphenylphosphine) palladium(ii)) ethanol solvate**

$C_{60}H_{30}F_{20}P_2Pd_2S_4$, $2C_2H_6O$
For complete entry see 85.40

86.C **1,2,3,4 - Tetrahydro - 1,4 - dioxo - naphthaleno(2,3 - c)(1′ - bis(triphenylphosphine)chloro - rhoda) cyclopentadiene monohydrate ethanol solvate**

$C_{60}H_{44}ClO_2P_2Rh$, H_2O , $0.5C_2H_6O$
For complete entry see 71.92

86.C **5,5,7,8,9,10 - Hexaphenyl - 6 - (triphenylphosphino) - 5,6 - dihydro - 5 - phospha - 6 - rhodabenzocyclo - octene toluene solvate**

$C_{64}H_{51}P_2Rh$, C_7H_8
For complete entry see 71.93

86.C **Hexa(acetato) - tris(triphenylphosphine) - oxotriruthenium**

$C_{66}H_{63}O_{13}P_3Ru_3$
For complete entry see 81.62

86.64 **Triphenylphosphine tris(o - diphenylphosphinophenyl)phosphine iridium(i) tetraphenylborate**

$C_{72}H_{57}IrP_5^+$, $C_{24}H_{20}B^-$
L.M.Venanzi, R.Spagna, L.Zambonelli *J. Chem. Soc. (D),* 1570, 1971
Residue 2 classified in 62

86.65 **Chloro - dioxo - bis(triphenylphosphine) rhodium dimer methylene chloride solvate**

$C_{72}H_{60}Cl_2O_4P_4Rh_2$, CH_2Cl_2
M.J.Bennett, P.B.Donaldson *J. Amer. Chem. Soc.*, **93,** 3307, 1971

86.66 **bis(bis(Tricyclohexylphosphine) nickel)dinitrogen**

$C_{72}H_{132}N_2Ni_2P_4$
P.W.Jolly, K.Jonas, C.Kruger, Y.-H.Tsay
J. Organometal. Chem., **33,** 109, 1971
Also classified in 82

86.67 **tetrakis(Triphenylphosphine) - di - μ - carbonyl - dirhodium dichloromethane solvate**

$C_{74}H_{60}O_2P_2Rh_2$, $2CH_2Cl_2$
C.B.Dammann, P.Singh, D.J.Hodgson *J. C. S. Chem. Comm.*, 586, 1972

86.68 **Diazido - μ - 1,2 - bis(diphenylphosphino)ethane - bis(1,2 - bis(diphenylphosphino)ethane) - dicopper(i)**

$C_{78}H_{72}Cu_2N_6P_6$
A.P.Gaughan, R.F.Ziolo, Z.Dori *Inorg. Chem.*, **10,** 2776, 1971

86.69 **Di - 2 - (5 - perfluoromethyltetrazolato) - μ - 1,2 - bis(diphenylphosphino) ethane - bis(1,2 - (diphenylphosphino) - ethane) dicopper(i)**

$C_{82}H_{72}Cu_2F_6N_8P_6$
A.P.Gaughan, K.S.Bowman, Z.Dori *Inorg. Chem.*, **11**, 601, 1972
Also classified in 83

86.C **Tetracopper - bis(triphenylphosphine - iridium - tetra(phenylacetylide))**

$C_{100}H_{70}Cu_4Ir_2P_2$
For complete entry see 71.94

86.70 **Triphenylphosphine - copper(i) hydride hexamer**

$C_{108}H_{96}Cu_6P_6$
S.A.Bezman, M.R.Churchill, J.A.Osborn, J.Wormald
J. Amer. Chem. Soc., **93**, 2063, 1971

86.71 **octakis(Tri - p - tolylphosphine) enneagold hexafluorophosphate**

$C_{168}H_{168}Au_9P_8^{3+}$, $3F_6P^-$
P.L.Bellon, F.Cariati, M.Manassero, L.Naldini, M.Sansoni
J. Chem. Soc. (D), 1423, 1971

FORMULA INDEX

C_1

$CBrCl_3Hg$	71.1	2
$CBrN_2O_4^-, K^+$	12.1	3
CBr_4, C_8H_{10}	60.1	2
ClN, C_5H_5N	60.65	2
CN_2^{2-}, Sr^{2+}	7.1	1
$CN_3O_6^-, Cs^+$	10.1	1
$CN_3O_6^-, I^+$	10.2	1
$CN_3O_6^-, K^+$	10.3	1
$CN_3O_6^-, Rb^+$	10.4	1
$CN_3O_6^-, H_5N_2^+$	12.2	3
$CO_5P^{3-}, 3Na^+, 6H_2O$	64.1+	4
$CHBrCl_2, C_4H_{10}O$	60.61	2
$CHBr_3$	5.1	4
$2CHBr_3, C_6H_{12}N_4$	60.1	4
$CHCl_3$	5.1	1
$CHCl_3, C_{16}H_{14}CuN_2O_2$	60.1	3
CHI_3	5.2	1
$CHI_3, 3S_8$	60.2	2
$CHI_3, C_4H_8O_2$	60.2	3
$CHI_3, C_4H_8S_2$	60.53	2
$CHI_3, C_6H_{12}N_4$	60.20	3
$CHI_3, 3C_9H_7N$	60.3	2
$2CHI_3, C_4H_8Se_2$	60.4	2
$CHN_2O_4^-, Rb^+$	12.1	1
CHN_4^-, Na^+, H_2O	32.1	1
CHO_2^-, Na^+	2.1	1
CHO_2^-, H_4N^+	2.2	1
$CHO_2^-, C_{11}H_{13}N_2O_2^+$	48.42	3
$2CHO_2^-, Ba^{2+}$	2.3	1
$2CHO_2^-, Ca^{2+}$	2.4	1
$2CHO_2^-, Mg^{2+}, 2H_2O$	2.5	1
$2CHO_2^-, Sr^{2+}$	2.6	1
$2CHO_2^-, Sr^{2+}, 2H_2O$	2.8	1
$2CHO_2^-, Sr^{2+}, 2H_2O$	2.1	3
$2CHO_2^-, Sr^{2+}, 2H_2O$	2.1	4
$6CHO_2^-, 16CH_4N_2S, 3Pb^{2+}$	8.2+	4
$CH_2NS_2^-, H_4N^+$	11.1	1
$CH_2N_4O_4^{2-}, 2K^+$	10.6	1
$CH_2N_5^-, H_5N_2^+$	32.2	1
CH_2O_2	1.1	1
CH_2O_2, CH_3NO	60.3	3
$CH_2O_6S_2^{2-}, 2K^+$	11.2	1
CH_2S_3	11.3	1
CH_3AsI_2	65.1	2
CH_3Cl	5.3	1
$CH_3CuN_3O_8$	84.1	2
CH_3NO	1.2	1
$CH_3N_2O^-, K^+$	9.1	1
$CH_3N_2S_2^-, H_5N_2^+$	9.2	1
$CH_3N_3O_9Sn$	69.1+	4
CH_3N_5, H_2O	32.3	1
$CH_3O_2S_2^-, Na^+, H_2O$	11.4	1
$CH_3O_3S^-, Cs^+$	11.5	1
$CH_3O_3S^-, Na^+, 2H_2O$	11.6	1
$CH_3O_3S^-, C_8H_{18}NO^+$	32.23	3
$2CH_3O_3S^-, C_{11}H_{26}N_2O^{2+}, H_2O$	38.10	3
$CH_3O_4P^{2-}, 2H_4N^+, 2H_2O$	46.1	4
$CH_3O_4S^-, C_{18}H_{23}N_2S^+$	41.18	3
$(CH_4CuN_2S^+)_n, nNO_3^-$	79.1	2
CH_4N_2O	8.1+	1
CH_4N_2O	10.7	1
$CH_4N_2O, Na^+, Cl^-, H_2O$	60.5	2
CH_4N_2O, H_3O_4P	8.6	1
CH_4N_2O, H_3O_4P	8.1	3
CH_4N_2O, H_4N^+, Br^-	60.6	2
CH_4N_2O, H_4N^+, Cl^-	60.7+	2
$CH_4N_2O, C_3H_2N_2O_3$	60.4	3
$CH_4N_2O, C_5H_4N_2O_4$	60.2	4
$CH_4N_2O, C_5H_{12}NO_2S^+, Cl^-$	60.68	2
$CH_4N_2O, C_6H_{12}O_6$	60.14	4
CH_4N_2O, C_7H_9N	60.100	2
$CH_4N_2O, C_8H_{12}N_2O_3$	60.27	3
$CH_4N_2O, C_{18}H_{24}O_2$	60.38	4
$2CH_4N_2O, C_2H_2O_4$	60.3	4
$CH_4N_2O_2$	8.7+	1
$CH_4N_2O_2$	10.8	1
$CH_4N_2O_2$	8.1	4
$CH_4N_2O_2S$	8.9	1
$CH_4N_2O_3, H_2O_2$	60.8	2
CH_4N_2S	8.10+	1
$CH_4N_2S, C_3H_2N_2O_3$	60.5	3
CD_4N_2S	8.14+	1
$2CH_4N_2S, C_5H_6N^+, Br^-$	60.6	4
$4CH_4N_2S, Cs^+, Cl^-, H_2O$	60.9	2
$4CH_4N_2S, Cs^+, F^-, 2H_2O$	60.10	2
$4CH_4N_2S, Tl^+, ClO_3^-$	60.6	3
$4CH_4N_2S, Tl^+, H_2O_4P^-$	60.11	2
$4CH_4N_2S, C_7H_5O_2^-, Tl^+$	60.98	2
$6CH_4N_2S, Pb^{2+}, 2ClO_4^-, 2H_2O$	8.16	1
$16CH_4N_2S, 6CHO_2^-, 3Pb^{2+}$	8.2	4
CH_4N_2Se	8.2+	3
$CH_4N_4NiS_4$	85.1	2
$CH_4N_4O_2$	8.17	1
$CH_4N_5^+, Cl^-$	9.1	4
CH_4O	5.4	1
CH_4O, C_2HgN_2	60.7	3
$CH_4O, C_{10}H_6BrNO_2$	60.32	3
$2CH_4O, Br_2$	60.12	2
$2CH_4O, H_4N_2$	60.13	2
$3CH_4O, Na^+, I^-$	60.14	2
$4CH_4O, H_4N_2$	60.15	2
CH_4O_2S	11.7	1
CH_5AgClN_3S	85.2	2

$(CH_5Cl_2CuN_3O)_n$	83.1+	4
$CH_5Cl_2N_3SZn$	85.3	2
CH_5N	3.1	1
CH_5NO_2S	4.1	1
$CH_5N_2O^+$, $CdCl_3^-$	8.18	1
$CH_5N_2O^+$, NO_3^-	8.19+	1
$CH_5N_2S^+$, NO_3^-	8.22+	1
$2CH_5N_3$, $C_8H_{12}Br_2O_8Rh_2$	81.56	2
$2CH_5N_3$, $C_8H_{12}Cl_2O_8Rh_2$	81.58	2
CH_5N_3S	8.4+	3
CH_6N^+, Br^-	3.2	1
CH_6N^+, Cl^-	3.3	1
CH_6N^+, ClO_4^-	3.4	1
CH_6N^+, Cl_3Ni^-	3.5	1
CH_6N^+, $H_{12}AlO_6^{3+}$, $2O_4S^{2-}$, $6H_2O$	3.6	1
CH_6N^+, $C_2H_7BNO^-$	62.6	2
CH_6N^+, $C_{12}H_{18}CuN_2O_2$, ClO_4^-	60.39	3
$2CH_6N^+$, $Br_2Cl_2Cu^{2-}$	3.1	3
CH_6NO^+, Cl^-	10.9	1
$CH_6N_3^+$, Cl^-	8.25	1
$CH_6N_3^+$, $H_{12}AlO_6^{3+}$, $2O_4S^{2-}$	8.26	1
$CH_6N_3^+$, $H_{12}CrO_6^{3+}$, $2O_4S^{2-}$	8.27	1
$CH_6N_3^+$, $C_8H_{11}N_2O_3^-$, $2H_2O$	43.4	4
$0.5CH_6N_3^+$, $C_5H_{14}N_3O_3P$, $0.5Cl^-$	8.9	4
$2CH_6N_3^+$, Ni^{2+}, $2O_4S^{2-}$, $6H_2O$	8.3	4
$2CH_6N_3^+$, Zn^{2+}, $2O_4S^{2-}$	8.6	3
$2CH_6N_3^+$, $C_{16}H_{28}Ce_2O_{18}^{2-}$, $2H_2O$	81.43	4
$3CH_6N_3^+$, C_4H_9NO, $3Cl^-$	60.16	2
$CH_6N_3O^+$, Cl^-	9.3	1
CH_6N_4S	8.28	1
$CH_7N_4^+$, Cl^-	8.29	1
$(CH_8CdN_2O_6S)_n$	79.2	2
$CH_8N_4S^{2+}$, $2Cl^-$, $2H_2O$	8.4	4
$CH_9N_6^+$, Cl^-	8.30	1

C_2

C_2AgO_4	81.2	2
$C_2Br_2O_2$	1.3	1
$C_2Br_2O_2$, $C_4H_8O_2$	60.17	2
C_2Br_4	5.5	1
C_2Br_4, $C_4H_4N_2$	60.38	2
C_2Br_6	5.6	1
C_2Br_6	5.2	4
$C_2Cl_2O_2$	1.4	1
$C_2Cl_2O_2$, $C_4H_8O_2$	60.18	2
$C_2Cl_4S_2$	39.1	1
C_2Cl_6	5.7	1
$C_2Cl_6S_3$	11.8	1
$C_2F_3O_2^-$, H_4N^+	2.9	1
$C_2F_3O_2^-$, $C_2HF_3O_2$, Cs^+	2.10	1
$C_2F_3O_2^-$, $C_2HF_3O_2$, K^+	2.11	1
$C_2F_3O_2^-$, $C_2HF_3O_2$, K^+	2.2	4
$C_2F_3O_2^-$, $C_2DF_3O_2$, K^+	2.3	4
C_2I_2, $C_4H_8O_2$	60.50	2
C_2I_2, $C_4H_8S_2$	60.19	2
C_2I_2, $C_4H_8Se_2$	60.20	2
C_2I_2, $C_6H_8O_2$	60.92	2
C_2I_4	5.8	1
C_2I_4, $C_4H_4N_2$	60.39	2
C_2I_4, $C_4H_8Se_2$	60.58	2
$C_2MoO_7^{2-}$, H_4N^+, Na^+, $2H_2O$	81.1	2
$C_2MoO_9^{2-}$, $2K^+$	81.1	3
C_2N_2	7.2	1
$C_2N_3O_4^-$, K^+	7.1	3
$C_2N_3O_4^-$, Rb^+	7.3	1
$C_2N_4O_8^{2-}$, $2K^+$	10.10+	1
$C_2NiS_6^{2-}$, $2C_{24}H_{20}As^+$	65.46	2
$C_2O_4^{2-}$, Ca^{2+}, H_2O	2.12	1
$C_2O_4^{2-}$, Ca^{2+}, $2H_2O$	2.13	1
$C_2O_4^{2-}$, $2K^+$, H_2O	2.15+	1
$C_2O_4^{2-}$, $2K^+$, H_2O	2.2	3
$C_2O_4^{2-}$, $2K^+$, H_2O_2	2.17	1
$C_2O_4^{2-}$, $2Li^+$	2.18	1
$C_2O_4^{2-}$, $2Li^+$, H_2O_2	2.3	3
$C_2O_4^{2-}$, $2Na^+$, H_2O_2	2.19	1
$C_2O_4^{2-}$, $2Rb^+$, H_2O	2.20	1
$C_2O_4^{2-}$, $2Rb^+$, H_2O_2	2.21	1
$C_2O_4^{2-}$, Sr^{2+}, $2.17H_2O$	2.22	1
$C_2O_4^{2-}$, $2H_4N^+$, H_2O	2.23+	1
$C_2O_4^{2-}$, $2H_4N^+$, H_2O_2	2.4	4
$C_2O_4^{2-}$, $C_2H_2O_4$, Ba^{2+}, $2H_2O$	2.5	3
$C_2HCl_3O_2$	1.5	1
$C_2HCl_3O_2$, C_5H_5NO	60.4	4
$C_2HF_3O_2$, $C_2F_3O_2^-$, Cs^+	2.10	1
$C_2HF_3O_2$, $C_2F_3O_2^-$, K^+	2.11	1
$C_2HF_3O_2$, $C_2F_3O_2^-$, K^+	2.2	4
$C_2DF_3O_2$, $C_2F_3O_2^-$, K^+	2.3	4
$C_2HNO_4^{2-}$, $2K^+$	2.26	1
$C_2HO_4^-$, K^+	2.27	1
$C_2HO_4^-$, K^+	2.6	3
$C_2HO_4^-$, K^+	2.5	4
$C_2HO_4^-$, Li^+, H_2O	2.7	3
$C_2HO_4^-$, Na^+	2.28	1
$C_2HO_4^-$, Na^+, H_2O	2.6	4
$C_2HO_4^-$, $H_5N_2^+$	2.29+	1
$C_2HO_4^-$, $C_2H_2O_4$, K^+, $2H_2O$	2.31	1
$C_2HO_4^-$, $C_2H_2O_4$, H_4N^+, $2H_2O$	2.32	1
C_2H_2	5.9	1
$C_2H_2B_{10}Cl_{10}$	62.1	2
$C_2H_2ClO_2^-$, $C_2H_3ClO_2$, H_4N^+	2.7	4
$C_2H_2Cl_2Hg$	71.2	2
$C_2H_2Cl_4N_2Ti$	83.3	4
$(C_2H_2CuO_2)_n$, $4nH_2O$	81.3+	2
$C_2H_2NO_3^-$, H_4N^+	2.33	1
$C_2H_2N_2OS_2$	41.1	1
$C_2H_2N_2S_3$	41.2+	1
$C_2H_2N_3O_5^-$, Rb^+	12.3	3
$C_2H_2N_4$	33.1	1
$C_2H_2O_4$	1.6+	1
$C_2H_2O_4$, $2H_2O$	1.9+	1
$C_2H_2O_4$, $2CH_4N_2O$	60.3	4
$C_2H_2O_4$, $C_2O_4^{2-}$, Ba^{2+}, $2H_2O$	2.5	3
$C_2H_2O_4$, $C_2HO_4^-$, K^+, $2H_2O$	2.31	1
$C_2H_2O_4$, $C_2HO_4^-$, H_4N^+, $2H_2O$	2.32	1
$C_2D_2O_4$, $2D_2O$	1.16+	1
$C_2H_2O_4S$	11.9	1
$C_2H_3BCl_3N$	7.2	3
$C_2H_3BF_3N$	7.4	1
$C_2H_3BF_3N$	7.3	3
$(C_2H_3BrCuN)_n$	83.4	4
$C_2H_3Br_4NORe^-$, $C_{24}H_{20}As^+$	83.1	2
$(C_2H_3ClCuN)_n$	83.2	2
$C_2H_3ClHgO_2$	71.3	2
$C_2H_3ClO_2$, $C_2H_2ClO_2^-$, H_4N^+	2.7	4
$C_2H_3Cl_2CuN_3$	83.3	2
$C_2H_3Cl_3$	5.3	4
$C_2H_3Cl_3O_2$	5.10	1
$(C_2H_3CuNO_4)_n$	81.1	4

$C_2H_3FO_2$	1.1	4
C_2H_3HgN	71.4+	2
C_2H_3N, Cl_5Sb	60.21	2
$2C_2H_3N$, Br_2	60.22	2
$C_2H_3N_3$	32.4+	1
$C_2H_3N_3S_2$	41.1	4
$C_2H_3N_4O_{10}V$	83.1	3
$C_2H_3O^+$, $AlCl_4^-$	12.1	4
$C_2H_3O^+$, Cl_6Sb^-	12.2	4
$C_2H_3O^+$, F_6Sb^-	12.3	1
$C_2H_3O_2^-$, Li^+, $2H_2O$	2.34+	1
$C_2H_3O_2^-$, Li^+, $2H_2O$	2.9	3
$C_2H_3O_2^-$, H_4N^+	2.36	1
$C_2H_3O_2^-$, H_4N^+, H_3N	2.37+	1
$C_2H_3O_2^-$, H_4N^+, $2H_3N$	2.39+	1
$C_2H_3O_2^-$, $H_5N_2^+$	2.10	3
$C_2H_3O_2^-$, $C_2H_4O_2$, K^+	2.8	4
$C_2H_3O_2^-$, $C_2H_4O_2$, Na^+	2.42	1
$C_2H_3O_2^-$, $C_2H_4O_2$, H_4N^+	2.11	3
$C_2H_3O_3^-$, Li^+, H_2O	2.43	1
$C_2H_3O_3^-$, $C_2H_4O_3$, K^+	2.44	1
$C_2H_3O_3^-$, $C_2H_4O_3$, Rb^+	2.45	1
$C_2H_2DO_3^-$, Li^+	2.46+	1
$C_2H_3O_4^-$, K^+	2.48	1
$(C_2H_4AgNO_2)_n$	82.1	4
$(C_2H_4AgNO_2)_{2n}$, nH_2O	82.2	4
$C_2H_4B_2Cl_4$	62.2	2
$C_2H_4B_{10}Cl_8$	62.3+	2
$C_2H_4BaO_6$	2.9	4
C_2H_4BrNO	1.1	3
$C_2H_4Br_3Pt^-$, K^+, H_2O	72.1	2
$(C_2H_4CdCl_2N_4)_n$	83.5	4
C_2H_4ClNO	1.21+	1
$C_2H_4ClPS_3$	64.1	3
$C_2H_4ClS_2Sb$	66.1	2
$C_2H_4Cl_2$	5.11+	1
$C_2H_4Cl_2N_6$	9.4	1
$C_2H_4Cl_3Pt^-$, K^+, H_2O	72.1	3
$C_2H_4Cl_3Pt^-$, K^+, H_2O	72.1	4
$C_2H_4Cl_3Pt^-$, K^+, H_2O	72.2	4
$(C_2H_4Cl_4Te)_n$	70.1	4
C_2H_4FNO	1.24	1
$C_2H_4FeO_6$	81.2+	3
$C_2H_4NO_2^-$, Ag^+	48.1	1
$2C_2H_4NO_2^-$, Ni^{2+}, $2H_2O$	48.2	1
$C_2H_4N_2NiS_4$	80.1	2
$C_2H_4N_2O_2$	1.25	1
$C_2H_4N_2O_2$	9.5	1
$C_2H_4N_2O_2$	10.13	1
$C_2H_4N_2S_2$	4.2	1
$C_2H_4N_3S_2^+$, Br^-	41.4	1
$C_2H_4N_3S_2^+$, Cl^-, $0.5H_2O$	41.5	1
$C_2H_4N_3S_2^+$, I^-	41.6	1
$C_2H_4N_3S_2^+$, I^-	41.2	4
$C_2H_4N_4$	7.5+	1
$C_2H_4N_4$	8.7	3
$C_2D_4N_4$	7.7	1
$C_2H_4N_4O_2$	9.6	1
$C_2H_4N_4O_4^{2-}$, $2Na^+$	9.7	1
C_2H_4O, $7.2H_2O$	61.1	2
$C_2H_4O_2$	1.26	1
$C_2H_4O_2$	1.2+	3
$C_2H_4O_2$, $C_2H_3O_2^-$, K^+	2.8	4
$C_2H_4O_2$, $C_2H_3O_2^-$, Na^+	2.42	1
$C_2H_4O_2$, $C_2H_3O_2^-$, H_4N^+	2.11	3
$C_2H_4O_2$, $C_6H_5N_2^+$, Cl^-	60.82	2
$C_2H_4O_2$, $C_{24}H_{40}O_4$	60.42	4
$C_2H_4O_3$	1.5+	3
$C_2H_4O_3$, $C_2H_3O_3^-$, K^+	2.44	1
$C_2H_4O_3$, $C_2H_3O_3^-$, Rb^+	2.45	1
$C_2H_4O_3Se_2$	39.2	1
$C_2H_4O_4S$	42.1	1
$C_2H_4O_6S_2^{2-}$, $2H_3O^+$	11.1	3
$C_2H_4O_8^{2-}$, $2Rb^+$	12.3	4
$C_2H_5AgN_2O_2^+$, NO_3^-	82.1	2
$C_2H_5BrN^+$, Br^-	3.7	1
$C_2H_5ClN^+$, Cl^-	3.8	1
$(C_2H_5IZn)_n$	71.6	2
C_2H_5NO	1.28	1
C_2H_5NO	1.2	4
$2C_2H_5NO$, Na^+, Br^-	60.23	2
C_2H_5NOS	4.1	4
$C_2H_5NO_2$	48.3+	1
$C_2H_5NO_2$	48.1	4
$C_2H_5NO_2$, $0.5H_2O$	10.1	3
$C_2H_5NO_2$, $C_2H_6NO_2^+$, Cl^-	48.6	1
$C_2H_5NO_2$, $C_2H_6NO_2^+$, NO_3^-	48.7+	1
$C_2H_5NO_2$, $2C_2H_6NO_2^+$, O_4S^{2-}	48.9	1
$C_2H_5NO_2$, $2C_2H_6NO_2^+$, O_4S^{2-}	48.2+	4
$2C_2H_5NO_2$, Ba^{2+}, $2Cl^-$, H_2O	48.1	3
$2C_2H_5NO_2$, Mn^{2+}, $2Cl^-$	48.2	3
$2C_2H_5NO_2$, Na^+, I^-, H_2O	48.4	4
$2C_2H_5NO_2$, Sr^{2+}, $2Cl^-$, $3H_2O$	48.3	3
C_2H_5NS	4.3	1
$C_2H_5N_3O_2$, $0.6H_2O$	8.31	1
$C_2H_5N_5$	32.6	1
$C_2H_5N_5S$	32.7	1
$C_2H_5N_5S$	32.1	3
$C_2H_5O^-$, $C_8H_7N_3O_7$, Cs^+	60.103	2
$C_2H_5O_2^+$, FO_3S^-	12.4	3
$C_2H_5O_2^+$, HO_4S^-	12.4	1
$C_2H_5O_4P^{2-}$, $2K^+$, $4H_2O$	46.2	4
$C_2H_5O_4S^-$, K^+	11.10	1
$2C_2H_5O_4S^-$, $C_{16}H_{18}N_4^{2+}$, $2H_2O$	8.12	4
$2C_2H_5O_4S^-$, $C_{16}H_{18}N_4O^{2+}$, $2H_2O$	8.13	4
$3C_2H_5O_4S^-$, $H_{18}ErO_9^{3+}$	11.11	1
$3C_2H_5O_4S^-$, $H_{18}O_9Pr^{3+}$	11.12	1
$3C_2H_5O_4S^-$, $H_{18}O_9Y^{3+}$	11.13	1
$C_2H_6Al_2Cl_4$	68.1	2
$C_2H_6As_2$	65.1	3
$C_2H_6BF_3OS$	62.5	2
$(C_2H_6Be_2)_n$	67.1	2
$(C_2H_6CaN_4O_4)_n$	67.1	4
$(C_2H_6CdCl_2N_2O)_n$	84.2	2
$C_2H_6CdCl_2N_2O_2$	84.3	2
$C_2H_6CdN_4O_4$	83.4	2
$C_2H_6CdN_4O_4$, H_2O	83.5	2
$(C_2H_6Cl_2CuN_2O)_n$	84.4	2
$(C_2H_6Cl_2CuOS)_n$	84.1	3
$(C_2H_6Cl_2Cu_2N_2)_n$	83.6	2
$C_2H_6Cl_2N_2P_2S_2$	64.1	2
$C_2H_6Cl_2Sn$	69.1+	3
$C_2H_6Cl_2Te$	70.1	2
$C_2H_6Cl_3Sn^-$, $C_{17}H_{17}ClN_3Sn^+$	69.32	2
$C_2H_6Cl_6N_2P_2$	64.2+	2
$C_2H_6Cl_8O_2Sb_2$	66.1	4
$(C_2H_6CoO_6)_n$	81.5	2
$C_2H_6CuN_2O_4$	81.6	2
$(C_2H_6CuO_6)_n$	81.7+	2
$C_2H_6F_2NP$	64.2	3
$(C_2H_6F_2O_6S_2Sn)_n$	69.2	4
$C_2H_6F_2Sn$	69.1	2

$C_2H_6F_6N_4P_4$ 64.3 3
$C_2H_6HgS_2$ 85.4 2
$C_2H_6I_2Te$ 70.2 4
$C_2H_6MnN_4O_4$, $2H_2O$ 83.7 2
$(C_2H_6MnO_6)_n$ 81.9+ 2
$C_2H_6NO^+$, C_2H_5NO , Br^- 1.7 3
$C_2H_6NO_2^+$, $C_2H_5NO_2$, Cl^- 48.6 1
$C_2H_6NO_2^+$, $C_2H_5NO_2$, NO_3^- 48.7+ 1
$C_2H_6N_2O_2$ 9.8 1
$C_2H_6N_2O_2$ 10.14 1
$C_2H_6N_2O_2$ 10.15+ 1
$C_2H_6N_2O_6Sn$ 69.3 3
$C_2H_6N_2S^+$, $C_7H_5ClO_2^-$ 8.32 1
$C_2H_6N_4O_4$ 10.17 1
$C_2H_6N_4O_4Zn$ 81.2 4
$(C_2H_6N_4O_4Zn)_n$ 83.2 3
$C_2H_6N_6S$ 32.2 3
$C_2H_6N_{10}$, H_2O 32.3+ 3
$(C_2H_6NiO_6)_n$ 81.11 2
$2C_2H_6O$, H_4N_2 60.24 2
C_2H_6OS 11.14+ 1
C_2H_6OS , $C_9H_{11}BrN_2O_6$ 47.4 1
$C_2H_6O_2PS_2^-$, K^+ 64.4 2
$2C_2H_6O_2PS_2^-$, $C_{36}H_{24}N_6Ni^{2+}$ 83.141 3
$C_2H_6O_2S$ 11.16 1
$C_2H_6O_2S$ 11.2 3
$C_2H_6O_4S_4$ 11.17 1
$C_2H_6O_4S_4Te$ 70.2 2
$C_2H_6P_2S_4$ 64.5 2
$C_2H_7AsO_2$ 65.2 2
$C_2H_7BNO^-$, CH_6N^+ 62.6 2
$C_2H_7Br_2NPt$ 72.3 2
$C_2H_7NO_3S$ 48.10+ 1
$C_2H_7NO_4S$ 4.4 1
$C_2H_7N_2O^+$, Cl^- 8.8+ 3
$C_2H_7N_2O^+$, NO_3^- 8.33 1
$2C_2H_7N_2S^+$, O_4S^{2-} 8.34 1
$C_2H_7N_{10}^+$, Br^- 32.5 3
$C_2H_7O_2P$ 64.6 2
$C_2H_8AgClN_4S_2$ 79.3 2
C_2H_8BN 62.1 3
$C_2H_8B_4$ 62.7+ 2
$C_2H_8Br_2N_4S_2Te$ 70.3 2
$(C_2H_8CaN_4O_5)_n$ 67.2 4
$C_2H_8CdCl_2N_4O_2$ 84.5 2
$C_2H_8CdCl_2N_4S_2$ 79.5 2
$C_2H_8CdN_4O_6S_2$ 79.6 2
$(C_2H_8ClCuN_4S_2)_n$ 79.8 3
$C_2H_8ClHgN_4S_2^+$, Cl^- 79.8 2
$C_2H_8Cl_2CoN_4S_2$ 79.2 3
$C_2H_8Cl_2CoN_4S_2$ 79.1 4
$C_2H_8Cl_2CuN_2$ 76.1 3
$C_2H_8Cl_2HgO_2$ 84.6 2
$C_2H_8Cl_2N_4PbS_2$ 79.9 2
$C_2H_8Cl_2N_4S_2Te$ 70.4 2
$C_2H_8Cl_2N_4S_2Zn$ 79.10 2
$C_2H_8HgI_2N_4S_2$ 79.11 2
$C_2H_8IN_4S_2^+$, I^- 8.5 4
$(C_2H_8LiN_4O_2^+)_n$, nI^- 67.3 3
$C_2H_8N^+$, Br^- 3.9 1
$C_2H_8N^+$, Cl^- 3.10 1
$C_2H_8N^+$, Cl_3Cu^- 3.11 1
$2C_2H_8N^+$, Cl_4Cu^{2-} 3.2 3
$2C_2H_8N^+$, Cl_4Cu^{2-} 3.3 3
$3C_2H_8N^+$, $BiBr_6^{3-}$ 3.12 1
$3C_2H_8N^+$, Cl^- , Cl_4Cu^{2-} 3.1 4
$C_2H_8NO^+$, ClO_4^- 3.5 3
$C_2H_8NO_2Se^+$, Cl^- 11.3 3
$C_2H_8NO_3P$ 64.7 2
$C_2H_8NO_4P$ 46.1+ 1
$C_2H_8N_2$ 3.2 4
$2C_2H_8N_2$, Li^+ , Br^- 60.25 2
$2C_2H_8N_2$, Li^+ , Cl^- 60.26+ 2
$3C_2H_8N_2$, Li^+ , I^- 3.3 4
$C_2H_8N_2O_6PdS_4^{2-}$, $C_4H_{16}N_4Pd^{2+}$ 76.14 3
$C_2H_8N_2S$ 4.5 1
$C_2H_8N_3^+$, NO_3^- 8.35 1
$C_2H_8N_4S_2^{2+}$, $2Br^-$, H_2O 8.36 1
$C_2H_8N_4S_2^{2+}$, $2Cl^-$ 8.6 4
$C_2H_8N_4S_2^{2+}$, $2I^-$, H_2O 8.37 1
$C_2H_8N_4Se_2^{2+}$, $2Cl^-$ 8.11 3
$C_2H_8N_6NiS_2$ 85.5 2
$C_2H_9B_{10}Br_3$ 62.9 2
$C_2H_{10}B_{10}Br_2$ 62.10 2
$C_2H_{10}B_{10}Br_2$ 62.11 2
$C_2H_{10}CrN_2O_5$, H_2O 76.1 2
$C_2H_{10}CuN_6O_2^{2+}$, $2Cl^-$ 83.8 2
$C_2H_{10}CuN_6S_2^{2+}$, $2NO_3^-$ 79.2 4
$C_2H_{10}CuN_6S_2^{2+}$, O_4S^{2-} 79.3 4
$C_2H_{10}N_2^{2+}$, $2Cl^-$ 3.13 1
$C_2H_{10}N_2^{2+}$, Cl_4Cu^{2-} 3.6 3
$C_2H_{10}N_2^{2+}$, O_4S^{2-} 3.15 1
$C_2H_{10}N_2^{2+}$, $C_{20}H_{30}Fe_2N_4O_{15}^{2-}$, $6H_2O$ 83.183 2
$2C_2H_{10}N_2^{2+}$, $2Br^-$, Br_4Cu^{2-} 3.4 4
$2C_2H_{10}N_2^{2+}$, $Cl_8Mo_2^{4-}$, $2H_2O$ 3.8 3
$C_2H_{10}N_4NiO_6$ 83.3 3
$C_2H_{10}N_6NiS_2^{2+}$, O_4S^{2-} 85.6 2
$C_2H_{10}N_6NiS_2^{2+}$, O_4S^{2-} , $3H_2O$ 85.7 2
$C_2H_{10}N_6O_2Zn^{2+}$, $2Cl^-$ 83.9 2
$C_2H_{11}B_{10}I$ 62.12 2
$C_2H_{11}CoN_6O_6$ 76.2+ 3
$C_2H_{12}CdCl_2N_8S_2$ 79.4 4
$C_2H_{12}Cl_2CrN_2O_2^+$, Cl^- 76.4 3
$C_2H_{13}B_5$ 62.13 2
$C_2H_{13}B_{10}I$ 62.14 2
$C_2H_{13}N_5P_2S_2$ 64.8 2
$C_2H_{14}B_2N_2$ 3.5 4
$C_2H_{14}B_4F_2NP$ 64.4 3
$C_2H_{14}N_6NiO_2S_2^{2+}$, $2NO_3^-$ 85.8 2
$C_2H_{14}N_8O_4Zn$ 83.10 2
$C_2H_{16}B_9N$ 62.15 2
$C_2H_{16}N_7P_2^+$, I^- 64.9 2
$C_2H_{18}B_{10}$ 62.16 2
$C_2H_{18}CoN_5O_2^{2+}$, Cl^- , ClO_4^- 81.12 2
$C_2H_{20}N_5Rh^{2+}$, $2Br^-$ 71.7 2
$C_2H_{21}N_5ORuS^{2+}$, $2F_6P^-$ 85.1 4

C_3

C_3BrN 7.8 1
C_3ClN 7.9 1
C_3Cl_6 20.1 1
$C_3Cl_6O_3$ 5.2 3
C_3IN 7.10 1
C_3N_{12} 33.2 1
$C_3O_2S_3$ 39.3 1
C_3HN 7.11 1
$C_3H_2F_2O_4$ 1.3 4

$C_3H_2N_2O_3$	32.8	1
$C_3H_2N_2O_3$, CH_4N_2O	60.4	3
$C_3H_2N_2O_3$, CH_4N_2S	60.5	3
$C_3H_2N_2S_2$	11.4	3
$C_3H_3AsN_2$	65.3	2
$C_3H_3BiO_6$	66.1	3
$C_3H_3BrO_2$	5.13	1
$C_3H_3FO_4$	1.8	3
$C_3H_3GdO_3$	81.13	2
$C_3H_3NOS_2$	41.7	1
$C_3H_3N_2O_6P_2^{3-}$, $1.5Ca^{2+}$, $6H_2O$	32.9	1
$C_3H_3N_3$	33.3+	1
$C_3H_3N_3O_3$	33.1+	3
$C_3H_3O_3^-$, Li^+	2.51	1
$C_3H_3O_3^-$, Na^+	2.52	1
$C_3H_3O_4^-$, K^+	2.12+	3
$(C_3H_3O_6Sc)_n$	81.14	2
$C_3H_3O_6Sn^-$, K^+	69.4	3
$C_3H_4Cl_2N_2O_2$	1.4	4
$C_3H_4Cl_2O_3S$	42.2	1
$C_3H_4N_2$	32.11	1
$C_3H_4N_2$	32.12	1
$C_3H_4N_2$	32.6	3
$C_3H_4N_2$	32.7+	3
$C_3H_4N_2$	32.1	4
$C_3H_4N_2O$	1.10	3
$C_3H_4N_2OS$	32.13	1
$C_3H_4N_3O_5^-$, K^+	12.4	4
$C_3H_4N_4NiOS_4$	83.11	2
$C_3H_4OS_3$	39.4	1
$C_3H_4O_2$	1.29+	1
$C_3H_4O_2S$	39.5	1
$C_3H_4O_3$	38.1	1
$C_3H_4O_4$	1.31	1
$C_3H_4O_5$	1.32	1
$C_3H_4O_6P^-$, $C_6H_{14}N^+$	46.3	1
$C_3H_5AlCl_4O$	68.1	4
C_3H_5NO	1.33	1
$C_3H_5NO_2$	40.1	4
$C_3H_5NO_4$	1.34	1
$C_3H_5NO_4$	1.35	1
$C_3H_5N_2^+$, $H_2O_4P^-$	32.2	4
$C_3H_5N_2S_2^+$, I^-	39.6	1
$C_3H_5O^+$, Cl_4Ga^-	12.5	4
$C_3H_5O^+$, Cl_6Sb^-	12.5	1
$C_3H_5OS_2^-$, K^+	11.18	1
$C_3H_5OS_2^-$, Na^+, $2H_2O$	11.5	3
$C_3H_6AgS_3^+$, ClO_4^-, H_2O	85.9	2
$C_3H_6AgS_3^+$, NO_3^-	85.10	2
$C_3H_6AgS_3^+$, NO_3^-, H_2O	85.11	2
C_3H_6AsN	65.4	2
$C_3H_6Cl_2HgS_3$	85.12	2
$(C_3H_6Cl_3Cu_3S_3)_n$	85.1	3
$C_3H_6Cl_3OZn^-$, $C_{52}H_{48}ClMoOP_4^+$, C_3H_6O	86.88	3
$C_3H_6Cl_5N_2P$	64.5	3
$(C_3H_6CoO_3)_n$	84.1	4
$C_3H_6N_2O_2$	1.11	3
$C_3H_6N_2S$	32.14	1
$C_3H_6N_4NiO_2S_4$	83.12	2
$C_3H_6N_4O_3$	8.39+	1
$C_3H_6N_6$	33.8	1
$C_3H_6N_6O_6$, $C_4H_8O_2S$	60.28	2
C_3H_6O, 2Br	60.29	2
C_3H_6O, $C_6H_6O_2$	60.88	2
$3C_3H_6O$, Na^+, I^-	60.30	2
$C_3H_6O_2$	1.36+	1
$C_3H_6O_3$	38.2+	1
$C_3H_6O_3S$	42.3	1
$C_3H_6O_7S_2^{2-}$, $2Na^+$, $4H_2O$	11.19	1
$C_3H_6S_3$	39.7+	1
$C_3H_6Se_3$	39.9	1
$(C_3H_7CdO_8P)_n$, $2nH_2O$	84.2	3
C_3H_7ClNO	1.12	3
$C_3H_7Cl_5NOSb$	66.2	2
C_3H_7NO	1.38	1
C_3H_7NO	10.18	1
C_3H_7NO, $C_{15}H_{13}I_2NO_4$	60.43	3
$2C_3H_7NO$, Na^+, ClO_4^-	60.31	2
$3C_3H_7NO$, Na^+, I^-	60.32	2
$4C_3H_7NO$, Li^+, Cl^-	60.33	2
$6C_3H_7NO$, $2K^+$, I^-, I_3^-	60.34	2
$C_3H_7NO_2$	1.39+	1
$C_3H_7NO_2$	48.12	1
$C_3H_7NO_2$	48.13	1
$C_3H_7NO_2$	48.14+	1
$C_3H_7NO_2$	48.5	4
$C_3H_7NO_2S$	48.16	1
$C_3H_7NO_2S$	48.6	4
$C_3H_7NO_3$	48.17	1
$C_3H_7NO_3$	48.7	4
$C_3H_7NO_3$, H_2O	48.8	4
$C_3H_7NO_5S$	48.18	1
$C_3H_7NO_5S$, H_2O	48.5	3
$C_3H_7NSe^+$, Br^-	42.4	1
$C_3H_7N_2^+$, Cl^-	32.15	1
$C_3H_7N_2O_2^+$, Cl^-	40.1	1
$C_3H_7N_5NiS_4$	83.6	4
$C_3H_7O_4P$	46.1	3
$C_3H_7O_4P$	46.2	3
$C_3H_7O_6P^{2-}$, $2Na^+$, $5H_2O$	46.5	1
$C_3H_7O_6P^{2-}$, $2Na^+$, $6H_2O$	46.3	4
$C_3H_8B_4Ga$	62.1	4
$C_3H_8NO_2^+$, Cl^-	48.19	1
$C_3H_8NO_2S^+$, Cl^-, H_2O	48.20	1
$C_3H_8NO_6P$	48.21	1
$C_3H_8NO_6P$	48.6	3
$C_3H_8NO_6P$, H_2O	48.7	3
$C_3H_8N_2O_3S$	8.7	4
$C_3H_8N_3O_2^+$, Br^-	1.41	1
$C_3H_8N_5^+$, Cl^-	32.16	1
$C_3H_8O_3$	5.14	1
$C_3H_9AlCl_3N$	68.2	2
$(C_3H_9AlS)_n$	68.1	3
$C_3H_9AsBr_3$	65.1	4
$C_3H_9AsCl_2$	65.2	4
$C_3H_9AsCl_3N$	3.9	3
$C_3H_9AsO_3$	65.3	4
$C_3H_9As_3Cl_3N_3$	65.5	2
$C_3H_9AuBr_3P$	86.1	2
$C_3H_9BBr_3N$	62.2	4
$C_3H_9BCl_3N$	62.17	2
$C_3H_9BCl_3N$	62.3	4
$C_3H_9BF_3N$	3.16	1
$C_3H_9BI_3N$	62.4	4
$C_3H_9B_3N_3$	62.18	2
$C_3H_9Br_2Sb$	66.4	2
$C_3H_9Cl_2Sb$	66.5	2
$C_3H_9Cl_2Sn^-$, $C_{15}H_{15}Mo_3S_4^+$	73.14	4
$C_3H_9Cl_4GeN$	69.3	4
$C_3H_9Cl_5OPSb$	66.6	2
$C_3H_9Cl_6Ge_3N_3$	69.4	4

C_3H_9FSn 69.2 2
$C_3H_9HgI_3S$ 85.13 2
$C_3H_9I_2Sb$ 66.3 2
C_3H_9In 68.3 2
C_3H_9N , ClI 60.35 2
C_3H_9N , I_2 60.36 2
C_3H_9N , O_2S 60.37 2
C_3H_9N , $6.5H_2O$ 61.1 3
C_3H_9N , $8H_2O$ 61.2 3
C_3H_9N , $10.25H_2O$ 61.2 2
C_3H_9NO 10.19 1
$C_3H_9NO_2S_2$ 4.6+ 1
$C_3H_9N_3O_2$ 3.6 4
$C_3H_9N_3O_2S$, $2H_2O$ 8.8 4
$C_3H_9N_6NiO_6$, H_2O 83.4 3
$C_3H_9N_6NiO_6^-$, $H_5N_2^+$, H_2O 83.13+ 2
$C_3H_9N_6O_6Zn^-$, K^+ 83.15 2
$C_3H_9OS^+$, BF_4^- 11.20 1
$C_3H_9OS^+$, ClO_4^- 11.21 1
$C_3H_9O_4ReSi$ 63.1 3
C_3H_9PS 64.6 3
$C_3H_9S^+$, I^- 11.22 1
$2C_3H_9S^+$, HgI_4^{2-} 11.23 1
$C_3H_9Se^+$, I^- 11.24 1
$C_3H_9Te^+$, $CH_3I_4Te^-$ 70.5 2
C_3H_9Tl 68.2 3
$C_3H_{10}AgN_7S_3$ 85.14 2
$(C_3H_{10}CdCl_2N_2)_n$ 83.7 4
$C_3H_{10}N^+$, Cl^- 3.17 1
$C_3H_{10}N^+$, I^- 3.10 3
$2C_3H_{10}N^+$, Cl_4Cu^{2-} 3.11 3
$2C_3H_{10}N^+$, Cl_4Mn^{2-} 3.12 3
$2C_3H_{10}N^+$, $Cl_{10}Cu_4^{2-}$ 3.13 3
$C_3H_{10}NO^+$, Cl^- 10.20+ 1
$C_3H_{10}OSn$ 69.3 2
$C_3H_{11}B_{10}Br_3$ 62.2 3
$C_3H_{11}NO_4Sn$ 69.5 4
$C_3H_{12}CdN_6O_4S_4$ 79.4 3
$C_3H_{12}Cl_2HgN_6S_3$ 79.12 2
$(C_3H_{12}CuN_6S_3^+)_n$, nCl^- 79.13 2
$C_3H_{12}GaN$ 68.4 2
$C_3H_{12}NO_9P_3$ 64.10 2
$C_3H_{12}N_2^{2+}$, $2Cl^-$ 3.18 1
$C_3H_{12}N_6O_3S_5Zn$, H_2O 79.14 2
$C_3H_{12}N_6O_4S_4Zn$ 79.15 2
$C_3H_{12}N_6Se_3^{2+}$, $2Br^-$, H_2O 8.12 3
$C_3H_{12}N_6Se_3^{2+}$, $2Cl^-$, H_2O 8.13 3
$(C_3H_{15}Ag_2Br_2N_9S_3)_n$ 79.16 2
$C_3H_{15}N_9NiS_3^{2+}$, $2NO_3^-$ 85.15 2
$C_3H_{15}N_9NiS_3^{2+}$, $2NO_3^-$, H_2O 85.16 2
$C_3H_{16}B_3N$ 62.19 2
$C_3H_{16}B_5BrSi$ 63.1 4
$C_3H_{18}CoN_5O_2^{2+}$, $2NO_3^-$ 82.2 2
$C_3H_{19}Na_2O_{12}P$ 67.3 4
$C_3H_{21}AlAsB_3$ 65.6 2
$C_3H_{21}AlB_3N$ 68.5+ 2
$C_3H_{21}B_9OS$ 62.5 4

C_4

$C_4As_4F_{12}$ 65.2 3
C_4BrN_3 7.1 4
C_4ClN_3 7.2 4
$C_4Cl_4N_4S_4$ 41.8 1
C_4Cl_8 20.2 1
$C_4Cl_{12}N_2W$ 83.8 4
$C_4Cl_{18}N_2W_2$ 83.5 3
$C_4CuO_8^{2-}$, $2H_4N^+$, $2H_2O$ 81.15 2
$C_4F_{12}P_4$ 64.11 2
C_4N_2 7.12 1
$(C_4N_2RbS_2^-)_n$, nRb^+ , nH_2O 67.4 3
$C_4N_3^-$, Ag^+ 7.13 1
$C_4N_3^-$, K^+ 7.14 1
$C_4N_3^-$, K^+ 7.3 4
$C_4N_3^-$, Na^+ 7.15 1
$C_4N_3^-$, H_4N^+ 7.16 1
$2C_4N_3^-$, Cu^{2+} 7.17 1
$C_4N_4NiS_4^{2-}$, $2C_{24}H_{20}As^+$ 65.47 2
$C_4NbO_{12}^{3-}$, $3H_4N^+$, H_2O 81.3 4
$C_4O_4^{2-}$, $2K^+$, H_2O 6.1 1
$C_4O_8Pd^{2-}$, $2K^+$, $4H_2O$ 81.17 2
$C_4O_8Pt^{2-}$, $2K^+$, $2H_2O$ 81.18 2
$C_4O_8Pt^{2-}$, $0.82Mg^{2+}$, $5.3H_2O$ 81.19 2
$C_4O_8Y^-$, $H_5O_2^+$, $2H_2O$ 81.4+ 4
$C_4O_{10}V^{3-}$, $3H_4N^+$, $2H_2O$ 81.9+ 4
$C_4H_2ClO_4^-$, K^+ 2.53 1
$C_4H_2Cl_4$ 5.4 4
$C_4H_2N_2O_3S^{2-}$, $2Na^+$, $6H_2O$ 44.1 3
$C_4H_2N_2O_4$ 43.1 1
$C_4H_2N_3O_4^-$, K^+ , $2H_2O$ 43.2 1
$C_4H_2N_3O_4^-$, Rb^+ 43.3 1
$2C_4H_2N_3O_4^-$, Sr^{2+} , $4H_2O$ 43.4 1
$C_4H_2O_3$ 1.42 1
$C_4H_2O_4$ 1.13 3
$C_4H_2O_4$ 20.1 3
$C_4H_2O_4$, $2H_2O$ 1.43 1
$C_4H_2O_4^{2-}$, $2K^+$, $2H_2O$ 2.15 3
$C_4H_2O_9Y^-$, H_4N^+ 81.20 2
$C_4H_3BrN_2O_2$ 50.1 1
$C_4H_3BrN_2O_2$ 44.2 3
$C_4H_3ClN_2O_2$ 44.3 3
$C_4H_3Cl_2N_3$ 44.1 1
$C_4H_3Cl_2N_3$ 44.2 1
$(C_4H_3ErO_9)_n$, $2nH_2O$ 81.4 3
$C_4H_3FN_2O$ 44.1 4
$C_4H_3FN_2O_2$, $C_4H_5N_3O$, H_2O 44.9 3
$C_4H_3FN_2O_2$, $C_5H_7N_3O$ 44.4 3
$C_4H_3FN_2O_2$, $C_7H_8N_4O$ 44.3 1
$C_4H_3N_2O_3^-$, H_4N^+ 43.5 1
$C_4H_3N_3O_4$, H_2O 44.4 1
$C_4D_3N_3O_4$, D_2O 43.6+ 1
$C_4H_3N_3O_5$ 43.8 1
$C_4H_3N_3O_5$, $3H_2O$ 43.9 1
$C_4H_3N_5O$ 40.2 4
$C_4H_3N_5O_2$, H_2O 44.5 1
$C_4H_3N_5O_2$, H_2O 44.6 3
$C_4H_3N_5S$ 41.3 4
$C_4H_3NbO_{11}^{2-}$, $2H^+$, $4H_2O$ 81.11 4
$C_4H_3O_4^-$, K^+ 2.55+ 1
$C_4H_3O_4^-$, Na^+ 2.16 3
$C_4H_3O_4^-$, Na^+ , $3H_2O$ 2.10 4
$C_4H_3O_4^-$, Rb^+ 2.17 3
$C_4H_3O_4^-$, $C_{16}H_{20}BrN_2^+$ 33.38 4
$2C_4H_3O_4^-$, $H_8CuO_4^{2+}$ 2.11 4
$2C_4H_3O_4^-$, $C_4H_4O_4$, $2K^+$ 2.18 3
$(C_4H_4AgN_2^+)_n$, nNO_3^- 83.16 2
$C_4H_4Ag_2N_2^{2+}$, $2NO_3^-$ 83.17 2
$C_4H_4ClNO_2$ 1.44 1
$C_4H_4ClN_3$ 33.1 4

$C_4H_4Cl_2O_3$	1.14	**3**
$C_4H_4Cl_4Ge_2$	69.4	**2**
$C_4H_4Cl_4O_2$	38.4	**1**
$C_4H_4Cl_{12}O_2Sb_2$	66.2	**4**
$C_4H_4CrO_{10}^-, K^+$	81.21	**2**
$(C_4H_4CuN_4O_6)_n$	83.6	**3**
$C_4H_4CuO_5$	81.12	**4**
$(C_4H_4CuO_6)_n, 3nH_2O$	81.13	**4**
$(C_4H_4CuO_6)_n, 3nH_2O$	81.14	**4**
$C_4H_4CuO_{10}^{2-}, 2Li^+, 4H_2O$	81.5	**3**
$C_4H_4Cu_2O_8, C_4H_8O_2$	81.22	**2**
$(C_4H_4Cu_2O_8)_n$	81.23	**2**
$C_4H_4F_6Si_3$	63.2	**4**
$C_4H_4Ge_2I_4$	69.5	**2**
$C_4H_4Mo_2O_{11}^{2-}, 2H_4N^+, 2.5H_2O$	81.24	**2**
$C_4H_4Mo_2O_{14}^{2-}, H_2BaO^{2+}, 2H_2O$	81.25	**2**
$C_4H_4N_2$	33.9	**1**
$C_4H_4N_2$	44.6	**1**
$C_4H_4N_2, C_2Br_4$	60.38	**2**
$C_4H_4N_2, C_2I_4$	60.39	**2**
$C_4H_4N_2O$	44.7	**3**
$C_4H_4N_2OS$	44.7	**1**
$C_4H_4N_2OSe$	44.8	**1**
$C_4H_4N_2O_2$	44.9+	**1**
$2C_4H_4N_2O_2, Cl_2Hg$	44.2	**4**
$C_4H_4N_2O_2S$	43.10	**1**
$C_4H_4N_2O_3$	43.11	**1**
$C_4H_4N_2O_3, 2H_2O$	43.12	**1**
$C_4H_4N_2O_4, H_2O$	43.13+	**1**
$C_4H_4N_2O_5$	43.16	**1**
$C_4H_4N_2O_5, 3H_2O$	43.17	**1**
$C_4H_4N_2S$	33.10	**1**
$C_4H_4N_2S_2$	11.25	**1**
$C_4H_4N_2S_2$	44.11	**1**
$C_4H_4N_2Se_2$	44.12	**1**
$C_4H_4N_3O_5^-, K^+$	12.6	**1**
$C_4H_4N_4$	7.18	**1**
$C_4H_4N_4$	7.19	**1**
$C_4H_4N_6, 0.5H_2O$	32.9	**3**
$C_4H_4N_6O, H_2O$	35.1+	**1**
$C_4H_4NiS_4$	85.2+	**3**
$C_4H_4NiS_4$	85.2	**4**
C_4H_4O	38.5+	**1**
$C_4H_4O_2$	38.7	**1**
$C_4H_4O_2S_2$	39.10	**1**
$C_4H_4O_3$	1.45+	**1**
$C_4H_4O_4$	1.49	**1**
$C_4H_4O_4$	1.47+	**1**
$C_4H_4O_4, 2C_4H_3O_4^-, 2K^+$	2.18	**3**
$C_4H_4O_5^{2-}, Ca^{2+}, 2H_2O$	2.59	**1**
$C_4H_4O_6, 2H_2O$	1.50	**1**
$C_4H_4O_6^{2-}, Ca^{2+}, 4H_2O$	2.60	**1**
$C_4H_4O_6^{2-}, K^+, Na^+, 4H_2O$	2.61+	**1**
$C_4H_4O_6^{2-}, 2K^+, 2H_2O$	2.63	**1**
$C_4H_4O_6^{2-}, Li^+, H_4N^+, H_2O$	2.64	**1**
$C_4H_4O_6^{2-}, Na^+, Rb^+$	2.65	**1**
$C_4H_4O_6^{2-}, 2Na^+, 2H_2O$	2.66	**1**
$C_4H_4O_6^{2-}, 2Rb^+, 2H_2O$	2.67	**1**
$C_4H_4O_6^{2-}, Sr^{2+}, 3H_2O$	2.68	**1**
$C_4H_4O_8Rh_2, 0.5H_2O$	81.26	**2**
$C_4H_4S_2$	39.11	**1**
$C_4H_4S_3$	39.12	**1**
$C_4H_4S_6$	39.13	**1**
$C_4H_5BrN_4$	44.13	**1**
$C_4H_5ClN_4$	44.14	**1**
$C_4H_5ClO_2$	1.51	**1**
$(C_4H_5CuINOS_2)_n$	85.3	**4**
$C_4H_5FO_5$	1.52	**1**
$C_4H_5NO_2$	1.53	**1**
$C_4H_5NO_3$	1.15	**3**
$C_4H_5N_3O$	44.15	**1**
$C_4H_5N_3O$	44.16	**1**
$C_4H_5N_3O, H_2O$	44.17	**1**
$C_4H_5N_3O, H_2O$	44.8	**3**
$C_4H_5N_3O, C_4H_3FN_2O_2, H_2O$	44.9	**3**
$C_4H_5N_3O, C_{10}H_{12}N_4O_5$	47.20	**3**
$C_4H_5N_3O, C_{13}H_9N, H_2O$	60.132	**2**
$C_4H_5N_3O_8$	1.16	**3**
$C_4H_5N_3S$	44.10	**3**
$C_4H_5O_2^-, 2C_4H_6O_2, Na^+$	60.8	**3**
$C_4H_5O_3^-, Na^+$	2.69	**1**
$C_4H_5O_4^-, Cs^+, H_2O$	2.12	**4**
$C_4H_5O_4^-, K^+$	2.13+	**4**
$C_4H_5O_4S_2Sb$	66.7	**2**
$C_4H_5O_6^-, K^+$	2.15	**4**
$C_4H_5O_6^-, H_4N^+$	2.70	**1**
C_4H_6	5.15	**1**
$C_4H_6BrNO_3$	1.17	**3**
$C_4H_6Br_2O_2$	38.8	**1**
$C_4H_6Br_2S_2$	39.14	**1**
$C_4H_6Br_4N_2Nb$	83.18	**2**
$C_4H_6Cl_2N_2Zn$	83.19	**2**
$C_4H_6Cl_2OS$	42.5	**1**
$C_4H_6Cl_2O_2$	38.9	**1**
$C_4H_6Cl_2O_2$	38.13	**1**
$C_4H_6Cl_2O_2$	38.10+	**1**
$C_4H_6Cl_2S_2$	39.15	**1**
$C_4H_6Cl_3NO$	32.17	**1**
$C_4H_6Cl_4Cu_2N_2$	83.20	**2**
$C_4H_6Cl_4N_2Sn$	69.6	**3**
$C_4H_6Cl_6Cu_3N_2$	83.21	**2**
$C_4H_6Cl_6Pt_2^{2-}, 2C_5H_{14}N^+$	72.3	**3**
$C_4H_6CuN_4O_6$	83.22	**2**
$C_4H_6CuN_6O_4^{2-}, 2K^+, 4H_2O$	83.23	**2**
$C_4H_6CuO_6$	81.27	**2**
$C_4H_6GeN_2$	69.6	**4**
$C_4H_6N_2O$	32.3	**4**
$C_4H_6N_2O_2$	33.11+	**1**
$C_4H_6N_2O_2$	44.11	**3**
$C_4H_6N_2O_2$	40.3	**4**
$C_4H_6N_2O_2, C_7H_9N_5$	44.55	**1**
$2C_4H_6N_2O_2, Cl_2Hg$	44.3	**4**
$C_4H_6N_2S$	41.4	**4**
$C_4H_6N_2S_2Sn$	69.7+	**3**
$C_4H_6N_2Si$	63.3	**4**
$C_4H_6N_2Sn$	69.7	**4**
$C_4H_6N_4NiO_4$	83.24+	**2**
$C_4H_6N_4O_3$	32.18	**1**
$C_4H_6N_4O_4Pt$	83.26	**2**
$C_4H_6O_2S_2$	39.16	**1**
$C_4H_6O_3S$	39.1	**4**
$C_4H_6O_4$	1.54	**1**
$C_4H_6O_4$	1.55	**1**
$C_4H_6O_4$	1.18	**3**
$C_4H_6O_4S$	1.56	**1**
$C_4H_6O_6$	1.60	**1**
$C_4H_6O_6$	1.57+	**1**
$C_4H_6O_6$	1.19	**3**
$C_4H_6O_6, H_2O$	1.61	**1**
$C_4H_6O_6, H_2O$	1.62+	**1**
$C_4H_7AsO_2S_2$	65.4	**4**
$(C_4H_7ErO_7)_n, nH_2O$	81.6	**3**

C_4H_7NO 20.3 1
$C_4H_7NO_2$ 32.11 3
$2C_4H_7NO_2$, Br^-, Na^+ 60.40 2
$2C_4H_7NO_2$, I^-, K^+ 60.41 2
$C_4H_7NO_3$ 48.23 1
$C_4H_7NO_3$ 48.8+ 3
$C_4H_7NO_4$ 48.24 1
$C_4H_7NO_4$ 48.25 1
$C_4H_7N_3O$ 48.26 1
$C_4H_7N_3S_2$ 32.12 3
$C_4H_7N_4O_4^-$, K^+ 2.71 1
$C_4H_7O^+$, Cl_6Sb^- 12.6 4
$C_4H_8BrS^+$, Br^- 39.1 3
$C_4H_8Br_2I_2S_2$ 39.2 3
$C_4H_8Br_2SSe$ 42.6 1
$C_4H_8Br_2Se$ 39.17 1
$C_4H_8CaO_7$ 2.16 4
$C_4H_8ClHgS^+$, Cl^- 85.17 2
$C_4H_8Cl_2CuN_2O_2$ 83.7 3
$C_4H_8Cl_2N_8Zn$ 83.8 3
$C_4H_8Cl_3O_2Pt^-$, $C_{24}H_{20}P^+$ 72.4 3
$C_4H_8Cl_3O_4P$ 64.7 3
$C_4H_8Cl_4Pd_2$ 72.4 2
$C_4H_8Cl_4Se_2$ 39.18 1
$C_4H_8Cl_8Si_4$ 63.1 2
$C_4H_8CoN_2O_8$, $2H_2O$ 81.15 4
$C_4H_8CuN_2O_4$, H_2O 82.4+ 2
$C_4H_8CuN_2O_4$, $2H_2O$ 82.3 2
$C_4H_8CuN_4S_2$ 76.1 4
$C_4H_8HgN_2O_2$ 83.9 3
$C_4H_8HgN_6S_4$ 79.17 2
$C_4H_8I_2STe$ 70.2 3
$C_4H_8I_2S_3$ 11.26+ 1
$C_4H_8NOS^+$, Cl^- 48.27+ 1
$C_4H_8NO_4^+$, Cl^- 48.29 1
$C_4H_8N_2NiO_2S_2$ 84.7 2
$C_4H_8N_2NiO_2S_2$ 85.18 2
$C_4H_8N_2NiO_2S_2$ 82.3 4
$C_4H_8N_2NiO_4$, $2H_2O$ 82.6 2
$C_4H_8N_2O$ 20.4 1
$C_4H_8N_2O_2$ 1.64 1
$C_4H_8N_2O_2$ 9.9 1
$C_4H_8N_2O_2$ 10.22 1
$C_4H_8N_2O_3$ 48.30 1
$C_4H_8N_2O_3$ 48.10 3
$C_4H_8N_2O_3$, Li^+, Br^- 48.9 4
$C_4H_8N_2O_3$, H_2O 48.31 1
$C_4H_8N_2O_3$, H_2O 48.10 4
$C_4D_8N_2O_3$ 48.11 3
$2C_4H_8N_2O_3$, Ca^{2+}, $2Cl^-$ 48.11 4
$C_4H_8N_2O_4Pt$ 82.7 2
$C_4H_8N_2O_8Zn$ 81.16 4
$C_4H_8N_2S$ 8.41 1
$C_4H_8N_2S$ 8.42 1
$C_4H_8N_3O^+$, $C_{10}H_{13}N_2O^+$, O_4S^{2-}, H_2O 35.28 1
$C_4H_8N_3O_5P^{2-}$, $2Na^+$, $4.5H_2O$ 64.12 2
$C_4H_8N_4$ 8.14 3
$C_4H_8N_4O_6$ 9.1 3
$C_4H_8N_6NiS_4$ 79.18 2
$C_4H_8N_6NiS_4$ 85.19 2
$C_4H_8N_6PdS_4$ 79.19 2
$C_4H_8N_8O_8$ 34.1+ 1
$C_4H_8N_8O_8$ 34.1 3
C_4H_8O, Br_2Hg 38.1 4
$8C_4H_8O$, $7.33H_2S$, $136H_2O$ 61.3 2
$2C_4H_8OS$, $C_6H_{10}N_2O_8$ 60.13 4
C_4H_8OSe, ClI 60.42 2
C_4H_8OSe, I_2 60.43 2
$C_4H_8O_2$ 1.65 1
$C_4H_8O_2$, Br_2 60.44 2
$C_4H_8O_2$, Cl_2Ge 60.9 3
$C_4H_8O_2$, Cl_2Hg 60.45 2
$C_4H_8O_2$, Li^+, Cl^- 60.46 2
$C_4H_8O_2$, Li^+, Cl^-, H_2O 60.47 2
$C_4H_8O_2$, N_2O_4 60.48 2
$C_4H_8O_2$, H_2O_4S 60.49 2
$C_4H_8O_2$, CHI_3 60.2 3
$C_4H_8O_2$, $C_2Br_2O_2$ 60.17 2
$C_4H_8O_2$, $C_2Cl_2O_2$ 60.18 2
$C_4H_8O_2$, C_2I_2 60.50 2
$C_4H_8O_2$, $C_4H_4Cu_2O_8$ 81.22 2
$C_4H_8O_2$, $2C_{10}H_{14}O_5V$ 60.34 3
$2C_4H_8O_2$, Br_2Hg 60.10 3
$3C_4H_8O_2$, Ag^+, ClO_4^- 60.51 2
$C_4H_8O_2S$ 11.28 1
$C_4H_8O_2S$, $C_3H_6N_6O_6$ 60.28 2
$C_4H_8O_2S_2$ 39.19 1
$C_4H_8O_5$ 38.14 1
$C_4H_8O_8$ 20.5 1
$(C_4H_8O_{10}Th)_n$, nH_2O 81.28 2
$C_4H_8O_{12}P_4^{4-}$, $2Ca^{2+}$, $10H_2O$ 64.3 4
$C_4H_8S_2$ 39.20 1
$C_4H_8S_2$, $2I_2$ 60.52 2
$C_4H_8S_2$, CHI_3 60.53 2
$C_4H_8S_2$, C_2I_2 60.19 2
$2C_4H_8S_2$, I_3Sb 60.54 2
C_4H_8Se, I_2 60.55+ 2
$C_4H_8Se_2$ 39.21 1
$C_4H_8Se_2$, $2I_2$ 60.57 2
$C_4H_8Se_2$, $2CHI_3$ 60.4 2
$C_4H_8Se_2$, C_2I_2 60.20 2
$C_4H_8Se_2$, C_2I_4 60.58 2
$C_4H_9AgNO^+$, I^- 83.27 2
C_4H_9Cl 5.16 1
C_4H_9GeN 69.6 2
C_4H_9NO, $3CH_6N_3^+$, $3Cl^-$ 60.16 2
C_4H_9NO, C_6H_5I 60.59 2
$C_4H_9NO_2$ 1.66+ 1
$C_4H_9NO_2$ 48.32 1
$C_4H_9NO_3$ 48.33 1
$C_4H_9NO_3S$ 48.34 1
C_4H_9NSSn 69.9 3
C_4H_9NSn 69.8 2
$C_4H_9N_2O_3^+$, Cl^-, H_2O 48.35 1
$C_4H_9N_3O_2$, H_2O 48.36+ 1
$C_4H_9O_2P$ 64.8 3
$C_4H_{10}Ag_2N_2O_4^{2+}$, $2NO_3^-$ 82.4+ 4
$C_4H_{10}Br_4O_2Ti^-$, $C_8H_{20}Br_2O_4Ti^+$ 84.27 2
$C_4H_{10}CdCl_2N_2O_2$ 83.28 2
$C_4H_{10}CdCl_2N_6O_4$ 84.8 2
$C_4H_{10}CdN_4O_4S_2$ 79.20 2
$(C_4H_{10}CdO_6)_n$ 81.17 4
$C_4H_{10}ClCuS_2$ 85.20 2
$C_4H_{10}ClHgO_3P$ 86.1 3
$C_4H_{10}ClHgS^+$, Cl^-, Cl_2Hg 85.21 2
$C_4H_{10}Cl_2N_4Pd$ 71.1 4
$C_4H_{10}Cl_2N_6O_4Zn$ 84.9 2
$C_4H_{10}Cl_3NPt$ 72.5 3
$C_4H_{10}Cl_3NPt$ 72.6 3
$(C_4H_{10}Cl_3NTi)_n$ 83.9 4
$C_4H_{10}Cl_6O_3S_2W_2$ 84.2 4

$C_4H_{10}Cl_8O_2Sb_2$ 66.8 **2**
$C_4H_{10}CuN_6O_4^{2+}$, $2Cl^-$ 84.10 **2**
$C_4H_{10}Fe_2N_4O_4S_2$ 85.22 **2**
$C_4H_{10}HgS_2$ 85.23 **2**
$4C_4H_{10}LiO^+$, $C_{12}H_{14}Cr_2^{4-}$ 71.12 **4**
$2C_4H_{10}NO^+$, $2C_{12}H_4N_4^-$, $C_{12}H_4N_4$ 40.4 **4**
$C_4H_{10}NO_6P$ 48.12 **3**
$C_4H_{10}NO_6P$ 48.12 **4**
$C_4H_{10}N_2$, $6H_2O$ 33.13 **1**
$0.5C_4H_{10}N_2$, Ag^+, I^- 60.60 **2**
$C_4H_{10}N_4NiS_2$ 85.4 **3**
$C_4H_{10}N_8NiS_4$ 83.30 **2**
$C_4H_{10}NiO_2S_2$ 85.24 **2**
$C_4H_{10}O$, $CHBrCl_2$ 60.61 **2**
$4C_4H_{10}O$, $Br_6Mg_4O_7$ 67.17 **2**
$C_4H_{10}O_4$ 45.1+ **1**
$2C_4H_{10}O_4P^-$, Ba^{2+} 46.6 **1**
$C_4H_{10}O_6S_3$ 11.29 **1**
$C_4H_{10}O_6Zn$ 81.29 **2**
$C_4H_{10}O_8Zn$ 81.30 **2**
$C_4H_{11}Cl_2NPt$ 72.5 **2**

$C_4H_{11}N$, $8.67H_2O$ 61.4 **2**
$C_4H_{11}N$, $9.75H_2O$ 61.5 **2**
$(C_4H_{11}NNdO_7^+)_n$, nCl^- 81.18 **4**
$C_4H_{11}NO_3S_2$ 4.8 **1**
$C_4H_{11}N_2O_2^+$, Br^- 1.68 **1**
$C_4H_{11}N_2O_2^+$, Cl^- 48.13+ **4**
$C_4H_{12}As^+$, Br^- 65.7 **2**
$C_4H_{12}As_2S_2$ 65.8 **2**
$(C_4H_{12}BLi)_n$ 62.3 **3**
$C_4H_{12}BN$ 62.4 **3**
$C_4H_{12}B_2Cl_4N_2$ 62.20 **2**
$C_4H_{12}B_2F_4N_2$ 62.21 **2**
$C_4H_{12}B_2N_8NiO_8$, $4H_2O$ 83.10 **4**
$C_4H_{12}B_4$ 62.22 **2**
$C_4H_{12}B_6$ 62.23 **2**
$C_4H_{12}B_{10}Br_4$ 62.24 **2**
$C_4H_{12}Br_4Pd_2S_2$ 85.26 **2**
$C_4H_{12}Br_4Pt_2S_2$ 85.27 **2**
$C_4H_{12}Cl_2CuN_8O_2$ 83.10 **3**
$C_4H_{12}Cl_2CuO_2S_2$ 84.3 **3**
$C_4H_{12}Cl_2N_2Pd$ 76.2 **4**
$C_4H_{12}Cl_2O_2PdS_2$ 85.28+ **2**
$C_4H_{12}Cl_3N_2Pt^+$, Cl^-, $0.5H_2O$ 72.3 **4**
$C_4H_{12}Cl_4O_6Re_2$ 81.7 **3**
$C_4H_{12}CuN_2O_4$ 81.31+ **2**
$C_4H_{12}Cu_2O_{10}$, $2H_2O$ 81.33 **2**
$C_4H_{12}F_4N_4P_4$ 64.9 **3**
$C_4H_{12}Ge_4S_6$ 69.10 **3**
$C_4H_{12}Li_4$ 67.5 **3**
$C_4H_{12}N^+$, $Ag_2I_3^-$ 3.19 **1**
$C_4H_{12}N^+$, $AlO_8Si_2^-$, H_2O 3.14 **3**
$C_4H_{12}N^+$, $AlO_{12}Si_5^-$ 3.15 **3**
$C_4H_{12}N^+$, Br_3Hg^- 3.20 **1**
$C_4H_{12}N^+$, Br_3Ni^- 3.21 **1**
$C_4H_{12}N^+$, $CeCl_6^-$ 3.22 **1**
$C_4H_{12}N^+$, Cl^-, $5Cl_2OSe$ 3.23 **1**
$C_4H_{12}N^+$, ClO_4^- 3.24 **1**
$C_4H_{12}N^+$, Cl_2I^- 3.25 **1**
$C_4H_{12}N^+$, Cl_3Mn^- 3.26 **1**
$C_4H_{12}N^+$, Cl_3Ni^- 3.27 **1**
$C_4H_{12}N^+$, Cs^+, $H_{13}B_{11}^{2-}$ 3.28 **1**
$C_4H_{12}N^+$, F^-, $4H_2O$ 61.6 **2**
$C_4H_{12}N^+$, HgI_3^- 3.29 **1**
$C_4H_{12}N^+$, HCl_2^- 3.16+ **3**
$C_4H_{12}N^+$, HO^-, $5H_2O$ 61.7 **2**
$C_4H_{12}N^+$, $C_4H_8B_3CrO_4^-$ 3.18 **3**
$C_4H_{12}N^+$, $C_4H_{16}B_{18}Br_6Co^-$ 3.31 **1**
$C_4H_{12}N^+$, $C_{14}H_{10}MnO_3^-$ 71.14 **3**
$C_4H_{12}N^+$, $C_{15}H_{30}N_3O_2S_6U^-$ 80.31 **2**
$C_4H_{12}N^+$, $C_{18}H_{13}O_4Si^-$ 3.32 **1**
$C_4H_{12}N^+$, $C_{18}H_{19}O_6Pb^-$ 69.30 **4**
$2C_4H_{12}N^+$, $13Ag^+$, $15I^-$ 3.19 **3**
$2C_4H_{12}N^+$, Cl_4Co^{2-} 3.33 **1**
$2C_4H_{12}N^+$, Cl_4Cu^{2-} 3.34 **1**
$2C_4H_{12}N^+$, Cl_4Ni^{2-} 3.35 **1**
$2C_4H_{12}N^+$, Cl_4Zn^{2-} 3.36 **1**
$2C_4H_{12}N^+$, $Cl_{18}Nb_6^{2-}$ 3.7 **4**
$2C_4H_{12}N^+$, O_4S^{2-}, $4H_2O$ 61.8 **2**
$2C_4H_{12}N^+$, $H_6B_6^{2-}$ 3.37 **1**
$2C_4H_{12}N^+$, $H_7B_{10}Cl_3^{2-}$ 3.8 **4**
$2C_4H_{12}N^+$, $H_{24}B_{20}Ni^{2-}$ 3.9 **4**
$2C_4H_{12}N^+$, $C_4N_4Pd^{2-}$ 3.10 **4**
$2C_4H_{12}N^+$, $C_6H_4Cu_2N_2O_8S_2^{2-}$ 3.38 **1**
$2C_4H_{12}N^+$, $C_8N_4NiS_4^{2-}$ 85.30 **2**
$2C_4H_{12}N^+$, $C_{10}H_{12}Cu_2N_2O_8S_2^{2-}$ 3.39 **1**
$2C_4H_{12}N^+$, $C_{12}N_6S_6V^{2-}$ 3.40 **1**
$2C_4H_{12}N^+$, $C_{17}Fe_6O_{16}^{2-}$ 71.35 **4**
$2C_4H_{12}N^+$, $C_{30}O_{30}Rh_{12}^{2-}$ 3.21 **3**
$3C_4H_{12}N^+$, $C_{16}O_{16}Rh_7^{3-}$ 3.41 **1**
$4C_4H_{12}N^+$, $Cl_{10}Cu_4O^{4-}$ 3.22 **3**
$C_4H_{12}N_2^{2+}$, $2Cl^-$, H_2O 33.14 **1**
$C_4H_{12}N_2^{2+}$, $2Cl_2I^-$ 60.62 **2**
$C_4H_{12}N_2^{2+}$, MoS_4^{2-} 33.2 **4**
$C_4H_{12}N_2NiO_6$ 83.31 **2**
$C_4H_{12}N_2O_2S$ 4.9 **1**
$C_4H_{12}N_2O_8PdS_2$ 85.31 **2**
$C_4H_{12}N_4$ 33.5 **3**
$C_4H_{12}N_4NiO_6$ 83.29 **2**
$C_4H_{12}NiO_4P_2S_4$ 85.32 **2**
$C_4H_{12}NiP_2S_4$ 85.33 **2**
$C_4H_{12}O_2Si$ 63.2 **2**
$C_4H_{12}O_4P_2S_4Te$ 70.6 **2**
$C_4H_{12}O_4Ti$ 84.11 **2**
$C_4H_{12}P_2S_2$ 64.13 **2**
$C_4H_{12}P_2S_2$ 64.10 **3**
$C_4H_{12}S_6Si_4$ 63.3 **2**
$C_4H_{12}S_6Sn_4$ 69.9 **2**
$C_4H_{12}Sb^+$, $C_{12}H_{36}AlO_4Si_4^-$ 68.21 **2**
$C_4H_{13}B_7$ 62.25 **2**
$C_4H_{13}Cl_3CrN_3$ 76.5 **3**
$C_4H_{13}CoN_6O_6$ 76.2 **2**
$C_4H_{13}MoN_3O_3$ 76.3 **2**
$C_4H_{14}B_8$ 62.5 **3**
$C_4H_{14}B_{10}Br_2$ 62.26+ **2**
$C_4H_{14}Cl_2MnN_2O_6$ 82.8 **2**
$C_4H_{14}CoO_8$ 81.34 **2**
$C_4H_{14}ErO_{10}^-$, $C_8H_{12}ErO_{12}^+$ 81.25 **4**
$C_4H_{14}N_2^{2+}$, $2Cl^-$ 3.42 **1**
$C_4H_{14}N_2O_2S_2^{2+}$, $2Cl^-$, H_2O 4.10 **1**
$C_4H_{14}N_4O_4S_6Te$ 70.7 **2**
$(C_4H_{14}N_4O_4Zn)_n$ 81.35 **2**
$C_4H_{14}N_{10}Ni^{2+}$, $2Cl^-$, $2H_2O$ 83.32 **2**
$C_4H_{14}NiO_8$ 81.36 **2**
$C_4H_{14}NiO_8$ 81.8 **3**
$C_4H_{15}B_7$ 62.6 **4**
$C_4H_{15}B_9$ 62.28 **2**
$C_4H_{16}AlB_9$ 62.7 **4**
$C_4H_{16}BN_2^+$, Cl^- 62.29 **2**
$C_4H_{16}B_2N_2$ 62.30 **2**
$C_4H_{16}B_{18}Br_6Co^-$, $C_4H_{12}N^+$ 3.31 **1**

$C_4H_{16}Br_2CoN_4^+, H_5O_2^+, 2Br^-$	76.4	2
$C_4H_{16}CdN_8S_4^{2+}, C_4N_4S_4Zn^{2-}$	79.21	2
$C_4H_{16}ClCoN_5O^+, ClO_4^-$	76.6	3
$C_4H_{16}ClN_4Ni^+, Cl^-$	76.6	2
$C_4H_{16}Cl_2CoN_4^+, Cl^-$	76.7	2
$C_4H_{16}Cl_2CoN_4^+, Cl^-, H_2O$	76.9+	2
$C_4H_{16}Cl_2CoN_4^+, Cl^-, H_2O$	76.3	4
$C_4H_{16}Cl_2CoN_4^+, 2Cl^-, H^+, 2H_2O$	76.8	2
$C_4H_{16}Cl_2CoN_4^+, 2Cl^-, H^+, 2H_2O$	76.7	3
$C_4H_{16}Cl_2CoN_4^+, NO_3^-$	76.11	2
$2C_4H_{16}Cl_2CoN_4^+, O_6S_6^{2-}, H_2O$	76.12	2
$C_4H_{16}Cl_2CoN_8S_4$	79.5	3
$C_4H_{16}Cl_2CoO_4$	84.12	2
$C_4H_{16}Cl_2CrN_4^+, 2Cl^-, H_5O_2^+$	76.13	2
$C_4H_{16}Cl_2N_4Pd$	76.14	2
$C_4H_{16}Cl_2N_4Pt^{2+}, 2Cl^-$	76.8	3
$C_4H_{16}Cl_2N_8NiS_4$	79.23	2
$C_4H_{16}CoN_6O_4^+, NO_3^-$	76.9	3
$C_4H_{16}CoN_6O_4^+, NO_3^-$	76.10	3
$C_4H_{16}CoN_8S_4^{2+}, 2NO_3^-$	79.6	3
$C_4H_{16}CoN_8S_4^{2+}, 2NO_3^-, H_2O$	79.7	3
$C_4H_{16}CoN_{10}^+, NO_3^-$	76.24	2
$C_4H_{16}CuN_4^{2+}, 2BF_4^-$	76.15	2
$C_4H_{16}CuN_4^{2+}, 2ClO_4^-$	76.16	2
$C_4H_{16}CuN_4^{2+}, 2NO_3^-$	76.17	2
$C_4H_{16}CuN_4^{2+}, O_3SSe^{2-}$	76.4	4
$C_4H_{16}CuN_4^{2+}, O_3S_2^{2-}$	76.5	4
$C_4H_{16}CuN_4^{2+}, CNS^-, ClO_4^-$	76.6	4
$C_4H_{16}CuN_4^{2+}, 2CNS^-$	76.18	2
$C_4H_{16}CuN_4^{2+}, C_4HgN_4S_4^{2-}$	76.19	2
$C_4H_{16}HgN_8S_4^{2+}, C_4CoN_4S_4^{2-}$	79.24	2
$C_4H_{16}N_3^{3+}, Cl^-, Cl_4Cu^{2-}$	3.11	4
$C_4H_{16}N_4Ni^{2+}, 2AgBr_2^-$	76.11	3
$C_4H_{16}N_4Ni^{2+}, 2AgI_2^-$	76.12	3
$nC_4H_{16}N_4Ni^{2+}, (AgBr_4^{2-})_n$	76.20	2
$C_4H_{16}N_4O_2Re^+, Cl^-$	76.13	3
$C_4H_{16}N_4Pd^{2+}, C_2H_8N_2O_6PdS_4^{2-}$	76.14	3
$C_4H_{16}N_5NiO_2^+, BF_4^-$	76.21	2
$C_4H_{16}N_5NiO_2^+, ClO_4^-$	76.22	2
$C_4H_{16}N_5NiO_2^+, I^-$	76.7	4
$C_4H_{16}N_6NiO_4$	76.15	3
$C_4H_{16}N_8PbS_4^{2+}, 2C_6H_2N_3O_7^-$	69.8	4
$C_4H_{16}N_8PdS_4^{2+}, 2Cl^-$	79.25	2
$C_4H_{16}N_8PdS_4^{2+}, 2Cl^-$	79.9	3
$C_4H_{16}N_8S_4, Tl^+, ClO_4^-$	68.7	2
$C_4H_{16}N_8S_4, Tl^+, NO_3^-$	68.8	2
$C_4H_{16}N_8S_4Te^{2+}, 2Cl^-$	70.8	2
$C_4H_{16}N_8S_4Te^{2+}, 2Cl^-, 2H_2O$	70.9	2
$C_4H_{16}N_8Se_4Te^{2+}, 2Cl^-$	70.3	4
$C_4H_{16}N_9Ru^+, F_6P^-$	76.16	3
$C_4H_{17}B_7$	62.31	2
$C_4H_{17}N_4O_2Re^{2+}, 2ClO_4^-$	76.25	2
$C_4H_{18}AlB_9$	62.8	4
$C_4H_{18}B_2P_2$	62.32	2
$C_4H_{18}B_{14}Co^-, C_8H_{20}N^+$	62.9	4
$C_4H_{18}ClCuN_4O^+, Cl^-$	76.26	2
$C_4H_{18}Co_2N_5O_2^{4+}, 4NO_3^-, H_2O$	76.17	3
$nC_4H_{18}CuN_4O^{2+}, (C_4Cu_2N_4^{2-})_n$	76.8	4
$C_4H_{18}CuO_{12}S_2$	84.4	3
$C_4H_{18}FeN_2O_8^{2+}, H_{12}FeO_6^{2+}, 2O_4S^{2-}$	82.9	2
$C_4H_{18}N_6Pt^{2+}, 2Cl^-, H_2O$	83.33	2
$C_4H_{19}Co_2N_5O_2^{4+}, 4NO_3^-, 2H_2O$	76.19	3
$C_4H_{20}N_4Pt^{2+}, 2Cl^-$	83.11	3
$C_4H_{20}N_4Pt^{2+}, Cl_4Pt^{2-}$	83.12	3
$C_4H_{22}B_{18}Co^-, Cs^+$	62.33	2
$C_4H_{22}B_{18}Cu^-, C_{19}H_{18}P^+$	64.56	2
$C_4H_{22}B_{18}Cu^{2-}, 2C_8H_{20}N^+$	62.34	2
$C_4H_{22}B_{18}Ni$	62.8	3
$C_4H_{22}B_{18}Ni^{2-}, C_8H_{18}N_2^{2+}$	37.5	3
$C_4H_{22}B_{20}$	62.35	2
$C_4H_{22}B_{20}N_2, C_2H_3N$	62.36	2
$C_4H_{22}CuN_2O_6^{2+}, O_4S^{2-}$	84.13	2
$C_4H_{22}Cu_2N_4O_2^{2+}, O_4S^{2-}, H_2O$	83.34	2
$C_4H_{24}B_8N_2$	62.37	2
$C_4H_{24}B_{10}S_2$	62.38	2
$C_4H_{26}B_{18}FeP_2$	62.39	2
$C_4H_{26}N_8NiO_3S_6^{2+}, O_3S_2^{2-}, H_2O$	79.26	2
$C_4H_{46}N_{14}O_2Ru_3^{6+}, 6Cl^-$	76.9	4

C_5

$C_5Cl_3N_2O_4^-, K^+$	20.1	4
$C_5F_{15}P_5$	64.14	2
$C_5O_5^{2-}, 2H_4N^+$	6.2	1
C_5HCl_4NO	33.3	4
$C_5HO_5^-, Rb^+$	6.3	1
$C_5HO_5^-, H_4N^+$	6.4	1
$C_5H_2FN_2O_4^-, Rb^+, H_2O$	44.19	1
$C_5H_3Br_2NO, Cl_2Hg$	84.14	2
$C_5H_3N_2O_4^-, H_4N^+, H_2O$	44.4	4
$C_5H_3N_3$	7.4	4
$C_5H_3N_4O_2^-, Na^+, 4H_2O$	44.12	3
$C_5H_3N_7, H_2O$	44.20	1
$C_5H_4BrN_2O_2, C_6H_7N_5$	44.21	1
$C_5H_4BrN_2O_3^-, C_5H_5BrN_2O_3, K^+$	44.5	4
$C_5H_4BrN_2O_3^-, C_5H_5BrN_2O_3, Rb^+$	44.6	4
C_5H_4ClNO	33.6	3
C_5H_4ClNO, C_5H_5NO	60.5	4
$(C_5H_4CuN_3)_n$	83.11	4
$C_5H_4N_2O_3$	33.15	1
$C_5H_4N_2O_3$	33.4	4
$2C_5H_4N_2O_3, H_4Cl_2CuO_2$	33.5	4
$C_5H_4N_2O_4, CH_4N_2O$	60.2	4
$C_5H_4N_4$	44.22+	1
$C_5H_4N_4O_2, 2H_2O$	44.7	4
$C_5H_4N_4O_3$	44.24	1
$C_5H_4N_4S, H_2O$	44.25+	1
$C_5H_4O_2S$	39.22	1
$C_5H_4O_2S$	39.23+	1
$C_5H_4O_2Se$	39.25	1
$C_5H_4O_2Te$	70.4	4
$C_5H_4O_3$	38.15	1
$C_5H_4S_2Se$	42.1	4
$C_5H_4S_3$	39.3	3
$C_5H_4Se_3$	39.2	4
$C_5H_5BrN_2O$	32.4	4
$C_5H_5BrN_2O_2, C_7H_8BrN_5$	44.13	3
$C_5H_5BrN_2O_2, C_7H_9N_5$	44.56	1
$C_5H_5BrN_2O_2, C_7H_9N_5$	44.14	3
$C_5H_5BrN_2O_3, C_5H_4BrN_2O_3^-, K^+$	44.15	3
$C_5H_5BrN_2O_3, C_5H_4BrN_2O_3^-, Rb^+$	44.16	3
C_5H_5ClIN	60.63	2
$C_5H_5CrNO_5$	83.35	2
$C_5H_5FN_2O_2, C_7H_9N_5$	44.57	1
$C_5H_5FN_2O_2, C_7H_9N_5$	44.58	1
$C_5H_5FN_2O_2, C_7H_9N_5$	44.17	3
$2C_5H_5IN_2O_2, C_7H_9N_5$	44.18	3
$2C_5H_5IN_2O_2, C_7H_{10}N_6$	44.19	3
$(C_5H_5In)_n$	68.9	2

C_5H_5N, Li^+, Cl^-	60.64	**2**
C_5H_5N, ClN	60.65	**2**
C_5H_5N, $C_6H_{14}O_6$	60.23	**3**
$2C_5H_5N$, Li^+, Cl^-, H_2O	60.66	**2**
C_5H_5NO	33.16	**1**
C_5H_5NO	10.2	**3**
C_5H_5NO, $C_2HCl_3O_2$	60.4	**4**
C_5H_5NO, C_5H_4ClNO	60.5	**4**
$C_5H_5NO_2$	38.16	**1**
C_5H_5NS	33.17	**1**
$C_5H_5N_3O$	33.18	**1**
$C_5H_5N_3O$	33.6+	**4**
$C_5H_5N_3O_2$	33.19	**1**
$C_5H_5N_4O^+$, Cl^-, H_2O	44.28	**1**
$C_5H_5N_5O$, H_2O	44.8	**4**
$C_5H_5N_5S$	44.20	**3**
C_5H_6	20.6	**1**
$C_5H_6BrN_3O$	32.13	**3**
$C_5H_6BrN_3O$, $C_7H_9N_5O$	44.59	**1**
$C_5H_6ClNO_3S$	40.5	**4**
$C_5H_6Cl_3N_5OZn$	83.14	**3**
$C_5H_6Cl_3N_5Zn$	83.15	**3**
$(C_5H_6Cl_4N_2Sn)_n$	69.10	**2**
$(C_5H_6CuO_8)_n$	84.15	**2**
$(C_5H_6CuO_8)_n$	84.17	**2**
$C_5H_6FN_3O$, $C_7H_9N_5O$	44.60	**1**
$C_5H_6FeN_4O_3$	83.36	**2**
$C_5H_6MnO_8$	84.16	**2**
$C_5H_6N^+$, AsF_4O^-	33.20	**1**
$C_5H_6N^+$, AsF_6^-	33.21	**1**
$C_5H_6N^+$, $BrCl_2^-$	33.7	**3**
$C_5H_6N^+$, Cl^-	33.22	**1**
$C_5H_6N^+$, Cl_4Sb^-	33.8	**3**
$C_5H_6N^+$, F_6P^-	33.24	**1**
$C_5H_6N^+$, F_6Sb^-	33.25	**1**
$C_5H_6N^+$, I_4Tl^-, C_5H_5N	33.9	**3**
$C_5H_6N^+$, NO_3^-	33.26	**1**
$C_5H_6N^+$, HBr_4Re^-	33.27+	**1**
$C_5H_6N^+$, $2CH_4N_2S$, Br^-	60.6	**4**
$C_5H_6N^+$, $C_4H_6CrN_6S_4^-$	33.29	**1**
$C_5H_6N^+$, $C_{24}H_{16}CrN_4O_4^-$	83.188	**2**
$C_5H_6N^+$, $C_{32}H_{18}CrN_6O_8^-$, $0.5C_5H_5N$	83.206	**2**
$2C_5H_6N^+$, $2H^+$, $Cl_8Re_2^{4-}$	33.30	**1**
$2C_5H_6N^+$, $H_2Mo_2O_{14}^{2-}$	33.8	**4**
$2C_5H_6N^+$, $H_4Mo_2O_{13}^{2-}$	33.9	**4**
$2C_5H_6N^+$, $C_{18}H_{12}O_6Si^{2-}$	63.10	**3**
$5C_5H_6N^+$, $Br_9Sb_2^{3-}$, $2Br^-$	33.10	**3**
$6C_5H_6N^+$, Br_6Sb^{3-}, $3Br_6Sb^-$	33.11	**3**
$C_5H_6NO^+$, Cl^-	10.24	**1**
$C_5H_6NO^+$, Cl^-	33.33	**1**
$C_5H_6NO^+$, Cl^-, H_2O	33.12	**3**
$C_5H_6NO_2^+$, Cl^-	33.13	**3**
$C_5H_6N_2OS$, $C_6H_7N_5$	44.21	**3**
$C_5H_6N_2O_2$	44.30	**1**
$C_5H_6N_2O_2$	44.31	**1**
$C_5H_6N_2O_2$, H_2O	44.32	**1**
$C_5H_6N_2O_2$, $C_6H_4O_2$	60.7	**4**
$C_5H_6N_2O_2$, $C_7H_9BrN_6$	44.27	**3**
$C_5H_6N_2O_3S$	41.9	**1**
$C_5H_6N_2S$	7.20	**1**
$C_5H_6N_4O_3$	44.9	**4**
$2C_5H_6N_5^+$, $2Cl^-$, H_2O	44.33+	**1**
$C_5H_6N_5O^+$, Cl^-, H_2O	44.35	**1**
$C_5H_6N_5O^+$, Cl^-, $2H_2O$	44.36	**1**
$2C_5H_6N_5O^+$, O_4S^{2-}, H_2O	44.22	**3**
$C_5H_6N_8O_{13}$	8.15	**3**
$C_5H_6O_2S$	39.26	**1**
$C_5H_6O_3$	38.17	**1**
$C_5H_6O_4$	20.2	**3**
$C_5H_6O_4$	20.3	**3**
$C_5H_6O_5$	1.69	**1**
$C_5H_7CuNO_5$, $2H_2O$	82.6	**4**
C_5H_7INO	32.14	**3**
$C_5H_7NOS_2$	41.1	**3**
$C_5H_7NO_2$	1.5	**4**
$C_5H_7NO_3$	48.15	**4**
$C_5H_7N_2O_2^+$, Br^-	44.38	**1**
$C_5H_7N_3O$	44.39	**1**
$C_5H_7N_3O$, $C_4H_3FN_2O_2$	44.4	**3**
$C_5H_7N_3O$, $C_7H_9N_5O$	44.61	**1**
$C_5H_7N_3OS$	32.5	**4**
$C_5H_7N_5O^{2+}$, O_4S^{2-}	44.10	**4**
$C_5H_7O_3^-$, Na^+	2.72	**1**
$C_5H_7O_4^-$, K^+	2.17	**4**
$C_5H_7O_4^-$, Rb^+	2.18	**4**
$C_5H_7O_4^-$, H_4N^+	2.19	**4**
$2C_5H_7S_2^+$, Cl_4Fe^{2-}	39.27+	**1**
C_5H_8ClNO	1.70	**1**
$C_5H_8ClNO_3$	48.13	**3**
$C_5H_8N_2OS$	11.6	**3**
$C_5H_8N_2OS$	44.11	**4**
$C_5H_8N_2O_2$	44.40	**1**
$C_5H_8N_2O_2$	44.41	**1**
$C_5H_8N_2O_2$, H_2O	32.15	**3**
$C_5H_8N_2O_4$	40.2+	**3**
$C_5H_8N_2S$	32.6+	**4**
$C_5H_8N_3O^+$, Br^-	44.42	**1**
$C_5H_8N_3O^+$, Cl^-	44.12	**4**
$C_5H_8N_4O_{12}$	45.3	**1**
$C_5H_8N_4S$	44.43	**1**
$C_5H_8O_2$	1.71	**1**
$C_5H_8O_2$	1.72	**1**
$C_5H_8O_2S$	39.29	**1**
$C_5H_8O_2S$	39.4	**3**
$C_5H_8O_3S$	39.30	**1**
$C_5H_8O_4$	1.73	**1**
$C_5H_8O_4$	1.74	**1**
$C_5H_9Br_2O_3P$	64.15	**2**
$C_5H_9Br_2O_3P$	64.4	**4**
$C_5H_9Cl_2NOPdS$	72.7	**3**
$C_5H_9Cl_3N_2O$	32.17	**3**
$C_5H_9CuNO_5$, H_2O	82.10	**2**
$C_5H_9NO_2$	48.38	**1**
$C_5H_9NO_3$	48.39+	**1**
$C_5H_9NO_4$	48.41	**1**
$C_5H_9NO_4$	48.14	**3**
$C_5H_9NO_4S_2$	4.11	**1**
$C_5H_9NO_5Zn$, H_2O	82.11	**2**
$C_5H_9N_2O_3S$	44.24	**3**
$(C_5H_9N_3Sn)_n$	69.9	**4**
$C_5H_9O_4P$	46.7	**1**
$C_5H_9O_4P$	46.8	**1**
$2C_5H_9O_6^-$, Ca^{2+}, $5H_2O$	45.4	**1**
$2C_5H_9O_6^-$, Sr^{2+}, $5H_2O$	45.5	**1**
$C_5H_9O_8P^{2-}$, Ba^{2+}, $5H_2O$	45.6	**1**
$C_5H_{10}B_{10}MoO_5^{2-}$, $2C_{16}H_{36}N^+$	62.9	**3**
$C_5H_{10}ClO_3P$	64.5	**4**
$C_5H_{10}NO_2^+$, Cl^-	48.42	**1**
$C_5H_{10}NO_4^+$, Cl^-	48.43	**1**
$C_5H_{10}NS_2^-$, Na^+	11.30	**1**
$C_5H_{10}NS_2^-$, Na^+, $3H_2O$	4.12	**1**

$C_5H_{10}N_2OS$	41.10	1
$C_5H_{10}N_2O_2$	1.75	1
$C_5H_{10}N_2O_2S$	4.2	4
$C_5H_{10}N_2O_3$	48.44	1
$C_5H_{10}N_2O_3$	48.15	3
$2C_5H_{10}N_2O_3S$, Na^+, I^-	60.67	2
$C_5H_{10}N_2O_4PbS$	69.11	2
$C_5H_{10}N_2S$	34.4	1
$C_5H_{10}O_2$	1.76	1
$C_5H_{10}O_4$	45.7	1
$C_5H_{10}O_5$	45.8	1
$C_5H_{10}O_5$	45.9	1
$C_5H_{10}O_5$	45.10	1
$C_5H_{10}O_5$	45.1	4
$C_5H_{10}S_5$	39.3+	4
$C_5H_{11}AgIN$	83.16	3
$C_5H_{11}BO_3$	62.40	2
$C_5H_{11}B_9O_3Re^-$, Cs^+	62.41	2
$C_5H_{11}Cl_2NO_2PdS$	82.1	3
$C_5H_{11}Cl_2NO_2PtS$	82.2	3
$C_5H_{11}Cl_2NO_2PtS$	82.3	3
$C_5H_{11}CuO_4^+$, $C_6N_3O_7^-$	77.1	2
$C_5H_{11}NO_2$	48.45	1
$C_5H_{11}NO_2$	48.16	3
$C_5H_{11}NO_2S$	41.11	1
$C_5H_{11}NO_2S$	48.46+	1
$C_5H_{11}N_2O_2^+$, ClO_4^-	48.48	1
$C_5H_{11}N_2O_3^+$, Cl^-	48.16	4
$C_5H_{11}N_2O_3^+$, Cl^-	48.17	4
$C_5H_{11}N_2O_3^+$, Cl^-, H_2O	48.49	1
$C_5H_{11}N_3^{2+}$, $2Cl^-$	59.1	1
$C_5H_{11}N_3^{2+}$, Cl_4Co^{2-}	32.8	4
$C_5H_{11}N_3^{2+}$, $2H_2O_4P^-$, H_2O	32.18+	3
$2C_5H_{11}N_3O_2$, $C_{10}H_{18}CoN_6O_4^-$, K^+	79.10	4
$C_5H_{11}O_4P$, H_2O	46.3	3
C_5H_{12}	5.17	1
$C_5H_{12}ClNS_2Sn$	69.11	3
$C_5H_{12}CuN_7S^+$, CNS^-	83.12	4
$C_5H_{12}LiN_2O_4^+$, Br^-, H_2O	67.6	3
$C_5H_{12}N^+$, Cl^-	33.34	1
$C_5H_{12}N^+$, ClO_4^-	3.44	1
$C_5H_{12}N^+$, $C_7H_4BrO_2^-$	33.10	4
$C_5H_{12}N^+$, $C_7H_4ClO_2^-$	33.11	4
$C_5H_{12}N^+$, $C_{40}H_{36}EuO_8^-$	77.28	3
$2C_5H_{12}N^+$, $As_4S_6^{2-}$	33.12	4
$2C_5H_{12}N^+$, $BiBr_5^{2-}$	33.35	1
$C_5H_{12}NO_2^+$, Br^-	48.50	1
$C_5H_{12}NO_2^+$, Cl^-	48.51	1
$C_5H_{12}NO_2^+$, Cl^-	48.52	1
$C_5H_{12}NO_2^+$, Cl^-	48.17	3
$C_5H_{12}NO_2^+$, Cl^-, H_2O	48.18	3
$C_5H_{12}NO_2S^+$, Cl^-, H_2O	48.18	4
$C_5H_{12}NO_2S^+$, CH_4N_2O, Cl^-	60.68	2
$C_5H_{12}N_2O_3S$	48.19	3
$C_5H_{12}O_2$	5.3	3
$C_5H_{12}O_4$	45.11+	1
$C_5H_{12}O_5$	45.14	1
$C_5H_{12}O_5$	45.15	1
$C_5H_{12}O_5$	45.16	1
$C_5H_{12}S_4$	11.31	1
$C_5H_{13}NO_4S$	59.2	1
$C_5H_{13}N_2O_2^+$. Br^-	48.20	3
$C_5H_{13}N_2O_2^+$, Br^-	48.19	4
$C_5H_{13}N_2O_2^+$, Cl^-	48.56	1
$C_5H_{13}N_2O_2^+$, Cl^-	48.21	3
$C_5H_{13}N_2O_3P$	64.6	4
$C_5H_{13}N_4NiS_2^+$, Cl^-	72.4	4
$(C_5H_{14}CuN_3O_2^+)_n$, $nCHO_2^-$	76.20	3
$2C_5H_{14}N^+$, $C_4H_6Cl_6Pt_2^{2-}$	72.3	3
$C_5H_{14}NO^+$, Cl^-	59.3	1
$C_5H_{14}NO^+$, Cl^-	59.1	4
$C_5H_{14}NO^+$, $C_4H_6CrN_6S_4^-$	59.4	1
$C_5H_{14}NO_6P$, H_2O	46.4	4
$C_5H_{14}N_3O_3P$, $0.5CH_6N_3^+$, $0.5Cl^-$	8.9	4
$C_5H_{14}N_8^{2+}$, $2Cl^-$, H_2O	8.43+	1
$C_5H_{15}As_5$	65.9	2
$C_5H_{16}ClN_5NiS$	76.27	2
$C_5H_{16}CoN_5O_3S_2$, $2H_2O$	76.28	2
$C_5H_{16}N_5NiS^+$, I^-	76.21	3
$C_5H_{16}N_6NiO_2S$	76.29	2
$(C_5H_{20}Cu_2N_{10}S_5^{2+})_n$, nO_4S^{2-}, $2nH_2O$	79.10	3
$C_5H_{21}B_{18}CoS_2$	71.2	4
$C_5H_{25}ClCoN_5^{2+}$, $2NO_3^-$	83.13	4

C_6

C_6BrCl_5	19.1	1
$C_6Br_2Cl_4$	19.2	1
$C_6Br_4O_2$	18.1	1
C_6Br_5Cl	19.3	1
C_6Br_6, $C_6H_2Br_4$	60.69	2
$C_6Cl_2O_4^{2-}$, $2H_4N^+$, H_2O	6.5	1
$C_6Cl_4O_2$	18.2+	1
$C_6Cl_4O_2$	18.1	4
$C_6Cl_4O_2$, $2C_9H_7NO$	60.107	2
$C_6Cl_4O_2$, $C_{10}H_{16}N_2$	60.122	2
$C_6Cl_4O_2$, $C_{12}H_{18}$	60.71	2
$C_6Cl_4O_2$, $2C_{16}H_{20}N_2$	60.33	4
$C_6Cl_4O_2$, $C_{18}H_{12}N_2O_2Pd$	60.70	2
C_6Cl_6	19.4	1
$C_6CoO_{12}^{3-}$, $C_{36}H_{24}N_6Ni^{2+}$, K^+, $2H_2O$	83.142	3
$C_6CrO_{12}^{3-}$, $3H_4N^+$	81.37	2
$C_6F_4O_2$, $C_{20}H_{12}$	60.152	2
C_6F_6, C_9H_{12}	60.21	4
C_6I_6	19.1+	3
$C_6MoN_6S_6^{3-}$, $3K^+$, $C_2H_4O_2$, H_2O	83.38	2
$C_6N_2O_8^{2-}$, $2H_4N^+$	6.6	1
$C_6N_2O_8^{2-}$, $2H_5O_2^+$, $2H_2O$	6.1	3
$C_6N_3O_7^-$, $C_5H_{11}CuO_4^+$	77.1	2
C_6N_4	7.21	1
C_6N_4	7.5	4
C_6N_4, $C_{10}H_8$	60.72	2
C_6N_4, $C_{10}H_{10}Fe$	60.119	2
C_6N_4, $C_{16}H_{10}$	60.142	2
C_6N_4, $C_{16}H_{10}$	60.44	3
C_6N_4, $C_{18}H_{20}$	60.37	4
C_6N_4, $8C_{20}H_{10}N_4$, $2CH_2Cl_2$	60.52	3
C_6N_4, $C_{20}H_{12}$	60.53	3
C_6N_4O	38.2+	4
$C_6N_4O_4$	21.1	3
$C_6N_6O_3$, $C_6H_{15}OP$	60.73	2
$C_6N_6O_3$, $C_6H_{15}O_4P$	60.15+	4
$C_6N_6O_6$	40.2+	1
$C_6N_6O_6$, $C_{12}H_{12}S_2$	60.74	2
$C_6N_6O_{12}$	15.1	1
$C_6N_9^{3-}$, $3Na^+$, $3H_2O$	33.36	1
$C_6NbO_{13}^{3-}$, $3H_4N^+$, H_2O	81.19	4
$C_6O_{12}Rh^{3-}$, $3K^+$, $4.5H_2O$	81.20	4

$C_6O_{12}Sb^{3-}$, $3K^+$, $4H_2O$	66.2	**3**
$C_6O_{12}Sb^{3-}$, $3H_4N^+$, $4H_2O$	66.3	**3**
$C_6O_{12}V^{3-}$, $3K^+$, $3H_2O$	81.39	**2**
$C_6O_{16}U_2{}^{2-}$, $2H_4N^+$	81.40	**2**
$C_6HCl_2N_3O_6$	15.2	**1**
C_6HCl_5O	17.1	**1**
C_6HCl_5O, C_6H_7N	60.17	**3**
$C_6HN_4O_6{}^-$, K^+, H_2O	40.6	**4**
$C_6HN_6O_4{}^-$, $C_{18}H_{16}OP^+$	64.43	**3**
$C_6H_2BrClO_2$	18.4	**1**
$C_6H_2Br_2Cl_2O$	21.1	**1**
$C_6H_2Br_2O_2$	18.5	**1**
$C_6H_2Br_4$	19.6+	**1**
$C_6H_2Br_4$, C_6Br_6	60.69	**2**
$C_6H_2ClN_3O_6$	15.3	**1**
$C_6H_2ClN_3O_6$	15.1+	**3**
$C_6H_2Cl_2N_2O$	21.2	**3**
$C_6H_2Cl_2N_2O_2$	40.7	**4**
$C_6H_2Cl_2O_2$	18.1	**3**
$C_6H_2Cl_2O_2$	18.2	**3**
$C_6H_2Cl_2O_2$	18.3	**3**
$C_6H_2Cl_2O_4$	18.8	**1**
$C_6H_2Cl_2O_4$, $2H_2O$	18.9	**1**
$C_6H_2Cl_4$	19.8	**1**
$C_6H_2Cl_4O_2$	17.2+	**1**
$C_6H_2Cl_4O_2$	21.2	**1**
$C_6H_2Cl_4O_2$, $C_{18}H_{15}AsO$	60.35	**4**
$C_6H_2Cl_5N$	16.1	**3**
$C_6H_2IN_3O_6$	15.4	**1**
$C_6H_2N_2O_8$, $6H_2O$	18.10	**1**
$C_6H_2N_2O_{10}S_2{}^{2-}$, $2K^+$, H_2O	15.3	**3**
$C_6H_2N_3O^-$, K^+	7.6	**4**
$C_6H_2N_3O_7{}^-$, K^+	15.5	**1**
$C_6H_2N_3O_7{}^-$, K^+	15.1	**4**
$C_6H_2N_3O_7{}^-$, H_4N^+	15.6	**1**
$C_6H_2N_3O_7{}^-$, $C_{10}H_{10}Fe^+$	73.25	**2**
$C_6H_2N_3O_7{}^-$, $C_{10}H_{13}N_2O^+$, H_2O	35.11	**4**
$2C_6H_2N_3O_7{}^-$, $C_4H_{16}N_8PbS_4{}^{2+}$	69.8	**4**
$C_6H_2N_3O_9S^-$, $H_5O_2{}^+$, $2H_2O$	15.2	**4**
$C_6H_2N_3O_9S$, $C_{13}H_{27}S^+$	11.8	**4**
$C_6H_2N_4O_4{}^{2-}$, $2Na^+$, $4H_2O$	35.3	**1**
$C_6H_2N_4O_4{}^{2-}$, $2Rb^+$, $2H_2O$	35.4	**1**
$C_6H_2N_4O_6$	40.8	**4**
$C_6H_2N_6O_4$, $C_{18}H_{15}OP$	60.75	**2**
$2C_6H_2N_6O_6$, $C_{18}H_{12}CuN_2O_2$	60.76	**2**
$C_6H_2O_4$, $2H_2O$	1.77	**1**
$C_6H_2O_4{}^{2-}$, $2K^+$	18.11	**1**
$C_6H_2O_8{}^{2-}$, $2K^+$	2.73	**1**
$C_6H_3BrN_2O_2$	40.9	**4**
$C_6H_3BrN_2O_4$	15.7	**1**
$C_6H_3BrN_2O_4$	15.5	**3**
$C_6H_3Br_2O_3S^-$, $H_7O_3{}^+$	11.1	**4**
$C_6H_3Br_3$	19.9	**1**
$C_6H_3ClN_2O_2$	40.4	**1**
$C_6H_3ClN_2O_4$	15.8	**1**
$C_6H_3ClN_2O_4$	15.6	**3**
$C_6H_3ClO_2$	18.4	**3**
$C_6H_3Cl_2O_3S^-$, $H_7O_3{}^+$	11.2	**4**
$C_6H_3Cl_3$	19.10+	**1**
$C_6H_3Cl_3$	19.1	**4**
$C_6H_3Cl_3O$, C_6H_7N	60.18	**3**
$C_6H_3Cl_9O_6{}^-$, $C_{10}H_{10}Fe^+$	73.3	**4**
$C_6H_3IN_2O_2$	40.10	**4**
$C_6H_3N_3$	20.7	**1**
$C_6H_3N_3O_6$	15.3	**4**
$C_6H_3N_3O_6$, C_6H_6IN	60.86	**2**
$C_6H_3N_3O_6$, $C_6H_9N_3$	60.12	**3**
$C_6H_3N_3O_6$, C_8H_7N	60.102	**2**
$C_6H_3N_3O_6$, C_9H_6OS	60.8	**4**
$C_6H_3N_3O_6$, C_9H_7NO	60.29	**3**
$C_6H_3N_3O_6$, C_9H_9N	60.108	**2**
$C_6H_3N_3O_6$, $C_9H_{18}N_6$	60.113	**2**
$C_6H_3N_3O_6$, $C_{10}H_8$	60.115+	**2**
$C_6H_3N_3O_6$, $C_{10}H_8CrO_4$	60.118	**2**
$C_6H_3N_3O_6$, $C_{12}H_9NS$	60.38	**3**
$C_6H_3N_3O_6$, $C_{14}H_{10}$	60.133+	**2**
$C_6H_3N_3O_6$, $C_{16}H_{10}$	60.143	**2**
$C_6H_3N_3O_6$, $C_{22}H_{28}CoN_2O_2$	60.9	**4**
$C_6H_3N_3O_6$, $C_{22}H_{28}CuN_2O_2$	60.10	**4**
$C_6H_3N_3O_6$, $C_{22}H_{28}N_2NiO_2$	60.11	**4**
$C_6H_3N_3O_7$, $C_{10}H_8BrN$	60.77	**2**
$C_6H_3N_5O_8$	16.1	**1**
C_6H_3P	64.7	**4**
$C_6H_4As_2Cl_2O$	65.10	**2**
C_6H_4BrCl	19.12	**1**
$C_6H_4BrNO_2$	15.9	**1**
$C_6H_4Br_3N$	16.2	**1**
$C_6H_4ClIO_2$	19.13	**1**
$C_6H_4ClNO_2$	15.10	**1**
$C_6H_4ClNO_2$	15.11	**1**
$C_6H_4ClNO_2$	18.12	**1**
$C_6H_4ClNO_4S$	11.7	**3**
$C_6H_4ClNO_5$	15.12	**1**
$C_6H_4Cl_2$	19.14+	**1**
$C_6H_4Cl_2$	19.2	**4**
$C_6H_4Cl_2N_2O_2$	18.13	**1**
$C_6H_4Cl_2N_4O_4$	40.5	**1**
$C_6H_4Cl_2O_2$	1.6	**4**
$C_6H_4Cl_3N$	16.1	**4**
$C_6H_4Cl_4N_2$	16.3	**1**
$C_6H_4CuN_3$	83.14	**4**
$C_6H_4Cu_2N_2O_8S_2{}^{2-}$, $2C_4H_{12}N^+$	3.38	**1**
$C_6H_4FO_7{}^{3-}$, Rb^+, H_4N^+, H^+, $2H_2O$	2.19	**3**
C_6H_4INO	10.25	**1**
$C_6H_4NO_3{}^-$, K^+, $0.5H_2O$	6.7	**1**
$C_6H_4N_2$	7.22	**1**
$C_6H_4N_2$	33.13	**4**
$C_6H_4N_2O$	40.6	**1**
$C_6H_4N_2O_2$	40.7	**1**
$C_6H_4N_2O_2$	40.8	**1**
$C_6H_4N_2O_2$	40.9	**1**
$C_6H_4N_2O_3S$	9.2	**3**
$C_6H_4N_2O_3S$	9.2	**4**
$C_6H_4N_2O_4$	15.15	**1**
$C_6H_4N_2O_4$	15.16	**1**
$C_6H_4N_2O_4S$, H_2O	9.11	**1**
$C_6H_4N_2O_8P^-$, $C_7H_{10}N^+$	46.5	**3**
$C_6H_4N_2S$	41.12	**1**
$C_6H_4N_2Se$	42.7	**1**
$C_6H_4N_4$	35.5	**1**
$C_6H_4N_4$	35.6	**1**
$C_6H_4N_4$	7.7+	**4**
$C_6H_4N_4{}^{2+}$, Cl_4Zn^{2-}	9.12	**1**
$C_6H_4N_4O_2$	9.13	**1**
$C_6H_4N_4O_2$, $1.5H_2O$	35.1	**4**
$C_6H_4O_2$	18.14+	**1**
$C_6H_4O_2$, $C_5H_6N_2O_2$	60.7	**4**
$C_6H_4O_2$, C_6H_5BrO	60.78	**2**
$C_6H_4O_2$, $2C_6H_5BrO$	60.79	**2**
$C_6H_4O_2$, C_6H_5ClO	60.80	**2**
$C_6H_4O_2$, $2C_6H_5ClO$	60.81	**2**
$C_6H_4O_2$, $C_6H_6O_2$	60.13	**3**

$C_6H_4O_2$, $2C_6H_6O_2$ 17.17 1
$2C_6H_4O_2$, $C_6H_6O_3$ 60.12 4
$C_6H_4O_4S$ 42.1 3
$C_6H_4O_5$ 38.18 1
$C_6H_4O_5$ 38.19 1
$C_6H_4O_6$, $2H_2O$ 18.16 1
$C_6H_4O_6^{2-}$, $2K^+$ 2.20 4
$C_6H_4O_7^{2-}$, Ca^{2+}, $4H_2O$ 2.22 3
$C_6H_4O_7^{2-}$, Ca^{2+}, $4H_2O$ 2.21 4
$C_6H_4S_2$ 39.31 1
$C_6H_4S_4$ 39.5 4
$C_6H_4Se_2$ 39.32 1
C_6H_5BrHg 71.8 2
C_6H_5BrMg, $2C_4H_{10}O$ 67.14 2
C_6H_5BrO, $C_6H_4O_2$ 60.78 2
C_6H_5BrO, $C_{19}H_{22}O_2$ 60.14 3
C_6H_5BrO, $C_{19}H_{26}O_2$ 60.39 4
C_6H_5BrO, $C_{19}H_{28}O_2$ 60.151 2
C_6H_5BrO, $C_{19}H_{28}O_2$ 60.15 3
$2C_6H_5BrO$, $C_6H_4O_2$ 60.79 2
C_6H_5BrOS 39.5 3
$C_6H_5Br_2NO$ 33.37 1
C_6H_5Cl 19.3 4
$C_6H_5ClN_2O_2$ 16.4 1
C_6H_5ClO 17.4 1
C_6H_5ClO 17.1 4
C_6H_5ClO, $C_6H_4O_2$ 60.80 2
$2C_6H_5ClO$, $C_6H_4O_2$ 60.81 2
$C_6H_5ClO_2Se$ 11.32 1
$C_6H_5Cl_2I$ 19.18 1
$C_6H_5Cl_2N$ 16.5 1
$C_6H_5Cl_5$ 21.3 1
$C_6H_5CrN_3O_3$ 73.1 3
$C_6H_5F_2N_3O_5S_3$ 11.8 3
$C_6H_5FeIO_3$ 72.7 2
$C_6H_5NO_2$ 15.17 1
$C_6H_5NO_2$ 33.38 1
$C_6H_5NO_3$ 17.5+ 1
$C_6H_5NO_3$ 33.14 4
$C_6H_5NO_3$, $C_{16}H_{14}CuN_2O_2$ 60.16 3
$C_6H_5N_2^+$, Br_3^- 9.14 1
$C_6H_5N_2^+$, $Br_3Cu_2^-$ 9.15 1
$C_6H_5N_2^+$, Cl^- 9.16 1
$C_6H_5N_2^+$, $C_2H_4O_2$, Cl^- 60.82 2
$C_6H_5N_5O_6$ 15.18 1
$2C_6H_5O_3S^-$, H_2MgO^{2+} 11.33 1
$2C_6H_5O_3S^-$, $H_{12}O_6Zn^{2+}$ 11.34 1
$C_6H_5O_4P$ 46.5 4
$C_6H_5O_4P^{2-}$, Ba^{2+}, $1.5H_2O$ 46.9 1
$C_6H_5O_4P^{2-}$, $2K^+$, $1.5H_2O$ 46.10 1
$C_6H_5O_6^-$, K^+ 2.74 1
$C_6H_5O_6^-$, K^+ 2.22 4
$2C_6H_5O_7^{3-}$, $3Mg^{2+}$, $10H_2O$ 2.77 1
C_6H_6 19.19+ 1
C_6H_6 19.3+ 3
C_6H_6 19.4 4
C_6H_6, Al_2Br_6 60.83 2
C_6H_6, Br_2 60.84 2
C_6H_6, N_2O_4 60.85 2
$2C_6H_6$, $C_4H_6CdN_6Ni$ 61.3 3
$(C_6H_6Ag^+)_n$, $(ClO_4^-)_n$ 74.1 2
$C_6H_6AgAlCl_4$ 74.2 2
$C_6H_6AlCl_4Cu$ 74.3 2
$C_6H_6Al_2Cl_8Sn$, C_6H_6 69.12 3
$C_6H_6AsO_2S^-$, $C_6H_{16}N^+$ 65.5 4
$C_6H_6BBrO_2$ 62.42 2
$C_6H_6Br_2N_4$ 35.7 1
C_6H_6ClN 16.6+ 1
$C_6H_6ClN_3$ 44.44 1
$C_6H_6Cl_2N_2O_4S_2$ 32.20 3
$C_6H_6Cl_4Ir_2O_4$ 71.9 2
$C_6H_6Cl_5Sb$ 66.9 2
$C_6H_6Cl_6$ 21.4 1
$C_6H_6Cl_6$ 21.5 1
$C_6H_6Cl_6$ 21.3 3
$C_6H_6Cl_6$ 21.4 3
$C_6H_6Cl_6FeOSi_2$ 73.2 3
$C_6H_6F_2N_2O_4S_2$ 44.13 4
C_6H_6IN, $C_6H_3N_3O_6$ 60.86 2
$C_6H_6MoS_6$ 85.34 2
$C_6H_6NO_2^+$, Cl^- 33.40 1
$C_6H_6NO_6^{3-}$, $3Na^+$, H_2O 2.75 1
$C_6H_6N_2O$ 33.41 1
$C_6H_6N_2O$ 33.42 1
$C_6H_6N_2O_2$ 16.10 1
$C_6H_2D_4N_2O_2$ 15.8 3
$2C_6H_6N_2O_2$, Cd_2Cl_4, $2H_2O$ 60.87 2
$4C_6H_6N_2O_2$, $C_{12}H_{28}N_2^{2+}$, $2Cl^-$ 60.40 3
$C_6H_6N_2S$ 33.43 1
$C_6H_6N_2S$ 33.15 4
$C_6H_6N_4O_4$ 40.10 1
$C_6H_6N_4S$, H_2O 44.45 1
$C_6H_6N_6O_6$ 15.19 1
$(C_6H_6N_6Sn)_n$ 69.10 4
$C_6H_6Nd_2O_{15}$, $4H_2O$ 81.41 2
C_6H_6O 17.7+ 1
C_6H_6O, $0.5H_2O$ 17.10 1
$3C_6H_6O$, $C_6H_{12}N_4$ 60.21 3
$C_6H_6O_2$ 17.13 1
$C_6H_6O_2$ 17.11+ 1
$C_6H_6O_2$ 17.14+ 1
$C_6H_6O_2$ 5.5 4
$C_6H_6O_2$ 17.2 4
$C_6H_6O_2$, O_2S 61.9 2
$C_6H_6O_2$, C_3H_6O 60.88 2
$C_6H_6O_2$, $C_6H_4O_2$ 60.13 3
$C_6H_6O_2$, $C_{17}H_{22}N_4O_6^{2+}$, $2Br^-$ 60.47 3
$1.5C_6H_6O_2$, $C_{13}H_{13}N_2O_2^+$, Br^- 60.30 4
$2C_6H_6O_2$, $C_6H_4O_2$ 17.17 1
$C_6H_6O_2S$ 39.6 3
$C_6H_6O_2S$ 39.7 3
$C_6H_6O_2Se$ 11.35 1
$C_6H_6O_3$ 17.18 1
$C_6H_6O_3$ 17.3 4
$C_6H_6O_3$, $2H_2O$ 17.19 1
$C_6H_6O_3$, $2C_6H_4O_2$ 60.12 4
$C_6H_6O_3S_2$ 39.8 3
$C_6H_6O_4$ 20.8 1
$C_6H_6O_4^{2-}$, $2C_6H_8O_4$, $2Na^+$ 20.9 1
$C_6H_6O_4P^-$, $C_6H_{14}N^+$ 46.6 3
$C_6H_6O_7^{2-}$, Li^+, H_4N^+, H_2O 2.76 1
$C_6H_6O_{24}P_6^{12-}$, $12Na^+$, $38H_2O$ 46.6 4
$C_6H_6Sn^{2+}$, $2AlCl_4^-$ 69.11 4
$C_6H_6U^{3+}$, $3AlCl_4^-$ 74.1 4
$C_6H_7AsO_3$ 65.12+ 2
$2C_6H_7BrN^+$, F_6Si^{2-}, $2H_2O$ 16.2 4
$C_6H_7BrN_2O_2$ 44.14+ 4
C_6H_7N, Cl_3Sb 60.89 2
C_6H_7N, I_2 60.90 2
C_6H_7N, C_6HCl_5O 60.17 3
C_6H_7N, $C_6H_3Cl_3O$ 60.18 3
$2C_6H_7N$, Cl_3Sb 60.19 3

C_6H_7NO 17.20 1
$C_6H_7NO_2$ 33.16 4
$C_6H_7NO_3S$ 16.11 1
$C_6H_7NO_3S$ 16.12 1
$C_6H_7NO_3S$, H_2O 16.13 1
$C_6H_7NO_6^{2-}$, Ca^{2+}, $2H_2O$ 2.23 4
$C_6H_7N_2O_2S^-$, Na^+, H_2O 16.14 1
$C_6H_7N_3O$ 33.17 4
$C_6H_7N_3O_3$ 44.46 1
$C_6H_7N_3S_2$ 39.9 3
$C_6H_7N_5$ 44.47 1
$C_6H_7N_5$, 2HBr 44.48 1
$C_6H_7N_5$, $C_5H_4BrN_2O_2$ 44.21 1
$C_6H_7N_5$, $C_5H_6N_2OS$ 44.21 3
$C_6H_7N_5$, $C_6H_8N_2O_2$ 44.51 1
$C_6H_7O_3$ 38.20 1
$C_6H_7O_6^-$, Na^+ 45.17 1
$C_6H_7O_7^-$, K^+ 2.78+ 1
$C_6H_7O_7^-$, Li^+ 2.80 1
$C_6H_7O_7^-$, Na^+ 2.81 1
$C_6H_7O_7^-$, Rb^+ 2.82 1
$2C_6H_7O_9S^-$, Ba^{2+}, $2H_2O$ 45.2 4
$C_6H_8AgN_4^+$, NO_3^- 83.15+ 4
$C_6H_8AsNO_3$ 65.14 2
$(C_6H_8Br_2N_2Ni)_n$ 83.40 2
$C_6H_8Br_4O_4$ 38.21 1
$(C_6H_8CdN_6Ni)_n$, $2nC_6H_6$ 61.4 3
$(C_6H_8Cl_2HgO_2)_n$ 60.91 2
$C_6H_8Cl_2N_4Zn$ 83.41 2
$C_6H_8Cl_4$ 21.6 1
$C_6H_8CrN_2O_8^-$, $C_6H_{16}CrN_4O_4^+$, $2H_2O$ 76.24 3
$C_6H_8N^+$, $BiBr_4^-$ 33.44 1
$C_6H_8N^+$, BiI_4^- 33.45 1
$C_6H_8N^+$, Br^- 16.15 1
$C_6H_8N^+$, Cl^- 16.16 1
$C_6H_8N^+$, $C_{18}H_{15}Br_3PZn^-$ 86.10 4
$2C_6H_8N^+$, Br_6Sb^-, Br_3^- 33.46 1
$C_6H_8NO^+$, Cl^- 17.21 1
$C_6H_8NO^+$, Cl^- 33.39 1
$C_6H_8NO_2^+$, Cl^- 17.4 4
$C_6H_8N_2$ 9.17 1
$C_6H_8N_2O$, $2H_2O$ 44.49 1
$C_6H_8N_2O_2$ 40.11 1
$C_6H_8N_2O_2$ 44.50 1
$C_6H_8N_2O_2$, $C_6H_7N_5$ 44.51 1
$2C_6H_8N_2O_2$, $C_7H_{10}N_6$, H_2O 44.25 3
$C_6H_8N_2O_2S$ 16.17+ 1
$C_6H_8N_2O_2S$ 16.3 4
$C_6H_8N_2O_2S$, H_2O 16.22 1
$C_6H_8N_2O_2S$, $C_9H_9N_3O_2S_2$ 60.30 3
$C_6H_8N_2O_3$ 43.1 4
$C_6H_8N_2O_4$ 43.2 4
$C_6H_8N_5O^+$, Br^- 44.52 1
$C_6H_8O_2$ 21.7+ 1
$C_6H_8O_2$, C_2I_2 60.92 2
$C_6H_8O_2S$ 39.6 4
$C_6H_8O_4$ 1.78 1
$C_6H_8O_4$ 20.11 1
$C_6H_8O_4$ 20.4 3
$C_6H_8O_4$ 20.5 3
$C_6H_8O_4$ 20.6 3
$C_6H_8O_4$ 20.7 3
$C_6H_8O_4$ 20.2 4
$C_6H_8O_4$ 45.3 4
$C_6H_8O_4^{2-}$, $C_6H_{18}N_2^{2+}$ 2.83+ 1
$2C_6H_8O_4$, $C_6H_6O_4^{2-}$, $2Na^+$ 20.9 1
$C_6H_8O_4S_2$ 39.35 1
$C_6H_8O_4Se_2$ 39.36 1
$C_6H_8O_6$ 45.18 1
$C_6H_8O_6$ 45.19+ 1
$C_6H_8O_6$ 45.2 3
$C_6H_8O_7$ 1.79 1
$C_6H_8O_7$, H_2O 1.7 4
$C_6H_9ClN_2O_4$ 48.22 3
$C_6H_9CuNO_5$, $2H_2O$ 82.13 2
$(C_6H_9EuO_9)_n$ 81.21 4
$C_6H_9GdO_9$, $0.5H_2O$ 81.10 3
C_6H_9NO 1.80 1
$C_6H_9NO_2$, H_2O 48.20 4
$C_6H_9NO_3$ 40.4 3
$C_6H_9NO_6$ 1.81 1
$C_6H_9N_2^+$, Cl^- 9.18 1
$C_6H_9N_3$, $C_6H_3N_3O_6$ 60.12 3
$C_6H_9N_3O^{2+}$, $2Cl^-$ 33.47+ 1
$C_6H_9N_3O_2$ 48.23 3
$C_6H_9N_3O_2$ 48.21 4
$C_6H_9N_4O_2S_2^{2+}$, $2Cl^-$ 48.58 1
$C_6H_9O_3PS$ 64.16 2
$C_6H_9O_7^-$, K^+, $2H_2O$ 45.22 1
$C_6H_9O_7^-$, Rb^+, $2H_2O$ 45.23 1
$2C_6H_9O_7^-$, Ca^{2+}, $2H_2O$ 45.24 1
$C_6H_9O_8U^-$, Na^+ 81.42 2
C_6H_{10} 20.13 1
$C_6H_{10}BrCl$ 21.9 1
$C_6H_{10}Cl_2$ 21.10 1
$C_6H_{10}Cl_2N_2Pd$ 83.42 2
$C_6H_{10}Cl_2Pd_2$ 72.8+ 2
$C_6H_{10}Cl_3HgN_3O_2$ 82.4 3
$C_6H_{10}Cr$ 72.12 2
$C_6H_{10}Mo_2N_2O_8S_2^{2-}$, $2Na^+$, $5H_2O$ 82.14 2
$C_6H_{10}NNdO_8$, H_2O 81.43 2
$C_6H_{10}NO_2^+$, Cl^-, H_2O 48.22 4
$C_6H_{10}NS^+$, Br^- 41.5 4
$C_6H_{10}N_2^{2+}$, $2Cl^-$ 16.23 1
$C_6H_{10}N_2^{2+}$, $2Cl^-$ 33.49 1
$C_6H_{10}N_2O_2$ 21.11 1
$C_6H_{10}N_2O_2$ 33.17 3
$C_6H_{10}N_2O_2$ 33.15+ 3
$C_6H_{10}N_2O_2$ 33.18+ 3
$C_6H_{10}N_2O_2$ 48.23 4
$C_6H_{10}N_2O_8$, $2C_4H_8OS$ 60.13 4
$C_6H_{10}N_3O_2^+$, Cl^-, H_2O 48.59 1
$C_6H_{10}N_3O_2^+$, Cl^-, H_2O 48.24 4
$C_6H_{10}N_3O_2^+$, Cl^-, $2H_2O$ 48.25 3
$C_6H_{10}N_3O_2^+$, $H_2O_4P^-$, H_3O_4P 48.25 4
$C_6H_{10}N_4$, ClI 60.93+ 2
$C_6H_{10}N_4NiS_4$ 85.35 2
$C_6H_{10}NiO_2S_4$ 80.3 2
$C_6H_{10}O_2PbS_4$ 69.12 2
$C_6H_{10}O_2S_4$ 11.9 3
$C_6H_{10}O_2S_4Te$ 70.10 2
$C_6H_{10}O_2S_4Zn$ 80.4 2
$C_6H_{10}O_4$ 1.86 1
$C_6H_{10}O_4$ 1.82+ 1
$C_6H_{10}O_4$ 38.22 1
$C_6H_{10}O_4$ 38.5 4
$C_6H_{10}O_4S$ 1.20 3
$C_6H_{10}O_5$ 45.3 3
$C_6H_{10}O_6$ 45.25 1
$C_6H_{10}O_6$ 45.4 3
$C_6H_{10}O_6$ 45.5 3

$C_6H_{10}S_3$	39.37	1
$C_6H_{10}S_4$	39.38	1
$C_6H_{10}S_4$	39.7	4
$(C_6H_{11}CaNO_8)_n$	67.4	4
$(C_6H_{11}CeO_7)_n$	81.44	2
$C_6H_{11}Cl_2NOS$	4.3	4
$C_6H_{11}NO_2$, H_2O	48.26	4
$C_6H_{11}NO_4S_2$	4.13	1
$C_6H_{11}N_3^{2+}$, $2Cl^-$, H_2O	35.8	1
$C_6H_{11}N_3O_4$	48.60	1
$C_6H_{11}N_3O_4$, Ca^{2+}, $2Cl^-$, $3H_2O$	60.95	2
$C_6H_{11}N_3O_4$, Li^+, Br^-	48.27	4
$C_6H_{11}N_4O_2^+$, Br^-	59.5	1
$C_6H_{11}N_4O_2^+$, Br^-	59.1	3
$C_6H_{11}O_4P$	64.17	2
$2C_6H_{11}O_6^-$, Ca^{2+}	45.27	1
$2C_6H_{11}O_6^-$, Sr^{2+}	45.28	1
$C_6H_{11}O_7^-$, K^+	45.29	1
$C_6H_{11}O_9P^{2-}$, $2K^+$, $2H_2O$	45.30	1
$C_6H_{11}O_9P^{2-}$, $2K^+$, $2H_2O$	45.6	3
$C_6H_{12}AgS_6^+$, NO_3^-	85.36	2
$C_6H_{12}BNO_3$	62.10	4
$C_6H_{12}Br_2N_4S_2Te$	70.11	2
$C_6H_{12}Br_2O_4$	45.7	3
$C_6H_{12}CdCl_2N_4S_2$	85.37	2
$(C_6H_{12}Ce_2O_{18})_n$, $4nH_2O$	81.11	3
$(C_6H_{12}ClCuN_3O_5)_{2n}$, nH_2O	82.15	2
$C_6H_{12}Cl_2O_4$	45.8	3
$C_6H_{12}Cl_4Te$	70.5	4
$C_6H_{12}CoN_3OS_4$	80.5	2
$C_6H_{12}CoN_3OS_4$	80.1+	4
$C_6H_{12}CrN_3O_6$, H_2O	82.7	4
$C_6H_{12}CuN_2O_4$	82.17	2
$C_6H_{12}CuN_2O_4$	82.5	3
$C_6H_{12}CuN_2O_6$	82.18	2
$C_6H_{12}CuO_7$, $0.5H_2O$	81.45	2
$C_6H_{12}Cu_2N_4O_{10}$	79.11	3
$C_6H_{12}FeN_3OS_4$	80.1+	3
$C_6H_{12}HgI_2N_2S_4$	85.4	4
$C_6H_{12}I_2N_4S_2Te$	70.12	2
$(C_6H_{12}La_2O_{18})_n$, $4nH_2O$	81.12	3
$C_6H_{12}LiN_2^+$, $C_7H_7^-$	67.7	3
$C_6H_{12}MoNO_5^-$, H^+	83.17	3
$C_6H_{12}NO_2^+$, Br^-	20.14	1
$C_6H_{12}NO_3S^+$, Cl^-, H_2O	41.13	1
$C_6H_{12}NO_4^+$, Cl^-	59.2	4
$C_6H_{12}N_2$	37.1+	1
$C_6H_{12}N_2O_2$	1.87	1
$C_6H_{12}N_2O_2$	1.88	1
$C_6H_{12}N_2O_2$	1.21	3
$C_6H_{12}N_2O_3$	48.26	3
$C_6H_{12}N_2O_4S_2$	48.61	1
$C_6H_{12}N_2O_4S_2$	48.27	3
$(C_6H_{12}N_2O_6Zn)_n$	82.6	3
$C_6H_{12}N_2S_4$	4.14	1
$C_6H_{12}N_3PS$	64.11	3
$C_6H_{12}N_4$	37.4+	1
$C_6H_{12}N_4$, $6H_2O$	61.10	2
$C_6H_{12}N_4$, $H_{12}MgO_6^{2+}$, $2Cl^-$, $4H_2O$	37.10	1
$C_6H_{12}N_4$, $2CHBr_3$	60.1	4
$C_6H_{12}N_4$, CHI_3	60.20	3
$C_6H_{12}N_4$, $3C_6H_6O$	60.21	3
$2C_6H_{12}N_4$, Ca^{2+}, $2Br^-$, $10H_2O$	60.97	2
$2C_6H_{12}N_4$, $H_{12}MgO_6^{2+}$, $Cr_2O_7^{2-}$	37.1	4
$C_6H_{12}N_8NiO_8^{2-}$, $2K^+$, $4H_2O$	83.43	2
$(C_6H_{12}Nd_2O_{18})_n$, $4nH_2O$	81.13	3
$(C_6H_{12}Nd_2O_{18})_{2n}$, $9nH_2O$	81.14	3
$C_6H_{12}NiO_6S_2$	81.22	4
$(C_6H_{12}O_2SSn)_n$	69.12	4
$C_6H_{12}O_2Se$	42.9	1
$C_6H_{12}O_2Si$	63.4	2
$C_6H_{12}O_3S$	42.10	1
$C_6H_{12}O_3S_2$, $0.5H_2O$	45.4	4
$C_6H_{12}O_4S$	45.31	1
$C_6H_{12}O_4S$	45.10	3
$C_6H_{12}O_4S$	42.2	4
$C_6H_{12}O_4S$	45.5	4
$C_6H_{12}O_4S$	45.6	4
$C_6H_{12}O_5$	45.32	1
$C_6H_{12}O_5$	45.33	1
$C_6H_{12}O_5$, H_2O	45.34	1
$C_6H_{12}O_5$, H_2O	45.7	4
$C_6H_{12}O_6$	21.12	1
$C_6H_{12}O_6$	45.35	1
$C_6H_{12}O_6$	45.38	1
$C_6H_{12}O_6$	45.36+	1
$C_6H_{12}O_6$	45.11	3
$C_6H_{12}O_6$	45.12	3
$C_6H_{12}O_6$	45.13	3
$C_6H_{12}O_6$	21.1	4
$C_6H_{12}O_6$, H_2O	45.39	1
$C_6H_{12}O_6$, H_2O	45.14	3
$C_6H_{12}O_6$, $2H_2O$	21.13	1
$C_6H_{12}O_6$, CH_4N_2O	60.14	4
$(C_6H_{12}O_{18}Pr_2)_n$, $4nH_2O$	81.15	3
$C_6H_{12}O_{18}Ti_2$, $4H_2O$	81.46	2
$C_6H_{12}S_3$	39.39	1
$C_6H_{12}Se_3$	39.40	1
$C_6H_{13}BO_8$, H_2O	45.8	4
$C_6H_{13}CuN_3O_4$, $4H_2O$	76.22	3
$C_6H_{13}NO_2$	1.89+	1
$C_6H_{13}NO_2$	48.62	1
$C_6H_{13}NO_2$	48.28	4
$C_6H_{13}NO_2$, $C_6H_{14}NO_2^+$, Cl^-	48.29	4
$C_6H_{13}N_2O_3^+$, Cl^-	48.63	1
C_6H_{14}	5.18	1
$C_6H_{14}AgN_{10}^+$, NO_3^-	83.44	2
$C_6H_{14}BN_3$	62.43	2
$(C_6H_{14}Ba_2CuO_{16})_n$	81.47	2
$C_6H_{14}ClNO_3Zn$	83.17	4
$C_6H_{14}Cl_2MoN_2O_4$	84.18	2
$C_6H_{14}CoN_4O_4^+$, Br^-	82.7	3
$C_6H_{14}CuN_4S_2$	85.5	3
$C_6H_{14}CuO_8$	81.48	2
$C_6H_{14}KO_3^+$, $C_{16}H_{16}Ce^-$	75.13	4
$C_6H_{14}N^+$, Cl^-	21.14+	1
$C_6H_{14}N^+$, $C_3H_4O_6P^-$	46.3	1
$C_6H_{14}N^+$, $C_6H_6O_4P^-$	46.6	3
$C_6H_{14}NO_2^+$, Br^-	48.64	1
$C_6H_{14}NO_2^+$, Br^-, H_2O	48.65	1
$C_6H_{14}NO_2^+$, I^-	48.28	3
$C_6H_{14}NO_4^+$, Cl^-	45.40	1
$C_6H_{14}NO_5^+$, Br^-	45.41	1
$C_6H_{14}NO_5^+$, Br^-	45.15	3
$C_6H_{14}NO_5^+$, Cl^-	45.43	1
$C_6H_{14}NO_5^+$, Cl^-	45.16	3
$C_6H_{14}N_2O_4S^{2+}$, $2Cl^-$	48.30	4
$C_6H_{14}N_2O_4S_2^{2+}$, $2Br^-$	48.66+	1
$C_6H_{14}N_2O_4S_2^{2+}$, $2Br^-$, $2H_2O$	48.31	4
$C_6H_{14}N_2O_4S_2^{2+}$, $2Cl^-$	48.68	1
$C_6H_{14}N_2S_2$	41.14	1
$C_6H_{14}N_3O_3^+$, Cl^-	48.32+	4

$(C_6H_{14}N_3O_6Zn^+)_{2n}$, nO_4S^{2-}, $4nH_2O$	82.8	3
$C_6H_{14}N_4O_2$, $2H_2O$	48.69	1
$C_6H_{14}N_4O_4S_2Zn$	79.27	2
$C_6H_{14}N_5O^+$, Cl^-	40.11	4
$C_6H_{14}NiO_8$	81.16	3
$C_6H_{14}O_2$, $6H_2O$	61.5	3
$C_6H_{14}O_6$	45.46	1
$C_6H_{14}O_6$	45.44+	1
$C_6H_{14}O_6$	45.17	3
$C_6H_{14}O_6$	45.9	4
$C_6H_{14}O_6$	45.12	4
$C_6H_{14}O_6$	45.10+	4
$C_6H_{14}O_6$, C_5H_5N	60.23	3
$C_6H_{15}BN_4$	37.11	1
$C_6H_{15}NS_2Sn$	69.13+	3
$C_6H_{15}N_2O_2^+$, Br^-	3.24	3
$C_6H_{15}N_2O_2^+$, Cl^-, $2H_2O$	48.70	1
$C_6H_{15}N_3$, $3H_2O$	33.50	1
$C_6H_{15}N_4O_2^+$, Br^-, H_2O	48.71	1
$C_6H_{15}N_4O_2^+$, Cl^-, H_2O	48.29	3
$C_6H_{15}N_4O_2^+$, $H_2O_4P^-$, H_2O	48.30	3
$C_6H_{15}OP$, $C_6N_6O_3$	60.73	2
$C_6H_{15}O_4P$, $C_6N_6O_3$	60.15+	4
$C_6H_{15}PS$	64.18	2
$C_6H_{15}PSe$	64.19	2
$C_6H_{16}BO_3P$	64.12	3
$C_6H_{16}B_2N_2$	62.11	4
$C_6H_{16}Cl_2CuN_{10}$, H_2O	83.45	2
$C_6H_{16}Cl_{10}Cu_5O_2$	84.19+	2
$C_6H_{16}CoN_6^+$, Cl^-	76.23	3
$C_6H_{16}CoN_6^+$, Cl^-, H_2O	76.10	4
$C_6H_{16}CrN_4O_4^+$, $C_6H_8CrN_2O_8^-$, $2H_2O$	76.24	3
$C_6H_{16}CuN_2O_6$, $4H_2O$	82.21	2
$C_6H_{16}CuN_{10}^{2+}$, $2Cl^-$, H_2O	79.12	3
$C_6H_{16}CuN_{10}^{2+}$, O_4S^{2-}, H_2O	83.46	2
$C_6H_{16}N^+$, Cl^-	3.45	1
$C_6H_{16}N^+$, $C_6H_6AsO_2S^-$	65.5	4
$C_6H_{16}N^+$, $C_9H_{10}N_2O_8P^-$	47.1	1
$C_6H_{16}N^+$, $C_{11}HFe_3O_{11}^-$	3.46	1
$C_6H_{16}N^+$, $C_{12}H_4N_4^-$, $C_{12}H_4N_4$	3.25	3
$C_6H_{16}N^+$, $C_{18}H_{12}O_6P^-$	46.7	4
$2C_6H_{16}N^+$, $H_{18}B_{20}^{2-}$	3.26	3
$3C_6H_{16}N^+$, $H_{18}B_{20}NO^{3-}$	3.27	3
$C_6H_{16}N_2$	3.49	1
$C_6H_{16}N_2^{2+}$, $2Br^-$	34.5	1
$C_6H_{16}N_2^{2+}$, $2Cl^-$	21.16	1
$C_6H_{16}N_2NiO_6$	82.19	2
$C_6H_{16}N_2NiO_8$	82.22	2
$C_6H_{16}N_2O_3^{2+}$, $2Cl^-$	21.17	1
$C_6H_{16}N_2O_4S_2$	4.4	4
$C_6H_{16}N_4NiO_4$	76.31	2
$C_6H_{16}N_4O_2^{2+}$, $2I^-$	48.72	1
$C_6H_{16}N_6NiS_2$	76.32	2
$C_6H_{16}N_8^{2+}$, $2Cl^-$, $2H_2O$	9.3	4
$C_6H_{16}N_{10}Ni^{2+}$, $2Cl^-$, H_2O	79.13+	3
$C_6H_{16}N_{10}Ni^{2+}$, $2Cl^-$, H_2O	79.5	4
$C_6H_{17}Cl_4NO_4Zn_2$	83.18	4
$C_6H_{18}Al_2$	68.10	2
$C_6H_{18}Al_2$	68.3	3
$C_6H_{18}Al_2$	68.2	4
$C_6H_{18}Al_2Br_4O_2$	68.11	2
$C_6H_{18}As_2Br_4Pd_2$	86.2	2
$C_6H_{18}As_2Cl_4Pt_2$	86.2	3
$C_6H_{18}As_4N_6$	65.15	2
$C_6H_{18}B_2Cl_4N_2$	3.28	3
$C_6H_{18}Br_3N_2Ti$	83.19	4
$(C_6H_{18}CaO_9^{2+})_n$, $2nBr^-$	67.5	4
$C_6H_{18}ClN_4Zn^+$, $C_{24}H_{20}B^-$	76.25	3
$C_6H_{18}Cl_2N_2OV$	83.47	2
$C_6H_{18}Cl_2O_2S_2Sn$	69.15	3
$C_6H_{18}Cl_2P_2Pt$	86.4	2
$C_6H_{18}Cl_3CrN_2$	83.20	4
$C_6H_{18}Cl_3N_2V$	83.18	3
$C_6H_{18}Cl_3N_6P_3$	64.8	4
$C_6H_{18}Cl_8N_6P_4$	64.20	2
$C_6H_{18}CoN_2O_8P_2$	84.21	2
$C_6H_{18}CoO_6P_3S_6$	85.5	4
$C_6H_{18}Ga_3N_3$	68.3	4
$C_6H_{18}IN_4Zn^+$, I^-	76.26	3
$C_6H_{18}N_2^{2+}$, $2Cl^-$	3.50	1
$C_6H_{18}N_2^{2+}$, $2I^-$	3.51	1
$C_6H_{18}N_2^{2+}$, $C_6H_8O_4^{2-}$	2.83+	1
$C_6H_{18}N_2Ti^{3+}$, $3Br^-$	83.49	2
$C_6H_{18}N_3NiO_3^-$, $C_6H_{21}CoN_3O_3^{3+}$, $2I^-$	83.21	4
$C_6H_{18}N_3O_6P_3$	64.21	2
$C_6H_{18}N_4O_4W$	71.3	4
$C_6H_{18}N_6P_2$	64.9	4
$C_6H_{18}OSb_2^{2+}$, $2ClO_4^-$	66.3	4
$C_6H_{18}OSb_2^{2+}$, $2N_3^-$	66.4	4
$C_6H_{18}O_3Si_3$	63.5	2
$C_6H_{18}O_8V_4$	84.22	2
$C_6H_{19}AlN_2$	68.12	2
$C_6H_{19}N_3NdO_8^{3+}$, $3Cl^-$, H_2O	82.9	3
$C_6H_{20}B_{20}O_2$	62.11	3
$(C_6H_{20}CaO_{10}^{2+})_n$, $2nBr^-$, nH_2O	67.6	4
$C_6H_{20}ClCoN_4O^{2+}$, $2ClO_4^-$	76.27	3
$C_6H_{20}ClCoN_6^{2+}$, $2Cl^-$, H_2O	76.11	4
$C_6H_{20}ClCoN_6O_4$	76.34	2
$C_6H_{20}Cl_2CoN_4^+$, $2Cl^-$, $H_5O_2^+$	76.35	2
$C_6H_{20}Cl_2CoN_4^+$, $2Cl^-$, $H_5O_2^+$, $2H_2O$	76.28	3
$C_6H_{20}Cl_2CuN_4O_8$	76.29	3
$C_6H_{20}CoN_6O_4^+$, Cl^-	76.30	3
$C_6H_{20}CoN_6O_6^+$, NO_3^-	76.31	3
$C_6H_{20}CuN_4^{2+}$, $Cr_2O_7^{2-}$	76.32	3
$C_6H_{20}CuN_4^{2+}$, $2NO_2^-$	76.33	3
$C_6H_{20}CuN_4^{2+}$, $2NO_3^-$	83.50	2
$C_6H_{20}CuN_4^{2+}$, $2NO_3^-$	76.34	3
$C_6H_{20}CuN_4^{2+}$, $2NO_3^-$	76.35	3
$C_6H_{20}CuN_4^{2+}$, O_4S^{2-}, H_2O	76.36	3
$C_6H_{20}CuN_4^{2+}$, O_4Se^{2-}, H_2O	76.37	3
$C_6H_{20}CuN_4^{2+}$, $2CNS^-$	76.12	4
$C_6H_{20}MoN_3O_7P$	84.3	4
$C_6H_{21}AlN_2$	68.13	2
$C_6H_{21}B_3N_6$	62.12	4
$C_6H_{21}ClCoN_5^{2+}$, $2NO_3^-$	76.38	3
$C_6H_{21}CoN_3O_3^{3+}$, $C_6H_{18}N_3NiO_3^-$, $2I^-$	83.21	4
$C_6H_{21}CrN_{15}^{3+}$, $3C_{10}H_{15}O_4S^-$	79.16	3
$C_6H_{21}N_4^{3+}$, $3Cl^-$	3.52+	1
$C_6H_{24}B_3N_3$	62.44	2
$C_6H_{24}B_3P_3$	64.22	2
$C_6H_{24}CdN_6^{2+}$, $O_3S_2^{2-}$	76.39	3
$C_6H_{24}Cd_2N_{12}O_8S_8$	79.17	3
$3C_6H_{24}CoN_6^+$, $H_{12}NaO_6^+$, $4Cl^-$	76.36	2
$C_6H_{24}CoN_6^{3+}$, $3Br^-$, H_2O	76.37	2
$C_6H_{24}CoN_6^{3+}$, $3Cl^-$, H_2O	76.38	2
$C_6H_{24}CoN_6^{3+}$, $C_5CrN_6O^{3-}$, $2H_2O$	76.40	3
$2C_6H_{24}CoN_6^{3+}$, $CdCl_6^{4-}$, $2Cl^-$, $2H_2O$	76.13	4

$2C_6H_{24}CoN_6^{3+}$, $Cl_8Cu_2^{4-}$, $2Cl^-$, $2H_2O$	76.41	3
$2C_6H_{24}CoN_6^{3+}$, $3HO_4P^{2-}$, $9H_2O$	76.14	4
$C_6H_{24}CrN_6^{3+}$, $C_5N_5Ni^{3-}$, $1.5H_2O$	76.39	2
$C_6H_{24}CrN_6^{3+}$, $C_6CoN_6^{3-}$, $6H_2O$	76.40	2
$C_6H_{24}CuN_6^{2+}$, O_4S^{2-}	76.42	3
$C_6H_{24}Cu_2N_{12}S_6^{2+}$, $2BF_4^-$	79.6	4
$C_6H_{24}Cu_2N_{12}S_6^{2+}$, $2ClO_4^-$	79.7	4
$C_6H_{24}N_4NiO_2^{2+}$, $2ClO_4^-$	76.43	3
$C_6H_{24}N_4NiO_2^{2+}$, $2NO_3^-$	83.51	2
$C_6H_{24}N_4NiO_2^{2+}$, $C_9H_{30}N_6Ni^{2+}$, $4Cl^-$, H_2O	83.33	4
$C_6H_{24}N_6Ni^{2+}$, K^+, $3CNSe^-$	76.41	2
$C_6H_{24}N_6Ni^{2+}$, $2NO_3^-$	76.42	2
$C_6H_{24}N_6Ni^{2+}$, O_4S^{2-}	76.44	3
$3C_6H_{24}N_6Ni^{2+}$, $O_{15}Si_6^{6-}$, $26H_2O$	76.43	2
$C_6H_{24}N_6Ru^{3+}$, $3Cl^-$, $3.5H_2O$	76.45	3
$C_6H_{24}N_{12}NiS_6^{2+}$, $2Br^-$	79.28	2
$C_6H_{24}N_{12}O_6Ti^{3+}$, $3ClO_4^-$	79.8	4
$C_6H_{24}N_{12}O_6Ti^{3+}$, $3I^-$	79.29	2
$C_6H_{24}N_{12}O_6Ti^{3+}$, $3I^-$	79.19	3
$C_6H_{24}N_{12}O_6Zn^{2+}$, $2NO_3^-$	84.4	4
$C_6H_{24}N_{12}PbS_6^{2+}$, $2ClO_4^-$	69.13	4
$C_6H_{24}N_{12}S_6Te_2^{4+}$, $4HF_2^-$	70.13	2
$C_6H_{26}N_{12}O_{12}U_2^{2+}$, I_4^{2-}	79.30	2
$C_6H_{32}B_{26}Co_2^{2-}$, $2Cs^+$, H_2O	62.12	3

C_7

$C_7HF_2MnO_5$	71.10	2
$C_7HF_6O_4Rh$	77.2	2
$C_7H_2Br_3N$	7.9	4
$C_7H_2Cl_3N$	7.23	1
$C_7H_2Cl_3N$	7.10	4
$C_7H_2N_5O_{10}^-$, Rb^+	12.7	4
$C_7H_3Br_2N_3$	9.19	1
$C_7H_3Br_5$	19.5	4
$C_7H_3FeNO_4$	72.13	2
$C_7H_3NO_4^{2-}$, Ca^{2+}, $3H_2O$	33.51	1
$C_7H_4BrN_3$	9.4	4
C_7H_4ClNOS	41.15	1
$C_7H_4ClNO_4$	13.1	1
C_7H_4ClNS	41.6	4
$C_7H_4ClN_2O_4^-$, K^+	12.8	4
$C_7H_4ClN_3$	9.20	1
$C_7H_4ClO_2^-$, $C_7H_5ClO_2$, K^+	14.1	1
$C_7H_4CrO_3S$	75.1	2
$(C_7H_4CuN_3)_n$	83.19	3
C_7H_4IN	7.24	1
$C_7H_4NO_4^-$, $C_7H_5NO_4$, K^+	14.3	1
$C_7H_4NO_4^-$, $C_7H_5NO_4$, Rb^+	14.2	1
$C_7H_4N_2O_6$, $C_{12}H_9NS$	60.130	2
$C_7H_4O_5^{2-}$, Ca^{2+}, $3H_2O$	38.23	1
$C_7H_5AlCl_4O$	68.14	2
$C_7H_5BrO_2$	13.2	1
$C_7H_5BrO_2$	13.3	1
$C_7H_5Br_2NO_2$	13.4+	1
$C_7H_5Br_3FeO_2Sn$	73.3	3
C_7H_5ClO	22.1	4
$C_7H_5ClO_2$	13.6	1
$C_7H_5ClO_2$	13.7	1
$C_7H_5ClO_2$, $C_7H_4ClO_2^-$, K^+	14.1	1
$C_7H_5ClO_2^-$, $C_2H_6N_2S^+$	8.32	1
$C_7H_5ClO_3$, $C_7H_8N_4O_2$	60.24	3
$C_7H_5ClO_3$, $C_8H_{10}N_4O_2$	60.104	2
$2C_7H_5ClO_3$, $C_7H_8N_5O_2$	60.17	4
$C_7H_5Cl_2CoHgO_2$	73.3	2
$C_7H_5Cl_2NO_2$	13.8	1
$C_7H_5Cl_3FeO_2Sn$	73.4	3
$C_7H_5Cl_4O_2Sb$	66.4	3
$C_7H_5Cl_6OSb$	66.10	2
$C_7H_5FO_2$	13.9	1
$C_7H_5IO_3$	42.8	1
C_7H_5NOS	41.2	3
$C_7H_5NOS_2$	39.41	1
$C_7H_5NO_3$	15.20+	1
$C_7H_5NO_3S$	41.16+	1
$C_7H_5NO_4$	13.10	1
$C_7H_5NO_4$	13.11+	1
$C_7H_5NO_4$	13.1	4
$C_7H_5NO_4$, $C_7H_4NO_4^-$, K^+	14.3	1
$C_7H_5NO_4$, $C_7H_4NO_4^-$, Rb^+	14.2	1
$C_7H_5NO_5$	13.13	1
$C_7H_5NO_5$	13.1	3
$C_7H_5NS_2$	41.18	1
$C_7H_5NS_2$	41.7	4
$C_7H_5N_2O_4^-$, K^+	12.5	3
$C_7H_5N_3O$	22.2	4
$C_7H_5N_5O_8$	16.24	1
$C_7H_5O_2^-$, Na^+	22.2	1
$C_7H_5O_2^-$, $4CH_4N_2S$, Tl^+	60.98	2
$C_7H_5O_2^-$, $C_7H_6O_2$, K^+	14.4	1
$C_7H_5O_2^-$, $C_7H_6O_2$, H_4N^+, H_2O	14.5	1
$C_7H_5O_2^-$, $C_{16}H_{36}N^+$, $39.5H_2O$	61.14	2
$C_7H_5O_3$, $C_{10}H_{15}N_2$	60.22	4
$C_7H_5O_3^-$, $C_7H_6O_3$, K^+, H_2O	14.6	1
$C_7H_5O_5S^-$, H_4N^+	13.15	1
$C_7H_5O_6S^-$, $H_7O_3^+$	13.2	3
$C_7H_5O_6S^-$, $H_7O_3^+$	13.2	4
$C_7H_6BF_3FeO_5S$	72.9	3
C_7H_6BrNO	13.16	1
C_7H_6BrNO	13.17	1
$C_7H_6BrN_3O$	9.5+	4
$C_7H_6BrN_4O$, $C_7H_8BrN_5$	44.26	3
C_7H_6ClNO	19.21	1
C_7H_6ClNO	19.22	1
C_7H_6ClNO	10.3	3
$C_7H_6ClNO_2$	18.17	1
$C_7H_6ClNO_2$	19.23	1
C_7H_6FNO	13.18	1
C_7H_6FNO	13.19	1
$C_7H_6FeO_3$	72.14	2
$C_7H_6NO_2^-$, K^+, $3H_2O$	14.7	1
$C_7H_6NO_2^-$, Na^+, $5H_2O$	14.8	1
$C_7H_6N_2O_2$	40.12	4
$C_7H_6N_2O_2S_2$, $0.75H_2O$	39.11	3
$C_7H_6N_2O_3$	15.9	3
$C_7H_6N_2S$	16.25	1
$C_7H_6N_4$	32.21	3
$C_7H_6O_2$	13.20	1
$C_7H_6O_2$	22.1	3
$C_7H_6O_2$, $C_7H_5O_2^-$, K^+	14.4	1
$C_7H_6O_2$, $C_7H_5O_2^-$, H_4N^+, H_2O	14.5	1
$C_7H_6O_2S$	39.42+	1
$C_7H_6O_3$	13.21+	1
$C_7H_6O_3$, $C_7H_5O_3^-$, K^+, H_2O	14.6	1
$C_7H_6O_3S$	42.11	1
$C_7H_7^+$, $C_8F_{12}NiS_4^-$	22.2	3
$C_7H_7^-$, $C_6H_{12}LiN_2^+$	67.7	3

$C_7H_7BrN_2O_2$	16.2	3
C_7H_7BrO	17.23	1
$C_7H_7BrO_4$	1.22	3
$C_7H_7Cl_2NO$	33.18	4
$(C_7H_7Cl_3CuO_4)_n$	77.1	3
$C_7H_7IrO_4$	77.2	3
C_7H_7NO	13.23	1
$2C_7H_7NO, HI_3$	60.99	2
$C_7H_7NO_2$	13.28	1
$C_7H_7NO_2$	13.29	1
$C_7H_7NO_2$	13.24+	1
$C_7H_7NO_2$	13.26+	1
$C_7H_7NO_2$	38.2	3
$C_7H_7NO_2$	15.4	4
$C_7H_7NO_2^+, Br^-$	58.1	3
$C_7H_7NO_3$	10.26	1
$C_7H_7NO_3$	13.30	1
$C_7H_7NO_3S$	15.22	1
$C_7H_7NO_7V, 2H_2O$	81.49	2
$C_7H_7N_2O^+, Cl_4Fe^-$	9.22	1
$C_7H_7N_2O_2S_2^-, Na^+, 2.5H_2O$	39.44	1
$C_7H_7N_5NiOS_4$	85.38	2
$C_7H_7O_2^+, Cl^-$	22.3	1
$C_7H_7O_3S^-, H_3O^+$	11.3	4
$C_7H_7O_3S^-, C_{10}H_{18}N^+$	37.5	4
$2C_7H_7O_3S^-, H_{12}O_6Zn^{2+}$	11.36	1
$C_7H_7O_4Rh$	77.3	2
$C_7H_7O_4Rh$	77.3	3
$C_7H_7O_4S^-, K^+$	11.37	1
$C_7H_8Ag_2^{2+}, 2NO_3^-$	75.2	2
$C_7H_8BrN_5, C_5H_5BrN_2O_2$	44.13	3
$C_7H_8BrN_5, C_7H_6BrN_4O$	44.26	3
$2C_7H_8BrN_5, C_{12}H_{12}N_2O_3$	60.131	2
$C_7H_8Cl_2O$	27.1	4
$C_7H_8Cl_2Pd$	75.3+	2
$2C_7H_8N^+, Cl_6Sn^{2-}$	3.54	1
$C_7H_8NO^+, Cl^-$	33.52	1
$C_7H_8N_2O$	13.3	4
$C_7H_8N_2O_3$	44.53	1
$C_7H_8N_2S$	33.20	3
$C_7H_8N_4O, C_4H_3FN_2O_2$	44.3	1
$C_7H_8N_4O_2, H_2O$	44.54	1
$C_7H_8N_4O_2, C_7H_5ClO_3$	60.24	3
$C_7H_8N_4O_2, C_9H_9N_3O_2S_2$	60.31	3
$C_7H_8N_4S$	33.21	3
$C_7H_8N_4Se$	33.19	4
$C_7H_8N_5O_2, 2C_7H_5ClO_3$	60.17	4
C_7H_8O	17.24	1
C_7H_8O	17.3	3
C_7H_8O	17.5	4
C_7H_8OS	38.24	1
$C_7H_8OS_2$	39.45	1
$C_7H_8O_2S$	11.38	1
$C_7H_8O_4$	1.91+	1
$C_7H_8S_3$	39.46+	1
$C_7H_9AsFeO_4$	65.6	4
$C_7H_9BrN_2STe$	70.14	2
$C_7H_9BrN_6, C_5H_6N_2O_2$	44.27	3
$C_7H_9BrN_6, C_7H_{10}N_2O_2$	44.28	3
$C_7H_9ClN_2S_2Te$	70.15	2
$C_7H_9CuNO_7$	83.52	2
$C_7H_9FeO_5Sb$	66.5	4
$C_7H_9MnN_2O_5$	71.2	3
C_7H_9N	16.27	1
C_7H_9N, CH_4N_2O	60.100	2
$C_7H_9NO_2S$	16.28	1
$C_7H_9NO_3S$	16.3	3
$C_7H_9N_2O^+, I^-$	33.54	1
$C_7H_9N_2O_2S_2^+, Br^-$	39.48	1
$C_7H_9N_3O_2, 2H_2O$	35.2	4
$C_7H_9N_4O_9^-, Rb^+$	12.6	3
$C_7H_9N_5, C_4H_6N_2O_2$	44.55	1
$C_7H_9N_5, C_5H_5BrN_2O_2$	44.56	1
$C_7H_9N_5, C_5H_5BrN_2O_2$	44.14	3
$C_7H_9N_5, C_5H_5FN_2O_2$	44.57	1
$C_7H_9N_5, C_5H_5FN_2O_2$	44.58	1
$C_7H_9N_5, C_5H_5FN_2O_2$	44.17	3
$C_7H_9N_5, 2C_5H_5IN_2O_2$	44.18	3
$C_7H_9N_5, C_8H_{12}N_2O_3$	60.18	4
$C_7H_9N_5, C_{10}H_{13}BrN_2O_3$	60.25	3
$C_7H_9N_5O, C_5H_6BrN_3O$	44.59	1
$C_7H_9N_5O, C_5H_6FN_3O$	44.60	1
$C_7H_9N_5O, C_5H_7N_3O$	44.61	1
$C_7H_9O_2^+, Br^-, H_2O$	38.25	1
$(C_7H_{10}Br_2CuN_2)_n$	83.22	4
$C_7H_{10}ClNO_3S$	42.2	3
$(C_7H_{10}Cl_2CuN_2)_n$	83.23	4
$C_7H_{10}N^+, Br_4Fe^-$	33.20	4
$C_7H_{10}N^+, Cl^-$	16.29	1
$C_7H_{10}N^+, HF_2^-$	16.4	4
$C_7H_{10}N^+, C_6H_4N_2O_8P^-$	46.5	3
$2C_7H_{10}N^+, Cl_6Re^{2-}$	16.30	1
$2C_7H_{10}N^+, F_6Si^{2-}, 2H_2O$	16.5	4
$2C_7H_{10}NO^+, O_4S^{2-}$	16.4	3
$C_7H_{10}N_2$	7.25	1
$C_7H_{10}N_2NiO$	72.15	2
$C_7H_{10}N_2O_2$	44.62	1
$C_7H_{10}N_2O_2, C_7H_9BrN_6$	44.28	3
$C_7H_{10}N_4O_2S, H_2O$	8.16	3
$C_7H_{10}N_6, 2C_5H_5IN_2O_2$	44.19	3
$C_7H_{10}N_6, 2C_6H_8N_2O_2, H_2O$	44.25	3
$C_7H_{10}O_2S$	39.12	3
$C_7H_{10}O_4$	20.3	4
$(C_7H_{11}NO_8Sr)_n$	67.7	4
$C_7H_{11}N_3O_2S$	4.5	4
$C_7H_{11}O_3^-, Na^+$	2.85	1
$C_7H_{12}Cl_2O_4$	45.47	1
$C_7H_{12}Cl_2O_4$	45.18	3
$C_7H_{12}CoN_4O_5$	83.20	3
$C_7H_{12}CuN_4O_4, 1.5H_2O$	82.23	2
$C_7H_{12}NOS^+, Br^-$	41.4	3
$C_7H_{12}N_4OS$	35.9	1
$C_7H_{12}O_4$	1.93	1
$C_7H_{12}O_4$	1.97	1
$C_7H_{12}O_4$	1.94+	1
$C_7H_{12}O_4$	1.8	4
$C_7H_{12}O_6, H_2O$	45.19	3
$C_7H_{13}BrO_5$	45.48	1
$C_7H_{13}ClHgO$	71.11	2
$C_7H_{13}ClHgO$	71.12	2
$C_7H_{13}ClN_2O_3PtS, H_2O$	82.10+	3
$C_7H_{13}ClO_5$	45.20	3
$C_7H_{13}CrN_3O_3$	76.44	2
$C_7H_{13}N_2O_2^+, Br^-$	32.21	1
$(C_7H_{14}AuNS_2)_n$	80.2	2
$C_7H_{14}N^+, Cl^-$	35.10	1
$4C_7H_{14}N^+, Br_6Sb^{3-}, Br_6Sb^-, 2Br_2$	37.1	3
$C_7H_{14}NO_2^+, Cl^-$	48.34	4
$C_7H_{14}NS_2Tl$	68.4	3
$C_7H_{14}N_2O_2$	1.98	1
$C_7H_{14}N_4NiO_2, 3H_2O$	76.15	4
$C_7H_{14}O_6$	45.49	1

$C_7H_{14}O_6$	45.21	3
$C_7H_{14}O_6$	45.22	3
$C_7H_{14}O_6$, H_2O	45.23	3
$C_7H_{14}O_7$	45.24	3
$C_7H_{14}O_7$, H_2O	45.25	3
$C_7H_{15}NS_3$	11.39	1
$C_7H_{15}PS_2$	64.23	2
C_7H_{16}	5.19	1
$C_7H_{16}B_9Fe$	73.5	2
$C_7H_{16}CoN_4O_3^+$, Cl^-, H_2O	81.17	3
$C_7H_{16}N^+$, Br^-	34.6	1
$C_7H_{16}NOS^+$, Br^-	3.29	3
$C_7H_{16}NOS^+$, Br^-	3.30	3
$C_7H_{16}NOS^+$, ClO_4^-	41.19+	1
$C_7H_{16}NOSe^+$, I^-	3.31	3
$C_7H_{16}NO_2^+$, Br^-	59.6+	1
$C_7H_{16}NO_2^+$, Cl^-	3.32	3
$2C_7H_{16}N_2^{2+}$, $Cl_8Ni_2^{4-}$	37.2	3
$C_7H_{16}N_3O_3^+$, Cl^-	48.35	4
$C_7H_{16}O$, $0.5H_2O$	5.20	1
$C_7H_{16}O$, $6C_{18}H_{20}O_2$	61.2	4
$C_7H_{16}O_4P$	46.11	1
$C_7H_{16}O_6S_3$	11.40	1
$C_7H_{17}Cl_3N_2NiO$	83.21	3
$C_7H_{17}N_2O_2^+$, Br^-	3.33	3
$C_7H_{18}CuN_5S^+$, CNS^-	76.46	2
$C_7H_{18}CuN_5S^+$, CNS^-	76.46	3
$C_7H_{18}N^+$, Cl^-	3.55	1
$(C_7H_{18}N_2Sn)_n$	69.16	3
$C_7H_{18}N_5SZn^+$, CNS^-	76.47	3
$C_7H_{20}CoN_4O_3^+$, ClO_4^-	83.24	4
$C_7H_{20}CuN_5S^+$, ClO_4^-	76.48	3
$C_7H_{22}CoN_5O_2^{2+}$, $2I^-$, $2H_2O$	82.24	2
$C_7H_{22}N_3^{3+}$, $3Cl^-$	59.8	1
$2C_7H_{22}N_3^{3+}$, $3HO_4P^{2-}$, $6H_2O$	59.9	1

C_8

$C_8Ag_2F_{14}O_4$	84.23	2
$C_8AuF_{12}S_4^-$, $C_{18}H_{15}ClP^+$	64.52	2
$C_8AuN_4S_4^-$, $C_{18}H_{36}AuN_2S_4^+$	80.12	4
$C_8Cl_2N_2O_2$, $C_{10}H_{12}O_4$	60.33	3
$C_8Cl_4O_3$	13.3	3
C_8Cl_8	20.8	3
$C_8CoF_{12}O_8^{2-}$, $2C_{24}H_{20}As^+$	81.50	2
$C_8CoN_4S_4^{2-}$, $2C_{16}H_{36}N^+$	85.39	2
$C_8CuN_4S_4^-$, $C_{16}H_{36}N^+$	85.40	2
$C_8F_6FeO_4$	72.16	2
C_8F_{12}	31.1	1
$C_8F_{12}NiS_4$, $C_{20}H_{12}$	60.54	3
$C_8F_{12}NiS_4^-$, $C_7H_7^+$	22.2	3
$C_8F_{24}N_8O_4S_4$	10.1	4
$C_8FeN_4S_4^-$, $C_{16}H_{36}N^+$	85.41	2
$C_8FeN_5OS_4^{2-}$, $2C_8H_{20}N^+$	85.42	2
$C_8N_4NiS_4^-$, $C_{19}H_{18}P^+$	85.43	2
$C_8N_4NiS_4^{2-}$, $2C_4H_{12}N^+$	85.30	2
C_8N_4S	39.49	1
$C_8N_4S_2$	39.51	1
$(C_8O_{16}Th^{4-})_n$, $4nK^+$, $4nH_2O$	81.51	2
$C_8O_{16}Zr^{4-}$, $4Na^+$, $3H_2O$	81.52	2
$C_8HBrN_4O_5S$	39.50	1
$C_8H_2F_8FeO_4$	71.13	2
$C_8H_2N_4$, $C_{12}H_{18}$	60.101	2
$C_8H_3O_9^-$, Rb^+	38.26	1
$C_8H_4Cl_2O_2$	13.4	4
$C_8H_4CuN_6O_8$, $4H_2O$	83.53	2
$C_8H_4Cu_2O_{12}^{4-}$, $4Na^+$, $10H_2O$	81.18	3
$C_8H_4F_3N_5O_8$	16.5	3
$C_8H_4FeO_8$	72.17	2
$C_8H_4FeO_8$	72.18	2
$(C_8H_4Hg_2O_4)_n$	81.53	2
$(C_8H_4I_2NiO_2)_n$	71.3	3
$C_8H_4NO_3^-$, $C_8H_7NO_2$, K^+	10.2	4
$C_8H_4N_2Se_2$	11.10	3
$C_8H_4N_4$	20.15	1
$C_8H_4O_2$	27.2	3
$C_8H_4O_4$	38.27	1
$C_8H_4O_4$	38.28	1
$C_8H_4O_{12}Sb_2^{2-}$, $2K^+$, $3H_2O$	66.5	3
$C_8H_4O_{12}Sb_2^{2-}$, $2K^+$, $3H_2O$	66.6	4
$C_8H_4O_{12}Sb_2^{2-}$, $2H_4N^+$, $3H_2O$	66.11	2
$C_8H_4O_{12}Sb_2^{2-}$, $2C_8H_{20}CoN_6S_2^+$, $4H_2O$	76.52	3
$C_8H_4O_{14}V_2^{4-}$, $4Na^+$, $12H_2O$	81.54	2
$C_8H_4O_{14}V_2^{4-}$, $4H_4N^+$, $2H_2O$	81.55	2
$C_8H_5BrN_2O_2$	40.12	1
$C_8H_5ClMoO_3$	73.6	2
$C_8H_5ClN_4$	9.23	1
$C_8H_5F_5IORh$	73.7	2
C_8H_5I, C_4H_9NO	60.59	2
$C_8H_5MnO_3$	73.8	2
$C_8H_5NO_2$	35.11	1
$C_8H_5NO_2$	35.3	4
$C_8H_5N_3$	33.55	1
$C_8H_5O_4^-$, K^+	14.9	1
$C_8H_5O_4^-$, H_4N^+	14.10	1
C_8H_6	5.21	1
$C_8H_6Br_2$	27.1	1
$C_8H_6Br_2N_2O_5$	28.1	1
$C_8H_6Br_4$	19.24	1
$C_8H_6ClNO_3$	18.18	1
$C_8H_6ClNO_3$	18.19	1
$C_8H_6Cl_2$	27.2	1
$C_8H_6Cl_2N_4O_2$	35.4	4
$C_8H_6CuO_5$	81.23	4
$C_8H_6Fe_2N_2O_6$	83.22	3
$C_8H_6IN_3$	16.6	3
$C_8H_6I_2$	27.3	1
$C_8H_6N_2$	35.5	4
$C_8H_6N_2$, $2H_2O$	35.12	1
$C_8H_6N_2$, $2H_2O$	35.13+	1
$C_8H_6N_2O_4S_2$	44.16	4
$C_8H_6N_2O_6$	13.32	1
$C_8H_6N_2O_6$	27.3	3
$C_8H_6N_4O_4^{2-}$, $2Rb^+$, $2H_2O$	35.15	1
$C_8H_6N_4O_8$, $2H_2O$	43.18	1
C_8H_6OS	39.52	1
$C_8H_6O_2S_2$	39.13	3
$C_8H_6O_3$	38.29	1
$C_8H_6O_4$	13.33	1
$C_8H_6O_4$	13.34+	1
$C_8H_6O_4$	13.5	4
$C_8H_6O_4$	38.6	4
$C_8H_6O_6$	1.23	3
$C_8H_6S_2$	39.53	1
$C_8H_6S_2$	39.54	1
$C_8H_6S_2$	39.55	1
$C_8H_6S_2$	39.14	3
$C_8H_7BrN_4O_5$	43.1	3

C_8H_7BrO 19.6 4
$C_8H_7BrO_2$ 22.4 1
$C_8H_7BrO_2$ 22.5 1
C_8H_7ClO 19.25 1
$C_8H_7ClO_5$ 38.7 4
$C_8H_7Cl_2NO_4$, H_2O 48.73 1
$C_8H_7CrNO_3$ 75.1 4
$C_8H_7CrNO_5$ 71.14 2
C_8H_7N, $C_6H_3N_3O_6$ 60.102 2
$C_8H_7NO_2$, $C_8H_4NO_3^-$, K^+ 10.2 4
$C_8H_7NO_2$, $C_8H_6NO_2^-$, K^+ 10.3 4
$C_8H_7NO_3$, H_2O 13.4+ 3
$C_8H_7NS_2$ 41.21 1
$C_8H_7NS_2$ 41.22 1
$C_8H_7NS_2$ 41.23 1
$C_8H_7N_3O_6$ 15.5 4
$C_8H_7N_3O_7$ 17.25 1
$C_8H_7N_3O_7$, $C_2H_5O^-$, Cs^+ 60.103 2
$C_8H_7N_4O_7^-$, K^+ 40.5 3
$C_8H_7N_5O$ 44.29 3
$C_8H_7O_2^-$, $C_8H_8O_2$, K^+ 2.86+ 1
$C_8H_7O_3^-$, Na^+, $0.5H_2O$ 2.24 4
$C_8H_7O_3^-$, $C_8H_8O_3$, K^+ 14.11 1
C_8H_8 31.2 1
$C_8H_8Ag^+$, NO_3^- 75.5 2
C_8H_8BrNO 16.31 1
$(C_8H_8ClCu)_n$ 75.6 2
C_8H_8ClNO 16.32 1
C_8H_8ClNO 19.26 1
$C_8H_8ClNO_3$ 18.20 1
$C_8H_8ClNO_3$ 18.21 1
$(C_8H_8Cl_2CoN_4)_n$ 83.25 4
$C_8H_8Cl_2HgN_4O_4$ 84.6 3
$C_8H_8Cl_2N_2O_2$ 16.33 1
$C_8H_8Cl_2N_2O_2$ 18.2 4
$C_8H_8Cl_2N_8Pd$ 83.26 4
$C_8H_8Cl_2O_4$ 1.24 3
$C_8H_8Cl_2Pd$ 75.7 2
$(C_8H_8CuN_4^+)_n$, $nClO_4^-$ 83.54 2
$C_8H_8CuN_6O_{10}$, $2H_2O$ 83.27 4
$C_8H_8N_2O_4$ 1.99 1
$C_8H_8N_2S$ 41.5 3
$C_8H_8N_3O_8^-$, K^+, $2H_2O$ 15.23 1
$C_8H_8N_4O_4$ 44.31 3
$C_8H_8N_4O_4$ 44.17 4
$C_8H_8N_4O_6$ 16.34 1
$C_8H_8O_2$ 13.36 1
$C_8H_8O_2$ 18.22 1
$C_8H_8O_2$ 18.25 1
$C_8H_8O_2$ 18.23+ 1
$C_8H_8O_2$ 17.4 3
$C_8H_8O_2$ 13.6 4
$C_8H_8O_2$, $C_8H_7O_2^-$, K^+ 2.86+ 1
$C_8H_8O_3$ 13.37 1
$C_8H_8O_3$ 13.6 3
$C_8H_8O_3$, $C_8H_7O_3^-$, K^+ 14.11 1
$C_8H_8O_3S$ 42.12 1
$C_8H_8O_5$ 38.8 4
$C_8H_8O_{12}Sb_2^{2-}$, $2H_4N^+$, $4H_2O$ 66.12 2
$C_8H_8Pd_2S_8$ 85.6 4
$C_8H_8Pt_2S_8$ 85.7 4
$C_8H_8S_2$ 39.56 1
$C_8H_8S_2$ 39.15 3
$C_8H_9F_6N_5$ 32.9 4
$C_8H_9MnO_5Sn$ 69.14 2
C_8H_9NO 13.38 1
C_8H_9NO 16.35 1
C_8H_9NOS 16.36 1
$C_8H_9NO_2$ 13.39+ 1
$C_8H_9NO_2$ 13.7 4
$C_8H_9PS_3$ 42.3 4
C_8H_{10} 20.16 1
C_8H_{10}, CBr_4 60.1 2
$C_8H_{10}Ag^+$, NO_3^- 31.1 4
$C_8H_{10}Br_2O_4$ 20.17 1
$C_8H_{10}Br_2O_4$ 20.18 1
$(C_8H_{10}Cl_2CuN_6)_n$ 83.23 3
$C_8H_{10}Cl_2CuN_6O_2$ 83.28 4
$C_8H_{10}Cl_2O_6Pd_2$, $3H_2O$ 72.10 3
$(C_8H_{10}CuN_2O_4)_n$ 81.19 3
$C_8H_{10}CuN_4O_3$, $1.5H_2O$ 82.25 2
$C_8H_{10}CuN_4O_5^{2-}$, $2Na^+$, $10H_2O$ 82.26 2
$C_8H_{10}NO_5PS$ 64.15 3
$C_8H_{10}N_2O_2$ 16.37 1
$C_8H_{10}N_2O_2$ 16.7 3
$C_8H_{10}N_2O_2Se$ 42.3 3
$C_8H_{10}N_2O_3$ 33.22 3
$2C_8H_{10}N_3^+$, Cl_4Zn^{2-} 9.7 4
$C_8H_{10}N_4NiO_5^{2-}$, $2Na^+$, $8H_2O$ 82.27 2
$C_8H_{10}N_4O_2$, H_2O 44.64 1
$C_8H_{10}N_4O_2$, $C_7H_5ClO_3$ 60.104 2
$C_8H_{10}N_4O_2$, $2C_8H_{12}N_2O_3$ 60.28 3
$C_8H_{10}N_4O_2$, $2C_8H_{12}N_2O_3$ 60.19 4
$C_8H_{10}O$ 17.26 1
$C_8H_{10}O$ 17.27 1
$C_8H_{10}O$ 17.28 1
$C_8H_{10}O$ 17.6 4
$C_8H_{10}O$ 17.7 4
$C_8H_{10}OS$ 11.11 3
$C_8H_{10}OS$ 39.8 4
$C_8H_{10}O_2$ 17.29 1
$C_8H_{10}O_4$ 1.100 1
$C_8H_{10}O_4$ 1.101 1
$C_8H_{10}O_4$ 1.102 1
$C_8H_{10}O_4^{2-}$, $2C_8H_{12}O_4$, $2K^+$ 21.2 4
$C_8H_{10}Pd$ 72.19 2
$C_8H_{11}Cl_2N_2^+$, Br^-, $2H_2O$ 32.10 4
$C_8H_{11}CoN_3O_8$, K^+, $1.5H_2O$ 76.16 4
$2C_8H_{11}IN^+$, Cl^-, I_3^- 60.105+ 2
$C_8H_{11}NO_2$ 48.31 3
$C_8H_{11}N_2O_3^-$, K^+ 43.20 1
$C_8H_{11}N_2O_3^-$, Na^+ 43.21 1
$C_8H_{11}N_2O_3^-$, Na^+ 43.3 4
$C_8H_{11}N_2O_3^-$, $CH_6N_3^+$, $2H_2O$ 43.4 4
$2C_8H_{11}N_2O_3^-$, Ca^{2+}, $3H_2O$ 43.5 4
$C_8H_{11}N_2O_6P$, $2H_2O$ 46.12 1
$C_8H_{11}N_2S^+$, Br^- 33.56 1
$C_8H_{11}N_2S^+$, Cl^- 33.57 1
$C_8H_{11}N_3O_3$, H_2O 48.36 4
$C_8H_{11}S^+$, ClO_4^- 11.41 1
$(C_8H_{12}BaO_5)_n$ 67.8 4
$C_8H_{12}Br_2O_4$ 38.9 4
$C_8H_{12}Br_2O_8Rh_2$, $2CH_5N_3$ 81.56 2
$C_8H_{12}Br_2O_8Rh_2^{2-}$, $2Br^-$, $4H_4N^+$ 81.57 2
$C_8H_{12}Br_4$ 22.6 1
$C_8H_{12}CdN_6S_4$ 79.33 2
$C_8H_{12}ClN_5O^+$, Cl^- 32.22 3
$C_8H_{12}ClN_6O^+$, Cl^- 32.22 1
$C_8H_{12}ClN_6O^+$, Cl^- 32.11 4
$C_8H_{12}Cl_2FeO_4S_4^+$, Cl_4Fe^- 84.24 2
$C_8H_{12}Cl_2O_8Rh_2$, $2CH_5N_3$ 81.58 2
$(C_8H_{12}Cl_4Hg_2S_2)_n$ 85.44 2

$C_8H_{12}CoN_2O_8^-$, Na^+, $2H_2O$ 81.24 4
$C_8H_{12}CoN_4O_6^-$, H_4N^+, $2H_2O$ 82.28 2
$2C_8H_{12}CoN_4O_6^-$, Ba^{2+}, xH_2O 82.12 3
$2C_8H_{12}CoN_4O_6^-$, $H_{12}CoO_6^{2+}$, $12H_2O$ 82.13 3
$2C_8H_{12}CoN_4O_6^-$, $H_{12}CoO_6^{2+}$, $6H_2O$ 82.14 3
$C_8H_{12}CuN_4O_6^{2-}$, $2K^+$, $6H_2O$ 82.29 2
$C_8H_{12}ErO_{12}^+$, $C_4H_{14}ErO_{10}^-$ 81.25 4
$C_8H_{12}Mo_2O_8$ 81.59 2
$C_8H_{12}N^+$, Cl^- 3.56 1
$C_8H_{12}N^+$, $C_{10}H_7MnO_5^-$ 3.57 1
$C_8H_{12}N^+$, $C_{11}H_7CrO_5^-$ 3.58 1
$C_8H_{12}N^+$, $C_{11}H_7CrO_5^-$ 3.59 1
$C_8H_{12}NO_2$ 37.3 3
$C_8H_{12}NO_2^+$, Cl^- 59.10 1
$C_8H_{12}NO_3^+$, Cl^- 59.11 1
$C_8H_{12}NO_3^+$, Cl^- 59.12 1
$C_8H_{12}NO_3^+$, Cl^- 59.3 4
$C_8H_{12}N_2$ 33.58 1
$C_8H_{12}N_2^+$, Br^- 16.39 1
$C_8H_{12}N_2O_2S$ 43.22 1
$C_8H_{12}N_2O_3$ 43.25 1
$C_8H_{12}N_2O_3$ 43.23+ 1
$C_8H_{12}N_2O_3$ 43.6 4
$C_8H_{12}N_2O_3$, CH_4N_2O 60.27 3
$C_8H_{12}N_2O_3$, $C_7H_9N_5$ 60.18 4
$C_8H_{12}N_2O_3$, $C_{13}H_{17}N_3O$ 60.31 4
$2C_8H_{12}N_2O_3$, $C_8H_{10}N_4O_2$ 60.28 3
$2C_8H_{12}N_2O_3$, $C_8H_{10}N_4O_2$ 60.19 4
$C_8H_{12}N_4$ 9.8 4
$C_8H_{12}N_4NiO_6^{2-}$, $2Na^+$, $8H_2O$ 82.30 2
$C_8H_{12}N_4NiO_6^{2-}$, $2Na^+$, $9H_2O$ 82.31 2
$C_8H_{12}N_5O^+$, Br^- 35.1 3
$(C_8H_{12}N_6NiS_4)_n$ 79.31 2
$C_8H_{12}N_6S_2Se_2Te$ 70.6 4
$C_8H_{12}N_6S_4Te$ 70.7 4
$C_8H_{12}O_2$ 20.19 1
$C_8H_{12}O_4$ 1.103 1
$C_8H_{12}O_4$ 21.18 1
$C_8H_{12}O_4$ 21.5 3
$C_8H_{12}O_4$ 21.6 3
$C_8H_{12}O_4$ 21.7 3
$C_8H_{12}O_4$ 21.3 4
$C_8H_{12}O_8Pb$ 69.15 2
$(C_8H_{12}O_8U)_n$ 81.60 2
$C_8H_{12}Rh^+$, Cl^- 72.20 2
$C_8H_{12}S_8V$ 85.8 4
$C_8H_{13}Cl_3O_3S_2SnW$ 85.6 3
$C_8H_{13}N_2O_2^+$, Br^-, H_2O 20.20 1
$C_8H_{13}N_4S_2Te^+$, Cl^- 70.16 2
$C_8H_{13}O_3^-$, Na^+ 2.89 1
$C_8H_{14}ClN$, H_2O 37.4 3
$C_8H_{14}ClNSn$ 69.16 2
$C_8H_{14}Cl_2$ 22.3 3
$C_8H_{14}Cl_2Pd_2$ 72.21 2
$C_8H_{14}CoN_4O_6^+$, ClO_4^- 82.15 3
$C_8H_{14}CuN_4O_4$ 83.56 2
$C_8H_{14}CuN_4O_4$ 83.24 3
$C_8H_{14}CuN_6OS_2$ 85.46 2
$C_8H_{14}CuN_{10}O_2^{2+}$, $2ClO_4^-$ 83.25 3
$C_8H_{14}FeN_2P_4S_4$ 85.9 4
$C_8H_{14}NO^+$, Cl^- 40.14 1
$C_8H_{14}NO_2^+$, Br^- 31.2 4
$C_8H_{14}NO_2^+$, Br^- 35.6 4
$C_8H_{14}N_2NiS_2$ 85.47 2
$C_8H_{14}N_2O_2$ 48.33 3
$C_8H_{14}N_2O_5P^+$, Cl^- 33.23 3
$C_8H_{14}N_4NiO_4$ 83.57 2
$C_8H_{14}N_4O_4Pd$ 83.58+ 2
$C_8H_{14}N_4O_4Pt$ 83.60 2
$C_8H_{14}N_6NiS_2$ 85.7 3
$C_8H_{14}Ni$ 72.22 2
$C_8H_{14}O_2S_2$ 39.57 1
$C_8H_{14}O_2S_2$ 39.9 4
$C_8H_{14}O_4$ 1.104 1
$C_8H_{14}O_4$ 1.105 1
$C_8H_{14}Pt$ 71.4 3
$C_8H_{14}Pt$ 71.4 4
$C_8H_{15}NO$ 58.1 1
$C_8H_{15}NO_2$ 21.20 1
$C_8H_{15}NO_6$ 45.51 1
$C_8H_{15}N_5O_2^{2+}$, $2Br^-$, $3H_2O$ 50.1 3
$2C_8H_{15}O_2^-$, Sr^{2+}, xH_2O 2.90 1
$C_8H_{16}Ag_2N_4O_6^{2+}$, $2NO_3^-$ 82.8 4
$C_8H_{16}Br_2HgO_4$ 84.5 4
$C_8H_{16}CaCdO_{10}$, $6H_2O$ 81.61 2
$C_8H_{16}CaCuO_{10}$, $6H_2O$ 81.62 2
$C_8H_{16}CaF_4O_8P_2$ 67.2 2
$C_8H_{16}ClOP$ 64.16 3
$C_8H_{16}Cl_2CuN_2O_2S$ 85.8 3
$C_8H_{16}Cl_2CuO_4S_2$ 84.7 3
$C_8H_{16}Cl_2HgO_2S_2$ 85.48 2
$C_8H_{16}Cl_2Rh_2$ 72.23 2
$C_8H_{16}Cl_4MnO_8P_2$ 84.25 2
$C_8H_{16}Cl_4S_2Sn$ 69.17 2
$C_8H_{16}Cl_8O_4Ti_2$ 84.26 2
$C_8H_{16}Cr_2O_{10}$ 81.63 2
$C_8H_{16}Cr_2O_{10}$ 81.26 4
$C_8H_{16}Cu_2O_{10}$ 81.64 2
$C_8H_{16}Cu_2O_{10}$ 81.27 4
$C_8H_{16}Ge_2$ 69.18 2
$C_8H_{16}Hg_2O_2$ 71.15 2
$C_8H_{16}Mo_2N_2O_6S_4$ 82.16 3
$C_8H_{16}N^+$, Br^- 37.12 1
$C_8H_{16}N^+$, $C_{11}O_4^-$ 35.7 4
$C_8H_{16}NO_2^+$, Br^- 21.21 1
$C_8H_{16}NO_2^+$, Br^- 21.22 1
$C_8H_{16}NO_2^+$, Br^-, H_2O 48.37 4
$C_8H_{16}NO_2^+$, Cl^- 21.23 1
$C_8H_{16}NO_2^+$, I^- 20.9 3
$C_8H_{16}N_2^+$, ClO_4^- 32.23 1
$C_8H_{16}N_2O^{2+}$, $2Cl^-$ 58.2 1
$C_8H_{16}N_2O_2$ 1.106 1
$C_8H_{16}N_2O_2$ 32.24 1
$C_8H_{16}N_6OS_2$ 4.15 1
$(C_8H_{16}N_8NiPd)_n$ 76.49 3
$2C_8H_{16}NaO_2^+$, $C_{24}H_{28}Al_2^{2-}$ 68.21 3
$C_8H_{16}Ni_2S_6$ 85.49+ 2
$C_8H_{16}O_2$ 20.21 1
$C_8H_{16}O_4$ 38.30 1
$C_8H_{16}O_5S$ 45.52 1
$C_8H_{16}O_6S$ 45.13 4
$C_8H_{16}O_{10}Rh_2$ 81.28 4
$C_8H_{16}P_2S_2$ 64.24 2
$C_8H_{17}CoN_5O_4$ 83.26 3
$C_8H_{17}GeNO_3$ 69.17 3
$C_8H_{17}NO_2Si$ 63.2 3
$C_8H_{17}N_2O_3^+$, Br^- 48.34 3
$C_8H_{17}O_2P$, H_2O 64.25 2
$C_8H_{17}O_3P$ 64.17 3
C_8H_{18} 5.22 1
$C_8H_{18}ClN_6NiS_2^+$, Cl^-, H_2O 85.9 3

$C_8H_{18}Cl_2N_2P_2$ 64.10 **4**
$C_8H_{18}Cl_2PdSe_2$ 85.10 **3**
$C_8H_{18}Cr_2N_4O_{10}$ 82.9 **4**
$C_8H_{18}CuO_8$ 81.66 **2**
$C_8H_{18}CuO_8$ 81.22 **3**
$C_8H_{18}HgS_2$ 85.51 **2**
$C_8H_{18}NO^+$, Br^- 58.3 **1**
$C_8H_{18}NO^+$, $CH_3O_3S^-$ 32.23 **3**
$C_8H_{18}NOS^+$, I^- 3.34 **3**
$C_8H_{18}NO_2^+$, I^- 59.13 **1**
$C_8H_{18}NO_2^+$, I^- 59.14 **1**
$C_8H_{18}NO_2^+$, I^- 38.3 **3**
$C_8H_{18}N_2^{2+}$, $C_4H_{22}B_{18}Ni^{2-}$ 37.5 **3**
$C_8H_{18}N_2O_{11}Zn$ 82.32 **2**
$C_8H_{18}N_6NiS_2$ 76.47 **2**
$C_8H_{18}N_6NiS_2$ 76.50 **3**
$C_8H_{18}N_6NiS_2$ 76.51 **3**
$C_8H_{18}N_6Pt^{2+}$, $2Cl^-$, $2H_2O$ 83.29 **4**
$C_8H_{18}O_2$, $4H_2O$ 5.6 **4**
$C_8H_{18}O_5Sn_2$ 69.20 **2**
$C_8H_{19}AsO_2$ 65.3 **3**
$C_8H_{19}N_2O_2P$ 64.11 **4**
$C_8H_{20}AlLi$ 68.15 **2**
$C_8H_{20}Au_2Br_2$ 71.16 **2**
$C_8H_{20}B_2F_4N_2$ 3.35 **3**
$(C_8H_{20}BrCuP_2)_n$ 86.3 **3**
$C_8H_{20}Br_2MgO_2$ 67.3 **2**
$C_8H_{20}Br_2O_4Ti^+$, $C_4H_{10}Br_4O_2Ti^-$ 84.27 **2**
$C_8H_{20}CdCl_2S_2$ 85.52 **2**
$C_8H_{20}ClCuN_4S_4$ 85.53 **2**
$C_8H_{20}Cl_2N_4NiO_4$ 83.30 **4**
$C_8H_{20}Cl_2PdSe_2$ 85.11 **3**
$C_8H_{20}Cl_4O_4Ti_2$ 84.28 **2**
$C_8H_{20}Cl_5Re_2S_4$ 85.54 **2**
$C_8H_{20}CoN_6O_4^+$, NO_3^- 83.61 **2**
$2C_8H_{20}CoN_6S_2^+$, $C_8H_4O_{12}Sb_2^{2-}$, $4H_2O$ 76.52 **3**
$C_8H_{20}CoS_4^{2+}$, $2ClO_4^-$ 85.55 **2**
$C_8H_{20}CuN_2O_6$ 83.62 **2**
$C_8H_{20}CuN_4^{2+}$, $2NO_3^-$ 76.53 **3**
$C_8H_{20}I_2N_4Rh^+$, I^- 83.28 **3**
$C_8H_{20}Li_4$ 67.4 **2**
$C_8H_{20}N^+$, Br_6Sb^- 3.36 **3**
$C_8H_{20}N^+$, Cl_4In^- 3.60 **1**
$C_8H_{20}N^+$, I_3^- 3.61+ **1**
$C_8H_{20}N^+$, $H_2Br_4O_2Re^-$ 3.63 **1**
$C_8H_{20}N^+$, $C_4Br_3O_4W^-$ 3.12 **4**
$C_8H_{20}N^+$, $C_4HCr_2O_4^-$ 3.64 **1**
$C_8H_{20}N^+$, $C_4H_{18}B_{14}Co^-$ 62.9 **4**
$C_8H_{20}N^+$, $C_8Br_2Co_2InO_8^-$ 3.13 **4**
$C_8H_{20}N^+$, $C_9H_{15}O_3PbS_6^-$ 69.15 **4**
$C_8H_{20}N^+$, $C_9H_{26}B_{17}CoN^-$ 62.13 **4**
$C_8H_{20}N^+$, $C_{10}HCr_2O_{10}^-$ 3.37 **3**
$C_8H_{20}N^+$, $C_{16}H_{16}N_2NiS_2^-$ 85.27 **3**
$C_8H_{20}N^+$, $C_{20}H_{40}EuN_4S_8^-$ 80.12 **3**
$C_8H_{20}N^+$, $C_{20}H_{40}N_4NpS_8^-$ 80.13+ **3**
$C_8H_{20}N^+$, $C_{28}H_{20}O_7V^{2-}$, Na^+, $2C_3H_8O$ 81.52 **3**
$2C_8H_{20}N^+$, $Br_6Pt_2^{2-}$ 3.65 **1**
$2C_8H_{20}N^+$, ClO_4Ti^{2-} 3.66 **1**
$2C_8H_{20}N^+$, Cl_4Ni^{2-} 3.67 **1**
$2C_8H_{20}N^+$, Cl_5In^{2-} 3.68 **1**
$2C_8H_{20}N^+$, Cl_5OPa^{2-} 3.14 **4**
$2C_8H_{20}N^+$, $C_4H_{22}B_{18}Cu^{2-}$ 62.34 **2**
$2C_8H_{20}N^+$, $C_8FeN_5OS_4^{2-}$ 85.42 **2**
$2C_8H_{20}N^+$, $C_{14}Mo_2Ni_4O_{14}^{2-}$ 3.15 **4**
$2C_8H_{20}N^+$, $C_{16}Ni_3O_{16}W_2^{2-}$ 3.16 **4**
$3C_8H_{20}N^+$, $C_8H_{42}B_{34}Co_3^{3-}$ 62.15 **3**
$3C_8H_{20}N^+$, $C_{12}InN_6S_6^{3-}$ 68.7 **4**
$4C_8H_{20}N^+$, $C_8N_8S_8U^{4-}$ 3.17 **4**
$C_8H_{20}NO_6P$ 46.13 **1**
$C_8H_{20}NO_6P$, $CdCl_2$, $3H_2O$ 46.14 **1**
$C_8H_{20}N_2NiO_6$, $2H_2O$ 82.33 **2**
$C_8H_{20}N_2NiS_2$ 85.56 **2**
$C_8H_{20}N_4NiS_4^{2+}$, $2Br^-$ 85.10 **4**
$C_8H_{20}NiO_4P_2S_4$ 85.57+ **2**
$C_8H_{20}NiP_2S_4$ 85.60 **2**
$C_8H_{20}Ni_2S_8$ 85.12 **3**
$C_8H_{20}O_4P_2PbS_4$ 69.14 **4**
$C_8H_{20}O_4P_2S_4Zn$ 85.61 **2**
$C_8H_{20}P_2S_2$ 64.26 **2**
$C_8H_{20}P_2S_2Se_2$ 64.27 **2**
$C_8H_{20}P_2S_2Se_2Te$ 64.18 **3**
$C_8H_{20}P_2Se_5$ 64.19 **3**
$C_8H_{22}B_{18}Fe_2O_4^{2-}$, $2Cs^+$, C_3H_6O, H_2O 62.13 **3**
$C_8H_{22}Be_2^{2-}$, $2C_4H_{10}NaO^+$ 67.5 **2**
$C_8H_{22}CoN_5O_2^{2+}$, $2Cl^-$, H_2O 76.17 **4**
$2C_8H_{22}CoN_5O_2^{2+}$, $4I^-$, H_2O 76.18 **4**
$C_8H_{22}CoN_6O_4^+$, Br^- 76.19 **4**
$C_8H_{22}CoN_6O_4^+$, Br^- 83.31 **4**
$C_8H_{22}CoN_6O_4^+$, ClO_4^- 76.54 **3**
$C_8H_{22}CoN_6O_4^+$, ClO_4^- 76.20 **4**
$C_8H_{22}CoN_6O_4^+$, ClO_4^- 76.21 **4**
$C_8H_{22}CuN_{10}O_4^{2+}$, $2Cl^-$ 83.29 **3**
$C_8H_{22}N_{10}NiO_4^{2+}$, $2Cl^-$, $2H_2O$ 83.30 **3**
$C_8H_{23}ClCoN_5^{2+}$, Cl^-, ClO_4^- 76.55 **3**
$C_8H_{23}ClCoN_5^{2+}$, $2ClO_4^-$ 76.48 **2**
$C_8H_{23}ClCoN_5^{2+}$, $2ClO_4^-$ 76.49 **2**
$C_8H_{23}ClCoN_5^{2+}$, Cl_4Zn^{2-} 76.22 **4**
$C_8H_{23}CoN_8^{2+}$, $2NO_3^-$, H_2O 76.23 **4**
$C_8H_{24}Al_2Mg$ 67.8 **3**
$C_8H_{24}Al_2N_2$ 68.5 **3**
$C_8H_{24}Al_3Br_5O_6Si_4$ 68.16 **2**
$C_8H_{24}Br_2N_4Ni$ 76.56 **3**
$C_8H_{24}Cl_2FeO_4S_4^+$, Cl_4Fe^- 84.29 **2**
$C_8H_{24}CuN_4^{2+}$, $2CNS^-$ 76.57 **3**
$C_8H_{24}Cu_4N_{12}$ 83.65 **2**
$C_8H_{24}I_2InO_4S_4^+$, I_4In^- 68.6 **3**
$C_8H_{24}LaN_3O_{13}S_4$ 84.6 **4**
$C_8H_{24}N_2NiP_4S_4$ 85.14 **3**
$C_8H_{24}N_4NiO_6^{2+}$, $2Cl^-$ 84.7 **4**
$C_8H_{24}N_4Ni_3S_4^{2+}$, $2Cl^-$ 83.31 **3**
$C_8H_{24}N_4O_8P_4$ 64.12 **4**
$C_8H_{24}N_4P_4$ 64.29 **2**
$C_8H_{24}N_6NiO_4$ 76.50 **2**
$C_8H_{24}O_2Si_4$ 63.6 **2**
$C_8H_{24}O_4Si_4$ 63.7 **2**
$C_8H_{24}O_4Zn_4$ 71.17 **2**
$C_8H_{24}O_6Si_5$ 63.8 **2**
$C_8H_{24}O_8Si_8$ 63.9 **2**
$C_8H_{24}O_{12}Si_8$ 63.10 **2**
$C_8H_{25}ClCoN_5^{2+}$, Cl_4Zn^{2-} 76.58 **3**
$C_8H_{25}ClCoN_5^{2+}$, $2I^-$, H_2O 76.24 **4**
$C_8H_{25}Cl_3CuN_4P_4$ 83.32 **3**
$2C_8H_{25}N_4P_4^+$, Cl_4Co^{2-} 64.20 **3**
$C_8H_{26}CoN_6^{3+}$, $3Br^-$ 76.25 **4**
$C_8H_{26}CuN_6^{2+}$, $2Br^-$, H_2O 76.59 **3**
$C_8H_{26}CuN_6^{2+}$, $2NO_3^-$ 76.51 **2**
$C_8H_{26}N_6Ni^{2+}$, $2Cl^-$, H_2O 76.60 **3**
$C_8H_{26}N_6Zn^{2+}$, $2NO_3^-$ 76.26 **4**
$C_8H_{28}Au_4O_4$ 71.18 **2**

$C_8H_{28}Br_2N_4Pt^{2+}$, $C_8H_{28}N_4Pt^{2+}$, $4Br^-$	83.67	**2**
$C_8H_{28}Cl_2N_4Pt^{2+}$, $C_8H_{28}N_4Pt^{2+}$, $4Cl^-$, $4H_2O$	83.68	**2**
$C_8H_{28}N_4Pt^{2+}$, Cl_4Pt^{2-}	83.33	**3**
$C_8H_{28}N_4Pt^{2+}$, $C_8H_{28}Br_2N_4Pt^{2+}$, $4Br^-$	83.67	**2**
$C_8H_{28}N_4Pt^{2+}$, $C_8H_{28}Cl_2N_4Pt^{2+}$, $4Cl^-$, $4H_2O$	83.68	**2**
$C_8H_{28}N_4Si_4$	63.11	**2**
$C_8H_{30}B_{18}Cr^-$, Cs^+, H_2O	71.5	**4**
$C_8H_{30}B_{18}Ni$	62.14	**3**
$C_8H_{32}B_4P_4$	62.45	**2**
$C_8H_{32}Cl_2N_8Ni_2^{2+}$, $2ClO_4^-$	76.27	**4**
$C_8H_{34}Co_2N_9O_2^{4+}$, $4NO_3^-$, H_2O	76.28	**4**
$C_8H_{34}Co_2N_9O_4S^{3+}$, $3Br^-$	76.29	**4**
$C_8H_{34}Co_2N_{10}O_2^{4+}$, $4NO_3^-$	76.54+	**2**
$C_8H_{35}Co_2N_9O^{4+}$, $4NO_3^-$, H_2O	76.30	**4**
$C_8H_{35}Co_2N_9O_2^{4+}$, $4NO_3^-$, $2H_2O$	76.56	**2**
$C_8H_{42}B_{34}Co_3^{3-}$, $3C_8H_{20}N^+$	62.15	**3**

C_9

$C_9FN_6^-$, $C_{24}H_{20}As^+$	65.40	**2**
$C_9F_8FeO_3$	75.8	**2**
$C_9H_2N_5^-$, K^+	7.11	**4**
$C_9H_4N_3O_2^-$, K^+	7.26	**1**
$C_9H_4O_3$	27.4	**1**
$C_9H_4S_2^+$, Br^-	39.16	**3**
$C_9H_5BrN_4O_4$	32.25	**1**
C_9H_5BrO	19.27	**1**
$C_9H_5BrO_3$, H_2O	38.32	**1**
$C_9H_5ClN_4O_4$	32.24	**3**
C_9H_5ClO	19.28	**1**
$C_9H_5CoF_6S_2$	73.9	**2**
$C_9H_5CoN_2S_2$	73.10	**2**
$C_9H_5NbO_4$	73.11	**2**
$C_9H_5O_4V$	73.12	**2**
C_9H_6ClN	35.16	**1**
C_9H_6ClN	35.17	**1**
C_9H_6ClN	35.18	**1**
$C_9H_6CrO_3$	74.4+	**2**
$C_9H_6CrO_{12}^{3-}$, $C_9H_{30}CoN_6^{3+}$, $3H_2O$	76.32	**4**
$C_9H_6Fe_2N_2O_7$	83.69	**2**
$C_9H_6INO_4S$	35.2	**3**
$C_9H_6NO_2S^-$, Na^+	41.24	**1**
$C_9H_6N_2O_3$	8.45	**1**
C_9H_6OS	39.58	**1**
C_9H_6OS, $C_6H_3N_3O_6$	60.8	**4**
$C_9H_6O_2$	1.107	**1**
$C_9H_6O_2$	38.33+	**1**
$C_9H_6O_3$, H_2O	38.35	**1**
$C_9H_6O_4$	27.5	**1**
$C_9H_6O_6$	13.41	**1**
$C_9H_6O_6$, $2H_2O$	13.8	**4**
$C_9H_6S_2$	39.17	**3**
$C_9H_7BrN_2O_2$	40.15	**1**
$C_9H_7BrN_2O_2$	40.16	**1**
$C_9H_7Br_3NNi^-$, $C_{16}H_{36}N^+$	3.80	**1**
$C_9H_7Cl_2HgNO$	84.30	**2**
$C_9H_7F_2NO_2$	10.5	**3**
$C_9H_7MnO_3$	75.1	**3**
$3C_9H_7N$, CHI_3	60.3	**2**
C_9H_7NO, $C_6H_3N_3O_6$	60.29	**3**
$2C_9H_7NO$, $C_6Cl_4O_2$	60.107	**2**
$C_9H_7NO_2$	40.17	**1**
$C_9H_7NO_2$	40.18	**1**
$C_9H_7NO_2$	40.6	**3**
$C_9H_7NO_2$	36.1	**4**
$C_9H_7O_2^-$, $C_9H_8O_2$, H_4N^+	2.91	**1**
$C_9H_7O_4^-$, $C_9H_8O_4$, K^+	14.13+	**1**
$C_9H_7O_4^-$, $C_9H_8O_4$, Rb^+	14.12	**1**
$C_9H_7S_2^+$, Cl^-, $0.79H_2O$	39.10	**4**
$C_9H_7S_2^+$, I^-	39.59	**1**
$C_9H_7S_2^+$, I^-	39.60	**1**
$C_9H_7S_2^+$, CNS^-	39.18	**3**
$C_9H_8Ag_2^{2+}$, $2ClO_4^-$	75.2	**3**
C_9H_8BrNO	13.7	**3**
$C_9H_8BrN_3O$	32.25	**3**
C_9H_8ClNO	1.25	**3**
C_9H_8ClNO	19.7	**4**
$C_9H_8FeO_4$	71.5	**3**
$C_9H_8FeO_5$	72.11	**3**
$C_9H_8N^+$, Cl^-	35.19	**1**
$C_9H_8N^+$, Cl^-, H_2O	35.20	**1**
$C_9H_8N^+$, $C_{12}H_4N_4^-$, $C_{12}H_4N_4$	35.4	**3**
$2C_9H_8N^+$, $CdCl_4^{2-}$	35.5	**3**
$2C_9H_8N^+$, $C_{10}N_6^{2-}$	60.20	**4**
$C_9H_8NO^+$, Cl_3OSe^-	35.22	**1**
$C_9H_8N_2OS$	41.25	**1**
$C_9H_8N_2OS$	41.6	**3**
$C_9H_8N_2O_2$	35.23	**1**
$C_9H_8N_2O_6$	13.9	**4**
$C_9H_8N_4O_7$	16.8	**3**
$C_9H_8O_2$	22.7	**1**
$C_9H_8O_2$, $C_9H_7O_2^-$, H_4N^+	2.91	**1**
$C_9H_8O_3$	1.108	**1**
$C_9H_8O_3$	52.1	**1**
$C_9H_8O_3$	17.5	**3**
$C_9H_8O_3$	31.1	**3**
$C_9H_8O_3S$	39.19	**3**
$C_9H_8O_4$	13.42	**1**
$C_9H_8O_4$	52.2	**1**
$C_9H_8O_4$	13.10	**4**
$C_9H_8O_4$, $C_9H_7O_4^-$, K^+	14.13+	**1**
$C_9H_8O_4$, $C_9H_7O_4^-$, Rb^+	14.12	**1**
$C_9H_8O_6$	38.4	**3**
$C_9H_8S_2$	39.20	**3**
$C_9H_9BrO_3$, $C_{10}H_{12}O_4$	1.119	**1**
$C_9H_9Br_2NO_3$	50.2	**3**
$C_9H_9ClN_2O_3$	15.6	**4**
$C_9H_9ClO_3$	59.2	**3**
$C_9H_9Cl_3$	19.29	**1**
$C_9H_9Cl_3O$	27.4	**3**
$C_9H_9Cl_3O_3$	38.36	**1**
$C_9H_9CuNO_4$, $0.5H_2O$	78.1	**2**
$C_9H_9CuNO_4$, $4H_2O$	78.2	**2**
$C_9H_9F_4Rh$	71.6	**4**
$C_9H_9I_2NO_3$, $2H_2O$	48.76	**1**
C_9H_9N, $C_6H_3N_3O_6$	60.108	**2**
$C_9H_9NO_2S$	33.21	**4**
$C_9H_9NO_3$	13.9	**3**
$C_9H_9NO_3$	35.8	**4**
$C_9H_9NO_3$	48.38	**4**
$C_9H_9NO_3S$	42.4	**4**
$C_9H_9N_2OS$	41.8	**4**
$C_9H_9N_2O_7PS^-$, $C_6H_{16}N^+$	47.1	**3**
$C_9H_9N_3O_2S_2$	41.7	**3**
$C_9H_9N_3O_2S_2$	41.9+	**4**
$C_9H_9N_3O_2S_2$, $C_6H_8N_2O_2S$	60.30	**3**
$C_9H_9N_3O_2S_2$, $C_7H_8N_4O_2$	60.31	**3**

$C_9H_9N_4O^+, Br^-$ 44.32 **3**
$C_9H_9O_4^-, K^+, 2H_2O$ 52.3 **1**
$C_9H_{10}BrNO$ 13.43 **1**
$C_9H_{10}Cl_4O_2$ 20.10 **3**
$C_9H_{10}MoO_2S$ 72.12 **3**
$C_9H_{10}NOSe$ 8.46 **1**
$C_9H_{10}NO_6S^-, K^+, 2H_2O$ 48.35 **3**
$C_9H_{10}N_2O_8P^-, C_6H_{16}N^+$ 47.1 **1**
$C_9H_{10}N_4O_5, 0.5H_2O$ 44.18 **4**
$C_9H_{10}O_2$ 13.44 **1**
$C_9H_{10}O_2$ 13.11 **4**
$C_9H_{10}O_4$ 18.3 **4**
$C_9H_{11}BrN_2O_5$ 47.2 **1**
$C_9H_{11}BrN_2O_6$ 47.3 **1**
$C_9H_{11}BrN_2O_6$ 47.3 **3**
$C_9H_{11}BrN_2O_6, C_2H_6OS$ 47.4 **1**
$C_9H_{11}BrN_2O_6, C_{10}H_{13}N_5O_4, H_2O$ 47.22 **1**
$C_9H_{11}ClN_2O_5$ 47.1 **4**
$C_9H_{11}ClN_2O_6$ 47.4 **3**
$C_9H_{11}FN_2O_5$ 47.5 **1**
$C_9H_{11}F_4O_2Rh$ 72.13 **3**
$C_9H_{11}IN_2O_5$ 47.6 **1**
$C_9H_{11}IN_2O_6$ 47.5 **3**
$C_9H_{11}NO$ 16.40 **1**
$C_9H_{11}NO$ 16.41 **1**
$C_9H_{11}NO$ 20.11 **3**
$C_9H_{11}NO_2$ 15.24 **1**
$C_9H_{11}NO_2$ 10.7 **3**
$C_9H_{11}NO_2$ 13.12 **4**
$C_9H_{11}NO_2$ 16.6 **4**
$C_9H_{11}NO_3$ 18.26 **1**
$C_9H_{11}NO_3$ 48.39 **4**
$C_9H_{11}NO_4$ 48.36 **3**
$C_9H_{11}NO_4$ 48.40 **4**
$C_9H_{11}NO_6$ 50.3 **3**
$C_9H_{11}NS$ 4.16 **1**
$C_9H_{11}N_2^+, C_{12}H_4N_4^-$ 35.9 **4**
$C_9H_{11}N_2O_3^+, Br^-, H_2O$ 33.24 **3**
$C_9H_{11}N_2O_9P^{2-}, Ba^{2+}, 8.9H_2O$ 47.7 **1**
$C_9H_{11}N_7O_9P^{2-}, 2Na^+, 4H_2O$ 47.2 **4**
$C_9H_{11}N_3OS_4$ 41.26 **1**
$2C_9H_{11}N_3OS_4, I_2$ 60.109 **2**
$C_9H_{11}N_3O_2$ 9.4 **3**
$C_9H_{11}N_3O_7P^-, Na^+, 2H_2O$ 47.6 **3**
$C_9H_{11}N_5S$ 44.19 **4**
$C_9H_{11}O_4P$ 46.15 **1**
C_9H_{12} 23.1 **1**
C_9H_{12}, C_6F_6 60.21 **4**
$C_9H_{12}Ag_3^{3+}, 3NO_3^-$ 75.9 **2**
$C_9H_{12}BrN_3O_4, C_{10}H_{13}N_5O_4$ 47.23 **1**
$C_9H_{12}Cl_2O$ 38.10 **4**
$C_9H_{12}CoN_3O_6$ 82.34 **2**
$C_9H_{12}FeO$ 72.14 **3**
$C_9H_{12}NO_2^+, Cl^-$ 48.41 **4**
$C_9H_{12}NO_2^+, C_{12}H_8N_2O_8P^-$ 13.13 **4**
$C_9H_{12}NO_3^+, Cl^-, H_2O$ 3.18 **4**
$C_9H_{12}NO_3^+, I^-$ 48.79 **1**
$C_9H_{12}NO_4^+, Cl^-$ 48.38 **3**
$C_9H_{12}N_2$ 22.8 **1**
$C_9H_{12}N_2OS_3$ 41.27 **1**
$C_9H_{12}N_2O_4S_2, H_2O$ 47.7 **3**
$C_9H_{12}N_2O_4S_2, H_2O$ 47.3 **4**
$C_9H_{12}N_2O_5S$ 47.8 **3**
$C_9H_{12}N_2O_5S$ 47.4 **4**
$C_9H_{12}N_2O_5S, H_2O$ 47.5 **4**
$C_9H_{12}N_2O_5S, 1.5H_2O$ 47.9 **3**
$C_9H_{12}N_2O_6$ 40.7 **3**
$C_9H_{12}N_2O_6$ 47.10 **3**
$C_9H_{12}N_2O_6, H_2O$ 47.11 **3**
$C_9H_{12}N_2S$ 33.25 **3**
$C_9H_{12}N_4O$ 44.65 **1**
$C_9H_{12}N_4O_3$ 44.66+ **1**
$C_9H_{12}N_4O_3, C_{16}H_{10}$ 60.110 **2**
$2C_9H_{12}N_4O_3, C_{20}H_{12}$ 60.111 **2**
$2C_9H_{12}N_4O_3, C_{24}H_{12}$ 60.112 **2**
$C_9H_{12}O_2$ 27.2 **4**
$C_9H_{12}O_2S_2$ 39.61 **1**
$C_9H_{12}O_4S_3$ 39.62 **1**
$C_9H_{13}BrO$ 52.4 **1**
$C_9H_{13}ClO$ 31.3 **1**
$C_9H_{13}ClO$ 52.5 **1**
$C_9H_{13}ClO_4$ 38.11 **4**
$C_9H_{13}NO_2S_2$ 11.4 **4**
$C_9H_{13}NO_3^-, K^+, 0.33H_2O$ 12.7 **3**
$C_9H_{13}N_2O_3^-, Na^+$ 43.7 **4**
$C_9H_{13}N_2O_6$ 47.6 **4**
$C_9H_{13}N_3O_4S, 2H_2O$ 47.12 **3**
$C_9H_{13}N_3O_5$ 47.9 **1**
$C_9H_{13}N_3O_5$ 37.2 **4**
$C_9H_{13}N_3O_6$ 37.3 **4**
$C_9H_{13}Rh$ 72.24 **2**
$C_9H_{14}BrN_3O$ 32.26 **3**
$C_9H_{14}Br_2$ 31.2 **3**
$C_9H_{14}CuN_4O_4, H_2O$ 82.35 **2**
$C_9H_{14}IN_3O_4$ 45.26 **3**
$2C_9H_{14}N^+, O_4S^{2-}$ 3.19 **4**
$4C_9H_{14}N^+, C_{24}Cu_8N_{12}S_{12}^{4-}$ 3.69 **1**
$C_9H_{14}NO^+, C_{10}H_{14}BrO_4S^-$ 16.7 **4**
$C_9H_{14}NO_5P$ 33.22 **4**
$C_9H_{14}NO_7P$ 33.26 **3**
$C_9H_{14}N_2O$ 33.60 **1**
$C_9H_{14}N_2O_6, 0.5H_2O$ 47.7+ **4**
$C_9H_{14}N_3O^+, H_2O_4P^-$ 33.27 **3**
$C_9H_{14}N_3O_4^+, Cl^-$ 47.13 **3**
$C_9H_{14}N_3O_5^+, Cl^-$ 47.14 **3**
$C_9H_{14}N_3O_7P, H_2O$ 47.9+ **4**
$C_9H_{14}N_3O_8P$ 47.10+ **1**
$C_9H_{14}N_5^+, Cl^-$ 35.24 **1**
$C_9H_{14}O_3$ 38.37 **1**
$C_9H_{14}O_4S_2$ 39.63 **1**
$C_9H_{14}O_5$ 1.110 **1**
$C_9H_{14}O_5$ 1.9 **4**
$C_9H_{15}AsO_3S_6$ 65.16 **2**
$C_9H_{15}CoO_3S_6$ 80.7 **2**
$C_9H_{15}Cr$ 72.15 **3**
$C_9H_{15}CrO_3S_6$ 80.2 **4**
$C_9H_{15}CuN_5O_5, H_2O$ 82.36 **2**
$C_9H_{15}FeO_3S_6$ 80.3+ **3**
$C_9H_{15}N_2O_2$ 32.12 **4**
$C_9H_{15}O_2Rh$ 72.16 **3**
$C_9H_{15}O_3PbS_6^-, C_8H_{20}N^+$ 69.15 **4**
$C_9H_{15}O_3S_6Sb$ 66.13 **2**
$C_9H_{15}O_4^-, Rb^+, 1.5H_2O$ 2.92 **1**
$C_9H_{16}N_2$ 33.28 **3**
$C_9H_{16}O, Cl_2Hg$ 23.1 **4**
$C_9H_{16}O, 6C_{18}H_{20}OS$ 61.1 **4**
$C_9H_{16}O_4$ 1.113 **1**
$C_9H_{16}O_4$ 1.111+ **1**
$C_9H_{17}BBrN_2^+, F_6P^-$ 62.16 **3**
$C_9H_{17}NS$ 34.1 **4**
$C_9H_{17}N_6Pt^-, C_{24}H_{20}B^+$ 71.7 **4**
$C_9H_{17}OS_2^+, ClO_4^-$ 39.21 **3**

$C_9H_{17}O_{10}S^-$, Rb^+	45.54	1
$C_9H_{18}AuBr_2NS_2$	80.8	2
$C_9H_{18}Br_2CuNS_2$	80.9	2
$C_9H_{18}Cl_2N_2^{2+}$, $2ClO_4^-$	37.4	4
$C_9H_{18}N^+$, I^-	35.25	1
$C_9H_{18}N^+$, $C_6H_4BrO_3S^-$	3.38	3
$C_9H_{18}NO^+$, Cl^-	33.29	3
$C_9H_{18}NO_2$	12.8	1
$C_9H_{18}NO_2$	12.8	3
$C_9H_{18}NO_2^+$, Br^-	48.42	4
$C_9H_{18}NO_6^+$, I^-	45.55	1
$C_9H_{18}N_2O_2$	48.80	1
$C_9H_{18}N_2O_2$	1.26+	3
$C_9H_{18}N_2O_2S$	48.43	4
$C_9H_{18}N_3^+$, ClO_4^-	20.4	4
$C_9H_{18}N_6$	33.23	4
$C_9H_{18}N_6$, $C_6H_3N_3O_6$	60.113	2
$C_9H_{18}O_3$	1.115	1
$C_9H_{18}O_6$	38.5	3
$C_9H_{19}N_2O_3^+$, Br^-	48.81	1
$C_9H_{20}CoN_4O_3^+$, ClO_4^-	83.34	3
$C_9H_{20}N^+$, Br^-	23.2	1
$C_9H_{20}NO_2^+$, I^-	38.38	1
$C_9H_{20}NO_2^+$, I^-	3.41	3
$C_9H_{20}NO_2^+$, I^-	3.39+	3
$C_9H_{20}N_2O$	9.25	1
$C_9H_{21}BrCoN_5^{2+}$, $2NO_3^-$	76.61	3
$C_9H_{21}N_4^{3+}$, $3Cl^-$, $0.5H_2O$	36.2	1
$C_9H_{23}Cl_2CoN_3$	76.62	3
$C_9H_{23}CoN_5O_4^+$, ClO_4^-	76.63	3
$C_9H_{24}B_3N_3$	62.46	2
$C_9H_{24}BrCoN_4^+$, Br^-, $0.5C_2H_6O$	83.32	4
$(C_9H_{24}CaN_3O_6^{2+})_n$, $2nCl^-$	67.9	4
$C_9H_{24}ClCuN_6S_3$	79.20	3
$C_9H_{24}CoN_6O_3^+$, ClO_4^-	82.37	2
$C_9H_{26}B_{17}CoN^-$, $C_8H_{20}N^+$	62.13	4
$C_9H_{27}CuP_3S_3^+$, ClO_4^-	85.15	3
$C_9H_{30}Al_3N_3$	68.7+	3
$C_9H_{30}CoN_6^{3+}$, $3Br^-$	76.58	2
$C_9H_{30}CoN_6^{3+}$, $3Br^-$	76.31	4
$C_9H_{30}CoN_6^{3+}$, $3Br^-$, H_2O	76.64	3
$C_9H_{30}CoN_6^{3+}$, $C_9H_6CrO_{12}^{3-}$, $3H_2O$	76.32	4
$C_9H_{30}CrN_6^{3+}$, $C_5N_5Ni^{3-}$, $2H_2O$	76.65	3
$C_9H_{30}CuN_6^{2+}$, $2Br^-$, $2H_2O$	76.66	3
$C_9H_{30}N_6Ni^{2+}$, $C_6H_{24}N_4NiO_2^{2+}$, $4Cl^-$, H_2O	83.33	4
$(C_9H_{36}Cu_4N_{18}S_9^{4+})_n$, $4nNO_3^-$	79.9	4

C_{10}

$C_{10}Cl_8$	20.22	1
$C_{10}Cl_8$	24.1	1
$C_{10}Cl_{10}Ru$	73.1	4
$C_{10}Cl_{12}O_3S$	31.4	1
$C_{10}F_8$	24.1	4
$C_{10}F_{24}Ni_2O_3P_4S_2$	86.4	3
$C_{10}N_6^{2-}$, Ba^{2+}, $6H_2O$	67.6	2
$C_{10}N_6^{2-}$, Ca^{2+}, $6H_2O$	67.7	2
$C_{10}N_6^{2-}$, $2K^+$	12.9	4
$C_{10}N_6^{2-}$, $2C_9H_8N^+$	60.20	4
$C_{10}H_2N_4$, $C_{10}H_8$	60.117	2
$C_{10}H_2N_4$, $C_{10}H_{16}N_2$	60.123	2
$C_{10}H_2N_4$, $C_{14}H_{10}$	60.32	4
$C_{10}H_2N_4$, $C_{18}H_{12}N_2O_2Pd$	60.148	2
$2C_{10}H_2N_4$, $C_{18}H_{12}CuN_2O_2$	60.146+	2
$C_{10}H_2O_6$, $C_{14}H_{10}$	60.135	2
$C_{10}H_2O_6$, $C_{16}H_{10}$	60.45+	3
$C_{10}H_2O_6$, $C_{20}H_{12}$	60.154	2
$C_{10}H_4Br_2Cl_2$	24.2	1
$C_{10}H_4Br_2O_2$	25.1	1
$C_{10}H_4Br_4$	24.3	1
$C_{10}H_4Br_5NO$	32.28	1
$C_{10}H_4Cl_2O_2$	25.2	1
$C_{10}H_4Cl_4$	24.4+	1
$C_{10}H_4Cl_5NO_2$	27.3	4
$C_{10}H_4N_4O_8$	24.6	1
$C_{10}H_5BrN_2O$	40.13	4
$C_{10}H_5BrO_2$	25.3	1
$C_{10}H_5BrO_2$	25.1	3
$C_{10}H_5Br_2N$	36.3	1
$C_{10}H_5ClN_2$	7.4	3
$C_{10}H_5ClO_2$	25.2	3
$C_{10}H_5ClO_3$	25.4	1
$C_{10}H_5Cl_2FN_2O_2$	32.29	1
$C_{10}H_5IO_2$	25.1	4
$C_{10}H_5IO_3$	25.2	4
$C_{10}H_6BrNO_2$	25.5	1
$C_{10}H_6BrNO_2$	25.3	4
$C_{10}H_6BrNO_2$, H_2O	25.3	3
$C_{10}H_6BrNO_2$, CH_4O	60.32	3
$C_{10}H_6BrNO_3S_2$	39.22	3
$C_{10}H_6BrNO_4$	13.45	1
$C_{10}H_6ClNO_2$	25.6	1
$C_{10}H_6Cl_2$	24.7	1
$C_{10}H_6Cl_2$	24.8	1
$C_{10}H_6Cl_2N_2O_2$	50.2	1
$C_{10}H_6Cl_2O$	20.23	1
$C_{10}H_6CrO_4$	75.3	3
$C_{10}H_6FeN_6^{2-}$, $C_8H_{18}N_2^{2+}$	83.34	4
$C_{10}H_6FeO_4$	75.10	2
$C_{10}H_6N_2$	22.9	1
$C_{10}H_6N_2$	7.12+	4
$C_{10}H_6N_2O_2$	24.2	4
$C_{10}H_6N_2O_3S_2$	39.23	3
$C_{10}H_6N_2O_4$	24.9+	1
$C_{10}H_6N_2O_4$	24.11+	1
$C_{10}H_6O_2$	20.24	1
$C_{10}H_6O_2$	25.7	1
$C_{10}H_6O_3$	25.4	4
$C_{10}H_6O_4$	25.8+	1
$C_{10}H_6O_4$	25.5+	4
$C_{10}H_6O_8$, $2H_2O$	13.14	4
$C_{10}H_6O_{10}Os_2$	81.29	4
$C_{10}H_7BrCl_2O_3$	1.29	3
$C_{10}H_7BrF_2O_3$	1.30	3
$C_{10}H_7BrO_3S$	42.5	4
$C_{10}H_7HgI$	71.19	2
$C_{10}H_7MnO_5^-$, $C_8H_{12}N^+$	3.57	1
$C_{10}H_7MoO_3^+$, BF_4^-	75.11	2
$C_{10}H_7NO_2$	25.13	1
$C_{10}H_7NO_2$, $0.5H_2O$	25.4	3
$C_{10}H_7N_3S$	41.8	3
$C_{10}H_8$	24.15	1
$C_{10}H_8$	24.13+	1
$C_{10}H_8$	27.6+	1
$C_{10}H_8$	24.1	3
$C_{10}H_8$, $2Cl_3Sb$	60.114	2
$C_{10}H_8$, C_6N_4	60.72	2
$C_{10}H_8$, $C_6H_3N_3O_6$	60.115+	2

$C_{10}H_8, C_{10}H_2N_4$ 60.117 2
$C_{10}D_8$ 24.16 1
$C_{10}H_8AsCl_3N_2$ 65.7 4
$C_{10}H_8BrN, C_6H_3N_3O_7$ 60.77 2
$C_{10}H_8BrNO_2$ 1.116 1
$C_{10}H_8BrNO_2$ 40.19 1
$C_{10}H_8BrN_3O_2$ 32.13 4
$C_{10}H_8Br_2MoN_2O_2$ 83.35 3
$C_{10}H_8Cl_2FeO_4S_2$ 73.14 2
$C_{10}H_8Cl_2O_6$ 13.15+ 4
$C_{10}H_8Cl_3NO_3S$ 17.30 1
$C_{10}H_8Cl_4$ 24.17 1
$C_{10}H_8CrN_2O_5$ 83.71+ 2
$C_{10}H_8CrO_4, C_6H_3N_3O_6$ 60.118 2
$C_{10}H_8CuN_4O_4$ 83.36 3
$C_{10}H_8CuN_{10}, 4H_2O$ 83.73 2
$C_{10}H_8F_4GeNi, CH_3NO_2$ 69.16 4
$C_{10}H_8F_4N_2Si$ 63.4 4
$C_{10}H_8F_4N_2Sn, CH_3NO_2$ 69.17 4
$C_{10}H_8MoO_3$ 75.12 2
$C_{10}H_8MoO_5$ 71.6 3
$C_{10}H_8NO_3S^-, Na^+, 4H_2O$ 24.18 1
$C_{10}H_8N_2$ 33.61 1
$C_{10}H_8N_2O_4$ 35.26 1
$C_{10}H_8N_4O_2$ 9.26 1
$C_{10}H_8O$ 24.19 1
$C_{10}H_8O$ 24.20 1
$C_{10}H_8O$ 38.39 1
$C_{10}H_8O_2$ 24.21 1
$1.5C_{10}H_8O_2, C_{11}H_9N_4O_2^+, Br^-, H_2O$ 60.35 3
$2C_{10}H_8O_2, C_{13}H_{12}N_4O_2$ 60.27 4
$2C_{10}H_8O_2, C_{13}H_{12}N_4O_2$ 60.28 4
$2C_{10}H_8O_2, C_{13}H_{12}N_4O_2, 3H_2O$ 60.29 4
$C_{10}H_8O_2S$ 13.46 1
$C_{10}H_8O_3V$ 75.13 2
$C_{10}H_8O_4$ 38.40 1
$C_{10}H_8O_4$ 38.41 1
$2C_{10}H_8O_4P^-, Ca^{2+}, 3H_2O$ 46.16 1
$C_{10}H_8O_7Ru_3$ 75.2 4
$C_{10}H_9BF_2O_2$ 62.14 4
$C_{10}H_9BrN_4O_2S$ 44.68 1
$C_{10}H_9BrO$ 5.7 4
$C_{10}H_9BrO_2$ 1.117 1
$C_{10}H_9BrO_2$ 1.118 1
$C_{10}H_9BrO_2$ 38.42 1
$C_{10}H_9Br_3O_2$ 22.3 4
$C_{10}H_9ClN_4$ 36.4 1
$C_{10}H_9Cl_3O$ 31.5 1
$C_{10}H_9CrNO_3$ 74.6 2
$C_{10}H_9N$ 27.8 1
$C_{10}H_9NO_2$ 35.27 1
$C_{10}H_9NO_2$ 40.20 1
$C_{10}H_9NO_4S$ 41.9 3
$C_{10}H_9NO_4S, H_2O$ 35.7 3
$2C_{10}H_9N_2^+, Cl_8I_6Mo_6^{2-}$ 33.30 3
$2C_{10}H_9N_2^+, Cl_{14}Mo_6^{2-}$ 33.31+ 3
$C_{10}H_9N_2S_2^+, HO^-$ 41.28 1
$C_{10}H_9N_5O_6$ 32.14 4
$C_{10}H_{10}$ 28.2 1
$C_{10}H_{10}$ 31.6+ 1
$C_{10}H_{10}Ag^+, BF_4^-, H_2O$ 75.15 2
$C_{10}H_{10}AgN_2^+, NO_3^-, H_2O$ 83.35 4
$C_{10}H_{10}Be$ 67.9 3
$C_{10}H_{10}Be$ 67.10 4
$C_{10}H_{10}BrNO_4S$ 41.11 4
$C_{10}H_{10}Br_2CuN_2$ 83.37 3
$C_{10}H_{10}Br_2O_2$ 22.4 4
$C_{10}H_{10}Br_2O_2$ 22.5 4
$C_{10}H_{10}CdCl_2N_2$ 83.74 2
$C_{10}H_{10}CdCl_2N_2$ 83.38 3
$C_{10}H_{10}CdN_8$ 83.39 3
$C_{10}H_{10}ClCrN_2O_2$ 73.16 2
$C_{10}H_{10}Cl_2$ 23.1 3
$C_{10}H_{10}Cl_2CoN_2$ 83.75 2
$C_{10}H_{10}Cl_2CuN_2$ 83.76+ 2
$C_{10}H_{10}Cl_2N_2OSe$ 11.42 1
$C_{10}H_{10}Cl_2N_2Zn$ 83.78 2
$C_{10}H_{10}Cl_2O_4$ 59.15 1
$C_{10}H_{10}Cl_2Pt$ 73.17 2
$C_{10}H_{10}Cl_4Cu_2N_2O_2$ 84.31 2
$C_{10}H_{10}Cl_4Cu_2N_2O_2$ 84.32 2
$C_{10}H_{10}Cl_4Cu_2N_2O_2$ 84.8 3
$C_{10}H_{10}Cl_4OTi_2$ 73.18 2
$C_{10}H_{10}Co^+, ClO_4^-$ 73.19 2
$C_{10}H_{10}Cr$ 73.20 2
$C_{10}H_{10}CuN_4$ 83.36 4
$C_{10}H_{10}CuN_4O_6, 0.5C_5H_5N$ 83.37 4
$C_{10}H_{10}CuN_8$ 83.40 3
$C_{10}H_{10}F_2Zr$ 73.2 4
$C_{10}H_{10}F_4N_2Si$ 63.12 2
$C_{10}H_{10}Fe$ 73.21+ 2
$C_{10}H_{10}Fe, C_6N_4$ 60.119 2
$C_{10}H_{10}Fe^+, I_3^-$ 73.24 2
$C_{10}H_{10}Fe^+, C_6H_2N_3O_7^-$ 73.25 2
$C_{10}H_{10}Fe^+, C_6H_3Cl_9O_6^-$ 73.3 4
$C_{10}H_{10}Fe_2O_6S_2$ 85.62 2
$C_{10}H_{10}INO_3S$ 40.21 1
$C_{10}H_{10}IN_2^+, I_3^-, 2I_2$ 60.120 2
$C_{10}H_{10}IRu^+, I_3^-$ 73.4 4
$C_{10}H_{10}I_2Zr$ 73.5 4
$C_{10}H_{10}Mn_2N_4O_5$ 73.6 4
$C_{10}H_{10}MoN_2O_2S_4$ 80.5 3
$C_{10}H_{10}MoO_3$ 73.26 2
$C_{10}H_{10}Mo_2O_2S_2$ 73.27 2
$C_{10}H_{10}NO^+, H_2AsO_4^-$ 35.8 3
$C_{10}H_{10}N_2^{2+}, 2Br^-$ 33.62 1
$C_{10}H_{10}N_2^{2+}, Br_4Co^{2-}$ 33.24 4
$C_{10}H_{10}N_2^{2+}, Cl_4OSe^{2-}$ 33.34 3
$C_{10}H_{10}N_2^{2+}, Cl_5Mn^{2-}$ 33.25 4
$C_{10}H_{10}N_4O_2^+, Br^-, 2H_2O$ 36.2 3
$C_{10}H_{10}N_4O_4S_2$ 44.33 3
$C_{10}H_{10}N_4O_6Zn$ 83.38 4
$C_{10}H_{10}N_8Zn$ 83.41 3
$C_{10}H_{10}O_2$ 20.25 1
$C_{10}H_{10}O_2$ 6.1 4
$C_{10}H_{10}O_2$ 23.2 4
$C_{10}H_{10}O_2$ 31.3 4
$C_{10}H_{10}O_3S$ 39.24 3
$C_{10}H_{10}O_3S$ 39.25 3
$C_{10}H_{10}O_4$ 31.4 3
$(C_{10}H_{10}Pb)_n$ 69.21 2
$C_{10}H_{10}Ru$ 73.28 2
$C_{10}H_{10}S_4W$ 73.7 4
$C_{10}H_{10}S_5Ti$ 73.8 4
$C_{10}H_{11}BrN_4O_5$ 47.15 3
$C_{10}H_{11}BrO_3$ 38.43 1
$C_{10}H_{11}ClO_2$ 38.6 3
$C_{10}H_{11}CrNO_6$ 71.8 4
$C_{10}H_{11}F_3N_2O_5$ 47.11 4
$C_{10}H_{11}NO_2$ 7.28 1
$C_{10}H_{11}NO_2S$ 41.29 1
$C_{10}H_{11}NO_4S_2$ 4.17 1

$C_{10}H_{11}N_3$	35.9	3
$C_{10}H_{11}N_3S_2$	8.17	3
$C_{10}H_{11}N_4O_4^-$, Na^+, H_2O	44.34	3
$C_{10}H_{11}N_4O_8P^{2-}$, Ba^{2+}, $6H_2O$	47.13	1
$C_{10}H_{11}N_4O_8P^{2-}$, $2Na^+$, $7.5H_2O$	47.14	1
$C_{10}H_{11}N_5O_6P$	47.15	1
$C_{10}H_{11}N_7$, $2H_2O$	44.20	4
$C_{10}H_{12}AgO^+$, BF_4^-	75.16	2
$C_{10}H_{12}AlN_2O_8^-$, K^+, $2H_2O$	68.9	3
$C_{10}H_{12}As_4Fe_2O_6$	86.5	2
$C_{10}H_{12}BrNO$	40.22	1
$C_{10}H_{12}BrNO_2$	52.6	1
$C_{10}H_{12}BrN_5O_3$, $C_{17}H_{20}N_4O_6$, $3H_2O$	60.34	4
$C_{10}H_{12}BrN_5O_5$, $2H_2O$	47.16	3
$C_{10}H_{12}ClNO$	10.27	1
$C_{10}H_{12}ClN_5$	32.27	3
$C_{10}H_{12}ClN_5O$, H_2O	32.30	1
$C_{10}H_{12}Cl_2$	19.8+	4
$C_{10}H_{12}Cl_2CuN_4O_4$	84.9	3
$C_{10}H_{12}Cl_6Cu_2N_{10}O_2$, $2H_2O$	83.39+	4
$C_{10}H_{12}Cl_8Cu_3N_{10}$, $4H_2O$	83.43	3
$C_{10}H_{12}CoN_2O_8^-$, H_4N^+, H_2O	76.59	2
$C_{10}H_{12}Cr_2N_4O_3$	73.6	3
$C_{10}H_{12}CuN_2O_8^{2-}$, $2K^+$, $3H_2O$	76.60	2
$(C_{10}H_{12}CuN_4^+)_n$, nCl_2Cu^-	83.79	2
$C_{10}H_{12}Cu_2N_2O_8S_2^{2-}$, $2C_4H_{12}N^+$	3.39	1
$C_{10}H_{12}FeN_6$	71.20	2
$C_{10}H_{12}FeN_6$, $4CHCl_3$	71.21	2
$C_{10}H_{12}Fe_2I_2O_6P_2$	86.6	2
$C_{10}H_{12}Mo$	73.29+	2
$C_{10}H_{12}Mo_2N_2O_{14}^{4-}$, $4Na^+$, $8H_2O$	76.61	2
$C_{10}H_{12}N_2OS_3$	41.30	1
$C_{10}H_{12}N_2O_3$	40.14	4
$C_{10}H_{12}N_2O_8^{2-}$, $2Rb^+$, $2H_2O$	2.23	3
$C_{10}H_{12}N_2O_8Sm^-$, Na^+, $8H_2O$	76.33	4
$(C_{10}H_{12}N_2O_8Sn_2)_n$, $2nH_2O$	69.18	4
$C_{10}H_{12}N_2O_8Tb^-$, Na^+	76.62	2
$C_{10}H_{12}N_2O_{10}V^{3-}$, $3Na^+$, $4H_2O$	76.34	4
$C_{10}H_{12}N_3O_2S_2^+$, Br^-, H_2O	41.31	1
$C_{10}H_{12}N_3O_8^-$, Cs^+	17.31	1
$C_{10}H_{12}N_3O_8^-$, K^+	17.32	1
$C_{10}H_{12}N_4O_4$	44.69	1
$C_{10}H_{12}N_4O_4$	35.10	4
$C_{10}H_{12}N_4O_4$, H_2O	44.70	1
$C_{10}H_{12}N_4O_4$, H_2O	44.21	4
$C_{10}H_{12}N_4O_4S$	47.16	1
$C_{10}H_{12}N_4O_5$	47.17	3
$C_{10}H_{12}N_4O_5$, $2H_2O$	47.18+	3
$C_{10}H_{12}N_4O_5$, $C_4H_5N_3O$	47.20	3
$C_{10}H_{12}N_4O_8P^-$, Na^+, $8H_2O$	47.21	3
$C_{10}H_{12}N_5O_5P$, H_2O	47.12	4
$C_{10}H_{12}N_5O_{10}P_2^{3-}$, $3Rb^+$, $3H_2O$	47.13	4
$2C_{10}H_{12}O$, $HClO_4$	60.121	2
$C_{10}H_{12}O_2$	18.27	1
$C_{10}H_{12}O_2$	20.26	1
$C_{10}H_{12}O_2$	28.3	1
$C_{10}H_{12}O_2$	13.12	3
$C_{10}H_{12}O_2$	13.13	3
$C_{10}H_{12}O_2$	23.3	3
$C_{10}H_{12}O_2$	31.5	3
$C_{10}H_{12}O_2$	22.6	4
$C_{10}H_{12}O_2S_4$	39.65	1
$C_{10}H_{12}O_3$	31.4	4
$C_{10}H_{12}O_4$	31.5	4
$C_{10}H_{12}O_4$, $C_8Cl_2N_2O_2$	60.33	3
$C_{10}H_{12}O_4$, $C_9H_9BrO_3$	1.119	1
$C_{10}H_{12}O_6$	21.24+	1
$C_{10}H_{12}S_2$	39.66	1
$C_{10}H_{13}As_2CrF_3O_4$	86.5	3
$C_{10}H_{13}As_2F_3MoO_4$	86.1	4
$C_{10}H_{13}Br$	19.30+	1
$C_{10}H_{13}BrN_2O_3$, $C_7H_9N_5$	60.25	3
$C_{10}H_{13}BrN_2O_4$	47.18	1
$C_{10}H_{13}BrO_3$	27.4	4
$C_{10}H_{13}BrO_4$	59.16	1
$C_{10}H_{13}Cl$	19.32	1
$C_{10}H_{13}ClNO$	40.23	1
$C_{10}H_{13}ClN_5^+$, I^-	44.71	1
$C_{10}H_{13}CuN_5O_6^{2-}$, $2Na^+$, $4.5H_2O$	82.17	3
$C_{10}H_{13}F_3MoO_4P_2$	86.7	3
$C_{10}H_{13}N_2^+$, Cl^-	35.10	3
$C_{10}H_{13}N_2O^+$, $C_4H_8N_3O^+$, O_4S^{2-}, H_2O	35.28	1
$C_{10}H_{13}N_2O^+$, $C_6H_2N_3O_7^-$, H_2O	35.11	4
$C_{10}H_{13}N_2O_8P^{2-}$, Ca^{2+}, $6H_2O$	47.19	1
$C_{10}H_{13}N_4O_7P$, $3H_2O$	47.20	1
$C_{10}H_{13}N_5O_3$, H_2O	47.21	1
$C_{10}H_{13}N_5O_4$	47.22	3
$C_{10}H_{13}N_5O_4$, $C_9H_{11}BrN_2O_6$, H_2O	47.22	1
$C_{10}H_{13}N_5O_4$, $C_9H_{12}BrN_3O_4$	47.23	1
$2C_{10}H_{13}N_5O_4$, $C_{62}H_{86}N_{12}O_{16}$, $12H_2O$	60.43+	4
$C_{10}H_{13}N_5O_4S$, H_2O	47.14	4
$C_{10}H_{13}N_5O_5$, $2H_2O$	47.23	3
$C_{10}H_{14}BeO_4$	67.8	2
$C_{10}H_{14}BeO_4$	67.11	4
$C_{10}H_{14}BrO_4S^-$, $C_9H_{14}NO^+$	16.7	4
$C_{10}H_{14}Br_2O_2$	23.4	3
$C_{10}H_{14}ClFeO_4$	77.4	3
$C_{10}H_{14}ClO_4Pt^-$, K^+	77.4	2
$C_{10}H_{14}Cl_2FeGe$	72.5+	4
$C_{10}H_{14}Cl_2O_3Pt$	75.17	2
$C_{10}H_{14}Cl_2O_4Re$	77.5	2
$C_{10}H_{14}Cl_6O_4Ti$	77.1	4
$C_{10}H_{14}CoN_4O_4$	81.30	4
$C_{10}H_{14}CoN_6O_4S_2^-$, H_4N^+, $3H_2O$	83.80	2
$C_{10}H_{14}CoS_4$	85.63	2
$C_{10}H_{14}CuN_2O_2^{2+}$, O_4S^{2-}	83.81	2
$C_{10}H_{14}CuO_4$	77.5	3
$C_{10}H_{14}FeN_2O_9^-$, Li^+, $2H_2O$	76.63	2
$C_{10}H_{14}FeN_2O_9^-$, Rb^+, H_2O	76.64	2
$C_{10}H_{14}MnN_2O_9^{2-}$, $2Li^+$, $4H_2O$	76.35	4
$C_{10}H_{14}N_2NiO_8$	83.41	4
$C_{10}H_{14}N_2O$	3.20	4
$C_{10}H_{14}N_2O_5$	47.25	1
$C_{10}H_{14}N_2O_5S$	47.24	3
$C_{10}H_{14}N_2O_6$, $0.5H_2O$	47.26	1
$C_{10}H_{14}N_2O_8^{2-}$, $2K^+$, H_2O	2.93	1
$C_{10}H_{14}N_2O_8^{2-}$, $2Rb^+$, $2H_2O$	2.94	1
$C_{10}H_{14}N_2O_9Sn$	69.19	4
$C_{10}H_{14}N_2O_{10}V^-$, H_4N^+, $3H_2O$	76.36	4
$C_{10}H_{14}N_2S$	41.12	4
$C_{10}H_{14}N_4O_2$	40.15	4
$C_{10}H_{14}N_5O_4^+$, Br^-, H_2O	50.3	1
$C_{10}H_{14}N_5O_6P$, C_2H_6O	47.15	4
$C_{10}H_{14}N_5O_7P$, H_2O	47.27	1
$C_{10}H_{14}N_5O_7P$, $2H_2O$	47.28	1
$C_{10}H_{14}N_5O_{13}P_3^{2-}$, $2Na^+$, $3H_2O$	47.25	3
$C_{10}H_{14}N_6O_3$, H_2O	47.26	3
$C_{10}H_{14}Ni_2$	72.17	3
$C_{10}H_{14}O_2S$	11.43	1
$C_{10}H_{14}O_4$	1.120+	1

$C_{10}H_{14}O_4$	5.23	**1**
$C_{10}H_{14}O_4$	31.6	**3**
$C_{10}H_{14}O_4$	38.12	**4**
$C_{10}H_{14}O_4Pd$	77.2	**4**
$C_{10}H_{14}O_4S_2$	11.5	**4**
$C_{10}H_{14}O_4S_3$	11.12	**3**
$C_{10}H_{14}O_5V$	77.6+	**2**
$2C_{10}H_{14}O_5V$, $C_4H_8O_2$	60.34	**3**
$C_{10}H_{14}O_8$	1.122	**1**
$C_{10}H_{15}BrClNO$	52.7	**1**
$C_{10}H_{15}BrO$	52.8	**1**
$C_{10}H_{15}BrO$	52.10	**1**
$C_{10}H_{15}BrO$	52.11	**1**
$C_{10}H_{15}BrO$	27.5	**3**
$C_{10}H_{15}BrO$	52.1	**3**
$C_{10}H_{15}Br_2N_3Zn$	83.44	**3**
$C_{10}H_{15}Br_3O$	52.12	**1**
$C_{10}H_{15}CeO_2^{2-}$, $2K^+$, H_2O	81.67	**2**
$C_{10}H_{15}ClO$	27.6	**3**
$C_{10}H_{15}CoN_5^+$, ClO_4^-	72.25	**2**
$C_{10}H_{15}CrN_2O_9$	76.66	**2**
$C_{10}H_{15}CuN_6O_3^+$, ClO_4^-	82.38	**2**
$C_{10}H_{15}FeN_2O_9$	76.37	**4**
$C_{10}H_{15}GaN_2O_9$	68.4	**4**
$2C_{10}H_{15}MnN_2O_9^-$, $H_8MnO_4^{2+}$, $4H_2O$	76.65	**2**
$C_{10}H_{15}NO$, $C_{20}H_{28}CuN_2O_2$, $2H_2O$	83.182	**2**
$C_{10}H_{15}NO_4$	59.17+	**1**
$C_{10}H_{15}NO_4$, H_2O	59.19	**1**
$C_{10}H_{15}NO_5Zn$, $2H_2O$	81.68	**2**
$C_{10}H_{15}N_2$, $C_7H_5O_3$	60.22	**4**
$2C_{10}H_{15}N_2NiO_9^-$, Ni^{2+} , $8H_2O$	76.67	**3**
$3C_{10}H_{15}O_4S^-$, $C_6H_{21}CrN_{15}^{3+}$	79.16	**3**
$C_{10}H_{16}$	31.9+	**1**
$C_{10}H_{16}Ag^+$, NO_3^-	75.4	**3**
$C_{10}H_{16}Ag_4O_4^{4+}$, $4ClO_4^-$	74.1	**3**
$C_{10}H_{16}As_2Cl_4OW$	86.8	**3**
$C_{10}H_{16}BrNO_2$	31.12	**1**
$C_{10}H_{16}BrNO_2$	52.13	**1**
$(C_{10}H_{16}Br_2MnO_4)_n$	84.10	**3**
$C_{10}H_{16}Br_2NiO_4$	77.3	**4**
$C_{10}H_{16}Br_2O$	21.26	**1**
$C_{10}H_{16}Br_2O$	52.14	**1**
$C_{10}H_{16}Cl_2N_2O_8Pd$, $5H_2O$	76.68	**3**
$C_{10}H_{16}Cl_2O$	21.4	**4**
$C_{10}H_{16}Cl_2Pt$	75.18	**2**
$C_{10}H_{16}CoN_2O_8^-$, Na^+ , $2H_2O$	82.10	**4**
$C_{10}H_{16}CuN_2O_6$, $4H_2O$	81.23	**3**
$C_{10}H_{16}CuN_2O_9$	76.69	**3**
$C_{10}H_{16}Cu_2N_2O_{10}$, $2H_2O$	76.70	**3**
$2C_{10}H_{16}N^+$, Cl_4Cu^{2-}	3.70	**1**
$C_{10}H_{16}NO^+$, Cl^-	58.4	**1**
$C_{10}H_{16}NO^+$, Cl^-	58.2	**3**
$C_{10}H_{16}NO_9S_2^-$, K^+ , H_2O	45.27	**3**
$C_{10}H_{16}N_2$, I	16.42	**1**
$C_{10}H_{16}N_2$, $C_6Cl_4O_2$	60.122	**2**
$C_{10}H_{16}N_2$, $C_{10}H_2N_4$	60.123	**2**
$C_{10}H_{16}N_2$, $C_{12}H_4N_4$	60.124	**2**
$C_{10}H_{16}N_2$, $2C_{12}H_4N_4$	60.127	**2**
$C_{10}H_{16}N_2^{2+}$, $2I^-$	58.3	**3**
$C_{10}H_{16}N_2NiO_9$	76.69	**2**
$C_{10}H_{16}N_2O_2S_4Se$	40.9	**3**
$C_{10}H_{16}N_2O_2S_4Te$	70.3	**3**
$C_{10}H_{16}N_2O_3$	43.8	**4**
$C_{10}H_{16}N_2O_4Pd$	82.18	**3**
$C_{10}H_{16}N_2O_5$	47.27	**3**
$C_{10}H_{16}N_2O_8$	1.123	**1**
$C_{10}H_{16}N_2O_8$	1.10	**4**
$C_{10}H_{16}N_2O_8^{2-}$, $H_{12}MgO_6^{2+}$	2.95	**1**
$C_{10}H_{16}N_4NiO_6$	83.45	**3**
$C_{10}H_{16}N_6S_2Se_2Te$	70.8	**4**
$C_{10}H_{16}O_2$	52.15	**1**
$C_{10}H_{16}O_2$	52.16	**1**
$C_{10}H_{16}O_3$	21.5	**4**
$C_{10}H_{16}O_3S$	52.17	**1**
$C_{10}H_{16}O_4$	22.11	**1**
$C_{10}H_{16}O_4$	22.12	**1**
$C_{10}H_{16}O_4$	21.8	**3**
$C_{10}H_{16}O_5Zn$	77.8+	**2**
$C_{10}H_{17}BrO$	21.27	**1**
$C_{10}H_{17}ClO$	21.6	**4**
$C_{10}H_{17}NO_3$, Na^+ , Br^- , H_2O	37.6	**3**
$C_{10}H_{17}N_2^+$, ClO_4^-	16.8+	**4**
$C_{10}H_{17}N_3O_6S$	48.82	**1**
$C_{10}H_{17}N_3O_6S$	48.39	**3**
$C_{10}H_{18}Br_2$	23.4	**1**
$C_{10}H_{18}Br_2Ni_2O_4$	72.27	**2**
$(C_{10}H_{18}CdO_2S_4)_n$	80.10	**2**
$C_{10}H_{18}ClNO_3S$	39.11	**4**
$C_{10}H_{18}Cl_2Pd_2$	72.28	**2**
$C_{10}H_{18}CoN_4O_4S_2$	79.21	**3**
$C_{10}H_{18}CoN_6O_4^-$, $2C_5H_{11}N_3O_2$, K^+	79.10	**4**
$C_{10}H_{18}CoO_6$	77.11	**2**
$C_{10}H_{18}CuN_2O_7S_2$, $2H_2O$	81.24	**3**
$C_{10}H_{18}CuN_4O_4$	83.82	**2**
$C_{10}H_{18}CuN_4O_6$, H_2O	82.39	**2**
$C_{10}H_{18}CuN_6^{2+}$, $2BF_4^-$	83.46	**3**
$C_{10}H_{18}CuN_6^{2+}$, $2ClO_4^-$	83.47	**3**
$C_{10}H_{18}KLaN_2O_{11}$, $5H_2O$	76.71	**2**
$C_{10}H_{18}MnO_6$	77.12	**2**
$C_{10}H_{18}MnO_6$	77.6	**3**
$C_{10}H_{18}N^+$, $C_7H_7O_3S^-$	37.5	**4**
$C_{10}H_{18}N_2NiO_5$, $2H_2O$	82.11	**4**
$C_{10}H_{18}N_2O_5$	40.10	**3**
$C_{10}H_{18}N_4NiO_4$	83.83	**2**
$C_{10}H_{18}N_4O_6S_2$, $2H_2O$	48.83	**1**
$C_{10}H_{18}NiO_6$	77.13	**2**
$C_{10}H_{18}O_2$	52.1	**4**
$C_{10}H_{18}O_2PbS_4$	69.22	**2**
$C_{10}H_{18}O_2Si_2$	63.13	**2**
$C_{10}H_{18}O_4$	1.124+	**1**
$C_{10}H_{18}O_4$	1.31	**3**
$C_{10}H_{18}O_4$	1.11	**4**
$C_{10}H_{18}O_7U$	77.10	**2**
$C_{10}H_{19}NO_8$, $3H_2O$	45.28	**3**
$C_{10}H_{19}O_2^-$, K^+	2.96	**1**
$C_{10}H_{20}AsBrN_2S_4$	65.4	**3**
$C_{10}H_{20}CdN_2S_4$	80.11	**2**
$C_{10}H_{20}ClFeN_2S_4$	80.6	**3**
$C_{10}H_{20}ClN_6NiO_5^+$, ClO_4^-	83.48	**3**
$C_{10}H_{20}Cl_2CoN_2O_2$	83.49	**3**
$C_{10}H_{20}CuN_2O_4S_2$	82.41	**2**
$C_{10}H_{20}CuN_2O_6$	82.42	**2**
$C_{10}H_{20}CuN_2S_4$	80.13+	**2**
$C_{10}H_{20}CuN_2Se_4$	80.7	**3**
$C_{10}H_{20}FeIN_2S_4$	80.3	**4**
$C_{10}H_{20}FeN_3OS_4$	80.16	**2**
$C_{10}H_{20}Mo_2N_2O_8S_2$	82.19	**3**
$C_{10}H_{20}N^+$, Cl^-	52.18	**1**
$C_{10}H_{20}N^+$, Cl^-	37.6	**4**
$C_{10}H_{20}N^+$, I^-	35.29	**1**
$C_{10}H_{20}NO^+$, Br^-	37.7	**3**
$C_{10}H_{20}N_2NiS_4$	80.17	**2**

$C_{10}H_{20}N_2NiSe_4$ 80.8 3
$C_{10}H_{20}N_2O_2$ 1.126 1
$C_{10}H_{20}N_2O_2$ 3.71 1
$C_{10}H_{20}N_2O_4Pd$, H_2O 82.20 3
$C_{10}H_{20}N_2PbS_4$ 69.20 4
$C_{10}H_{20}N_2PdS_4$ 80.4 4
$C_{10}H_{20}N_2PtS_4$ 80.19 2
$C_{10}H_{20}N_2S_4$ 4.18 1
$C_{10}H_{20}N_2S_4Se$ 39.26 3
$C_{10}H_{20}N_2S_4Te$ 70.9 4
$C_{10}H_{20}N_2S_4Zn$ 80.21 2
$C_{10}H_{20}N_2Se_4Zn$ 80.9 3
$C_{10}H_{20}N_3ReS_4$ 80.5 4
$C_{10}H_{20}NiO_6^{2+}$, $2ClO_4^-$ 77.4 4
$C_{10}H_{20}O_2$ 23.5+ 3
$C_{10}H_{20}O_4S_2$ 45.14 4
$C_{10}H_{20}P_2S_2$ 64.21 3
$C_{10}H_{21}AgNPS$ 85.64 2
$C_{10}H_{21}LaN_2O_{12}$, $3H_2O$ 76.72 2
$C_{10}H_{21}MgO_6$ 67.9 2
$C_{10}H_{21}NO$ 1.127 1
$C_{10}H_{21}OP$ 64.13 4
$C_{10}H_{22}AlN$ 68.5 4
$C_{10}H_{22}Cl_2N_2O_2Pt$ 83.50 3
$C_{10}H_{22}N^+$, Cl^-, $1.5H_2O$ 23.5 1
$C_{10}H_{22}NO_4^+$, I^- 45.29 3
$C_{10}H_{22}N_2O_4S_2^{2+}$, $2Cl^-$ 48.44 4
$C_{10}H_{23}Cl_2CoN_4O_2$ 83.51 3
$C_{10}H_{23}N_4NiO_2^+$, Cl^-, H_2O 83.84 2
$C_{10}H_{23}N_4NiO_2^+$, Cl^-, H_2O 83.42 4
$C_{10}H_{23}N_4O_2Pt^+$, Cl^-, H_2O 83.52 3
$C_{10}H_{24}Br_4N_4S_2Te$ 70.17 2
$C_{10}H_{24}ClN_2Ni^+$, Cl^- 83.85 2
$C_{10}H_{24}ClN_4Ni^+$, Cl^- 83.86 2
$C_{10}H_{24}Cl_2CuN_4S_2$ 79.22 3
$C_{10}H_{24}Cl_2N_4O_3V$ 79.23 3
$C_{10}H_{24}Cl_2N_4S_2Zn$ 79.11 4
$C_{10}H_{24}Cl_4N_4S_2Te$ 70.18 2
$C_{10}H_{24}CuN_4^{2+}$, $2NO_3^-$, $0.5H_2O$ 83.87 2
$C_{10}H_{24}CuN_4O_4^{2+}$, $2Cl^-$, $2H_2O$ 82.21 3
$C_{10}H_{24}N^+$, Cl^- 3.72 1
$C_{10}H_{24}N_2^{2+}$, $2Cl^-$ 23.6+ 1
$C_{10}H_{24}N_2^{2+}$, $2Cl^-$, $2H_2O$ 23.8 1
$C_{10}H_{25}BrMgO_2$ 67.10 2
$C_{10}H_{26}Al_2O_2$ 68.18 2
$C_{10}H_{26}N_2^{2+}$, $2Br^-$, $2H_2O$ 3.21 4
$C_{10}H_{28}CoN_6^{3+}$, $C_6CoN_6^{3-}$, $2H_2O$ 76.71 3
$C_{10}H_{28}N_4Zn_2$ 76.72+ 3
$C_{10}H_{30}Al_4Cl_4N_6$ 68.6 4
$C_{10}H_{30}I_2Rh_2S_3$ 85.65 2
$C_{10}H_{30}N_2Si_4$ 63.14 2
$C_{10}H_{30}N_4^{4+}$, $4Cl^-$ 59.20 1
$C_{10}H_{30}N_4^{4+}$, $2HO_4P^{2-}$, $6H_2O$ 59.21 1
$C_{10}H_{30}N_5Nb$ 83.53 3
$C_{10}H_{32}N_{10}Ni_2S_2^{2+}$, $2ClO_4^-$ 76.70 2
$C_{10}H_{32}N_{10}Ni_2S_2^{2+}$, $2I^-$ 76.74 2
$C_{10}H_{33}N_6^{5+}$, $5Cl^-$ 3.73 1
$C_{10}H_{45}N_5Si_5$ 63.15 2

C_{11}

$C_{11}H_3Co_3O_9$ 71.22 2
$C_{11}H_5CoFeHgO_6$ 73.31 2
$C_{11}H_5F_7MoO_3$ 73.32 2
$C_{11}H_5O_4ReS_2$ 85.11 4
$C_{11}H_6BrN_3O_2$ 40.16 4
$C_{11}H_6Br_2O$ 27.5 4
$C_{11}H_6Fe_3N_2O_9$ 83.88 2
$C_{11}H_7ClN_4O_2S$ 42.4 3
$C_{11}H_7ClO$ 27.6 4
$C_{11}H_7ClO_2$ 25.15 1
$C_{11}H_7CrO_5^-$, $C_8H_{12}N^+$ 3.58 1
$C_{11}H_7CrO_5^-$, $C_8H_{12}N^+$ 3.59 1
$C_{11}H_7FeO_3$ 71.9 4
$C_{11}H_7NO$ 36.2 4
$C_{11}H_8Br_2O_3$ 1.128 1
$C_{11}H_8ClN_3O_2S$ 41.10 3
$C_{11}H_8CrO_4$ 74.2 3
$C_{11}H_8CrO_5$ 74.7 2
$C_{11}H_8CrO_5$ 74.2 4
$C_{11}H_8F_2$ 31.6 4
$C_{11}H_8FeO_3$ 75.19 2
$C_{11}H_8Fe_2N_2O_6$ 83.43 4
$C_{11}H_8MoO_3$ 75.20 2
$C_{11}H_8NO$ 36.5 1
$C_{11}H_8N_2O_3S_2$ 59.4+ 4
$2C_{11}H_8N_4O_2$, $1.5CHO_2^-$, $1.5Ag^+$, xH_2O 36.3 4
$C_{11}H_8O$ 27.10 1
$C_{11}H_8O_2$ 24.22 1
$C_{11}H_8O_2$ 24.23 1
$C_{11}H_8O_2$ 28.4 1
$C_{11}H_8O_3$ 25.16 1
$C_{11}H_8O_3$ 24.3 4
$C_{11}H_8O_3Ru$ 75.5 3
$C_{11}H_9BrO_3$ 38.8 3
$C_{11}H_9ClN_2O_3$ 40.25 1
$C_{11}H_9ClO$ 20.12 3
$C_{11}H_9FeNO_4$ 75.6 3
$C_{11}H_9FeNO_5$ 75.7 3
$C_{11}H_9FeO_3^+$, BF_4^- 75.21 2
$C_{11}H_9NO_2$ 25.17 1
$C_{11}H_9NO_2$ 40.17 4
$C_{11}H_9NO_2$, H_2O 25.5 3
$C_{11}H_9NO_2$, $0.25H_2O$ 25.18 1
$C_{11}H_9N_4O_2^+$, Br^-, $2H_2O$ 36.6 1
$C_{11}H_9N_4O_2^+$, Br^-, $2H_2O$ 36.3 3
$C_{11}H_9N_4O_2^+$, $1.5C_{10}H_8O_2$, Br^-, H_2O 60.35 3
$C_{11}H_9O_3Re$ 73.33 2
$C_{11}H_9O_5S^-$, Na^+ 11.44 1
$C_{11}H_{10}CoN_4^{2+}$, $2Cl^-$ 83.89 2
$C_{11}H_{10}CrO_3$ 75.22 2
$C_{11}H_{10}FeO_4S$ 71.8 3
$C_{11}H_{10}N_2$ 27.11 1
$C_{11}H_{10}N_2O_2S$ 41.32 1
$C_{11}H_{10}N_2O_3$ 40.26 1
$C_{11}H_{10}N_2O_3$ 43.26 1
$C_{11}H_{10}O_2$ 24.24 1
$C_{11}H_{10}O_2S$ 27.7 3
$C_{11}H_{10}O_6$ 31.7 4
$C_{11}H_{11}BrN_2O$ 32.31 1
$C_{11}H_{11}IO_5$ 27.7 4
$C_{11}H_{11}NO_2$ 40.11 3
$C_{11}H_{11}P$ 64.22 3
$C_{11}H_{12}As_2F_6MoO_4$ 86.9 3
$C_{11}H_{12}BrNO$ 1.12 4
$C_{11}H_{12}NO^+$, Cl^- 24.25 1
$C_{11}H_{12}NS^+$, Br^- 41.11 3
$C_{11}H_{12}N_2$ 35.11 3

$C_{11}H_{12}N_2O$	32.28	3
$C_{11}H_{12}N_2O$	32.15	4
$C_{11}H_{12}N_2O_3$	48.40	3
$C_{11}H_{12}N_2O_3$	48.45	4
$C_{11}H_{12}OS_2$	11.6	4
$C_{11}H_{12}O_2$	20.13	3
$C_{11}H_{12}O_2$	29.1	4
$C_{11}H_{12}O_3S$	39.27	3
$C_{11}H_{12}O_4$	1.32	3
$C_{11}H_{12}O_6$	38.9	3
$C_{11}H_{13}ClO_2S$	20.5	4
$C_{11}H_{13}CuNO_4$	78.1	4
$C_{11}H_{13}I_2NO_3$	48.41	3
$C_{11}H_{13}MoNO$	71.10	4
$C_{11}H_{13}N_2O_2^+$, Br^-	48.84+	1
$C_{11}H_{13}N_2O_2^+$, Cl_2^-	48.86	1
$C_{11}H_{13}N_2O_2^+$, CHO_2^-	48.42	3
$C_{11}H_{13}OP$	64.14	4
$C_{11}H_{13}O_3ReSi$	73.9	4
$C_{11}H_{14}AgP$	72.29	2
$C_{11}H_{14}AuN$	71.23	2
$C_{11}H_{14}CoN_2O_8^-$, K^+, $2H_2O$	81.31	4
$C_{11}H_{14}CrNO_5$	71.24	2
$C_{11}H_{14}CuP$	72.30	2
$C_{11}H_{14}N_2O_3$	48.43	3
$C_{11}H_{14}N_2O_3S$	33.35	3
$C_{11}H_{14}N_2O_3S$	41.13	4
$C_{11}H_{14}O$	27.8	4
$C_{11}H_{14}O_2$	31.8	3
$C_{11}H_{15}BrN_2O_4$	45.57+	1
$C_{11}H_{15}BrN_2O_4$	45.59+	1
$C_{11}H_{15}BrN_3O_7^+$, Br^-	58.1	4
$C_{11}H_{15}BrO_7$	45.15	4
$C_{11}H_{15}Cl$	19.33	1
$C_{11}H_{15}NO$	52.19	1
$C_{11}H_{15}NO_2S$	39.12	4
$C_{11}H_{15}NO_3$	48.44	3
$C_{11}H_{15}N_2O_3^+$, Cl^-, H_2O	48.46	4
$C_{11}H_{13}N_2O_4^+$, Cl^-, H_2O	48.87	1
$C_{11}H_{15}N_3$	37.15	1
$C_{11}H_{15}N_5S$	44.22	4
$C_{11}H_{15}O_2Tl$	68.19	2
$C_{11}H_{15}O_3P$	64.23	3
$C_{11}H_{16}BrN_3O_{10}S$, H_2O	50.4	1
$C_{11}H_{16}Br_2$	28.1	3
$C_{11}H_{16}ClN_5^+$, Cl^-	8.10	4
$C_{11}H_{16}NOSe^+$, Cl^-	3.22	4
$C_{11}H_{16}N_2O_3$	43.27	1
$C_{11}H_{16}N_2O_5S$	47.28	3
$C_{11}H_{16}N_4NiO_4$	83.44	4
$C_{11}H_{16}N_5O_3^+$, Br^-	50.5	1
$C_{11}H_{16}N_5O_6P$, C_2H_6O	47.16	4
$C_{11}H_{16}O_4S$	31.9	3
$C_{11}H_{16}P^+$, I^-	64.31	2
$C_{11}H_{17}ClN_5^+$, Br^-	8.47	1
$C_{11}H_{17}ClN_5^+$, Cl^-	8.48	1
$C_{11}H_{17}NO_5$	45.16	4
$C_{11}H_{17}N_2^+$, Br^-	58.4	3
$C_{11}H_{17}N_2O_2^+$, Cl_3Ge^-, $0.5H_2O$	58.6	1
$2C_{11}H_{17}N_3S_4Zn$, C_6H_6	80.22	2
$C_{11}H_{17}N_6O_3^+$, I^-, H_2O	47.17	4
$2C_{11}H_{18}NO_3^+$, O_4S^{2-}, $2H_2O$	59.3	3
$C_{11}H_{18}N_2O_2$	48.47	4
$C_{11}H_{18}N_2O_3$	43.28+	1
$C_{11}H_{18}N_2O_3$	43.4+	3
$C_{11}H_{18}N_3O_8^+$, Br^-	58.7	1
$C_{11}H_{18}N_3O_8^+$, Br^-	58.2	4
$C_{11}H_{18}O$	31.10	3
$C_{11}H_{19}NO_2S_4$	58.5	3
$C_{11}H_{19}N_2^+$, BF_4^-	35.12	4
$C_{11}H_{19}N_2^+$, ClO_4^-	32.29	3
$C_{11}H_{20}CuP$	73.7	3
$C_{11}H_{20}N_2$	32.16	4
$C_{11}H_{20}O_2$	21.7	4
$C_{11}H_{20}O_4$	1.129	1
$C_{11}H_{21}BrO_2$	1.130	1
$C_{11}H_{21}N_4Ni^+$, ClO_4^-	83.45	4
$C_{11}H_{22}O_2S$	1.131	1
$C_{11}H_{23}BrO$	5.8	4
$C_{11}H_{23}MoN_4O_6P$	83.46	4
$C_{11}H_{23}O_4P$	64.24	3
$C_{11}H_{24}I_3Zn^-$, $C_{11}H_{25}S^+$	11.45	1
$C_{11}H_{24}N^+$, Cl^-	34.7	1
$C_{11}H_{24}NO_2^+$, Br^-, $0.5H_2O$	1.132	1
$C_{11}H_{25}S^+$, $C_{11}H_{24}I_3Zn^-$	11.45	1
$C_{11}H_{26}CoN_5O_2^{2+}$, Cl_4Zn^{2-}	76.74	3
$C_{11}H_{26}CoN_5O_2^{2+}$, $2I^-$, $2H_2O$	76.75	3
$C_{11}H_{26}N_2O^{2+}$, $2CH_3O_3S^-$, H_2O	38.10	3
$C_{11}H_{27}As_3Br_2Ni$	86.7	2
$C_{11}H_{28}N_2^{2+}$, $2I^-$, $0.25H_2O$	3.74	1

C_{12}

$C_{12}Cl_{12}$	28.5	1
$C_{12}Cl_{12}$	31.13	1
$C_{12}Cl_{14}$	22.13	1
$C_{12}Co_2F_5O_7$	73.34	2
$C_{12}Co_2F_6O_6$	75.23	2
$C_{12}F_{10}$	19.34	1
$C_{12}F_{10}Hg$	71.25	2
$C_{12}F_{10}Hg$	71.9	3
$C_{12}F_{12}FeO_4$	73.35	2
$C_{12}F_{18}MoSe_6$	85.12	4
$C_{12}FeN_6S_6^{2-}$, $2C_{24}H_{20}As^+$	85.16	3
$C_{12}InN_6S_6^{3-}$, $3C_8H_{20}N^+$	68.7	4
$C_{12}MoN_6S_6^{2-}$, $2C_{24}H_{20}As^+$	85.17	3
$C_{12}N_6S_6V^{2-}$, $2C_4H_{12}N^+$	3.40	1
$C_{12}HF_{10}OTl$	68.10	3
$C_{12}H_2Cl_6O_2$	59.22	1
$C_{12}H_4ClO_6Re$	71.10	3
$C_{12}H_4Cl_2O_3$	24.4	4
$C_{12}H_4Cl_4N_2$	36.7	1
$C_{12}H_4Cl_4N_2$	36.8	1
$C_{12}H_4Cl_4O_2$	38.13	4
$C_{12}H_4N_2O_7$	24.5	4
$C_{12}H_4N_4$	7.29	1
$C_{12}H_4N_4$, $C_{10}H_{16}N_2$	60.124	2
$C_{12}H_4N_4$, $2C_{12}H_4N_4^-$, $2Cs^+$	7.30	1
$C_{12}H_4N_4$, $C_{12}H_8N_2$	60.23	4
$C_{12}H_4N_4$, $C_{12}H_{18}$	60.36	3
$C_{12}H_4N_4$, $C_{13}H_{11}NS$	60.24	4
$C_{12}H_4N_4$, $C_{14}H_{10}$	60.125	2
$C_{12}H_4N_4$, $C_{18}H_{12}CuN_2O_2$	60.126	2
$C_{12}H_4N_4^-$, Rb^+	12.10	4
$C_{12}H_4N_4^-$, $C_6H_{16}N^+$, $C_{12}H_4N_4$	3.25	3
$C_{12}H_4N_4^-$, $C_9H_8N^+$, $C_{12}H_4N_4$	35.4	3
$C_{12}H_4N_4^-$, $C_9H_{11}N_2^+$	35.9	4
$C_{12}H_4N_4^-$, $C_{12}H_{14}N^+$, $C_{12}H_4N_4$	60.25	4
$C_{12}H_4N_4^-$, $C_{13}H_{11}N_2^+$	36.23	1

$C_{12}H_4N_4^-$, $C_{14}H_{16}Cr^+$ 74.16 2
$C_{12}H_4N_4^-$, $C_{14}H_{16}Cr^+$, $C_{12}H_4N_4$ 60.140 2
$C_{12}H_4N_4^-$, $C_{19}H_{18}As^+$, $C_{12}H_4N_4$ 65.9 4
$C_{12}H_4N_4^-$, $C_{19}H_{18}P^+$, $C_{12}H_4N_4$ 64.24 4
$C_{12}H_4N_4^-$, $C_{21}H_{21}N_2S_2^+$, $C_{12}H_4N_4$ 60.40 4
$C_{12}H_4N_4^-$, $C_{22}H_{26}ClN_2^+$, $C_{12}H_4N_4$ 60.41 4
$2C_{12}H_4N_4^-$, $2C_4H_{10}NO^+$, $C_{12}H_4N_4$ 40.4 4
$2C_{12}H_4N_4$, $C_{10}H_{16}N_2$ 60.127 2
$C_{12}H_4N_8O_{12}$ 9.5 3
$C_{12}H_5Cl_7O_4$ 31.14 1
$C_{12}H_5FeMnO_7$ 73.36 2
$C_{12}H_6BrNO_3$ 24.6 4
$C_{12}H_6Cl_2O_2$ 38.14 4
$C_{12}H_6Fe_2O_6$ 72.32 2
$C_{12}H_6Fe_2O_6$ 73.38 2
$C_{12}H_6Fe_2O_8$ 72.31 2
$C_{12}H_6N_2S$ 41.14 4
$C_{12}H_6O_2$ 28.6 1
$C_{12}H_6O_3$ 24.7 4
$C_{12}H_6O_{12}$ 13.47 1
$C_{12}H_8$ 28.7 1
$(C_{12}H_8AgClO_4)_n$ 74.3 4
$C_{12}H_8AsO_4^-$, K^+ 65.17 2
$C_{12}H_8As_2S$ 65.5 3
$C_{12}H_8BrCl$ 28.8 1
$C_{12}H_8Br_2$ 28.9 1
$C_{12}H_8Br_2$ 28.1 4
$C_{12}H_8Br_2N_2$ 9.27+ 1
$C_{12}H_8Br_2N_2O_2$ 10.28 1
$C_{12}H_8Br_2O$ 17.33 1
$C_{12}H_8Br_2O_2S$ 11.46 1
$C_{12}H_8Br_2O_2S_2$ 11.47 1
$C_{12}H_8Br_2S$ 11.48 1
$C_{12}H_8Br_2S_2$ 11.49 1
$C_{12}H_8Cl_2$ 28.10 1
$C_{12}H_8Cl_2$ 28.2 4
$C_{12}H_8Cl_2CoN_2^+$, Cl^-, $3H_2O$ 83.91 2
$C_{12}H_8Cl_2I_2Te$ 70.19 2
$C_{12}H_8Cl_2N_2$ 9.29 1
$C_{12}H_8Cl_2N_2Zn$ 83.92 2
$C_{12}H_8Cl_2O_2S$ 11.50+ 1
$C_{12}H_8Cl_2Se_2$ 11.52 1
$C_{12}H_8Cl_2Te_2$ 70.20 2
$C_{12}H_8Cl_3N_2Tl$ 68.8 4
$C_{12}H_8Cl_4S$ 11.13 3
$C_{12}H_8CoN_2O_4$ 83.103 2
$C_{12}H_8CrN_2O_5$ 83.93 2
$C_{12}H_8CuN_2O_4$, K^+, CNS^- 60.37 3
$C_{12}H_8CuN_2O_4$, $2H_2O$ 81.25 3
$C_{12}H_8FeO_4$ 72.7 4
$C_{12}H_8Fe_2O_8$ 72.33 2
$C_{12}H_8HgN_2^{2+}$, $2NO_3^-$ 83.94 2
$C_{12}H_8I_2OTe$ 70.10 4
$C_{12}H_8N_2$ 36.9 1
$C_{12}H_8N_2$ 36.10 1
$C_{12}H_8N_2$ 36.4 4
$C_{12}H_8N_2$ 36.5 4
$C_{12}H_8N_2$, I_2 60.128 2
$C_{12}H_8N_2$, $C_{12}H_4N_4$ 60.23 4
$C_{12}H_8N_2$, $C_{12}H_8N_2O$ 36.6 4
$C_{12}H_8N_2$, $C_{12}H_8N_2O$ 36.7 4
$C_{12}H_8N_2$, $C_{12}H_8N_2O$ 36.8 4
$C_{12}H_8N_2NbO_6^-$, K^+, $3H_2O$ 83.47 4
$C_{12}H_8N_2NbO_6^-$, K^+, $3H_2O$, H_2O_2 83.48 4
$C_{12}H_8N_2O$ 36.11 1
$C_{12}H_8N_2O$ 36.9+ 4
$C_{12}H_8N_2O_2$ 33.63 1
$C_{12}H_8N_2O_2$ 36.12+ 1
$C_{12}H_8N_2O_2$ 33.36 3
$C_{12}H_8N_2O_4$ 15.25 1
$C_{12}H_8N_2O_4$ 36.14+ 1
$C_{12}H_8N_2O_4S_2$ 11.14 3
$C_{12}H_8N_2O_4S_2$ 11.15 3
$C_{12}H_8N_2O_6$ 28.11 1
$C_{12}H_8N_2O_8P^-$, $C_9H_{12}NO_2^+$ 13.13 4
$C_{12}H_8N_2O_8P^-$, $C_{13}H_{21}N_2O_2^+$ 3.47 3
$C_{12}H_8N_2O_8P^-$, $C_{18}H_{23}N_2O_2^+$, H_2O 16.17 4
$C_{12}H_8N_4$ 36.18 1
$C_{12}H_8N_4$ 21.9 3
$C_{12}H_8O$ 38.15 4
$C_{12}H_8OS$ 42.13 1
$C_{12}H_8O_2S$ 39.67 1
$C_{12}H_8O_2S_2$ 39.68 1
$C_{12}H_8O_2S_2$ 39.28 3
$C_{12}H_8O_4Te$ 70.21 2
$C_{12}H_8O_6Os_2$ 72.34 2
$C_{12}H_8S$ 39.29 3
$C_{12}H_8S_2$ 39.69 1
$C_{12}H_8S_2$ 39.70+ 1
$C_{12}H_8Se$ 39.30 3
$C_{12}H_8Se_2$ 39.72 1
$C_{12}H_9AsBrN$ 65.18 2
$C_{12}H_9AsClN$ 65.19+ 2
$C_{12}H_9BrClNO_2S$ 16.43 1
$C_{12}H_9Br_2N_3$ 9.30 1
$C_{12}H_9Br_2N_3$ 9.31 1
$C_{12}H_9ClN_2O_2$ 32.30 3
$C_{12}H_9Cl_2N$ 16.44 1
$C_{12}H_9Cl_2NO$ 33.26 4
$C_{12}H_9Cl_2O_4P$ 46.17 1
$C_{12}H_9Cl_3$ 31.8 4
$C_{12}H_9N$ 36.19 1
$C_{12}H_9N$ 36.4 3
$C_{12}H_9NS$ 41.33 1
$C_{12}H_9NS$, $C_6H_3N_3O_6$ 60.38 3
$C_{12}H_9NS$, $C_7H_4N_2O_6$ 60.130 2
$C_{12}H_9N_3O_4$ 33.64 1
$C_{12}H_{10}$ 19.35+ 1
$C_{12}H_{10}$ 28.12 1
$C_{12}H_{10}Ag^+$, ClO_4^- 75.8 3
$C_{12}H_{10}AsBr$ 65.21 2
$C_{12}H_{10}AsCl$ 65.22 2
$C_{12}H_{10}AsI$ 65.23 2
$C_{12}H_{10}As_2S_3$ 65.8 4
$C_{12}H_{10}BrNO_2S$ 16.45 1
$C_{12}H_{10}BrNO_2S$ 34.8 1
$C_{12}H_{10}BrN_3$ 9.32+ 1
$C_{12}H_{10}Br_2Se$ 11.53 1
$C_{12}H_{10}Br_2Te$ 70.22 2
$C_{12}H_{10}Cl_2N_2$ 16.46 1
$C_{12}H_{10}Cl_2N_2$ 16.47 1
$C_{12}H_{10}Cl_2Se$ 11.54 1
$C_{12}H_{10}Cl_2Sn$ 69.21 4
$C_{12}H_{10}Cl_4N_3P_3$ 64.32 2
$C_{12}H_{10}Cl_8N_6P_6$ 64.15 4
$C_{12}H_{10}CoN_4S_2$ 83.95 2
$C_{12}H_{10}CrO_5$ 74.4 4
$C_{12}H_{10}CuN_4O_4$ 84.33 2
$C_{12}H_{10}F_4N_3P_3$ 64.25 3
$C_{12}H_{10}FeO_2$ 73.39 2
$C_{12}H_{10}FeO_4$ 73.8 3
$C_{12}H_{10}Hg$ 71.26 2

$C_{12}H_{10}HgN_2O_3S$ 85.18 3
$C_{12}H_{10}I^+$, BF_4^- 19.38 1
$C_{12}H_{10}I^+$, NO_3^- 19.10 4
$C_{12}H_{10}NO_4S_2^-$, Na^+ 11.16 3
$C_{12}H_{10}N_2$ 9.34 1
$C_{12}H_{10}N_2$ 9.35+ 1
$C_{12}H_{10}N_2$ 22.14 1
$C_{12}H_{10}N_2$ 9.9 4
$C_{12}H_{10}N_2^{2+}$, $2Br^-$, H_2O 36.11 4
$C_{12}H_{10}N_2O_2$ 33.65 1
$C_{12}H_{10}N_2O_2$ 9.6 3
$C_{12}H_{10}N_2O_5$ 35.30 1
$C_{12}H_{10}N_4NiO_2$, $2H_2O$ 83.96+ 2
$C_{12}H_{10}O$ 31.11 3
$C_{12}H_{10}OS$ 11.55 1
$C_{12}H_{10}O_2$ 25.19 1
$C_{12}H_{10}O_2$ 28.13 1
$C_{12}H_{10}O_2$ 31.15 1
$C_{12}H_{10}O_2$ 17.8 4
$C_{12}H_{10}O_4$ 17.35+ 1
$C_{12}H_{10}O_4S_2$ 11.7 4
$C_{12}H_{10}O_4S_2Se$ 11.56 1
$C_{12}H_{10}O_4S_2W$ 85.19 3
$C_{12}H_{10}O_4S_3$ 11.57 1
$C_{12}H_{10}O_4S_4Te$ 70.11 4
$C_{12}H_{10}S_2$ 11.58 1
$C_{12}H_{10}S_3$ 39.13 4
$C_{12}H_{10}Se_2$ 11.59 1
$C_{12}H_{10}Sn$, $0.33C_8H_{10}$ 69.23 2
$C_{12}H_{11}Br$ 24.26 1
$C_{12}H_{11}BrO_2$ 31.12 3
$C_{12}H_{11}N$, $C_{13}H_{10}O$ 60.26 4
$C_{12}H_{11}NO_3Ru$ 75.3 4
$C_{12}H_{11}NO_4S_2$ 11.17 3
$C_{12}H_{11}NS_4$ 39.14 4
$C_{12}H_{11}N_3$ 9.10 4
$C_{12}H_{11}O_3^+$, Br^- 59.23 1
$C_{12}H_{12}$ 24.27 1
$C_{12}H_{12}Al_2Cl_8Pd_2$ 74.3 3
$C_{12}H_{12}Al_4Cl_{14}Pd_2$ 74.4 3
$C_{12}H_{12}BrNO$ 35.31 1
$C_{12}H_{12}BrNO_8S$ 45.61 1
$C_{12}H_{12}Br_2O_5$ 59.4 3
$C_{12}H_{12}Br_6$ 19.39 1
$C_{12}H_{12}ClN_3$ 35.13 4
$C_{12}H_{12}Cl_2N_2O_4$ 50.1 4
$C_{12}H_{12}Cl_2O_2$ 38.11 3
$C_{12}H_{12}Cl_2Sn_2^{2+}$, $2AlCl_4^-$ 69.22 4
$C_{12}H_{12}Cl_3OSb$ 66.14 2
$C_{12}H_{12}Cl_4$ 28.14 1
$C_{12}H_{12}Co_2O_4$ 72.35 2
$C_{12}H_{12}Cr$ 74.9+ 2
$C_{12}H_{12}CrO_5$ 74.5 3
$C_{12}H_{12}CuN_2O_6$ 81.32 4
$(C_{12}H_{12}Cu_2N_8)_n$ 83.39 2
$C_{12}H_{12}FeN_2O_5$ 75.4 4
$C_{12}H_{12}Fe_2O_8P_2$ 86.8 2
$C_{12}H_{12}Fe_2O_8Sn_2$ 69.23 4
$C_{12}H_{12}GdO_{15}^{3-}$, $2ClO_4^-$, $5Na^+$, $6H_2O$ 81.69 2
$C_{12}H_{12}MoN_2O_5$ 71.27 2
$C_{12}H_{12}MoN_8$ 71.11 4
$C_{12}H_{12}N_2^{2+}$, $2Br^-$, H_2O 36.20 1
$C_{12}H_{12}N_2NiO_6$, $2H_2O$ 81.33+ 4
$C_{12}H_{12}N_2O_2$ 33.27 4
$C_{12}H_{12}N_2O_2S$ 11.18 3
$C_{12}H_{12}N_2O_2S$ 16.10 4
$C_{12}H_{12}N_2O_3$, $2C_7H_8BrN_5$ 60.131 2
$C_{12}H_{12}N_2O_4S_2$ 41.15 4
$C_{12}H_{12}N_2O_6Zn$, $2H_2O$ 81.26 3
$C_{12}H_{12}N_2O_6Zn$, $2H_2O$ 81.35 4
$C_{12}H_{12}N_2O_{12}Zr^{2-}$, $2K^+$, H_2O 81.70 2
$C_{12}H_{12}N_2S_2$ 11.61 1
$C_{12}H_{12}N_2S_2$ 11.19 3
$C_{12}H_{12}N_4Ni$ 83.99 2
$C_{12}H_{12}N_4O_8$, $2H_2O$ 44.23 4
$C_{12}H_{12}NdO_{15}^{3-}$, $2ClO_4^-$, $5Na^+$, $6H_2O$ 81.72 2
$C_{12}H_{12}NdO_{15}^{3-}$, $3Na^+$, $6H_2O$ 81.71 2
$C_{12}H_{12}NdO_{15}^{3-}$, $5Na^+$, $2ClO_4^-$, $6H_2O$ 81.27 3
$C_{12}H_{12}O_2$ 31.13 3
$C_{12}H_{12}O_4$ 31.9 4
$C_{12}H_{12}O_{12}$ 45.17 4
$C_{12}H_{12}O_{15}Yb^{3-}$, $5Na^+$, $2ClO_4^-$, $6H_2O$ 81.28 3
$C_{12}H_{12}S_2$, $C_6N_6O_6$ 60.74 2
$C_{12}H_{12}V$ 75.24 2
$C_{12}H_{13}Cl_2N_2NbO_2$ 83.57 3
$C_{12}H_{13}FeNO$ 73.9 3
$C_{12}H_{13}NO$ 35.14 4
$C_{12}H_{13}NO_4$ 35.32 1
$C_{12}H_{14}BrNO_2S$ 27.9 4
$C_{12}H_{14}Br_2N_2Zn$ 83.49 4
$(C_{12}H_{14}CdN_4OS_2)_n$ 83.50 4
$C_{12}H_{14}ClNO_2$ 40.12 3
$C_{12}H_{14}ClN_4^+$, Br^- 44.36 3
$C_{12}H_{14}Cl_2CuN_2$ 83.100 2
$C_{12}H_{14}Cl_2CuN_2$ 83.101 2
$C_{12}H_{14}Cl_2CuN_2O_2$ 84.11 3
$C_{12}H_{14}Cl_2N_2Zn$ 83.51 4
$C_{12}H_{14}Cr_2^{4-}$, $4C_4H_{10}LiO^+$ 71.12 4
$C_{12}H_{14}CuN_4O_6$ 83.52+ 4
$C_{12}H_{14}HgN_4O_4$ 83.60 3
$C_{12}H_{14}INO_2S$ 35.33 1
$C_{12}H_{14}N^+$, I^- 35.12 3
$C_{12}H_{14}N^+$, $C_{12}H_4N_4^-$, $C_{12}H_4N_4$ 60.25 4
$C_{12}H_{14}N_2^{2+}$, $2Br^-$ 33.28 4
$C_{12}H_{14}N_2^{2+}$, $2Cl^-$ 33.29 4
$C_{12}H_{14}N_2^{2+}$, Cl_4Co^{2-} 33.38 3
$C_{12}H_{14}N_2^{2+}$, Cl_4Cu^{2-} 33.66 1
$C_{12}H_{14}N_2^{2+}$, Cl_4Pd^{2-} 33.39 3
$C_{12}H_{14}N_2^{2+}$, $2I^-$ 33.30 4
$C_{12}H_{14}N_2^{2+}$, $C_4N_4Ni^{2-}$ 33.67 1
$nC_{12}H_{14}N_2^{2+}$, $(Cl_2Cu^-)_{2n}$ 33.40 3
$C_{12}H_{14}N_2O_2$ 37.16 1
$C_{12}H_{14}N_2O_4$, H_2O 48.48 4
$C_{12}H_{14}N_4$ 32.31 3
$C_{12}H_{14}N_4O_2$ 32.32 1
$C_{12}H_{14}N_4O_8S_2$ 48.45 3
$C_{12}H_{14}N_6NiS_2$ 85.20 3
$C_{12}H_{14}O_3$ 31.10 4
$C_{12}H_{14}O_4$ 13.48 1
$C_{12}H_{15}NO_2$ 58.6 3
$C_{12}H_{15}NO_3$ 37.17 1
$C_{12}H_{15}N_5O_5$ 47.30 3
$(C_{12}H_{15}NaO_4)_n$ 67.11 3
$C_{12}H_{16}$ 22.4 3
$(C_{12}H_{16}AgN_4^+)_n$, $nClO_4^-$ 83.61 3
$C_{12}H_{16}B_2CoO_2$ 75.5 4
$C_{12}H_{16}BrFO_7$ 45.31 3
$C_{12}H_{16}Br_2N_8Ni$ 83.102 2

$C_{12}H_{16}CdN_6O_4$, H_2O	82.45	2
$C_{12}H_{16}Cl_2N_2O_2Sn$	69.24	2
$C_{12}H_{16}Cl_2N_5O^+$, Cl^-	33.31	4
$C_{12}H_{16}CoN_6O_4$, H_2O	82.46	2
$C_{12}H_{16}CoN_6O_4$, $2H_2O$	82.22	3
$C_{12}H_{16}CuN_4O_4^{2+}$, $2Cl^-$	83.62	3
$C_{12}H_{16}CuN_8^{2+}$, $2I^-$	83.104	2
$C_{12}H_{16}CuN_8O_4^{2+}$, $2Cl^-$, $3H_2O$	83.54	4
$C_{12}H_{16}Cu_2N_6O_8^{2-}$, $2Na^+$, $2H_2O$	82.47	2
$C_{12}H_{16}MoNS^+$, I^-	73.40	2
$C_{12}H_{16}Mo_2N_6O_8$, $3H_2O$	82.23	3
$C_{12}H_{16}NO_2^+$, Br^-	3.43	3
$C_{12}H_{16}NO_2^+$, Cl^-	3.44	3
$C_{12}H_{16}NO_3^+$, Br^-	58.3	4
$C_{12}H_{16}NO_4^+$, Br^-, $2H_2O$	48.88	1
$C_{12}H_{16}N_2O_3$	43.6	3
$C_{12}H_{16}N_2O_3$	43.7	3
$C_{12}H_{16}N_2O_5S$	47.18	4
$C_{12}H_{16}N_2O_5Si$	63.3	3
$C_{12}H_{16}N_2O_8Zn$	83.55	4
$C_{12}H_{16}N_4NiO_4^{2+}$, $2Cl^-$	83.105	2
$C_{12}H_{16}N_4OS$, $2H_2O$	59.5	3
$C_{12}H_{16}N_4O_4$	44.72	1
$C_{12}H_{16}N_6NiO_4$, H_2O	82.48	2
$C_{12}H_{16}N_6O_4Zn$, $2H_2O$	82.49	2
$C_{12}H_{16}N_6O_4Zn$, $2H_2O$	82.12	4
$C_{12}H_{16}N_6O_4Zn$, $5H_2O$	82.50	2
$C_{12}H_{16}N_8Ni^{2+}$, $2Cl^-$	83.106	2
$C_{12}H_{16}N_8NiO_2^{2+}$, $2NO_3^-$	83.107	2
$C_{12}H_{16}O_2S_2$	39.32	3
$C_{12}H_{16}O_6$	38.12	3
$C_{12}H_{16}O_8$	20.6	4
$C_{12}H_{16}O_8$	45.18	4
$C_{12}H_{17}BrN_2O_5$	45.63+	1
$C_{12}H_{17}BrN_2O_5$	45.33	3
$C_{12}H_{17}CuN_3O_3$	76.38	4
$C_{12}H_{17}IP^+$, I^-	64.16	4
$C_{12}H_{17}NO_2$	37.7	4
$C_{12}H_{17}NO_3$	32.17	4
$C_{12}H_{17}NO_3Si$	63.16	2
$C_{12}H_{17}NO_4S$	41.16	4
$C_{12}H_{17}N_4OS^+$, Cl^-, H_2O	41.17	4
$C_{12}H_{17}N_5O_5$	47.19	4
$C_{12}H_{18}$	19.40	1
$C_{12}H_{18}$	23.9	1
$C_{12}H_{18}$	19.6+	3
$C_{12}H_{18}$	20.14	3
$C_{12}H_{18}$, $C_6Cl_4O_2$	60.71	2
$C_{12}H_{18}$, $C_8H_2N_4$	60.101	2
$C_{12}H_{18}$, $C_{12}H_4N_4$	60.36	3
$C_{12}H_{18}Ag_2^{2+}$, $2NO_3^-$	75.9	3
$C_{12}H_{18}B_2CoF_2N_6O_6$	83.56	4
$C_{12}H_{18}B_2CoF_2N_6O_6^+$, BF_4^-	83.57	4
$C_{12}H_{18}Be_4O_{13}$	67.11	2
$C_{12}H_{18}Cl_2O_6Pd_2$	72.18	3
$C_{12}H_{18}Cl_2Ru$	72.37	2
$(C_{12}H_{18}Cl_6Cu_3N_2O_4)_n$	84.34	2
$C_{12}H_{18}CoN_3O_3$	77.5	4
$C_{12}H_{18}CuN_2O_2$	83.112	2
$C_{12}H_{18}CuN_2O_2$, $0.5H_2O$	83.109	2
$C_{12}H_{18}CuN_2O_2$, CH_6N^+, ClO_4^-	60.39	3
$C_{12}H_{18}CuO_4$	77.14	2
$C_{12}H_{18}CuO_6$	77.15+	2
$C_{12}H_{18}FeN_6^{2+}$, $2Cl^-$, $3H_2O$	71.28	2
$C_{12}H_{18}Fe_2Ge_3O_6$	69.20	3
$C_{12}H_{18}N_2O_3V$	76.76	3
$C_{12}H_{18}N_4OS^{2+}$, $2Cl^-$, H_2O	41.34	1
$C_{12}H_{18}N_4O_7P_2S$, $4H_2O$	46.18	1
$C_{12}H_{18}N_6O_6$, $0.5H_2O$	48.89	1
$C_{12}H_{18}O_2$	21.8	4
$C_{12}H_{18}O_4$	20.7	4
$C_{12}H_{18}O_4P_2$	64.26+	3
$C_{12}H_{18}O_{12}Pd_3$, $0.5H_2O$	81.29	3
$C_{12}H_{18}O_{13}Zn_4$	81.74	2
$C_{12}H_{19}BrO$	27.12	1
$C_{12}H_{19}Cl_2NPt$	72.38	2
$C_{12}H_{19}Cl_2NPt$	72.19	3
$C_{12}H_{19}Cl_2NPt$	72.8	4
$C_{12}H_{19}Co$	72.39	2
$C_{12}H_{19}NO_3$	21.10	3
$C_{12}H_{19}N_2O_2^+$, Br^-	3.45	3
$C_{12}H_{19}N_2O_2^+$, Br^-	16.11	4
$C_{12}H_{19}N_4O_7P_2S^+$, Cl^-	41.18	4
$C_{12}H_{20}$	21.28	1
$C_{12}H_{20}AgN_9O_3$	83.58	4
$C_{12}H_{20}Br_2O$	20.28	1
$C_{12}H_{20}Br_2O$	23.7	3
$C_{12}H_{20}Cl_2Rh_2$	72.40	2
$C_{12}H_{20}Cl_4Pt_4$	71.13	4
$C_{12}H_{20}Cr_2$	72.20	3
$C_{12}H_{20}CuN_2O_3$	77.17	2
$C_{12}H_{20}CuN_2O_4$	83.113	2
$C_{12}H_{20}Cu_2O_8$	81.30	3
$C_{12}H_{20}Mo_2$	72.9	4
$C_{12}H_{20}Mo_2N_6O_6S_2$, $3H_2O$	82.13	4
$C_{12}H_{20}Mo_2O_7S_8$	80.23	2
$C_{12}H_{20}NO^+$, ClO_4^-	58.4	4
$C_{12}H_{20}N_2O_3$	43.9	4
$C_{12}H_{20}N_2S_3$	33.41	3
$C_{12}H_{20}N_4O_4$	48.46	3
$C_{12}H_{20}O_2S$	39.73	1
$C_{12}H_{20}O_4$	38.46	1
$C_{12}H_{20}O_6$	38.47	1
$C_{12}H_{20}O_6$	45.34	3
$C_{12}H_{20}S_4$	39.15	4
$C_{12}H_{20}S_6$	39.33	3
$C_{12}H_{21}Cl_2NOPd_2$	72.10	4
$C_{12}H_{21}Cl_2O_6Re_2^+$, O_4Re^-	81.31	3
$C_{12}H_{22}Br_2N_2O_2$	1.133	1
$C_{12}H_{22}Cl_2N_2O_2Pd$	83.114	2
$C_{12}H_{22}Cl_4O_2Rh_2$, CH_4O	71.12	3
$C_{12}H_{22}CuN_6O_6^{2+}$, $2NO_3^-$	82.51	2
$C_{12}H_{22}N_2O_2$	34.9	1
$C_{12}H_{22}N_2O_2$	21.11	3
$C_{12}H_{22}N_2O_2$	34.2	3
$C_{12}H_{22}N_2O_4$	33.68	1
$C_{12}H_{22}O_4$	1.134	1
$C_{12}H_{22}O_6$	21.12	3
$C_{12}H_{22}O_{11}$	45.69	1
$C_{12}H_{22}O_{11}$	45.65+	1
$C_{12}H_{22}O_{11}$	45.19	4
$C_{12}H_{22}O_{11}$, Na^+, Br^-, $2H_2O$	45.70	1
$C_{12}H_{22}O_{11}$, H_2O	45.35	3
$C_{12}H_{22}O_{11}$, H_2O	45.36	3
$C_{12}H_{22}O_{11}$, H_2O	45.20	4
$C_{12}H_{22}O_{11}$, $2H_2O$	45.21	4
$C_{12}H_{23}N_3O_3^{2+}$, $2Cl^-$	58.8	1
$C_{12}H_{24}$	23.10	1
$C_{12}H_{24}Cl_2MnN_4$, $2H_2O$	83.115	2
$C_{12}H_{24}Cl_2N_8NiS_4$	79.34+	2
$C_{12}H_{24}Cr_3O_{16}^+$, Cl^-, $5H_2O$	81.75	2
$C_{12}H_{24}Cr_3O_{16}^+$, Cl^-, $6H_2O$	81.32	3

$(C_{12}H_{24}CuN_2O_4)_n, 2nH_2O$ 81.33 **3**
$C_{12}H_{24}CuN_2O_8$ 82.24 **3**
$C_{12}H_{24}Cu_2O_8$ 81.76 **2**
$C_{12}H_{24}Fe_3O_{16}^+, ClO_4^-, 2C_2H_4O_2$ 81.34 **3**
$C_{12}H_{24}N^+, Br^-$ 52.2 **3**
$C_{12}H_{24}NO^+, I^-$ 58.5 **4**
$C_{12}H_{24}N_2O_2$ 1.135+ **1**
$C_{12}H_{24}N_4O_2S_4Sn_4$ 69.21 **3**
$C_{12}H_{24}N_4S_8Zn_2$ 80.24 **2**
$C_{12}H_{24}N_6S_2Se_2Te$ 70.12 **4**
$C_{12}H_{24}O_2$ 1.137 **1**
$C_{12}H_{24}O_4$ 38.16 **4**
$C_{12}H_{24}Pd_3S_9$ 85.66 **2**
$C_{12}H_{25}N_3Nd_2O_{16}^{2+}, 2Cl^-, 3H_2O$ 82.52 **2**
$C_{12}H_{26}BrNbO_2S_4$ 80.6 **4**
$C_{12}H_{26}ClNbO_2S_4$ 80.7 **4**
$C_{12}H_{26}CuN_2O_5$ 82.53 **2**
$C_{12}H_{26}KN_2O_4^+, CNS^-$ 67.12 **4**
$C_{12}H_{26}N_2O_2$ 1.138 **1**
$C_{12}H_{26}N_2O_2$ 34.10 **1**
$C_{12}H_{26}N_2O_4$ 40.18 **4**
$C_{12}H_{27}S^+, F^-, 20H_2O$ 61.11 **2**
$C_{12}H_{27}S^+, F^-, 23H_2O$ 61.12 **2**
$(C_{12}H_{28}AgP_2PtS_2^+)_n, nNO_3^-$ 85.67 **2**
$C_{12}H_{28}BrCuN_6$ 83.116 **2**
$C_{12}H_{28}Br_2Cu_2N_2O_2$ 83.59 **4**
$C_{12}H_{28}CuN_4^{2+}, 2NO_3^-$ 83.117 **2**
$(C_{12}H_{28}HgO_4P_2S_4)_n$ 85.21 **3**
$C_{12}H_{28}N^+, Br^-$ 3.75 **1**
$C_{12}H_{28}N^+, C_{20}H_8CoN_6S_4^-$ 85.35 **3**
$C_{12}H_{28}N_2^{2+}, 2Br^-$ 34.11 **1**
$C_{12}H_{28}N_2^{2+}, 4C_6H_6N_2O_2, 2Cl^-$ 60.40 **3**
$C_{12}H_{28}N_4Ni^{2+}, 2NO_3^-$ 83.118 **2**
$C_{12}H_{28}N_4O_2P_2$ 64.34 **2**
$C_{12}H_{28}O_4P_2PbS_4$ 69.24 **4**
$C_{12}H_{28}O_4P_2S_4$ 64.28 **3**
$C_{12}H_{28}Si_4$ 63.17 **2**
$C_{12}H_{29}Cl_2CoN_3$ 76.75 **2**
$C_{12}H_{29}Cl_2N_7P_3^+, Cl^-$ 64.29 **3**
$C_{12}H_{29}N_4O_4PS^+, HO_4P^-, 3H_2O$ 46.20 **1**
$C_{12}H_{30}Al_2F^-, K^+$ 68.20 **2**
$C_{12}H_{30}Al_3N_3$ 68.11 **3**
$C_{12}H_{30}B_3N_3$ 62.47 **2**
$C_{12}H_{30}BrCuN_4^+, Br^-$ 83.119 **2**
$C_{12}H_{30}BrFeN_4^+, Br^-$ 76.76 **2**
$C_{12}H_{30}BrMnN_4^+, Br^-$ 76.77 **2**
$C_{12}H_{30}BrN_4Ni^+, Br^-$ 83.120 **2**
$C_{12}H_{30}BrN_4Zn^+, Br^-$ 76.78 **2**
$C_{12}H_{30}Br_2CoN_4$ 76.79 **2**
$C_{12}H_{30}Br_2NiP_2$ 86.9 **2**
$C_{12}H_{30}Br_2P_2Pt$ 86.10 **2**
$C_{12}H_{30}Cl_2P_2Pt$ 86.11 **2**
$C_{12}H_{30}Cl_6P_2Re_2$ 86.12 **2**
$C_{12}H_{30}CoN_6O_3^{3+}, 3ClO_4^-, H_2O$ 82.54 **2**
$C_{12}H_{30}InO_6P_3S_6$ 68.12 **3**
$C_{12}H_{30}N_2^{2+}, 2Br^-, 2H_2O$ 3.77 **1**
$C_{12}H_{30}N_2^{2+}, 2Cl^-, 2H_2O$ 3.78 **1**
$C_{12}H_{30}N_2NiO_6^{2+}, 2NO_3^-$ 83.60 **4**
$C_{12}H_{30}N_2O_{16}P_2U$ 84.35 **2**
$C_{12}H_{30}N_3O_6P_3$ 64.30 **3**
$C_{12}H_{30}N_3O_6P_3$ 64.17 **4**
$C_{12}H_{30}N_{18}Ni_3O_6^{6+}, 6NO_3^-, 2H_2O$ 83.63 **3**
$C_{12}H_{30}O_6P_3S_6V$ 85.22 **3**
$C_{12}H_{31}BrP_2Pt$ 86.13 **2**
$(C_{12}H_{32}Al_2MgO_6)_n$ 67.12 **2**
$C_{12}H_{32}Mg_2N_4$ 67.12 **3**
$C_{12}H_{32}N_8S_4Te^{2+}, 2Br^-$ 70.13 **4**
$C_{12}H_{32}N_8S_4Te^{2+}, 2Cl^-$ 70.14 **4**
$(C_{12}H_{34}CaO_{15}^{2+})_n, 2nBr^-, 3nH_2O$ 67.13 **4**
$C_{12}H_{34}Cu_2N_4O_2^{2+}, 2Br^-$ 76.77 **3**
$C_{12}H_{34}N_6Ni^{2+}, 2ClO_4^-$ 83.64 **3**
$C_{12}H_{36}AlO_4Si_4^-, C_4H_{12}Sb^+$ 68.21 **2**
$C_{12}H_{36}As_4Cu_4I_4$ 86.14 **2**
$C_{12}H_{36}Be_3N_6$ 67.13 **3**
$C_{12}H_{36}Cl_4Pt_4$ 71.29 **2**
$C_{12}H_{36}Co_3N_6O_6^{2+}, 2C_2H_3O_2^-$ 83.65 **3**
$C_{12}H_{36}CrN_4OSi_3$ 83.61 **4**
$C_{12}H_{36}Cs_4O_4Si_4$ 63.4 **3**
$C_{12}H_{36}CuN_4O_2^{2+}, 2NO_3^-$ 76.78 **3**
$C_{12}H_{36}K_4O_4Si_4$ 63.5 **3**
$C_{12}H_{36}N_6W$ 83.123 **2**
$C_{12}H_{36}O_4Rb_4Si_4$ 63.6 **3**
$C_{12}H_{36}O_6Si_8$ 63.18 **2**
$C_{12}H_{40}O_4Pt_4$ 71.30+ **2**
$C_{12}H_{41}B_9P_2PtS$ 86.10 **3**
$C_{12}H_{42}N_6Pt_2^{2+}, 2I^-$ 76.80 **2**
$C_{12}H_{54}Co_4N_{12}O_6^{6+}, 3O_6S_2^{2-}, 8H_2O$ 76.39 **4**
$C_{12}H_{54}Cr_4N_{12}O_6^{6+}, 6N_3^-, 4H_2O$ 76.81 **2**

C_{13}

$C_{13}Cl_{11}$ 12.11 **4**
$C_{13}H_3N_6^-, H_4N^+$ 7.6 **3**
$C_{13}H_4BrN_3O_7, C_{14}H_{10}$ 60.42 **3**
$C_{13}H_4BrN_3O_7, C_{20}H_{16}$ 60.55 **3**
$C_{13}H_4BrN_3O_7, C_{26}H_{16}$ 60.156 **2**
$C_{13}H_5Co_2NO_4Pt$ 83.124 **2**
$C_{13}H_5MnMoO_8$ 73.41 **2**
$C_{13}H_5MoN_4S_4^-, C_{24}H_{20}P^+$ 64.52 **3**
$C_{13}H_6Cl_2N_2O_2$ 31.11 **4**
$C_{13}H_7BrN_2$ 9.7 **3**
$C_{13}H_7BrO$ 28.2 **3**
$C_{13}H_8Br_2O$ 19.41 **1**
$C_{13}H_8Br_2O_4S_2$ 11.20 **3**
$C_{13}H_8Br_4GeN_2O_3W$ 83.66 **3**
$C_{13}H_8CrO_3$ 74.13 **2**
$C_{13}H_8CrO_5S$ 71.14 **4**
$C_{13}H_8CrO_6$ 71.32 **2**
$C_{13}H_8FeO_5$ 75.10 **3**
$C_{13}H_8Fe_2O_5$ 75.26 **2**
$C_{13}H_8Fe_2O_6$ 75.6 **4**
$C_{13}H_8N_2$ 28.15 **1**
$C_{13}H_8N_2O_3$ 36.21 **1**
$C_{13}H_8O$ 28.3 **4**
$C_{13}H_9BrN_2O_4$ 35.13 **3**
$C_{13}H_9Br_3O$ 20.29 **1**
$C_{13}H_9ClO_2$ 13.49 **1**
$C_{13}H_9ClO_2$ 13.50 **1**
$C_{13}H_9Cl_2N$ 16.12 **4**
$C_{13}H_9Cl_2N$ 16.13 **4**
$C_{13}H_9CrNO_3$ 74.14 **2**
$C_{13}H_9FeO_4$ 72.21 **3**
$C_{13}H_9Fe_3NO_{10}Si$ 63.5 **4**
$C_{13}H_9IO_2$ 13.14 **3**
$C_{13}H_9N$ 36.22 **1**
$C_{13}H_9N$ 36.5 **3**
$C_{13}H_9N, C_4H_5N_3O, H_2O$ 60.132 **2**
$C_{13}H_9NO$ 36.6 **3**
$C_{13}H_9NOS$ 41.13 **3**

$C_{13}H_9NO_2$	33.69	1
$C_{13}H_9N_3S$	9.38	1
$C_{13}H_{10}$	28.16	1
$C_{13}H_{10}BrN$	33.42	3
$C_{13}H_{10}BrN$	16.14	4
$C_{13}H_{10}BrNO$	16.48	1
$C_{13}H_{10}BrNO$	16.49	1
$C_{13}H_{10}Br_2S_2$	11.21	3
$C_{13}H_{10}ClNO$	16.50	1
$C_{13}H_{10}ClNO$	16.51	1
$C_{13}H_{10}ClNO$	16.52+	1
$C_{13}H_{10}Cl_2FeO_2Sn$	69.22	3
$C_{13}H_{10}Cl_2HgN_4S$	85.13	4
$C_{13}H_{10}Cl_2O_2$	17.38	1
$C_{13}H_{10}FeN_2$	73.10	3
$C_{13}H_{10}FeO_3$	72.41	2
$C_{13}H_{10}N_2O_4$	50.6	1
$C_{13}H_{10}N_2O_4$	35.14+	3
$C_{13}H_{10}N_2O_4$	35.15	4
$C_{13}H_{10}N_2O_4$	35.16	4
$C_{13}H_{10}N_4S$	32.33	1
$C_{13}H_{10}N_4S$	41.35	1
$C_{13}H_{10}N_4S^+$, HO^-	32.32	3
$C_{13}H_{10}O$	19.42+	1
$C_{13}H_{10}O$, $C_{12}H_{11}N$	60.26	4
$C_{13}H_{10}O_2S$	39.74	1
$C_{13}H_{10}O_2S$, H_2O	17.39	1
$C_{13}H_{10}O_3Rh_2$	73.42	2
$C_{13}H_{11}Cl_2N_2O^+$, I^-, H_2O	33.43	3
$C_{13}H_{11}FeN$	73.10	4
$C_{13}H_{11}I_2IrN_2O_3$	71.33	2
$C_{13}H_{11}N$	16.9	3
$C_{13}H_{11}NO_2$	37.8	4
$C_{13}H_{11}NO_2S_2$	13.51	1
$C_{13}H_{11}NO_3$	10.29	1
$C_{13}H_{11}NS$	39.34	3
$C_{13}H_{11}NS$	41.19	4
$C_{13}H_{11}NS$, $C_{12}H_4N_4$	60.24	4
$C_{13}H_{11}N_2^+$, $C_{12}H_4N_4^-$	36.23	1
$C_{13}H_{11}O_2P$	64.36	2
$C_{13}H_{12}BrNO_2$	33.70	1
$C_{13}H_{12}BrNO_2S$	16.54	1
$C_{13}H_{12}Br_2N_2SZn$	83.62	4
$C_{13}H_{12}ClNO_3S$	39.16	4
$C_{13}H_{12}ClN_3O_2S$	41.20	4
$C_{13}H_{12}Cl_2CuN_2S$	85.14	4
$C_{13}H_{12}CrO_3$	75.11	3
$C_{13}H_{12}Cu_3S_6$	85.69	2
$C_{13}H_{12}FeO$	73.43	2
$C_{13}H_{12}FeO_2$	75.12	3
$C_{13}H_{12}FeO_3$	75.13	3
$C_{13}H_{12}N_2O_4S$	9.8	3
$C_{13}H_{12}N_4O_2$, $2C_{10}H_8O_2$	60.27	4
$C_{13}H_{12}N_4O_2$, $2C_{10}H_8O_2$	60.28	4
$C_{13}H_{12}N_4O_2$, $2C_{10}H_8O_2$, $3H_2O$	60.29	4
$C_{13}H_{13}Br_2NO_5$	21.13	3
$C_{13}H_{13}Br_2N_5O_2$	58.7	3
$C_{13}H_{13}FNP$	64.31	3
$C_{13}H_{13}FeO_2^-$, $C_{20}H_{25}N_2O_2^+$, H_2O	73.44	2
$C_{13}H_{13}N_2O_2^+$, $1.5C_6H_6O_2$, Br^-	60.30	4
$C_{13}H_{13}N_4O_2^+$, I^-	36.7	3
$C_{13}H_{13}OPS$	64.38	2
$C_{13}H_{13}OPSe$	64.39	2
$C_{13}H_{13}P$	64.32	3
$C_{13}H_{13}Ti$	75.14	3
$C_{13}H_{14}Cl_2N_2O_2$, $2H_2O$	33.71	1
$C_{13}H_{14}Cl_2Ti$	73.11	4
$C_{13}H_{14}INO_2$	33.72	1
$C_{13}H_{14}MoN_2O_5^+$, F_6P^-	71.15	4
$C_{13}H_{14}N_2O$	33.73	1
$C_{13}H_{14}N_2O_3$	43.9	3
$C_{13}H_{14}N_2O_4S_2$	50.7	1
$C_{13}H_{14}O_5$	59.6	4
$C_{13}H_{15}CuN_3O_4$, $2H_2O$	82.25	3
$C_{13}H_{15}FeNO_2$	75.15	3
$C_{13}H_{15}NO_2$	37.9	4
$C_{13}H_{15}N_3O_3$, $2H_2O$	48.90	1
$C_{13}H_{15}N_5O_3S$, $0.33H_2O$	47.32	3
$C_{13}H_{16}As_2FeO_3$	86.15	2
$C_{13}H_{16}Cl_2N_2Pt$	71.16	4
$C_{13}H_{16}Cl_2N_2Pt$	83.63	4
$C_{13}H_{16}Cl_4N_2Pt$, $CHCl_3$	71.17	4
$C_{13}H_{16}CuN_2O_4$	82.55	2
$C_{13}H_{16}FeO$	75.7	4
$C_{13}H_{16}INO_2S$	35.17	4
$C_{13}H_{16}NO_2^+$, Br^-, $2H_2O$	58.9	1
$C_{13}H_{16}N_2O_2$	40.13	3
$C_{13}H_{16}N_4O_4$	44.37	3
$C_{13}H_{16}N_5O_3^+$, I^-	47.29	1
$C_{13}H_{16}N_6Ni_2O_2S_2$, $2C_3H_7NO$	85.15	4
$C_{13}H_{16}O$	52.20	1
$C_{13}H_{14}D_2O$	52.21	1
$C_{13}H_{17}BrO_7$	38.13	3
$C_{13}H_{17}NO_4$, H_2O	48.49	4
$C_{13}H_{17}N_3O$, $C_8H_{12}N_2O_3$	60.31	4
$C_{13}H_{17}N_3O_4$, H_2O	48.91	1
$C_{13}H_{18}Cl_2O_5$	45.71	1
$2C_{13}H_{18}FeN^+$, Cl_4Zn^{2-}, H_2O	71.18	4
$C_{13}H_{18}N_2O_2S$	41.14	3
$C_{13}H_{18}N_2O_3$	43.30	1
$C_{13}H_{18}O_2Pd$	75.27	2
$C_{13}H_{18}O_3S$	21.9+	4
$C_{13}H_{19}ClO_2S$	39.17	4
$(C_{13}H_{19}CuNO_4)_n$	81.35	3
$C_{13}H_{19}NO_2PbS_4$	69.27	2
$C_{13}H_{19}NO_5$, $0.5H_2O$	37.8	3
$C_{13}H_{19}N_2^+$, ClO_4^-	32.18	4
$C_{13}H_{19}N_2O^+$, Cl^-	35.16	3
$2C_{13}H_{19}N_2OS^+$, Cl_4Co^{2-}	41.21	4
$2C_{13}H_{19}N_2OS^+$, Cl_4Cu^{2-}	41.22	4
$C_{13}H_{19}Re$	73.45	2
$C_{13}H_{20}Cl_2NPPt$	71.19	4
$C_{13}H_{20}N^+$, Br^-	32.34	1
$C_{13}H_{20}N^+$, I^-	32.19	4
$C_{13}H_{20}NO_3^+$, Br^-	58.6	4
$C_{13}H_{20}NO_5^+$, ClO_4^-	31.14	3
$C_{13}H_{20}N_2OS_2$	8.18	3
$C_{13}H_{20}NiO_2$	75.28	2
$C_{13}H_{20}O_8$	45.72	1
$C_{13}H_{21}^+$, $AlCl_4^-$	12.9	1
$C_{13}H_{21}Cl_2NPt$	72.11	4
$C_{13}H_{21}NO_3$	3.23	4
$C_{13}H_{21}N_2O_2^+$, Cl^-	3.24	4
$C_{13}H_{21}N_2O_2^+$, $C_{12}H_8N_2O_8P^-$	3.47	3
$C_{13}H_{21}N_6OP$	32.35	1
$C_{13}H_{21}N_6OP$	32.36	1
$C_{13}H_{21}N_6OP$	32.37	1
$C_{13}H_{22}Br_2N_2NiO$	83.127	2
$C_{13}H_{22}NO^+$, Br^-	3.48	3
$C_{13}H_{22}NO_4^+$, Br^-	50.4	3
$C_{13}H_{22}N_3O^+$, Br^-	58.7	4
$C_{13}H_{22}N_3O_4$	48.92	1

$C_{13}H_{23}FeN_7^{2+}$, $2F_6P^-$	71.20	4
$C_{13}H_{23}NO$, $C_{13}H_{24}NO^+$, Cl^-	58.8	4
$C_{13}H_{24}N^+$, Br^-	52.2	4
$C_{13}H_{24}N_2O$	8.11	4
$C_{13}H_{24}O_3$	1.139	1
$C_{13}H_{24}O_4$	1.140	1
$C_{13}H_{24}O_5S$	45.37	3
$C_{13}H_{24}O_{11}$, H_2O	45.73	1
$C_{13}H_{24}O_{11}$, CH_4O	45.38	3
$C_{13}H_{25}BrN^+$, Br^-	52.3	4
$C_{13}H_{25}NNiO_4P_2S_4$	85.70	2
$C_{13}H_{25}NS$	34.2	4
$C_{13}H_{26}N^+$, Cl^-	58.9	3
$C_{13}H_{26}N_2^{2+}$, $2I^-$	36.25	1
$C_{13}H_{27}BrMgO_3$	67.14	3
$C_{13}H_{27}S^+$, $C_6H_2N_3O_9S^-$	11.8	4
$C_{13}H_{28}N^+$, I^-	52.22	1
$C_{13}H_{30}ClOP_2Pt^+$, F_5Si^-	86.16	2

C_{14}

$C_{14}Cl_{12}$	22.7	4
$C_{14}H_5CoF_{12}O$	73.46	2
$C_{14}H_6Br_2O_2$	26.3	1
$C_{14}H_6Br_2O_2$	26.1+	1
$C_{14}H_6Cl_2O_2$	26.4+	1
$C_{14}H_6F_2O_2$	26.6	1
$C_{14}H_6Fe_2O_6$	71.13	3
$C_{14}H_6Fe_4O_{12}S_3$	85.71	2
$C_{14}H_6I_2O_2$	26.7+	1
$C_{14}H_6N_2O_8$	26.9	1
$C_{14}H_6O_8$	59.24	1
$C_{14}H_7BrO_2$	26.1	3
$C_{14}H_7ClO_2$	26.2	3
$C_{14}H_7O_{12}Re_3Sn$	69.25	4
$C_{14}H_8AgN_2O_8$, H_2O	81.36	3
$C_{14}H_8AgN_2O_8$, $2H_2O$	81.36	4
$C_{14}H_8BrCl$	26.10	1
$C_{14}H_8BrN_5O_2$, $0.5C_7H_8$	37.18	1
$C_{14}H_8Br_2$	26.11+	1
$C_{14}H_8Br_2F_2$	5.24	1
$C_{14}H_8Br_2O$	26.13	1
$C_{14}H_8Br_2O_3$	13.52	1
$C_{14}H_8Br_2O_4$	13.17	4
$C_{14}H_8ClNO_2$	35.36	1
$C_{14}H_8Cl_2$	26.14	1
$C_{14}H_8Cl_2$	26.15	1
$C_{14}H_8Cl_2N_2O_2$	40.19	4
$C_{14}H_8Cl_2O_3$	13.53	1
$C_{14}H_8Cl_2O_4$	13.54	1
$C_{14}H_8Cl_2O_4$	13.15	3
$C_{14}H_8Cl_4$	26.16	1
$C_{14}H_8Fe_2O_6$	75.29	2
$C_{14}H_8I_2O_4$	13.18	4
$C_{14}H_8N_2NiO_8$, $3H_2O$	81.37	4
$C_{14}H_8N_2O_4$	26.17	1
$C_{14}H_8N_2O_8S_2$, H_2O	11.62	1
$C_{14}H_8O_2$	26.18+	1
$C_{14}H_8O_4$	26.25	1
$C_{14}H_8O_4$	26.28	1
$C_{14}H_8O_4$	26.29	1
$C_{14}H_8O_4$	26.26+	1
$C_{14}H_8O_4S$	42.5	3
$C_{14}H_8O_4S_2$	39.18	4
$C_{14}H_8O_6Ru_2$	75.30	2
$C_{14}H_8S_2$	39.19	4
$C_{14}H_9BrO_2$	38.48	1
$C_{14}H_9ClO_3$	22.8	4
$C_{14}H_9Cl_5$	19.11	4
$C_{14}H_9FeN_2$	73.12	4
$C_{14}H_9N$	36.12	4
$C_{14}H_9NO_2$	26.30	1
$C_{14}H_9NO_2$	40.27	1
$C_{14}H_{10}$	5.25+	1
$C_{14}H_{10}$	23.11	1
$C_{14}H_{10}$	26.31+	1
$C_{14}H_{10}$	28.17+	1
$C_{14}H_{10}$	26.3+	3
$C_{14}H_{10}$	28.3+	3
$C_{14}H_{10}$	28.4	4
$C_{14}H_{10}$	29.2	4
$C_{14}H_{10}$, $2Cl_3Sb$	28.5	4
$C_{14}H_{10}$, $C_6H_3N_3O_6$	60.133+	2
$C_{14}H_{10}$, $C_{10}H_2N_4$	60.32	4
$C_{14}H_{10}$, $C_{10}H_2O_6$	60.135	2
$C_{14}H_{10}$, $C_{12}H_4N_4$	60.125	2
$C_{14}H_{10}$, $C_{13}H_4BrN_3O_7$	60.42	3
$C_{14}H_{10}$, $C_{18}H_{18}Si$	60.36	4
$C_{14}H_{10}BrN_5O$	44.24	4
$C_{14}H_{10}Br_2N_2$	9.11	4
$C_{14}H_{10}CaO_6$	67.14	4
$C_{14}H_{10}ClNO_3$	13.16	3
$C_{14}H_{10}Cl_2CuN_2O_4$	78.3	2
$C_{14}H_{10}Cl_2FeHg_2N_2O_4$	83.67	3
$C_{14}H_{10}Cl_2Fe_2Ge$	73.47	2
$C_{14}H_{10}Cl_2Fe_2O_4Sn$	73.48	2
$C_{14}H_{10}Cl_2N_2$	36.8	3
$C_{14}H_{10}Cl_2N_2Pt$	71.21	4
$C_{14}H_{10}CrO_3$	75.31	2
$C_{14}H_{10}CuN_6O_3$	83.64	4
$C_{14}H_{10}CuO_4$	78.4+	2
$C_{14}H_{10}CuO_4$	84.36	2
$C_{14}H_{10}CuO_4$, $3H_2O$	81.77	2
$C_{14}H_{10}F_4$	19.44	1
$C_{14}H_{10}Fe_2N_2O_4Sn$	73.49	2
$C_{14}H_{10}Fe_2O_4$	73.51	2
$C_{14}H_{10}Fe_2O_4$	73.12	3
$C_{14}H_{10}Fe_2O_4$	73.13	3
$C_{14}H_{10}Fe_2O_6$	71.34	2
$C_{14}H_{10}Fe_2O_6$	75.32	2
$C_{14}H_{10}I_4O_4Re_2$	81.37	3
$C_{14}H_{10}MnO_3^-$, $C_4H_{12}N^+$	71.14	3
$C_{14}H_{10}N_2O_2$	32.38	1
$C_{14}H_{10}N_2O_2$	35.18	4
$C_{14}H_{10}N_2O_4$	15.10	3
$C_{14}H_{10}N_2O_{15}Ti_2^{2-}$, $2K^+$, $3H_2O$	81.38	3
$C_{14}H_{10}N_2S$	41.36	1
$C_{14}H_{10}N_4$	33.32	4
$C_{14}H_{10}NiS_4$	85.16	4
$C_{14}H_{10}NiS_6$	85.17	4
$C_{14}H_{10}O$	26.33	1
$C_{14}H_{10}O_2$	19.45	1
$C_{14}H_{10}O_2$	38.49	1
$C_{14}H_{10}O_2S_2$	11.22	3
$C_{14}H_{10}O_3$	13.17	3
$C_{14}H_{10}O_3W$	73.53	2
$C_{14}H_{10}O_4$	13.55	1
$C_{14}H_{10}O_4Ru_2$	73.54	2
$C_{14}H_{10}PdS_4$	85.72	2

$C_{14}H_{10}S_4Zn$ 85.18 **4**
$C_{14}H_{10}S_6Zn$ 85.19 **4**
$C_{14}H_{11}BrO$ 19.46 **1**
$C_{14}H_{11}ClO$ 19.47 **1**
$C_{14}H_{11}Cl_3MoN_2O_3Sn$ 69.23 **3**
$C_{14}H_{11}FO$ 19.48 **1**
$C_{14}H_{11}NO_2$ 16.10 **3**
$C_{14}H_{12}$ 19.49 **1**
$C_{14}H_{12}$ 29.1 **1**
$C_{14}H_{12}$ 22.9 **4**
$C_{14}H_{12}$, $4Cl_3Sb$ 60.136 **2**
$C_{14}H_{12}Ag_4O^{4+}$, $4ClO_4^-$ 74.6 **3**
$C_{14}H_{12}As_2Co_2F_4O_6$ 86.11 **3**
$C_{14}H_{12}As_2F_4Fe_2O_6$ 75.33 **2**
$C_{14}H_{12}BrClO_2$ 27.8 **3**
$C_{14}H_{12}BrNO_3$ 38.17 **4**
$C_{14}H_{12}Br_2OS$ 11.9 **4**
$C_{14}H_{12}ClN_3O$ 35.17 **3**
$2C_{14}H_{12}Cl_2N_2Ni$, $2CHCl_3$ 83.128 **2**
$C_{14}H_{12}Cl_2N_2Zn$ 83.68 **3**
$C_{14}H_{12}Cl_2O_2$ 17.40 **1**
$C_{14}H_{12}CoS_4^-$, $C_{19}H_{18}As^+$, $0.5C_2H_6O$ 85.73 **2**
$C_{14}H_{12}CrO_3S$ 75.8 **4**
$C_{14}H_{12}CuN_2O_2$ 78.7 **2**
$C_{14}H_{12}CuN_2O_4$ 78.8 **2**
$C_{14}H_{12}N^+$, I^- 36.9 **3**
$C_{14}H_{12}N_2$ 9.9 **3**
$C_{14}H_{12}N_2NiO_2$ 78.9 **2**
$C_{14}H_{12}N_2NiO_4$ 78.10 **2**
$C_{14}H_{12}N_2OSe$ 8.49 **1**
$C_{14}H_{12}N_2O_2$ 36.26 **1**
$C_{14}H_{12}N_2O_2$ 9.10 **3**
$C_{14}H_{12}N_2O_2$ 16.11 **3**
$C_{14}H_{12}N_2O_4Pd$ 78.1 **3**
$C_{14}H_{12}N_2O_5U$ 83.65 **4**
$C_{14}H_{12}N_2S_4$ 11.10 **4**
$C_{14}H_{12}N_4O_4Pd$ 83.66 **4**
$C_{14}H_{12}N_4S_3$ 39.35 **3**
$C_{14}H_{12}OS$ 39.75 **1**
$C_{14}H_{12}O_2$ 31.12 **4**
$C_{14}H_{12}O_3$ 59.6 **3**
$C_{14}H_{12}O_4$ 38.50 **1**
$C_{14}H_{12}O_7$ 25.20 **1**
$C_{14}H_{12}S_2$ 39.20 **4**
$C_{14}H_{13}BrN_4O_2$, H_2O 36.10 **3**
$C_{14}H_{13}BrN_4O_3$, H_2O 43.31 **1**
$C_{14}H_{13}BrO_2$ 52.23 **1**
$C_{14}H_{13}ClO$ 31.16 **1**
$C_{14}H_{13}Cl_2NO_3$ 50.5 **3**
$C_{14}H_{13}HgNO_3$ 71.22 **4**
$C_{14}H_{13}MnO_2$ 75.34 **2**
$C_{14}H_{13}NO$ 1.141 **1**
$C_{14}H_{13}N_4O_2^+$, I^-, H_2O 36.11 **3**
$C_{14}H_{14}$ 19.50 **1**
$C_{14}H_{14}$ 19.51+ **1**
$C_{14}H_{14}$, $2Cl_3Sb$ 60.137 **2**
$C_{14}H_{14}$, $4Cl_3Sb$ 60.138 **2**
$C_{14}H_{14}Ag^+$, ClO_4^- 74.7 **3**
$C_{14}H_{14}As_2$ 65.24 **2**
$C_{14}H_{14}As_2Br_2$ 65.25 **2**
$C_{14}H_{14}As_2I_2$ 65.26 **2**
$C_{14}H_{14}BrNO_2S$ 16.55 **1**
$C_{14}H_{14}BrNO_2S$ 16.56 **1**
$C_{14}H_{14}Br_2O_4$, $5C_4H_8O_2$ 31.16 **3**
$C_{14}H_{14}Br_2Se$ 11.63 **1**
$C_{14}H_{14}ClN_2O_4Tl$ 68.22 **2**
$C_{14}H_{14}Cl_2CoN_2$ 83.130+ **2**
$C_{14}H_{14}Cl_2CoN_6O_2$ 83.67 **4**
$C_{14}H_{14}Cl_2CuN_2$ 83.132 **2**
$C_{14}H_{14}Cl_2CuN_2O$ 83.69 **3**
$C_{14}H_{14}Cl_2HgN_2O_3$, $0.5H_2O$ 84.37 **2**
$C_{14}H_{14}Cl_2Se$ 11.64 **1**
$C_{14}H_{14}Cl_4$ 20.30 **1**
$C_{14}H_{14}CrO_3$ 75.9 **4**
$C_{14}H_{14}CuO_8$, $2H_2O$ 81.78 **2**
$C_{14}H_{14}FeO$ 73.55 **2**
$C_{14}H_{14}FeO_2$ 73.14 **3**
$C_{14}H_{14}GdN_7O_9$ 83.70 **3**
$C_{14}H_{14}Hg$ 71.15 **3**
$C_{14}H_{14}NO_3$ 17.41 **1**
$C_{14}H_{14}N_2$ 9.40+ **1**
$C_{14}H_{14}N_2O_3$ 9.11+ **3**
$C_{14}H_{14}N_2S$ 16.12 **3**
$C_{14}H_{14}N_4O_2$ 36.13 **4**
$C_{14}H_{14}N_6Ni$ 83.71 **3**
$C_{14}H_{14}NiO_6$ 78.11 **2**
$C_{14}H_{14}O_2$ 17.42 **1**
$C_{14}H_{14}O_2PS_2^-$, K^+ 64.33 **3**
$C_{14}H_{14}O_2Ru$ 73.57 **2**
$C_{14}H_{14}O_3$ 38.18 **4**
$C_{14}H_{14}O_4S_2$ 39.21 **4**
$C_{14}H_{14}O_4S_4Te$ 70.24 **2**
$C_{14}H_{14}O_4Se_2$ 39.22 **4**
$C_{14}H_{14}O_8Zn$ 78.12 **2**
$C_{14}H_{14}S$ 11.65 **1**
$C_{14}H_{14}S$, I_2 60.139 **2**
$C_{14}H_{14}S_2$ 11.66 **1**
$C_{14}H_{14}S_2$ 11.25 **3**
$C_{14}H_{14}S_2$ 11.23+ **3**
$C_{14}H_{14}S_2$ 11.11 **4**
$C_{14}H_{14}Se$ 11.67 **1**
$C_{14}H_{14}Te$ 70.25 **2**
$C_{14}H_{15}BrO_3$ 53.1 **1**
$C_{14}H_{15}BrO_3$ 59.7 **3**
$C_{14}H_{15}BrO_3S$ 31.17 **1**
$C_{14}H_{15}BrO_7$ 59.25 **1**
$C_{14}H_{15}ClN_4O_5$ 44.25 **4**
$C_{14}H_{15}NO_2$ 21.14 **3**
$C_{14}H_{15}NO_4S$ 11.12 **4**
$C_{14}H_{15}N_3$ 9.42 **1**
$C_{14}H_{15}O_4P$ 46.21 **1**
$C_{14}H_{16}Ag^+$, NO_3^- 75.35 **2**
$C_{14}H_{16}Ag^+$, NO_3^- 31.13 **4**
$C_{14}H_{16}BrNO_3$ 21.29 **1**
$C_{14}H_{16}Cr^+$, I^- 74.15 **2**
$C_{14}H_{16}Cr^+$, $C_{12}H_4N_4^-$ 74.16 **2**
$C_{14}H_{16}Cr^+$, $C_{12}H_4N_4^-$, $C_{12}H_4N_4$ 60.140 **2**
$C_{14}H_{16}CuN_2O_4$, C_5H_5N 81.79 **2**
$C_{14}H_{16}CuN_4O_2^{2+}$, $2ClO_4^-$ 83.68 **4**
$C_{14}H_{16}CuN_4O_2^{2+}$, $Cl_5Cu_3^{2-}$ 83.72 **3**
$C_{14}H_{16}F_4N_2P_2$ 64.41 **2**
$C_{14}H_{16}Fe_2O_2S_2^+$, BF_4^- 73.15 **3**
$C_{14}H_{16}N^+$, ClO_4^- 33.74 **1**
$C_{14}H_{16}NOP$ 64.42 **2**
$C_{14}H_{16}N_2$ 16.57 **1**
$C_{14}H_{16}N_2$ 16.58 **1**
$C_{14}H_{16}N_2^{2+}$, $2Br^-$, H_2O, $0.33HBr$ 34.3 **3**
$C_{14}H_{16}N_2^{2+}$, $C_4N_4Ni^{2-}$ 33.44 **3**
$C_{14}H_{16}N_2O_6Zn$ 81.38 **4**
$C_{14}H_{16}O_8$ 13.18 **3**
$C_{14}H_{16}P_2S_2$ 64.43 **2**
$C_{14}H_{16}S_5$ 39.23 **4**

$C_{14}H_{17}BrO_2S$	11.26	**3**
$C_{14}H_{17}Br_2N$	36.14	**4**
$C_{14}H_{17}Br_2N_3Ni$	83.73	**3**
$C_{14}H_{17}CoO$	73.58	**2**
$C_{14}H_{18}$	31.14	**4**
$C_{14}H_{18}Br_2CuN_2$	83.69	**4**
$C_{14}H_{18}ClNOPt$	71.23	**4**
$C_{14}H_{18}Cl_2CoN_2$	83.134	**2**
$C_{14}H_{18}Cl_2CuN_2$	83.74	**3**
$C_{14}H_{18}Cl_2CuN_2$	83.70	**4**
$C_{14}H_{18}Cl_2CuN_2O_2$	84.38	**2**
$C_{14}H_{18}Cl_2N_2O_2Zn$	84.39	**2**
$C_{14}H_{18}Cl_2N_2Zn$	83.135	**2**
$C_{14}H_{18}CoI_2N_2$	83.136	**2**
$C_{14}H_{18}FeNO$	73.59	**2**
$C_{14}H_{18}NO^+$, I^-	37.10	**4**
$C_{14}H_{18}NO_2^+$, I^-	58.10	**1**
$C_{14}H_{18}N_2$	24.8	**4**
$C_{14}H_{18}N_2^{2+}$, $2Cl^-$	16.59	**1**
$C_{14}H_{18}N_2^{2+}$, $C_4N_4Ni^{2-}$, $3H_2O$	33.33	**4**
$C_{14}H_{18}N_4O_4$, $2H_2O$	44.38	**3**
$C_{14}H_{18}OP^+$, I^-	64.18	**4**
$C_{14}H_{18}O_3S$	31.15	**4**
$C_{14}H_{19}ClO_9$	45.39	**3**
$C_{14}H_{19}N_2^+$, $C_{15}H_3CuF_{18}O_6^-$	77.7	**4**
$C_{14}H_{19}N_2^+$, $C_{15}H_3F_{18}MgO_6^-$	67.17	**4**
$C_{14}H_{19}N_4O_3^+$, Br^-	44.39	**3**
$C_{14}H_{20}$	31.18	**1**
$C_{14}H_{20}As_2ClF_{12}ORh$	71.24	**4**
$C_{14}H_{20}Cl_2Cu_2N_2O_2$	83.71	**4**
$C_{14}H_{20}Cl_2NPt$	71.25+	**4**
$C_{14}H_{20}CuN_6O_6$	83.75	**3**
$C_{14}H_{20}F_6FeN_2S_6$	80.8	**4**
$2C_{14}H_{20}FeN_2O_4^-$, Ca^{2+}, $8H_2O$	83.137	**2**
$C_{14}H_{20}N_2^{2+}$, $2Br^-$	37.19	**1**
$C_{14}H_{20}N_2^{2+}$, $2Br^-$, $2H_2O$	37.11	**4**
$C_{14}H_{20}N_3S^+$, Cl^-	33.34	**4**
$C_{14}H_{20}N_4O_2$	37.20	**1**
$C_{14}H_{20}N_4O_4$	44.73	**1**
$C_{14}H_{20}N_4O_4$	44.40	**3**
$C_{14}H_{20}O_2S_3$	39.24	**4**
$C_{14}H_{21}BrO$	17.43	**1**
$C_{14}H_{21}BrO$	53.1	**3**
$C_{14}H_{21}BrO$	53.2	**3**
$C_{14}H_{21}BrO_3$	38.51	**1**
$C_{14}H_{21}ClO_7$	45.40	**3**
$C_{14}H_{21}NO$	33.35	**4**
$C_{14}H_{21}NO$	33.36	**4**
$C_{14}H_{21}NO$	33.37	**4**
$C_{14}H_{21}N_3S^+$, Cl^-	39.36	**3**
$C_{14}H_{21}OP$	64.34	**3**
$C_{14}H_{21}OP$	64.35	**3**
$C_{14}H_{21}OP$	64.19	**4**
$C_{14}H_{22}$	52.24	**1**
$C_{14}H_{22}Cl_4Cu_2N_2O_4S_2$	84.8	**4**
$C_{14}H_{22}Cr_2N_4O_2$	73.16	**3**
$C_{14}H_{22}Cr_2N_4O_2$	73.17	**3**
$C_{14}H_{22}FeN_8O_4$, $2CH_4O$	83.72	**4**
$C_{14}H_{22}Mo_2S_4$	73.18	**3**
$C_{14}H_{22}Mo_2S_4^+$, F_6P^-	73.19	**3**
$C_{14}H_{22}N^+$, ClO_4^-	33.45	**3**
$C_{14}H_{22}NO_3^+$, Cl^-	3.49	**3**
$C_{14}H_{22}N_2Ni_2O_4$	84.12	**3**
$C_{14}H_{22}N_2O_2$	40.14	**3**
$C_{14}H_{22}N_2S_4$	58.11	**1**
$C_{14}H_{22}N_6O_6$, $3H_2O$	48.47	**3**
$C_{14}H_{22}NaO_6^+$, I^-	38.15	**3**
$C_{14}H_{22}NiO_2S_2$	85.20	**4**
$C_{14}H_{22}O$	53.3	**3**
$C_{14}H_{22}O_7$	45.22	**4**
$C_{14}H_{23}ClO_2$	53.2	**1**
$C_{14}H_{23}N_2O^+$, AsF_6^-	16.15	**4**
$C_{14}H_{23}N_2O_2^+$, Br^-	3.50	**3**
$C_{14}H_{23}N_6OP$	32.39	**1**
$C_{14}H_{23}N_6OP$	32.40	**1**
$C_{14}H_{24}$	28.19	**1**
$C_{14}H_{24}$	31.16	**4**
$C_{14}H_{24}Br_2Pd_2$	75.36	**2**
$C_{14}H_{24}Cl_4Pt_2$, $2CCl_4$	72.22	**3**
$C_{14}H_{24}CuN_2S_4$	80.25	**2**
$C_{14}H_{24}CuO_8S_2$	84.40	**2**
$C_{14}H_{24}N_4O_4$	34.12	**1**
$C_{14}H_{24}NaO_6^+$, I^-	67.15	**4**
$C_{14}H_{24}OS$	39.37	**3**
$C_{14}H_{24}O_4$	38.52	**1**
$C_{14}H_{24}P_2^+$, Br^-	64.36	**3**
$C_{14}H_{25}Co_3O_4S_5$	85.74	**2**
$C_{14}H_{25}MoO_4P_5$	86.17	**2**
$C_{14}H_{26}Cl_2Pd_2$	72.42	**2**
$C_{14}H_{26}N_3O_9^+$, Br^-, H_2O	50.9	**1**
$C_{14}H_{27}BrO_4$	1.142	**1**
$C_{14}H_{27}Cl_2N_2Rh$	75.37	**2**
$C_{14}H_{28}CuN_2O_4$	82.26	**3**
$C_{14}H_{28}CuN_2S_4$	80.26	**2**
$C_{14}H_{28}N^+$, I^-	35.37	**1**
$C_{14}H_{28}N_2NiS_4$	80.27+	**2**
$C_{14}H_{28}N_2O_8^{2+}$, $2Br^-$, $5H_2O$	50.2	**4**
$C_{14}H_{28}N_6NiO_4Se_2$	83.138	**2**
$C_{14}H_{28}P_2^{2+}$, $2Br^-$	64.37	**3**
$C_{14}H_{29}NO$	1.143	**1**
$C_{14}H_{29}O_4P$	64.38	**3**
$C_{14}H_{30}Be_2N_2$	67.16	**4**
$C_{14}H_{30}ClN_4NiO^+$, Cl^-	83.73	**4**
$C_{14}H_{30}N_2O_4^{2+}$, $2ClO_4^-$	1.13	**4**
$C_{14}H_{30}N_2O_4^{2+}$, $2I^-$	3.51	**3**
$C_{14}H_{32}NiO_6$	77.6	**4**

C_{15}

$C_{15}H_3CuF_{18}O_6^-$, $C_{14}H_{19}N_2^+$	77.7	**4**
$C_{15}H_3F_{18}MgO_6^-$, $C_{14}H_{19}N_2^+$	67.17	**4**
$C_{15}H_6F_4FeO_3$	75.16	**3**
$C_{15}H_7F_5MoO_2$	71.16	**3**
$C_{15}H_8F_{12}NiS_4$	85.23	**3**
$C_{15}H_8Fe_2O_5$	75.38	**2**
$C_{15}H_8INO_4$, H_2O	35.38	**1**
$2C_{15}H_9BrO_2$, $0.5C_6H_6$	27.10	**4**
$C_{15}H_9Br_2N$	5.27	**1**
$C_{15}H_9ClN_4$	27.11	**4**
$C_{15}H_9N$	26.34	**1**
$C_{15}H_{10}BrNO$	40.20	**4**
$C_{15}H_{10}BrNO_3$	35.39	**1**
$C_{15}H_{10}BrNO_3$	15.11	**3**
$C_{15}H_{10}Br_2O_2$	19.54	**1**
$C_{15}H_{10}ClNO_3$	15.12	**3**
$C_{15}H_{10}Cl_2$	26.35	**1**
$C_{15}H_{10}Cl_2$	26.9	**3**
$C_{15}H_{10}Cl_2O_2$	19.55	**1**
$C_{15}H_{10}FeO_3$	72.23	**3**

$C_{15}H_{10}N_2S_3$	41.23	4
$C_{15}H_{10}N_4$	27.12	4
$C_{15}H_{10}O$	26.36	1
$C_{15}H_{10}O$	28.20	1
$C_{15}H_{10}O_5Re_2$	73.13	4
$C_{15}H_{10}O_9Ru$	71.27	4
$C_{15}H_{10}S$	20.8	4
$C_{15}H_{11}BrN_2O$	32.41	1
$C_{15}H_{11}BrO_2$	38.16	3
$C_{15}H_{11}ClN_2O_3$	40.15	3
$C_{15}H_{11}Cl_2CoN_3$	83.76	3
$C_{15}H_{11}Cl_2N_3Zn$	83.141+	2
$C_{15}H_{11}Cl_3GaN_3$	68.13	3
$C_{15}H_{11}I_4NO_4$, $5H_2O$	48.48	3
$C_{15}H_{11}N_3O_9S_2$	11.68	1
$C_{15}H_{11}N_5O_6$	35.18	3
$C_{15}H_{12}$	26.10	3
$C_{15}H_{12}$	26.1	4
$C_{15}H_{12}Br_2N_4O_3$	9.43	1
$C_{15}H_{12}Cl_4O_2$	38.19	4
$C_{15}H_{12}CuN_3O_2^+$, ClO_4^-	83.140	2
$C_{15}H_{12}Fe_2O_6$	71.28+	4
$C_{15}H_{12}Fe_2O_7$	71.17	3
$C_{15}H_{12}NO_2S^+$, Cl^-	41.24	4
$C_{15}H_{12}N_2OS$	41.25	4
$C_{15}H_{12}N_2O_2$	32.20	4
$C_{15}H_{12}N_4$	9.13	3
$C_{15}H_{12}N_4S_3$	41.26	4
$C_{15}H_{12}O$	19.8	3
$C_{15}H_{12}O$	26.2	4
$C_{15}H_{12}O_2$	19.56	1
$C_{15}H_{12}O_4S$	42.6	4
$C_{15}H_{12}O_5$	59.26+	1
$C_{15}H_{12}O_6$	59.8	3
$C_{15}H_{13}Br$	5.28	1
$C_{15}H_{13}FeO_2^+$, F_6Sb^-	71.18	3
$C_{15}H_{13}I_2NO_4$, C_3H_7NO	60.43	3
$C_{15}H_{13}I_3NO_4$	48.50	4
$C_{15}H_{13}NO_3$	38.17	3
$C_{15}H_{14}Br_2O_8$	59.7	4
$C_{15}H_{14}CrO_3S$	75.10	4
$C_{15}H_{14}MoO_2$	74.17	2
$C_{15}H_{14}N_2$	32.42	1
$C_{15}H_{14}N_2$	16.16	4
$C_{15}H_{14}OS$	39.25	4
$C_{15}H_{14}OS_2$	39.26	4
$C_{15}H_{14}O_3$	17.44	1
$C_{15}H_{15}BrO_5$	59.9	3
$C_{15}H_{15}BrO_6$	59.28	1
$C_{15}H_{15}BrO_6$	59.29	1
$C_{15}H_{15}BrO_6$	59.30	1
$C_{15}H_{15}ClU$	73.62	2
$C_{15}H_{15}Cl_2CoN_3OS$	83.74	4
$C_{15}H_{15}Ge_3O_9Ru_3$	69.26	4
$C_{15}H_{15}Mn_3N_4O_4$	73.63	2
$C_{15}H_{15}MoNO$	73.20	3
$C_{15}H_{15}Mo_3S_4^+$, $C_3H_9Cl_2Sn^-$	73.14	4
$C_{15}H_{15}N_2O^+$, I^-	33.46	3
$C_{15}H_{15}N_5O_6Zn$	83.77	3
$C_{15}H_{15}Ni_3S_2$	73.64	2
$C_{15}H_{15}O_3P$	64.20	4
$C_{15}H_{15}Sm$	73.65	2
$C_{15}H_{16}BrNO_2$	37.12	4
$C_{15}H_{16}BrO_3$	53.3	1
$C_{15}H_{16}CrO_3$	75.39	2
$C_{15}H_{16}N_2O_6$	27.9	3
$C_{15}H_{17}BrN_4O_2$	36.15	4
$C_{15}H_{17}BrO_3$	31.19	1
$C_{15}H_{17}BrO_3$	53.5	1
$C_{15}H_{17}BrO_3$	53.6	1
$C_{15}H_{17}BrO_3$	53.7	1
$C_{15}H_{17}BrO_3$	53.8	1
$C_{15}H_{17}BrO_4$	53.4	3
$C_{15}H_{17}BrO_5$, H_2O	53.9	1
$C_{15}H_{17}BrO_6$	59.31	1
$C_{15}H_{17}ClO_3$	53.5	3
$C_{15}H_{17}CrNO_6$	71.30	4
$C_{15}H_{17}N_4O_2^+$, I^-, H_2O	36.15	3
$C_{15}H_{17}O_2Rh$	74.8+	3
$C_{15}H_{18}$	29.2	1
$C_{15}H_{18}BrCuN_4O^+$, ClO_4^-	83.78	3
$C_{15}H_{18}BrNO_2$	37.21	1
$C_{15}H_{18}ClN_3O$	38.18	3
$C_{15}H_{18}CrO_3$	74.18	2
$C_{15}H_{18}O_6Rh_2$	72.43	2
$C_{15}H_{19}BrO_3$	53.1	4
$C_{15}H_{19}BrO_3S$	31.20	1
$C_{15}H_{19}ClN_3Pd^+$, Cl^-	83.75	4
$C_{15}H_{19}ClO$	21.11	4
$C_{15}H_{19}ClO_4Zr$	73.21+	3
$C_{15}H_{19}Cl_3N_3Ti$	83.76	4
$C_{15}H_{19}O_3Re$	75.40	2
$C_{15}H_{19}O_4Rh$	74.10	3
$C_{15}H_{19}O_4Rh$	74.5	4
$C_{15}H_{19}Ti$	72.44	2
$C_{15}H_{20}Ag_2O_2^{2+}$, $2NO_3^-$	53.7	3
$C_{15}H_{20}As_2FeI_2NiO$	73.15	4
$C_{15}H_{20}N^+$, Br^-	37.9	3
$C_{15}H_{20}N_2O_3$	40.21	4
$C_{15}H_{20}N_3O_9^+$, $0.5CH_4O$, I^-	58.12	1
$C_{15}H_{20}N_6O_7$, H_2O	44.26	4
$C_{15}H_{20}O_5$	53.8	3
$C_{15}H_{20}O_{12}$	45.41	3
$C_{15}H_{21}Ag_3NiO_6^{2+}$, $2NO_3^-$, H_2O	77.19	2
$C_{15}H_{21}BMoN_4O_2$	72.12	4
$C_{15}H_{21}BrO_5$, $0.5C_3H_6O$	50.3	4
$C_{15}H_{21}Br_2ClO_2$	53.2	4
$C_{15}H_{21}Br_2ClO_3$	59.8	4
$C_{15}H_{21}ClN_3O_5Pr^{2+}$, $2Cl^-$, $3H_2O$	83.77	4
$C_{15}H_{21}ClO_6Zr$	77.8	4
$C_{15}H_{21}CoO_6$	77.20	2
$C_{15}H_{21}CoO_6$	77.8	3
$C_{15}H_{21}CoO_6$, $C_{15}H_{21}AlO_6$	77.9	4
$C_{15}H_{21}CoO_6^-$, $C_{16}H_{36}N^+$	77.21	2
$C_{15}H_{21}CrO_6$	77.22	2
$C_{15}H_{21}CrO_6$	77.9	3
$C_{15}H_{21}FeO_6$	77.23	2
$C_{15}H_{21}FeS_6$	85.24	3
$C_{15}H_{21}MnO_6$	77.24	2
$C_{15}H_{21}N_2O_4^+$, Br^-	48.49	3
$C_{15}H_{21}N_3O_4$	48.51	4
$C_{15}H_{21}N_5O_5$	47.33	3
$C_{15}H_{21}O_6Rh$	77.25	2
$C_{15}H_{21}O_6V$	77.26+	2
$C_{15}H_{21}O_6V$	77.10	4
$C_{15}H_{21}O_6Yb$	77.28	2
$C_{15}H_{22}BrN_4Ni^+$, Br^-, H_2O	83.143	2
$C_{15}H_{22}Br_2O_4$, $0.5H_2O$	53.9	3
$C_{15}H_{22}NO_2^+$, Br^-	58.10	3
$C_{15}H_{22}O_5$	59.9	4
$C_{15}H_{22}O_{10}$	45.42	3
$C_{15}H_{23}Br$	53.10	3

$C_{15}H_{23}Br$ 53.11 3
$C_{15}H_{23}BrO$ 53.12 3
$C_{15}H_{23}BrO_3P$ 53.13 3
$C_{15}H_{23}ClO_3P$ 53.14 3
$C_{15}H_{23}OP$ 64.39 3
$C_{15}H_{23}O_2Rh$ 72.45 2
$C_{15}H_{23}O_7Yb$ 77.29 2
$C_{15}H_{23}O_7Yb$ 77.11 3
$C_{15}H_{23}O_7Yb$, $0.5C_6H_6$ 77.30 2
$C_{15}H_{24}Ag^+$, NO_3^- 53.15 3
$C_{15}H_{24}AgNO_3$ 53.10 1
$C_{15}H_{24}Ag_2^{2+}$, $2NO_3^-$ 53.11 1
$C_{15}H_{24}Br_2$ 28.5 3
$C_{15}H_{24}Br_2$ 31.18 3
$C_{15}H_{24}Br_2$ 53.3 4
$C_{15}H_{24}Cl_2CuN_6$ 83.79 3
$C_{15}H_{24}FeN_3S_6$ 80.9 4
$C_{15}H_{24}IN_2O_2$ 53.12 1
$C_{15}H_{24}N^+$, Br^- 58.9 4
$C_{15}H_{24}NO^+$, Br^- 58.11 3
$C_{15}H_{24}N_4O_2S_2$ 44.42 3
$C_{15}H_{24}O$ 17.45 1
$C_{15}H_{25}Br$ 53.16 3
$C_{15}H_{25}BrO$ 53.13 1
$C_{15}H_{25}Br_2$ 53.14 1
$C_{15}H_{25}Cl$ 53.15 1
$C_{15}H_{25}Cl$ 53.16 1
$C_{15}H_{25}Cl$ 53.17 1
$C_{15}H_{25}Cl$ 53.18 1
$C_{15}H_{25}Cl_2OPPt$ 71.38 2
$C_{15}H_{25}EuO_8$, H_2O 77.12 3
$C_{15}H_{25}HoO_8$, $2H_2O$ 77.11 4
$C_{15}H_{25}LaO_8$ 77.32 2
$C_{15}H_{25}N_2^+$, ClO_4^- 59.32 1
$C_{15}H_{25}N_6OP$ 32.43 1
$C_{15}H_{25}N_6OP$, $0.5H_2O$ 32.44 1
$C_{15}H_{25}NdO_8$ 77.12 4
$C_{15}H_{25}NdO_8$, H_2O 77.33 2
$C_{15}H_{25}O_8Y$, H_2O 77.34 2
$C_{15}H_{26}Br_2$ 53.19 1
$C_{15}H_{26}Cl_2$ 53.20 1
$C_{15}H_{26}N_2$, H_2O 58.13 1
$C_{15}H_{26}N_4Ni^{2+}$, $2ClO_4^-$ 83.80 3
$C_{15}H_{27}FeS_9$ 80.10 4
$C_{15}H_{27}N_2O^+$, ClO_4^- 58.14 1
$2C_{15}H_{27}N_2O^+$, H^+, $3ClO_4^-$ 58.15 1
$C_{15}H_{27}N_3O_9Pd_3$, $0.5C_6H_6$ 81.39 4
$C_{15}H_{28}NO^+$, Br^- 58.10 4
$C_{15}H_{28}NO^+$, Br^- 58.11 4
$C_{15}H_{28}NO_2^+$, I^- 27.13 4
$C_{15}H_{28}O_2$ 23.3 4
$C_{15}H_{29}N_2^+$, I^- 37.10 3
$C_{15}H_{30}AsN_3S_6$ 65.27 2
$C_{15}H_{30}B_3CrN_3O_3$ 75.11 4
$C_{15}H_{30}CoN_3S_6$ 80.29+ 2
$C_{15}H_{30}N_3O_2S_6U^-$, $C_4H_{12}N^+$ 80.31 2
$C_{15}H_{30}N_3RuS_6$ 80.32 2
$C_{15}H_{30}N_4ORuS_6$ 80.33 2
$C_{15}H_{31}O_4P$ 64.40 3
$C_{15}H_{36}CoN_6^{3+}$, $3Cl^-$, $4H_2O$ 76.40 4
$C_{15}H_{42}CoN_6^{3+}$, $3Cl^-$, H_2O 83.78 4
$C_{15}H_{45}As_5NiO_5^{2+}$, $2ClO_4^-$ 84.41 2
$C_{15}H_{45}N_3Si_6$ 63.7 3

C_{16}

$C_{16}As_2F_{10}FeO_4$ 86.2 4
$C_{16}Co_2F_{24}S_8$ 85.45 2
$C_{16}F_{24}Fe_2S_8^{2-}$, $2C_{16}H_{36}N^+$ 85.25 3
$C_{16}H_4N_6O_8$ 28.6 4
$C_{16}H_5Fe_3O_{11}Rh$ 73.16 4
$C_{16}H_5N_5O_6$ 28.21 1
$C_{16}H_6N_4$ 24.9 4
$C_{16}H_6N_4O_4$ 28.22 1
$C_{16}H_8Cl_2$ 29.3 1
$C_{16}H_8Cl_2N_2O_4$ 40.17 3
$C_{16}H_8Cl_4$ 27.13 1
$C_{16}H_8F_8$ 31.17 4
$C_{16}H_8FeN_4O_3$ 71.31 4
$C_{16}H_8Mn_2O_6$ 75.42 2
$C_{16}H_8Mo_2O_6$ 75.43+ 2
$C_{16}H_8O_2S_2$ 39.76+ 1
$C_{16}H_8O_2Se_2$ 39.78 1
$C_{16}H_8O_4$ 38.53 1
$C_{16}H_9ClN_2O$ 36.16 4
$(C_{16}H_9CuF_{12}NO_5)_n$ 77.13 4
$C_{16}H_{10}$ 29.4 1
$C_{16}H_{10}$ 29.5 1
$C_{16}H_{10}$ 29.3 4
$C_{16}H_{10}$, C_6N_4 60.142 2
$C_{16}H_{10}$, C_6N_4 60.44 3
$C_{16}H_{10}$, $C_6H_3N_3O_6$ 60.143 2
$C_{16}H_{10}$, $C_9H_{12}N_4O_3$ 60.110 2
$C_{16}H_{10}$, $C_{10}H_2O_6$ 60.45+ 3
$C_{16}H_{10}BrN_4V$ 73.24 3
$C_{16}H_{10}Br_4O_5$ 38.20 4
$C_{16}H_{10}Cl_2O_2$ 19.9 3
$C_{16}H_{10}Cl_4Co_2FeGe_2O_6$ 73.17 4
$C_{16}H_{10}Co_4O_{10}$ 72.46 2
$C_{16}H_{10}Fe_2O_6$ 72.25 3
$C_{16}H_{10}Fe_2O_6$ 72.26 3
$C_{16}H_{10}Fe_2O_6$ 75.17 3
$C_{16}H_{10}Mo_2O_6$ 73.66 2
$C_{16}H_{10}N_2$ 5.29 1
$C_{16}H_{10}N_2O_2$ 35.40 1
$C_{16}H_{10}N_2O_2$ 35.41 1
$C_{16}H_{10}N_2O_2$ 35.19+ 3
$C_{16}H_{10}O$ 31.21 1
$C_{16}H_{10}O_2$ 25.21 1
$C_{16}H_{10}O_2$ 13.19+ 4
$C_{16}H_{11}ClO_3$ 38.54 1
$C_{16}H_{11}Cl_2NO$ 35.19 4
$C_{16}H_{11}FeO_4P$ 86.12 3
$C_{16}H_{11}N_3O_3$ 9.44 1
$C_{16}H_{11}N_4^+$, Cl^-, H_2O 36.16 3
$C_{16}H_{12}$ 31.22 1
$C_{16}H_{12}$ 30.1 3
$C_{16}H_{12}$ 31.20 3
$C_{16}H_{12}As_2F_4I_2Mn_2O_8$ 86.3 4
$C_{16}H_{12}As_2F_4Mn_2O_8$ 86.4 4
$C_{16}H_{12}Br_2$ 29.4 4
$C_{16}H_{12}Cl_2$ 26.3 4
$C_{16}H_{12}Cl_2CuN_4$ 83.81 3
$C_{16}H_{12}CrO_3$ 75.46 2
$C_{16}H_{12}CuN_2O_6$ 84.42 2
$C_{16}H_{12}Fe_2O_6$ 75.12 4
$C_{16}H_{12}Fe_2O_8$ 75.18 3
$C_{16}H_{12}NO^+$, HO_4Se^- 50.4 4
$C_{16}H_{12}N_4OS$ 41.37 1

$C_{16}H_{12}OS_2$	39.27	4
$C_{16}H_{12}O_2$	19.10	3
$C_{16}H_{12}O_2$	26.4	4
$C_{16}H_{12}O_2$	27.14	4
$C_{16}H_{12}O_6$	50.6	3
$C_{16}H_{12}O_{12}Re_4S_4$	85.21	4
$C_{16}H_{13}BiN_2O_2S_2$	66.7	4
$C_{16}H_{13}Br$	26.37	1
$C_{16}H_{13}Br$	28.6	3
$C_{16}H_{13}ClN_2O$	35.20	4
$C_{16}H_{13}ClN_4O_4$	15.26	1
$C_{16}H_{13}MoN_3O_2S$	72.27	3
$C_{16}H_{13}NO_3$	1.33	3
$C_{16}H_{13}N_2O_2^+$, Br^-, $2H_2O$	58.12	3
$0.5C_{16}H_{14}$, $C_{19}H_{14}FeO_3$	60.51	3
$C_{16}H_{14}BrNO$	31.23	1
$C_{16}H_{14}BrN_5O_3$	37.23	1
$C_{16}H_{14}BrO_6$	38.21	4
$C_{16}H_{14}ClFeN_2O_2$	78.13	2
$C_{16}H_{14}ClFeN_2O_2$, CH_3NO_2	78.14	2
$C_{16}H_{14}Cl_2Cu_2N_2O_2$	78.2	3
$C_{16}H_{14}Cl_2N_2$	35.21	4
$C_{16}H_{14}Cl_2N_2O_2Ti$, C_4H_8O	78.2	4
$C_{16}H_{14}Cl_2O$	31.21	3
$C_{16}H_{14}Cl_2S_2$	11.69	1
$C_{16}H_{14}CoN_2O_2$, $CHCl_3$	78.15	2
$C_{16}H_{14}CrO_5$	74.11	3
$C_{16}H_{14}CuN_2O_2$	78.16	2
$C_{16}H_{14}CuN_2O_2$, $CHCl_3$	78.17	2
$C_{16}H_{14}CuN_2O_2$, $CHCl_3$	60.1	3
$C_{16}H_{14}CuN_2O_2$, $C_6H_5NO_3$	78.18	2
$C_{16}H_{14}CuN_2O_2$, $C_6H_5NO_3$	60.16	3
$2C_{16}H_{14}CuN_2O_2$, Na^+, ClO_4^-, C_8H_{10}	60.145	2
$C_{16}H_{14}CuO_{10}$	81.80	2
$C_{16}H_{14}F_2$	5.30	1
$C_{16}H_{14}MoS_2$	73.26	3
$C_{16}H_{14}N_2NiO_2$	78.3	4
$C_{16}H_{14}O$	38.55	1
$C_{16}H_{14}O_2$	17.6	3
$C_{16}H_{14}O_2$	6.2	4
$C_{16}H_{14}O_2S$	1.34	3
$C_{16}H_{14}S_3$	39.38	3
$C_{16}H_{15}Br$	5.31	1
$C_{16}H_{15}Br_2N_3O_2Si$, C_3H_6O	63.19	2
$C_{16}H_{15}Co$	73.67	2
$C_{16}H_{15}Co_3O_2$	73.27	3
$C_{16}H_{15}N$	37.11	3
$C_{16}H_{15}N_2O^+$, Br^-	32.45	1
$C_{16}H_{16}$	31.24	1
$C_{16}H_{16}$	31.25+	1
$C_{16}H_{16}$	19.11	3
$C_{16}H_{16}$	23.8+	3
$C_{16}H_{16}$	31.18	4
$C_{16}H_{16}Ag^+$, NO_3^-	75.47	2
$C_{16}H_{16}Br_2N_2$	28.7	3
$C_{16}H_{16}Ce^-$, $C_6H_{14}KO_3^+$	75.13	4
$C_{16}H_{16}Cl_2Cu_2N_2O_2$	78.4	4
$C_{16}H_{16}Cl_4CuN_2O_4$	81.40	4
$C_{16}H_{16}Cl_4Pd_2$	72.47	2
$C_{16}H_{16}CuN_2O_2$	78.19+	2
$C_{16}H_{16}CuN_2O_2$	78.5	4
$C_{16}H_{16}CuN_4O_8^{2-}$, $2K^+$	83.79	4
$C_{16}H_{16}CuN_4O_8^{2-}$, $2Li^+$	83.80	4
$C_{16}H_{16}Fe$	75.48	2
$C_{16}H_{16}Fe_2O_4Pb$	73.68	2
$C_{16}H_{16}Fe_2O_4Sn$	73.69	2
$C_{16}H_{16}HgN_2O_2S_2$	85.75	2
$C_{16}H_{16}HgO_2$	71.39	2
$C_{16}H_{16}IN$	37.13	4
$C_{16}H_{16}MoO_4S_2Ti$	73.28	3
$C_{16}H_{16}N_2NiO_2$	78.22+	2
$C_{16}H_{16}N_2NiO_6$	83.144	2
$C_{16}H_{16}N_2NiS_2^-$, $C_8H_{20}N^+$	85.27	3
$C_{16}H_{16}N_2O_2$	9.45	1
$C_{16}H_{16}N_2O_2Zn$	78.24	2
$C_{16}H_{16}N_2O_3Zn$	78.25	2
$C_{16}H_{16}N_2O_5S$	50.7	3
$C_{16}H_{16}NiO_4$	72.48	2
$C_{16}H_{16}OS_2$	39.39	3
$C_{16}H_{16}O_3$	17.46	1
$C_{16}H_{16}O_4$	27.14	1
$C_{16}H_{16}Th$	75.14	4
$C_{16}H_{16}Ti$	75.19	3
$C_{16}H_{16}U$	75.15	4
$C_{16}H_{17}BCo_3NO_{10}$	71.40	2
$C_{16}H_{17}BrN_4O_3$	36.17	4
$C_{16}H_{17}Mo_2O_4P$	73.70	2
$C_{16}H_{17}NOS$	41.27	4
$C_{16}H_{17}NO_3$	38.20	3
$C_{16}H_{17}N_2O_4S^-$, K^+	50.10	1
$C_{16}H_{17}N_2O_4S^-$, Na^+	50.11	1
$C_{16}H_{18}$	19.57	1
$C_{16}H_{18}ClN_3O$	36.18	4
$C_{16}H_{18}Cl_2CuN_2O_4$	81.40	3
$C_{16}H_{18}CrN_2O_4^+$, Cl^-	78.3	3
$C_{16}H_{18}CrO_4$	75.16	4
$C_{16}H_{18}CuO_8$	81.81	2
$C_{16}H_{18}Fe$	73.71	2
$C_{16}H_{18}NO_4^+$, Br^-, $2H_2O$	58.16	1
$C_{16}H_{18}N_2O_2$	9.14	3
$C_{16}H_{18}N_2O_3$	9.15	3
$C_{16}H_{18}N_2O_3$	58.12	4
$C_{16}H_{18}N_2O_4S$	50.12	1
$C_{16}H_{18}N_2O_5S$	50.13	1
$C_{16}H_{18}N_2O_6S$, CH_4O	50.8	3
$C_{16}H_{18}N_3S^+$, Cl^-, $5H_2O$	41.15	3
$C_{16}H_{18}N_4^{2+}$, $2C_2H_5O_4S^-$, $2H_2O$	8.12	4
$C_{16}H_{18}N_4O^{2+}$, $2C_2H_5O_4S^-$, $2H_2O$	8.13	4
$C_{16}H_{18}N_4O_3$	36.27	1
$C_{16}H_{18}N_6Ni$	71.20	3
$C_{16}H_{18}Ni_2$	72.13	4
$C_{16}H_{19}BrN_2O_2S$	33.75	1
$C_{16}H_{19}F_{12}N_3NiO$	71.32	4
$C_{16}H_{19}NO_2S_2$	11.13	4
$C_{16}H_{19}N_3O_4S$, $3H_2O$	50.14	1
$C_{16}H_{19}N_4O_2^+$, ClO_4^-	36.19	4
$C_{16}H_{20}Ag^+$, ClO_4^-	74.12	3
$C_{16}H_{20}Ag^+$, ClO_4^-	74.13	3
$C_{16}H_{20}Ag^+$, ClO_4^-	74.14	3
$C_{16}H_{20}As_2PdS_2$, C_5H_5N	86.5	4
$C_{16}H_{20}As_2PtS_2$	86.13	3
$C_{16}H_{20}BrN$	31.27	1
$C_{16}H_{20}BrNO_5$	50.15	1
$C_{16}H_{20}BrN_2^+$, $C_4H_3O_4^-$	33.38	4
$C_{16}H_{20}Cl_2Sn_2^{2+}$, $2AlCl_4^-$	69.27	4
$C_{16}H_{20}Co_2O_4$	72.29	3
$C_{16}H_{20}CuN_5S^+$, CNS^-	83.82+	3
$C_{16}H_{20}CuO_6$, $4H_2O$	81.41	4
$C_{16}H_{20}CuO_9$	81.42	4
$C_{16}H_{20}Fe$	73.72	2
$C_{16}H_{20}Fe_2Ge_2O_8$	69.29	2

$C_{16}H_{20}GeO_{10}, 2H_2O$ 69.30 **2**
$C_{16}H_{20}MoO_8P_2$ 86.6 **4**
$C_{16}H_{20}NO_2^+, Br^-$ 58.17 **1**
$C_{16}H_{20}NO_4^+, Br^-$ 58.18 **1**
$2C_{16}H_{20}N_2, C_6Cl_4O_2$ 60.33 **4**
$C_{16}H_{20}N_2NiO_4P_2S_4$ 85.28+ **3**
$C_{16}H_{20}N_3O_2^+, Br^-$ 51.1 **3**
$C_{16}H_{20}N_3O_8S^-, Na^+, 2H_2O$ 50.16 **1**
$C_{16}H_{20}NiP_2S_4$ 85.76 **2**
$C_{16}H_{20}O_2$ 31.28 **1**
$C_{16}H_{20}S_5$ 39.28 **4**
$C_{16}H_{21}Cl_3CuNOP$ 86.7 **4**
$C_{16}H_{21}NO$ 36.18 **3**
$C_{16}H_{21}N_3^+, ClO_3^-$ 12.12 **4**
$C_{16}H_{21}N_3^+, ClO_4^-$ 12.10 **1**
$C_{16}H_{21}N_3^+, I^-$ 12.11 **1**
$C_{16}H_{21}O_{11}^-, Rb^+, 2H_2O$ 52.25 **1**
$C_{16}H_{22}$ 31.19 **4**
$C_{16}H_{22}Al_2$ 68.14 **3**
$C_{16}H_{22}BrNO_3$ 58.19 **1**
$C_{16}H_{22}Br_3NiP_2, 0.5C_{16}H_{22}Br_2NiP_2, C_6H_6$ 86.14 **3**
$C_{16}H_{22}ClN_4NiS_2^+, ClO_4^-$ 85.30 **3**
$C_{16}H_{22}Cl_2P_2Pd$ 86.8 **4**
$C_{16}H_{22}Cl_4Hg_2N_2$ 83.145 **2**
$C_{16}H_{22}INO_2S$ 36.19 **3**
$C_{16}H_{22}I_2P_2Pd$ 86.18+ **2**
$C_{16}H_{22}NO_2^+, Cl^-$ 24.10 **4**
$C_{16}H_{22}NO_3^+, Br^-$ 58.20 **1**
$C_{16}H_{22}N_2$ 35.42 **1**
$C_{16}H_{22}N_2$ 35.22 **4**
$C_{16}H_{22}N_2O_2$ 18.4 **4**
$C_{16}H_{22}NiO_{10}$ 81.42 **3**
$C_{16}H_{22}O_8Si_2W_2$ 63.6 **4**
$C_{16}H_{23}Br$ 21.12 **4**
$C_{16}H_{23}IO_2S$ 52.26 **1**
$C_{16}H_{23}NO$ 52.27 **1**
$C_{16}H_{23}NO_5$ 58.13 **4**
$C_{16}H_{23}O_2S_2$ 39.29 **4**
$C_{16}H_{24}$ 22.10 **4**
$C_{16}H_{24}Br_2MnN_8$ 83.81 **4**
$C_{16}H_{24}Cl_2Cu_2$ 75.50 **2**
$C_{16}H_{24}Cl_2Rh_2$ 75.51 **2**
$C_{16}H_{24}Cl_3IrSn$ 75.52 **2**
$C_{16}H_{24}Cl_4Ni_2, C_6H_6$ 75.53 **2**
$C_{16}H_{24}CoN_5O_6$ 83.146 **2**
$C_{16}H_{24}CoN_9O_3^+, NO_3^-, 0.5C_2H_6O$ 83.82 **4**
$C_{16}H_{24}NO_2^+, Br^-$ 33.76 **1**
$C_{16}H_{24}NO_2^+, Cl^-$ 33.77 **1**
$C_{16}H_{24}NO_2^+, Cl^-$ 33.78 **1**
$C_{16}H_{24}Ni$ 75.54 **2**
$C_{16}H_{24}O_4$ 59.10 **4**
$C_{16}H_{24}O_4Pt_2$ 71.33 **4**
$C_{16}H_{25}AsN_2S_4$ 65.28 **2**
$C_{16}H_{25}Cl_2NPt$ 75.21 **3**
$C_{16}H_{25}Co$ 75.55 **2**
$(C_{16}H_{25}Mn_3O_{17})_n$ 81.43 **3**
$C_{16}H_{26}Br_2O_8Sn$ 69.31 **2**
$C_{16}H_{26}Br_2O_8Sn$ 69.25 **3**
$C_{16}H_{26}Cu_2N_2O_4$ 77.14 **3**
$C_{16}H_{26}NO^+, Br^-$ 58.14 **4**
$C_{16}H_{26}NO_2^+, Br^-$ 58.13 **3**
$C_{16}H_{27}NO_5$ 58.15 **4**
$C_{16}H_{27}N_2O_2^+, Br^-$ 58.21 **1**
$C_{16}H_{28}AuN_{16}^-, C_{24}H_{20}As^+$ 71.34 **4**
$C_{16}H_{28}CaN_4O_8^{2+}, 2Br^-$ 67.17 **3**
$C_{16}H_{28}Ce_2O_{18}^{2-}, 2CH_6N_3^+, 2H_2O$ 81.43 **4**
$C_{16}H_{28}Cu_2O_8$ 81.82 **2**
$C_{16}H_{28}Cu_2O_8$ 81.44 **4**
$C_{16}H_{28}NO^+, Br^-$ 58.16 **4**
$C_{16}H_{28}N_4Ni^{2+}, 2ClO_4^-$ 83.83 **4**
$C_{16}H_{28}N_4O_2$ 12.13 **4**
$C_{16}H_{28}N_{12}Ni_2O_4^{4+}, 4Cl^-, 2H_2O$ 83.84 **3**
$C_{16}H_{28}O_4$ 38.56 **1**
$C_{16}H_{28}O_8Re_2^{2+}, 2O_4Re^-$ 81.44 **3**
$C_{16}H_{28}O_8Ru_2^+, Cl^-$ 81.83 **2**
$C_{16}H_{30}Cl_2Pd_2$ 72.14 **4**
$C_{16}H_{30}Co_2S_{14}$ 85.22 **4**
$C_{16}H_{30}Fe_2S_{14}$ 85.31 **3**
$C_{16}H_{30}O_6Ru_2Sn_4$ 69.26 **3**
$C_{16}H_{31}O_2^-, K^+$ 2.97 **1**
$C_{16}H_{32}BeN_2$ 67.18 **4**
$C_{16}H_{32}Cl_2N_8NiS_4$ 79.36 **2**
$C_{16}H_{32}MgN_2$ 67.18 **3**
$C_{16}H_{32}N_4Ni^{2+}, 2ClO_4^-$ 83.85 **3**
$C_{16}H_{32}N_4Ni^{2+}, 2ClO_4^-$ 83.84 **4**
$C_{16}H_{34}O$ 5.32 **1**
$C_{16}H_{36}As_3NNiP^+, ClO_4^-$ 86.20 **2**
$C_{16}H_{36}N^+, Ag_3I_4^-$ 3.52 **3**
$C_{16}H_{36}N^+, F^-, 32.8H_2O$ 61.13 **2**
$C_{16}H_{36}N^+, C_2Cl_2O_2Rh^-$ 3.53 **3**
$C_{16}H_{36}N^+, C_7H_5O_2^-, 39.5H_2O$ 61.14 **2**
$C_{16}H_{36}N^+, C_8CuN_4S_4^-$ 85.40 **2**
$C_{16}H_{36}N^+, C_8FeN_4S_4^-$ 85.41 **2**
$C_{16}H_{36}N^+, C_9H_7Br_3NNi^-$ 3.80 **1**
$C_{16}H_{36}N^+, C_{15}IO_{15}Rh_6^-$ 3.54 **3**
$C_{16}H_{36}N^+, C_{15}H_{21}CoO_6^-$ 77.21 **2**
$C_{16}H_{36}N^+, C_{26}H_{15}F_{12}FeOPS_4^-$ 84.20 **3**
$C_{16}H_{36}N^+, C_{26}H_{20}CoN_2S_4^-$ 85.45 **3**
$2C_{16}H_{36}N^+, O_{19}W_6^{2-}$ 3.55 **3**
$2C_{16}H_{36}N^+, C_5H_{10}B_{10}MoO_5^{2-}$ 62.9 **3**
$2C_{16}H_{36}N^+, C_8CoN_4S_4^{2-}$ 85.39 **2**
$2C_{16}H_{36}N^+, C_{16}F_{24}Fe_2S_8^{2-}$ 85.25 **3**
$2C_{16}H_{36}N^+, C_{16}O_{16}Re_4^{2-}$ 3.82 **1**
$2C_{16}H_{36}N^+, C_{24}Cl_{16}Co_2S_8^{2-}$ 85.86 **2**
$3C_{16}H_{36}N^+, C_6ErN_6S_6^{3-}$ 3.83 **1**
$3C_{16}H_{36}N^+, C_8MoN_8^{3-}$ 3.56 **3**
$(C_{16}H_{36}O_2P_2S_2Zn)_n$ 85.32+ **3**
$C_{16}H_{36}O_4Ti_2$ 71.21 **3**
$C_{16}H_{38}N_2^{2+}, 2Br^-, 2H_2O$ 3.84 **1**
$C_{16}H_{40}Br_2Mg_2N_2$ 67.15 **2**
$C_{16}H_{40}Br_2Mg_2O_4$ 67.16 **2**
$C_{16}H_{40}CrN_2O_2Si_4$ 83.85 **4**
$C_{16}H_{40}Mo_2O_{11}P_4S_8, 2C_6H_4Cl_2$ 85.78 **2**
$C_{16}H_{40}P_4S_8Zn_2$ 85.34 **3**
$C_{16}H_{44}B_4N_8$ 62.48 **2**
$C_{16}H_{48}Be_4O_4Si_4$ 63.8 **3**
$C_{16}H_{48}CuN_8O_6P_4^{2+}, 2ClO_4^-$ 84.13 **3**
$C_{16}H_{48}Hg_4O_4Si_4$ 63.9 **3**
$C_{16}H_{48}N_8O_{16}P_8$ 64.45 **2**
$C_{16}H_{48}N_{12}P_4$ 64.46 **2**
$C_{16}H_{64}B_{20}Cd_2O_4$ 84.9 **4**

C_{17}

$C_{17}Fe_6O_{16}^{2-}, 2C_4H_{12}N^+$ 71.35 **4**
$C_{17}H_5Co_3GeO_{11}$ 69.27 **3**
$C_{17}H_5F_{18}Rh$ 74.19 **2**
$C_{17}H_8Fe_2O_5$ 75.22 **3**

$C_{17}H_8N_2O$	26.38	1
$C_{17}H_{10}BrNS_2$	41.38	1
$C_{17}H_{10}CrO_3$	74.20	2
$C_{17}H_{10}CrO_3$	74.21+	2
$C_{17}H_{10}MnNO_4$	71.36	4
$C_{17}H_{11}BrS_3$	39.40	3
$C_{17}H_{11}MnO_5$	72.15	4
$C_{17}H_{11}NO_2S_2$	39.41	3
$C_{17}H_{12}$	29.1	3
$C_{17}H_{12}As_2F_4Fe_3O_9$	71.37	4
$C_{17}H_{12}BrN$	24.11	4
$C_{17}H_{12}BrNO_2$	32.21	4
$C_{17}H_{12}BrN_3O_4$	36.20	3
$C_{17}H_{12}CrO_3$	74.23	2
$C_{17}H_{12}OS_2$	39.42	3
$C_{17}H_{12}O_2$	27.10	3
$C_{17}H_{12}O_6$	59.10	3
$C_{17}H_{12}O_6$	59.11	4
$C_{17}H_{12}O_6$, $CHCl_3$	59.11	3
$C_{17}H_{12}O_7$, $0.5C_4H_3BrS$	59.33	1
$C_{17}H_{12}O_7$, $0.5C_6H_5Br$	59.34	1
$C_{17}H_{12}O_7$, $0.5C_6H_6$	59.35	1
$C_{17}H_{12}S_3$	39.83	1
$C_{17}H_{12}S_3$	39.43	3
$C_{17}H_{12}S_3$	39.30	4
$C_{17}H_{13}ClN_4$, $C_{17}H_{14}ClN_4^+$, Br^-, $2C_2H_6O$	36.20	4
$C_{17}H_{14}$	29.7	1
$C_{17}H_{14}$	31.20	4
$C_{17}H_{14}CoO$	73.73	2
$C_{17}H_{14}CuN_3O_2^+$, BF_4^-	81.45	3
$C_{17}H_{14}CuN_3O_2^+$, F_6P^-	81.46	3
$C_{17}H_{14}NO_3S^-$, K^+	24.4	3
$C_{17}H_{14}N_4S_2$	41.39	1
$C_{17}H_{14}O_2Ti$	71.38	4
$C_{17}H_{14}O_6$	59.12	3
$C_{17}H_{15}As_2Co_3F_4O_7$	86.9	4
$C_{17}H_{15}BMoN_8O_2$	83.86	3
$C_{17}H_{15}BrN_2$	5.33	1
$C_{17}H_{15}BrO_2$	31.25	3
$C_{17}H_{15}BrO_3S$	52.28	1
$C_{17}H_{15}BrO_3S$	31.26	3
$C_{17}H_{15}Cl_3N_2O$	35.43	1
$C_{17}H_{15}FSi$	63.21	2
$C_{17}H_{15}N$	33.39	4
$C_{17}H_{15}Ni_3O_2$	73.74	2
$C_{17}H_{16}BrClO_6$	50.17	1
$C_{17}H_{16}Cl_2Cu_2N_2O_2$	78.4	3
$C_{17}H_{16}CuN_2O_2$, H_2O	78.26	2
$C_{17}H_{16}CuN_3O_3^+$, BF_4^-	81.47	3
$C_{17}H_{16}Fe_2O_5$	75.57	2
$C_{17}H_{16}Fe_2O_5$	75.58	2
$C_{17}H_{16}Ge$	69.28	3
$C_{17}H_{16}MoS_2$	73.75	2
$C_{17}H_{16}N_2O_3V$	78.5	3
$C_{17}H_{16}O_4$	26.5	4
$C_{17}H_{16}O_6$	59.12	4
$C_{17}H_{16}O_7$	59.13	3
$C_{17}H_{16}Si$	63.22	2
$C_{17}H_{17}BBrNO_4$	45.43	3
$C_{17}H_{17}ClN_3Sn^+$, $C_2H_6Cl_3Sn^-$	69.32	2
$C_{17}H_{17}NO$	35.23	4
$C_{17}H_{18}BrNO$	40.22	4
$C_{17}H_{18}BrNO_2$	13.19+	3
$C_{17}H_{18}CuN_2O_8P$	82.57	2
$C_{17}H_{18}N_2O_5S$	36.21	3
$C_{17}H_{19}BrN_4O_3$	36.22	3
$C_{17}H_{19}ClN_2S$	41.40	1
$C_{17}H_{19}NO_2$	21.13	4
$C_{17}H_{19}NO_3S_2$	37.12	3
$C_{17}H_{19}NO_5$	58.22	1
$C_{17}H_{19}N_3O_5$, $4H_2O$	59.13	4
$C_{17}H_{20}ClN_2OP$	64.21	4
$C_{17}H_{20}ClN_2S^+$, Cl^-	41.28	4
$C_{17}H_{20}NO_3^+$, Br^-, H_2O	58.23	1
$C_{17}H_{20}NO_3^+$, I^-, $2H_2O$	58.14	3
$C_{17}H_{20}N_4O_6$, $C_{10}H_{12}BrN_5O_3$, $3H_2O$	60.34	4
$C_{17}H_{20}N_8O_4$, H_2O	44.43	3
$C_{17}H_{21}BrO_2$	53.17	3
$C_{17}H_{21}BrO_5$	53.22	1
$C_{17}H_{21}BrO_5$	53.23	1
$C_{17}H_{21}BrO_5$	53.4	4
$C_{17}H_{21}IO_2$	52.29	1
$C_{17}H_{21}NO_4S$	11.14	4
$C_{17}H_{21}N_2S^+$, Br^-	41.41	1
$C_{17}H_{21}N_2S^+$, Cl^-	41.42	1
$C_{17}H_{21}N_4O_6^+$, Br^-, H_2O	59.15	3
$C_{17}H_{22}BrP$	64.48	2
$C_{17}H_{22}NO_2^+$, Br^-	51.2	3
$C_{17}H_{22}NO_4^+$, Br^-	58.24	1
$C_{17}H_{22}NO_4^+$, Br^-	3.25	4
$C_{17}H_{22}NO_4^+$, Br^-	3.26	4
$C_{17}H_{22}NO_4^+$, Cl^-	58.25	1
$C_{17}H_{22}N_2O_3$	51.1	1
$C_{17}H_{22}N_2O_6S$, H_2O	48.93	1
$C_{17}H_{22}N_2O_6S$, H_2O	48.52	4
$C_{17}H_{22}N_3^+$, ClO_4^-	16.60	1
$C_{17}H_{22}N_4O_6^{2+}$, $C_6H_6O_2$, $2Br^-$	60.47	3
$C_{17}H_{23}BrO_3$	59.16	3
$C_{17}H_{23}BrO_4S$	31.30	1
$C_{17}H_{23}BrO_5$	53.25	1
$C_{17}H_{23}CuN_3O_5$, $0.5C_4H_{10}O$, $4H_2O$	82.58	2
$C_{17}H_{23}FeN_7S_2^+$, ClO_4^-	83.150	2
$C_{17}H_{23}N^+$, I^-	3.85	1
$C_{17}H_{23}NO_2$	51.1	4
$C_{17}H_{24}NO^+$, Br^-, H_2O	58.15	3
$C_{17}H_{24}OS_2$	39.85	1
$C_{17}H_{24}O_2$	1.144	1
$C_{17}H_{24}O_2$	57.1	4
$C_{17}H_{24}O_2Pd$	75.23	3
$C_{17}H_{25}Al_2N$	68.15	3
$C_{17}H_{25}BrO_4$	53.18	3
$C_{17}H_{26}BrNO_5$	45.23	4
$C_{17}H_{26}N^+$, Cl^-	33.47	3
$C_{17}H_{26}NO_4^+$, Br^-, $0.5H_2O$	58.16	3
$C_{17}H_{26}O_3P$	52.4	4
$C_{17}H_{26}O_3S$	21.15	3
$C_{17}H_{26}O_3S$	21.16	3
$C_{17}H_{26}O_3S$	21.14	4
$C_{17}H_{27}Cl_2NPt$	72.30	3
$C_{17}H_{27}NNiP_2S_4$	83.87	3
$C_{17}H_{27}N_3O_6$, H_2O	48.53	4
$C_{17}H_{27}N_8O_5^+$, Br^-	50.18	1
$C_{17}H_{27}O_2PS$	52.5	4
$C_{17}H_{29}BrO$	20.9	4
$C_{17}H_{35}CoN_5S_3$, $0.5C_4H_{10}O$	83.88	3
$C_{17}H_{39}NNiOP_3^+$, BF_4^-	86.16	3
$C_{17}H_{49}N_3Si_7$	63.7	4

C_{18}

$C_{18}N_2O_2$ 10.30 **1**
$C_{18}HCo_5O_{15}$ 71.39 **4**
$C_{18}H_8Fe_2N_2O_6$ 83.89 **3**
$C_{18}H_9ClO_2$ 29.8 **1**
$C_{18}H_{10}$ 30.1 **1**
$C_{18}H_{10}Cl_2O_2$ 20.15 **3**
$C_{18}H_{10}Cr_2O_6$ 74.24+ **2**
$C_{18}H_{10}F_{12}O_8V_2$ 73.29 **3**
$C_{18}H_{10}Fe_2N_2O_6$ 83.152 **2**
$C_{18}H_{10}Fe_2O_8Rh_2$ 73.18 **4**
$C_{18}H_{10}Mn_2N_2O_8Pt$ 83.153 **2**
$C_{18}H_{12}$ 23.15 **1**
$C_{18}H_{12}$ 29.9 **1**
$C_{18}H_{12}$ 29.10 **1**
$C_{18}H_{12}$ 29.11 **1**
$C_{18}H_{12}$ 29.14 **1**
$C_{18}H_{12}$ 29.12+ **1**
$C_{18}H_{12}$ 30.2 **1**
$C_{18}H_{12}As_2F_4Fe_3O_{10}$ 86.17 **3**
$C_{18}H_{12}As_2F_4O_{10}Ru_3$ 86.18 **3**
$C_{18}H_{12}AuCl_3N_2$ 83.154 **2**
$C_{18}H_{12}BFFeN_6O_3P^+, BF_4^-, CH_2Cl_2$ 83.86 **4**
$C_{18}H_{12}BFN_6NiO_3P^+, BF_4^-$ 83.90 **3**
$C_{18}H_{12}B_3NO_6$ 62.17 **3**
$C_{18}H_{12}Br_2NiO_2S_2$ 85.23 **4**
$C_{18}H_{12}ClNO_3$ 24.28 **1**
$C_{18}H_{12}CuN_2O_2$ 83.155+ **2**
$C_{18}H_{12}CuN_2O_2, 2C_6H_2N_6O_6$ 60.76 **2**
$C_{18}H_{12}CuN_2O_2, 2C_{10}H_2N_4$ 60.146+ **2**
$C_{18}H_{12}CuN_2O_2, C_{12}H_4N_4$ 60.126 **2**
$C_{18}H_{12}F_3N$ 16.14 **3**
$C_{18}H_{12}Fe_2O_7$ 75.17+ **4**
$C_{18}H_{12}I_3N$ 16.15 **3**
$C_{18}H_{12}MoN_2O_4$ 84.10 **4**
$C_{18}H_{12}Mo_2O_6$ 75.25 **3**
$C_{18}H_{12}N_2O_2Pd$ 83.159 **2**
$C_{18}H_{12}N_2O_2Pd, C_6Cl_4O_2$ 60.70 **2**
$C_{18}H_{12}N_2O_2Pd, C_{10}H_2N_4$ 60.148 **2**
$C_{18}H_{12}N_2O_2Zn, 2H_2O$ 83.160 **2**
$C_{18}H_{12}N_2PbS_2$ 69.28 **4**
$C_{18}H_{12}N_2PdS_2$ 85.80 **2**
$C_{18}H_{12}N_3O_6P_3, 0.5C_6H_5Br$ 64.49 **2**
$C_{18}H_{12}N_3O_6P_3, 0.5C_6H_6$ 64.50 **2**
$C_{18}H_{12}N_3O_{10}P$ 46.7 **3**
$C_{18}H_{12}N_5O_6, C_6H_6$ 12.12 **1**
$C_{18}H_{12}O_2$ 31.21 **4**
$C_{18}H_{12}O_2S_2$ 39.31 **4**
$C_{18}H_{12}O_6P^-, C_6H_{16}N^+$ 46.7 **4**
$C_{18}H_{12}O_6Si^{2-}, 2C_5H_6N^+$ 63.10 **3**
$C_{18}H_{13}AgN_2O_2, C_5H_5N$ 83.161 **2**
$C_{18}H_{13}As$ 65.29 **2**
$C_{18}H_{13}FeNO_3$ 72.31 **3**
$C_{18}H_{13}MoN_3O_3$ 83.87 **4**
$C_{18}H_{13}O_4Si^-, C_4H_{12}N^+$ 3.32 **1**
$C_{18}H_{13}S_3$ 39.32 **4**
$C_{18}H_{14}$ 19.58 **1**
$C_{18}H_{14}$ 19.59 **1**
$C_{18}H_{14}$ 30.3 **1**
$C_{18}H_{14}$ 19.12+ **3**
$C_{18}H_{14}$ 19.12 **4**
$C_{18}H_{14}, C_{26}H_{28}N_6NiS_2$ 61.8 **3**
$C_{18}H_{14}BrOP$ 64.22 **4**
$C_{18}H_{14}BrP$ 64.51 **2**
$C_{18}H_{14}Br_2N_2$ 9.47 **1**
$C_{18}H_{14}ClOP$ 64.23 **4**
$C_{18}H_{14}CuO_8$ 81.84 **2**
$C_{18}H_{14}Fe$ 75.60 **2**
$C_{18}H_{14}Fe_2O_4$ 73.76+ **2**
$C_{18}H_{14}Fe_2O_4$ 71.24 **3**
$C_{18}H_{14}N_4$ 9.12 **4**
$C_{18}H_{14}Ru$ 73.78 **2**
$C_{18}H_{15}AsO, H_2O$ 65.30 **2**
$C_{18}H_{15}AsO, C_6H_2Cl_4O_2$ 60.35 **4**
$C_{18}H_{15}Bi$ 66.15+ **2**
$C_{18}H_{15}BiCl_2$ 66.17 **2**
$C_{18}H_{15}BrO_5$ 31.27 **3**
$C_{18}H_{15}Br_3PZn^-, C_6H_8N^+$ 86.10 **4**
$C_{18}H_{15}ClP^+, C_8AuF_{12}S_4^-$ 64.52 **2**
$C_{18}H_{15}ClSn$ 69.29 **4**
$C_{18}H_{15}Cl_2Sb$ 66.18 **2**
$C_{18}H_{15}FeN_6O_6$ 84.44 **2**
$C_{18}H_{15}Ga$ 68.16 **3**
$C_{18}H_{15}I_3NiP^-, C_{24}H_{20}As^+$ 65.42 **2**
$C_{18}H_{15}In$ 68.17 **3**
$C_{18}H_{15}N_4PS_3$ 64.41 **3**
$C_{18}H_{15}N_6O_8U^-, H_4N^+$ 84.45 **2**
$C_{18}H_{15}OP$ 64.42 **3**
$C_{18}H_{15}OP, Cl_2OSe$ 60.48 **3**
$C_{18}H_{15}OP, C_6H_2N_6O_4$ 60.75 **2**
$4C_{18}H_{15}OP, C_8H_4Mo_4N_4O_{16}$ 60.149 **2**
$5C_{18}H_{15}OP, I^-, Li^+$ 61.15 **2**
$C_{18}H_{15}O_3Rh_3$ 73.79 **2**
$C_{18}H_{15}O_3Rh_3$ 73.80+ **2**
$C_{18}H_{15}O_4P$ 46.23+ **1**
$C_{18}H_{15}P$ 64.53 **2**
$2C_{18}H_{15}PS, 3I_2$ 60.150 **2**
$C_{18}H_{16}$ 31.22 **4**
$C_{18}H_{16}Ag_2ClO_4^+, ClO_4^-$ 75.19 **4**
$C_{18}H_{16}AsBrO$ 65.31 **2**
$C_{18}H_{16}AsClO$ 65.32 **2**
$C_{18}H_{16}BrNO_2$ 31.31 **1**
$C_{18}H_{16}Cl_2Co_2O_4Sn$ 75.20 **4**
$C_{18}H_{16}Cl_2N_2O_2$ 31.23 **4**
$(C_{18}H_{16}CoN_3O_2)_n, nCH_4O$ 71.25 **3**
$C_{18}H_{16}Cu_3N_6O_5S, 16.3H_2O$ 83.88 **4**
$C_{18}H_{16}ErN_3O_8$ 81.45 **4**
$(C_{18}H_{16}LaN_3O_8)_n$ 81.46 **4**
$C_{18}H_{16}Mn_2O_4$ 72.51 **2**
$C_{18}H_{16}N_2O_4Zn$ 83.162 **2**
$C_{18}H_{16}N_4O_2$ 32.22 **4**
$C_{18}H_{16}OP^+, C_6HN_6O_4^-$ 64.43 **3**
$C_{18}H_{16}O_3$ 38.22 **4**
$C_{18}H_{17}BrO_3$ 59.38 **1**
$C_{18}H_{17}Cl$ 26.6 **4**
$C_{18}H_{17}ClN_2$ 20.10 **4**
$C_{18}H_{17}NO_3$ 24.29 **1**
$C_{18}H_{18}$ 23.16 **1**
$C_{18}H_{18}$ 28.23+ **1**
$C_{18}H_{18}Br_2$ 31.33 **1**
$C_{18}H_{18}ClNO$ 32.33 **3**
$C_{18}H_{18}Cl_2$ 5.34 **1**
$C_{18}H_{18}Cl_2Cu_2N_2O_2$ 78.6 **4**
$C_{18}H_{18}Cl_2Pd_2$ 72.32 **3**
$C_{18}H_{18}Co_2$ 75.61 **2**
$C_{18}H_{18}Fe_2O_7$ 75.27 **3**
$C_{18}H_{18}I_2N_2NiO_2$ 78.6 **3**
$C_{18}H_{18}N_2O_3$ 32.23 **4**
$C_{18}H_{18}N_2O_5$ 13.20 **4**
$C_{18}H_{18}N_2PS_2^+, ClO_4^-$ 64.54 **2**

$C_{18}H_{18}N_2S_2$ 41.16 **3**
$C_{18}H_{18}N_6$ 7.32 **1**
$C_{18}H_{18}N_6S_2$ 41.44 **1**
$C_{18}H_{18}O$ 21.17 **3**
$C_{18}H_{18}O_2$ 29.15 **1**
$C_{18}H_{18}O_2$ 17.9 **4**
$C_{18}H_{18}S_2$ 11.15 **4**
$C_{18}H_{18}S_3$ 39.33 **4**
$C_{18}H_{18}Si, C_{14}H_{10}$ 60.36 **4**
$C_{18}H_{19}BrO_3$ 31.28 **3**
$C_{18}H_{19}CoN_2O_2$ 71.40 **4**
$(C_{18}H_{19}HoN_3O_8{}^{3+})_n, nC_6CrN_6S_6{}^{3-}, 2nH_2O$ 81.47 **4**
$C_{18}H_{19}IO_8$ 59.17 **3**
$C_{18}H_{19}N_2{}^+, I^-$ 35.44 **1**
$C_{18}H_{19}N_3O_4U$ 78.7 **3**
$C_{18}H_{19}N_3O_6S, C_2H_4O_2, H_2O$ 50.9 **3**
$C_{18}H_{19}N_6NiO_3{}^{3+}, 3Br^-, H_2O$ 83.92 **3**
$C_{18}H_{19}O_6Pb^-, C_4H_{12}N^+$ 69.30 **4**
$C_{18}H_{20}$ 29.16 **1**
$C_{18}H_{20}$ 31.34 **1**
$C_{18}H_{20}$ 31.35 **1**
$C_{18}H_{20}$ 26.7 **4**
$C_{18}H_{20}, C_6N_4$ 60.37 **4**
$C_{18}H_{20}B_2CoN_{12}$ 83.93 **3**
$C_{18}H_{20}BrNO_{10}S$ 45.44 **3**
$C_{18}H_{20}Br_2N_2$ 33.40 **4**
$C_{18}H_{20}ClNO_2$ 31.24 **4**
$C_{18}H_{20}ClN_3O_2$ 36.21 **4**
$C_{18}H_{20}ClN_3O_6S_2, 0.65CH_2Br_2$ 58.27 **1**
$C_{18}H_{20}Cl_2Cu_2N_2O_2$ 78.7 **4**
$C_{18}H_{20}Cl_2N_2$ 9.17 **3**
$C_{18}H_{20}CuN_2O_2$ 78.27 **2**
$C_{18}H_{20}CuN_2O_2$ 78.28 **2**
$C_{18}H_{20}CuN_2O_2$ 78.8 **3**
$C_{18}H_{20}CuN_2O_4$ 78.29 **2**
$C_{18}H_{20}CuN_2O_4$ 78.9 **3**
$C_{18}H_{20}CuN_2O_4$ 82.14 **4**
$C_{18}H_{20}CuN_2O_6$ 82.28 **3**
$C_{18}H_{20}NO_3{}^+, Br^-$ 58.28 **1**
$C_{18}H_{20}N_2NiO_2$ 78.30 **2**
$C_{18}H_{20}N_2O_2Pd$ 78.31 **2**
$C_{18}H_{20}N_2O_2PdS_2$ 83.89 **4**
$C_{18}H_{20}N_2O_4Pd, 0.5H_2O$ 82.29 **3**
$C_{18}H_{20}N_2S_2$ 41.29 **4**
$C_{18}H_{20}N_4$ 31.29 **3**
$3C_{18}H_{20}OS, C_2H_6O$ 61.16 **2**
$6C_{18}H_{20}OS, C_9H_{16}O$ 61.1 **4**
$C_{18}H_{20}O_2$ 17.47 **1**
$C_{18}H_{20}O_2$ 17.7+ **3**
$C_{18}H_{20}O_2, 0.167CHCl_3$ 61.6 **3**
$C_{18}H_{20}O_2, CH_4O, H_2O$ 17.9 **3**
$C_{18}H_{20}O_2, C_2H_6O, H_2O$ 17.10 **4**
$C_{18}H_{20}O_2, 0.33C_2H_6O$ 61.7 **3**
$C_{18}H_{20}O_2, C_2H_6OS$ 17.11 **4**
$6C_{18}H_{20}O_2, C_7H_{16}O$ 61.2 **4**
$C_{18}H_{20}O_6, 2C_3H_7NO$ 31.25 **4**
$C_{18}H_{20}S_2$ 39.34 **4**
$C_{18}H_{20}S_3$ 39.44 **3**
$C_{18}H_{21}BrN_2OPd$ 71.26 **3**
$C_{18}H_{21}BrO_2$ 51.2+ **1**
$C_{18}H_{21}BrO_7S$ 45.45 **3**
$C_{18}H_{21}ClN_4Pd$ 71.41 **4**
$C_{18}H_{21}CuN_3{}^+, ClO_4{}^-$ 83.90 **4**
$C_{18}H_{21}NO, H_2O$ 12.14 **4**
$C_{18}H_{21}N_2O_2$ 40.29 **1**
$C_{18}H_{22}BrN_3$ 36.22 **4**
$C_{18}H_{22}Br_2O_2$ 51.2+ **4**
$C_{18}H_{22}Co_2O_4$ 72.52 **2**
$C_{18}H_{22}CrN_4O_2{}^+, I^-$ 78.8 **4**
$C_{18}H_{22}Cu_2N_2O_8$ 81.85+ **2**
$C_{18}H_{22}FeNO_3{}^+, Cl^-$ 72.33 **3**
$C_{18}H_{22}HgN_2$ 71.41 **2**
$C_{18}H_{22}NO_3{}^+, Br^-$ 53.26 **1**
$C_{18}H_{22}NO_3{}^+, Br^-, 2H_2O$ 58.29 **1**
$C_{18}H_{22}NS^+, Cl^-$ 39.35 **4**
$C_{18}H_{22}N_2O_2$ 37.14 **4**
$C_{18}H_{22}N_2O_2Si_2$ 63.11 **3**
$C_{18}H_{22}N_2S$ 41.17 **3**
$C_{18}H_{22}N_4O_2$ 36.23 **4**
$C_{18}H_{22}N_4O_4S_6Te$ 70.15 **4**
$C_{18}H_{22}N_4O_{10}S_2$ 47.30 **1**
$C_{18}H_{22}N_4O_{10}S_2$ 47.34 **3**
$C_{18}H_{22}Ni$ 75.62 **2**
$C_{18}H_{22}O_2$ 51.5 **3**
$C_{18}H_{22}O_2$ 51.6 **3**
$C_{18}H_{22}O_2$ 51.4+ **4**
$C_{18}H_{22}O_2S$ 11.27 **3**
$C_{18}H_{23}BrO_2, CH_4O$ 51.3 **1**
$C_{18}H_{23}BrO_3$ 52.5 **3**
$C_{18}H_{23}CoN_2O_2, H_2O$ 76.81 **3**
$C_{18}H_{23}CuF_{15}N_4O_4$ 76.41 **4**
$C_{18}H_{23}N_2O^+, Br^-$ 58.17 **3**
$C_{18}H_{23}N_2OS^+, C_4H_3O_4{}^-$ 41.30 **4**
$C_{18}H_{23}N_2O_2{}^+, C_{12}H_8N_2O_8P^-, H_2O$ 16.17 **4**
$C_{18}H_{23}N_2S^+, Cl^-$ 41.45 **1**
$C_{18}H_{23}N_2S^+, CH_3O_4S^-$ 41.18 **3**
$C_{18}H_{24}$ 24.30 **1**
$C_{18}H_{24}$ 31.30 **3**
$C_{18}H_{24}CdN_{12}{}^{2+}, 2NO_3{}^-$ 83.91 **4**
$C_{18}H_{24}CdN_{12}{}^{2+}, HO^-, NO_3{}^-, 4H_2O$ 83.92 **4**
$C_{18}H_{24}CdN_{12}{}^{2+}, CO_3{}^{2-}, 5H_2O$ 83.93 **4**
$C_{18}H_{24}Cl_2N_6NiP_2$ 86.21 **2**
$C_{18}H_{24}CoN_{12}{}^{2+}, 2NO_3{}^-$ 83.94+ **4**
$C_{18}H_{24}CoN_{12}{}^{2+}, CO_3{}^{2-}, 5H_2O$ 83.96 **4**
$C_{18}H_{24}CoN_{12}{}^{2+}, 2C_2H_3O_2{}^-, H_2O$ 83.97 **4**
$C_{18}H_{24}CuN_2O_4, 3H_2O$ 81.48 **4**
$C_{18}H_{24}NO_2{}^+, Br^-$ 51.7 **4**
$C_{18}H_{24}NO_3{}^+, I^-$ 58.30 **1**
$C_{18}H_{24}NO_5{}^+, Br^-$ 58.19 **3**
$C_{18}H_{24}N_2NiO_4P_2S_4$ 83.94 **3**
$C_{18}H_{24}N_2O_2Pt$ 83.164 **2**
$C_{18}H_{24}N_4O_2, 3H_2O$ 36.24 **4**
$C_{18}H_{24}N_{12}Ni^{2+}, 2NO_3{}^-$ 83.165 **2**
$C_{18}H_{24}N_{12}Ni^{2+}, 2NO_3{}^-$ 83.95 **3**
$C_{18}H_{24}N_{12}Zn^{2+}, 2Cl^-, 4H_2O$ 83.166 **2**
$C_{18}H_{24}NiO_2$ 75.63 **2**
$C_{18}H_{24}O_2, H_2O$ 51.4 **1**
$C_{18}H_{24}O_2, 0.5H_2O$ 51.8 **4**
$C_{18}H_{24}O_2, CH_4N_2O$ 60.38 **4**
$C_{18}H_{24}O_2, C_3H_8O$ 51.9 **4**
$C_{18}H_{24}O_3$ 51.5 **1**
$C_{18}H_{24}O_3$ 29.2 **3**
$C_{18}H_{24}O_3$ 51.10 **4**
$C_{18}H_{25}AlN_2O_2$ 68.9 **4**
$C_{18}H_{25}NO$ 58.31 **1**
$C_{18}H_{25}NOP^+, Cl_2Cu^-$ 64.44 **3**
$C_{18}H_{26}As_2Cl_3O_2Rh$ 86.19 **3**
$C_{18}H_{26}As_2I_2PdS_2$ 86.20 **3**
$C_{18}H_{26}BrNO_6, 0.5C_2H_6O$ 58.32 **1**
$C_{18}H_{26}CoN_4$ 83.98 **4**
$C_{18}H_{26}CuN_4$ 83.97+ **3**

$C_{18}H_{26}CuN_8O_6$	81.49	4
$C_{18}H_{26}FeN_8O_6$, $2H_2O$	83.167	2
$C_{18}H_{26}NO^+$, Br^-, H_2O	58.33	1
$C_{18}H_{26}N_4Ni$	83.99	3
$C_{18}H_{26}O_2$	51.7	3
$C_{18}H_{27}BrO_3$	31.26	4
$C_{18}H_{28}ClN_3^{2+}$, $2H_2O_4P^-$, H_2O	35.21	3
$C_{18}H_{28}Cl_2FeMoS_2$	73.19	4
$C_{18}H_{28}NO_3^+$, I^-	58.20	3
$2C_{18}H_{28}NO_6^+$, Cl_6Pt^{2-}, $2H_2O$	58.34	1
$C_{18}H_{28}N_{10}NiS_2$	83.99	4
$C_{18}H_{30}$	29.19	1
$C_{18}H_{30}$	29.20	1
$C_{18}H_{30}$	29.17+	1
$C_{18}H_{30}$	52.30	1
$C_{18}H_{30}$, $0.5CHCl_3$	61.17	2
$C_{18}H_{30}$, $0.39C_6H_{12}$	61.18	2
$C_{18}H_{30}$, $0.225C_7H_{16}$	61.19	2
$C_{18}H_{30}Ag_2^{2+}$, $2ClO_4^-$	72.53	2
$C_{18}H_{30}Al_2$	68.18	3
$C_{18}H_{30}BrNO$	21.31	1
$C_{18}H_{30}Br_2O$	54.1	1
$C_{18}H_{30}Ca_2O_{12}Sr$	67.18+	2
$C_{18}H_{30}ClCo_3F_9O_{16}$	81.50	4
$C_{18}H_{30}Cl_2$	5.35	1
$C_{18}H_{30}Cl_4O_2Sn_2$	69.34	2
$C_{18}H_{30}Co_2HgO_6P_2$	86.23	2
$C_{18}H_{30}N^+$, I^-	58.35	1
$C_{18}H_{30}NO_2S_2^+$, BF_4^-	39.45	3
$C_{18}H_{30}N_2NiO_4P_2S_4$	85.81	2
$C_{18}H_{31}Cl_4P_2RhSi$, $xHCl_3Si$	86.21	3
$C_{18}H_{32}$	21.15	4
$C_{18}H_{32}CoN_6^+$, ClO_4^-	83.100	3
$C_{18}H_{32}Cu_2N_4O_2^{2+}$, $2ClO_4^-$	83.100	4
$C_{18}H_{32}NO_3^+$, Br^-	53.19	3
$C_{18}H_{32}O_2$	28.8	3
$C_{18}H_{32}O_{16}$	45.24	4
$C_{18}H_{32}O_{16}$, $2H_2O$	45.25	4
$C_{18}H_{32}O_{16}$, $5H_2O$	45.47	3
$C_{18}H_{34}ClN_2O_5S^+$, Cl^-, H_2O	50.10	3
$C_{18}H_{34}O_2$	1.145	1
$C_{18}H_{34}O_2$	1.146	1
$C_{18}H_{34}O_8$	1.35	3
$C_{18}H_{35}BrO_2$	1.147	1
$C_{18}H_{35}N_2O_6S^+$, Cl^-, H_2O	50.19	1
$C_{18}H_{36}$	21.32	1
$C_{18}H_{36}AuN_2S_4^+$, $AgBr_2^-$	80.11	4
$C_{18}H_{36}AuN_2S_4^+$, $AuBr_2^-$	80.34	2
$C_{18}H_{36}AuN_2S_4^+$, Br^-	80.11	3
$C_{18}H_{36}AuN_2S_4^+$, $C_8AuN_4S_4^-$	80.12	4
$C_{18}H_{36}CsN_2O_6^+$, CNS^-, H_2O	67.19	3
$C_{18}H_{36}Cs_2N_2S_4$	67.20	3
$C_{18}H_{36}Cu_2N_{12}S_6^{2+}$, $2BF_4^-$	79.24	3
$C_{18}H_{36}KN_2O_6^+$, I^-	67.21	3
$C_{18}H_{36}N_2NaO_6^+$, I^-	67.22	3
$C_{18}H_{36}N_2O_6Rb^+$, CNS^-, H_2O	67.23	3
$C_{18}H_{36}N_4Ni^{2+}$, $2ClO_4^-$	83.101	4
$C_{18}H_{36}O_2$	1.148	1
$C_{18}H_{36}O_2$	1.36	3
$C_{18}H_{36}O_6Pt_2$	77.36	2
$C_{18}H_{38}N_4O_{11}^{2+}$, O_4S^{2-}, H_2O	50.20	1
$C_{18}H_{38}N_4O_{11}^{2+}$, O_4Se^{2-}, H_2O	50.21	1
$C_{18}H_{39}N_4NiO_2^+$, ClO_4^-	81.48	3
$C_{18}H_{40}N_2O_4Pt_2$	77.37	2
$C_{18}H_{42}Cl_4P_2Pt_2$	86.24	2
$C_{18}H_{42}CoN_6^{3+}$, $3Cl^-$, $5H_2O$	76.82	3
$C_{18}H_{42}CrN_3$	83.102	3
$C_{18}H_{48}Cu_2N_{12}S_6^{2+}$, $2BF_4^-$	79.25	3
$C_{18}H_{54}AlN_3Si_6$	63.12	3
$C_{18}H_{54}B_2N_4Si_6$	63.23	2
$C_{18}H_{54}Cl_3N_9O_3P_3Pr$	84.14	3
$C_{18}H_{54}FeN_3Si_6$	63.24	2
$C_{18}H_{54}Li_3N_3Si_6$	63.13	3

C_{19}

$C_{19}H_{10}Br_2O_6$	38.23	4
$C_{19}H_{10}Fe_2N_2O_7$	83.168+	2
$C_{19}H_{11}BrN_4O_7$	17.12	4
$C_{19}H_{11}BrN_4O_7$	17.13	4
$C_{19}H_{12}Br_2N_2$	35.45	1
$C_{19}H_{12}Br_2O$	19.15	3
$C_{19}H_{12}N_3O_6$	12.13	1
$C_{19}H_{12}O_2$	29.21	1
$C_{19}H_{12}O_2$	29.22	1
$C_{19}H_{12}O_6$	59.39	1
$C_{19}H_{13}BrOS_4$	39.46	3
$C_{19}H_{14}BrNO_9$, C_3H_7NO	50.22	1
$C_{19}H_{14}Br_2O$	20.31	1
$C_{19}H_{14}Br_2O_2$	17.14	4
$C_{19}H_{14}ClNO_9$, C_3H_7NO	50.23	1
$C_{19}H_{14}CrO_3$	75.28	3
$C_{19}H_{14}FeO_3$, $0.5C_{16}H_{14}$	60.51	3
$C_{19}H_{14}I_2O$	20.32	1
$C_{19}H_{15}^+$, ClO_4^-	12.14	1
$C_{19}H_{15}AuNP$	86.22	3
$C_{19}H_{15}Br$	19.60	1
$C_{19}H_{15}Br$	19.16	3
$C_{19}H_{15}BrOSi$	63.8	4
$C_{19}H_{15}Fe_2NO_{10}S$	72.54	2
$C_{19}H_{15}MoN_3O_2S$	71.27	3
$C_{19}H_{15}N$	28.9	3
$C_{19}H_{15}N_3O$	32.24	4
$C_{19}H_{16}$	19.13	4
$C_{19}H_{16}BrN_3O_6$	36.25	4
$C_{19}H_{16}ClNO_4$	35.24	4
$C_{19}H_{16}Cl_3NO_2S_2$	24.12	4
$C_{19}H_{16}FeO_2$	73.20	4
$C_{19}H_{16}FeO_3$	75.64	2
$C_{19}H_{16}FeO_3$	75.29	3
$C_{19}H_{16}N_4$	35.25	4
$C_{19}H_{16}O_3S$	39.88	1
$C_{19}H_{16}P^+$, Cl_4Fe^-	12.15	1
$C_{19}H_{17}NOS_2$	39.36	4
$C_{19}H_{17}NO_2$	25.6	3
$C_{19}H_{17}NO_2S$	25.7	4
$C_{19}H_{17}NO_2S_2$	11.28	3
$C_{19}H_{17}NS_3$	39.47	3
$C_{19}H_{17}P$	64.45	3
$C_{19}H_{18}As^+$, $C_{12}H_4N_4^-$, $C_{12}H_4N_4$	65.9	4
$C_{19}H_{18}As^+$, $C_{14}H_{12}CoS_4^-$, $0.5C_2H_6O$	85.73	2
$2C_{19}H_{18}As^+$, Cl_4Ni^{2-}	65.33	2
$C_{19}H_{18}CrO_4$	74.6+	4
$C_{19}H_{18}N_3^+$, ClO_4^-	12.15	4
$C_{19}H_{18}N_3O_4S_2^+$, Cl^-, H_2O	50.11	3
$C_{19}H_{18}OPdS$	72.34	3
$C_{19}H_{18}O_6$	59.40	1
$C_{19}H_{18}O_6$	59.14	4
$C_{19}H_{18}P^+$, $H_{24}B_{20}Zn^-$	64.46	3

$C_{19}H_{18}P^+$, $C_4H_{22}B_{18}Cu^-$ 64.56 2
$C_{19}H_{18}P^+$, $C_8N_4NiS_4^-$ 85.43 2
$C_{19}H_{18}P^+$, $C_{12}H_4N_4^-$, $C_{12}H_4N_4$ 64.24 4
$C_{19}H_{19}BrO$ 21.16 4
$C_{19}H_{19}BrO_5$ 59.41 1
$C_{19}H_{19}NO_4$, H_2O 58.17 4
$C_{19}H_{19}O_2Rh$ 73.21 4
$C_{19}H_{20}BrN_5O_2S$ 39.37 4
$C_{19}H_{20}BrN_5O_2S$ 39.38 4
$C_{19}H_{20}Co_6O_{11}S_5$ 85.82 2
$C_{19}H_{20}CuN_2O_2$ 78.10 3
$C_{19}H_{20}F_2NO_2^+$, Br^- 33.48 3
$C_{19}H_{21}CuNO_4$ 77.15 3
$C_{19}H_{21}NO_2$ 36.24 3
$C_{19}H_{21}NO_5$ 28.7 4
$C_{19}H_{22}ClN_3O_2$ 36.26 4
$C_{19}H_{22}N^+$, Br^- 36.27 4
$C_{19}H_{22}NO_4^+$, Br^- 58.21 3
$C_{19}H_{22}N_3PPdS_2$ 86.23 3
$C_{19}H_{22}O$ 38.57 1
$C_{19}H_{22}O_2$, C_6H_5BrO 60.14 3
$C_{19}H_{23}BMoN_4O_2$ 75.21 4
$C_{19}H_{23}BrO_3$ 54.2 1
$C_{19}H_{23}BrO_4$ 54.1 3
$C_{19}H_{23}BrO_{10}S$ 45.74 1
$C_{19}H_{23}N_2O_3^+$, Br^- 48.54 4
$C_{19}H_{23}N_7O_{12}P$ 47.20 4
$C_{19}H_{24}$ 28.10 3
$C_{19}H_{24}Br_2O_6$ 53.20 3
$C_{19}H_{24}ClNO_6$ 58.36 1
$C_{19}H_{24}Cl_2O_7$ 53.5 4
$C_{19}H_{24}FeNiO_3$ 75.30 3
$C_{19}H_{24}N^+$, Br^- 33.80 1
$C_{19}H_{24}N^+$, Br^- 32.34 3
$C_{19}H_{24}NNi$ 73.82 2
$C_{19}H_{24}NO_4^+$, Br^- 58.22 3
$C_{19}H_{24}NO_4^+$, I^- 58.23 3
$C_{19}H_{24}NO_5^+$, I^- 58.24 3
$C_{19}H_{24}NO_5^+$, I^-, CH_4O 58.25 3
$C_{19}H_{24}N_2NiO_3$ 78.32 2
$C_{19}H_{24}N_7O_{12}P$, $4H_2O$ 47.31 1
$C_{19}H_{24}O_2$ 51.8 3
$C_{19}H_{24}O_3$ 51.11 4
$C_{19}H_{25}BrO_2$ 27.11 3
$C_{19}H_{25}BrO_2$ 51.9 3
$C_{19}H_{25}Cl_2IrO_3S_2$ 71.28 3
$C_{19}H_{25}N_2^+$, Br^- 58.26 3
$C_{19}H_{25}N_2O_2^+$, I^- 58.27 3
$C_{19}H_{25}N_2S^+$, Cl^- 41.31 4
$C_{19}H_{25}O_2Rh$ 74.16 3
$C_{19}H_{26}NO_5^+$, I^- 58.28 3
$C_{19}H_{26}NO_7^+$, $C_{10}H_{14}BrO_4S^-$, H_2O 58.38 1
$C_{19}H_{26}O_2$ 20.16 3
$C_{19}H_{26}O_2$ 51.12 4
$C_{19}H_{26}O_2$, C_6H_5BrO 60.39 4
$C_{19}H_{26}O_3$ 51.10 3
$C_{19}H_{27}BrO_2$ 54.3 1
$C_{19}H_{27}BrO_3$ 51.13 4
$C_{19}H_{28}Co_2N_2O$ 73.33 3
$C_{19}H_{28}O_2$, C_6H_5BrO 60.151 2
$C_{19}H_{28}O_2$, C_6H_5BrO 60.15 3
$C_{19}H_{29}Al_2NO_2$ 68.19 3
$C_{19}H_{29}BrO_3$ 51.11 3
$C_{19}H_{29}ClO$, $0.5CH_4O$ 51.12 3
$C_{19}H_{29}NO$ 32.35 3
$C_{19}H_{30}Br_2$ 51.14 4
$C_{19}H_{30}NO^+$, Cl^- 32.25 4
$C_{19}H_{30}O_2$ 51.6 1
$C_{19}H_{30}O_2$ 51.15 4
$C_{19}H_{30}O_2$, H_2O 51.16 4
$C_{19}H_{31}Cl_2CrO_3$ 84.46 2
$C_{19}H_{31}KO_9S$ 67.19 4
$C_{19}H_{31}O_2Rh$ 72.35 3
$C_{19}H_{32}IMoO_2P$ 73.22 4
$C_{19}H_{32}NO_2^+$, CH_4O, Br^- 58.39 1
$C_{19}H_{32}O_2$ 51.17 4
$C_{19}H_{32}O_2$ 51.18 4
$C_{19}H_{36}O_2$ 1.149 1
$C_{19}H_{38}BaN_3O_7S^+$, CNS^- 67.20 4
$C_{19}H_{38}O_2$ 1.150 1
$C_{19}H_{38}O_2$ 1.37 3

C_{20}

$C_{20}Co_6O_{18}$ 71.42 4
$C_{20}H_4AmF_{24}O_8^-$, Cs^+ 77.16 3
$C_{20}H_4EuF_{24}O_8^-$, Cs^+ 77.17 3
$C_{20}H_4F_{24}O_8Y^-$, Cs^+ 77.39 2
$C_{20}H_8Br_4O_4$ 30.4 1
$C_{20}H_8CoN_6S_4^-$, $C_{12}H_{28}N^+$ 85.35 3
$C_{20}H_8F_{18}O_6Zr$ 73.34 3
$C_{20}H_{10}$ 5.36 1
$C_{20}H_{10}$ 30.2 3
$C_{20}H_{10}Cl_4$ 20.17 3
$C_{20}H_{10}Co_2O_6$ 72.56 2
$C_{20}H_{10}CrO_5$ 71.30 3
$C_{20}H_{10}CuF_{12}N_2O_4$ 77.18 3
$C_{20}H_{10}Fe_2O_6$ 72.57 2
$C_{20}H_{10}Fe_2O_6S$ 71.31 3
$C_{20}H_{10}Fe_2O_6S_2$ 85.83 2
$C_{20}H_{10}Fe_2O_8$ 71.32 3
$8C_{20}H_{10}N_4$, C_6N_4, $2CH_2Cl_2$ 60.52 3
$C_{20}H_{10}O_4$ 25.22 1
$C_{20}H_{11}Mn_2O_8P$ 86.25 2
$C_{20}H_{12}$ 30.5+ 1
$C_{20}H_{12}$, $C_6F_4O_2$ 60.152 2
$C_{20}H_{12}$, C_6N_4 60.53 3
$C_{20}H_{12}$, $C_8F_{12}NiS_4$ 60.54 3
$C_{20}H_{12}$, $2C_9H_{12}N_4O_3$ 60.111 2
$C_{20}H_{12}$, $C_{10}H_2O_6$ 60.154 2
$C_{20}H_{12}Br_2N_2O$ 40.30 1
$C_{20}H_{12}Cl_2Fe_2N_2O_8$ 83.103 3
$C_{20}H_{12}Fe_4O_{16}Sn_3$ 69.35 2
$C_{20}H_{12}IN_3$ 28.25 1
$C_{20}H_{12}N_2$ 36.29 1
$C_{20}H_{12}N_2O_2$ 36.30 1
$C_{20}H_{12}O_8Re_2Si$ 63.14 3
$C_{20}H_{13}Br$ 31.37 1
$C_{20}H_{13}Fe_2NO_6$ 83.170 2
$C_{20}H_{13}NSe$ 42.14 1
$C_{20}H_{14}$ 24.5 3
$C_{20}H_{14}$ 31.31 3
$C_{20}H_{14}$ 31.27 4
$C_{20}H_{14}BrN$ 16.16 3
$C_{20}H_{14}Br_2O_2$ 30.1 4
$C_{20}H_{14}Cr_2O_6$ 75.22 4
$C_{20}H_{14}N_4$ 49.1 1
$C_{20}H_{14}N_4$ 49.1 4
$C_{20}H_{15}CoNO_3P$ 86.11 4

$C_{20}H_{15}F_{12}O_2P$	64.25	4
$C_{20}H_{15}OP$	64.57	2
$C_{20}H_{15}PS$	64.58	2
$C_{20}H_{16}$	29.23	1
$C_{20}H_{16}$	29.24	1
$C_{20}H_{16}$	28.11	3
$C_{20}H_{16}$	29.5	4
$C_{20}H_{16}$, $C_{13}H_4BrN_3O_7$	60.55	3
$C_{20}H_{16}ClGaN_2O_2$	68.10	4
$C_{20}H_{16}Cl_2Cu_2N_{16}O_4^{2+}$, $2Cl^-$, $6H_2O$	83.104	3
$C_{20}H_{16}Cl_2Fe_2$	73.85	2
$C_{20}H_{16}Cl_2GaN_4^+$, Cl_4Ga^-	68.11	4
$C_{20}H_{16}CoN_5O_3^{2+}$, HO^-, NO_3^-, $4H_2O$	83.102	4
$C_{20}H_{16}CuIN_4^+$, I^-	83.171	2
$C_{20}H_{16}CuN_4^{2+}$, $2ClO_4^-$	83.103	4
$C_{20}H_{16}CuN_5O_2^+$, NO_3^-	83.105	3
$C_{20}H_{16}CuO_4$	77.40	2
$C_{20}H_{16}Fe$	75.31	3
$C_{20}H_{16}Fe_2$	73.35	3
$C_{20}H_{16}LaN_4^{3+}$, $3NO_3^-$	83.172	2
$C_{20}H_{16}N_2$	33.41	4
$C_{20}H_{16}N_2O_3V$	83.104	4
$C_{20}H_{16}N_4O_4$	35.26	4
$C_{20}H_{16}N_4Pd^{2+}$, $2NO_3^-$, H_2O	83.105+	4
$C_{20}H_{16}N_6Pd$	83.173	2
$C_{20}H_{16}N_7O_9Tb$	83.174	2
$C_{20}H_{16}O_2$	24.6	3
$C_{20}H_{16}O_4Ru_3$	75.66	2
$C_{20}H_{16}O_5$	1.38	3
$C_{20}H_{16}O_6$	59.15	4
$C_{20}H_{16}O_7$	38.24	4
$C_{20}H_{17}AuBr_3P$	71.43	4
$C_{20}H_{17}ClN_2O_3$	34.3	4
$C_{20}H_{17}Cl_3O_5$	50.25	1
$C_{20}H_{17}NO_5$	58.29	3
$C_{20}H_{17}N_4$	12.17	1
$2C_{20}H_{17}N_4^+$, $C_4CoN_4S_4^{2-}$	32.36	3
$C_{20}H_{18}$	19.61	1
$C_{20}H_{18}Br_2$	21.18	3
$C_{20}H_{18}Br_2N_2$	9.18	3
$C_{20}H_{18}Br_2N_4$	36.28	4
$C_{20}H_{18}Br_2N_4$	36.29	4
$C_{20}H_{18}Br_2N_4$	36.30	4
$C_{20}H_{18}ClN_3O_4$	9.19	3
$C_{20}H_{18}Cl_2$	5.37	1
$C_{20}H_{18}Cl_2$	5.38	1
$C_{20}H_{18}CuN_6^{2+}$, $2ClO_4^-$	83.107	3
$C_{20}H_{18}Cu_2N_4O_2^{2+}$, $2NO_3^-$	83.108	3
$C_{20}H_{18}Fe_2$	73.86+	2
$C_{20}H_{18}GeO$	69.36	2
$C_{20}H_{18}NO_5^+$, Br^-	58.18	4
$C_{20}H_{18}N_2O_2Sn$	69.37	2
$C_{20}H_{18}OSi$	63.15	3
$C_{20}H_{18}O_4Pd$	77.41	2
$C_{20}H_{18}O_5V$	77.42+	2
$C_{20}H_{18}O_6$, CCl_4	59.16	4
$C_{20}H_{18}O_7$	59.19	3
$C_{20}H_{19}BrN_2O$	37.24	1
$C_{20}H_{19}BrO_7$	59.42	1
$C_{20}H_{19}CuN_5^{2+}$, $2BF_4^-$	83.107	4
$C_{20}H_{19}I_2NiPS_2$	85.36	3
$C_{20}H_{19}NO_5$	58.41	1
$C_{20}H_{19}N_2O_2^+$, AsF_6^-	35.22	3
$C_{20}H_{19}N_2S^+$, I^-	41.46	1
$C_{20}H_{20}$	29.25	1
$C_{20}H_{20}BrNO_3$	58.30	3
$C_{20}H_{20}BrNO_4$	13.21	4
$C_{20}H_{20}Br_2N_2$	37.15	4
$C_{20}H_{20}Br_2N_4Ni$	83.175	2
$C_{20}H_{20}Br_4Cu_2N_4O_4$	84.11	4
$C_{20}H_{20}ClN$	28.26	1
$C_{20}H_{20}Cl_2CoN_4$	83.176	2
$C_{20}H_{20}Cl_2Cu_2N_{20}^{2+}$, $2Cl^-$, $6H_2O$	83.108	4
$C_{20}H_{20}Cl_2N_4Ni$	83.177	2
$C_{20}H_{20}Cl_2N_4Rh^+$, $HN_2O_6^-$	83.178	2
$C_{20}H_{20}Cl_4O_4Ti_4$	73.36	3
$C_{20}H_{20}Cl_6N_4W_2$, xC_3H_6O	83.109	4
$C_{20}H_{20}Co_4S_6$, $0.5CHCl_3$	73.37	3
$C_{20}H_{20}CuN_2O_2$	83.179	2
$C_{20}H_{20}CuN_4O_4^{2+}$, $2BF_4^-$	84.47	2
$C_{20}H_{20}CuN_4O_4^{2+}$, $2ClO_4^-$	84.48	2
$C_{20}H_{20}CuN_5O_3^+$, NO_3^-, H_2O	83.109	3
$C_{20}H_{20}Cu_2N_4O_7S$, $4H_2O$	83.110	3
$C_{20}H_{20}Cu_2N_8O_{12}$, C_5H_5N	83.111	3
$C_{20}H_{20}Cu_2N_8O_{16}$	84.49	2
$C_{20}H_{20}Fe_4S_4$	73.89+	2
$C_{20}H_{20}NO_2Re^+$, Cl^-, $2H_2O$	83.110	4
$C_{20}H_{20}N_2$	9.20	3
$C_{20}H_{20}N_5NiO_3^+$, NO_3^-	83.111	4
$C_{20}H_{20}Nb_2$	71.44	4
$C_{20}H_{20}Ni_5S_4$	73.91	2
$C_{20}H_{20}O_4$	59.20	3
$2C_{20}H_{20}P^+$, $CeN_5O_{15}^{2-}$	64.48	3
$C_{20}H_{20}Pt_2$	75.23	4
$C_{20}H_{20}Ti$	73.23	4
$C_{20}H_{20}Zr$	73.38	3
$C_{20}H_{21}BrO_6$	59.43	1
$C_{20}H_{21}O_2Sb$	66.19	2
$C_{20}H_{21}Rh_3$	73.92	2
$C_{20}H_{22}Al_2O_6W_2$	73.24	4
$C_{20}H_{22}BrNO_3$	13.22	4
$C_{20}H_{22}BrNO_4S_2$	39.39	4
$C_{20}H_{22}Br_2O_5$	59.44	1
$C_{20}H_{22}Cl_2Fe_2N_2O_4$	78.11	3
$C_{20}H_{22}Cl_4Cu_2N_2O_8$	81.49	3
$C_{20}H_{22}Fe_2N_2O_4$	73.39	3
$C_{20}H_{22}NO_4^+$, I^-	58.42	1
$C_{20}H_{22}NO_5^+$, Br^-	58.31	3
$C_{20}H_{22}N_2$	34.4	3
$C_{20}H_{22}N_2O_2$	36.25	3
$C_{20}H_{22}N_2O_3$	58.43	1
$C_{20}H_{22}N_4O$	32.26	4
$C_{20}H_{22}NiS_5$	85.37	3
$C_{20}H_{22}NiS_6$	85.38	3
$C_{20}H_{22}O_4Pt$	78.33	2
$C_{20}H_{23}BrO_4$	21.33	1
$C_{20}H_{23}BrO_4$	54.1	4
$C_{20}H_{23}BrO_4$	54.2	4
$C_{20}H_{23}BrO_6$	54.4	1
$C_{20}H_{23}CoN_2O_3$	78.12	3
$C_{20}H_{23}IN_2O_4$	59.45	1
$C_{20}H_{23}N_2O^+$, Br^-, C_2H_6O	58.32	3
$C_{20}H_{23}N_3NiO_2$	78.9	4
$C_{20}H_{24}BrNO$	21.17	4
$C_{20}H_{24}BrNO_2$	31.32	3
$C_{20}H_{24}Br_2N_2Ni$	76.83	3
$C_{20}H_{24}ClFeN_2O_2$	78.13	3
$C_{20}H_{24}Cl_2Cu_2N_2O_4$	78.10	4
$C_{20}H_{24}Cl_5NO_3$	37.25	1
$C_{20}H_{24}CoN_2O_4$	77.44	2
$C_{20}H_{24}CuN_2O_2$	78.34	2

$C_{20}H_{24}CuN_2O_2$ 78.35 **2**
$C_{20}H_{24}NO_4^+$, I^- 58.44 **1**
$C_{20}H_{24}NO_5^+$, Br^- 58.45 **1**
$C_{20}H_{24}NO_5^+$, Br^-, H_2O 36.26 **3**
$C_{20}H_{24}NS^+$, Cl^-, H_2O 39.40 **4**
$C_{20}H_{24}N_2NiO_2$ 78.36 **2**
$C_{20}H_{24}N_2NiO_4$ 77.45 **2**
$C_{20}H_{24}N_2NiO_6$ 77.46 **2**
$C_{20}H_{24}N_2O_2$ 58.19 **4**
$C_{20}H_{24}N_3O_6^+$, Br^- 48.94 **1**
$C_{20}H_{24}N_4NiO_2$, $6H_2O$ 76.42 **4**
$C_{20}H_{24}NaO_6^+$, Br^-, $2H_2O$ 38.21 **3**
$C_{20}H_{24}O_3$ 24.13 **4**
$C_{20}H_{24}O_3S$ 51.19 **4**
$C_{20}H_{24}O_4$ 54.2 **3**
$C_{20}H_{24}O_4S_8$ 39.41 **4**
$2C_{20}H_{24}O_6$, $0.55Rb^+$, $0.45Na^+$, CNS 38.22 **3**
$C_{20}H_{24}S_5$ 39.42 **4**
$C_{20}H_{25}BrO_7$ 54.3 **4**
$C_{20}H_{25}Co_5O_{10}S_5$ 85.84 **2**
$C_{20}H_{25}IO_3$ 31.38 **1**
$C_{20}H_{25}NO_3S$, CH_3NO_2 51.15 **3**
$C_{20}H_{26}BrNaO_7$, $C_{20}H_{28}NaO_8^+$, Br^-, $2H_2O$ 67.21 **4**
$C_{20}H_{26}Br_2O$ 59.46 **1**
$C_{20}H_{26}Br_2O_3$ 54.4 **4**
$C_{20}H_{26}CuN_4O_2$, $2H_2O$ 83.181 **2**
$C_{20}H_{26}HgN_4O_2$ 83.112 **4**
$C_{20}H_{26}N^+$, Cl^- 33.42 **4**
$C_{20}H_{26}NO_3^+$, Cl^- 58.20 **4**
$C_{20}H_{26}NO_4^+$, $C_6H_2BrN_2O_5^-$ 58.34 **3**
$C_{20}H_{26}N_3O_4S^+$, Br^-, H_2O 50.26 **1**
$C_{20}H_{26}N_4O_4S_6Te$ 70.4 **3**
$C_{20}H_{26}O_2$ 54.5 **1**
$C_{20}H_{27}BrO_3$ 51.7 **1**
$C_{20}H_{27}BrO_4$ 27.15 **4**
$C_{20}H_{27}IO_3$ 51.20 **4**
$C_{20}H_{27}N_2^+$, I^- 58.46 **1**
$C_{20}H_{27}N_2O^+$, Br^- 16.61 **1**
$C_{20}H_{27}N_2O^+$, Br^- 58.47 **1**
$C_{20}H_{27}N_2O_2^+$, I^- 58.48 **1**
$C_{20}H_{27}N_2O_2^+$, I^- 58.49 **1**
$C_{20}H_{27}N_2O_2P$ 64.26 **4**
$C_{20}H_{27}N_2O_3^+$, Br^- 58.35 **3**
$C_{20}H_{28}$ 31.33 **3**
$C_{20}H_{28}CeO_8$ 77.47 **2**
$C_{20}H_{28}CeO_8$ 77.19 **3**
$C_{20}H_{28}Cl_2O_9Ti_2$, $CHCl_3$ 77.48 **2**
$C_{20}H_{28}CuN_2O_2$, $0.67C_6H_6$ 61.21 **2**
$C_{20}H_{28}CuN_2O_2$, $C_{10}H_{15}NO$, $2H_2O$ 83.182 **2**
$C_{20}H_{28}NO_3^+$, Br^- 58.50 **1**
$C_{20}H_{28}N_2NiO_4P_2S_4$ 83.113 **4**
$C_{20}H_{28}O$ 54.5 **4**
$C_{20}H_{28}O$ 54.6+ **4**
$C_{20}H_{28}O_2$ 59.47 **1**
$C_{20}H_{28}O_3$ 51.21 **4**
$(C_{20}H_{28}O_4P_2Zn)_n$ 84.50+ **2**
$C_{20}H_{28}O_6$ 54.8 **4**
$C_{20}H_{28}O_8Th$ 77.49 **2**
$C_{20}H_{28}O_8U$ 77.20 **3**
$C_{20}H_{28}O_8Zr$ 77.50 **2**
$C_{20}H_{28}O_{10}Ti_2$, $2C_4H_8O_2$ 77.21 **3**
$C_{20}H_{29}BrO_3$ 51.22 **4**
$C_{20}H_{29}NO_7Yb$, H_2O 77.22 **3**
$C_{20}H_{29}N_2O^+$, I^- 58.36 **3**
$C_{20}H_{30}$ 31.39 **1**
$C_{20}H_{30}As_4Cl_2Ni^+$, Cl^- 86.26 **2**
$C_{20}H_{30}Br_2O_3$ 54.7 **1**
$C_{20}H_{30}Cl_3OP_2Re$ 86.27 **2**
$C_{20}H_{30}Co_2N_{10}^{4+}$, $4ClO_4^-$ 71.44 **2**
$C_{20}H_{30}Fe_2N_4O_{15}^{2-}$, $C_2H_{10}N_2^{2+}$, $6H_2O$ 83.183 **2**
$C_{20}H_{30}Mn_2O_8P_2$ 86.28 **2**
$C_{20}H_{30}NO_7Yb$ 77.23 **3**
$C_{20}H_{30}N_8P_2Pd$ 83.114 **4**
$C_{20}H_{30}O$ 59.21 **3**
$C_{20}H_{30}O$ 54.9 **4**
$C_{20}H_{30}O_2$ 28.8 **4**
$C_{20}H_{30}O_5$ 50.5 **4**
$C_{20}H_{32}$ 54.8 **1**
$C_{20}H_{32}$ 54.4 **3**
$C_{20}H_{32}As_2Cl_3N_2Rh$ 86.25 **3**
$C_{20}H_{32}As_2Cu_2I_2N_2$ 86.12 **4**
$C_{20}H_{32}As_4AuI_2^+$, I^- 86.26 **3**
$C_{20}H_{32}As_4Cl_2Co^+$, ClO_4^- 86.27 **3**
$C_{20}H_{32}As_4Cl_2Pt$ 86.31 **2**
$C_{20}H_{32}As_4Co^{2+}$, $2ClO_4^-$ 86.32 **2**
$C_{20}H_{32}As_4CoNO^{2+}$, $2ClO_4^-$ 86.13 **4**
$C_{20}H_{32}As_4I_2Ni$ 86.33 **2**
$C_{20}H_{32}As_4I_2Pd$ 86.34 **2**
$C_{20}H_{32}As_4I_2Pt$ 86.35 **2**
$C_{20}H_{32}Cl_4Ru_2$ 72.16 **4**
$C_{20}H_{32}Co_2O_{10}$ 77.51 **2**
$C_{20}H_{32}N_8Ni^{2+}$, $2F_6P^-$ 83.112 **3**
$C_{20}H_{33}BrO$ 51.23 **4**
$C_{20}H_{35}BrO_2$ 54.9 **1**
$C_{20}H_{35}N_3^{2+}$, $2ClO_4^-$ 58.51 **1**
$C_{20}H_{36}$, Ag^+, NO_3^- 23.4 **4**
$C_{20}H_{36}Ag^+$, NO_3^- 75.67 **2**
$C_{20}H_{36}AgO_{12}P_4^+$, ClO_4^- 86.36 **2**
$C_{20}H_{36}B_4N_8S_4$ 62.50 **2**
$C_{20}H_{38}Cl_4$ 5.39 **1**
$C_{20}H_{38}FeSi_4$ 73.94 **2**
$C_{20}H_{40}Cd_2N_4S_8$ 80.35+ **2**
$C_{20}H_{40}Cu_4N_4S_8$ 80.37 **2**
$C_{20}H_{40}EuN_4S_8^-$, $C_8H_{20}N^+$ 80.12 **3**
$C_{20}H_{40}N_4NpS_8^-$, $C_8H_{20}N^+$ 80.13 **3**
$C_{20}H_{40}N_4O_3Re_2S_8$ 80.13+ **4**
$C_{20}H_{40}N_4Pb_2S_8$ 69.38 **2**
$C_{20}H_{40}N_4S_8Sn$ 69.31 **4**
$C_{20}H_{40}N_4S_8Te$ 70.16 **4**
$C_{20}H_{40}N_4S_8Th$ 80.15 **3**
$C_{20}H_{40}N_4S_8Ti$ 80.15 **4**
$C_{20}H_{40}N_4S_8Zn_2$ 80.20 **2**
$C_{20}H_{40}O_2$ 1.151+ **1**
$C_{20}H_{42}Cl_2N_2P_2Pt_2S_2$ 86.28+ **3**
$C_{20}H_{42}N_2P_2PdS_2$ 86.38 **2**
$C_{20}H_{44}BaN_2O_9^{2+}$, $2CNS^-$ 67.22 **4**
$C_{20}H_{44}N^+$, F^-, $38H_2O$ 61.22 **2**
$C_{20}H_{47}Cl_2P_2Rh$ 86.14 **4**
$C_{20}H_{48}N_{12}O_4P_4W$ 64.27 **4**

C_{21}

$C_{21}Co_6O_{19}$ 71.45 **2**
$C_{21}H_7F_{24}O_2Rh$ 71.34 **3**
$C_{21}H_9N_3NdO_{12}^{3-}$, $3Na^+$, $15H_2O$ 81.51 **4**
$C_{21}H_9N_3O_{12}Yb^{3-}$, ClO_4^-, $4Na^+$, $10H_2O$ 81.52 **4**

$C_{21}H_{9}N_{3}O_{12}Yb^{3-}, 3Na^{+}, 14H_{2}O$	81.50	**3**
$C_{21}H_{9}N_{3}O_{12}Yb^{3-}, 3Na^{+}, 13H_{2}O$	81.53	**4**
$C_{21}H_{10}CoMnO_{9}Sn$	69.39	**2**
$C_{21}H_{10}Mo_{2}O_{11}Re_{2}S_{2}$	85.24	**4**
$C_{21}H_{11}Fe_{2}O_{6}^{+}, BF_{4}^{-}$	75.24	**4**
$C_{21}H_{12}N_{2}O_{3}$	36.27	**3**
$C_{21}H_{12}N_{3}$	36.31	**1**
$2C_{21}H_{12}N_{4}^{+}, Cl_{4}Cu^{2-}$	36.32	**1**
$C_{21}H_{13}BrN_{2}O_{3}$	40.18	**3**
$C_{21}H_{13}Co_{3}O_{6}$	71.35	**3**
$C_{21}H_{13}Fe_{2}NO_{6}$	72.17	**4**
$C_{21}H_{13}N$	36.33	**1**
$C_{21}H_{14}BrNO_{3}$	19.62	**1**
$C_{21}H_{14}BrN_{3}OS$	41.32	**4**
$C_{21}H_{15}^{+}, ClO_{4}^{-}$	12.18	**1**
$C_{21}H_{15}BrOSi$	63.9	**4**
$2C_{21}H_{15}ClFe_{3}O_{6}Sb^{+}, Cl_{4}Fe^{2-}, CH_{2}Cl_{2}$	73.25	**4**
$C_{21}H_{15}ClO_{6}Sn, CHCl_{3}$	69.29	**3**
$C_{21}H_{15}Cl_{2}N$	35.27	**4**
$C_{21}H_{15}CrS_{6}$	85.85	**2**
$C_{21}H_{15}FeO_{6}$	84.52	**2**
$C_{21}H_{15}MnNO_{4}P$	86.39	**2**
$C_{21}H_{15}N_{3}$	33.81	**1**
$C_{21}H_{16}$	30.8	**1**
$C_{21}H_{16}CrO_{5}$	75.68	**2**
$C_{21}H_{16}O$	23.5	**4**
$C_{21}H_{16}O_{7}Sn, 2.1H_{2}O, 0.45CH_{4}O$	69.30	**3**
$C_{21}H_{16}O_{9}Ru_{3}$	71.45	**4**
$C_{21}H_{17}N$	16.17	**3**
$C_{21}H_{17}N_{3}S_{2}Zn$	85.25	**4**
$C_{21}H_{18}$	29.6	**4**
$C_{21}H_{18}CuN_{7}S^{+}, C_{22}H_{18}CuN_{8}S_{2} . ClO_{4}^{-}$	83.117	**4**
$C_{21}H_{18}FeN_{3}O_{6}, 3H_{2}O$	84.53	**2**
$C_{21}H_{18}Fe_{2}O$	73.95	**2**
$C_{21}H_{18}Mo_{2}O_{6}$	75.69	**2**
$C_{21}H_{18}N_{2}$	9.13	**4**
$C_{21}H_{18}N_{2}$	9.14	**4**
$C_{21}H_{19}AuBr_{3}P$	71.46	**4**
$C_{21}H_{19}BrO_{7}S$	53.21	**3**
$C_{21}H_{19}BrS$	39.90	**1**
$C_{21}H_{19}ClN_{2}O_{4}$	35.28	**4**
$C_{21}H_{19}ClN_{2}O_{4}$	35.29	**4**
$C_{21}H_{19}ClN_{2}O_{4}$	40.23	**4**
$C_{21}H_{19}CoN_{3}O_{2}$	78.14	**3**
$C_{21}H_{19}N_{4}NiO_{5}^{+}, NO_{3}^{-}$	83.184	**2**
$C_{21}H_{20}BrNO_{7}S$	45.48	**3**
$C_{21}H_{20}BrO_{8}$	31.40	**1**
$C_{21}H_{20}ClPPd$	72.36	**3**
$C_{21}H_{20}Cl_{3}PPdSn, 0.4C_{3}H_{6}O$	72.37	**3**
$C_{21}H_{20}N_{3}^{+}, Br^{-}, H_{2}O$	36.31	**4**
$C_{21}H_{20}N_{3}^{+}, Br^{-}, CH_{4}O$	36.34	**1**
$C_{21}H_{21}As$	65.34	**2**
$C_{21}H_{21}BrO_{4}$	24.14	**4**
$C_{21}H_{21}ClNiPS_{3}^{+}, ClO_{4}^{-}$	86.31	**3**
$C_{21}H_{21}CoN_{2}O_{4}, 0.7H_{2}O$	77.14	**4**
$C_{21}H_{21}N_{2}S_{2}^{+}, Br^{-}$	41.47	**1**
$C_{21}H_{21}N_{2}S_{2}^{+}, Br^{-}, C_{2}H_{6}O$	41.48	**1**
$C_{21}H_{21}N_{2}S_{2}^{+}, C_{12}H_{4}N_{4}^{-}, C_{12}H_{4}N_{4}$	60.40	**4**
$C_{21}H_{21}OP$	64.28	**4**
$C_{21}H_{21}O_{4}P$	46.8	**3**
$C_{21}H_{21}P$	64.29	**4**
$C_{21}H_{21}PSe$	64.30	**4**
$C_{21}H_{22}BrNO_{4}$	53.22	**3**
$C_{21}H_{22}NO_{4}^{+}, I^{-}, 2H_{2}O$	58.21	**4**
$C_{21}H_{22}NO_{5}^{+}, Br^{-}$	58.52	**1**
$C_{21}H_{22}NO_{6}^{+}, I^{-}$	58.37	**3**
$C_{21}H_{22}NO_{6}^{+}, I^{-}$	58.22	**4**
$C_{21}H_{22}O_{2}$	21.18	**4**
$C_{21}H_{22}O_{2}$	21.19	**4**
$C_{21}H_{23}AgAs_{3}CoO_{4}$	86.32	**3**
$C_{21}H_{23}BrO_{3}$	24.15	**4**
$C_{21}H_{23}BrO_{4}$	38.58	**1**
$C_{21}H_{23}F_{2}NO_{2}$	33.43	**4**
$C_{21}H_{23}NO_{5}$	58.53	**1**
$C_{21}H_{23}NO_{5}V$	77.15	**4**
$C_{21}H_{23}N_{2}O_{2}^{+}, Br^{-}, 2H_{2}O$	58.54	**1**
$C_{21}H_{24}Br_{2}O_{7}$	59.17	**4**
$C_{21}H_{24}F_{2}NO_{2}^{+}, Cl^{-}$	33.44	**4**
$C_{21}H_{24}FeO_{4}S$	73.40	**3**
$C_{21}H_{24}NO_{3}^{+}, Br^{-}$	37.26	**1**
$C_{21}H_{24}N_{2}O_{2}^{2+}, O_{4}S^{2-}, 5H_{2}O$	58.55	**1**
$C_{21}H_{24}OSi_{2}$	63.25	**2**
$C_{21}H_{24}O_{9}$	53.6	**4**
$C_{21}H_{25}BrO_{5}, CH_{3}NO_{2}$	59.18	**4**
$C_{21}H_{25}N_{2}O^{+}, I^{-}$	58.56	**1**
$C_{21}H_{25}N_{2}O_{3}^{+}, I^{-}$	58.23	**4**
$C_{21}H_{25}N_{3}NiO_{2}$	78.15	**3**
$C_{21}H_{25}N_{5}O$	35.30	**4**
$C_{21}H_{26}BrNO$	53.27	**1**
$C_{21}H_{26}ClNO_{7}$	58.57	**1**
$C_{21}H_{26}ClN_{3}O_{2}$	36.32	**4**
$C_{21}H_{26}N^{+}, Br^{-}, H_{2}O$	35.46	**1**
$C_{21}H_{26}NO_{5}^{+}, Br^{-}$	58.58	**1**
$C_{21}H_{27}BrN_{2}O_{2}$	53.7	**4**
$C_{21}H_{27}BrO_{2}$	51.8	**1**
$C_{21}H_{27}BrO_{2}$	27.16	**4**
$C_{21}H_{27}ClO_{5}$	51.16	**3**
$C_{21}H_{27}NO_{7}, H_{2}O$	58.24	**4**
$C_{21}H_{27}N_{2}O_{2}^{+}, Br^{-}$	58.59	**1**
$C_{21}H_{27}N_{2}O_{3}^{+}, Cl^{-}$	58.25	**4**
$C_{21}H_{28}BrO_{3}$	51.9	**1**
$C_{21}H_{28}Br_{2}O_{2}$	59.23	**3**
$C_{21}H_{28}IN_{3}O_{2}$	53.28	**1**
$C_{21}H_{28}N^{+}, Cl^{-}$	33.45	**4**
$C_{21}H_{28}NO^{+}, Br^{-}$	58.60	**1**
$C_{21}H_{28}NO_{2}^{+}, I^{-}$	58.38	**3**
$C_{21}H_{28}NO_{4}^{+}, I^{-}$	58.39	**3**
$C_{21}H_{28}O_{2}$	17.50	**1**
$C_{21}H_{28}O_{2}$	51.17	**3**
$C_{21}H_{28}O_{2}$	17.15	**4**
$C_{21}H_{28}O_{5}$	51.24	**4**
$C_{21}H_{28}O_{5}$	51.25	**4**
$C_{21}H_{29}BrN_{2}O_{3}S$	53.29	**1**
$C_{21}H_{29}BrO_{2}$	51.10	**1**
$C_{21}H_{29}BrO_{3}$	51.11	**1**
$C_{21}H_{29}BrO_{3}S$	53.23	**3**
$C_{21}H_{29}BrO_{4}S$	53.24	**3**
$C_{21}H_{29}Cl_{2}N_{2}PPt$	71.37	**3**
$C_{21}H_{29}FO_{5}$	51.26	**4**
$C_{21}H_{29}FO_{5}$	51.27	**4**
$C_{21}H_{29}FO_{5}$	51.28	**4**
$C_{21}H_{29}N_{2}O_{4}^{+}, I^{-}$	58.62	**1**
$C_{21}H_{30}NO_{4}^{+}, I^{-}$	58.40	**3**
$C_{21}H_{30}NO_{4}^{+}, I^{-}$	58.41	**3**
$C_{21}H_{30}O_{2}$	51.29	**4**
$C_{21}H_{30}O_{3}$	51.18	**3**
$C_{21}H_{30}O_{3}$	51.30	**4**
$C_{21}H_{31}BrO_{3}$	51.12	**1**
$C_{21}H_{31}BrO_{4}$	51.13	**1**
$C_{21}H_{31}BrO_{4}$	51.14	**1**

$C_{21}H_{31}ClO_5$	54.10	**4**
$C_{21}H_{31}NO_2^+$, I^-	54.5	**3**
$C_{21}H_{31}N_2^+$, I^-	58.63	**1**
$C_{21}H_{33}BrHgO_5$	59.48	**1**
$C_{21}H_{33}NO^+$, I^-	58.64	**1**
$C_{21}H_{33}N_5O_3S$	44.76	**1**
$C_{21}H_{34}NO^+$, I^-	58.65	**1**
$C_{21}H_{34}O_3S$	21.20	**4**
$C_{21}H_{34}O_4$	31.28	**4**
$C_{21}H_{35}N_3$	58.66	**1**
$C_{21}H_{35}N_3S_6Te$	70.17	**4**
$C_{21}H_{36}Mo_3O_{13}$	77.53	**2**
$C_{21}H_{36}N_3^+$, I^-, $0.5CH_4O$	37.27	**1**
$C_{21}H_{36}O_3$	38.59	**1**
$C_{21}H_{40}N_8O_{12}$, $1.5H_2O_4Se$, $4H_2O$	50.27	**1**
$C_{21}H_{42}O_2$	1.153	**1**

C_{22}

$C_{22}Co_6O_{18}$	71.38	**3**
$C_{22}H_{10}$	5.40	**1**
$C_{22}H_{10}Fe_2O_8$	71.46	**2**
$C_{22}H_{10}O_2$	30.3	**3**
$C_{22}H_{12}$	30.9+	**1**
$C_{22}H_{12}O_2$	30.11	**1**
$C_{22}H_{12}S_2$	39.91	**1**
$C_{22}H_{12}S_2$	39.48	**3**
$C_{22}H_{14}$	30.12	**1**
$C_{22}H_{14}$	30.13	**1**
$C_{22}H_{14}$	30.14+	**1**
$C_{22}H_{14}O_9Ru_4$	75.32+	**3**
$C_{22}H_{15}AuCoO_4P$	86.40	**2**
$C_{22}H_{15}ClFe_2MoO_7Sn$	73.96	**2**
$C_{22}H_{15}ClMnO_4P$	86.15	**4**
$C_{22}H_{15}ClN_2O$	32.27	**4**
$C_{22}H_{16}$	30.16+	**1**
$C_{22}H_{16}$	24.7	**3**
$C_{22}H_{16}$	30.4	**3**
$C_{22}H_{16}$	30.2	**4**
$C_{22}H_{16}$	30.3	**4**
$C_{22}H_{16}AsF_4FeO_4P$	86.16	**4**
$C_{22}H_{16}FeN_6S_2$	83.115+	**4**
$C_{22}H_{16}O_{10}Ru_4$	71.39	**3**
$C_{22}H_{17}Cl$	31.34	**3**
$C_{22}H_{17}Co_3O_6$	71.40	**3**
$C_{22}H_{17}IN_2O_5$	58.67	**1**
$2C_{22}H_{17}N_2O_3^+$, Cl_4Hg^{2-}, $2H_2O$	58.68	**1**
$C_{22}H_{17}P$	64.31	**4**
$C_{22}H_{18}CuN_8S_2$, $C_{21}H_{18}CuN_7S^+$, ClO_4^-	83.117	**4**
$C_{22}H_{18}Fe$	73.98	**2**
$C_{22}H_{18}Fe$	73.41	**3**
$C_{22}H_{18}N_8NiS_2$, H_2O	83.113	**3**
$C_{22}H_{19}Br$	19.14	**4**
$C_{22}H_{19}BrOS$	39.43	**4**
$C_{22}H_{19}BrO_6$	50.13	**3**
$C_{22}H_{19}BrO_7$	53.25	**3**
$C_{22}H_{19}CuN_3O_6$	83.114	**3**
$C_{22}H_{20}Cr_2N_2O_2S_2$	73.99	**2**
$C_{22}H_{20}FeN_6S_2$	83.185	**2**
$C_{22}H_{20}N_2O_5U$	78.11	**4**
$C_{22}H_{20}N_2O_8S_2$	59.50	**1**
$C_{22}H_{20}N_6NiS_2$	83.186	**2**
$C_{22}H_{20}O_2$	31.42	**1**
$C_{22}H_{20}O_2$	28.9	**4**
$C_{22}H_{20}O_4$	29.26+	**1**
$C_{22}H_{20}O_7$	53.8	**4**
$C_{22}H_{20}Ti$	71.47	**4**
$C_{22}H_{21}CuNO_4$	77.54	**2**
$C_{22}H_{21}NO_{11}$	58.26	**4**
$C_{22}H_{21}O_3PS_4U$	85.26	**4**
$C_{22}H_{22}BrN_3O_8S$, $0.5C_6H_6$	50.28	**1**
$C_{22}H_{22}ClPPd$	72.59	**2**
$C_{22}H_{22}CoN_3O_3$	77.16	**4**
$C_{22}H_{22}CoN_3O_3$, CH_4O	78.16	**3**
$C_{22}H_{22}CuO_4$	77.55	**2**
$C_{22}H_{22}Cu_2N_2O_4$	83.90	**2**
$C_{22}H_{22}S_5$	39.49	**3**
$C_{22}H_{23}As_3CuMnO_5$	86.41	**2**
$C_{22}H_{23}NO_2$	24.32	**1**
$C_{22}H_{23}N_2^+$, Br^-	35.31	**4**
$C_{22}H_{23}N_2^+$, Br^-	35.32	**4**
$C_{22}H_{23}N_2O_7^+$, Br^-	50.14	**3**
$C_{22}H_{24}BrN_3O$	36.33	**4**
$C_{22}H_{24}Br_2O_6$	38.60	**1**
$C_{22}H_{24}ClN_2O_8^+$, Cl^-	50.29	**1**
$C_{22}H_{24}Cl_2N_4O_2Zn$	84.12	**4**
$C_{22}H_{24}FeO$	73.26	**4**
$C_{22}H_{24}NO_4^+$, I^-	58.69	**1**
$C_{22}H_{24}O_5Zn$	77.24	**3**
$C_{22}H_{25}BrN_2O_4$	48.55	**4**
$C_{22}H_{25}BrO_2$	31.29	**4**
$C_{22}H_{25}BrO_4$	53.32	**1**
$C_{22}H_{25}BrO_5$	53.9	**4**
$C_{22}H_{25}BrO_8$, $0.5C_6H_6$	53.34	**1**
$C_{22}H_{25}ClN_2O_4$	48.56	**4**
$C_{22}H_{25}NO_8^+$, Cl^-	50.6	**4**
$C_{22}H_{25}N_2O_9^+$, Cl^-	50.30	**1**
$C_{22}H_{25}O_6^+$, Cl_4Fe^-	59.51	**1**
$C_{22}H_{26}ClN_2^+$, $C_{12}H_4N_4^-$, $C_{12}H_4N_4$	60.41	**4**
$C_{22}H_{26}Cl_2NiP_2$	86.17	**4**
$C_{22}H_{26}N^+$, Br^-	32.28	**4**
$C_{22}H_{26}N_2O_2Pd$	78.38+	**2**
$C_{22}H_{26}N_4O$	32.29	**4**
$C_{22}H_{27}BrO_2$	53.35	**1**
$C_{22}H_{27}BrO_3$	51.19	**3**
$C_{22}H_{27}BrO_4$	53.27	**3**
$C_{22}H_{27}N_2O_3^+$, I^-	58.71	**1**
$C_{22}H_{27}N_2O_3^+$, I^-, H_2O	58.72	**1**
$C_{22}H_{27}N_2O_4^+$, Br^-, H_2O	58.73	**1**
$C_{22}H_{27}N_2O_6^+$, Br^-, C_2H_6O	58.42	**3**
$C_{22}H_{28}$	31.43	**1**
$C_{22}H_{28}BrN_3O_6S$	50.15	**3**
$C_{22}H_{28}Br_2O_4$	59.24	**3**
$C_{22}H_{28}CoN_2O_2$, $C_6H_3N_3O_6$	60.9	**4**
$C_{22}H_{28}CuN_2O_2$	78.40	**2**
$C_{22}H_{28}CuN_2O_2$	78.41	**2**
$C_{22}H_{28}CuN_2O_2$, $C_6H_3N_3O_6$	60.10	**4**
$C_{22}H_{28}N_2NiO_2$	78.42	**2**
$C_{22}H_{28}N_2NiO_2$	78.43	**2**
$C_{22}H_{28}N_2NiO_2$, $C_6H_3N_3O_6$	60.11	**4**
$C_{22}H_{28}N_2O_2Pd$	78.44	**2**
$C_{22}H_{28}N_2O_2Pd$	78.45	**2**
$C_{22}H_{28}N_4^{2+}$, $2Br^-$	58.74	**1**
$C_{22}H_{28}N_4^{2+}$, $2Br^-$, $2H_2O$	58.70	**1**
$C_{22}H_{28}N_4Ni$	71.48	**4**
$C_{22}H_{28}O_7S$	27.17	**4**
$C_{22}H_{29}BrO_3$	51.20	**3**
$C_{22}H_{29}BrO_6$	53.36	**1**

$C_{22}H_{29}ClO_7$ 59.52 **1**
$C_{22}H_{29}IO_6$ 54.10 **1**
$C_{22}H_{29}N_2O_4^+$, I^- 58.75 **1**
$C_{22}H_{29}N_2O_4^+$, CH_4O, Br^- 58.76 **1**
$C_{22}H_{29}N_3S_2$ 41.20 **3**
$C_{22}H_{30}Cl_2O_2Pt_2$ 75.71 **2**
$C_{22}H_{30}CoN_8O_6$ 83.115 **3**
$C_{22}H_{30}CuN_2O_2$, $2H_2O$ 83.187 **2**
$C_{22}H_{30}CuN_4O_2$ 78.12 **4**
$C_{22}H_{30}NO_5^+$, Br^-, H_2O 58.43 **3**
$C_{22}H_{30}NO_5^+$, I^- 58.77 **1**
$C_{22}H_{30}NO_5^+$, $C_2HO_4^-$ 58.27 **4**
$C_{22}H_{30}N_2O_8$ 36.29 **3**
$C_{22}H_{30}N_2O_8$ 37.13 **3**
$C_{22}H_{30}N_8NiO_6$ 83.118+ **4**
$C_{22}H_{30}N_8O_6Zn$ 83.116 **3**
$C_{22}H_{30}O_2$ 29.7 **4**
$C_{22}H_{30}O_3$ 51.21 **3**
$C_{22}H_{30}O_5$ 51.31 **4**
$C_{22}H_{31}As_2Br_3OPt$ 71.49 **4**
$C_{22}H_{31}BrO_3$ 51.22 **3**
$C_{22}H_{31}FO_3$ 51.23 **3**
$C_{22}H_{31}FO_3$ 51.24 **3**
$C_{22}H_{31}IO_5$ 54.11 **1**
$C_{22}H_{31}NO_2$, $0.5C_6H_6$ 58.28 **4**
$C_{22}H_{31}NO_5$ 58.78 **1**
$C_{22}H_{31}N_2O^+$, I^- 58.79 **1**
$C_{22}H_{31}N_7O_5^{2+}$, $2Cl^-$, $5H_2O$ 50.16 **3**
$C_{22}H_{32}IO_3$ 51.32 **4**
$C_{22}H_{32}NO_3^+$, I^- 58.44 **3**
$C_{22}H_{32}N_2P^+$, ClO_4^- 64.59 **2**
$C_{22}H_{32}O_2$ 51.25 **3**
$C_{22}H_{32}O_8$ 23.10 **3**
$C_{22}H_{32}O_8$ 27.18 **4**
$C_{22}H_{33}BrO_6S$ 59.53 **1**
$C_{22}H_{33}IO_3$ 51.26 **3**
$C_{22}H_{33}NO$ 32.30 **4**
$C_{22}H_{33}NO_2$ 58.45 **3**
$C_{22}H_{34}NO_5^+$, Br^-, H_2O 58.80 **1**
$C_{22}H_{36}NO_2^+$, Br^-, H_2O 58.81 **1**
$C_{22}H_{36}O_2Si$ 51.33 **4**
$C_{22}H_{36}O_7$ 54.6 **3**
$C_{22}H_{38}$ 19.63 **1**
$C_{22}H_{38}BaN_2O_7S_2$ 67.23 **4**
$C_{22}H_{38}NiO_2S_2$ 77.17 **4**
$C_{22}H_{38}NiO_4$ 77.56 **2**
$C_{22}H_{38}O_4Zn$ 77.57 **2**
$C_{22}H_{40}OsP_4$ 71.41 **3**
$C_{22}H_{40}P_4Ru$ 71.42 **3**
$C_{22}H_{42}Mo_2O_6P_4$ 86.43 **2**
$C_{22}H_{48}N_{10}NiS_6$ 79.26 **3**

C_{23}

$C_{23}H_4O_6Re$ 71.47 **2**
$C_{23}H_{10}Fe_3O_9$ 72.61 **2**
$C_{23}H_{15}As_2Co_2O_5P$ 86.33 **3**
$C_{23}H_{15}CrO_5P$ 86.34 **3**
$C_{23}H_{15}CrO_8P$ 86.35 **3**
$C_{23}H_{15}GeMnO_5$ 69.48 **2**
$C_{23}H_{15}MnO_5Sn$ 69.40 **2**
$C_{23}H_{16}BrNO$ 32.46+ **1**
$C_{23}H_{16}CoF_4O_3P$ 86.44 **2**
$C_{23}H_{16}S_3$ 39.44 **4**
$C_{23}H_{17}ClN_2O_2Ti$ 73.42 **3**
$C_{23}H_{19}BrO_7$, C_3H_7NO 50.31 **1**
$C_{23}H_{19}IO_2$ 24.16 **4**
$C_{23}H_{20}Cl_3GeNiP$, $0.5C_6H_6$ 73.43 **3**
$C_{23}H_{20}CuP$ 73.44 **3**
$C_{23}H_{20}FeIO_2P$ 72.62+ **2**
$C_{23}H_{21}NO_3S$ 11.16 **4**
$C_{23}H_{22}N_2OS_2$ 41.49 **1**
$C_{23}H_{23}BrO_7$ 59.54 **1**
$C_{23}H_{23}IO_8$ 45.75 **1**
$C_{23}H_{23}N_2^+$, Cl^-, H_2O 35.33 **4**
$C_{23}H_{24}BrNO_6$ 53.37 **1**
$C_{23}H_{24}Br_2N_4O_3S$ 59.55 **1**
$C_{23}H_{24}N_6Zn^{2+}$, $2ClO_4^-$ 83.117 **3**
$C_{23}H_{24}O_2Sn$ 69.41 **2**
$C_{23}H_{25}AuNPS_2$ 80.16 **4**
$C_{23}H_{25}BrO_5$ 53.38 **1**
$C_{23}H_{27}BrO_3$ 28.12 **3**
$C_{23}H_{28}FeO_4$ 72.38 **3**
$C_{23}H_{29}N_3O_2S_2$ 39.92 **1**
$C_{23}H_{29}N_5Ni^{2+}$, $2ClO_4^-$, CH_3NO_2 76.84 **3**
$C_{23}H_{29}O_5P$ 64.60+ **2**
$C_{23}H_{30}NO_6^+$, Br^-, $2H_2O$ 58.46 **3**
$C_{23}H_{30}O_6$ 51.34 **4**
$C_{23}H_{31}BrN_4O_7$ 48.95 **1**
$C_{23}H_{31}ClO_5$, CH_4O 51.35 **4**
$C_{23}H_{31}N_2O_4^+$, I^- 58.82 **1**
$C_{23}H_{32}ClN_3O^{2+}$, $2Cl^-$, $2H_2O$ 36.30 **3**
$C_{23}H_{32}O_3$ 51.28 **3**
$C_{23}H_{32}O_5$, C_6H_5BrO 51.36 **4**
$C_{23}H_{32}O_6$, $0.5H_2O$ 51.37 **4**
$C_{23}H_{33}BrO_5$ 51.17 **1**
$C_{23}H_{33}BrO_5S$ 28.10 **4**
$C_{23}H_{33}BrO_6$ 54.7 **3**
$C_{23}H_{33}N_2O^+$, I^- 58.47 **3**
$C_{23}H_{33}N_2O_2^+$, I^- 58.84 **1**
$C_{23}H_{33}N_2O_3^+$, I^- 58.85 **1**
$C_{23}H_{34}O_4$ 51.18 **1**
$C_{23}H_{35}BrO_9$ 27.15 **1**
$C_{23}H_{35}IO_3$ 51.30 **3**
$C_{23}H_{35}NO_2$ 58.29 **4**
$C_{23}H_{35}N_5O_4S$ 48.51 **3**
$C_{23}H_{35}N_5O_4Se$ 48.52 **3**
$C_{23}H_{36}NO_2^+$, I^- 37.28 **1**
$C_{23}H_{36}NO_2^+$, I^- 58.48 **3**
$C_{23}H_{36}NO_4^+$, Br^-, $2H_2O$ 58.86 **1**
$C_{23}H_{37}BrN_2$ 51.38 **4**
$C_{23}H_{37}ClCoN_5O_4$, CH_2Cl_2 83.120 **4**
$C_{23}H_{38}NO_6^+$, Br^- 58.49 **3**

C_{24}

$C_{24}Cl_{16}Co_2S_8^{2-}$, $2C_{16}H_{36}N^+$ 85.86 **2**
$C_{24}Cu_8N_{12}S_{12}^{4-}$, $4C_9H_{14}N^+$ 3.69 **1**
$C_{24}F_{20}P_4$ 64.49 **3**
$C_{24}H_3AsCo_6O_{18}$ 72.39 **3**
$C_{24}H_8N_8^-$, $C_{24}H_{20}P^+$ 64.66 **2**
$C_{24}H_{12}$ 30.18+ **1**
$C_{24}H_{12}$ 29.3+ **3**
$C_{24}H_{12}$, $2C_9H_{12}N_4O_3$ 60.112 **2**
$C_{24}H_{12}AsN_3O_6$ 65.35 **2**
$C_{24}H_{12}O_{14}Ru_6$ 71.48 **2**

$C_{24}H_{14}Br_2N_2O_2$	28.13	3
$C_{24}H_{15}BrMnO_4P$	71.50	4
$C_{24}H_{15}N_4O_5P$, C_6H_6	40.19	3
$C_{24}H_{15}N_6O_3P$	40.20	3
$C_{24}H_{15}P$	64.62	2
$C_{24}H_{16}$	31.35	3
$C_{24}H_{16}As_4$	65.6	3
$C_{24}H_{16}BrN_3O_4$	36.31	3
$(C_{24}H_{16}Cl_2HgO_2S_2)_n$	85.39	3
$C_{24}H_{16}Cl_6PtS_2$	85.27	4
$C_{24}H_{16}CrN_4O_4^-$, $C_5H_6N^+$	83.188	2
$C_{24}H_{16}Fe_2$	75.34	3
$C_{24}H_{16}N_2O_2$	40.32	1
$C_{24}H_{17}BrO_2$	31.44	1
$C_{24}H_{17}NO_3Si$	63.26	2
$C_{24}H_{18}$	19.64	1
$C_{24}H_{18}FeO_2$	73.101	2
$C_{24}H_{18}MnO_5P$	86.36	3
$C_{24}H_{18}N_2P^+$, Cl^-	64.32	4
$C_{24}H_{18}O_4$	21.21	4
$C_{24}H_{18}O_6$, C_6H_5Br	24.33	1
$C_{24}H_{19}BrNP$	64.50	3
$C_{24}H_{19}P_3$	64.63	2
$C_{24}H_{20}$	31.45	1
$C_{24}H_{20}Ag_2Cl_2O_8$	74.8	4
$C_{24}H_{20}As^+$, F_5Si^-	65.7	3
$C_{24}H_{20}As^+$, $FeN_4O_{12}^-$	65.8	3
$C_{24}H_{20}As^+$, I^-	65.36	2
$C_{24}H_{20}As^+$, I_3^-	65.10	4
$C_{24}H_{20}As^+$, $HN_2O_6^-$	65.37	2
$C_{24}H_{20}As^+$, $H_2Br_4MoO_2^-$	65.38	2
$C_{24}H_{20}As^+$, $H_4Cl_4O_2Ru^-$, H_2O	65.39	2
$C_{24}H_{20}As^+$, $C_2H_3Br_4NORe^-$	83.1	2
$C_{24}H_{20}As^+$, $C_9FN_6^-$	65.40	2
$C_{24}H_{20}As^+$, $C_{12}H_2O_{12}Re_3^-$	65.41	2
$C_{24}H_{20}As^+$, $C_{16}H_{28}AuN_{16}^-$	71.34	4
$C_{24}H_{20}As^+$, $C_{18}H_{15}I_3NiP^-$	65.42	2
$2C_{24}H_{20}As^+$, $Cl_{11}Re_3^{2-}$, H_2O	65.43	2
$2C_{24}H_{20}As^+$, $CoN_4O_{12}^{2-}$	65.44	2
$2C_{24}H_{20}As^+$, FeN_{15}^{2-}	65.45	2
$2C_{24}H_{20}As^+$, $MnN_4O_{12}^{2-}$	65.9	3
$2C_{24}H_{20}As^+$, $N_{18}Pd_2^{2-}$	65.11	4
$2C_{24}H_{20}As^+$, $C_2NiS_6^{2-}$	65.46	2
$2C_{24}H_{20}As^+$, $C_4N_4NiS_4^{2-}$	65.47	2
$2C_{24}H_{20}As^+$, $C_5N_5NbOS_5^{2-}$	65.10	3
$2C_{24}H_{20}As^+$, $C_8CoF_{12}O_8^{2-}$	81.50	2
$2C_{24}H_{20}As^+$, $C_{12}FeN_6S_6^{2-}$	85.16	3
$2C_{24}H_{20}As^+$, $C_{12}MoN_6S_6^{2-}$	85.17	3
$C_{24}H_{20}As_2O$	65.48	2
$C_{24}H_{20}B^+$, $C_9H_{17}N_6Pt^-$	71.7	4
$C_{24}H_{20}B^-$, Rb^+	62.51	2
$C_{24}H_{20}B^-$, $C_6H_{18}ClN_4Zn^+$	76.25	3
$C_{24}H_{20}B^-$, $C_{26}H_{42}BrN_2NiP^+$	83.126	4
$C_{24}H_{20}B^-$, $C_{54}H_{42}ClCoP_4^+$	86.56	4
$C_{24}H_{20}B^-$, $C_{58}H_{48}NOP_4Ru^+$, C_3H_6O	86.63	4
$C_{24}H_{20}B^-$, $C_{72}H_{57}IrP_5^+$	86.64	4
$4C_{24}H_{20}B^-$, $xC_{72}H_{62}N_4P_4Pt_2^{2+}$, $yC_{72}H_{64}N_2P_4Pt_2^{2+}$	86.98	3
$C_{24}H_{20}B_2I_4P_2$	62.15	4
$C_{24}H_{20}BrNO_5$	59.19	4
$C_{24}H_{20}ClNO_2$	36.32	3
$C_{24}H_{20}Cl_2N_3P_3$	64.64	2
$C_{24}H_{20}Cl_4N_4P_4$	64.65	2
$C_{24}H_{20}Cl_4N_4P_4$	64.51	3
$C_{24}H_{20}Cl_4O_4Ti_2$	84.54	2
$(C_{24}H_{20}Cl_4Pb_2)_n$	69.42	2
$(C_{24}H_{20}CoO_2P_2S_2)_n$	85.40	3
$C_{24}H_{20}CuN_2O_2$	78.46	2
$C_{24}H_{20}Cu_2N_6$	83.180	2
$C_{24}H_{20}F_3NiP$	71.43	3
$C_{24}H_{20}Fe_2O_2S_2$	73.102	2
$C_{24}H_{20}Fe_2O_4^+$, F_6P^-	73.27	4
$C_{24}H_{20}Fe_2O_4Sn$	69.31	3
$C_{24}H_{20}Fe_4O_4$	73.28	4
$C_{24}H_{20}Ge$	69.32+	4
$C_{24}H_{20}NS_2^+$, I^-	41.21	3
$C_{24}H_{20}N_2S_2$	41.33	4
$C_{24}H_{20}N_6O_2Pd$	83.189	2
$C_{24}H_{20}N_8O_8Zr$	84.16	3
$C_{24}H_{20}NiP_2S_4$	85.88	2
$C_{24}H_{20}Ni_2$	72.64	2
$C_{24}H_{20}P^+$, $C_4H_8Cl_3O_2Pt^-$	72.4	3
$C_{24}H_{20}P^+$, $C_{13}H_5MoN_4S_4^-$	64.52	3
$C_{24}H_{20}P^+$, $C_{24}H_8N_8^-$	64.66	2
$C_{24}H_{20}Se_4$	39.45	4
$C_{24}H_{20}Si$	63.10+	4
$C_{24}H_{20}Sn$	69.32	3
$C_{24}H_{20}Sn$	69.34	4
$C_{24}H_{21}BrPRh$	72.18	4
$C_{24}H_{21}FeS_7$	85.89	2
$C_{24}H_{21}InS_6$	68.20	3
$C_{24}H_{21}NO_2$	27.12	3
$C_{24}H_{21}OSb$	66.20	2
$C_{24}H_{22}BrNO_5$	58.50	3
$C_{24}H_{22}Fe_2O_4$	73.104	2
$C_{24}H_{22}Fe_2O_4$	73.105	2
$C_{24}H_{22}N_2^{2+}$, $2I^-$	33.46	4
$2C_{24}H_{22}N_4NiO_4$, I_2	60.155	2
$C_{24}H_{22}Nb_2O_{10}$	72.40	3
$C_{24}H_{22}O_7$	59.25	3
$C_{24}H_{23}N_3O_4S$	33.47	4
$C_{24}H_{24}As_4Co_4F_8O_8$	86.18	4
$C_{24}H_{24}As_4F_8O_8Ru_3$	86.37	3
$C_{24}H_{24}FeN_3S_6$	80.17	4
$C_{24}H_{24}Fe_2$	73.106	2
$C_{24}H_{24}IN_3O_5$, C_2H_6O	58.87	1
$C_{24}H_{24}N_2O_4$	35.34	4
$C_{24}H_{24}N_3P_2^+$, Cl^-	64.67	2
$C_{24}H_{24}N_6Ni^{2+}$, $2ClO_4^-$	83.121	4
$C_{24}H_{24}N_6Zn^{2+}$, $2ClO_4^-$	83.190	2
$C_{24}H_{24}O_2$	30.20	1
$C_{24}H_{24}Ti_2$	75.72	2
$C_{24}H_{25}OP$	64.68	2
$C_{24}H_{25}O_3P$	64.33	4
$C_{24}H_{26}$	29.8	4
$C_{24}H_{26}ClNO_6$	38.25	4
$C_{24}H_{26}Cl_2$	20.11	4
$C_{24}H_{26}Fe$	73.45	3
$C_{24}H_{26}Fe_2$	73.107	2
$C_{24}H_{26}N_2O_8S_2$	59.56	1
$C_{24}H_{27}As$	65.49	2
$C_{24}H_{27}FeN_7^{2+}$, $2BF_4^-$	83.118	3
$C_{24}H_{28}Al_2^{2-}$, $2C_8H_{16}NaO_2^+$	68.21	3
$C_{24}H_{28}Cl_4Cu_2N_4O_4$	84.17	3
$C_{24}H_{28}Cl_6Cu_4N_4O$, xH_2O	83.119	3
$C_{24}H_{28}N_8Ni^{2+}$, $2ClO_4^-$	83.120	3
$C_{24}H_{28}N_8O_6$, $3H_2O$	37.16	4
$C_{24}H_{30}As_4ClPd^+$, ClO_4^-, C_6H_6	86.45	2
$C_{24}H_{30}Fe_2O_6$	72.65	2
$C_{24}H_{30}N_2O_2Pt_2$	83.191	2
$C_{24}H_{30}Na_2O_8$	67.24	3
$C_{24}H_{31}BrO_5$	53.28	3

$C_{24}H_{31}BrO_8$ 54.12 **1**
$C_{24}H_{31}N_2O_5^+, I^-$ 58.51 **3**
$C_{24}H_{31}N_2O_6^+, I^-$ 58.88+ **1**
$C_{24}H_{32}Ag^+, ClO_4^-$ 74.9 **4**
$C_{24}H_{32}CuN_2O_2$ 78.18 **3**
$C_{24}H_{32}NO_3^+, Br^-$ 58.52 **3**
$C_{24}H_{32}NO_6^+, I^-$ 58.30 **4**
$C_{24}H_{32}N_2NiO_2$ 78.47 **2**
$C_{24}H_{32}N_2O_2Pd$ 78.48 **2**
$C_{24}H_{32}Nb_2NiS_4^{2+}, 2BF_4^-, 2H_2O$ 73.29 **4**
$C_{24}H_{32}O_2Pd$ 71.51 **4**
$C_{24}H_{32}O_4S$ 51.39 **4**
$C_{24}H_{32}U$ 75.25 **4**
$C_{24}H_{33}B_3N_3O_3$ 62.18 **3**
$C_{24}H_{33}BrO_8$ 54.13 **1**
$C_{24}H_{33}Cl_2MoOP_3$ 86.19 **4**
$C_{24}H_{34}Cl_2Pt_2$ 75.35 **3**
$C_{24}H_{34}N_4NiO_2$ 78.49 **2**
$C_{24}H_{35}As_2Br_3OPt$ 86.46 **2**
$C_{24}H_{35}IO_4$ 54.14 **1**
$C_{24}H_{36}Ag_2^{2+}, 2NO_3^-, 2H_2O$ 53.29 **3**
$C_{24}H_{36}BrNO_2$ 52.6 **3**
$C_{24}H_{36}Cl_2CoN_4^+, ClO_4^-$ 76.85 **3**
$C_{24}H_{36}Cl_6Pt_3Sn_2$ 69.43 **2**
$C_{24}H_{36}CoN_7^{2+}, 2BF_4^-$ 83.121 **3**
$C_{24}H_{36}CuN_7^{2+}, 2BF_4^-$ 83.122 **3**
$C_{24}H_{36}FeN_7^{2+}, 2BF_4^-$ 83.123 **3**
$C_{24}H_{36}Fe_2O_4$ 72.41 **3**
$C_{24}H_{36}MnN_7^{2+}, 2BF_4^-$ 83.124 **3**
$C_{24}H_{36}NO_4^+, I^-$ 58.90 **1**
$C_{24}H_{36}NO_8^+, I^-, 3H_2O$ 58.91 **1**
$C_{24}H_{36}N_7Ni^{2+}, 2F_6P^-$ 83.125 **3**
$C_{24}H_{36}N_7Zn^{2+}, 2BF_4^-$ 83.126 **3**
$C_{24}H_{36}Ru$ 74.10 **4**
$C_{24}H_{39}ClN_2P_2Pd$ 71.44 **3**
$C_{24}H_{40}NO_6^+, I^-, H_2O$ 58.92 **1**
$C_{24}H_{40}O_4, C_2H_4O_2$ 60.42 **4**
$C_{24}H_{42}Cl_2Cu_2$ 75.36 **3**
$C_{24}H_{42}N_2O_6$ 48.53 **3**
$C_{24}H_{42}O_8Zn_7$ 71.45 **3**
$C_{24}H_{43}NO_5$ 59.57 **1**
$C_{24}H_{44}AgNO_3$ 75.37 **3**
$C_{24}H_{44}N_4O_4$ 34.5 **3**
$C_{24}H_{44}P_4$ 64.69 **2**
$C_{24}H_{48}O_4Pt_2$ 77.58 **2**
$C_{24}H_{48}O_4Pt_2$ 71.46 **3**
$C_{24}H_{49}I$ 5.4 **3**
$C_{24}H_{52}CuN_4O_8^{2+}, 2ClO_4^-$ 81.51 **3**
$C_{24}H_{56}Cd_2O_8P_4S_8$ 85.41 **3**
$C_{24}H_{56}Cr_2Li_4O_4$ 71.47 **3**
$C_{24}H_{56}Cu_4O_8P_4S_8$ 85.28 **4**
$C_{24}H_{56}O_8P_4S_8Zn_2$ 85.42 **3**
$C_{24}H_{60}Cl_4Pt_4$ 71.48 **3**
$C_{24}H_{60}Ni_6O_{12}S_{12}$ 85.25 **2**
$C_{24}H_{60}Ni_6O_{12}S_{12}$ 85.43 **3**
$C_{24}H_{62}Nb_2Si_6$ 71.52 **4**
$C_{24}H_{66}Mo_2Si_6$ 71.53 **4**
$2C_{24}H_{72}ClCoN_{18}P_6^+, Cl_6Co_2^{2-}, 2CHCl_3$ 83.122 **4**
$C_{24}H_{72}ClCuN_{18}P_6^+, Cl_2Cu^-$ 83.127 **3**
$C_{24}H_{72}CoN_{12}O_9P_6^{2+}, 2ClO_4^-$ 84.18 **3**
$C_{24}H_{72}CuN_{12}O_9P_6^{2+}, 2ClO_4^-$ 84.19 **3**
$C_{24}H_{72}MgN_{12}O_9P_6^{2+}, 2ClO_4^-$ 67.25 **3**
$C_{24}H_{72}N_{18}P_6$ 64.70 **2**

C_{25}

$C_{25}H_{11}CdMn_2N_3O_{10}$ 83.123 **4**
$C_{25}H_{15}BrO_7$ 59.20 **4**
$C_{25}H_{16}$ 30.4 **4**
$C_{25}H_{16}Fe_3O_9$ 71.54 **4**
$C_{25}H_{16}Fe_3O_9, 0.5C_6H_6$ 71.55 **4**
$C_{25}H_{17}MoO_4P$ 72.66 **2**
$C_{25}H_{18}Cl_2NO_2Rh$ 83.128 **3**
$C_{25}H_{18}Cl_3N_3O_4$ 24.17 **4**
$C_{25}H_{18}FeO_2$ 75.38 **3**
$C_{25}H_{18}FeO_3$ 71.49 **3**
$C_{25}H_{18}Fe_2O_8$ 71.50 **3**
$C_{25}H_{18}Fe_2O_8$ 71.51 **3**
$C_{25}H_{19}MoO_3P$ 75.39 **3**
$C_{25}H_{20}$ 19.65 **1**
$C_{25}H_{20}IMoO_2P$ 73.46 **3**
$C_{25}H_{20}MnO_4P$ 71.56 **4**
$C_{25}H_{20}OS$ 39.93 **1**
$C_{25}H_{20}S_4$ 11.17 **4**
$C_{25}H_{21}CrO_5P$ 86.47 **2**
$C_{25}H_{21}CrO_5P$ 71.52 **3**
$C_{25}H_{22}CoN_5O_2$ 71.57 **4**
$C_{25}H_{22}NO_2PPd$ 71.58 **4**
$C_{25}H_{22}NO_2PS$ 64.34 **4**
$C_{25}H_{22}O_{10}, CH_4O$ 59.26 **3**
$C_{25}H_{22}Sn$ 69.35 **4**
$C_{25}H_{23}IO_8$ 59.58 **1**
$C_{25}H_{23}OSb$ 66.21 **2**
$C_{25}H_{23}O_2P$ 64.54 **3**
$C_{25}H_{23}P$ 64.55 **3**
$C_{25}H_{24}BrNO_4$ 35.23 **3**
$C_{25}H_{24}Co_2$ 73.108 **2**
$C_{25}H_{26}BrN_3O_5$ 36.34 **4**
$C_{25}H_{26}BrN_5O_{13}, 7H_2O$ 59.21 **4**
$C_{25}H_{27}BrO_3$ 51.20 **1**
$C_{25}H_{29}BrNO_5$ 58.93 **1**
$C_{25}H_{29}BrO_4S$ 51.40 **4**
$C_{25}H_{29}BrO_4S$ 51.41 **4**
$C_{25}H_{29}BrO_8, 0.8CHCl_3$ 54.15 **1**
$C_{25}H_{30}NO_3^+, I^-$ 58.94 **1**
$C_{25}H_{30}N_5Ni^+, Cl^-, xH_2O, yCH_4O$ 49.2 **1**
$C_{25}H_{30}N_5Ni^+, Cl^-, xCH_4O$ 49.2 **4**
$C_{25}H_{31}BrO_2$ 51.31 **3**
$C_{25}H_{31}BrO_4S$ 51.42 **4**
$C_{25}H_{31}N_3O_8S$ 50.7 **4**
$C_{25}H_{32}N_3O_4^+, Br^-, H_2O$ 58.53 **3**
$C_{25}H_{32}N_3O_4^+, I^-, H_2O$ 58.54 **3**
$C_{25}H_{33}BrO_8$ 50.32 **1**
$C_{25}H_{33}O_2$ 12.10 **3**
$C_{25}H_{34}Br_2Cl_2O_8S_2$ 50.33 **1**
$C_{25}H_{34}F_3P_2Pt^+, F_6Sb^-$ 71.59 **4**
$C_{25}H_{37}IrP_2$ 75.26 **4**
$C_{25}H_{38}BrNO_3$ 58.31 **4**
$C_{25}H_{38}Br_2O_3$ 51.21 **1**
$C_{25}H_{38}Br_2O_3$ 51.22 **1**
$C_{25}H_{44}N_2^{2+}, 2Br^-$ 58.55 **3**
$C_{25}H_{44}N_2^{2+}, 2I^-$ 58.56 **3**
$C_{25}H_{46}Br_2O_5$ 1.39 **3**
$C_{25}H_{46}N_{13}O_{10}^{3+}, 2Br^-, Cl^-, 3H_2O$ 50.8 **4**
$C_{25}H_{48}O_5S_2$ 1.154 **1**
$C_{25}H_{50}N_5Nb$ 83.129 **3**
$C_{25}H_{60}S_5Zn_5$ 71.50 **2**
$C_{25}H_{62}Si_7$ 63.16 **3**

C_{26}

$C_{26}H_{14}N_8$	49.1	3
$C_{26}H_{14}N_8Ni$	49.3	1
$C_{26}H_{15}AsF_{12}FeS_4$	85.44	3
$C_{26}H_{15}Br$	30.5	4
$C_{26}H_{15}F_{12}FeOPS_4^-$, $C_{16}H_{36}N^+$	84.20	3
$C_{26}H_{16}$	28.27	1
$C_{26}H_{16}$	30.21	1
$C_{26}H_{16}$	24.18	4
$C_{26}H_{16}$, $C_{13}H_4BrN_3O_7$	60.156	2
$C_{26}H_{16}CuN_2O_4$	83.130	3
$C_{26}H_{16}HgN_6S_2$	83.124	4
$C_{26}H_{16}N_2O_4Pd$	84.57	2
$C_{26}H_{16}O_2$	38.61+	1
$C_{26}H_{18}CuN_2O_2$	78.50	2
$C_{26}H_{18}N_2O_2$	26.39	1
$C_{26}H_{18}N_2O_2$	26.40	1
$C_{26}H_{18}N_2S_2$	41.34	4
$C_{26}H_{20}AuO_3PW$	73.109	2
$C_{26}H_{20}ClOP$	64.72	2
$C_{26}H_{20}CoN_2S_4^-$, $C_{16}H_{36}N^+$	85.45	3
$C_{26}H_{20}CrO_4S_2W$	73.47	3
$C_{26}H_{20}CuN_2O_2$	78.51	2
$C_{26}H_{20}Fe_2O_8S_2Sn$	73.110	2
$C_{26}H_{20}IOP$	64.73	2
$C_{26}H_{20}MoO_4S_2W$	73.48	3
$C_{26}H_{20}N_2NiO_2$	78.52	2
$C_{26}H_{20}N_2OS$	41.23	3
$C_{26}H_{20}N_{10}NiS_2$	83.131	3
$C_{26}H_{20}Ni$	75.27	4
$C_{26}H_{20}O_4S_2W_2$	73.49	3
$C_{26}H_{20}P_2$	64.74	2
$C_{26}H_{21}N_4O_2Rh$	71.53	3
$C_{26}H_{22}CuN_2O_2$	83.192	2
$C_{26}H_{22}HgN_8S_2$, $2C_5H_5N$	85.90	2
$C_{26}H_{22}N_8NiS_2$	85.46	3
$C_{26}H_{22}N_8S_2Zn$	85.29+	4
$C_{26}H_{22}Se_2$	11.70	1
$C_{26}H_{23}CoN_2O_4$, $1.5H_2O$	78.13	4
$C_{26}H_{23}N_3P_2$	64.56	3
$C_{26}H_{23}O_2PS$	64.75	2
$(C_{26}H_{24}Cl_2CuO_2P_2)_n$	84.13	4
$C_{26}H_{25}Cl_2NP_2Pd$	86.48	2
$C_{26}H_{25}IO_8$	53.30	3
$C_{26}H_{26}Cl_2CoNOP$	86.20	4
$C_{26}H_{26}CuN_2O_7$	81.54	4
$C_{26}H_{26}Cu_2N_2O_8$	81.87	2
$C_{26}H_{26}O_2$	31.46	1
$C_{26}H_{27}N_2O_{11}$	50.17	3
$C_{26}H_{28}BrNO_7S$	58.32	4
$C_{26}H_{28}CoP$, $0.5C_6H_6$	72.67	2
$C_{26}H_{28}NO_9^+$, Br^-	58.57	3
$C_{26}H_{28}N_2O_{11}$	50.9	4
$C_{26}H_{28}N_6NiS_2$	83.125	4
$C_{26}H_{28}N_6NiS_2$, $C_{18}H_{14}$	61.8	3
$C_{26}H_{29}ClN_4O_4PRh$	83.132	3
$C_{26}H_{30}NO^+$, Br^-	17.10	3
$C_{26}H_{30}N_4Ni$	83.133	3
$C_{26}H_{30}O_4$	31.30	4
$C_{26}H_{30}O_8$	31.31	4
$C_{26}H_{31}BrO_3$	51.32	3
$C_{26}H_{31}BrO_3$	51.33	3
$C_{26}H_{31}BrO_3$	51.43+	4
$C_{26}H_{31}BrO_5$	51.24	1
$C_{26}H_{31}BrO_5S$	54.11	4
$C_{26}H_{31}N_2O_5^+$, Br^-	58.33	4
$C_{26}H_{32}K_2N_2O_8S_2$	67.24	4
$C_{26}H_{32}NO_5^+$, Br^-	58.34	4
$C_{26}H_{33}BrO_4S$	51.25	1
$C_{26}H_{33}BrO_7$	59.27	3
$C_{26}H_{33}N_2NiP_3$	86.39	3
$C_{26}H_{34}BrNO_2$	51.45	4
$C_{26}H_{34}MnN_4O_8$, $4H_2O$	82.56	2
$C_{26}H_{34}O_8$	54.8	3
$C_{26}H_{35}ClF_4N_2P_3Rh$	86.21	4
$C_{26}H_{36}Cl_2N_4NiO_2$	78.53	2
$C_{26}H_{36}O_8$	54.12	4
$C_{26}H_{37}BrN_2O_5S$	54.13	4
$C_{26}H_{37}BrO_3S$	54.14	4
$C_{26}H_{38}B_2Cl_2N_2$	62.19	3
$C_{26}H_{38}Br_2O_8$	54.9	3
$C_{26}H_{39}BrO_5$	55.1	1
$C_{26}H_{40}Al_2Fe_2O_4$	73.50	3
$C_{26}H_{40}Cl_4O_2Rh_4$	71.54	3
$C_{26}H_{40}O_4Pd_2$	71.60	4
$C_{26}H_{42}BrN_2NiP^+$, $C_{24}H_{20}B^-$	83.126	4
$C_{26}H_{42}Mn_2N_2O_8$	77.18	4
$C_{26}H_{142}FeN_2O_9$	71.61	4

C_{27}

$C_{27}H_{15}Fe_2O_9PPt$	86.22	4
$C_{27}H_{18}$	30.6	4
$C_{27}H_{18}N_2$	36.35+	1
$C_{27}H_{19}Fe_2O_6P$	72.68	2
$C_{27}H_{19}N_3O_5U$	83.193	2
$C_{27}H_{19}N_3O_5U$, $CHCl_3$	83.194	2
$C_{27}H_{20}NOP$, CH_4O	64.35	4
$C_{27}H_{21}ClU$	75.28	4
$C_{27}H_{21}Sm$	75.42	3
$C_{27}H_{22}As_2CrO_2$	74.17	3
$C_{27}H_{23}MoO_3P$	71.51	2
$C_{27}H_{26}BrNO_6$	58.58	3
$C_{27}H_{26}N_4P_2$	64.57	3
$C_{27}H_{27}BrO_7$	56.1	3
$C_{27}H_{27}BrO_{10}$, $2C_2H_6O$	54.17	1
$C_{27}H_{27}NO_5$	35.35	4
$C_{27}H_{27}O_2Rh$	72.19	4
$C_{27}H_{29}N_2P$	64.36	4
$C_{27}H_{29}N_3O_4S_2$	36.35	4
$C_{27}H_{29}N_8^+$, I^-, C_3H_8O	36.33	3
$C_{27}H_{31}BrO_3$	51.26	1
$C_{27}H_{31}BrO_3$	30.5	3
$C_{27}H_{31}BrO_4$	31.32	4
$C_{27}H_{31}BrO_7$	53.10	4
$C_{27}H_{32}NO_5^+$, CH_4O, Br^-	58.100	1
$C_{27}H_{33}BrO_3$	51.46	4
$C_{27}H_{33}BrO_6$	54.18	1
$C_{27}H_{33}IO_5$	54.10	3
$C_{27}H_{34}NO_6^+$, I^-	58.101	1
$C_{27}H_{34}N_4O_8^{2+}$, $C_4H_{10}O_2$, $2Br^-$, H_2O	50.34	1
$C_{27}H_{34}N_5Ni^+$, ClO_4^-, $0.8C_3H_6O_2$	49.3	4
$C_{27}H_{34}N_5Pd^+$, ClO_4^-, $0.6C_3H_6O$	49.4	4
$C_{27}H_{34}N_5Pt^+$, ClO_4^-	49.5	4
$C_{27}H_{35}LiN_2$	67.26	3
$C_{27}H_{36}N_5NiO^+$, ClO_4^-	49.6	4
$C_{27}H_{37}BrO_3$	51.47	4

$C_{27}H_{37}BrO_8$	54.11	3
$C_{27}H_{37}Cl_3NOP_2Re$	83.195	2
$C_{27}H_{37}IO$	30.6	3
$C_{27}H_{37}IO_4$	51.35	3
$C_{27}H_{38}NO_7^+$, Br^-	58.102	1
$C_{27}H_{38}NO_7^+$, Br^-	58.59	3
$C_{27}H_{39}NO$	13.23	4
$C_{27}H_{41}NO_3$	58.103	1
$C_{27}H_{44}Br_2$	51.27	1
$C_{27}H_{44}NO_2^+$, I^-	58.104	1
$C_{27}H_{44}NO_2^+$, I^-, CH_4O	58.105	1
$C_{27}H_{44}NO_8^+$, I^-, $2H_2O$	58.106	1
$C_{27}H_{44}O_6$	51.28	1
$C_{27}H_{45}Cl_2NO$	51.36	3
$C_{27}H_{45}Cl_2NO$, $0.5C_3H_6O$	36.36	4
$C_{27}H_{45}I$	51.29	1
$C_{27}H_{46}BrCl$	51.30	1
$C_{27}H_{46}BrCl$	51.31	1
$C_{27}H_{46}Br_2$	51.32	1
$C_{27}H_{46}Cl_2$	51.33	1
$C_{27}H_{46}Cl_2$	51.34	1
$C_{27}H_{46}NO^+$, I^-	58.107	1
$C_{27}H_{46}NO_2^+$, Br^-	58.108	1
$C_{27}H_{46}NO_2^+$, I^-	58.109	1
$C_{27}H_{48}NO_3^+$, I^-, CH_4O	51.48	4
$C_{27}H_{54}FeN_3S_6$	80.40	2
$C_{27}H_{54}MoN_4OS_6$	80.17	3
$C_{27}H_{54}N_3NiS_6^+$, Br^-	80.18	3
$C_{27}H_{54}N_3NiSe_6^+$, Br^-	80.18	4
$C_{27}H_{63}Cl_3O_3P_3Rh$	86.40	3

C_{28}

$C_{28}H_{12}Cl_2N_2O_2Ti$	83.196	2
$C_{28}H_{12}N_2O_2$	36.37	1
$C_{28}H_{14}N_2O_4$	36.38	1
$C_{28}H_{16}$	30.22	1
$C_{28}H_{16}Br_2O_4$	13.56	1
$C_{28}H_{16}Cl_2O_4$	13.57	1
$C_{28}H_{16}N_2O_4$	26.11	3
$C_{28}H_{16}O_2$	28.28	1
$C_{28}H_{18}$	31.36	3
$C_{28}H_{18}Co_3O_8P$	71.55	3
$C_{28}H_{18}FeO_4$	75.75	2
$C_{28}H_{18}N_2$	35.36	4
$C_{28}H_{18}O_2$	28.29	1
$C_{28}H_{18}O_4$	26.41	1
$C_{28}H_{19}BrN_2O_3$	9.49	1
$C_{28}H_{20}$	31.47	1
$C_{28}H_{20}Cl^+$, Cl_5Sn^-	12.19	1
$C_{28}H_{20}Cl_2O_8Re_2$, $2CHCl_3$	81.88	2
$C_{28}H_{20}N_2O_2$	40.33	1
$C_{28}H_{20}N_4Ni^{2+}$, $2BF_4^-$	83.134	3
$C_{28}H_{20}N_4Ni^{2+}$, $2I^-$, H_2O	83.135	3
$C_{28}H_{20}N_6Ni$	83.127	4
$C_{28}H_{20}NiS_4$	85.91	2
$C_{28}H_{20}Ni_2O_4P_2$	86.41	3
$C_{28}H_{20}O_7V^{2-}$, $C_8H_{20}N^+$, Na^+, $2C_3H_8O$	81.52	3
$C_{28}H_{20}S_8V$	85.31	4
$C_{28}H_{21}NO_2PRh$	83.128	4
$C_{28}H_{22}$	5.41	1
$C_{28}H_{22}CoN_2O_2$	78.19	3
$C_{28}H_{22}CrN_3O_4$	83.197	2
$C_{28}H_{22}Mo_2O_6$	75.76	2
$C_{28}H_{22}N_3^+$, ClO_4^-	36.37	4
$C_{28}H_{23}BrO_{13}$	59.28	3
$C_{28}H_{23}FeOPS$	71.56	3
$C_{28}H_{23}FeO_2P$	72.42	3
$C_{28}H_{24}$	20.34	1
$C_{28}H_{24}Cl_4N_4Ni_2$, $2CHCl_3$	83.136	3
$C_{28}H_{24}Ge_2$	69.45	2
$C_{28}H_{24}N_2O_4Zn$	78.20	3
$C_{28}H_{24}N_2P_2PdS_2$	86.42	3
$C_{28}H_{25}ClSiZr$	73.30	4
$C_{28}H_{26}O_4Sn_2$	69.33	3
$C_{28}H_{28}Br_2NNiP_2$	86.49	2
$C_{28}H_{28}Cl_2NiOP_2$	86.43	3
$C_{28}H_{28}Hf$	71.62	4
$C_{28}H_{28}I_2Sn_2$	69.34	3
$C_{28}H_{28}Sn$	69.36	4
$C_{28}H_{28}Ti$	71.63+	4
$C_{28}H_{28}Zr$	71.57	3
$C_{28}H_{29}Br_2NNiP_2$	86.44	3
$C_{28}H_{30}NP_2^+$, I^-	64.76	2
$C_{28}H_{31}BrO_5$	51.35	1
$C_{28}H_{31}ClP_2Pt$	86.50	2
$C_{28}H_{31}IO_6$	56.1	1
$C_{28}H_{32}Cl_4Cu_4$	75.77	2
$C_{28}H_{33}BrO_5S$	50.35	1
$C_{28}H_{33}ClIrOP_2^+$, BF_4^-, CH_2Cl_2	71.58	3
$C_{28}H_{33}IO_9$	56.2	1
$C_{28}H_{34}O_3$	17.11	3
$C_{28}H_{35}AsN_2O_3S_4U$	80.19	3
$C_{28}H_{35}AsN_2O_3Se_4U$	85.32	4
$C_{28}H_{35}N_2O_3PS_4U$	80.20	3
$C_{28}H_{36}NO_4^+$, Br^-	58.110	1
$C_{28}H_{36}N_4Ni^{2+}$, $2ClO_4^-$	83.198	2
$C_{28}H_{36}N_4Ni^{2+}$, $2ClO_4^-$	83.199	2
$C_{28}H_{36}N_8P_4$	64.37	4
$C_{28}H_{36}O_2$	30.7	4
$C_{28}H_{37}Cl_3NOP_2Re$	83.200	2
$C_{28}H_{38}BrN_5O_8$, $0.75C_4H_8O_2$, H_2O	48.57	4
$C_{28}H_{38}NO_{12}$, $1.5H_2O$	52.6	4
$C_{28}H_{38}O_7$, $2H_2O$	51.49	4
$C_{28}H_{39}BrO_5$	55.2	1
$C_{28}H_{39}BrO_9$	54.12	3
$C_{28}H_{39}BrO_{16}$	52.31	1
$C_{28}H_{40}Al_2Ti_2$	73.111	2
$C_{28}H_{40}KO_{10}^+$, I^-	67.25	4
$C_{28}H_{40}NiP_2$	86.51+	2
$C_{28}H_{40}O_{10}$	38.26	4
$C_{28}H_{41}BrO_4$	54.20	1
$C_{28}H_{42}AsN_5NiS_2$	76.86	3
$C_{28}H_{42}Br_2O_9$	54.13	3
$C_{28}H_{44}Br_2O$	51.37+	3
$C_{28}H_{44}Cu_4N_4O_8$	77.25	3
$C_{28}H_{46}NO_3^+$, Br^-	58.111	1
$C_{28}H_{46}NO_3^+$, Br^-	58.60	3
$C_{28}H_{48}N_8NiS_4^{2+}$, $2I^-$	79.12	4
$C_{28}H_{49}NiP$	71.65	4
$C_{28}H_{52}$, Ag^+, NO_3^-	23.6	4
$C_{28}H_{55}BrO$	51.50	4
$C_{28}H_{55}BrO$	51.51	4
$C_{28}H_{56}Mo_2N_4S_8$	71.66	4
$C_{28}H_{58}Cl_6Mg_4O_6$	67.26	4
$C_{28}H_{64}N_6Ni_2O_2^{2+}$, $2ClO_4^-$	76.87	3
$C_{28}H_{72}O_{16}Ti_4$	84.58	2
$C_{28}H_{72}O_{16}Ti_4$	84.22	3

C_{29}

$C_{29}H_{15}Fe_3O_{11}P$, $C_{29}H_{15}Fe_3O_{11}P$ 86.53 **2**
$C_{29}H_{18}Br_2O_2S_2$ 39.46 **4**
$C_{29}H_{18}Fe_2N_2O_6$ 83.137 **3**
$C_{29}H_{20}Br$ 20.12 **4**
$C_{29}H_{20}F_5NiP$ 73.112 **2**
$C_{29}H_{20}F_6FeN_2O_2P_2$ 86.45 **3**
$C_{29}H_{21}BrO_{11}$ 59.60 **1**
$C_{29}H_{22}$ 20.18 **3**
$C_{29}H_{22}MoO_4P_2$ 86.46 **3**
$C_{29}H_{24}Br_2MoO_3P_2$, C_3H_6O 86.23 **4**
$C_{29}H_{25}NiP$ 73.113 **2**
$C_{29}H_{25}O_3P$ 64.38 **4**
$C_{29}H_{28}Br_2O_6$ 53.39 **1**
$C_{29}H_{30}ClIrO_3P_2$ 86.24 **4**
$C_{29}H_{32}BrClO_6S$ 50.18 **3**
$C_{29}H_{32}BrNO_9$ 59.61 **1**
$C_{29}H_{33}BrClO_6S$ 50.10 **4**
$C_{29}H_{33}Cl_3NP_2Re$ 86.47 **3**
$C_{29}H_{34}BrFO_5$ 51.39 **3**
$C_{29}H_{34}CoN_7$ 49.2 **3**
$C_{29}H_{34}N_4$ 49.3 **3**
$C_{29}H_{37}AgNO_5^+$, BF_4^- , $2C_3H_8O$ 50.19 **3**
$C_{29}H_{39}BrN_2O_5S$ 51.36 **1**
$C_{29}H_{39}NO_9^+$, I^- 58.35 **4**
$C_{29}H_{43}IO_4$ 51.37 **1**
$C_{29}H_{44}Br_2O_4$ 51.38 **1**
$C_{29}H_{45}BrO_4$ 51.39 **1**
$C_{29}H_{45}IO_4$ 51.40 **1**
$C_{29}H_{47}NO_5$, CH_4O 58.61 **3**
$C_{29}H_{48}O_2S$ 51.41 **1**

C_{30-34}

$C_{30}Co_8O_{24}$ 71.59 **3**
$C_{30}Co_8O_{24}$, $0.5C_6H_8$ 71.60 **3**
$C_{30}H_{14}$ 30.23 **1**
$C_{30}H_{14}O_2$ 30.24 **1**
$C_{30}H_{14}O_2$ 30.25 **1**
$C_{30}H_{15}AuClF_{10}P$ 71.52 **2**
$C_{30}H_{16}$ 30.26 **1**
$C_{30}H_{16}$ 30.8 **4**
$C_{30}H_{16}Fe_4O_{10}$, $C_2H_4Cl_2$ 75.43 **3**
$C_{30}H_{18}Br_2O_7$ 59.62 **1**
$C_{30}H_{18}Cl_2$ 29.28 **1**
$C_{30}H_{18}Fe_2O_6$ 72.69 **2**
$C_{30}H_{18}O_2$ 30.27 **1**
$C_{30}H_{20}$ 31.37+ **3**
$C_{30}H_{20}Br_2$ 31.39 **3**
$C_{30}H_{20}F_{12}O_2S$ 11.18 **4**
$C_{30}H_{20}Ni_2O_6P_2$ 86.55 **2**
$C_{30}H_{20}O_2$ 31.48 **1**
$C_{30}H_{20}S_2$ 39.47 **4**
$C_{30}H_{22}Br_2O_{10}$, $2CH_4O$, H_2O 59.29 **3**
$C_{30}H_{22}Cl_2$ 20.13 **4**
$C_{30}H_{22}CuO_4$ 77.59 **2**
$C_{30}H_{22}CuO_4$ 77.26 **3**
$C_{30}H_{22}O_2PdS_2$ 85.92 **2**
$C_{30}H_{22}O_4Pd$ 77.60 **2**
$C_{30}H_{24}CrO_4P_2$ 86.25 **4**
$C_{30}H_{24}HgN_4^{2+}$, $2ClO_4^-$ 71.67 **4**
$C_{30}H_{24}N_6V$ 83.202 **2**
$C_{30}H_{25}BrNO_4P$ 64.78 **2**
$C_{30}H_{25}FeOP$ 71.61 **3**
$C_{30}H_{25}MoNO_4P_2$ 86.56 **2**
$C_{30}H_{25}Ni_2O_5S_4$ 85.47 **3**
$C_{30}H_{25}P$ 64.79 **2**
$C_{30}H_{25}P_5$ 64.80 **2**
$C_{30}H_{25}Sb$ 66.22+ **2**
$C_{30}H_{26}^{2+}$, $2ClO_4^-$ 27.16 **1**
$C_{30}H_{26}Br_2Sn$ 69.35 **3**
$C_{30}H_{26}Co_2O_4Sn$ 75.29 **4**
$C_{30}H_{26}Fe_3$ 73.115 **2**
$C_{30}H_{26}Ni_2O_5S_4$ 84.60 **2**
$C_{30}H_{27}IO_{14}$ 50.36 **1**
$C_{30}H_{28}CoN_6S_2$ 83.129 **4**
$C_{30}H_{28}CuN_2O_2$ 78.56 **2**
$C_{30}H_{28}CuN_2O_2$ 78.57 **2**
$C_{30}H_{28}CuN_2O_2$ 78.21 **3**
$C_{30}H_{28}FeN_4S_2^{2+}$, $2ClO_4^-$, CH_4O 83.138 **3**
$C_{30}H_{28}O_4^{2+}$, $2Cl_2I^-$ 12.20 **1**
$C_{30}H_{28}O_6$ 59.22 **4**
$C_{30}H_{29}O_7Y$ 77.61 **2**
$C_{30}H_{30}Ag^+$, BF_4^- 75.78 **2**
$C_{30}H_{30}Br_2N_4O_4$ 59.64 **1**
$C_{30}H_{30}FeN_6^{2+}$, $C_{13}F_5O_{13}^{2-}$ 83.203 **2**
$C_{30}H_{30}N_8Ni$ 83.204 **2**
$C_{30}H_{31}BrNiP_2$ 72.43 **3**
$C_{30}H_{32}Al_2N_2O_2$ 68.12 **4**
$C_{30}H_{32}CuN_2O_6$ 84.23 **3**
$C_{30}H_{32}F_{21}LuO_7$ 77.19 **4**
$C_{30}H_{32}I_2NiO_2P_2$ 86.48 **3**
$C_{30}H_{33}BrF_4O_3$ 51.40 **3**
$C_{30}H_{34}CuN_4$ 49.4 **3**
$C_{30}H_{35}CoN_5O_5^{2+}$, $2ClO_4^-$ 84.24 **3**
$C_{30}H_{36}O_2Sn$ 69.47 **2**
$C_{30}H_{36}O_4$ 59.65 **1**
$C_{30}H_{37}BrO_{12}$, $2CH_4O$ 59.30 **3**
$C_{30}H_{38}BO_6P_2Rh$ 86.57 **2**
$C_{30}H_{38}ClIO_8$ 59.66 **1**
$C_{30}H_{38}Cl_2N_2P_3Re$ 86.49 **3**
$C_{30}H_{39}BrO_4$ 54.22 **1**
$C_{30}H_{39}IO_8$ 56.3 **1**
$C_{30}H_{40}CoN_7O^+$, Br^- , $xC_4H_8O_2$, yCH_4O 49.5 **1**
$C_{30}H_{40}NO_{13}^+$, I^- , $0.5CH_4O$ 58.36 **4**
$C_{30}H_{41}BrO_4S$ 59.23 **4**
$C_{30}H_{41}CoN_7O^+$, ClO_4^- 49.6 **1**
$C_{30}H_{41}IO_4$ 59.67 **1**
$C_{30}H_{42}Ni_3O_{12}$ 77.62 **2**
$C_{30}H_{42}O_{12}Zn_3$ 77.63 **2**
$C_{30}H_{44}BrNO_3$ 51.52 **4**
$C_{30}H_{44}BrN_3O_3$ 50.37 **1**
$C_{30}H_{44}Br_2O_2$ 51.42 **1**
$C_{30}H_{44}Br_2O_3$ 56.2 **3**
$C_{30}H_{44}Co_3O_{13}$ 77.64 **2**
$C_{30}H_{44}N_2O_2S^{2+}$, $2Br^-$, $4H_2O$ 58.112 **1**
$C_{30}H_{45}Cl_2MoOP_3$ 86.26 **4**
$C_{30}H_{45}Cl_2NP_3Re$ 86.58 **2**
$C_{30}H_{45}Cl_3P_3Ru^-$, $C_{60}H_{90}Cl_3P_6Ru_2^+$ 86.117 **2**
$C_{30}H_{45}Cl_9P_3Re_3$ 86.59 **2**
$C_{30}H_{45}NiO_{15}P_5^{2+}$, $2ClO_4^-$ 86.27 **4**
$C_{30}H_{46}GeOP_2Pt$ 71.68 **4**
$C_{30}H_{47}BrO$ 56.4 **1**
$C_{30}H_{47}BrO_7$, C_6H_6 56.3 **3**
$C_{30}H_{48}O_4$ 56.4 **3**
$C_{30}H_{49}Br$ 51.41 **3**

$C_{30}H_{50}CrO_4$	53.40	1
$C_{30}H_{50}CrO_4$	53.11	4
$C_{30}H_{50}Fe_2N_{10}O_3^{4+}$, $4ClO_4^-$	83.205	2
$C_{30}H_{50}NO_{11}^+$, Br^-, xC_2H_6O	50.38	1
$C_{30}H_{50}NO_{11}^+$, Br^-, $0.6C_2H_6O$	50.20	3
$C_{30}H_{50}O$	56.6	1
$C_{30}H_{50}O$	56.1	4
$C_{30}H_{50}O_6$	59.24	4
$C_{30}H_{51}BrO_3$	56.7	1
$C_{30}H_{51}CoN_2PSi_4$	83.130	4
$C_{30}H_{52}$	56.5	3
$C_{30}H_{52}$	56.6	3
$C_{30}H_{54}Co_2O_6P_2$	86.60+	2
$C_{30}H_{54}IMoN_6^+$, I^-	71.69	4
$C_{30}H_{60}Ag_6N_6S_{12}$	80.41	2
$C_{31}H_{20}Cl_2Fe_2O_8P_2Pd_2$, C_7H_8	86.62	2
$C_{31}H_{20}FeO_3$	75.79	2
$C_{31}H_{20}Fe_2Ge_2O_7$	69.36	3
$C_{31}H_{21}BrO_{12}$	59.31	3
$C_{31}H_{21}F_{12}PRu$	71.70	4
$C_{31}H_{25}BrO_2Si$	63.12	4
$C_{31}H_{25}ClN_2Ni$, C_5H_5N	75.30	4
$C_{31}H_{25}ClN_3O_4P$	64.59	3
$C_{31}H_{25}FeO_2P$	71.62	3
$C_{31}H_{25}NO_2Sn$	69.37	4
$C_{31}H_{27}NO_9Th$	84.25	3
$C_{31}H_{30}Cl_2N_2PRh$, $0.61CHCl_3$	71.63	3
$C_{31}H_{31}BrO_{10}$	54.15	4
$C_{31}H_{32}BN$	62.16	4
$C_{31}H_{32}Br_2O_{10}$	59.32	3
$C_{31}H_{35}BrO_9$	54.16	4
$C_{31}H_{35}INNiOP_2^+$, I^-	86.52	3
$C_{31}H_{37}IO_{12}$	59.25	4
$C_{31}H_{38}BrNO_6$	58.113	1
$C_{31}H_{38}BrNO_6$	58.63	3
$C_{31}H_{38}BrN_3O_8S$	50.21	3
$C_{31}H_{38}O_{11}$	54.14	3
$C_{31}H_{43}IO_5$	56.9	1
$C_{31}H_{45}ClO_5$	51.53	4
$C_{31}H_{45}CuN_3O_2$	78.58	2
$C_{31}H_{48}O_3$	56.2	4
$C_{31}H_{51}BrO_2$	56.7	3
$C_{31}H_{52}O_4$	51.42	3
$C_{32}H_{14}$	30.28	1
$C_{32}H_{16}$	30.29	1
$C_{32}H_{16}Cl_2N_8Sn$	49.7	4
$C_{32}H_{16}CuN_8$	49.7	1
$C_{32}H_{16}F_{12}O_8PrS_4^-$, H_4N^+, H_2O	77.65	2
$C_{32}H_{16}MgN_8$, $2C_5H_5N$, H_2O	49.8	4
$C_{32}H_{16}N_8Ni$	49.8	1
$C_{32}H_{16}N_8Pt$	49.9+	1
$C_{32}H_{16}N_8Sn$	49.5	3
$C_{32}H_{18}CrN_6O_8^-$, $C_5H_6N^+$, $0.5C_5H_5N$	83.206	2
$C_{32}H_{18}MgN_8O$, $2C_5H_5N$	49.6	3
$C_{32}H_{18}N_8$	49.12+	1
$C_{32}H_{18}N_8$	49.7	3
$C_{32}H_{20}FeO_4$	72.20	4
$C_{32}H_{22}CuN_4O_2$	83.207	2
$C_{32}H_{24}$	31.49	1
$C_{32}H_{24}Cu_2F_{12}O_8$	74.11	4
$C_{32}H_{24}FeN_8^{2+}$, $2ClO_4^-$	83.139	3
$C_{32}H_{25}BrO_2Si$	63.13	4
$C_{32}H_{28}CoI_2N_4$	71.64	3
$C_{32}H_{28}Co_2N_4O_4$	78.59	2
$C_{32}H_{28}Co_2N_4O_4$	78.14	4
$C_{32}H_{28}Fe_2N_4O_5$, CH_2Cl_2	78.22+	3
$C_{32}H_{28}Fe_2N_4O_5$, $2C_5H_5N$	78.24	3
$C_{32}H_{28}Ni_2S_8$	85.93	2
$C_{32}H_{28}S_8V$	85.48	3
$C_{32}H_{29}ClMoOP_2$	73.51+	3
$C_{32}H_{32}N_{12}Ni_2$	83.131	4
$C_{32}H_{32}Si_2$	63.28	2
$C_{32}H_{34}N_4OV$, $C_2H_4Cl_2$	49.14	1
$C_{32}H_{35}N_2NiOP_2S^+$, F_6P^-	86.53	3
$C_{32}H_{36}N_4Ni$	49.15	1
$C_{32}H_{37}BrO_{10}$	54.24	1
$C_{32}H_{38}BrCoN_2P_2^+$, F_6P^-	76.43	4
$C_{32}H_{38}BrN_2NiP_2^+$, Br^-, $0.5C_4H_{10}O$	76.44	4
$C_{32}H_{38}N_4$	49.9	4
$C_{32}H_{39}BrO_{11}$	56.10	1
$C_{32}H_{40}Cl_2P_4Ru$	86.63	2
$C_{32}H_{40}O_8$	54.15	3
$C_{32}H_{40}P_4Pd^{2+}$, $2Cl^-$, C_2H_6O	86.54	3
$C_{32}H_{41}N_2O_{12}^+$, Br^-	35.37	4
$C_{32}H_{42}Ag_2O_6$, $2H_2O$	81.55	4
$C_{32}H_{43}BrO_2$	55.3	1
$C_{32}H_{44}ClN_2P_4Re$	86.55	3
$C_{32}H_{45}N_2NiO_6P_3$	86.56	3
$C_{32}H_{45}N_3Si_5$	63.29	2
$C_{32}H_{48}Ce_2Cl_2O_4$	75.31	4
$C_{32}H_{48}O_3$	56.3	4
$C_{32}H_{48}O_6$	56.8	3
$C_{32}H_{50}NO_5^+$, Br^-	58.114	1
$C_{32}H_{50}NO_8^+$, I^-, $2C_3H_6O$	58.37	4
$C_{32}H_{51}IO_2$	56.11	1
$C_{32}H_{52}Br_2O_3$	51.43	1
$C_{32}H_{52}Br_2O_7S$	1.14	4
$C_{32}H_{52}IO_2$	56.12	1
$C_{32}H_{52}NO_3^+$, I^-	58.115	1
$C_{32}H_{53}IO_2$	51.44	1
$C_{32}H_{53}IO_2$	56.13	1
$C_{32}H_{54}Co_2N_2O_8$	77.20	4
$C_{32}H_{54}O_3$	56.4	4
$C_{32}H_{56}O_8S_{16}Zn_4$	80.19	4
$C_{32}H_{80}Cd_{10}O_{16}S_{16}^{4+}$, $2O_4S^{2-}$, $4H_2O$	85.94	2
$C_{32}H_{80}O_{16}Ti_4$	84.61	2
$C_{32}H_{80}S_8Zn_8$	71.54	2
$C_{33}H_{24}FeO_9$	73.116	2
$C_{33}H_{30}Fe_3$	73.53	3
$C_{33}H_{31}INO_6$, C_3H_6O	54.25	1
$C_{33}H_{33}Br_3O_{12}$	54.17	4
$C_{33}H_{33}CrNO_3P_2$	86.57	3
$C_{33}H_{33}Fe_3O_9P_3$	86.28	4
$C_{33}H_{35}PPtS_4$	85.49	3
$C_{33}H_{36}BiCl_3N_{12}O_9S_3$	66.8	4
$C_{33}H_{37}IO_{11}S$, C_3H_6O	50.39	1
$C_{33}H_{38}I_2O_{12}$	59.26	4
$C_{33}H_{38}N_4O_2$	49.16	1
$C_{33}H_{39}BrO_{15}$	51.54	4
$C_{33}H_{40}BrN_3O_7$	50.22	3
$C_{33}H_{40}N_2O_9$	58.116	1
$C_{33}H_{47}IN_2O_6$	40.24	4
$C_{33}H_{51}BrO_4$	56.14	1
$C_{33}H_{51}IO_4$	56.15	1
$C_{33}H_{51}NO_6$	56.16	1
$C_{33}H_{52}NO_4^+$, I^-, C_3H_6O	58.64	3
$C_{33}H_{57}ErO_6$	77.21	4
$C_{33}H_{59}DyO_7$	77.22	4
$C_{33}H_{62}O_6$	1.155	1
$C_{34}H_{16}O_2$	30.30	1
$C_{34}H_{16}O_2$	30.31	1

$C_{34}H_{18}$ 30.32 1
$C_{34}H_{20}$ 30.9 4
$C_{34}H_{21}Br_2N_7O_3S_4$, $0.5C_7H_8O$ 50.40 1
$C_{34}H_{24}$ 24.19 4
$C_{34}H_{25}NbO$ 72.72 2
$C_{34}H_{25}OTa$ 72.73 2
$C_{34}H_{26}N_4NiO_2$ 83.208 2
$C_{34}H_{28}O_4$ 21.22 4
$C_{34}H_{28}O_4$ 28.11 4
$C_{34}H_{28}O_4$ 28.12 4
$C_{34}H_{30}Br_2O_8$ 54.26 1
$C_{34}H_{30}Co_2P_2$ 73.117 2
$C_{34}H_{30}CrN_4O_2^+$, I^-, H_2O 71.65 3
$C_{34}H_{30}N_2NiP_2S_4$ 85.50 3
$C_{34}H_{30}Ni_2P_2$ 73.118 2
$C_{34}H_{32}ClFeN_4O_4$ 49.17 1
$C_{34}H_{33}N_2O_2PPdS_2$ 85.33 4
$C_{34}H_{36}N_4NiO_2$, $0.5C_2H_4Cl_2$ 49.10 4
$C_{34}H_{38}Al_2N_2O_4$ 68.28 2
$C_{34}H_{38}INO_{12}$, C_3H_6O 56.17 1
$C_{34}H_{38}IO_9$ 56.18 1
$C_{34}H_{38}NO_{10}^+$, I^- 58.38 4
$C_{34}H_{46}Cu_2N_6O_{10}$, $8H_2O$, $C_4H_{10}O$ 82.15 4
$C_{34}H_{47}BrO_3$ 51.55 4
$C_{34}H_{51}IO_6$ 56.19 1
$C_{34}H_{53}O_8^-$, Ag^+, C_3H_6O 50.23 3
$2C_{34}H_{53}O_8^-$, Ba^{2+}, H_2O 50.24 3
$C_{34}H_{68}$ 23.17 1

C_{35-39}

$C_{35}H_{20}Fe_3N_4O_9$ 83.209 2
$C_{35}H_{21}N_5$ 49.11 4
$C_{35}H_{24}BrNO$ 40.25 4
$C_{35}H_{24}Br_2N_2$ 35.24 3
$C_{35}H_{24}I_2N_2$ 34.6 3
$C_{35}H_{25}O_2V$ 75.46 3
$C_{35}H_{26}BrO_3P$ 64.39 4
$C_{35}H_{34}Fe_3O$ 73.54 3
$C_{35}H_{39}IrP_2$ 75.32 4
$C_{35}H_{41}Br_2ClO_4$ 51.56 4
$C_{35}H_{46}BrClO_{13}$, C_2H_6O 54.18 4
$C_{35}H_{46}INO_4$ 51.45 1
$C_{35}H_{46}INO_4$ 51.46 1
$C_{35}H_{46}INO_4$ 51.43 3
$C_{35}H_{47}BrO_2$ 51.47 1
$C_{35}H_{60}N_2O_4^{2+}$, $2Br^-$, CH_2Cl_2, H_2O 51.44 3
$C_{36}H_{20}Br_4N_4O_4W$, C_6H_6 83.132 4
$C_{36}H_{20}Fe_3O_8$ 72.74+ 2
$C_{36}H_{20}O_8Os_3$ 72.21 4
$C_{36}H_{24}$ 31.40+ 3
$C_{36}H_{24}FeN_6^{2+}$, $C_8H_4O_{12}Sb_2^{2-}$, $8H_2O$ 83.210 2
$C_{36}H_{24}N_3O_6P_3$ 64.40 4
$C_{36}H_{24}N_6Ni^{2+}$, $2C_2H_6O_2PS_2^-$ 83.141 3
$C_{36}H_{24}N_6Ni^{2+}$, $2C_5MnO_5^-$ 83.133 4
$C_{36}H_{24}N_6Ni^{2+}$, $C_6CoO_{12}^{3-}$, K^+, $2H_2O$ 83.142 3
$C_{36}H_{26}Br_2CrN_8O_6^-$, $C_5H_6N^+$, $3H_2O$ 83.211 2
$C_{36}H_{27}N^+$, ClO_4^- 21.19 3
$C_{36}H_{28}Fe_2N_2O_6$ 83.212 2
$C_{36}H_{28}NiP_2Se_2$ 85.51 3
$C_{36}H_{30}Al_2$ 68.29+ 2
$C_{36}H_{30}As_2Cl_2HgO_2$ 84.62 2
$C_{36}H_{30}As_2Cl_4Hg_2O_2$ 84.63 2
$C_{36}H_{30}As_6$ 65.11 2
$C_{36}H_{30}Bi_2O^{2+}$, $2ClO_4^-$ 66.9 4
$C_{36}H_{30}Br_2NiP_2$ 86.65 2
$C_{36}H_{30}CdCl_2P_2$ 86.58 3
$C_{36}H_{30}CeN_4O_{14}P_2$ 84.26 3
$C_{36}H_{30}ClN_2O_2P_2Ru^+$, F_6P^-, C_6H_6 86.29 4
$C_{36}H_{30}Cl_2IrNOP_2$ 86.60 3
$C_{36}H_{30}Cl_2NP_2Re$ 86.66 2
$C_{36}H_{30}Cl_2NiP_2$ 86.67 2
$C_{36}H_{30}Cl_3HgNOOsP_2$ 86.61 3
$C_{36}H_{30}Cl_3InP_2$ 86.68 2
$C_{36}H_{30}Cl_4Hg_2P_2Se_2$ 85.52 3
$C_{36}H_{30}CoN_9$, C_7H_8 83.213 2
$C_{36}H_{30}CrO_4Si_2$ 63.17 3
$C_{36}H_{30}CuNO_3P_2$ 86.62 3
$C_{36}H_{30}CuP_2^+$, H_4B^- 86.69 2
$C_{36}H_{30}Fe_2N_4O_6$ 83.214 2
$C_{36}H_{30}IrN_2O_2P_2^+$, ClO_4^- 86.63 3
$C_{36}H_{30}Ir_2N_2O_3P_2$, C_6H_6 86.70 2
$C_{36}H_{30}NP_2^+$, $C_{10}Cr_2IO_{10}^-$ 64.62 3
$2C_{36}H_{30}NP_2^+$, $C_{10}Cr_2O_{10}^{2-}$, CH_2Cl_2 64.63 3
$2C_{36}H_{30}NP_2^+$, $C_{10}Mo_2O_{10}^{2-}$, CH_2Cl_2 64.64 3
$2C_{36}H_{30}NP_2^+$, $C_{16}Mo_2Ni_3O_{16}^{2-}$ 64.41 4
$2C_{36}H_{30}NP_2^+$, $C_{16}Ni_3O_{16}W_2^{2-}$ 64.42 4
$C_{36}H_{30}N_3O_6P_3$ 64.65 3
$C_{36}H_{30}N_3P_3$ 64.81 2
$C_{36}H_{30}N_4NiOP_2$ 86.30 4
$C_{36}H_{30}N_6OSb_2$ 66.10 4
$C_{36}H_{30}N_{10}NiS_2$ 83.143 3
$C_{36}H_{30}O_2P_2Pt$, $2CHCl_3$ 86.31 4
$C_{36}H_{30}O_2P_2Pt$, $1.5C_6H_6$ 86.71 2
$C_{36}H_{30}O_2P_2Pt$, $2C_7H_8$ 86.64 3
$C_{36}H_{30}P_3Pt$ 86.72 2
$C_{36}H_{30}P_6$ 64.82+ 2
$C_{36}H_{30}Pb_2$ 69.38 4
$2C_{36}H_{31}As_2O_2^+$, $Br_6Hg_2^{2-}$ 65.51 2
$C_{36}H_{31}BrSn$ 69.37 3
$C_{36}H_{31}ClSn$ 69.38 3
$C_{36}H_{31}N_2O_3OsP_2^+$, F_6P^- 86.65 3
$C_{36}H_{32}AsBrO_2^+$, Br_3^- 65.11 3
$C_{36}H_{32}La_2N_6O_{16}$ 81.56 4
$C_{36}H_{32}N_6O_{16}Pr_2$ 81.57 4
$C_{36}H_{32}N_6O_{16}Sm_2$ 81.58 4
$C_{36}H_{33}Br_2CoP_3$ 86.73 2
$C_{36}H_{33}Cl_3NOsP_2$ 86.66 3
$C_{36}H_{34}Br_2N_2O_{12}$, C_3H_6O 35.38 4
$C_{36}H_{35}Cl_2CuO_2P_2$ 84.14 4
$C_{36}H_{36}N_4NiO_6$, $0.5C_6H_6$ 49.18 1
$C_{36}H_{37}ClNOP_3Rh^+$, F_6P^- 86.32 4
$C_{36}H_{37}ClP_3Rh$ 86.33 4
$C_{36}H_{38}B_3CuP_2$ 86.67 3
$C_{36}H_{38}N_4O_5$ 49.12 4
$C_{36}H_{39}AlO_{12}W_3$ 73.31 4
$C_{36}H_{40}Cu_2N_4O_{16}$, $4H_2O$ 81.59 4
$C_{36}H_{40}N_4O_4Ti$ 78.61 2
$C_{36}H_{40}N_4O_4Zr$ 78.25 3
$C_{36}H_{41}Cl_3N_4O_8Pd$ 83.144 3
$C_{36}H_{42}BrNO_{17}$, H_2O 58.39 4
$C_{36}H_{42}N_{12}Ni_3O_{12}$, C_6H_6 83.215 2
$C_{36}H_{44}Cl_2N_4Sn$ 49.9 3
$C_{36}H_{46}N_2$ 21.20 3
$C_{36}H_{47}N_6^{3+}$, $3I^-$ 58.118 1
$C_{36}H_{51}CrO_6$ 77.27 3
$C_{36}H_{51}FO_2$ 51.57 4

$C_{36}H_{54}F_{10}P_2PtS_2$ 85.53 3
$C_{36}H_{61}AgO_{11}$, $2H_2O$ 50.25 3
$C_{36}H_{62}O_{11}$, H_2O 50.11 4
$C_{36}H_{65}Br_3O_6$ 1.156 1
$C_{36}H_{66}Cl_2NiP_2$ 86.74 2
$C_{36}H_{66}CuNO_3P_2$ 86.34 4
$C_{36}H_{72}AlLiN_4$ 68.13 4
$C_{36}H_{74}$ 5.42+ 1
$C_{36}H_{84}Pd_6S_{12}$ 85.95 2
$C_{37}H_{24}O_7Os_3P_2$ 71.71 4
$C_{37}H_{27}AsO_2$ 65.12 4
$C_{37}H_{30}ClIrNO_2P_2^+$, BF_4^- 86.75 2
$C_{37}H_{30}ClIrO_3P_2$ 86.76 2
$C_{37}H_{30}ClIrO_3P_2S$ 86.77 2
$C_{37}H_{30}ClO_3P_2RhS$ 86.68 3
$C_{37}H_{30}ClP_2RhS$ 86.78 2
$C_{37}H_{30}FP_2PtS_2^+$, HF_2^- 85.34 4
$C_{37}H_{30}IIrNO_2P_2^+$, BF_4^-, C_6H_6 86.69 3
$C_{37}H_{30}IIrO_3P_2$, CH_2Cl_2 86.79 2
$C_{37}H_{30}O_3P_2Pt$ 86.80 2
$C_{37}H_{30}P_2$ 64.43 4
$C_{37}H_{30}P_2PdS_2$ 86.81 2
$C_{37}H_{30}P_2PtS_2$ 86.70 3
$C_{37}H_{33}IIrNOP_2$ 71.66 3
$C_{37}H_{33}IO_2P_2PtS$ 71.72 4
$C_{37}H_{33}I_2P_2Rh$, C_6H_6 71.55 2
$C_{37}H_{34}F_6FeP_2Sn$ 73.32 4
$C_{37}H_{35}MoN_2O_3P_3$ 86.71 3
$C_{37}H_{37}BrF_2N_2O_5$ 51.45 3
$C_{37}H_{40}BrNO_7$, C_3H_8O 50.12 4
$C_{37}H_{40}BrNO_7$, $0.5C_3H_8O$ 50.42 1
$C_{37}H_{42}N_4O_7^{2+}$, $2Br^-$ 58.65 3
$C_{37}H_{43}BrO_8$, $0.5C_4H_8O_2$ 51.48 1
$C_{37}H_{43}FeN_4O_5$ 49.19 1
$C_{37}H_{44}O_5P_4Pt_4$ 86.72 3
$C_{37}H_{51}BrClN_3O_{10}$ 58.40 4
$C_{37}H_{51}IO_{10}S$ 59.33 3
$C_{37}H_{52}BrNO$, C_2H_6O 56.10 3
$C_{37}H_{53}BrO_4$, $0.5H_2O$ 56.20 1
$C_{37}H_{55}BrO_3$ 56.5 4
$C_{37}H_{50}O_8$, C_6H_6 51.49 1
$C_{37}H_{68}NO_{13}^+$, I^-, $2H_2O$ 50.43 1
$C_{38}H_{18}$ 30.33 1
$C_{38}H_{24}O_8Os_3P_2$ 71.73 4
$C_{38}H_{26}Cl_5F_5NiP_2$ 71.74 4
$C_{38}H_{26}F_{10}NiP_2$ 71.75 4
$C_{38}H_{28}^{2+}$, $2Cl_6Sb^-$ 12.11 3
$C_{38}H_{30}ClF_3P_2Pt$ 72.22 4
$C_{38}H_{30}ClF_4P_2Rh$ 86.84 2
$C_{38}H_{30}ClIrO_2P_2$, C_6H_6 86.73 3
$C_{38}H_{30}Cl_2F_2P_2Pt$ 72.23 4
$C_{38}H_{30}Cl_4P_2Pt$ 72.24 4
$C_{38}H_{30}MnNO_3P_2$ 86.85 2
$C_{38}H_{30}NO_3OsP_2^+$, ClO_4^-, CH_2Cl_2 86.35 4
$C_{38}H_{30}N_4O_5STh$, C_2H_6OS 83.145 3
$C_{38}H_{31}IrO_2P_2$ 86.74 3
$C_{38}H_{31}N_9Zn$, $0.5C_6H_{15}N$ 49.13 4
$C_{38}H_{32}Fe_2O_{10}S_2Sn_2$ 73.33 4
$C_{38}H_{33}Cl_3NP_2Re$ 86.75 3
$C_{38}H_{34}Br_2NiP_2$ 86.76 3
$C_{38}H_{34}Br_2O_2$ 20.14 4
$C_{38}H_{34}F_6FeOP_2Sn$ 73.55 3
$C_{38}H_{34}NiP_2$ 72.77 2
$C_{38}H_{34}NiP_2$ 72.25 4
$C_{38}H_{34}P_2Pt$ 72.26 4
$C_{38}H_{35}NO_2Si_3$ 63.30 2
$C_{38}H_{41}N_2O_7^+$, Br^- 58.41 4
$C_{38}H_{42}Co_2N_6O_8$ 78.26 3
$C_{38}H_{46}CuN_4O_8$ 83.146 3
$C_{38}H_{46}N_4O_8Pd$ 83.147 3
$C_{38}H_{52}CoP_2$ 86.86 2
$C_{38}H_{53}BrO_5$ 56.6 4
$C_{38}H_{56}Cl_2HgO_4$ 84.64 2
$C_{38}H_{95}O_{24}Ti_7$ 84.15 4
$C_{39}H_{27}Fe_2N_2O_5P$ 86.36 4
$C_{39}H_{30}CoGeO_3P$ 69.39 3
$C_{39}H_{30}F_4IIrOP_2$ 86.87 2
$C_{39}H_{30}F_6NiOP_2$ 71.76 4
$C_{39}H_{30}O_3OsP_2$ 86.88 2
$C_{39}H_{33}Br_2P_3Pt$ 86.37 4
$C_{39}H_{34}IP_2Rh$ 72.78 2
$C_{39}H_{34}N_8OSi_2$ 49.14 4
$C_{39}H_{34}P_2Pt$ 72.27 4
$C_{39}H_{35}Cl_5O_3P_2Re_2$ 81.89 2
$C_{39}H_{35}IP_2Pd_2$, C_6H_6 72.28 4
$C_{39}H_{36}O_3P_2Pt$ 86.89 2
$C_{39}H_{39}BrF_2N_2O_6$, $2C_4H_{10}O$ 51.46 3
$C_{39}H_{39}BrO_9S$ 59.68 1
$(C_{39}H_{42}N_6Na^+)_n$, nCl^- 67.27+ 4
$C_{39}H_{44}Br_2N_4O_{10}$ 59.27 4
$C_{39}H_{51}Br_2NO_{11}$, CH_4O 58.120 1
$C_{39}H_{51}Br_2NO_{11}$, C_2H_6O 59.28 4
$C_{39}H_{53}BrO_7$ 51.50 1
$C_{39}H_{55}IO_5$ 51.58 4
$C_{39}H_{64}LuNO_6$ 84.16 4
$C_{39}H_{74}O_6$ 1.157 1

C_{40-49}

$C_{40}H_{20}$ 30.34 1
$C_{40}H_{20}Co_4F_6O_{10}P_2$ 72.29 4
$C_{40}H_{28}Cu_2N_4O_4$, $2CHCl_3$ 78.27 3
$C_{40}H_{30}AuMnO_7P_2$ 86.90 2
$C_{40}H_{30}Au_2F_6P_2$ 71.77 4
$C_{40}H_{30}BrN_3O_3$ 52.32 1
$C_{40}H_{30}Cl_2N_2P_2Pt$ 72.30 4
$C_{40}H_{30}CrO_{10}P_2$ 86.38 4
$C_{40}H_{30}F_6OP_2$ 64.84 2
$C_{40}H_{30}MnO_4PSn$ 86.91 2
$C_{40}H_{30}N_2O_2S$ 35.39 4
$C_{40}H_{30}N_2P_2Pt$ 71.78 4
$C_{40}H_{30}Nb_2O_2$ 72.48 3
$C_{40}H_{31}F_6N_3P_2Pt$ 86.92 2
$C_{40}H_{32}Al_2N_4O_5$ 68.22 3
$C_{40}H_{32}Mn_2N_8O_2^{3+}$, $3ClO_4^-$, $3H_2O$ 83.134 4
$C_{40}H_{32}N_2P_2Pt$ 72.49 3
$C_{40}H_{32}N_8^{6+}$, $6Cl^-$, H_2O 49.21 1
$C_{40}H_{34}HgN_4$ 71.79 4
$C_{40}H_{34}O_3P_2Pd$ 71.80 4
$C_{40}H_{36}Cu_2I_2N_{12}^{2+}$, I^-, ClO_4^- 83.135 4
$C_{40}H_{36}EuO_8^-$, $C_5H_{12}N^+$ 77.28 3
$C_{40}H_{37}As_2Cl_2Rh$ 72.50 3
$C_{40}H_{37}EuO_8$, $C_4H_{11}N$ 77.29 3
$C_{40}H_{37}GdO_8$, $C_5H_{12}N$ 77.30 3
$C_{40}H_{40}LaN_8O_8^{3+}$, $3ClO_4^-$ 84.17 4
$C_{40}H_{41}ClPtSi$ 71.81 4
$C_{40}H_{44}N_4O^{2+}$, $2I^-$ 58.42 4
$C_{40}H_{44}N_4O_2^{2+}$, $2I^-$ 58.121 1
$C_{40}H_{44}Pt_2S_8$ 85.35 4

$C_{40}H_{46}Co_2N_6O_6$, C_7H_8 78.15 **4**
$C_{40}H_{48}Fe_2N_4O_5$ 78.16 **4**
$C_{40}H_{48}N_4O_2^{2+}$, $2I^-$, xH_2O 58.122 **1**
$C_{40}H_{50}N_2O_3$ 51.47 **3**
$C_{40}H_{50}O_2$ 57.1 **1**
$C_{40}H_{52}O_2$ 57.2 **1**
$C_{40}H_{54}$ 57.3 **1**
$C_{40}H_{56}$ 57.4 **1**
$C_{40}H_{56}Co_4O_{16}$ 77.67 **2**
$C_{40}H_{56}Cu_4N_4$ 71.82 **4**
$C_{40}H_{56}Ni_4O_{16}$ 78.28 **3**
$C_{40}H_{58}$ 57.5 **1**
$C_{40}H_{58}Cl_4N_4O_4Rh_2$ 71.83 **4**
$C_{40}H_{61}CoO_8P_4$ 86.77 **3**
$C_{40}H_{62}FeO_8P_4$ 86.39+ **4**
$C_{40}H_{64}BO_{14}^-$, Rb^+, $2CH_4O$ 50.13 **4**
$C_{40}H_{64}O_{12}$, CNS^-, K^+ 50.44 **1**
$C_{40}H_{67}AgO_{10}$ 50.26 **3**
$C_{40}H_{67}O_{10}Tl$, H_2O 50.14 **4**
$C_{40}H_{67}O_{11}^-$, Ag^+ 50.45 **1**
$C_{40}H_{67}O_{11}^-$, Ag^+ 50.27 **3**
$C_{40}H_{100}Nb_8O_{30}$ 84.65 **2**
$C_{41}H_{28}BrNO$, C_2H_6O 33.82 **1**
$C_{41}H_{30}F_6OP_2RuS_2$ 85.54 **3**
$C_{41}H_{33}FeIO_6P_2$ 73.119 **2**
$C_{41}H_{33}IrN_2OP_2$ 86.79 **3**
$C_{41}H_{33}N_2NiP_3$ 86.80 **3**
$C_{41}H_{33}N_2NiP_3$, $0.33H_2O$ 86.81 **3**
$C_{41}H_{33}N_2NiP_3$, CH_4O 86.82 **3**
$C_{41}H_{34}CuF_3O_2P_2$ 86.83 **3**
$C_{41}H_{35}FeIO_6P_2$ 73.120 **2**
$C_{41}H_{36}O_8$ 5.9 **4**
$C_{41}H_{37}Cl_2O_2P_2Re$ 77.68 **2**
$C_{41}H_{38}P_2Pt$ 75.33 **4**
$C_{41}H_{48}N_4O^{2+}$, $2I^-$, H_2O 58.67 **3**
$C_{41}H_{48}N_4O_4$, CH_4O 58.123 **1**
$C_{41}H_{49}N_5Zn$ 49.15 **4**
$C_{41}H_{52}O_{17}$ 56.11 **3**
$C_{41}H_{58}FeN_9O_{20}$, $4H_2O$ 82.59 **2**
$C_{42}H_{18}$ 30.35 **1**
$C_{42}H_{26}N_4$, C_3H_6O 36.34 **3**
$C_{42}H_{30}$ 19.67 **1**
$C_{42}H_{30}CoN_6^{3+}$, $3I^-$, $3H_2O$ 83.148 **3**
$C_{42}H_{30}Co_2O_{12}P_2$ 86.94 **2**
$C_{42}H_{30}F_{12}N_2P_2Pt$, $0.4CH_2Cl_2$ 71.84 **4**
$C_{42}H_{30}N_4P_2Pt$ 72.51 **3**
$C_{42}H_{30}Ni_3S_{12}$ 85.96 **2**
$C_{42}H_{30}ReS_6$ 85.97 **2**
$C_{42}H_{30}S_6V$ 85.98 **2**
$C_{42}H_{33}BrCrMnO_8P_3$ 86.41 **4**
$C_{42}H_{34}FeO_2P_2$ 72.52 **3**
$C_{42}H_{35}P_3PdS_4$ 85.36 **4**
$C_{42}H_{38}P_2Pt$ 75.34 **4**
$C_{42}H_{39}Br_2P_2Pt$ 86.42 **4**
$C_{42}H_{39}Br_2P_3Pd$ 86.43 **4**
$C_{42}H_{40}Cl_3O_5P_2Re_2$ 81.53 **3**
$C_{42}H_{40}N_8O_4Ru_2^{2+}$, $2BF_4^-$ 81.54 **3**
$C_{42}H_{40}P_2Ru$, C_7H_8 72.31 **4**
$C_{42}H_{42}ClCoNP_3^+$, F_6P^- 86.84 **3**
$C_{42}H_{42}CoINP_3^+$, I^- 86.95 **2**
$C_{42}H_{42}CoI_2NP_3$ 86.44 **4**
$C_{42}H_{42}INNiP_3^+$, I^- 86.85 **3**
$C_{42}H_{46}N_4O_5^{2+}$, $2Br^-$, $2H_2O$ 58.43 **4**
$C_{42}H_{47}Br_3O_8$ 59.69 **1**
$C_{42}H_{48}NNiP_2Si_2$ 83.136 **4**
$C_{42}H_{48}N_{12}Ni_4O_6P_4$ 86.97 **2**
$C_{42}H_{59}IO_{14}S$ 54.19 **4**
$C_{42}H_{84}Ag_6N_6O_6S_6$ 80.20 **4**
$C_{42}H_{84}Ag_6N_6S_{12}$ 80.21 **3**
$C_{42}H_{84}Co_3O_{24}P_6$ 84.66 **2**
$C_{42}H_{84}Cu_6N_6O_6S_6$ 80.22 **3**
$C_{43}H_{30}BrIrN_4OP_2$ 86.98 **2**
$C_{43}H_{30}OW$ 72.32 **4**
$C_{43}H_{33}AsO$ 65.13 **4**
$C_{43}H_{33}OP$ 64.44 **4**
$C_{43}H_{34}AsO^+$, ClO_4^- 65.14 **4**
$C_{43}H_{34}ClFIrN_2OP_2^+$, BF_4^-, C_3H_6O 71.85 **4**
$C_{43}H_{35}As_3BrNiS^+$, ClO_4^-, C_6H_5Cl 85.37 **4**
$C_{43}H_{35}MoNO_2P_2$, $0.5CH_2Cl_2$ 73.56 **3**
$C_{43}H_{37}MnN_4O_2$ 77.31 **3**
$C_{43}H_{40}P_2Pt$ 75.35 **4**
$C_{43}H_{50}N_4O_2$ 49.16 **4**
$C_{43}H_{57}IO_6S$ 51.48 **3**
$C_{43}H_{67}EuN_2O_6$ 77.23 **4**
$0.54C_{44}H_{28}AgN_4$, $0.46C_{44}H_{30}N_4$ 49.17 **4**
$C_{44}H_{28}CuN_4$ 49.22 **1**
$C_{44}H_{28}N_4Pd$ 49.23 **1**
$C_{44}H_{28}N_4Zn$, H_2O 49.24 **1**
$C_{44}H_{30}ClF_{12}RhSb_2$, CH_2Cl_2 71.68 **3**
$C_{44}H_{30}Cu_2N_2O_4$ 78.29 **3**
$C_{44}H_{30}F_2$ 27.17 **1**
$C_{44}H_{30}F_{12}Ir_2N_2O_2P_2$ 71.86 **4**
$C_{44}H_{30}FeN_{10}$, $2C_6H_7N$ 49.18 **4**
$C_{44}H_{30}MgN_4O$ 49.10 **3**
$C_{44}H_{30}N_4$ 49.25+ **1**
$C_{44}H_{30}N_4^{2+}$, Cl^-, Cl_4Fe^- 49.27 **1**
$C_{44}H_{31}FeN_4O_2$ 49.28+ **1**
$C_{44}H_{32}Cl_4F_{24}Rh_4$ 71.87 **4**
$C_{44}H_{32}N_4^{2+}$, $2Cl^-$ 49.30 **1**
$C_{44}H_{32}N_4O_2Zn$ 49.31 **1**
$C_{44}H_{39}N_2NiP_3$ 86.99 **2**
$C_{44}H_{39}N_2NiP_3$ 86.86 **3**
$C_{44}H_{42}Cl_2O_6P_2Rh_2$ 75.47 **3**
$C_{44}H_{42}HgN_4$ 71.88 **4**
$C_{44}H_{42}O_{14}P_2U_2$ 81.55 **3**
$C_{44}H_{48}Br_2O_6$ 56.12 **3**
$C_{44}H_{48}Cl_4Hg_2O_4S_4$ 84.67 **2**
$C_{44}H_{48}Cr_2O_{12}P_2$ 77.69 **2**
$C_{44}H_{56}Br_2O_6$, $2C_2H_6O$ 56.13 **3**
$C_{44}H_{63}N_2O_2^+$, I^-, C_2H_6O 58.44 **4**
$C_{45}H_{29}N_9Zn$ 49.11 **3**
$C_{45}H_{30}FeN_3O_{12}^-$, Na^+, $3CCl_4$, $3CH_4O$ 83.137 **4**
$C_{45}H_{30}FeN_3O_{12}^-$, Na^+, $2C_3H_6O$ 83.138 **4**
$C_{45}H_{30}O_9Os_3P_2$ 71.89 **4**
$C_{45}H_{33}EuN_9^{3+}$, $3ClO_4^-$ 83.217 **2**
$C_{45}H_{35}HoO_7$ 77.32 **3**
$C_{45}H_{51}IN_2O_{14}$, C_2H_6OS, H_2O 50.28 **3**
$C_{45}H_{53}IN_2O_{13}$, C_3H_6O, $5H_2O$ 50.48 **1**
$C_{45}H_{60}Br_2O_5$ 51.49 **3**
$C_{45}H_{71}HoN_2O_6$ 77.24 **4**
$C_{45}H_{77}AgO_{13}$ 50.15 **4**
$C_{46}H_{28}N_4O_2Ru$ 49.19 **4**
$C_{46}H_{30}Ir_4O_{10}P_2$ 86.100 **2**
$C_{46}H_{38}N_2P_2RuS_2$ 85.38 **4**
$C_{46}H_{58}ClCoN_6O_{13}$, C_3H_6O, $2H_2O$ 49.32 **1**
$C_{46}H_{66}CoN_{11}O_9$, $11H_2O$ 49.20 **4**
$C_{47}H_{59}N_4O_{10}^+$, I^-, $2H_2O$ 58.125 **1**
$C_{47}H_{76}KO_{14}$ 50.16 **4**
$C_{48}H_{28}Fe_2O_4$ 75.48 **3**
$C_{48}H_{30}$, C_6H_4Cl 31.42 **3**
$C_{48}H_{30}Cu_4F_{18}N_4O_{14}$ 81.60 **4**

$C_{48}H_{30}N_8O_4P_2$ 40.21 **3**
$C_{48}H_{32}Br_2O_2$, $1.5C_6H_6$ 17.51 **1**
$C_{48}H_{35}NbO$ 72.53 **3**
$C_{48}H_{36}$ 27.19 **4**
$C_{48}H_{40}Al_4N_4$ 68.14 **4**
$C_{48}H_{40}As_4N_4$ 65.12 **3**
$C_{48}H_{40}I_2O_{16}$ 59.70 **1**
$C_{48}H_{40}N_4P_4$ 64.67 **3**
$C_{48}H_{40}N_{12}Ni_2$ 83.218 **2**
$C_{48}H_{40}OP_2Pt$ 71.90 **4**
$C_{48}H_{44}Cu_4N_8O_8^{4+}$, $4NO_3^-$ 78.17 **4**
$C_{48}H_{48}N_4O_8P_2Rh_2$, H_2O 81.61 **4**
$C_{48}H_{144}Cl_5P_4Ru_2$ 86.101 **2**
$C_{49}H_{22}As_2F_{20}Hg_2$ 71.91 **4**
$C_{49}H_{31}IrN_8OP_2$, $0.5C_6H_6$ 86.45 **4**
$C_{49}H_{38}F_2IrN_4OP_2^+$, BF_4^-, C_6H_6 83.139 **4**
$C_{49}H_{40}Br_4Pb_4$ 69.40 **3**
$C_{49}H_{74}INO_{18}$, $3C_4H_8O$, H_2O 50.17 **4**

C_{50-99}

$C_{50}H_{33}ClN_4Rh$, H_2O 49.33 **1**
$C_{50}H_{35}MnO_4Sn$ 73.121 **2**
$C_{50}H_{36}FeN_8^+$, Cl^- 49.12 **3**
$C_{50}H_{36}FeN_8^+$, Cl^-, $4CH_4O$ 49.21 **4**
$C_{50}H_{38}Al_2Br_2N_2$, $2C_6H_6$ 68.32 **2**
$C_{50}H_{40}N_2P_2$ 64.69 **3**
$C_{50}H_{40}N_2P_2$ 64.45 **4**
$C_{50}H_{40}P_2Pt$ 72.83 **2**
$C_{50}H_{46}F_{30}Ni_6O_{24}$, C_7H_8 77.33 **3**
$C_{50}H_{75}Cl_4P_5Ru_2$, $xC_6H_{12}O_2$ 86.102 **2**
$C_{50}H_{92}NiP_4$ 86.46 **4**
$C_{52}H_{40}Fe_4O_6P_2$ 73.122 **2**
$C_{52}H_{44}As_4Br_2MoO_2$ 86.47 **4**
$C_{52}H_{44}As_4Cl_2O_2Rh_2$ 86.87 **3**
$C_{52}H_{44}CoO_2P_4^+$, BF_4^-, $2C_6H_6$ 86.48 **4**
$C_{52}H_{48}ClCoP_4^+$, Cl_3Sn^- 86.49 **4**
$C_{52}H_{48}ClCoP_4^+$, Cl_3Sn^-, C_6H_5Cl 86.50 **4**
$C_{52}H_{48}ClMoOP_4^+$, $C_3H_6Cl_3OZn^-$, C_3H_6O 86.88 **3**
$C_{52}H_{48}IrO_2P_2^+$, F_6P^- 86.89 **3**
$C_{52}H_{48}IrP_4S_2^+$, Cl^-, C_2H_3N 86.51 **4**
$C_{52}H_{48}MoN_4P_4$ 86.52 **4**
$C_{52}H_{48}O_2P_4Rh^+$, F_6P^- 86.90 **3**
$C_{52}H_{48}P_4Rh^+$, ClO_4^- 86.91 **3**
$C_{52}H_{52}As_4ClCoO_8^+$, ClO_4^- 84.68 **2**
$C_{52}H_{52}IrP_4^+$, BF_4^-, C_6H_{12} 86.53 **4**
$C_{52}H_{58}N_8O_8P_2Rh_2$, C_3H_8O, H_2O 86.54 **4**
$C_{52}H_{59}BBrNO_{17}$, CH_2Cl_2 50.18 **4**
$C_{52}H_{60}I_2O_{14}$ 59.29 **4**
$C_{53}H_{44}As_4Br_2O_3W$ 86.55 **4**
$C_{53}H_{48}IrOP_4^+$, Cl^- 86.104 **2**
$C_{54}H_{42}As_4Br_2Hg$, CH_2Cl_2 86.105 **2**
$C_{54}H_{42}As_4Br_2Ru$ 86.106 **2**
$C_{54}H_{42}ClCoP_4^+$, $C_{24}H_{20}B^-$ 86.56 **4**
$C_{54}H_{43}ClIrO_9P_3$ 71.70 **3**
$C_{54}H_{45}ClP_3Rh$ 86.108 **2**
$C_{54}H_{45}Cl_2Cu_2P_3$ 86.57 **4**
$C_{54}H_{45}Cl_2Cu_2P_3$, C_6H_6 86.92 **3**
$C_{54}H_{45}Cl_2P_3Ru$ 86.109 **2**
$C_{54}H_{45}IrNOP_3$ 86.58 **4**
$C_{54}H_{46}ClP_3Ru$, C_6H_6 86.110 **2**
$C_{54}H_{46}CoN_2P_3$ 86.93 **3**
$C_{54}H_{46}CoN_2P_3$, $C_4H_{10}O$ 86.112 **2**
$C_{54}H_{46}IrNOP_3^+$, ClO_4^- 86.59 **4**
$C_{54}H_{46}NOP_3Ru$ 86.60 **4**
$C_{54}H_{50}FeN_6$ 49.22 **4**
$C_{54}H_{52}Cu_2N_2P_4S_2$ 86.61 **4**
$C_{54}H_{54}Co_2N_6O_{12}$, $2C_2H_6O_5$, $2H_2O$ 83.150 **3**
$C_{54}H_{90}N_6O_{18}$ 50.19 **4**
$C_{55}H_{45}OP_3Pt$ 86.94 **3**
$C_{55}H_{45}OP_3Pt_2S$ 86.95 **3**
$C_{55}H_{46}BrOOsP_3$ 86.83 **2**
$C_{55}H_{46}OP_3Rh$ 86.114 **2**
$C_{56}H_{40}$ 22.16+ **1**
$C_{56}H_{45}O_2P_3Pt_2S$ 86.115 **2**
$C_{56}H_{48}Cl_3N_{16}Ni^+$, Cl^-, $4C_3H_6O$ 83.219 **2**
$C_{56}H_{70}Cu_2P_4$ 71.71 **3**
$C_{57}H_{49}O_2P_3Ru$ 86.116 **2**
$C_{58}H_{40}F_{12}P_4Rh^+$, $C_2Cl_2O_2Rh^-$ 86.62 **4**
$C_{58}H_{48}NOP_4Ru^+$, $C_{24}H_{20}B^-$, C_3H_6O 86.63 **4**
$C_{58}H_{52}Br_3NO_{16}$ 50.30 **3**
$C_{59}H_{48}Ir_2O_2P_2S_6$ 85.39 **4**
$C_{60}H_{30}F_{20}P_2Pd_2S_4$ 85.55 **3**
$C_{60}H_{30}F_{20}P_2Pd_2S_4$, $2C_2H_6O$ 85.40 **4**
$C_{60}H_{40}Br_2Mo_2O_4$ 75.49 **3**
$C_{60}H_{44}ClO_2P_2Rh$, H_2O, $0.5C_2H_6O$ 71.92 **4**
$C_{60}H_{45}N_6O_6P_3$, $0.5C_7H_8$ 64.70 **3**
$C_{60}H_{50}Cl_2O_2Pd_2$ 75.80 **2**
$C_{60}H_{50}Cl_2O_2Pd_2$ 75.81 **2**
$C_{60}H_{52}O_{10}Ti_2$ 84.18 **4**
$C_{60}H_{64}F_{42}O_{14}Pr_2$, $2H_2O$ 77.34 **3**
$C_{60}H_{90}Cl_3P_6Ru_2^+$, $C_{30}H_{45}Cl_3P_3Ru^-$ 86.117 **2**
$C_{62}H_{57}P_4Re$ 86.118 **2**
$C_{62}H_{86}N_{12}O_{16}$, $2C_{10}H_{13}N_5O_4$, $12H_2O$ 60.43+ **4**
$C_{63}H_{45}Ir_4O_9P_3$ 86.119 **2**
$C_{63}H_{88}CoN_{14}O_{14}P$, $18H_2O$ 49.13 **3**
$C_{63}H_{88}CoN_{14}O_{14}P$, $20H_2O$ 49.23 **4**
$C_{63}H_{88}CoN_{14}O_{14}P$, $22H_2O$ 49.34 **1**
$C_{63}H_{89}CoN_{14}O_{17}P_2$, $15H_2O$ 49.14 **3**
$C_{64}H_{32}N_{16}U$ 49.15 **3**
$C_{64}H_{31}P_2Rh$, C_7H_8 71.93 **4**
$C_{64}H_{52}Co_2F_8N_8O_{14}$, $2CHCl_3$, $C_5H_{11}N$ 78.30 **3**
$C_{66}H_{60}Br_2O_{12}$, $2C_6H_6$ 59.34 **3**
$C_{66}H_{63}O_{13}P_3Ru_3$ 81.62 **4**
$C_{66}H_{72}CaN_{12}O_6^{2+}$, $2ClO_4^-$ 67.20 **2**
$C_{66}H_{72}MgN_{12}O_6^{2+}$, $2ClO_4^-$ 67.21 **2**
$C_{66}H_{72}N_{12}O_6Pb^{2+}$, $2ClO_4^-$ 69.49 **2**
$C_{66}H_{72}N_{12}O_6Y^{3+}$, $3I^-$ 84.28 **3**
$C_{66}H_{114}O_{12}Pr_2$ 77.35 **3**
$C_{68}H_{50}Nb_2O_2$ 72.84 **2**
$C_{69}H_{84}N_{18}O_{18}S_5$ 50.31 **3**
$C_{72}H_{57}IrP_5^+$, $C_{24}H_{20}B^-$ 86.64 **4**
$C_{72}H_{60}Cl_2O_4P_4Rh_2$, CH_2Cl_2 86.65 **4**
$C_{72}H_{60}Cl_6Cu_4O_5P_4$ 84.69 **2**
$C_{72}H_{60}Cu_2N_6P_4$ 86.96 **3**
$C_{72}H_{60}N_{18}OZn_4$, C_6H_6 83.151 **3**
$C_{72}H_{61}AsP_3Rh$, $0.5C_6H_6$ 86.97 **3**
$C_{72}H_{61}P_4Rh$ 86.120 **2**
$xC_{72}H_{62}N_4P_4Pt_2^{2+}$, $4C_{24}H_{20}B^-$, $yC_{72}H_{64}N_2P_4Pt_2^{2+}$ 86.98 **3**
$C_{72}H_{100}CoN_{18}O_{17}P$, $17H_2O$ 49.35 **1**
$C_{72}H_{120}O_{60}$, $3.8C_2H_3O_2^-$, $3.8K^+$, $19.4H_2O$ 45.76 **1**
$C_{72}H_{132}N_2Ni_2P_4$ 86.66 **4**
$C_{74}H_{42}Mn_2N_{18}O$, $2C_5H_5N$ 49.36 **1**
$C_{74}H_{60}O_2P_2Rh_2$, $2CH_2Cl_2$ 86.67 **4**

$C_{75}H_{66}NaO_6P_6^+$, Br^-, $3H_2O$ 67.27 **3**
$C_{78}H_{72}Cu_2N_6P_6$ 86.68 **4**
$C_{82}H_{72}Cu_2F_6N_8P_6$ 86.69 **4**
$C_{84}H_{84}Br_4O_{16}$ 30.36 **1**
$C_{88}H_{56}Fe_2N_8O$ 49.24 **4**

$C_{100-149}$

$C_{100}H_{70}Cu_4Ir_2P_2$ 71.94 **4**
$C_{108}H_{96}Cu_6P_6$ 86.70 **4**
$C_{126}H_{84}Au_{11}Cl_{21}I_3P_7$ 86.99 **3**

$C_{150-199}$

$C_{168}H_{168}Au_9P_8^{3+}$, $3F_6P^-$ 86.71 **4**

Ag

Vol. 1 53.10 **Vol. 2** 72.29, 72.53, 74.1, 74.2, 75.2, 75.5, 75.9, 75.15, 75.16, 75.35, 75.47, 75.67, 75.78, 76.20, 77.19, 79.3, 79.16, 80.41, 81.2, 82.1, 83.16, 83.17, 83.27, 83.44, 83.161, 84.23, 85.2, 85.9, 85.10, 85.11, 85.14, 85.36, 85.64, 85.67, 86.36 **Vol. 3** 50.19, 53.7, 53.15, 53.29, 74.1, 74.6, 74.7, 74.12, 74.13, 74.14, 75.2, 75.4, 75.8, 75.9, 75.37, 76.11, 76.12, 80.21, 81.36, 83.16, 83.61, 86.32 **Vol. 4** 23.4, 23.6, 31.1, 31.13, 74.3, 74.8, 74.9, 75.19, 80.11, 80.20, 81.36, 81.55, 82.1, 82.2, 82.4, 82.5, 82.8, 83.15, 83.16, 83.35, 83.58

Am

Vol. 3 77.16

Au

Vol. 2 64.52, 71.16, 71.18, 71.23, 71.52, 73.109, 80.2, 80.8, 80.34, 83.154, 86.1, 86.40, 86.90 **Vol. 3** 80.11, 86.22, 86.26, 86.99 **Vol. 4** 71.34, 71.43, 71.46, 71.77, 80.11, 80.12, 80.16, 86.71

Cd

Vol. 2 79.2, 79.5, 79.6, 79.20, 79.21, 79.33, 80.10, 80.11, 80.35, 80.36, 81.61, 82.45, 83.4, 83.5, 83.28, 83.74, 84.2, 84.3, 84.5, 84.8, 85.37, 85.52, 85.94 **Vol. 3** 61.4, 76.39, 79.4, 79.17, 83.38, 83.39, 84.2, 85.41, 86.58 **Vol. 4** 76.13, 79.4, 81.17, 83.5, 83.7, 83.50, 83.91, 83.92, 83.93, 83.123, 84.9

Ce

Vol. 2 77.47, 81.44, 81.67 **Vol. 3** 77.19, 81.11, 84.26 **Vol. 4** 75.13, 75.31, 81.43

Co

Vol. 1 3.31 **Vol. 2** 62.33, 71.22, 71.40, 71.44, 71.45, 72.25, 72.35, 72.39, 72.46, 72.52, 72.56, 72.67, 73.3, 73.9, 73.10, 73.19, 73.31, 73.34, 73.46, 73.58, 73.67, 73.73, 73.108, 73.117, 75.23, 75.55, 75.61, 76.2, 76.4, 76.7, 76.8, 76.9, 76.10, 76.11, 76.12, 76.24, 76.28, 76.34, 76.35, 76.36, 76.37, 76.38, 76.40, 76.48, 76.49, 76.54, 76.55, 76.56, 76.58, 76.59, 76.75, 76.79, 77.11, 77.20, 77.21, 77.44, 77.51, 77.64, 77.67, 78.15, 78.59, 79.24, 80.5, 80.7, 80.29, 80.30, 81.5, 81.12, 81.34, 81.50, 82.2, 82.24, 82.28, 82.34, 82.37, 82.46, 82.54, 83.61, 83.75, 83.80, 83.89, 83.91, 83.95, 83.103, 83.124, 83.130, 83.131, 83.134, 83.136, 83.146, 83.176, 83.213, 84.12, 84.21, 84.66, 84.68, 85.39, 85.45, 85.55, 85.63, 85.73, 85.74, 85.82, 85.84, 85.86, 86.23, 86.32, 86.40, 86.44, 86.60, 86.61, 86.73, 86.86, 86.94, 86.95, 86.112 **Vol. 3** 62.12, 69.39, 71.25, 71.35, 71.38, 71.40, 71.55, 71.59, 71.60, 71.64, 72.29, 72.39, 73.27, 73.33, 73.37, 76.2, 76.3, 76.6, 76.7, 76.9, 76.10, 76.17, 76.19, 76.23, 76.27, 76.28, 76.30, 76.31, 76.38, 76.40, 76.41, 76.52, 76.54, 76.55, 76.58, 76.61, 76.62, 76.63, 76.64, 76.71, 76.74, 76.75, 76.81, 76.82, 76.85, 77.8, 78.12, 78.14, 78.16, 78.19, 78.26, 78.30, 79.2, 79.5, 79.6, 79.7, 79.21, 81.17, 82.7, 82.12, 82.13, 82.14, 82.15, 82.22, 83.20, 83.26, 83.34, 83.49, 83.51, 83.65, 83.76, 83.88, 83.93, 83.100, 83.115, 83.121, 83.142, 83.148, 83.150, 84.18, 84.24, 85.35, 85.40, 85.45, 86.11, 86.27, 86.32, 86.33, 86.77, 86.84, 86.93 **Vol. 4** 60.9, 71.2, 71.39, 71.40, 71.42, 71.57, 72.29, 73.17, 75.5, 75.20, 75.29, 76.3, 76.10, 76.11, 76.13, 76.14, 76.16, 76.17, 76.18, 76.19, 76.20, 76.21, 76.22, 76.23, 76.24, 76.25, 76.28, 76.29, 76.30, 76.31, 76.32, 76.39, 76.40, 76.43, 77.5, 77.9, 77.14, 77.16, 77.20, 78.13, 78.14, 78.15, 79.1, 79.10, 80.1, 81.15, 81.24, 81.30, 81.31, 81.50, 82.10, 83.13, 83.21, 83.24, 83.25, 83.31, 83.32, 83.56, 83.57, 83.67, 83.74, 83.78, 83.82, 83.94, 83.95, 83.96, 83.97, 83.98, 83.102, 83.120, 83.122, 83.129, 83.130, 84.1, 85.5, 85.22, 86.9, 86.11, 86.13, 86.18, 86.20, 86.44, 86.48, 86.49, 86.50, 86.56

Cr

Vol. 1 3.58, 3.59 **Vol. 2** 60.118, 60.140, 71.14, 71.24, 71.32, 72.12, 73.16, 73.20, 73.99, 74.4, 74.5, 74.6, 74.7, 74.9, 74.10, 74.11, 74.12, 74.13, 74.14, 74.15, 74.16, 74.18, 74.20, 74.21, 74.22, 74.23, 74.24, 74.25, 75.1, 75.22, 75.31, 75.39, 75.46, 75.68, 76.1, 76.13, 76.39, 76.40, 76.44, 76.66, 76.81, 77.22, 77.69, 81.21, 81.37, 81.63, 81.75, 83.35, 83.71, 83.72, 83.93, 83.188, 83.197, 83.206, 83.211, 84.46, 85.85, 86.47 **Vol. 3** 71.30, 71.47, 71.52, 71.65, 72.15, 72.20, 73.1, 73.6, 73.16, 73.17, 73.47, 74.2, 74.5, 74.11, 74.17, 75.3, 75.11, 75.28, 76.4, 76.5, 76.24, 76.32, 76.40, 76.65, 77.9, 77.27, 78.3, 79.16, 81.32, 83.102, 86.5, 86.34, 86.35, 86.57 **Vol. 4** 71.5, 71.8, 71.12, 71.14, 71.30, 74.2, 74.4, 74.6, 74.7, 75.1, 75.8, 75.9, 75.10, 75.11, 75.16, 75.22, 76.32, 78.8, 80.2, 81.26, 81.47, 82.7, 82.9, 83.20, 83.61, 83.85, 86.25, 86.38, 86.41

Cu

Vol. 1 3.38, 3.39, 3.69 **Vol. 2** 60.76, 60.126, 60.145, 60.146, 60.147, 61.21, 62.34, 72.30, 74.3, 75.6, 75.50, 75.77, 76.15, 76.16, 76.17, 76.18, 76.19, 76.26, 76.46, 76.51, 76.60, 77.1, 77.14, 77.15, 77.16, 77.17, 77.40, 77.54, 77.55, 77.59, 78.1, 78.2, 78.3, 78.4, 78.5, 78.6, 78.7, 78.8, 78.16, 78.17, 78.18, 78.19, 78.20, 78.21, 78.26, 78.27, 78.28, 78.29, 78.34, 78.35, 78.40, 78.41, 78.46, 78.50, 78.51, 78.56, 78.57, 78.58, 79.1, 79.13, 80.9, 80.13, 80.14, 80.15, 80.25, 80.26, 80.37, 81.3, 81.4, 81.6, 81.7, 81.8, 81.15, 81.22, 81.23, 81.27, 81.31, 81.32, 81.33, 81.45, 81.47, 81.48, 81.62, 81.64, 81.66, 81.76, 81.77, 81.78, 81.79, 81.80, 81.81, 81.82, 81.84, 81.85, 81.86, 81.87, 82.3, 82.4, 82.5, 82.10, 82.13, 82.15, 82.17, 82.18, 82.21, 82.23, 82.25, 82.26, 82.29, 82.35, 82.36, 82.38, 82.39, 82.41, 82.42, 82.47, 82.51, 82.53, 82.55, 82.57, 82.58, 83.2, 83.3, 83.6, 83.8, 83.20, 83.21, 83.22, 83.23, 83.34, 83.39, 83.45, 83.46, 83.50, 83.52, 83.53, 83.54, 83.56, 83.62, 83.65, 83.73, 83.76, 83.77, 83.79, 83.81, 83.82, 83.87, 83.90, 83.100, 83.101, 83.104, 83.109, 83.112, 83.113, 83.116, 83.117, 83.119, 83.132, 83.140, 83.155, 83.156, 83.157, 83.158, 83.171, 83.179, 83.180, 83.181, 83.182, 83.187, 83.192, 83.207, 84.1, 84.4, 84.10, 84.13, 84.15, 84.17, 84.19, 84.20, 84.31, 84.32, 84.33, 84.34, 84.36, 84.38, 84.40, 84.42, 84.47, 84.48, 84.49, 84.69, 85.20, 85.40, 85.46, 85.53, 85.69, 86.14, 86.41, 86.69 **Vol. 3** 60.1, 60.16, 60.37, 60.39, 71.71, 73.7, 73.44, 75.36, 76.1, 76.20, 76.22, 76.29, 76.32, 76.33, 76.34, 76.35, 76.36, 76.37, 76.41, 76.42, 76.46, 76.48, 76.53, 76.57, 76.59, 76.66, 76.69, 76.70, 76.77, 76.78, 77.1, 77.5, 77.14, 77.15, 77.18, 77.25, 77.26, 78.2, 78.4, 78.8, 78.9, 78.10, 78.18, 78.21, 78.27, 78.29, 79.8, 79.10, 79.11, 79.12, 79.20, 79.22, 79.24, 79.25, 80.7, 80.22, 81.5, 81.18, 81.19, 81.22, 81.23, 81.24, 81.25, 81.30, 81.33, 81.35, 81.40, 81.45, 81.46, 81.47, 81.49, 81.51, 82.5, 82.17, 82.21, 82.24, 82.25, 82.26, 82.28, 83.6, 83.7, 83.10, 83.19, 83.23, 83.24, 83.25, 83.29, 83.32, 83.36, 83.37, 83.40, 83.43, 83.46, 83.47, 83.62, 83.69, 83.72, 83.74, 83.75, 83.78, 83.79, 83.81, 83.82, 83.83, 83.97, 83.98, 83.104, 83.105, 83.107, 83.108, 83.109, 83.110, 83.111, 83.114, 83.119, 83.122, 83.127, 83.130, 83.146, 84.1, 84.3, 84.4, 84.7, 84.8, 84.9, 84.11, 84.13, 84.17, 84.19, 84.23, 85.1, 85.5, 85.8, 85.15, 86.3, 86.62, 86.67, 86.83, 86.92, 86.96 **Vol. 4** 60.10, 71.82, 71.94, 74.11, 76.1, 76.4, 76.5, 76.6, 76.8, 76.12, 76.38, 76.41, 77.7, 77.13, 78.1, 78.4, 78.5, 78.6, 78.7, 78.10, 78.12, 78.17, 79.2, 79.3, 79.6, 79.7, 79.9, 81.1, 81.12, 81.13, 81.14, 81.23, 81.27, 81.32, 81.40, 81.41, 81.42, 81.44, 81.48, 81.49, 81.54, 81.59, 81.60, 82.6, 82.14, 82.15, 83.1, 83.2, 83.4, 83.11, 83.12, 83.14, 83.22, 83.23, 83.27, 83.28, 83.36, 83.37, 83.39, 83.40, 83.52, 83.53, 83.54, 83.59, 83.64, 83.68, 83.69, 83.70, 83.71, 83.79, 83.80, 83.88, 83.90, 83.100, 83.103, 83.107, 83.108, 83.117, 83.135, 84.8, 84.11, 84.13, 84.14, 85.3, 85.14, 85.28, 86.7, 86.12, 86.34, 86.57, 86.61, 86.68, 86.69, 86.70

Dy

Vol. 4 77.22

Er

Vol. 3 81.4, 81.6 **Vol. 4** 77.21, 81.25, 81.45

Eu

Vol. 2 83.217 **Vol. 3** 77.12, 77.17, 77.28, 77.29, 80.12 **Vol. 4** 77.23, 81.21

Fe

Vol. 2 60.119, 62.39, 63.24, 69.35, 71.13, 71.20, 71.21, 71.28, 71.34, 71.46, 72.7, 72.13, 72.14, 72.16, 72.17, 72.18, 72.31, 72.32, 72.33, 72.41, 72.54, 72.57, 72.61, 72.62, 72.63, 72.65, 72.68, 72.69, 72.74, 72.75, 73.5, 73.14, 73.21, 73.22, 73.23, 73.24, 73.25, 73.31, 73.35, 73.36, 73.38, 73.39, 73.43, 73.44, 73.47, 73.48, 73.49, 73.51, 73.55, 73.59, 73.68, 73.69, 73.71, 73.72, 73.76, 73.77, 73.85, 73.86, 73.87, 73.89, 73.90, 73.94, 73.95, 73.96, 73.98, 73.101, 73.102, 73.104, 73.105, 73.106, 73.107, 73.110, 73.115, 73.116, 73.119, 73.120, 73.122, 75.8, 75.10, 75.19, 75.21, 75.26, 75.29, 75.32, 75.33, 75.38, 75.48, 75.57, 75.58, 75.60, 75.64, 75.75, 75.79, 76.63, 76.64, 76.76, 77.23, 78.13, 78.14, 80.16, 80.40, 82.9, 82.59, 83.36, 83.69, 83.88, 83.137, 83.150, 83.152, 83.167, 83.168, 83.169, 83.170, 83.183, 83.185, 83.203, 83.205, 83.209, 83.210, 83.212, 83.214, 84.24, 84.29, 84.44, 84.52, 84.53, 85.22, 85.41, 85.42, 85.62, 85.71, 85.83, 85.89, 86.5, 86.6, 86.8, 86.15, 86.53, 86.62 **Vol. 3** 60.51, 69.22, 69.31, 71.5, 71.8, 71.13, 71.17, 71.18, 71.24, 71.31, 71.32, 71.49, 71.50, 71.51, 71.56, 71.61, 71.62, 72.9, 72.11, 72.14, 72.21, 72.23, 72.25, 72.26, 72.31, 72.33, 72.38, 72.41, 72.42, 72.52, 73.2, 73.3, 73.4, 73.8, 73.9, 73.10, 73.12, 73.13, 73.14, 73.15, 73.35, 73.39, 73.40, 73.41, 73.45, 73.50, 73.53, 73.54, 73.55, 75.6, 75.7, 75.10, 75.12, 75.13, 75.15, 75.16, 75.17, 75.18, 75.22, 75.27, 75.29, 75.30, 75.31, 75.34, 75.38, 75.43, 75.48, 77.4, 78.11, 78.13, 78.22, 78.23, 78.24, 80.1, 80.2, 80.3, 80.4, 80.6, 81.2, 81.3, 81.34, 83.22, 83.67, 83.89, 83.103, 83.118, 83.123, 83.137, 83.138, 83.139, 84.20, 85.16, 85.24, 85.25, 85.31, 85.44, 86.12, 86.17, 86.45
Vol. 4 63.5, 71.9, 71.18, 71.20, 71.28, 71.29, 71.31, 71.35, 71.37, 71.54, 71.55, 71.61, 72.5, 72.6, 72.7, 72.17, 72.20, 73.3, 73.10, 73.12, 73.15, 73.16, 73.17, 73.18, 73.19, 73.20, 73.25, 73.26, 73.27, 73.28, 73.32, 73.33, 75.4, 75.6, 75.7, 75.12, 75.17, 75.18, 75.24, 76.37, 78.16, 80.3, 80.8, 80.9, 80.10, 80.17, 83.34, 83.43, 83.72, 83.86, 83.115, 83.116, 83.137, 83.138, 85.9, 86.2, 86.16, 86.22, 86.28, 86.36, 86.39, 86.40

Gd

Vol. 2 81.13, 81.69 **Vol. 3** 77.30, 81.10, 83.70

Hf

Vol. 4 71.62

Hg

Vol. 2 60.91, 71.1, 71.2, 71.3, 71.4, 71.5, 71.8, 71.11, 71.12, 71.15, 71.19, 71.25, 71.26, 71.39, 71.41, 73.3, 73.31, 76.19, 79.8, 79.11, 79.12, 79.17, 79.24, 81.53, 83.94, 83.145, 84.6, 84.30, 84.37, 84.62, 84.63, 84.64, 84.67, 85.4, 85.12, 85.13, 85.17, 85.21, 85.23, 85.44, 85.48, 85.51, 85.75, 85.90, 86.23, 86.105 **Vol. 3** 63.9, 71.9, 71.15, 82.4, 83.9, 83.60, 83.67, 84.6, 85.18, 85.21, 85.39, 85.52, 86.1, 86.61 **Vol. 4** 71.22, 71.67, 71.79, 71.88, 71.91, 83.112, 83.124, 84.5, 85.4, 85.13

Ho

Vol. 3 77.32 **Vol. 4** 77.11, 77.24, 81.47

Ir

Vol. 2 71.9, 71.33, 75.52, 86.70, 86.75, 86.76, 86.77, 86.79, 86.87, 86.98, 86.100, 86.104, 86.119 **Vol. 3** 71.28, 71.58, 71.66, 71.70, 77.2, 86.60, 86.63, 86.69, 86.73, 86.74, 86.79, 86.89 **Vol. 4** 71.85, 71.86, 71.94, 75.26, 75.32, 83.139, 85.39, 86.24, 86.45, 86.51, 86.53, 86.58, 86.59, 86.64

La

Vol. 2 76.71, 76.72, 77.32, 83.172
Vol. 3 81.12 **Vol. 4** 81.46, 81.56, 84.6, 84.17

Lu

Vol. 4 77.19, 84.16

Mn

Vol. 1 3.57 **Vol. 2** 69.14, 69.40, 71.10, 72.51, 73.8, 73.36, 73.41, 73.63, 73.121, 75.34, 75.42, 76.65, 76.77, 77.12, 77.24, 81.9, 81.10, 82.8, 82.56, 83.7, 83.115, 83.153, 84.16, 84.25, 86.25, 86.28, 86.39, 86.41, 86.85, 86.90, 86.91 **Vol. 3** 71.2, 71.14, 75.1, 77.6, 77.31, 81.43, 83.124, 84.10, 86.36 **Vol. 4** 71.36, 71.50, 71.56, 72.15, 73.6, 76.35, 77.18, 83.81, 83.123, 83.133, 83.134, 86.3, 86.4, 86.15, 86.41

Mo

Vol. 2 71.27, 71.51, 72.66, 73.6, 73.26, 73.27, 73.29, 73.30, 73.32, 73.40, 73.41, 73.66, 73.70, 73.75, 73.96, 74.17, 75.11, 75.12, 75.20, 75.43, 75.44, 75.45, 75.69, 75.76, 76.3, 76.61, 77.53, 80.23, 81.1, 81.24, 81.25, 81.59, 82.14, 83.38, 84.18, 85.34, 85.78, 86.17, 86.43, 86.56 **Vol. 3** 64.52, 69.23, 71.6, 71.16, 71.27, 72.12, 72.27, 73.18, 73.19, 73.20, 73.26, 73.28, 73.46, 73.48, 73.51, 73.52, 73.56, 75.25, 75.39, 75.49, 80.5, 80.17, 81.1, 82.16, 82.19, 82.23, 83.17, 83.35, 83.86, 85.17, 86.7, 86.9, 86.46, 86.71, 86.88 **Vol. 4** 71.10, 71.11, 71.15, 71.53, 71.66, 71.69, 72.9, 72.12, 73.14, 73.19, 73.22, 75.21, 82.13, 83.46, 83.87, 84.3, 84.10, 85.12, 85.24, 86.1, 86.6, 86.19, 86.23, 86.26, 86.47, 86.52

Nb

Vol. 2 72.72, 72.84, 73.11, 83.18, 84.65
Vol. 3 72.40, 72.48, 72.53, 83.53, 83.57, 83.129 **Vol. 4** 71.44, 71.52, 73.29, 80.6, 80.7, 81.3, 81.11, 81.19, 83.47, 83.48

Nd

Vol. 2 77.33, 81.41, 81.43, 81.71, 81.72, 82.52
Vol. 3 81.13, 81.14, 81.27, 82.9 **Vol. 4** 77.12, 81.18, 81.51

Ni

Vol. 1 3.80 **Vol. 2** 60.155, 65.42, 65.46, 65.47, 72.15, 72.22, 72.27, 72.48, 72.64, 72.77, 73.64, 73.74, 73.82, 73.91, 73.112, 73.113, 73.118, 75.28, 75.53, 75.54, 75.62, 75.63, 76.6, 76.20, 76.21, 76.22, 76.27, 76.29, 76.31, 76.32, 76.39, 76.41, 76.42, 76.43, 76.47, 76.50, 76.69, 76.70, 76.74, 77.13, 77.19, 77.45, 77.46, 77.56, 77.62, 78.9, 78.10, 78.11, 78.22, 78.23, 78.30, 78.32, 78.36, 78.42, 78.43, 78.47, 78.49, 78.52, 78.53, 79.18, 79.23, 79.26, 79.28, 79.31, 79.34, 79.35, 79.36, 80.1, 80.3, 80.17, 80.27, 80.28, 81.11, 81.36, 82.6, 82.19, 82.22, 82.27, 82.30, 82.31, 82.33, 82.48, 83.11, 83.12, 83.13, 83.14, 83.24, 83.25, 83.29, 83.30, 83.31, 83.32, 83.40, 83.43, 83.51, 83.57, 83.83, 83.84, 83.85, 83.86, 83.96, 83.97, 83.99, 83.102, 83.105, 83.106, 83.107, 83.118, 83.120, 83.127, 83.128, 83.138, 83.143, 83.144, 83.165, 83.175, 83.177, 83.184, 83.186, 83.198, 83.199, 83.204, 83.208, 83.215, 83.218, 83.219, 84.7, 84.41, 84.60, 85.1, 85.5, 85.6, 85.7, 85.8, 85.15, 85.16, 85.18, 85.19, 85.24, 85.25, 85.30, 85.32, 85.33, 85.35, 85.38, 85.43, 85.47, 85.49, 85.50, 85.56, 85.57, 85.58, 85.59, 85.60, 85.70, 85.76, 85.81, 85.88, 85.91, 85.93, 85.96, 86.7, 86.9, 86.20, 86.21, 86.26, 86.33, 86.49, 86.51, 86.52, 86.55, 86.65, 86.67, 86.74, 86.97, 86.99 **Vol. 3** 22.2, 60.54, 61.4, 61.8, 71.3, 71.20, 71.43, 72.17, 72.43, 73.43, 75.30, 76.11, 76.12, 76.15, 76.21, 76.43, 76.44, 76.49, 76.50, 76.51, 76.56, 76.60, 76.65, 76.67, 76.83, 76.84, 76.86, 76.87, 77.33, 78.6, 78.15, 78.28, 79.13, 79.14, 79.26, 80.8, 80.18, 81.8, 81.16, 81.42, 81.48, 83.3, 83.4, 83.21, 83.30, 83.31, 83.45, 83.48, 83.63, 83.64, 83.71, 83.73, 83.80, 83.84, 83.85, 83.87, 83.90, 83.92, 83.94, 83.95, 83.99, 83.112, 83.113, 83.120, 83.125, 83.131, 83.133, 83.134, 83.135, 83.136, 83.141, 83.142, 83.143, 84.12, 85.2, 85.3, 85.4, 85.7, 85.9, 85.12, 85.14, 85.20, 85.23, 85.27, 85.28, 85.29, 85.30, 85.36, 85.37, 85.38, 85.43, 85.46, 85.47, 85.50, 85.51, 86.4, 86.14, 86.16, 86.31, 86.39, 86.41, 86.43, 86.44, 86.48, 86.52, 86.53, 86.56, 86.76, 86.80, 86.81, 86.82, 86.85, 86.86 **Vol. 4** 60.11, 71.32, 71.48, 71.65, 71.74, 71.75, 71.76, 72.4, 72.13, 72.25, 73.15, 73.29, 75.27, 75.30, 76.7, 76.15, 76.27, 76.42, 76.44, 77.3, 77.4, 77.6, 77.17, 78.3, 78.9, 79.5, 79.12, 80.18, 81.22, 81.33, 81.34, 81.37, 82.3, 82.11, 83.6, 83.10, 83.21, 83.30, 83.33, 83.41, 83.42, 83.44, 83.45, 83.60, 83.73, 83.83, 83.84, 83.99, 83.101, 83.111, 83.113, 83.118, 83.119, 83.121, 83.125, 83.126, 83.127, 83.131, 83.133, 83.136, 84.7, 85.2, 85.10, 85.15, 85.16, 85.17, 85.20, 85.23, 85.37, 86.17, 86.27, 86.30, 86.46, 86.66

Np

Vol. 3 80.13

Os

Vol. 2 72.34, 86.83, 86.88 **Vol. 3** 71.41, 86.61, 86.65, 86.66 **Vol. 4** 71.71, 71.73, 71.89, 72.21, 81.29, 86.35

Pd

Vol. 2 60.70, 60.148, 72.4, 72.8, 72.9, 72.10, 72.11, 72.19, 72.21, 72.28, 72.42, 72.47, 72.59, 75.3, 75.4, 75.7, 75.27, 75.36, 75.80, 75.81, 76.14, 77.41, 77.60, 78.31, 78.38, 78.39, 78.44, 78.45, 78.48, 79.19, 79.25, 81.17, 83.42, 83.58, 83.59, 83.114, 83.159, 83.173, 83.189, 84.57, 85.26, 85.28, 85.29, 85.31, 85.66, 85.72, 85.80, 85.92, 85.95, 86.2, 86.18, 86.19, 86.34, 86.38, 86.45, 86.48, 86.62, 86.81 **Vol. 3** 71.26, 71.44, 72.7, 72.10, 72.18, 72.32, 72.34, 72.36, 72.37, 74.3, 74.4, 75.23, 76.14, 76.49, 76.68, 78.1, 79.9, 81.29, 82.1, 82.18, 82.20, 82.29, 83.144, 83.147, 85.10, 85.11, 85.55, 86.20, 86.23, 86.42, 86.54 **Vol. 4** 71.1, 71.41, 71.51, 71.58, 71.60, 71.80, 72.10, 72.14, 72.28, 76.2, 77.2, 80.4, 81.39, 83.26, 83.66, 83.75, 83.89, 83.105, 83.106, 83.114, 85.6, 85.33, 85.36, 85.40, 86.5, 86.8, 86.43

Pr

Vol. 2 77.65 **Vol. 3** 77.34, 77.35, 81.15, 84.14 **Vol. 4** 81.57, 83.77

Pt

Vol. 2 69.43, 71.29, 71.30, 71.31, 71.38, 72.1, 72.3, 72.5, 72.38, 72.83, 73.17, 75.17, 75.18, 75.71, 76.80, 77.4, 77.36, 77.37, 77.58, 78.33, 80.19, 81.18, 81.19, 82.7, 83.26, 83.33, 83.60, 83.67, 83.68, 83.124, 83.153, 83.164, 83.191, 85.27, 85.67, 86.4, 86.10, 86.11, 86.13, 86.16, 86.24, 86.31, 86.35, 86.46, 86.50, 86.71, 86.72, 86.80, 86.89, 86.92, 86.115 **Vol. 3** 71.4, 71.37, 71.46, 71.48, 72.1, 72.3, 72.4, 72.5, 72.6, 72.19, 72.22, 72.30, 72.49, 72.51, 75.21, 75.35, 76.8, 82.2, 82.3, 82.10, 82.11, 83.11, 83.12, 83.33, 83.50, 83.52, 85.49, 85.53, 86.2, 86.10, 86.13, 86.28, 86.29, 86.64, 86.70, 86.72, 86.94, 86.95, 86.98 **Vol. 4** 71.4, 71.7, 71.13, 71.16, 71.17, 71.19, 71.21, 71.23, 71.25, 71.26, 71.33, 71.49, 71.59, 71.68, 71.72, 71.78, 71.81, 71.84, 71.90, 72.1, 72.2, 72.3, 72.8, 72.11, 72.22, 72.23, 72.24, 72.26, 72.27, 72.30, 75.23, 75.33, 75.34, 75.35, 83.29, 83.63, 85.7, 85.27, 85.34, 85.35, 86.22, 86.31, 86.37, 86.42

Re

Vol. 2 62.41, 71.47, 73.33, 73.45, 75.40, 76.25, 77.5, 77.68, 81.88, 81.89, 83.1, 83.195, 83.200, 85.54, 85.97, 86.12, 86.27, 86.58, 86.59, 86.66, 86.118 **Vol. 3** 71.10, 76.13, 81.7, 81.31, 81.37, 81.44, 81.53, 86.47, 86.49, 86.55, 86.75 **Vol. 4** 73.9, 73.13, 80.5, 80.13, 80.14, 83.110, 85.11, 85.21, 85.24

Rh

Vol. 2 71.7, 71.55, 72.20, 72.23, 72.24, 72.40, 72.43, 72.45, 72.78, 73.7, 73.42, 73.79, 73.80, 73.81, 73.92, 74.19, 75.37, 75.51, 77.2, 77.3, 77.25, 81.26, 81.56, 81.57, 81.58, 83.178, 85.65, 86.57, 86.78, 86.84, 86.108, 86.114, 86.120 **Vol. 3** 71.12, 71.34, 71.53, 71.54, 71.63, 71.68, 72.13, 72.16, 72.35, 72.50, 74.8, 74.9, 74.10, 74.16, 75.47, 77.3, 83.28, 83.128, 83.132, 86.19, 86.21, 86.25, 86.40, 86.68, 86.87, 86.90, 86.91, 86.97 **Vol. 4** 71.6, 71.24, 71.83, 71.87, 71.92, 71.93, 72.18, 72.19, 73.16, 73.18, 73.21, 74.5, 81.20, 81.28, 81.61, 83.128, 86.14, 86.21, 86.32, 86.33, 86.54, 86.62, 86.65, 86.67

Ru

Vol. 2 71.48, 72.37, 73.28, 73.54, 73.57, 73.78, 75.30, 75.66, 80.32, 80.33, 81.83, 86.63, 86.101, 86.102, 86.106, 86.109, 86.110, 86.116, 86.117 **Vol. 3** 71.39, 71.42, 75.5, 75.32, 75.33, 76.16, 76.45, 81.54, 85.54, 86.18, 86.37 **Vol. 4** 71.27, 71.45, 71.70, 72.16, 72.31, 73.1, 73.4, 74.10, 75.2, 75.3, 76.9, 81.62, 85.1, 85.38, 86.29, 86.60, 86.63

Sc

Vol. 2 81.14

Sm

Vol. 2 73.65 **Vol. 3** 75.42 **Vol. 4** 76.33, 81.58

Ta

Vol. 2 72.73

Tb

Vol. 2 76.62, 83.174

Th

Vol. 2 77.49, 81.28, 81.51 **Vol. 3** 80.15, 83.145, 84.25 **Vol. 4** 75.14

Ti

Vol. 2 72.44, 73.18, 73.111, 75.72, 77.48, 78.61, 79.29, 81.46, 83.49, 83.196, 84.11, 84.26, 84.27, 84.28, 84.54, 84.58, 84.61 **Vol. 3** 71.21, 73.28, 73.36, 73.42, 75.14, 75.19, 77.21, 79.19, 81.38, 84.22 **Vol. 4** 71.38, 71.47, 71.63, 71.64, 73.8, 73.11, 73.23, 77.1, 78.2, 79.8, 80.15, 83.3, 83.9, 83.19, 83.76, 84.15, 84.18

U

Vol. 2 73.62, 77.10, 79.30, 80.31, 81.40, 81.42, 81.60, 83.193, 83.194, 84.35, 84.45
Vol. 3 77.20, 78.7, 80.19, 80.20, 81.55
Vol. 4 74.1, 75.15, 75.25, 75.28, 78.11, 83.65, 85.26, 85.32

V

Vol. 1 3.40 **Vol. 2** 73.12, 75.13, 75.24, 77.6, 77.7, 77.26, 77.27, 77.42, 77.43, 81.39, 81.49, 81.54, 81.55, 83.47, 83.202, 84.22, 85.98 **Vol. 3** 60.34, 73.24, 73.29, 75.46, 76.76, 78.5, 79.23, 81.52, 83.1, 83.18, 85.22, 85.48 **Vol. 4** 76.34, 76.36, 77.10, 77.15, 81.9, 81.10, 83.104, 85.8, 85.31

W

Vol. 2 73.53, 73.109, 83.123 **Vol. 3** 73.47, 73.48, 73.49, 83.5, 83.66, 85.6, 85.19, 86.8 **Vol. 4** 64.27, 71.3, 72.32, 73.7, 73.24, 73.31, 83.8, 83.109, 83.132, 84.2, 86.55

Y

Vol. 2 77.34, 77.39, 77.61, 81.20
Vol. 3 84.28 **Vol. 4** 81.4, 81.5, 81.6, 81.7, 81.8

Yb

Vol. 2 77.28, 77.29, 77.30 **Vol. 3** 77.11, 77.22, 77.23, 81.28, 81.50 **Vol. 4** 81.52, 81.53

Zn

Vol. 1 11.45 **Vol. 2** 71.6, 71.17, 71.50, 71.54, 76.78, 77.8, 77.9, 77.57, 77.63, 78.12, 78.24, 78.25, 79.10, 79.14, 79.15, 79.21, 79.27, 80.4, 80.20, 80.21, 80.22, 80.24, 81.29, 81.30, 81.35, 81.68, 81.74, 82.11, 82.32, 82.49, 82.50, 83.9, 83.10, 83.15, 83.19, 83.41, 83.78, 83.92, 83.135, 83.139, 83.141, 83.160, 83.162, 83.166, 83.190, 84.9, 84.39, 84.50, 84.51, 85.3, 85.61 **Vol. 3** 71.45, 76.25, 76.26, 76.47, 76.58, 76.72, 76.73, 76.74, 77.24, 78.20, 80.9, 81.26, 82.6, 82.8, 83.2, 83.8, 83.14, 83.15, 83.41, 83.44, 83.68, 83.77, 83.116, 83.117, 83.126, 83.151, 85.32, 85.33, 85.34, 85.42, 86.88 **Vol. 4** 71.18, 76.22, 76.26, 79.11, 80.19, 81.2, 81.16, 81.35, 81.38, 82.12, 83.17, 83.18, 83.38, 83.49, 83.51, 83.55, 83.62, 84.4, 84.12, 85.18, 85.19, 85.25, 85.29, 85.30, 86.10

Zr

Vol. 2 77.50, 81.52, 81.70 **Vol. 3** 71.57, 73.21, 73.22, 73.34, 73.38, 78.25, 84.16 **Vol. 4** 73.2, 73.5, 73.30, 77.8

AUTHOR INDEX

Aarskog, J. **3** 8.12, 8.13
Aasen, A. **4** 33.21
Aava, U. **3** 67.20, 80.22
Abdul'minev, I.K. **4** 81.45, 81.57
Abdullaev, G.K. **2** 84.13
Abe, J. **1** 50.31 **3** 50.20, 59.1
Abel, E.W. **2** 85.25
Abel, H. **1** 54.24
Abel, J.E. **4** 15.3
Abidi, S. **4** 35.12
Ablov, A.B. **2** 83.135
Ablov, A.V. **2** 66.11, 81.32, 81.76, 81.82, 81.87, 83.91
Ablov, A.W. **2** 60.87
Abola, J. **4** 47.8, 47.12, 47.15
Abowitz, G. **1** 32.5
Abraham, D.J. **1** 32.22 **3** 32.22, 47.10, 58.23, 59.26 **4** 37.4, 44.9, 44.18, 44.25, 50.2
Abraham, N.A. **3** 51.7
Abrahams, S.C. **1** 11.29, 11.50, 11.55, 11.65, 11.67, 13.54, 24.13 **2** 70.25, 73.30
Abrahamsson, S. **1** 1.131, 1.146, 1.147, 5.32, 16.34, 41.28, 46.13, 50.13, 53.32, 59.62, 59.69 **3** 5.4 **4** 39.1, 39.16
Abramson, H.N. **1** 51.1
Acheson, R.M. **4** 35.34
Acland, C.B. **4** 82.1, 82.2, 82.8, 83.15
Acton, N. **2** 75.35 **3** 75.34
Adam, G. **1** 58.104, 58.105 **4** 51.48, 51.56
Adam, V.C. **3** 72.3, 86.88
Adams, M.J. **3** 82.4
Adamson, G.W. **2** 67.5, 71.50, 71.54 **3** 63.7, 64.31 **4** 71.4
Adamson, M.G. **2** 65.39
Adcock, S. **3** 27.4
Addamiano, A. **1** 9.28
Adeoye, S.A. **1** 59.66
Adley, A.D. **4** 63.4, 69.16, 69.17
Adman, E. **1** 16.30, 20.9 **2** 60.119 **3** 20.6, 44.31 **4** 44.16
Admiraal, L.J. **1** 58.11 **2** 83.130 **3** 58.5
Adolf, W. **3** 54.8, 54.15 **4** 54.12
Agawa, T. **3** 41.23 **4** 35.39
Agre, V.M. **2** 80.11, 80.36 **4** 69.28
Agrell, I. **3** 83.39, 83.40, 83.41
Ahmed, F.R. **1** 1.14, 16.61, 24.14, 29.14, 33.76, 33.77, 33.78, 35.46, 36.22, 38.57, 58.19, 58.29, 58.41, 58.53, 58.60 **2** 64.32, 64.64, 64.81 **3** 26.4, 64.56, 64.57 **4** 33.35, 33.36, 33.37, 51.45, 64.8
Ahmed, N.A. **4** 33.32
Ahmed, N.A.K. **1** 2.29
Aihara, A. **3** 21.9
Aime, S. **3** 25.4
Ainalem, I.-B. **2** 83.71, 83.72
Akhmed, N.A. **4** 17.8, 69.34
Akhtar, F. **2** 83.104 **4** 83.82
Akhtar, M.N. **2** 81.51 **3** 78.7
Akimoto, H. **1** 24.33
Akiyama, T. **3** 56.3
Akopyan, Z.A. **1** 15.1, 24.11, 24.12 **2** 83.172 **3** 40.2, 64.48 **4** 84.17
Al-Shamma, A.A. **4** 54.10
Albano, V. **1** 40.25, 40.26 **2** 60.149, 86.72, 86.74, 86.100, 86.118, 86.119
Albano, V.G. **1** 3.41 **2** 71.33 **3** 3.21, 3.54, 86.94, 86.99 **4** 86.57, 86.58
Albert, R. **3** 39.1
Albertsson, J. **2** 81.41, 81.69, 81.71, 81.72, 82.52 **3** 81.27, 81.28, 81.50 **4** 81.51, 81.52, 81.53
Albinati, A. **4** 35.21, 71.60, 76.26
Albrecht, G. **2** 72.12, 83.202
Alcala, R. **4** 13.5
Alcock, N.W. **1** 39.65 **2** 69.41, 69.47, 73.45, 81.40, 83.208, 86.102 **4** 38.23, 69.30
Alden, R.A. **1** 31.39, 52.24, 52.30, 59.65
Alderman, P.R.H. **2** 72.5, 80.5
Aleby, S. **1** 1.150, 1.152, 1.153 **4** 39.45
Aleksandrov, G.G. **2** 72.73, 72.84 **3** 73.29, 74.8, 74.9, 74.10, 74.16, 75.46 **4** 69.34, 71.38, 73.21, 74.5
Alexander, E. **1** 31.31
Alexander, L.E. **1** 1.97, 7.25, 34.9, 43.14 **2** 63.11, 63.13, 78.12 **3** 34.2
Alexander, R. **2** 75.2
Ali, S.M. **1** 2.24
Alison, J.M.C. **4** 85.36
Allcock, H.R. **4** 46.7, 64.40
Alleaume, M. **1** 13.26, 16.14, 16.18, 16.19, 16.22 **3** 8.16, 50.26 **4** 13.3, 16.3, 16.10, 50.14
Allegra, G. **1** 23.9, 29.17, 29.18, 29.19, 38.27, 38.28, 38.41 **2** 61.17, 61.18, 61.19, 68.1, 68.20, 71.45, 72.20, 72.39, 72.43, 72.45, 72.67, 73.18, 74.5, 74.24, 74.25, 75.13, 75.48 **3** 1.31, 21.6, 21.8, 21.10, 39.1, 62.16, 74.3, 74.4 **4** 71.63, 72.16
Allen, C.W. **3** 64.25
Allen, D.W. **2** 86.99 **3** 65.5, 86.80
Allen, F.H. **1** 53.13 **3** 10.5, 27.2, 27.3, 35.15, 51.49, 52.1, 53.15, 56.10, 73.41, 86.40 **4** 36.27, 52.6, 56.2, 69.2
Allentoff, N. **1** 9.7
Alleyne, A.B. **1** 38.17
Allison, A.J. **3** 54.2

Allison, G.B. **3** 35.18 **4** 81.49
Allman, R. **4** 29.3
Allmann, R. **1** 17.51, 33.82, 34.12, 41.44 **2** 60.105, 60.106, 64.54, 64.59 **4** 75.4
Almlof, J. **4** 60.5
Almodovar, I. **2** 81.3, 81.4, 81.8, 81.10
Alsop, P.A. **3** 85.46
Altenburg, H. **4** 36.36
Altenkirk, B. **3** 58.13, 58.52
Altona, C. **1** 38.8, 38.9, 38.10, 38.11, 38.12, 38.13, 42.3, 42.5 **4** 21.20, 27.2, 31.15
Alver, E. **2** 64.31 **3** 64.8
Alvino, G. **1** 1.133
Amakasu, O. **1** 58.7, 58.12
Amano, Y. **2** 61.21, 83.187
Amanov, A.Z. **2** 80.19
Ambady, G. **4** 48.51, 58.25
Ambady, G.K. **1** 2.60, 2.66, 2.68
Ambats, I. **1** 40.32
Amendola, A. **1** 39.18
Amin, A. **4** 17.8
Amirthalingam, V. **1** 2.35, 41.9 **2** 67.8 **3** 41.1 **4** 53.11
Amisimov, K.N. **2** 72.72
Amit, A. **1** 31.6
Amit, A.G. **1** 9.27
Amma, B.L. **4** 75.19
Amma, E.L. **2** 68.3, 68.10, 74.2, 74.3, 79.1, 79.3, 79.19, 79.28, 83.16, 85.19, 85.56, 86.4, 86.10, 86.11, 86.51 **3** 69.12, 74.1, 74.6, 74.7, 74.11, 74.12, 74.13, 74.14, 75.2, 75.8, 79.5, 79.6, 79.7, 79.8, 79.9, 79.20, 79.22, 79.24, 79.25 **4** 41.2, 69.11, 69.22, 69.27, 74.3, 74.8, 74.9, 74.11, 79.6, 79.9, 85.10, 85.27, 86.24, 86.48
Ammon, H.L. **1** 25.22, 27.14, 35.44, 37.20, 39.90 **3** 11.18, 15.1, 15.2, 20.11, 28.9, 36.27, 39.6, 39.28, 39.32 **4** 8.10, 11.9, 33.26, 33.31, 33.39, 35.19, 35.27, 37.9, 39.6, 39.43
Amzel, L.M. **3** 76.14, 79.26
Anagnostopoulos, A. **2** 83.103
Anan'eva, N.N. **4** 76.35
Ananthakrishnan, N. **1** 48.66
Anca, R. **1** 13.44
Andersen, E.K. **1** 6.5, 6.6, 18.8, 18.9, 18.10, 39.13
Andersen, P. **1** 7.14, 7.15, 12.13 **2** 60.61, 64.16
Anderson, B. **3** 50.31
Anderson, B.F. **3** 54.4
Anderson, D.N. **3** 3.2 **4** 3.4
Anderson, J.S. **3** 83.11
Anderson, O.P. **3** 40.9 **4** 70.13, 70.14
Anderson, W.A. **4** 86.34
Ando, K. **3** 50.18
Ando, O. **1** 48.52
Ando, T. **1** 17.40, 17.42
Andre, D. **4** 19.3, 19.4
Andre, J.M. **4** 38.15
Andreetti, G.D. **1** 16.31, 35.8, 39.41, 41.21 **2** 79.14, 79.15, 79.27 **3** 8.5, 61.8, 76.47, 81.45, 81.46, 81.47 **4** 9.5, 9.6, 26.5, 35.2, 39.11, 76.12, 83.7, 83.12, 83.33, 83.125, 83.129
Andresen, O. **1** 9.14
Andrew, J.E. **3** 78.28, 83.84
Andrianov, V.G. **2** 62.12, 62.14, 71.34, 72.62, 72.63, 73.119, 73.120 **3** 71.56 **4** 16.1, 72.5, 72.6
Andrianov, V.I. **1** 32.17 **3** 27.12, 33.28, 64.30
Aneja, R. **3** 58.39
Angeli, F.d' **1** 33.75
Anisimov, K.N. **2** 71.47, 73.68 **3** 71.56
Ansell, G.B. **2** 60.60, 64.21, 83.27, 85.33 **3** 21.1, 32.27, 33.5, 83.16 **4** 32.6, 32.7, 38.17, 64.12, 83.114
Anthonsen, T. **4** 56.1
Antosz, F.J. **4** 50.18
Antsyshkina, A.S. **1** 8.6 **2** 76.6, 76.27, 81.5, 81.14, 81.26, 81.28, 83.138, 83.175, 83.186 **3** 76.15, 76.21 **4** 76.7, 76.27
Antti, B.-M. **4** 83.93
Antti, C.-J. **4** 83.16
Anzenhofer, K. **2** 62.18, 81.79 **3** 31.31, 72.50, 81.34 **4** 77.4
Aoki, K. **3** 48.30
Aoki, T. **3** 72.20
Aota, K. **4** 53.1
Apgar, P.A. **4** 31.2
Arai, G. **1** 58.47
Archer, E.M. **1** 19.13, 19.18
Archer, R.D. **4** 83.132
Arcovito, G. **3** 9.10
Argay, G. **4** 9.8
Argos, P. **3** 33.47
Arigone, D. **1** 54.7
Arima, K. **3** 50.18
Arita, M. **4** 54.3
Ariyaratne, J.K.P. **3** 71.5, 71.6
Armstrong, R.A. **2** 81.27, 81.45, 81.48, 81.66, 81.81
Armstrong, V.S. **2** 75.22
Arnott, S. **1** 56.2
Arora, S. **3** 34.4
Arora, S.K. **1** 13.8 **3** 47.12, 50.16 **4** 11.3, 41.14, 50.1
Arrington, D.E. **3** 11.8
Arrington, W.S. **2** 84.45
Artaud, M. **3** 43.7
Arthur, P. **1** 7.30
Arutnyan, E.G. **2** 81.28
Arutunian, E.G. **1** 8.6
Arutyunyan, A.V. **4** 86.36
Asahara, T. **3** 1.23
Asahi, K. **1** 50.4
Asahi, Y. **3** 44.42
Asai, M. **1** 48.73 **3** 50.13 **4** 50.6
Ase, K. **3** 70.4 **4** 70.6, 70.7, 70.8, 70.11, 70.12, 70.15
Asher, J.D.M. **1** 53.6, 53.25, 58.48, 59.70
Ashida, T. **1** 3.13, 3.42, 11.37, 13.3, 20.30, 22.9, 27.13, 31.27, 33.63, 33.65, 33.69, 38.42, 48.18, 48.52, 48.56, 48.63, 48.84, 48.86, 48.88, 48.95, 56.12, 58.42 **2** 63.21, 63.22, 78.1, 78.2, 82.13, 82.29 **3** 33.36, 59.15 **4** 44.14, 44.15, 48.33, 48.35, 48.57, 49.13, 49.18, 54.7, 67.9
Ashworth, R.S. **2** 85.9, 85.10, 85.11, 85.36
Askew, W.B. **2** 71.40

Aslanov, L.A. **2** 77.28, 77.33, 81.24 **3** 77.11, 77.12, 77.22, 77.29, 77.30, 82.9 **4** 77.11, 77.12, 81.45, 81.57
Aslanov, L.P. **3** 77.28
Asperger, R.G. **3** 76.85
Astakhova, I.S. **2** 71.47 **3** 62.2, 71.10, 71.49, 75.38
Atoji, M. **1** 3.1, 5.7 **2** 76.42
Atovmyan, L.O. **1** 10.2, 10.3, 13.43, 32.17 **2** 60.140, 74.16, 81.1, 83.78 **3** 12.1, 13.7, 69.17, 78.22, 83.17 **4** 60.40, 60.41, 81.10, 84.10
Atwood, J.L. **2** 67.12, 68.18 **3** 67.8, 67.13, 68.11, 75.42 **4** 67.18, 68.5
Aubagnac, J.-L. **4** 32.18
Aubry, A. **4** 48.23, 48.43
Aurivillius, B. **1** 5.13 **4** 39.22
Auvray, P. **4** 7.12
Avdeef, A. **3** 80.18, 85.38 **4** 75.14, 75.15
Avitabile, G. **1** 1.69, 1.110 **3** 1.32, 22.4, 83.86 **4** 1.9, 15.6, 16.6, 50.17, 72.12
Avoyan, R.L. **1** 28.8, 28.9, 28.10, 29.28 **2** 73.87
Axelrod, M. **1** 52.26
Ayer, W.A. **3** 58.13, 58.52 **4** 58.14
Ayerst, E.M. **1** 1.25
Aynehchi, Y. **3** 53.25
Ayres, F.D. **2** 83.40
Ayyar, R.R. **1** 48.20, 48.85
Azarnia, N. **3** 45.2 **4** 45.9, 45.12
Azzoni, G. **2** 84.2

Baba, S. **4** 71.58, 71.80
Babeau, A. **3** 3.50
Babeshkina, G.K. **3** 76.13
Babushkina, T.A. **1** 16.3, 19.1, 19.2, 19.3 **2** 71.1 **3** 16.1
Bacon, G.E. **1** 2.86, 2.87, 11.51, 17.15, 19.19, 19.20
Baddley, W.H. **4** 71.78, 71.90
Badley, E.M. **2** 71.38
Baenziger, N.C. **1** 6.2, 6.3, 6.4, 12.9, 12.20 **2** 60.93, 60.94, 72.4, 72.47, 73.17, 75.2, 75.3, 75.4, 75.6, 75.18, 75.77 **3** 11.13, 83.8
Baerlocher, C. **3** 3.14, 3.15
Baert, F. **4** 21.13
Bagga, M.M. **2** 83.214 **3** 71.50, 71.51
Baggio, S. **2** 76.28 **3** 76.14, 79.17, 79.26 **4** 38.6
Baikie, P.E. **2** 71.14, 75.31, 75.46, 83.152, 83.170, 83.209, 83.214
Bailey, A.S. **2** 60.75, 60.76 **3** 35.18, 40.19
Bailey, M. **1** 3.71, 13.34, 13.35, 13.48, 26.5, 26.9, 26.39, 26.40, 36.38
Bailey, M.F. **2** 74.4, 74.18, 75.1 **3** 81.48, 83.121, 83.122, 83.123, 83.124, 83.125, 83.126 **4** 83.83, 83.84
Bailey, N.A. **1** 23.11, 23.15, 38.39 **2** 65.6, 68.5, 68.6, 71.9, 73.35, 75.23, 77.2, 77.3, 83.85, 83.182, 86.18, 86.19 **3** 71.3, 83.75, 83.82, 83.83, 85.7, 85.20 **4** 12.14, 28.11, 35.23, 71.57, 78.13, 83.131
Baillie, A.C. **4** 32.10
Bain, V.A. **2** 63.12
Baird, H.W. **2** 73.9
Baird, W. **2** 73.11
Baird, W.C. **2** 75.50
Bakakin, V.V. **3** 76.39 **4** 76.4, 76.5
Baker Junior, W.A. **3** 78.11 **4** 81.56, 81.58
Baker-Hawkes, M.J. **2** 85.86
Baker, C. **3** 39.2
Baker, D.J. **2** 85.49
Baker, E.N. **2** 78.7, 78.17, 78.18, 78.27 **3** 60.1, 60.16, 60.39
Baker, R.J. **3** 83.72
Baker, R.W. **2** 71.52, 86.120 **3** 83.67, 84.28, 86.97
Balaban, A.T. **1** 33.74
Balasubramanian, R. **2** 76.24
Balch, A.L. **3** 84.20 **4** 71.7, 71.20
Balchin, A.A. **1** 45.24
Baldwin, J.E. **1** 37.18, 37.24, 39.61, 41.38 **3** 20.12
Balkenhol, W.G. **3** 59.32
Ball, R. **3** 69.27
Ball, R.D. **2** 76.26
Bally, R. **2** 65.28, 80.15 **3** 64.15, 65.4 **4** 41.8, 41.25
Balta, E.Ya. **2** 81.28
Balyaeva, K.F. **3** 82.9
Ban, Y. **3** 58.36
Bando, S. **4** 48.57, 67.9
Bandoli, G. **3** 39.39, 40.15, 64.42, 69.33, 72.49, 72.51, 81.55 **4** 39.27, 40.19, 71.78, 78.11, 83.65
Bang, W.B. **1** 23.1
Baraniak, E. **3** 83.20
Barassin, J. **1** 1.116
Barber, P.G. **3** 9.11 **4** 13.20
Barclay, G.A. **2** 76.34, 77.15, 81.23, 81.85, 83.90, 83.113, 83.171, 85.50 **3** 76.30, 83.92
Bardi, R. **1** 4.16, 39.45 **2** 83.56
Bardos, T.J. **1** 47.30 **4** 36.18
Barendregt, F. **3** 3.11
Barieau, R.E. **1** 48.17
Barker, D.L. **1** 44.15
Barlex, D.M. **3** 71.34
Barlow, M. **1** 11.20, 39.17, 47.31
Barnes, W.H. **1** 33.76, 33.77, 33.78, 36.22, 58.13, 58.25, 58.29 **2** 64.32, 64.64, 64.81
Barnes, W.M. **4** 11.8
Barnet, M.T. **3** 82.12, 82.13, 82.14, 82.15
Barnett, A.J. **4** 39.14
Barnett, B. **4** 71.65
Barnett, B.L. **3** 27.9, 30.1, 31.29, 71.13 **4** 63.5
Barnett, W.E. **4** 42.5
Barnhart, D.M. **2** 69.10 **3** 78.21, 83.61
Barnighausen, H. **1** 40.12 **4** 9.1
Barr, R.E. **3** 33.47
Barrans, Y. **1** 3.78, 19.25, 52.4, 52.5 **3** 3.24, 3.33, 3.45, 3.50 **4** 3.21
Barrett, A.N. **1** 46.12
Barrett, E.K. **1** 35.23 **3** 28.13, 36.33 **4** 32.26, 32.29
Barrett, J.H. **1** 34.8
Barrick, J.C. **4** 11.10
Barriola, A. **4** 38.6
Barrow, M.J. **3** 75.3, 75.28, 83.103 **4** 72.15, 75.9, 81.41, 81.54

Bart, J.C.J. 1 19.67, 41.16, 57.1, 57.2 2 63.3, 64.69, 64.74 3 64.31, 64.32, 64.45 4 26.1, 26.2, 26.4, 71.4
Bartindale, G.W.R. 2 71.28
Bartlett, M. 3 77.31, 86.83
Bartlett, M.W. 4 83.120
Bartolozzo, G. 3 64.42
Barton, D.H.R. 3 58.34
Barton, T.J. 3 32.20 4 60.36
Barve, J.V. 4 15.4
Basolo, F. 4 71.72
Bassi, I.W. 2 72.53, 75.48 4 71.63
Basson, S.S. 1 33.67 3 33.44 4 33.33
Bastian, J.M. 4 39.35
Bateman, L.R. 3 71.54
Bates, J.W. 4 60.3
Bates, R.B. 1 39.4 3 36.3, 53.1, 53.2, 53.19, 53.27
Battaglia, A. 3 40.15 4 40.19
Battelle, L. 1 42.6
Batterham, T.J. 3 35.1
Battersby, A.R. 4 52.6, 58.30
Battiste, M.A. 4 38.19
Battisti, A. 4 39.8
Battistini, G.G. 2 79.18, 83.8, 83.9
Bau, R. 1 3.82 2 65.41 4 27.3, 72.32
Baudour, J. 1 19.30, 19.31
Baudour, J.L. 3 32.30 4 19.12
Bauer, H. 4 32.22
Baughman, R.H. 2 60.59
Baxter, R.M. 3 39.16
Baxter, W.J. 4 59.25, 68.8
Bayer, W. 1 3.72
Beagley, B. 1 2.18, 2.33, 2.51
Beal, J.K. 4 54.10
Beale, J.P. 3 45.37, 45.41, 45.42, 86.13, 86.20 4 3.23, 3.25, 3.26, 86.5
Beall, H. 2 62.11
Beall, R. 3 3.49, 39.4
Bear, C.A. 3 59.28, 60.47, 78.2, 78.4, 78.27
Beard, D.W. 3 49.1, 49.7
Beard, L.N. 1 32.9
Beasley, G.H. 4 56.3
Beattie, I.R. 2 69.17
Beauchamp, A.L. 2 66.20, 66.23 4 83.124
Beaute, C. 3 1.7
Bechtel, F. 4 27.10
Becka, L.N. 1 37.4, 37.5, 37.6, 37.7, 37.8, 37.9 2 76.28 3 76.14, 79.26
Becker, J.W. 3 48.36
Becker, K.A. 1 20.31 2 76.7
Becker, M.R. 1 31.3
Becker, P. 4 17.3
Becker, T.L. 3 38.16
Beckett, R. 2 85.63 3 85.24 4 83.88
Beddoes, R.L. 3 75.11, 75.13 4 78.2
Bednowitz, A. 1 31.4
Bednowitz, A.L. 1 1.48, 1.90, 30.27, 40.11 3 13.4, 13.5, 20.3 4 69.35
Beer, R.J.S. 3 39.40, 42.3 4 39.14
Beevers, C.A. 1 1.59, 21.13, 30.1, 45.30, 45.34, 45.70, 58.54 4 45.20, 45.21
Begley, M.J. 3 59.23 4 59.18
Beineke, T.A. 1 27.17, 33.2 2 64.15 3 63.16, 83.107
Beintema, J. 1 24.27
Beisler, J.A. 3 51.48 4 31.25
Bekoe, A. 1 45.2
Bekoe, D.A. 1 7.21, 59.66 2 67.6, 67.7
Bel'skaya, N.P. 3 68.9
Belford, R. 3 71.39
Belford, R.L. 2 77.7, 77.40, 77.41, 77.42, 77.43, 77.55, 81.49, 81.54 3 77.24, 81.18, 81.52
Belinskij, I. 3 76.33
Belitskus, D. 1 1.115
Bell, J.D. 1 41.33 2 82.23, 82.36, 82.38 4 76.16
Bellingen, I.van 3 60.17, 60.18
Bello, C.di 1 34.4, 35.9
Bellobono, I.R. 3 31.25
Bellon, P. 2 60.149, 86.118 4 86.58
Bellon, P.L. 1 3.41, 40.25, 40.26 2 71.33, 86.72, 86.74, 86.100, 86.119 3 3.21, 3.54, 86.22, 86.94, 86.99 4 86.57, 86.71
Belluco, U. 3 72.49 4 71.78
Belov, N.V. 2 66.11, 66.12
Belyaeva, K.F. 2 81.43
Benci, P. 3 37.7 4 37.2, 37.3
Bencze, W.L. 4 24.16
Benedetti, E. 1 1.93 3 1.31, 20.4, 21.5, 21.6, 21.7, 21.8, 21.10, 33.16, 33.19, 39.1, 62.16, 72.19 4 1.8, 48.7, 71.25, 71.26, 72.8
Bennett, A. 1 30.26
Bennett, I. 3 48.25
Bennett, J. 3 86.1
Bennett, M.A. 2 86.46 3 72.52 4 71.43, 71.46, 71.49, 71.56, 75.35, 86.21
Bennett, M.J. 2 66.20, 66.23, 73.26, 73.39, 75.39, 75.66, 77.39, 77.63, 81.83, 81.88, 84.24, 84.29, 85.29, 85.54, 86.28, 86.97 3 69.27 4 21.22, 28.12, 63.6, 86.25, 86.65
Benno, R.H. 3 69.10
Benschop, H.P. 4 52.5
Bentley, G.A. 2 82.57 3 86.61
Bentrude, W.G. 3 64.17
Beran, G. 3 68.13, 86.42
Berends, E.A. 4 82.10
Bergin, R. 1 59.10, 59.11 3 58.2 4 3.18, 3.19
Bergman, J. 4 39.16
Bergman, J.G. 2 65.44, 81.50
Berkebile, C.A. 1 14.14
Berking, B. 1 21.31 3 51.36 4 36.36, 43.3, 43.5, 43.7, 45.3, 45.21
Berlin, Yu. 4 50.13
Berliner, L.J. 3 12.8
Berman, H. 1 8.8
Berman, H.M. 1 45.44, 45.46, 45.49 3 32.11, 32.23, 45.2, 45.5, 45.47 4 2.22, 26.6, 36.32, 47.20
Bernal, I. 1 39.69 2 80.29, 85.41 3 11.14, 11.15, 80.17, 83.139, 84.20, 85.16, 85.25, 85.44, 85.54 4 33.25, 39.34, 39.41, 73.7, 73.8, 73.11
Bernard, M.-J. 2 83.2
Bernard, M.J. 4 83.4
Bernard, W.H. 3 76.77
Bernauer, K. 3 58.67
Berndt, A.F. 2 73.8
Bernotat-Wulf, H. 3 59.9

Bernstein, J. **4** 12.9, 16.12, 16.13, 31.17, 31.18, 60.37
Bernstein, T. **2** 73.24
Bersted, B.H. **2** 73.106, 81.49
Berta, D.A. **3** 69.1, 79.9
Berthold, H.J. **1** 11.8
Berthou, J. **1** 9.47, 31.12, 43.20, 43.21 **2** 60.28 **3** 9.18, 9.20, 11.20, 11.21, 32.7 **4** 8.8
Berti, G. **4** 21.12
Bertini, I. **4** 83.126
Bertinotti, A. **1** 16.27
Bertinotti, F. **1** 13.30, 33.1
Bertrand, G. **4** 2.9
Bertrand, J.A. **1** 46.11, 53.28, 59.5 **2** 84.69, 86.73 **3** 3.22, 77.14, 77.25, 78.11, 83.65, 84.12 **4** 32.10, 77.20, 78.10, 83.21, 84.14
Berzjak, A. **1** 8.45
Besancon, J. **3** 74.2, 74.5 **4** 74.2, 74.4, 74.6, 74.7
Beschastnov, A.S. **3** 71.56
Betkouski, M. **4** 38.10
Betzner, G.L. **2** 76.25, 77.68
Beurskens, P. **1** 13.13
Beurskens, P.T. **1** 1.34 **2** 61.12, 80.8, 80.9, 80.34 **3** 80.11 **4** 79.10, 80.4, 80.11, 80.12, 80.16, 80.18, 85.3, 85.4
Bevan, C.W.L. **1** 58.73 **3** 56.11
Bevan, J.A. **2** 78.6, 83.157
Beyer, H. **1** 39.1
Beyer, L. **4** 85.23
Bezjak, A. **2** 81.60
Bezman, S.A. **4** 71.94, 75.26, 75.32, 86.70
Bezzi, S. **1** 19.14, 39.45 **2** 83.56 **4** 19.2
Bhalla, V.K. **3** 58.39
Bhandary, K.K. **4** 84.6
Bhat, H.B. **1** 59.68
Bhattacharyya, S.C. **1** 48.19
Biagini Cingi, M. **4** 84.12
Biagini, S. **1** 1.46, 40.9, 40.17, 40.18 **3** 40.6, 76.26, 76.60, 83.64
Bianchi, A. **3** 86.53, 86.84 **4** 76.43, 76.44
Bianco, V.D. **2** 86.74
Bichan, D. **4** 39.46
Bick, I.R.C. **3** 58.6
Bickley, D.G. **4** 77.1
Bideau, J.-P. **1** 43.22, 43.26, 43.30 **3** 3.33, 43.6, 43.7, 43.9 **4** 43.8
Biedl, A. **3** 45.28
Bierlein, T.K. **1** 10.18
Bierrum, A.M. **3** 71.5, 71.6
Bierstedt, R. **1** 44.46
Biffar, B. **4** 33.47
Bigliardi, G. **2** 81.35, 83.10, 83.15
Bigoli, F. **2** 83.4, 83.5 **3** 83.2, 83.3 **4** 79.4, 81.2, 83.50
Bihan, M.S.le **2** 67.3
Bijvoet, J.M. **1** 2.63, 2.65, 2.67, 2.70, 21.5, 48.65, 58.55
Bilton, M.S. **4** 69.3
Binas, H. **2** 60.21
Binnie, W.P. **1** 3.49, 3.50
Biondi, C. **1** 7.17
Birchall, T. **2** 73.122 **3** 83.137
Bird, M.J. **4** 81.44
Bird, P.H. **2** 65.6, 65.41, 68.5, 68.6, 75.40, 75.42, 75.43, 75.69, 75.76 **3** 58.17, 72.39, 75.32, 75.33, 75.43 **4** 63.4, 69.16, 69.17, 77.1
Birkett-Clews, C.J. **1** 19.59
Birnbaum, G.I. **1** 54.15, 58.112, 59.32 **3** 58.49 **4** 44.23, 75.17
Birnbaum, K.B. **1** 54.18 **3** 36.24 **4** 58.20, 58.24, 58.27, 75.18
Birrell, G.B. **3** 3.6
Biryukov, B.P. **2** 69.39, 73.41, 73.49, 73.68, 73.69 **3** 69.31
Bissell, E.C. **1** 3.69 **3** 49.1, 49.7 **4** 46.7
Biswas, A.B. **1** 2.52, 2.69, 2.72, 2.85, 2.89, 48.30 **2** 82.19
Biswas, S.G. **1** 39.58
Bjamer, K. **1** 45.59, 54.20, 54.22
Bjorvatten, T. **1** 7.8, 7.9 **2** 60.2, 60.3, 60.4, 60.53, 60.54 **3** 60.2
Bkouche-Waksman, I. **2** 84.12
Blaauw, H.J.A. **2** 80.34
Black, K.T. **4** 63.13
Black, M. **2** 86.24
Blackmer, G.L. **4** 76.16
Blackmore, T. **4** 71.70
Blackmore, W.R. **1** 11.65, 11.67 **2** 70.25
Blackstone, M. **2** 77.59
Blake, A.B. **2** 80.23 **3** 78.28, 83.84
Blakeslee, A.E. **2** 84.23
Bland, W.J. **2** 86.92
Blank, G.E. **4** 46.6, 59.5
Blayden, H.E. **3** 69.6
Blaylock, B.T. **4** 16.14, 24.11
Bleidelis, Y.Y. **3** 69.17
Blessing, R.H. **4** 32.2, 48.25
Block, S. **1** 39.42, 39.43 **2** 62.35, 83.92 **3** 19.4
Blom, E.A. **2** 69.24
Bloom, A. **4** 36.14, 49.22
Bloomer, A.C. **3** 33.22
Bloor, J.H. **2** 60.93, 60.94
Blot, J. **4** 19.8, 19.9
Blount, J.F. **1** 3.46, 41.33, 52.29 **2** 72.61, 82.24, 82.25, 83.54 **3** 9.19, 35.17, 35.23, 36.31, 64.67, 65.12, 82.17 **4** 50.15, 58.3, 58.6
Blum, J.J. **1** 32.8
Blumburg, H. **3** 35.24
Blundell, T.L. **2** 68.22, 69.48, 86.40, 86.41, 86.45 **3** 86.32 **4** 86.56
Bo, I. **4** 9.4
Boccalon, G. **1** 33.75
Bocelli, G. **4** 26.5, 83.125
Bock, C.M. **1** 20.5 **3** 1.16, 3.5, 20.1, 62.11
Bodner, R.L. **4** 83.58
Bodor, G. **1** 1.90
Boelema, E. **1** 39.26
Boer, F.P. **1** 3.32, 11.9, 12.3, 42.1, 50.32 **2** 62.7, 62.22, 63.16, 63.26 **3** 9.1, 20.8, 20.10, 32.26, 38.18, 42.1, 63.2, 63.3, 63.10, 63.11, 69.35, 69.37, 69.38 **4** 27.1, 31.12, 32.12, 33.3, 33.18, 38.13, 38.14, 46.5, 69.18, 69.19, 75.20, 75.29
Boer, J.J.de **3** 31.31, 72.22, 72.35, 72.50, 81.34 **4** 75.33

Boer, J.L.de **1** 16.42 **2** 60.122, 86.78 **4** 16.8, 16.9
Boeyens, J.C.A. **1** 8.16, 30.4 **2** 60.9, 60.10, 60.11, 60.40, 60.41, 60.98, 60.135, 60.154, 68.7, 68.8 **3** 12.7, 60.6, 67.17, 77.34, 77.35 **4** 77.17, 77.19, 77.21, 77.22
Bois d'Enghien-Peteau, M. **2** 60.109
Bois d'Enghien, H. **1** 5.27
Bois, C. **1** 17.24 **3** 17.3 **4** 17.5
Bok, L.D.C. **1** 33.67 **3** 33.44 **4** 33.33
Bokii, G.B. **2** 72.1, 72.3, 81.1 **4** 72.1
Bokii, N.G. **1** 9.22 **2** 69.4, 69.5, 69.18, 69.45 **3** 60.9, 64.30, 73.22, 73.38 **4** 69.29, 83.128, 86.36
Bokiy, N.K. **2** 63.28
Boldrini, P. **1** 8.18 **2** 79.2, 79.20, 83.8, 83.9 **3** 79.6, 79.7, 79.9 **4** 79.6, 85.10
Bolhuis, F.van **3** 39.17
Bolton, N. **1** 43.1
Bolton, W. **1** 27.4, 30.30, 30.31, 43.8, 43.11, 43.13
Bombieri, G. **2** 69.21, 73.19, 77.10, 78.28, 78.35 **3** 71.26, 72.51, 80.19, 80.20, 81.55, 83.45, 86.19, 86.25 **4** 71.23, 71.41, 85.26, 85.32, 86.12
Bommel, A.J.van **1** 2.65, 2.70
Bonamartini, A.C. **4** 41.21, 41.22, 83.64
Bonamico, M. **1** 3.70, 33.70, 33.72 **2** 61.13, 61.14, 68.11, 68.16, 80.13, 80.17, 80.20, 84.42, 84.60, 85.72, 85.85, 85.93, 85.96 **3** 9.10, 68.20, 80.7, 80.8, 80.9, 83.114, 85.48 **4** 79.11, 85.17, 85.18, 85.19, 85.31
Bonati, F. **2** 77.2, 77.3 **3** 86.98
Bonds Junior, W.D. **4** 83.132, 86.51
Bonnemere, C. **1** 59.55
Bonnet, J.-J. **3** 83.46, 83.47, 83.48
Bonnet, J.J. **4** 32.8
Bonnett, R. **3** 49.13
Bonnettt, R. **4** 32.17
Bonsma, G.F. **1** 58.28
Boonstra, E.G. **1** 15.25 **3** 36.32 **4** 32.14
Bootsma, G.A. **1** 1.60, 1.62, 1.63
Bordner, J. **3** 9.8, 24.6, 27.11, 28.12, 30.5, 31.32, 38.17 **4** 22.10, 24.5, 27.16, 31.31, 37.15, 51.20, 64.32
Bordwell, F.G. **3** 11.26
Borgen, B. **1** 7.10
Borisov, S.V. **3** 76.39 **4** 76.4, 76.5
Borkhoven, C. **1** 58.55
Born, L. **2** 64.34
Borovyak, T.E. **4** 73.10
Bortin, O. **3** 76.9, 76.10, 76.61
Boschi, T. **3** 71.26
Bosman, W.P.J.H. **4** 80.16, 85.3
Bosnich, B. **2** 83.86
Bottcher, B. **3** 63.8, 63.13
Boulton, A.J. **4** 32.22
Bour, J.J. **4** 79.10
Boutin, H.P. **3** 34.1
Bovio, B. **1** 36.7, 36.8, 36.29 **4** 32.27, 36.16, 36.25, 40.16
Bowen, B.J. **3** 31.2, 83.23
Bowler, S.J. **3** 83.75
Bowman, K. **2** 80.31 **4** 83.72
Bowman, K.S. **4** 86.69
Boyce, C.B. **1** 58.100
Boyd, D.R. **4** 40.14
Boyko, E.R. **1** 29.2 **2** 78.29
Boyum, K. **4** 70.6, 70.7
Bracher, B.H. **1** 1.39, 1.40 **3** 10.1
Bradford, C.W. **4** 71.71, 71.73, 71.89
Bradley, D.C. **2** 63.24, 78.61, 83.123, 84.65, 85.4, 85.23 **3** 78.25, 83.102 **4** 80.15, 83.61, 83.85, 83.130, 83.136
Bradsher, C.K. **4** 37.10
Brahde, L.B. **1** 39.38
Braibanti, A. **1** 8.28, 9.2 **2** 79.5, 81.35, 83.4, 83.5, 83.7, 83.10, 83.14, 83.15 **3** 83.2, 83.3, 83.4 **4** 8.4, 67.1, 67.2, 79.4, 81.2, 81.16, 83.50
Brain, E.G. **4** 24.13, 24.14, 24.15
Branchi, G. **2** 69.11, 85.3
Branden, C. **1** 2.59 **2** 83.166, 85.20
Branden, C.-I. **2** 66.2, 66.6, 84.62, 84.63, 85.17, 85.21 **3** 33.43
Brandl, F. **1** 28.24, 28.25 **3** 54.8, 54.15 **4** 54.8, 54.16, 71.92
Brandon, J.K. **1** 11.5
Brasen, S. **3** 31.16
Brassy, C. **4** 60.26
Bratek-Wiewiorowska, M.D. **1** 59.32
Braterman, P.S. **3** 73.47, 73.48, 73.49
Bratholdt, J.S. **4** 39.25
Brathovde, J.R. **1** 1.127 **2** 77.24
Bratton, W.K. **2** 81.88 **3** 81.37
Brauer, D.J. **3** 68.1, 68.21
Braun, P.B. **3** 21.20, 31.8, 51.8, 51.10, 51.17, 51.18, 51.21, 51.23, 51.24, 51.25
Braun, R.L. **2** 78.43, 78.47, 78.48
Bravic, G. **1** 43.26, 59.39 **3** 25.3 **4** 25.3, 27.10, 38.22
Bray, L.S. **3** 72.41
Breazeale, J.D. **2** 78.11
Breece, J.L. **3** 78.11 **4** 78.10
Breen, J.J. **4** 64.14
Bregman, J. **1** 16.51, 16.52, 16.53, 23.16
Brehm, L. **1** 40.22 **3** 1.22, 15.9 **4** 40.3
Bremner, J.B. **3** 58.6 **4** 44.16
Breneman, G.L. **3** 71.2
Brenic, J.V. **3** 3.8
Brennan, T. **2** 80.29 **4** 26.7
Brennan, T.F. **3** 3.39, 60.30, 60.31, 80.17 **4** 47.19, 60.17
Brenner, S. **1** 20.25
Brenner, S.A. **1** 1.74, 35.28
Breton-Lacombe, M. **1** 25.1, 25.19
Breton, M. **1** 25.15
Brettle, R. **1** 33.80
Brewis, S. **3** 56.8
Brice, M.D. **1** 52.17 **3** 71.35, 71.40, 71.55 **4** 71.42
Bridson, J.N. **4** 35.34
Bright, A. **3** 71.12 **4** 71.83
Bright, D. **2** 83.195, 83.200, 83.212 **3** 38.22, 67.11, 67.24, 86.47, 86.66 **4** 72.20, 75.33
Bright, W.M. **3** 11.2
Brill, R. **1** 1.9
Brimacombe, J.S. **1** 45.71 **3** 45.26 **4** 45.16
Briner, R.C. **1** 58.100
Bringeland, R. **1** 11.25
Brink-Shoemaker, C. **1** 18.12, 49.34
Briscoe, O.V. **1** 41.33

Brisse, F. **3** 58.45, 58.60
Britnell, D. **4** 84.2
Britton, D. **1** 7.13, 7.24, 39.49, 40.4 **2** 65.3, 69.6, 69.8, 71.18 **3** 11.4 **4** 7.1, 7.2, 7.3, 7.4, 7.9, 7.10, 40.7, 40.9, 40.10, 40.12, 63.3, 69.6, 69.7
Britts, K. **1** 4.18, 20.17, 20.18, 20.23, 20.25, 32.3, 35.27, 37.15, 39.57, 46.20, 52.27
Broadhurst, M.J. **4** 27.11, 27.12, 35.14
Broadley, J.S. **1** 1.55
Broch, N.C. **2** 68.14
Brocklehurst, K. **3** 1.33
Brockway, L.O. **1** 19.40
Brodherr, N. **1** 51.49 **4** 27.19
Broggi, R. **2** 61.17, 61.19
Brook, A.G. **2** 71.11, 71.12 **4** 63.12
Brook, P.R. **3** 27.8
Brooks, E.H. **3** 69.27
Brooks, J.J. **3** 67.26
Broomhead, J.M. **1** 11.33, 11.34, 44.33, 44.35
Brossi, A. **4** 58.3, 58.6
Brot, F.E. **3** 31.10
Browall, K.W. **4** 85.2, 85.6, 85.7
Brown, B.W. **2** 76.18, 76.32, 78.3
Brown, C.J. **1** 1.47, 2.83, 8.47, 8.48, 9.36, 9.40, 9.43, 13.25, 13.34, 13.35, 16.16, 16.35, 17.12, 17.20, 19.45, 19.57, 24.18, 26.9, 26.39, 26.40, 31.24, 32.38, 34.10, 38.1, 45.33, 45.66, 49.7, 49.10, 49.11
Brown, D. **3** 80.12, 80.13, 80.15 **4** 3.14
Brown, D.B. **3** 83.50
Brown, D.H. **3** 83.62
Brown, D.R. **1** 33.80
Brown, D.S. **1** 3.68 **2** 60.116, 60.133, 60.134, 76.15, 84.47, 84.48, 86.15
Brown, G.F. **3** 85.17
Brown, G.M. **1** 3.8, 13.39, 13.40, 44.26, 45.35, 45.69 **2** 83.198, 83.199 **3** 16.14, 16.15, 21.19, 45.19, 58.21, 58.31, 73.9 **4** 45.21, 73.1, 81.27
Brown, I. **1** 39.50
Brown, I.D. **1** 11.5 **2** 76.25, 77.68, 83.6, 83.180
Brown, J.N. **3** 8.8, 34.6, 64.37 **4** 37.13, 37.14, 51.1, 51.7, 81.59
Brown, K. **4** 59.22
Brown, R. **1** 30.27
Brown, R.N. **1** 1.44
Brown, S.H. **1** 58.63
Brown, T.L. **2** 68.15
Brown, W.A.C. **1** 31.30, 50.17, 56.17 **2** 75.68
Brownlee, G.S. **4** 69.1
Brubaker, G.R. **3** 79.16
Bruce, M.I. **3** 71.39 **4** 71.36, 71.70, 71.94
Bruckner, S. **2** 78.59 **3** 76.81
Brueckner, D.A. **1** 52.13
Brufani, M. **1** 35.3, 35.4, 35.7, 35.12, 35.13, 35.14, 35.15, 50.36, 50.48 **3** 50.28 **4** 1.11, 35.26, 54.19
Bruijnsvoort, A.van **1** 19.63
Bruins, D. **3** 76.76
Brun, G. **4** 64.3
Brun, L. **2** 66.2, 84.26
Brunel, Y. **1** 31.12
Brunn, E. **3** 64.59
Brunton, G.D. **3** 81.4 **4** 81.4, 81.5, 81.6, 81.7, 81.8
Brusentsev, F.A. **2** 72.3
Brushmiller, J.G. **4** 82.10
Brusset, H. **1** 17.27 **2** 84.6, 84.12 **4** 17.3, 17.6, 17.7
Bruun, T. **4** 56.1
Bryan, J.G.H. **3** 45.26
Bryan, R.F. **1** 2.91, 12.15, 12.19, 13.37, 23.2, 44.42, 44.48 **2** 69.14, 69.40, 73.31, 73.110, 73.121, 82.49, 83.62, 85.83, 86.23, 86.60, 86.91 **3** 15.3, 44.36, 44.39, 62.13, 69.22, 73.3, 73.4, 73.12, 73.13, 85.47 **4** 16.14, 24.11, 53.10, 54.4, 58.36, 58.37, 58.40, 59.29, 62.1, 69.21, 73.33, 82.7
Bryant, M.W.R. **1** 11.58, 11.66 **3** 11.19 **4** 11.15
Bryden, J.H. **1** 8.17, 8.29, 8.33, 9.4, 9.6, 10.6, 11.32, 11.35, 32.2, 32.6, 32.16 **3** 32.21 **4** 15.5
Bryson, W.C. **4** 38.24
Bryukhova, E.V. **2** 71.1
Brzechffa, M. **3** 31.42
Bua, E. **1** 19.14
Buchanan, G.L. **1** 20.20, 38.58
Buchardt, O. **1** 40.30
Buchi, G. **1** 53.40 **4** 59.27
Buck, K.W. **1** 42.10
Buckingham, D.A. **2** 76.48, 76.49, 82.54 **3** 82.12, 82.13, 82.14, 82.15 **4** 76.11, 76.17, 76.18
Buckles, R.E. **1** 12.20 **3** 11.13
Bugg, C. **1** 7.26, 33.55
Bugg, C.E. **1** 47.11 **3** 44.2, 44.3, 44.5, 44.20, 47.15, 47.16, 47.19, 47.23 **4** 35.11, 36.31, 44.8, 44.20, 47.14, 64.36, 67.5, 67.6, 67.13
Bukowska-Strzyzewska, M. **2** 81.7, 81.22, 81.31
Bullen, G.J. **2** 62.43, 64.21, 64.46, 64.65, 77.11, 77.62 **3** 62.4, 62.17, 64.51 **4** 33.23, 62.15, 62.16, 64.12, 64.37
Bullitt, J.G. **4** 81.29
Bullough, R.K. **1** 4.17
Buraway, A. **2** 71.16
Burbank, R.D. **1** 5.3 **3** 83.50
Burckhalter, J.H. **1** 51.1
Burger, K. **4** 32.9
Burgi, H.B. **1** 22.11 **3** 16.9, 16.10, 16.11
Burgner II, J.W. **3** 32.11, 32.23
Burke, M.E. **1** 36.5, 36.18
Burke, T.S. **4** 77.6
Burkhardt, L.A. **1** 8.17
Burkholder, P.R. **3** 58.7
Burki, H. **1** 44.5, 51.27
Burlingame, A.L. **3** 53.25
Burlitch, J.M. **4** 73.31
Burnell, R.H. **3** 58.17
Burnett, R.M. **4** 21.5
Burns, D.M. **1** 28.16, 29.11
Burns, J.H. **2** 65.9 **3** 75.42, 77.16, 77.17 **4** 75.28
Burow, N.E. **1** 2.78, 2.79
Burr, A.H. **1** 16.49 **2** 64.65
Bursh, T. **4** 85.6, 85.7
Burzlaff, H. **1** 36.25, 39.63 **3** 39.15, 39.34
Busch, D.H. **4** 85.25

Busetta, B. 1 36.34, 41.41, 43.22, 51.4 3 17.7, 17.9, 36.30, 51.5 4 8.12, 8.13, 13.8, 17.9, 17.10, 19.14, 51.4, 51.5, 51.8, 51.9, 51.12, 51.16, 51.18, 51.21
Busetti, V. 1 1.51, 38.2, 38.3, 39.8, 39.39 2 69.42 3 16.12, 32.12 4 39.4
Bush, M.A. 1 3.57, 3.58, 3.59 2 66.1, 73.47, 86.17 3 38.15, 38.21, 73.1, 73.16, 73.17, 73.46, 73.52 4 10.2, 10.3, 38.26, 67.15, 67.21, 67.25, 73.2, 73.5, 76.41
Bushnell, G.W. 2 86.15
Busing, W.R. 1 8.21 2 83.199 3 8.1
Bussemeier, B. 4 71.65
Butler, I.S. 4 86.41
Butler, K.R. 3 83.142 4 76.32, 81.24
Butler, W.M. 4 71.1, 71.7
Butman, L.A. 3 77.11, 77.29, 77.30
Buzell, A. 1 7.4
Bycroft, B.W. 4 50.8
Bye, E. 3 48.42
Byram, S.K. 3 68.3
Byrn, S.R. 4 13.15, 13.16, 20.11, 20.13
Bywater, R.P. 3 1.33

Cady, H.H. 1 15.19, 16.24, 34.1, 34.2, 40.2
Cady, R.T. 2 83.162
Caglioti, L. 3 36.21 4 71.41
Caira, M.R. 4 77.15
Cairns, M.A. 3 71.39
Calabrese, J.C. 1 13.45 3 1.29, 1.30, 3.38 4 63.1
Calas, M.-R. 1 43.10 4 41.31
Calderon, J.L. 3 73.20 4 73.6, 73.23, 75.21
Calhoun, H.P. 4 64.27
Calleri, M. 1 10.13, 13.53, 40.15, 40.16, 46.17 2 83.25 3 35.22, 59.4
Calleri, M.B. 3 65.11
Calligaris, M. 1 21.29 2 78.59 3 76.81, 78.14, 78.26, 83.150, 85.32, 85.33, 85.34, 85.40 4 71.40, 77.14
Callow, R.K. 1 51.38, 51.39, 51.40
Calvarin, G. 4 73.20
Calvert, L.D. 1 1.35
Calvin, M. 3 49.6 4 49.8, 49.12
Calvo, C. 3 8.2, 8.3, 81.31, 81.44 4 11.10, 58.16, 71.51, 83.110
Calzolari, L.C. 3 79.10
Cam, Y. 3 41.18
Cambie, R.C. 1 54.2 4 59.22
Camellini, M. 2 83.7
Camellini, M.T. 3 83.2 4 8.4, 79.4, 81.2, 81.15, 81.16
Camerman, A. 1 29.5, 30.6, 33.74, 45.61, 45.74, 58.79 2 65.19 3 24.4, 32.35, 44.40, 52.6 4 13.21, 32.20, 32.25, 35.20, 48.50
Camerman, N. 1 44.69, 44.73, 45.61, 47.6, 58.46, 58.79 2 65.1, 65.4, 65.8 3 32.35, 44.40 4 32.20, 32.25, 35.20, 48.50
Cameron, A.F. 1 31.20 3 28.5, 31.18, 31.27, 31.28, 33.35, 51.44, 53.17, 59.16, 83.77, 83.111, 86.58 4 3.6, 3.20, 11.4, 31.24, 41.13, 58.30, 64.34, 83.37, 83.38, 83.41, 83.52, 83.53
Cameron, J.U. 4 65.7
Cameron, T.S. 2 60.73, 60.75 3 40.19, 40.20, 40.21, 64.43, 64.70, 73.47, 73.48, 73.49 4 24.2, 60.15, 60.16, 64.21, 64.28, 64.29, 64.30, 71.15, 73.19
Camp, U.de la 3 1.19, 11.11
Campbell, B.S. 4 64.20
Campbell, J. 3 77.21
Campbell, J.C. 3 45.31
Campbell, R.B. 1 29.12, 30.14, 30.15
Campbell, R.V.M. 4 59.18
Campsteyn, H. 4 51.28, 51.29
Candeloro, S. 4 83.137, 83.138
Candlin, R. 2 82.45 3 82.22
Canepa, F.G. 1 3.74
Cannas, M. 1 1.46, 40.7, 40.8, 40.9, 40.17, 40.18 3 40.6, 76.26, 76.48, 76.60, 83.64 4 76.6
Cannilo, E. 2 83.81
Cannon, J.F. 4 40.14
Cano, F.H. 3 13.13 4 71.11
Cantrell, J.S. 1 59.22 3 38.16
Canty, A.J. 4 71.79, 71.88, 71.91
Capacchi, L. 1 11.1 2 85.35
Capacchi, L.C. 2 79.16, 85.14 4 66.8
Capiomont, A. 3 37.3 4 12.13
Carazzolo, G. 1 38.2, 39.8, 39.9, 39.39
Cardin, D.J. 3 71.37
Carfagna, P.D. 2 86.51
Carfagno, P. 3 79.6, 79.7
Carhart, R.E. 3 21.17
Cariati, F. 2 86.80 4 86.71
Carlisle, C.H. 1 33.10, 45.24, 46.18, 51.29, 56.13, 56.15 3 9.12, 9.15 4 39.9
Carlstrom, D. 1 8.39, 33.54, 37.26, 59.10, 59.11 3 35.16 4 3.19
Carmichael, J.W. 2 77.55 3 83.11 4 83.29
Carnduff, J. 1 30.20
Carney, R.W. 4 33.47
Caron, A. 1 8.5, 10.19, 10.20, 48.59
Carpenter, C. 2 75.4
Carpenter, C.L. 2 72.47, 73.17
Carpenter, D.C. 1 38.30
Carpenter, G.B. 1 7.11 2 79.34, 79.35
Carr, G. 3 83.74
Carrabine, J.A. 3 84.6 4 44.2, 44.3, 83.28, 83.40
Carrai, G. 2 65.16
Carreck, P.W. 4 83.25
Carrell, H.L. 2 62.32 3 2.19, 23.3, 31.30 4 2.20, 2.22, 26.6, 36.26
Carruthers, J.R. 2 81.27, 81.45, 81.48, 81.66, 81.81 3 59.33, 81.22 4 2.11, 33.42, 33.45, 81.12, 81.13, 81.14, 81.23
Carstensen-Oeser, E. 1 28.23 2 60.77 4 35.36
Carta, G. 3 76.48 4 76.6
Carter, D. 2 83.140
Carter, J.C. 2 62.6
Carter, O.L. 2 60.118, 73.16, 73.44, 74.6, 74.7, 74.14
Carter, V.B. 4 7.9, 7.10
Carter, W. 2 73.122
Carthy, A.J. 3 78.5
Carty, A.J. 2 73.122 3 68.13, 83.128, 83.137 4 71.61, 72.29, 83.106, 86.34
Carty, P. 2 86.70

Casagrande, G. 3 74.3
Casagrande, G.T. 3 74.4
Casalone, G. 1 19.50 3 15.10, 19.11, 39.35
Casalone, G.L. 1 5.28, 5.31 2 60.103
Casciato, C.A. 3 2.22 4 2.20, 2.21
Case, J. 3 51.7
Case, J.A.M. 2 72.65, 86.94
Casey, A.T. 3 83.110
Casini, G. 1 35.3 4 35.26
Casinovi, C.G. 3 59.33
Caspi, E. 4 51.55
Cassady, J.M. 1 53.34 3 53.25
Cassidy, F. 4 24.13, 24.14, 24.15
Cassol, A. 3 80.19, 80.20
Castain, W. 1 9.8
Castellano, E. 3 54.14, 60.29
Castellano, E.E. 3 25.6 4 60.9, 60.10, 60.11
Catesby, C.G.C. 2 60.6
Caticha-Ellis, S. 1 13.54
Cattalini, L. 4 71.41, 83.65
Caughey, W.S. 1 49.18, 49.19
Caughlan, C.N. 1 39.93, 46.5, 46.7, 46.9, 46.10, 46.16, 46.24, 53.1, 53.3 2 64.25, 64.42, 69.10, 77.48, 84.22, 84.54, 84.58 3 46.2, 46.7, 53.4, 53.18, 64.17, 76.44, 77.21, 83.61, 83.70, 84.26 4 31.13, 53.1, 64.16, 64.19, 64.25, 64.39, 79.5, 84.15, 86.11
Caughman, H.D. 3 64.44 4 86.7
Caulton, K.G. 2 81.83 4 86.54
Cavalca, L. 1 16.31, 39.41, 41.15 2 79.2, 79.5, 79.14, 79.15, 79.27, 79.33, 83.28, 84.5, 84.8, 85.3, 85.5, 85.37 3 16.4, 41.2, 61.8, 79.4, 79.10 4 9.5, 9.6, 26.5, 35.2, 39.11, 66.8, 76.12, 81.1, 83.7, 83.33
Cavan, M.P. 3 30.4
Cavelier, C. 1 43.21
Cefola, M. 2 77.65 3 33.7
Celotti, G. 3 1.36
Cenini, S. 2 86.89 3 86.98
Cerrini, S. 3 32.36, 58.15, 59.33 4 36.34, 54.19
Cesari, M. 3 71.25, 78.12, 78.16 4 74.1
Cesario, M. 4 58.7
Cesur, A.F. 1 17.21, 53.18
Cetinkaya, B. 3 71.37
Chacko, K.K. 4 48.26, 48.34, 48.37
Chaiwasie, S. 2 73.6
Chamberlin, J.W. 1 50.45
Chan, L.Y.Y. 3 73.6 4 70.2, 86.3, 86.4
Chan, N. 4 83.29
Chandrasekaran, K.S. 1 45.43
Chandrasekaran, R. 1 16.23, 20.14, 44.45, 45.43, 48.81, 48.85
Chandrasekharan, G. 1 20.14
Chandrasekharan, R. 3 45.15, 45.16
Chaney, M.O. 1 48.58 3 29.2, 38.20, 48.27, 48.28, 48.45 4 51.10
Chang, F.C. 3 56.2
Chang, G. 3 86.40
Chang, K. 3 84.1, 84.3, 84.7
Chang, S.C. 2 83.97 3 81.32
Chanon, M. 4 41.5
Chao-Fa, L. 1 3.54
Chao, G.Y. 2 60.52, 60.57, 70.19
Chao, K. 4 67.10
Chao, K.-J. 3 67.9
Chapelle, A. 3 58.17
Chapman, O.L. 3 38.8 4 36.14, 59.16
Charbonneau, G. 1 19.30, 19.32
Charbonneau, G.-P. 1 19.29, 19.33
Charlton, R.J. 2 83.154
Charlton, T.L. 1 15.9
Chasar, D.W. 1 39.74
Chasseau, D. 3 25.3, 25.5, 60.32 4 25.3, 35.9
Chastain Junior, R.V. 3 37.6
Chastain, B. 3 44.2, 44.3, 44.5
Chastain, R.V. 2 78.52, 83.52 3 8.14
Chasteen, N.D. 3 77.24, 81.18, 81.52
Chatani, Y. 1 1.30 3 9.11
Chatt, J. 2 71.38 3 86.72
Chatzidakis, A. 2 64.68
Chaudhuri, B. 1 9.38, 17.50 4 49.11
Chawdhury, S.A. 1 16.46, 16.57 3 11.27
Che'ng, W. 2 76.25, 77.68
Checchi, C.de 1 39.40, 42.9
Cheeda, G.B. 3 44.32
Cheer, C.J. 3 21.15, 21.16, 53.27
Cheeseman, T.P. 2 78.40, 78.50, 83.192
Chen, B.M.L. 4 49.1
Chen, I.Y. 4 28.10
Chen, S.M. 1 1.123
Chenery, D.H. 3 47.25
Cheng-Chung Lin, A 1 48.48
Cheng, J.C. 1 32.29 3 50.8
Cheng, P.-T. 3 3.43, 3.44, 86.64 4 72.25, 72.26, 86.31
Cheng, P.T. 3 81.54
Cherepenskii-Malov, V.D. 4 64.17
Cherepinskii-Malov, V.D. 3 64.30
Cherkasova, T.G. 4 83.128
Chernyshev, E.A. 4 63.8
Chesick, J.P. 4 41.7
Chesnut, D.B. 1 20.4 4 64.24, 65.9
Chetkina, L.A. 1 5.24, 5.30, 26.1, 26.2, 26.4, 26.6, 26.7, 26.8 4 24.4, 24.6, 24.7
Cheung, K.K. 1 54.25, 59.32, 59.33, 59.34, 59.35 2 65.51, 79.8, 85.44 3 8.18, 59.16, 86.46, 86.71 4 75.23
Chevrier, B. 1 12.5 2 66.10
Chheda, G.B. 3 44.29
Chiari, G. 3 59.4
Chiaroni, A. 3 16.2, 16.7 4 51.23
Chiba, A. 1 48.56
Chiba, T. 1 20.1
Chidambaram, R. 1 17.3 3 2.2 4 81.27
Chidester, C.G. 4 27.9, 34.3, 35.28, 35.29, 36.20, 36.33, 40.23, 45.23, 50.3, 51.46, 51.55
Chieh, P.C. 1 37.27 3 1.10, 1.11, 33.27, 63.15, 69.32, 84.23 4 63.11, 69.32, 69.33, 78.12, 83.106
Chiesi Villa, A. 1 39.32
Chiesi, A. 3 76.1, 83.10 4 79.12
Chih, C. 3 67.9
Childs, J.D. 2 60.37
Chin-You, N. 4 58.16
Chini, P. 3 86.72
Chioccola, G. 2 63.29, 64.84, 86.101 4 71.63
Chion, B. 3 37.3 4 12.13
Chipman, D.M. 3 71.2
Chisholm, N.H. 2 83.123
Chiu, C.C. 1 1.128, 45.75
Chiu, Y.H. 3 46.1, 64.26, 64.27

Chlenov, I.E. **3** 40.2
Choi, C.S. **3** 34.1 **4** 15.3
Chomyn, A. **4** 26.6
Chothia, C. **1** 59.13, 59.14 **3** 20.9
Chou, K.-D. **1** 11.19
Chow, Y.M. **3** 69.7, 69.21 **4** 63.3, 69.6, 69.7, 69.9, 69.10
Christensen, A.T. **1** 7.20, 16.2, 51.26 **3** 27.7, 51.31, 51.40, 51.45, 51.46
Christensen, J.J. **4** 67.23
Christian, S.D. **2** 60.37 **4** 4.4
Christofferson, G.D. **2** 70.1, 70.22
Christoph, G.G. **3** 16.8 **4** 83.34
Chu, S. **1** 13.13
Chu, S.-C. **3** 13.1
Chu, S.C. **1** 58.101
Chu, S.S.C. **1** 18.2, 45.37, 45.41, 45.67, 45.73 **4** 39.40
Chui, K.M. **4** 31.7, 41.27, 86.37, 86.42, 86.43
Chupakhina, R.A. **4** 81.45
Churchill, M.R. **1** 3.82 **2** 65.41, 71.13, 71.51, 72.27, 72.41, 73.7, 73.10, 73.32, 73.67, 73.73, 73.77, 73.112, 73.113, 74.19, 75.8, 75.27, 75.38, 75.40, 75.42, 75.43, 75.69, 75.76 **3** 62.14, 62.15, 64.52, 71.8, 71.16, 71.18, 71.24, 71.43, 71.60, 72.9, 72.43, 73.35, 73.40, 75.1, 75.22, 75.31, 75.32, 75.33, 75.43, 83.90, 83.93, 85.14 **4** 62.7, 62.8, 62.13, 71.2, 71.35, 71.74, 71.75, 71.94, 73.16, 73.18, 75.2, 75.26, 75.32, 83.86, 85.9, 86.70
Cian, A.de **3** 60.51, 72.21, 72.31 **4** 72.7
Ciana, A. **3** 85.32, 85.33, 85.40
Ciani, G. **2** 60.149 **4** 86.57
Ciani, G.F. **1** 3.41
Cid-Dresdner, H. **1** 50.30
Ciechanowicz, M. **3** 86.74
Cingi, M.B. **2** 81.80 **3** 81.19 **4** 76.38, 83.55
Clapp, R.C. **1** 41.10 **3** 45.44
Clardy, J. **3** 58.24, 58.50 **4** 27.11, 27.12, 35.14, 36.14, 36.22, 39.17, 60.36
Clardy, J.C. **3** 13.15, 31.9, 32.20, 38.8, 56.2, 58.19 **4** 13.17, 13.18, 59.16, 64.4, 64.31
Clark, G.L. **1** 50.11
Clark, G.R. **1** 54.2 **2** 75.11, 77.17, 78.27, 83.109 **3** 38.11, 39.36, 54.1, 78.8, 86.23 **4** 33.34, 58.41, 86.35, 86.53
Clark, H.C. **2** 69.2, 86.16
Clark, H.G. **1** 7.25
Clark, J.R. **1** 2.8, 2.34, 2.61
Clark, N.H. **3** 62.4
Clark, R.J.H. **3** 86.19, 86.25
Clarke, D.G. **3** 59.23
Clarke, P.T. **2** 62.50
Clarke, T.A. **4** 75.22
Clastre, J. **1** 17.11, 19.16, 26.34, 41.42, 48.50, 48.51 **3** 3.24
Clausen, A. **3** 76.51
Clauss, K. **4** 40.5, 44.13
Claverini, R. **4** 1.8
Clearfield, A. **3** 83.139, 85.54 **4** 35.5
Clegg, H.J. **3** 81.25
Clegg, W. **4** 83.123
Clemens, J. **4** 71.84, 71.86
Clemente, D.A. **1** 33.75 **3** 39.39, 40.15, 64.42, 69.33, 72.49 **4** 39.27, 40.19, 71.78, 78.11, 83.65
Clews, C.J.B. **1** 44.1, 44.2, 44.13, 44.14, 44.43, 44.44 **3** 19.12, 19.13, 19.14
Clippard, P.H. **4** 62.2, 62.3, 62.4
Clunie, J.S. **1** 53.17
Coates, E. **2** 77.2, 77.3
Coates, R.M. **1** 53.35
Coburn, R.A. **3** 45.44
Cocco, G. **1** 2.12
Cochran, T.G. **3** 52.6 **4** 13.21, 44.9, 44.18, 44.25, 50.2
Cochran, W. **1** 2.55, 13.22, 44.1, 44.2, 44.13, 44.34, 44.44, 45.70, 48.44
Coda, A. **1** 36.13
Codding, P.W. **4** 49.9
Cody, V. **3** 48.41, 60.43, 69.34 **4** 51.2, 51.3, 51.42, 51.47, 51.50, 51.51
Coetzer, J. **3** 75.47, 79.23, 85.8 **4** 54.17, 59.25, 77.17
Coggon, P. **1** 53.7 **3** 3.48, 50.1, 53.28, 53.29, 54.7, 58.20, 59.7, 59.17, 59.34, 64.22, 68.12, 78.3, 78.23, 86.1 **4** 28.10, 51.40, 51.41, 59.19, 59.26, 81.55
Coghi, L. **2** 83.28, 83.46, 84.2 **3** 79.14, 83.10 **4** 83.5, 83.12
Cohen, G.H. **1** 49.24, 49.29 **2** 83.137
Cohen, M.D. **3** 1.38
Coiro, V.M. **4** 8.11
Cojazzi, G. **3** 32.12
Colapietro, M. **1** 4.12 **2** 65.27, 80.16 **3** 32.36, 72.4 **4** 80.15
Cole, F.E. **3** 11.22, 48.12, 48.13, 48.39, 58.1 **4** 33.22, 48.12, 48.18
Cole, S.J. **2** 82.50
Coleman, J.M. **2** 73.117, 73.118, 85.71
Coletta, M. **1** 33.75
Coll, J.C. **4** 37.6
Collamati, I. **2** 84.42
Colleter, J.-C. **1** 33.43
Colleter, J.C. **1** 33.56, 33.57 **3** 33.25
Collier, J.E. **4** 64.11, 64.26
Collier, M.R. **4** 71.81
Collin, R.L. **1** 7.12 **3** 64.33
Collins, D.M. **1** 58.44 **3** 49.11, 49.12, 69.29, 69.30 **4** 49.21, 49.24, 76.36
Collins, E. **2** 65.7
Collins, R.K. **4** 83.76
Collman, J.P. **4** 86.20
Colman, P.M. **3** 60.4, 60.5
Colombo, A. **1** 29.17, 29.18, 38.27, 38.28, 38.41 **2** 61.17, 61.19, 75.48 **4** 72.16
Colson, J. **3** 21.3, 21.4
Colton, R. **3** 74.17
Colton, R.H. **1** 2.43 **3** 60.36
Comberton, G. **4** 8.12, 17.11, 51.12
Comer, F.W. **1** 53.38
Conde, A. **4** 33.19
Connelly, N.G. **3** 73.15, 73.18, 73.19
Conner, S.H. **1** 33.21, 33.24, 33.25
Connor, J.A. **2** 71.24
Constant, G. **4** 83.3
Conti, F. **2** 86.89 **3** 86.98
Cook, C.D. **3** 86.64 **4** 72.25, 72.26, 86.31
Cook, D.S. **1** 46.18
Cook, R.E. **3** 39.45, 45.29, 58.29 **4** 39.47
Cook, W.J. **4** 67.5, 67.6, 67.13
Cooke, J. **3** 64.52, 85.14

Cooke, P.W. **2** 60.86
Coomes, R.M. **1** 58.52
Cooper, A. **1** 51.5, 51.7, 51.9, 51.10, 51.12, 51.13, 51.14, 51.17, 51.21, 51.22, 51.41, 51.50 **2** 60.151, 77.15, 82.53, 84.64 **3** 17.8, 51.12, 51.16 **4** 51.15, 51.17
Cooper, M.J. **3** 37.6
Cooper, R.D.G. **3** 50.8
Cooper, W.F. **4** 39.5
Copeland, R.F. **1** 3.44, 33.21, 33.24, 33.25
Copeland, V.C. **4** 83.22, 83.23
Coppens, P. **1** 1.18, 1.19, 15.20, 15.21, 17.5, 17.6, 33.3, 33.4, 33.5 **2** 64.4 **3** 33.1, 33.4, 33.7, 60.44 **4** 7.5, 22.9, 33.4, 39.5, 76.15
Coppola, F. **1** 33.70, 33.72
Coppola, J. **1** 58.47
Coppola, J.C. **3** 46.5, 46.6, 47.25, 47.33, 51.42, 53.12, 53.13, 53.14, 65.5 **4** 8.9, 23.2
Corao, E. **3** 79.17
Corbett, M. **2** 83.213, 83.218 **3** 83.151
Corbridge, D.E.C. **1** 24.18 **2** 83.141
Corden, B.J. **3** 3.56
Cordes, A.W. **1** 35.22 **3** 33.34, 83.11 **4** 64.9, 65.5, 65.8, 83.29
Corey, E.J. **3** 40.13
Corey, E.R. **2** 64.41, 64.67, 73.48, 73.96, 83.65, 84.18 **3** 69.34
Corey, R.B. **1** 48.33, 48.60
Corfield, P.W.R. **1** 28.15, 45.18 **2** 71.23, 72.29, 72.30, 76.39, 86.16, 86.58 **3** 11.26, 21.17, 64.6, 65.7, 71.71, 77.23, 85.15, 86.16 **4** 45.15, 83.31
Corney, D.J. **4** 33.23
Corradi, A. **1** 41.21
Corradi, A.B. **4** 59.28
Corradini, P. **1** 1.86, 1.133, 44.53 **2** 73.111, 74.24 **3** 20.4, 21.5, 21.7, 33.16, 33.19, 72.19
Costello, W.R. **2** 85.12
Costin, C.R. **1** 1.122
Cosulich, D.B. **1** 59.50 **3** 21.13, 50.2 **4** 25.7, 41.24
Coton, F.A. **2** 86.97
Cotrait, M. **1** 2.93, 2.94 **3** 2.23 **4** 1.10, 24.10, 48.46
Cotton, F.A. **1** 3.63 **2** 65.44, 65.47, 66.20, 66.23, 71.44, 72.25, 73.39, 73.63, 74.9, 74.17, 75.30, 75.32, 75.39, 75.57, 75.58, 75.66, 76.3, 76.44, 77.39, 77.51, 77.56, 77.57, 77.61, 77.63, 77.64, 77.67, 80.23, 81.25, 81.50, 81.83, 81.88, 81.89, 82.49, 83.1, 84.21, 84.24, 84.29, 84.66, 85.29, 85.54, 85.55, 86.12, 86.59 **3** 3.8, 11.16, 11.17, 73.20, 73.44, 75.5, 77.33, 81.37, 81.53, 83.132, 83.133, 86.57 **4** 71.10, 72.9, 73.6, 73.23, 75.3, 75.6, 75.12, 75.21, 81.26, 81.28, 81.29, 81.62, 86.25, 86.54
Couch, D.A. **3** 83.147
Coucouvanis, D. **1** 3.69 **2** 85.89 **3** 85.31
Coulter, C.L. **1** 11.21, 31.22, 47.1 **3** 47.4, 47.6, 49.14
Countryman, R. **3** 49.12 **4** 3.17, 49.21, 71.32, 71.76, 76.34
Courseille, C. **1** 25.21 **3** 25.1, 25.2, 36.30, 51.5 **4** 8.12, 8.13, 13.8, 17.9, 17.10, 19.14, 25.2, 51.4, 51.5, 51.9, 51.12, 51.16, 51.18, 51.21
Court, W.A. **4** 58.40
Courtois, A. **4** 2.9
Couvillion, J. **1** 3.55
Coville, N.J. **4** 86.41
Cow, J.W. **2** 64.67
Cox, A. **4** 71.45
Cox, D.A. **4** 56.3
Cox, D.E. **3** 28.3, 28.4
Cox, E.G. **1** 1.8, 9.8, 39.31, 39.70, 45.11, 45.13 **2** 76.80, 83.141, 85.22 **3** 19.3
Cox, G. **1** 45.33
Cox, G.W. **1** 1.12
Cox, J.R. **1** 46.11
Cox, J.W. **2** 64.41
Cox, M.M. **2** 83.197
Cox, M.R. **1** 2.92 **4** 53.10
Cox, P.J. **4** 24.13, 24.14, 24.15
Coxon, J.M. **1** 52.17
Coyle, B.A. **3** 84.24
Cradwick, E.M. **3** 83.66
Cradwick, M.E. **3** 83.12, 83.33 **4** 76.42
Cradwick, P.D. **3** 53.5, 76.50 **4** 3.13, 25.4, 25.5, 25.6, 37.12, 53.7, 76.42
Craig, D.C. **4** 83.113
Craik, A.R.M. **3** 71.53
Cramer, R. **4** 71.6
Cramer, R.E. **4** 77.23
Crane, R.I. **4** 53.3
Cras, J.A. **2** 80.8, 80.9, 80.34 **3** 80.11 **4** 80.4, 80.11, 80.18, 85.4
Craven, B. **1** 1.149
Craven, B.M. **1** 1.12, 19.8, 43.5, 43.6, 43.7, 43.9, 43.15, 43.23, 43.24, 43.25, 43.27, 43.28, 43.29, 44.4, 58.56, 58.85, 59.30, 59.31 **2** 83.67, 83.68 **3** 43.4, 43.5, 44.8, 60.27, 60.28, 83.115, 83.116 **4** 43.1, 43.2, 43.3, 43.4, 43.6, 43.9, 46.4, 60.19, 60.42
Credali, L. **1** 39.40, 42.9
Creed, D. **4** 31.30
Creitz, T.C. **2** 83.32
Croatto, U. **1** 19.14 **2** 69.21 **3** 64.42 **4** 78.11, 85.26, 85.32
Crociani, B. **3** 71.26
Crombie, L. **3** 59.23 **4** 59.18
Cromer, D.T. **1** 33.58, 34.1, 34.2, 40.2 **2** 81.3, 81.4 **3** 83.19 **4** 33.5, 76:8, 83.11, 83.14
Cross, A.D. **1** 51.25
Cross, J.H. **3** 73.51 **4** 73.22
Cross, R.J. **4** 75.23
Crosse, B.C. **2** 85.25
Crossing, P.F. **4** 76.31
Crow, J.P. **4** 86.3, 86.4
Crowder, M.M. **1** 10.26
Crowfoot, D. **1** 51.29
Cruickshank, D.W.J. **1** 1.14, 1.55, 2.9, 19.44, 19.52, 24.14, 24.15, 37.4, 37.5, 37.6, 37.7, 37.8, 37.9, 45.13, 49.34, 52.18, 59.18 **3** 19.3, 26.4, 26.5 **4** 22.2, 46.2, 71.11, 78.2
Crump, D.B. **4** 71.59
Crump, D.R. **1** 54.2
Cruse, W.B.T. **4** 54.10, 65.4
Cser, F. **1** 52.3
Cucka, P. **1** 33.19
Cullen, D. **4** 49.19
Cullen, D.L. **3** 49.9, 76.42, 83.29, 83.30 **4** 49.15

Cullen, W.R. **2** 65.10, 65.48 **3** 86.5, 86.7 **4** 86.3, 86.4
Cunningham, J.A. **2** 77.29, 77.30, 77.34 **3** 3.56, 85.51
Cuperus, A.J. **4** 27.14
Cupper, G.L. **3** 72.14, 75.15
Curran, R. **3** 85.51
Currie, M. **1** 2.32 **3** 2.12, 35.8 **4** 2.8, 2.13, 2.14, 49.3, 49.4, 49.5
Curry, J.D. **4** 66.7
Curry, N.A. **1** 2.32, 2.86, 2.87, 11.51, 17.15, 19.19, 19.20
Curti, R. **1** 36.7, 36.8, 36.11, 36.13
Curtin, D.Y. **1** 9.49 **3** 9.6 **4** 13.15, 13.16, 17.12, 17.13
Curtis, N.F. **3** 81.48 **4** 83.101
Curtis, R.M. **1** 8.35, 39.11
Cusatis, C. **1** 43.27
Cutfield, J.F. **2** 82.55
Cyrot, E. **3** 30.6
Czerinski, E.W. **4** 50.16

D'Addario, A. **4** 49.14
D'yachenko, O.A. **3** 78.22
Dabrowiak, J.C. **4** 83.31
Dahan, F. **4** 37.1
Dahl, L.F. **1** 3.46, 3.64 **2** 71.22, 72.9, 72.46, 72.48, 72.61, 73.11, 73.27, 73.58, 73.64, 73.70, 73.90, 73.91, 73.116, 73.117, 73.118, 74.4, 74.18, 75.1, 75.62, 75.63, 75.75, 75.80, 75.81, 83.203, 84.15, 84.16, 84.17, 85.25, 85.62, 85.71, 85.74, 85.82, 85.84, 86.20 **3** 3.37, 50.7, 50.9, 50.11, 64.62, 64.63, 64.64, 64.67, 65.12, 71.54, 72.12, 73.15, 73.18, 73.19, 73.27, 73.37, 75.30, 75.48, 83.31, 86.33, 86.72 **4** 3.15, 3.16, 22.7, 63.1, 64.41, 64.42, 65.11, 71.34, 73.14, 73.25, 73.27, 73.28, 85.24
Dahl, S. **4** 23.1
Dahl, T. **1** 5.5, 21.10 **2** 60.38, 60.39, 60.58, 60.65 **3** 60.20 **4** 60.1, 60.21
Dahlen, B. **4** 39.16, 64.28, 64.29, 64.30
Dahm, D. **3** 58.24, 58.50
Dahm, D.J. **2** 86.53
Dahne, S. **1** 18.11
Dakin, T.W. **1** 35.26
Dale, D. **2** 83.204 **4** 49.20
Daley, L. **3** 83.60
Dalley, N.K. **3** 23.10 **4** 67.23
Dalton, R.F. **3** 69.14
Daltrozzo, E. **4** 36.37
Daly, J. **4** 40.14
Daly, J.J. **1** 2.75, 4.13 **2** 63.3, 63.19, 63.29, 64.5, 64.10, 64.53, 64.57, 64.58, 64.63, 64.80, 64.82, 64.83, 64.84, 84.46, 86.101 **3** 63.7, 64.32, 64.49, 64.55, 65.1, 71.65 **4** 64.6, 71.4
Daly, J.W. **3** 58.9
Dalzell, B.C. **4** 81.20
Damiani, A. **1** 33.81 **2** 60.110, 60.111, 60.112
Damm, E. **2** 60.17, 60.18
Dammann, C.B. **4** 86.67
Dammeier, B. **4** 13.22, 35.33
Danford, M.D. **3** 77.16, 77.17
Dangoumau, J. **3** 3.45 **4** 24.10
Daniels, P. **4** 59.12
Danielsen, J. **1** 38.36, 59.28 **2** 84.25 **3** 39.7
Dansted, E. **1** 52.17
Dapporto, P. **3** 76.87, 83.88, 86.48, 86.52, 86.85 **4** 83.126
Daran, J.C. **4** 83.3
Darden Junior, W.J. **4** 21.21
Dargay, J.M. **4** 2.22
Darlow, S.F. **1** 2.55, 2.56, 2.73, 13.47
Darnin, C. **1** 10.28
Das, I.M. **4** 49.11
Dashevii, V.G. **1** 15.1
Datt, I.D. **1** 7.7
Dauben, W.G. **4** 56.3
Dauphin, G. **1** 16.55
Davey, G. **3** 76.20, 81.40, 81.49 **4** 81.40
Davidkov, K. **4** 33.21
Davidson, A.G.H. **3** 48.25
Davie, A.W. **1** 56.2
Davies, A.G. **3** 69.2
Davies, D.R. **1** 1.64, 32.8
Davies, G.R. **2** 72.28, 86.6, 86.52 **3** 71.57, 72.30, 73.28, 73.47, 73.48, 73.49, 80.1, 80.2 **4** 69.36, 71.62, 71.64
Davies, J.E. **3** 78.13 **4** 78.16
Davies, V.H. **4** 53.10
Davies, W.O. **1** 46.23
Davis, B.R. **2** 86.112 **3** 76.16, 86.55, 86.93 **4** 39.34, 73.7, 73.11
Davis, M. **4** 41.6
Davis, P.H. **3** 79.19
Davis, R.E. **1** 45.52, 50.19, 59.38 **2** 72.68, 73.76 **3** 20.15, 27.9, 30.1, 31.29, 71.13, 72.14, 72.23, 72.25, 72.26, 72.41, 72.42, 75.10, 75.12, 75.15, 75.39, 83.131 **4** 71.70, 71.84, 75.24
Davison, A. **2** 73.39
Davison, B.E. **3** 45.48
Davydova, M.A. **1** 24.2, 24.3, 24.5 **4** 24.6
Dawson, B. **1** 48.29, 48.43
Dawson, I.M. **1** 11.26
Dawson, M.I. **3** 27.11
Day, A.C. **3** 13.18
Day, F.H. **4** 37.10
Day, V.W. **2** 78.45 **3** 84.25 **4** 49.24
De Camp, W.H. **3** 32.28 **4** 33.35, 33.36, 33.37
Dean, C. **1** 19.8
Debaerdemaeker, T. **3** 48.44
Debaerdemaeker, T.D.J. **4** 51.6
Debbaudt, M. **4** 83.39
DeBoer, B.G. **1** 3.31 **3** 83.133 **4** 73.6, 73.23, 75.6, 81.26, 81.28
DeCamp, W.H. **4** 32.3
Decap, J. **1** 13.26, 16.18, 16.19, 16.22
Declerq, J.P. **3** 67.6 **4** 32.21, 51.25, 51.30, 51.31, 51.34, 83.39
Degeilh, R. **1** 33.12 **2** 84.44
Degreve, Y. **2** 72.57
Dehand, J. **2** 83.124, 83.153
Dehli, J. **3** 23.7
Dehnicke, K. **2** 73.80
DeIasi, R. **4** 78.14
Dejace, J. **1** 1.21, 19.58
DeJarnette, F. **4** 51.2, 51.3

DeJarnette, F.E. **4** 51.47, 51.50, 51.51
Dekaprilevich, M.O. **2** 77.33 **4** 77.12
Del Pra, A. **1** 3.85, 33.75, 34.4, 35.9, 38.3, 39.9 **2** 69.42 **4** 39.3
DeLacy, T.P. **4** 17.15, 19.11
Delaplane, R.G. **1** 1.13, 1.20 **3** 33.1
Delbaere, L.T.J. **3** 73.7, 82.23
Delettre, J. **3** 13.17
Della Casa de Marcano, D.P. **3** 54.14
Dellaca, R.J. **1** 26.35 **3** 33.30, 33.31, 33.32, 71.35, 71.38, 71.40, 71.59 **4** 71.39, 76.17, 76.18
DeLuke, A.B. **4** 47.13
Demalde, A. **4** 28.5
DeMarco, P.V. **3** 50.8
Dempf, D. **4** 75.25
Dempsey, J.N. **2** 72.4
Denne, W.A. **4** 1.2, 16.2, 16.4, 16.5, 41.6, 58.9, 58.10, 58.11, 58.35
Dennis, D. **4** 59.4
Denniston, M.L. **4** 62.1
Derieg, M.E. **3** 9.19, 35.17
Derissen, J.L. **1** 48.25 **3** 1.18
Derry, J.E. **1** 36.20 **3** 34.3 **4** 22.6, 36.11
Desiderato, R. **1** 7.16, 11.28, 35.26, 35.30 **4** 33.16, 36.1
DeSimone, R.E. **4** 86.10
Dessy, G. **1** 3.70 **2** 68.11, 68.16, 80.13, 80.17, 84.42, 84.60, 85.72, 85.85, 85.93, 85.96 **3** 68.20, 80.7, 80.8, 80.9, 83.114, 85.48 **4** 79.11, 85.17, 85.18, 85.19, 85.31
DeStevens, G. **4** 33.47
Destro, R. **1** 17.25, 17.31, 17.32, 52.1 **2** 60.103 **3** 31.1, 31.25 **4** 31.4
Desvergne, J.-P. **3** 26.9
DeTitta, G.T. **4** 46.4, 60.42
Deuschl, H. **1** 32.4 **2** 74.21
DeVille, T.E. **1** 21.33 **4** 54.1, 54.2
Dewar, R. **1** 42.11, 42.12 **3** 62.18, 83.80
Dewar, R.B.K. **2** 75.26
Dexter, D.D. **4** 3.22, 3.24
Deyrieux, R. **3** 81.2, 81.3
Deyrup, J. **4** 38.10
Dhaneshwar, N.N. **4** 13.7
Diamand, R.D. **2** 77.1 **4** 59.7
Diana, G. **1** 38.59 **2** 75.37
Dias, H.W. **1** 8.42
Dichmann, K. **4** 39.46
Dichmann, K.S. **3** 36.18, 36.25, 60.34
Dickens, B. **2** 75.19, 75.29 **3** 12.2
Dickerson, R.E. **1** 33.51 **2** 68.15 **3** 9.8, 24.6, 27.11, 28.12, 30.5, 31.32 **4** 27.16
Dickinson, C. **1** 12.6, 15.2, 16.1, 24.6 **3** 11.18, 16.5 **4** 32.14
Dickson, R.S. **4** 71.48
Dideberg, O. **4** 38.15, 51.28, 51.29, 51.39
Dierks, H. **2** 75.54, 75.72
Dieterich, D.A. **3** 9.6
Dietrich, H. **1** 10.30 **2** 67.4, 72.22, 72.77, 75.54, 75.72 **3** 20.14, 72.15, 75.19
Dijk, B.van **3** 11.23
Dijkstra, A. **2** 82.17
Dik-Edixhoven, L. **3** 33.46
Dikareva, L.M. **2** 81.56, 81.57, 81.58
Dimroth, K. **2** 64.68
Dittmar, G. **3** 63.9
Dixon, K.R. **2** 86.16
Djerassi, C. **1** 58.63 **3** 51.41, 58.35, 58.42 **4** 58.33
Dobinson, G.C. **2** 83.178 **3** 86.98
Dobler, M. **1** 22.12, 31.15, 37.12, 50.36, 53.40 **3** 40.7 **4** 49.6
Dodge, R. **4** 60.13
Dodge, R.P. **2** 72.34, 72.74, 72.75, 75.10, 75.79, 77.6 **3** 76.25 **4** 21.21, 38.11
Dodson, G.G. **1** 58.34 **4** 76.42
Doedens, R.J. **2** 73.70, 73.116, 83.36, 83.69, 83.88, 83.203, 86.25, 86.58, 86.66, 86.79 **3** 79.11, 81.35, 83.22, 83.89 **4** 71.36, 81.42, 81.60, 83.43
Doherty, D.G. **4** 48.55, 48.56
Doherty, G.O.P. **1** 59.45
Dollase, W.A. **1** 39.51 **2** 74.9
Domenicano, A. **1** 4.12 **2** 65.27, 80.16, 80.32, 80.33, 80.35, 85.9, 85.10, 85.11, 85.36, 85.69 **3** 85.1
Domiano, P. **1** 16.31, 35.8, 41.21 **2** 79.2, 79.14, 79.18, 79.27, 85.37 **3** 8.5 **4** 79.1, 83.55
Dominguez, X.A. **4** 56.4, 59.14
Dominianni, S.J. **3** 38.20
Donaldson, D.M. **1** 24.30, 30.28
Donaldson, P.B. **4** 86.65
Donaruma, L.G. **1** 30.27
Dondoni, A. **3** 39.39, 40.15 **4** 39.27, 40.19
Donohue, J. **1** 8.5, 8.17, 10.19, 10.20, 11.27, 15.12, 16.10, 31.11, 48.13, 48.17, 48.23, 48.33, 48.40, 48.59, 50.29 **2** 62.32, 64.11 **3** 1.24, 2.6, 11.25, 23.3, 31.30, 40.11, 44.6, 60.8, 65.2, 86.4 **4** 5.2, 24.1, 37.8, 41.7, 51.14, 52.4, 52.5
Doomes, E. **3** 11.26
Doorakian, G.A. **3** 69.35
Doorne, W.van **4** 64.9
Dorfelt, C. **2** 69.9
Dori, Z. **1** 3.40 **2** 76.75, 80.31, 83.116, 85.73, 85.86 **3** 85.27, 86.96 **4** 82.13, 83.72, 86.61, 86.68, 86.69
Dorignac-Calas, M.-R. **4** 41.28
Dornberg, M.L. **1** 2.74, 2.80, 2.81
Dorset, D. **3** 1.39
Doskotch, R.W. **4** 54.10
Dougherty, T.A. **2** 83.18
Dougill, M.W. **1** 1.8, 1.54 **2** 64.29
Douglas, B. **1** 58.100, 58.101
Douglas, P.G. **3** 86.49
Douglas, W.E. **4** 73.29
Douwes, H. **2** 64.4
Dow, J. **1** 47.15 **3** 48.29
Downie, T.C. **1** 14.5 **3** 81.8 **4** 33.15, 41.1
Downs, G.L. **2** 84.15, 84.17
Doyle, J.R. **2** 72.47, 73.17, 75.2, 75.3, 75.4, 75.6, 75.18, 75.77
Doyne, T.H. **1** 56.6 **2** 82.32 **3** 86.36
Drager, M. **3** 67.4
Dragonette, K.S. **1** 5.41, 8.41, 35.28
Drake, J.E. **2** 83.47
Draper, M.D. **3** 33.48
Dreele, R.B.von **3** 50.17 **4** 50.9, 77.9
Dreissig, W. **2** 72.77 **4** 64.22, 64.23
Drenth, W. **1** 19.61

Drew, M.G.B. **1** 24.32, 31.43, 54.8 **2** 76.21, 76.31, 76.50, 81.46, 82.34, 83.103, 83.219, 86.21 **3** 33.9, 81.36, 82.16, 82.19, 83.5, 86.8, 86.75 **4** 3.12, 64.33, 81.36, 83.8, 83.75, 83.76, 84.2, 86.23, 86.47, 86.55
Drew, R.E. **4** 69.5
Dreyfus, H. **3** 53.12, 53.13, 53.14
Driel, C.A.A.van **1** 1.70
Driver, R. **2** 82.23, 82.36, 82.38
Drummond, J. **2** 65.45 **3** 65.9 **4** 78.17
Dryer, H.B. **2** 60.67
Duax, W. **4** 51.2, 51.3
Duax, W.L. **3** 48.41, 51.16, 60.14, 60.15, 60.43 **4** 11.12, 11.14, 50.19, 51.24, 51.26, 51.27, 51.35, 51.36, 51.42, 51.47, 51.50, 51.51, 60.38, 60.39
Dubey, R.J. **3** 1.1
Duchamp, D.J. **1** 13.41 **3** 50.6, 50.10 **4** 27.9, 34.3, 35.28, 35.29, 36.20, 36.33, 40.23, 45.23, 50.3, 50.7, 51.46, 51.55
Duckworth, V.F. **2** 77.54, 83.100, 83.101 **3** 86.26, 86.49 **4** 23.5
Ducruix, A. **4** 21.16, 45.14
Duerst, R.W. **1** 51.47
Duesler, E.M. **4** 71.50
Duesler, E.N. **4** 76.14
Duffield, A.M. **1** 59.41
Duffin, B. **1** 32.42 **2** 83.22, 84.1 **3** 11.28, 35.2, 35.7
Duke, A.J. **3** 27.8
Duke, J.R.C. **1** 1.25 **3** 27.8, 32.3, 32.4, 32.5
Duke, R.P. **4** 33.42, 33.45
Dukefos, T. **1** 45.63
Dullforce, T.A. **1** 3.57, 3.58, 3.59 **4** 53.4, 53.9
Dulova, V.G. **2** 63.28
Dumbleton, J.H. **1** 2.97
Dumke, W.L. **1** 19.54
Dunaj-Jurco, M. **2** 83.30 **4** 76.1
Dunitz, J.D. **1** 1.11, 1.43, 1.77, 21.16, 21.18, 22.11, 22.12, 23.2, 23.4, 23.6, 23.7, 23.8, 23.10, 31.15, 34.6, 34.7, 34.11, 37.12, 38.37, 38.49, 46.21, 48.15, 49.2, 50.29, 50.44, 51.45, 53.40 **2** 73.22, 75.12, 75.53, 75.67, 83.6, 83.75, 83.77, 83.180 **3** 16.9, 16.10, 16.11, 23.5, 23.6, 31.6, 40.7, 83.85 **4** 23.3, 23.4, 23.6, 31.14, 31.16, 49.2, 49.3, 49.4, 49.5, 49.6, 50.11, 50.13, 83.120
Dunken, H. **1** 13.9, 33.20
Dunlop, J.H. **2** 82.34, 82.37
Dunlop, R.S. **2** 73.59
Dunn, J.B.R. **4** 60.28
Dunn, R.M. **4** 2.24
Dunn, R.M.A. **4** 24.2
Dunne, T.G. **2** 71.44, 72.25
Dunphy, R.F. **4** 31.26
Dupont, L. **4** 38.15, 51.28, 51.29, 51.39
Dupuy Junior, A.E. **4** 32.30
Durant, F. **2** 60.25, 60.26, 60.46, 60.47, 60.64, 60.66 **3** 48.44 **4** 48.49, 59.2
Duranti, D. **1** 35.13
Durcanska, E. **4** 79.7
Durovic, S. **3** 83.37
Dusausoy, Y. **3** 2.5, 74.2, 74.5 **4** 73.26, 74.2, 74.4, 74.6, 74.7, 75.8, 75.10
Dutkey, S.D. **4** 11.9
Dutta, S.N. **2** 64.26
Duvall, L.A. **4** 67.28
Dvoryankin, V.F. **1** 8.10
Dwek, R.A. **3** 45.31
Dwivedi, G.L. **4** 83.67
Dwyer, M. **3** 76.38
Dyer, G. **2** 86.105 **4** 85.37
Dyer, J.R. **1** 59.5 **4** 32.10
Dyke, M. **1** 10.11

Eaborn, C. **4** 71.81
Earle, K.M. **3** 48.38
Earley, J.E. **4** 76.9
Eck, C. **4** 53.3
Eck, J. **4** 4.1
Eckstein, F. **3** 47.1
Eddy, L.P. **3** 83.29
Edmond, E. **3** 49.13 **4** 49.23
Edmonds, J. **3** 7.6
Edmonds, J.W. **3** 19.6, 19.7 **4** 9.3, 39.5
Edmunds, I.G. **1** 17.39
Edwards, I.A.S. **3** 3.35, 30.3
Edwards, J.O. **3** 83.28
Edwards, W.T. **2** 75.30, 75.32
Eeles, W.T. **1** 13.46, 58.106
Effinger, J. **1** 41.46
Eger, C. **3** 60.14, 60.15 **4** 51.35, 51.36, 60.39
Eglington, G. **1** 39.56
Egmond, J.v. **3** 22.3
Ehrenberg, M. **1** 1.45, 28.29, 29.15, 31.47, 31.48, 31.49
Ehrlich, H.W.W. **1** 28.12, 30.1, 32.12 **2** 86.27
Eian, G.L. **4** 36.14
Eichelberger, H.R. **4** 81.59
Eichhorn, E.L. **1** 9.26, 33.15, 44.46
Eick, H.A. **3** 73.21
Eijck, B.P.van **1** 1.32
Eiland, P.F. **2** 73.21
Eilermann, L. **1** 19.63
Einspahr, H. **3** 2.6, 11.25, 86.4 **4** 24.8
Einspahr, H.M. **3** 1.24
Einstein, F. **2** 70.5
Einstein, F.W.B. **1** 3.60, 3.68 **2** 69.32, 71.10, 75.33, 83.139, 86.32 **3** 68.6, 73.6, 73.55, 83.1 **4** 68.7, 69.5, 70.2, 71.37, 71.85, 73.32, 83.139, 86.3, 86.4, 86.9, 86.16, 86.18, 86.62
Einstein, J.R. **1** 44.72 **3** 44.34, 85.18 **4** 30.2, 45.2, 47.4, 48.55, 48.56
Eisenberg, R. **2** 76.75, 85.30, 85.73, 85.86, 85.97, 85.98, 86.50 **3** 3.56, 83.28, 85.27, 85.35, 85.45, 85.51, 86.31, 86.96 **4** 73.15, 85.12, 85.16, 85.39, 86.29, 86.60, 86.63
Eisenhuth, W. **2** 65.5, 65.15
Eisner, U. **3** 36.29, 37.13
Eiss, R. **2** 77.63, 77.64, 84.66 **3** 72.11, 75.5, 81.53, 83.148
Elcombe, M.M. **1** 8.12, 8.13, 8.14, 8.15
Elder, M. **3** 63.14, 69.20, 69.23, 69.36, 73.34, 83.146, 85.6 **4** 73.17
Elder, R.C. **2** 73.63, 76.3, 77.44, 77.45, 77.51, 77.67, 83.94
Eley, D.D. **2** 60.83
Elguero, J. **1** 9.47 **3** 32.7 **4** 32.18
Elix, J.A. **3** 36.16
Eller-Pandraud, H.von **1** 35.41

Eller, G.von 1 16.29
Eller, H.von 1 39.76, 39.78 3 35.19
Eller, P.G. 3 64.6, 85.15, 86.16
Elliott, I.W. 4 37.15
Elliott, N. 4 33.25
Ellison, R.D. 1 2.53 3 1.6
Elmes, P.S. 4 86.2
Elmore, N.F. 3 33.35
Emerson, G.F. 2 73.77
Emerson, K. 2 84.41 3 76.44, 83.81, 83.147
Emerson, M.T. 1 53.3 3 53.18 4 53.1
Emmert, D.E. 4 51.46
Endeman, H.J. 1 48.25
Enemark, J.H. 2 62.36, 64.52, 85.45, 86.39, 86.85, 86.112 3 76.40 4 71.1, 71.7, 71.20, 80.1, 86.13, 86.30
Engebretson, G.R. 1 19.55 2 75.24
Engel, D.W. 4 54.12
Engel, J.F. 3 64.22 4 86.17
Engel, M.R. 4 59.16
Engmann, R. 1 7.25 2 63.13
Entwistle, R.F. 1 29.7, 30.11
Enwall, E. 3 83.1
Enwall, E.L. 3 83.81 4 4.4, 51.57, 51.58, 82.14
Epstein, E.F. 3 75.30, 75.48, 84.20, 85.25, 85.44, 85.54 4 73.8
Eraker, J. 1 20.13
Erasmus, C.S. 3 77.35 4 77.22
Ercolani, C. 2 84.42
Erdtman, H. 4 59.12
Erichsen, H.H. 1 9.23
Erickson, J.L. 3 33.5
Eriks, K. 1 11.14, 12.14 2 83.38 3 46.8 4 12.15, 81.20
Ermer, O. 3 23.5, 23.6, 31.6 4 23.4, 23.6, 31.14, 31.16
Ern, V. 4 36.12
Erskine, G.J. 2 86.46
Escande, A. 4 32.4
Escobar, C. 1 41.42
Eser, H. 4 23.3, 23.4, 23.6
Esperas, S. 4 70.16, 70.17
Estienne, J. 4 71.66, 81.50
Estlin, J.A. 1 4.18, 35.32, 39.57, 58.22 3 11.3, 31.34
Evans, B.E. 4 40.2, 41.3
Evans, D.A. 3 30.5
Evans, H. 3 49.13
Evans, J.A. 3 71.34, 72.13, 72.16 4 71.83, 71.87, 85.34
Evans, J.M. 2 60.75
Evans, M. 4 71.27
Evers, H. 1 28.1
Evertsson, B. 2 82.51
Evonkova, E.V. 2 83.19
Evrard, G. 3 20.18 4 20.12, 24.19, 59.2
Exner, O. 3 10.7

Faber, D.H. 4 21.20
Fabiani, C. 4 70.9
Fackler Junior, J.P. 3 80.18, 85.37, 85.38 4 85.35
Fackler, J.P. 1 3.69
Fagherazzi, G. 2 72.39, 72.53 4 46.1
Fair, C.K. 4 65.5, 83.29
Faithful, B.D. 2 65.37
Falgueirettes, J. 1 9.45, 32.25 3 2.9, 27.5, 27.6, 32.24 4 9.11, 32.13, 63.9
Falkenberg, G. 3 35.16
Fallani, G. 4 83.126
Faller, J.W. 2 73.39
Fanfani, L. 1 38.43 4 35.10, 70.4, 83.49, 85.8
Fankuchen, I. 1 1.1, 40.11, 44.68 2 63.7
Farag, M.S. 1 19.64 4 17.8
Fares, V. 2 84.60, 85.93, 85.96 3 68.20, 85.48 4 79.11, 85.17, 85.18, 85.19, 85.31
Farina, M. 1 29.19 2 61.17, 61.19
Farney, R.F. 1 53.35
Farnham, P. 4 86.20
Farnham, W.B. 4 52.4, 52.5
Farnsworth, N.R. 3 59.26
Farrier, D.S. 3 58.20
Farris, F.F. 4 60.35
Fasiska, E.J. 3 38.10
Faught, J.B. 3 64.25
Faure, R. 4 81.38
Fava, G. 1 9.3, 32.15, 39.23, 39.25 2 69.11, 79.9, 79.33, 84.5, 84.8, 84.9, 85.5
Fawcett, J.K. 1 28.7, 30.18, 45.55, 51.35, 51.43 2 65.24 3 20.17, 23.1, 36.18, 36.25, 46.5, 46.6, 58.47, 68.3
Fay, R.C. 4 77.8
Fayos, J. 3 13.2 4 63.7, 83.9
Fealey, T. 4 76.9
Featherman, S.I. 4 64.13
Fechtig, B. 4 33.47
Fedder, W. 4 83.10
Fedeli, W. 1 35.3, 35.4, 35.7, 35.12, 35.14, 35.15, 38.37, 50.48 3 33.45, 50.28, 58.15, 59.33 4 1.11, 32.22, 33.42, 33.45, 35.26, 36.34, 54.19
Fedorov, Yu.G. 4 24.7
Fedutin, D.N. 4 60.40
Feher, F. 3 39.33 4 39.7
Fehlhammer, W.P. 4 65.11, 71.34, 73.27
Fehlmann, M. 1 25.20, 58.122 2 82.54 3 41.5
Fehn, J. 4 32.9
Feil, D. 1 8.19, 8.20, 8.22, 8.23 2 61.22 4 60.3
Felix, D. 1 11.45
Feltham, R.D. 4 80.1, 86.13
Fenical, W. 4 53.2, 59.8
Fenimore, C.P. 1 28.27
Fenn, R.H. 1 11.23 2 73.6, 81.39, 85.13 3 71.27, 72.27, 73.51, 83.35, 85.53, 85.55 4 46.3, 73.22, 85.40
Fennessey, J.P. 2 71.51, 73.10, 73.32 3 58.26, 71.18, 85.14
Fenton, D.E. 4 67.24, 76.41

Ferguson, G. 1 13.1, 13.2, 13.6, 19.27, 19.28, 22.6, 29.8, 31.20, 31.38, 37.28, 38.58, 52.7, 53.12, 54.3, 54.15, 54.18, 54.20, 54.22, 58.68, 59.45, 59.54, 59.58, 59.67 2 65.30, 65.31, 65.32, 65.51, 66.16, 66.17, 73.102, 74.22, 74.23 3 31.27, 31.28, 39.38, 51.26, 51.30, 51.44, 53.17, 53.24, 58.54, 59.16, 59.27, 60.42, 60.55, 65.11, 71.50, 71.51, 86.58 4 35.8, 51.13, 51.22, 51.32, 58.4, 58.5, 64.44, 65.12, 65.13, 65.14, 66.3, 66.4, 66.9, 66.10, 85.1
Ferguson, G.L. 3 3.6 4 3.11
Ferguson, R.B. 3 73.7
Fergusson, J.E. 3 83.144, 83.147
Fernando, Q. 1 32.33 2 83.140, 85.47, 85.58, 85.60, 85.70, 85.81 3 32.32, 68.22, 77.15, 83.87, 83.94, 83.141, 85.28, 85.29 4 59.6, 68.10, 83.104, 85.13
Ferrari, A. 2 81.35, 83.10
Ferrari, M. 1 11.1 2 79.16, 85.14 3 79.10
Ferrari, M.B. 4 66.8
Ferraris, G. 1 10.13, 13.53, 40.15, 40.16 2 83.25, 83.26 3 40.17 4 72.21
Ferrier, R.P. 1 29.21, 29.22
Ferrier, W.G. 1 29.16, 45.36, 45.39, 46.2
Ferris, J.P. 1 58.100
Field, D.S. 3 73.13
Figgis, B.N. 2 81.75 4 79.8
Filappakis, S.E. 1 1.80
Filippakis, S.E. 1 1.100, 7.22, 38.29, 39.42, 39.43
Filippini, G. 3 31.1 4 22.2, 31.4, 31.10
Filippova, T.V. 3 76.70
Finch, N. 3 58.51 4 24.16
Fink, R. 3 31.13
Fink, W. 2 63.19, 63.30
Finnegan, W.G. 2 60.60, 83.27
Fischer Junior, R.G. 3 80.18
Fischer, E.O. 2 71.14, 73.20 4 74.10
Fischer, J. 2 84.3
Fischer, M.S. 3 48.17, 49.6 4 49.8, 49.12
Fischer, W. 2 64.68
Fischinger, A.J. 2 81.30
Fischmann, E. 1 18.12, 18.17, 18.18, 18.19
Fisher, C.D. 2 60.121
Fisher, C.G. 3 82.6
Fisher, J. 4 51.15, 51.17
Fisher, M.S. 1 20.11
Fitzgerald, A. 4 64.16, 64.19
Fitzwater, D.R. 1 11.11, 11.12, 11.13, 46.8
Flaig, B.D. 4 8.1, 44.7
Fleischer, E. 3 32.14, 32.15, 83.80
Fleischer, E.B. 1 19.43, 31.2, 42.11, 42.12, 49.1, 49.15, 49.21, 49.22, 49.23, 49.27, 49.28, 49.30, 49.31, 49.33, 52.26 2 73.55, 75.26, 81.12, 83.143, 83.150, 83.184, 83.205 3 16.8, 51.7, 62.18, 83.34, 83.117, 83.134, 83.135, 83.138 4 49.24, 83.121
Fleming, J.E. 1 39.7 2 83.161, 83.193, 84.35
Fletcher, R.O.W. 1 3.6
Fletcher, S.R. 4 71.3, 80.5, 80.14, 85.38
Fletterick, R.J. 3 48.26 4 71.9
Flilippini, G. 1 52.1
Flippen, J. 1 41.39
Flippen, J.L. 3 8.17, 61.6, 61.7 4 31.5, 34.1, 34.2, 37.16, 41.16, 41.33, 44.26, 51.37, 61.2
Flood, M.T. 2 76.81 4 86.39, 86.40
Florencio, F. 3 13.12 4 13.11
Florian, L.R. 2 84.18
Floyd, J.C. 1 59.5
Flurry, R.L. 1 3.44
Flynn, J.J. 1 3.32, 11.9, 42.1, 50.32 2 63.26 3 20.10, 42.1, 63.10 4 21.18, 21.19, 69.18, 69.19, 75.20, 75.29
Fodor, G. 3 37.6 4 35.12
Folkers, J.B. 1 3.67
Follner, H. 3 2.7 4 2.6, 83.17, 83.18
Folting, K. 1 19.21, 19.22, 19.26, 32.13 2 83.197
Fontal, B. 2 65.41
Forder, R.A. 3 69.8, 69.9, 69.16 4 10.1
Foresti Serantoni, A. 4 9.13
Foresti Serantoni, E. 4 9.14
Foresti-Serantoni, E. 4 64.1, 64.2
Foresti, E. 3 28.10, 32.13, 36.21
Fornies-Marquina, J.M. 4 13.8, 17.9, 19.14, 51.16
Forrest, J.G. 2 81.27, 81.45, 81.48, 81.55, 81.66, 81.81
Forrest, K.P. 3 83.111, 86.58 4 75.23, 83.37
Forrester, A.R. 2 73.59
Forrester, J.D. 2 82.59, 85.39, 85.40
Forsellini, E. 2 78.35 3 71.26, 72.51, 80.19, 80.20, 83.45 4 71.23, 71.41, 85.26, 85.32, 86.12
Forst, D. 2 64.2
Forster, R. 1 59.6
Forsythe, G.D. 3 53.1, 53.2
Fosheim, K. 2 70.8, 70.9
Foss, O. 1 8.36, 8.37, 11.4, 11.25, 39.16, 39.35, 39.36, 41.6 2 70.2, 70.3, 70.4, 70.7, 70.8, 70.9, 70.11, 70.12, 70.13, 70.14, 70.15, 70.16, 70.24, 76.12 4 70.6, 70.7, 70.8, 70.12
Foster, M.S. 3 86.33
Foster, R. 1 40.29
Fourme, R. 1 5.1, 38.5, 38.6 4 19.3, 19.4
Foust, A.S. 2 60.155 3 86.33 4 73.13
Fowell, D.T. 3 56.6
Fowles, G.W.A. 2 81.46, 84.27 3 83.5 4 84.2
Fowlie, A.D. 3 76.5
Fowweather, F. 1 16.58, 16.59
Fox, M.D. 2 78.42
Fox, M.R. 2 78.23, 78.36
Foxman, B.M. 2 81.89, 86.12 3 81.53 4 76.11, 83.13
Francis, J.N. 3 62.15 4 72.22, 72.23, 72.24, 72.30
Frank, G.W. 3 47.24, 47.28, 58.1 4 30.5, 30.6
Frank, J.A.K. 3 71.31
Frank, R.W. 3 42.2
Franks, W.A. 2 82.18, 82.58 4 82.15
Franzini, M. 2 80.3
Frasci, A. 1 1.121
Fraser-Reid, B. 3 38.11
Fraser, A.R. 3 72.39 4 63.4, 69.16, 69.17
Fraser, K.A. 2 80.22, 82.25, 82.48
Frasson, E. 2 68.9, 73.19, 77.10, 78.22, 78.31, 78.44, 83.56, 83.59, 83.60, 83.82, 83.83, 85.64, 86.38 4 19.2
Fraterman, H.A. 4 21.17

Fratini, A.V. **1** 16.60, 31.45, 31.46, 52.27, 58.31, 58.33 **3** 60.52 **4** 28.9
Fray, R.C. **4** 77.9
Fredericks, R.J. **3** 86.36
Freedman, H.H. **1** 22.17 **3** 69.35
Freedman, L.D. **4** 64.32
Freeman, G.R. **3** 16.14, 16.15, 21.19, 44.2, 44.3, 44.5 **4** 36.1
Freeman, H.C. **1** 8.31, 39.27, 41.33 **2** 82.5, 82.6, 82.7, 82.15, 82.23, 82.24, 82.25, 82.26, 82.27, 82.30, 82.31, 82.35, 82.36, 82.38, 82.39, 82.47, 82.54, 83.23, 83.31, 83.54, 83.173, 84.10 **3** 48.11, 76.27, 76.74, 76.75, 82.2, 82.3, 82.10, 82.11, 82.12, 82.13, 82.14, 82.15, 82.17, 83.20 **4** 82.1, 82.2, 82.8, 83.15
Freer, S.T. **1** 48.27, 48.28
Fregerslev, S. **1** 58.6
Frehel, D. **3** 37.6
Freiburg, C. **2** 78.19
Freiser, H. **3** 83.143 **4** 85.30
Frenz, B.A. **4** 75.12, 83.133
Frew, D. **3** 39.40
Frey, A.J. **1** 58.87
Frey, M. **3** 60.7, 60.10 **4** 38.1, 84.5
Frey, M.N. **4** 48.8
Fridrichsons, J. **1** 48.93, 50.7, 51.44, 58.23, 58.27, 58.32, 58.65, 58.77, 58.81, 58.94, 58.118 **4** 83.66
Fried, J. **3** 51.7
Friedel, M.K. **3** 49.5
Friedlander, P.H. **1** 20.19, 29.9, 29.23, 30.24
Friedman, L.B. **2** 62.13, 62.36
Fries, D.C. **3** 45.35, 48.35, 85.37, 85.38
Fritchie Junior, C.J. **1** 3.28 **2** 69.35 **3** 35.24, 36.2, 60.35, 60.38, 69.10 **4** 36.3, 60.27, 60.28, 60.29, 71.86
Fritchie, C.J. **1** 7.30, 27.11, 32.41, 36.6, 36.23, 52.7 **2** 60.130, 73.89, 85.43
Fritz, G. **2** 63.17
Fritz, H. **3** 37.9
Fritz, H.P. **2** 85.65
Frost, G.H. **2** 83.217
Frost, R. **2** 81.12
Frye, R.S. **3** 33.11
Fryer, R.I. **3** 9.19, 35.17
Fuchs, B. **4** 38.5, 38.9
Fuganti, C. **3** 58.28 **4** 58.26
Fuji, K. **1** 52.25 **3** 17.5
Fujii, K. **1** 1.108
Fujimaki, H. **4** 78.1
Fujimoto, Y. **4** 48.20, 48.22
Fujita, J. **4** 83.78
Fujita, T. **4** 53.10
Fujiwara, H. **3** 27.4
Fujiwara, T. **1** 56.20, 58.93 **3** 33.26, 35.10, 44.12, 47.32, 48.40, 56.13 **4** 44.14, 44.15, 56.5
Fukazawa, Y. **1** 31.16, 58.111 **4** 22.3, 22.4, 22.5, 27.15
Fukuyo, M. **1** 13.12 **2** 65.18, 65.20 **3** 36.4
Fulke, J.W. **1** 59.67
Funakoshi, K. **4** 48.33, 48.35
Furberg, S. **1** 11.56, 38.22, 44.40, 45.4, 45.5, 45.6, 45.7, 45.57, 45.58, 45.59, 45.60, 45.64, 47.9, 47.15 **3** 44.7, 44.10, 45.33 **4** 44.1
Furlani, C. **3** 85.22 **4** 85.33
Furnas Junior, T.C. **3** 49.1, 49.7
Furst, A. **4** 31.28
Furue, K. **3** 69.11
Furukawa, K. **1** 22.4, 22.5
Furusaki, A. **1** 1.89, 1.108, 13.36, 22.13, 28.5, 28.14, 31.13, 50.31, 50.38, 58.35, 58.120, 59.25 **3** 17.5, 33.42, 36.4, 50.5, 50.20, 53.20, 72.20, 81.24 **4** 58.1, 58.2, 81.32, 81.33, 81.35

Gabe, E.J. **1** 2.46, 2.47, 2.76, 3.2, 4.15, 35.46, 52.22, 53.9, 58.25 **2** 85.46 **4** 26.3, 38.25
Gadet, A. **4** 81.30, 83.97
Gadret, M. **1** 33.43, 33.56, 33.57 **3** 33.20, 33.25
Gaetani, A. **1** 35.8 **3** 41.2, 76.1
Gafner, G. **1** 19.6, 19.7, 24.1, 24.4, 58.11 **2** 60.69, 83.130, 86.57 **3** 41.7, 58.5, 75.47 **4** 41.9, 41.10, 59.25, 68.8
Gagnaux, P. **2** 60.50
Gainsford, G.J. **3** 33.30, 33.31, 33.32, 71.71 **4** 71.71, 71.73, 71.89, 73.24
Gainullina, R.S. **3** 76.13
Galeffi, C. **3** 59.33
Galesic, N. **4** 81.11
Galigne, J.-L. **1** 9.45, 32.25 **3** 2.9, 9.14, 32.24 **4** 63.9
Galigne, J.L. **4** 2.1, 9.11
Gallagher, R.T. **3** 59.28
Galles, C.J. **3** 71.2
Gallezot, P. **1** 44.41
Ganis, P. **1** 1.78, 1.86, 1.93, 1.97, 1.110, 21.24, 21.25, 38.49, 38.59 **2** 72.38, 73.18, 75.67 **3** 1.32, 22.4, 75.36, 75.37, 83.86 **4** 1.9, 15.6, 16.6, 50.17, 72.12, 83.90
Gantzel, P.K. **1** 11.21, 31.35 **2** 67.6, 67.7
Garaj, J. **2** 81.6, 83.30 **4** 76.1
Garbassi, F. **4** 46.1
Garbuglio, C. **1** 1.51, 39.62
Garcia-Blanco, S. **1** 13.44, 59.46 **3** 13.13 **4** 13.11, 77.10
Gardia, C. **1** 3.78
Gardner, A.P. **4** 78.8
Gardner, J.A.F. **1** 59.61
Garner, C.D. **3** 69.3
Gartland, G.L. **3** 43.4, 43.5, 60.27, 60.28 **4** 43.9, 60.19
Garton, G. **2** 86.67
Garza, A. **4** 71.70
Gash, A.G. **4** 69.11, 69.22, 69.27, 74.11
Gasparri, G.F. **1** 41.15 **2** 79.2, 79.16, 79.18, 79.20, 79.26, 79.27, 79.31, 80.1, 83.8, 83.9, 85.2, 85.14, 85.35 **3** 8.5, 16.4, 79.10 **4** 66.8
Gasparrini, F. **4** 71.41
Gassmann, J. **1** 49.16, 54.24
Gastaldi, L. **4** 83.89, 85.33
Gatehouse, B.M. **2** 86.5 **3** 45.21, 45.22, 45.23, 78.13 **4** 24.3, 43.1, 43.2, 71.79, 71.88, 71.91, 78.8, 78.16
Gattow, G. **1** 11.3 **3** 67.4
Gaughan, A.P. **3** 86.96 **4** 82.13, 83.72, 86.61, 86.68, 86.69

Gaultier, J. **1** 11.44, 24.21, 24.24, 24.25, 25.3, 25.4, 25.5, 25.6, 25.7, 25.13, 25.16, 25.17, 25.18, 25.21, 26.3, 38.32, 38.35, 59.39 **3** 25.1, 25.2, 25.4, 25.5, 26.9, 60.32 **4** 21.4, 25.1, 27.10, 35.9, 38.22
Gauthier, J. **4** 41.30
Gavezzotti, A. **3** 19.11 **4** 31.20
Gaw, H. **4** 81.37
Gebala, A.E. **3** 83.117 **4** 83.121
Gebreyesus, T. **4** 58.33
Gee, R.J.D. **4** 71.68
Gehrz, R.C. **4** 40.10
Geise, H.J. **1** 42.3, 46.15, 51.8, 51.30, 51.31, 51.32, 51.33, 51.34 **3** 51.37, 51.38
Geissman, T.A. **3** 53.18
Gelfano, E.M. **3** 17.4
Geller, S. **1** 3.16 **3** 3.19
Gemenden, C.W. **3** 58.51
Gemmel, K.W. **3** 53.23
Gemmell, K.W. **3** 58.67
Genet, F. **1** 3.45, 33.49, 35.19, 35.20 **2** 84.14 **4** 7.12
Gent, A.R. **4** 71.15
Gent, P.A. **1** 45.71
Gentile, P.S. **3** 79.6, 79.7
Genzurg, S.L. **3** 17.4
Geoffre, S. **1** 26.3 **4** 25.2, 51.9
George, J.W. **2** 70.17, 70.18
Gerdil, R. **1** 44.32 **4** 31.14, 31.16
Gerken, V.A. **1** 3.29
Gerkin, R.M. **3** 40.18
Gerloch, M. **2** 73.29, 73.46, 78.13, 78.14, 83.89 **3** 78.24
Germain, G. **1** 10.14, 10.15, 10.16, 41.26, 41.27, 41.30, 41.46 **3** 20.18, 48.15, 48.44, 60.17, 60.18 **4** 32.23, 33.43, 33.44, 40.17, 48.49, 51.25, 51.30, 51.31, 51.34
Gerteis, R.L. **2** 68.15
Gervasio, G. **4** 72.21
Geue, R.J. **4** 83.24
Ghilardi, C.A. **3** 78.18, 86.44, 86.53 **4** 76.43, 76.44, 83.56, 83.57
Ghose, S. **1** 43.12
Giacomello, G. **1** 1.133, 13.30, 33.1, 33.70, 33.72, 35.3, 35.4, 35.7, 35.12, 35.13, 35.14, 35.15, 44.53, 50.48 **3** 50.28, 58.15
Giacomello, P. **4** 8.11
Giannini, U. **2** 72.39
Giarda, L. **4** 46.1
Gibbons, C.S. **1** 31.44 **3** 58.64 **4** 31.1, 31.3, 71.18
Gibson, C.S. **2** 71.16
Gibson, D.T. **1** 11.29
Gibson, J.G. **2** 83.85
Gibson, J.W. **1** 37.17 **3** 48.47 **4** 44.21
Gibson, K.H. **4** 52.6
Gibson, R.E. **3** 36.27, 39.32
Gieren, A. **1** 51.49 **3** 39.33, 49.15 **4** 32.9, 36.37, 71.92
Giering, W.P. **2** 73.77
Giese Junior, R.F. **3** 21.3, 21.4
Giesen, M.T. **4** 21.6
Giesen, W.van de **4** 84.4
Giglio, E. **1** 33.81, 44.67, 59.8, 59.20 **2** 60.110, 60.111, 60.112 **4** 8.11
Gignere, P.A. **2** 60.8
Gil, E.R.de **3** 72.12
Gilardi, R. **3** 64.59
Gilardi, R.D. **1** 58.31, 58.33 **3** 51.28, 58.63 **4** 9.12, 34.1, 34.2, 37.16, 40.15, 42.6, 51.37, 54.5, 54.6
Gilchrist, A.B. **4** 71.85, 83.139
Gill, N.S. **3** 83.119
Gillard, J. **3** 58.6
Gillard, R.D. **2** 77.1, 81.39, 82.28, 82.34, 82.37 **3** 76.63, 82.5 **4** 83.63
Gilli, G. **4** 78.2
Gillier-Pandraud, H. **1** 17.7, 17.8, 17.27, 17.28 **2** 84.12 **3** 1.7 **4** 3.2, 3.3, 17.3, 17.6, 17.7
Gillier, H. **1** 43.2, 43.3
Gillot, R.J.J.H. **1** 39.31
Gillum, W.O. **2** 83.190
Gilmore, C.J. **3** 3.52, 71.64 **4** 58.40, 69.23, 71.77
Ginderow, D. **4** 69.12
Ginsberg, A.P. **2** 73.30
Ginsberg, S.L. **3** 40.2
Giordano, F. **1** 21.29 **2** 64.6, 84.50, 84.51 **3** 33.23
Gipson, R.M. **1** 27.15
Giraldi, G. **1** 9.3, 39.23, 39.25 **2** 84.9
Girard, C. **4** 65.6, 66.5
Girling, R.L. **2** 79.19, 85.56 **3** 45.10, 79.20, 79.24, 79.25 **4** 45.4, 45.5, 45.6
Gitany, R. **3** 75.34
Giudice, F.L. **2** 72.39
Giuliano, V. **3** 75.37
Giuseppetti, G. **2** 83.81
Gladkova, V.F. **4** 9.10
Glanville, J.O. **2** 72.83
Glass, G.E. **2** 71.18
Glasser, L.S.D. **3** 76.24, 85.52
Glazer, A.M. **4** 36.5, 36.6, 36.7, 36.8, 36.9, 36.10
Glehn, M.von **1** 45.27 **3** 36.7, 36.10 **4** 36.15
Glen, G.L. **2** 81.52
Glick, M.D. **1** 3.83, 49.24, 49.29 **2** 75.63, 76.61, 78.45, 84.15, 84.16, 84.17 **3** 30.4, 39.45, 45.29, 58.29, 64.67, 65.12, 68.18, 84.14 **4** 24.12, 39.47, 81.46, 81.47, 81.56, 81.58, 83.77
Glidewell, C. **4** 63.10
Glinskaya, E.A. **2** 85.59
Glockling, F. **3** 73.43
Glockner, P. **3** 75.17
Gloss, R.A. **1** 4.5
Glotter, E. **4** 51.49
Glusker, J.P. **1** 1.79, 2.74, 2.76, 2.78, 2.79, 2.80, 2.81, 4.15, 44.20, 48.91 **2** 85.46 **3** 2.19, 2.22 **4** 2.20, 2.21, 2.22, 26.3, 26.6, 36.21, 36.32
Gluskova, V.P. **1** 11.30 **2** 62.29, 62.42, 69.20
Gluyas, R.E. **3** 15.1, 15.2
Go, K.C. **4** 32.11, 51.53
Go, K.T. **1** 34.5, 41.19, 41.20 **3** 54.9
Goaman, L.C.G. **1** 21.26, 21.27
Gobillon, Y. **2** 60.23, 60.30, 60.32, 60.46
Godfrey, J.M. **3** 49.13
Goebel, C.V. **4** 81.42
Goebel, P. **1** 23.1
Goedken, V. **4** 83.34, 85.25

Goedkoop, J.A. **1** 1.31
Goel, R.G. **4** 66.3, 66.4, 66.9, 66.10
Goggin, P.L. **2** 85.27
Gogoi, B.N. **3** 11.7
Gol'der, G.A. **1** 5.24, 5.30, 7.23, 15.3, 26.1, 26.2, 26.7, 26.8, 39.77 **3** 26.1, 26.2, 35.20 **4** 24.4, 24.6, 24.7
Golankiewicz, K. **3** 44.37
Gold, K. **2** 72.41 **3** 62.14, 71.60, 75.32, 75.33, 83.93 **4** 62.13, 71.2
Goldberg, I. **4** 8.2, 38.5, 38.9, 39.18, 39.19, 60.23, 69.13
Goldberg, S.Z. **4** 71.50
Goldfarb, T.D. **2** 63.15
Goldish, E. **1** 10.19, 16.10 **4** 12.9
Goldman, M. **1** 1.5
Goldschmidt, G.H. **1** 35.11
Goldschmied, E. **2** 76.34 **3** 76.30, 83.76
Goldstein, M. **4** 83.25
Goldstein, P. **1** 22.8, 32.5, 32.35, 32.36, 32.37, 32.39, 32.40, 32.43, 32.44 **2** 62.45, 64.66, 76.55
Golic, L. **1** 2.10, 2.11, 2.45 **4** 48.29, 60.4
Gollnick, K. **3** 28.5, 31.18
Golomb, M.L. **2** 82.7 **3** 82.2, 82.3, 82.10, 82.11
Golovastikov, N.I. **2** 66.11, 66.12
Golovina, N.I. **1** 10.2, 10.3 **3** 12.1
Gomes de Mesquita, A.H. **1** 11.61, 12.14
Good, M.L. **4** 81.59
Good, R.H. **4** 35.8
Goodacre, G.W. **2** 64.24 **3** 64.1, 64.10, 64.21 **4** 42.3
Goodall, B.L. **4** 71.36
Goodall, D.C. **2** 85.49, 86.105
Gooden, E.L. **3** 37.6
Goodfellow, R.J. **2** 85.27
Goodgame, D.M.L. **1** 3.38, 3.39 **2** 76.21, 76.50, 83.104, 83.215 **3** 83.43
Goodman, M. **4** 15.6, 16.6
Goodman, S.H. **1** 31.11
Goodwin, F.H. **1** 45.72
Goodwin, T.H. **1** 13.20, 17.29, 30.24, 45.11
Gopal, H. **4** 31.5
Gopalakrishna, E.M. **1** 15.10, 51.10, 51.12, 51.21, 51.22 **2** 60.151, 84.64 **3** 15.5, 15.6
Gorbel, C.V. **2** 75.7
Gordeeva, G.A. **2** 81.24
Gordon-Gray, C.G. **4** 38.21
Gordon, J.T. **1** 56.6
Gore, E.S. **4** 83.31
Gorres, B.T. **3** 60.33
Gosling, K. **3** 68.7, 68.8
Goto, K. **1** 58.36, 58.57
Gottardi, G. **2** 65.16, 66.13
Gottlicher, S. **1** 37.19 **2** 60.77, 84.53 **3** 35.5, 37.9 **4** 37.11
Gough, T.E. **3** 77.18
Gougoutas, J.Z. **1** 17.30 **3** 13.15, 40.13, 51.39, 58.30 **4** 13.17, 13.18, 35.37, 59.27
Gould, E.S. **1** 32.24, 39.2, 39.18
Gould, R.O. **2** 85.24 **3** 85.43 **4** 45.21, 85.36
Goursolle, M. **3** 33.20, 33.25
Graaf, R.A.G.de **4** 21.6, 27.2
Grabowski, M. **3** 39.8
Graddon, D.P. **2** 78.6, 83.101, 83.157
Graeber, E.J. **1** 3.26 **3** 9.5
Graham, A.J. **2** 81.39 **3** 71.27, 72.27
Graham, J. **4** 79.8
Graham, W.A.G. **3** 69.27 **4** 73.13
Grainger, C.T. **1** 9.44 **4** 3.23, 21.14
Gram, F. **1** 9.20
Gramaccioli, C.M. **1** 17.25, 17.31, 17.32, 52.1 **2** 60.103, 82.10, 82.11 **3** 31.1, 31.25 **4** 31.4, 31.6, 31.21
Gramacciolo, C.M. **4** 31.22
Grampel, J.C.van de **4** 27.14
Grandjean, D. **1** 41.14
Granoff, B. **2** 75.34, 77.21
Grant, D.F. **1** 13.32, 21.26, 21.27, 52.22, 53.18 **2** 68.2 **4** 17.4, 53.11
Grant, I.J. **1** 56.1, 58.74
Grant, W.K. **1** 28.4
Gray, H.B. **1** 3.40 **2** 76.75, 76.81, 85.73, 85.86, 85.98, 86.26 **3** 85.27, 86.77 **4** 73.3, 73.4, 86.39, 86.40
Gray, R.T. **1** 52.8
Graybeal, J.D. **3** 69.1
Graziani, R. **2** 78.28, 78.35 **3** 72.49, 72.51, 80.19, 80.20, 81.55, 83.45, 85.19, 86.19, 86.25 **4** 71.23, 71.41, 71.92, 85.26, 85.32, 86.12
Grdenic, D. **1** 8.45 **2** 71.15, 77.47, 77.49, 81.60 **3** 66.5, 83.9 **4** 83.137, 83.138
Greaves, M.L. **3** 47.6, 49.14
Green, C.D. **1** 53.26 **2** 85.58 **3** 53.19
Green, D.W. **1** 44.30
Green, E.A. **3** 47.10
Green, M. **1** 52.26 **3** 71.39 **4** 71.84, 71.86
Green, M.J. **3** 51.7
Green, M.L.H. **3** 71.5, 71.6, 73.47, 73.48, 73.49
Greenberg, B. **1** 9.11, 20.15
Greene, M.L.H. **4** 73.29
Greene, P.T. **3** 62.13, 69.22, 73.4, 73.12, 73.13, 83.18, 85.47, 86.43 **4** 62.1, 69.21, 82.7, 83.19, 83.20
Greenwood, N.N. **3** 64.46 **4** 84.9
Gregory, U.A. **3** 71.41, 71.42, 86.28, 86.29, 86.88
Grenthe, I. **3** 81.6, 81.10 **4** 81.21, 81.25
Grethe, G. **1** 41.29 **3** 35.23
Gribova, E.A. **1** 39.77 **3** 35.20
Grice, N. **4** 75.24
Grieb, R. **2** 83.188, 83.206
Griffe, M. **2** 60.47 **3** 48.44
Griffith, E.A.H. **3** 75.2, 75.8, 79.22 **4** 74.9, 75.19
Griffith, R.L. **2** 81.2
Griffiths, A. **3** 9.7, 28.2 **4** 83.87
Grigg, R. **3** 49.4
Grigor'eva, N.V. **1** 7.3, 10.1, 10.4, 12.1 **3** 7.1, 12.3, 12.5, 12.6 **4** 12.4, 12.7
Grim, S.O. **2** 72.83 **3** 86.34, 86.35 **4** 86.38
Grimes, R.N. **4** 62.1
Grimvall, S. **1** 14.12
Groen, M.B. **3** 39.48
Gronbaek, R. **1** 3.52 **2** 85.7
Groot, A.de **1** 35.42
Grosse, G. **2** 76.7
Grostic, M.F. **4** 27.9

Groth, P. **1** 1.3, 1.4, 21.8, 21.11, 21.20, 21.21, 38.46, 38.50, 38.52, 38.56, 45.32 **2** 60.12, 60.48, 60.91, 60.92 **3** 21.12, 21.14, 21.18, 23.4, 23.7, 33.17, 38.5, 38.9, 41.9, 48.46, 86.54 **4** 23.1, 33.21, 38.16, 67.19
Grout, R.J. **4** 35.25
Groves, D. **3** 62.3
Grundtvig, F. **4** 39.10
Grundy, K.R. **4** 86.35
Grunze, H. **2** 67.2
Grutzmacher, H.-F. **3** 63.4, 63.5, 63.6
Gsell, R. **2** 83.32
Guastini, C. **2** 81.80 **3** 81.19 **4** 76.38, 79.2, 79.3, 83.55, 84.12
Gubin, S.P. **3** 71.4, 74.9 **4** 73.10, 74.5
Gubler, B. **1** 53.40
Guggenberger, L.J. **1** 38.54 **2** 67.10, 69.43, 71.40 **3** 3.18, 62.9, 86.10 **4** 3.9, 36.12, 71.6, 71.44, 86.39, 86.40
Guggisberg, A. **3** 58.67
Gugielmetti, L. **1** 54.7
Guha, S. **3** 48.21, 82.21
Guilard, R. **4** 75.8, 75.10
Guilhem, J. **1** 26.27, 26.29 **3** 58.27 **4** 51.38, 58.7
Guin, H.W. **1** 27.15
Gum, P. **1** 35.27
Gupta, M.P. **1** 1.50 **3** 2.15, 2.16, 2.17, 2.18, 13.6 **4** 2.10, 2.16, 13.10, 19.6
Gupta, N.P. **1** 1.50
Gurr, G.E. **1** 45.22, 45.23 **2** 83.179 **3** 60.33
Gurskaya, G.V. **3** 17.4 **4** 48.41
Gusejnova, M.K. **2** 81.5, 81.14, 83.13
Gusev, A.I. **2** 72.72, 72.73, 72.84 **3** 40.2, 72.40, 72.48, 72.53, 75.46 **4** 63.8, 64.17, 73.10
Guss, J.M. **2** 82.6, 82.27, 82.30, 82.31, 82.39, 83.31 **3** 71.70 **4** 71.71, 71.73, 71.82, 71.89
Guss, M. **4** 75.22
Gustaffson, B. **3** 3.47
Guttman, L.J. **4** 59.29
Guttormson, R. **3** 54.2
Gwinup, P.D. **4** 65.8

Haagensen, C.O. **1** 3.73, 38.36
Haas, D.J. **1** 1.74, 2.31, 8.25, 56.10 **2** 60.16, 60.31, 60.33, 60.68
Haase, W. **1** 3.66, 33.20 **2** 84.28
Habermehl, G. **1** 28.23, 37.19 **2** 60.77 **3** 37.9, 58.9 **4** 37.11
Hacker, M.J. **4** 85.34
Hackert, M.L. **3** 3.36, 45.4
Hadicke, E. **3** 71.14 **4** 5.5
Hadinec, I. **3** 53.7
Hady, S.A. **3** 2.10
Hadzi, D. **4** 2.2, 2.3, 60.4
Hagihara, H. **2** 69.12, 69.22, 69.27, 80.4, 85.61 **4** 69.20
Hagihara, N. **2** 86.71, 86.81
Hahn, T. **1** 48.6
Hai-Fu Fan, A **1** 48.48
Haigh, J.M. **4** 77.15
Haight, H.L. **2** 75.2, 75.77
Haines, L.M. **2** 86.57
Hair, N.J. **3** 33.35 **4** 3.6, 3.20, 11.4, 41.13, 64.34
Haisa, M. **4** 33.10, 33.11
Hale, P.K. **3** 76.41
Hale, R.L. **3** 51.41
Haley, J. **3** 83.120 **4** 83.99
Hall, D. **1** 10.7, 26.26, 50.14, 58.34 **2** 76.26, 77.16, 77.17, 78.4, 78.5, 78.7, 78.16, 78.17, 78.18, 78.21, 78.25, 78.27, 78.29, 78.40, 78.41, 78.46, 78.50, 78.58, 82.55, 83.67, 83.68, 83.109, 83.112, 83.192, 83.194 **3** 45.39, 54.4, 60.1, 60.16, 60.39, 69.20, 69.23, 76.50, 78.8, 78.29, 83.12, 83.33, 83.66, 85.6 **4** 25.4, 25.5, 25.6, 59.22, 76.42
Hall, D.DeW. **3** 79.2
Hall, E.A. **2** 85.19 **3** 74.1, 74.6, 74.7, 74.12, 74.14
Hall, G.S. **2** 83.99
Hall, L.H. **2** 62.35 **3** 58.21, 73.9
Hall, M.C. **3** 86.91
Hall, S.R. **1** 16.11, 16.12, 56.19, 58.41, 58.53, 58.78
Hallak, N. **4** 38.21
Halpern, J. **2** 83.94 **4** 81.61
Halsall, T.G. **3** 54.14, 56.8, 56.11
Haltiwanger, R.C. **4** 54.4, 58.40
Haluska, R.J. **1** 34.8 **4** 36.22
Ham, J.T. **3** 45.38
Ham, P.J. **3** 56.12
Hamanaka, T. **4** 54.7
Hamasaki, T. **4** 59.20
Hamburger, G. **1** 11.53, 11.54
Hamelin, M. **1** 43.4 **2** 83.53 **4** 83.27
Hamer, G. **3** 60.34
Hamer, J. **4** 40.7
Hamilton, J. **3** 48.48
Hamilton, J.A. **1** 20.20, 48.76, 53.22, 56.1, 58.76, 59.55 **4** 58.34
Hamilton, W.C. **1** 1.28, 8.43, 8.44, 10.22, 14.14, 15.22, 28.15, 30.27 **2** 63.15, 64.22, 64.60, 64.61, 66.19, 66.21, 69.1, 77.65, 85.41 **3** 3.39, 19.6, 19.7, 64.69, 72.1, 83.51 **4** 9.3, 45.10, 45.11, 48.5, 48.8, 48.10, 48.29, 64.45, 76.33, 83.42, 83.132
Hammer, C.F. **1** 51.42
Hammer, H. **1** 45.32
Hamor, M.J. **1** 49.19, 49.26 **2** 76.63
Hamor, T.A. **1** 20.20, 35.6, 36.20, 42.10, 45.71, 49.18, 49.19, 49.26, 52.13, 54.13, 56.1, 58.70, 58.74, 58.76, 59.64 **2** 82.56, 84.52 **3** 34.3, 45.26 **4** 22.1, 22.6, 36.11, 42.2
Hampson, G.C. **1** 9.34 **2** 71.16
Hampton, C.R.S.M. **4** 86.62
Han, K. **1** 3.51
Hanazaki, F. **1** 21.22, 21.23
Handa, R. **4** 40.11
Handy, L.B. **1** 3.64 **3** 3.37, 64.62, 64.63, 64.64
Haner, B.A. **1** 51.17
Hanic, F. **1** 37.11, 53.19, 59.12 **2** 74.20, 81.78, 81.86, 83.117, 83.118 **3** 73.54 **4** 79.7
Hanicova, K. **2** 81.86
Hannah, R.B. **1** 7.12
Hannaway, C. **2** 73.102 **3** 28.5, 31.18, 51.44 **4** 58.30

Hanouskova, D. **4** 31.32
Hansen, F. **3** 8.4
Hansen, H.N. **4** 45.20
Hansen, L.K. **3** 39.3
Hansen, P.J. **2** 73.36
Hanson, A.W. **1** 17.41, 29.4, 29.25, 29.26, 29.27, 30.26, 31.21, 31.34, 36.3, 36.14, 36.15, 36.16, 36.17, 38.55, 50.6, 58.20, 58.60 **2** 60.102, 60.108, 60.115, 60.124, 60.127, 60.143, 60.152 **3** 31.20, 37.11, 39.44 **4** 16.15, 39.33, 62.14
Hanson, J.C. **1** 51.1 **3** 30.2, 51.6, 83.49 **4** 45.19, 62.2, 62.3, 62.4
Hanson, J.R. **3** 51.9 **4** 54.11, 54.14
Hansson, E. **2** 81.41 **3** 81.14
Hansson, I. **2** 66.7
Hanzawa, Y. **3** 36.20
Haque, F. **4** 72.15
Haque, M.-ul- **1** 39.93, 46.5, 46.10, 53.1, 53.2, 53.23 **2** 64.42, 69.10 **3** 46.2, 46.7, 53.4, 53.18, 64.16, 64.17, 64.34, 64.35, 64.39, 76.44, 83.61, 83.70, 84.26 **4** 53.1, 64.25
Harada, H. **1** 50.37, 58.86
Harada, S. **3** 50.22
Harada, Y. **1** 13.16
Harayama, T. **1** 58.93
Hardcastle, G.A. **1** 50.26
Hardgrove Junior, G.L. **4** 39.25
Hardgrove, G.L. **1** 27.1, 27.2, 27.3, 51.47 **2** 73.28 **4** 40.9
Harding, M.M. **1** 1.147, 48.16 **2** 80.22, 82.45, 82.46, 82.48, 82.50, 85.90 **3** 48.25, 82.22, 85.43
Hardy, A.D.U. **3** 73.46, 73.52
Hardy, R. **1** 45.72
Hare, C.R. **2** 81.61, 81.62, 84.40, 85.31 **3** 82.20, 82.29
Hargis, J.H. **3** 64.17
Hargreaves, A. **1** 8.9, 11.36, 16.46, 16.57, 16.59, 17.50, 19.36, 24.19, 24.20 **3** 11.7, 11.27
Hargreaves, R.N. **3** 71.46, 71.48
Harkema, S. **1** 8.19, 8.20 **4** 60.3
Harker, D. **2** 63.8
Harle, E. **1** 54.24
Harley-Mason, J. **4** 53.5
Harlow, R.L. **3** 78.1
Harnik, E. **1** 28.28
Harper, F.R. **2** 65.25, 65.26
Harreld, C.S. **4** 69.31
Harris, C.B. **2** 65.47
Harris, C.M. **2** 83.154
Harris, D.R. **1** 8.25, 47.5, 50.43, 58.52, 58.73 **2** 60.16
Harris, G.S. **2** 65.51
Harris, P.M. **3** 15.1, 15.2
Harrison, H.R. **3** 49.3, 56.8, 56.11
Harrison, I.T. **1** 55.3
Harrison, P.G. **4** 69.37
Harrison, P.M. **1** 48.61 **2** 83.182
Harrison, R.W. **2** 69.36
Harrison, W. **3** 81.8, 86.11, 86.45 **4** 33.15, 41.1, 48.38, 68.3, 73.9, 81.17, 83.122, 85.21
Hart, F.A. **2** 83.217 **3** 83.70, 84.26
Hart, H. **2** 62.23
Hartke, K. **3** 39.15
Hartman, A. **1** 20.7
Hartmann, G. **2** 64.1, 64.20
Hartshorn, M.P. **1** 52.17
Hartsuck, J.A. **1** 54.26
Hartung, H. **2** 85.76 **3** 86.3
Haschemeyer, A.E.V. **1** 47.22, 47.23
Hasegawa, M. **4** 33.41
Hasekamp, R.C.D.E. **1** 42.2
Haser, R. **1** 18.4, 18.5
Hashimoto, M. **1** 50.2
Hassanali, A. **1** 59.23
Hassel, O. **1** 1.3, 1.4, 5.5, 7.10, 21.6, 21.8, 21.9, 21.10, 21.21, 38.22 **2** 60.3, 60.12, 60.17, 60.18, 60.29, 60.35, 60.38, 60.39, 60.44, 60.45, 60.48, 60.49, 60.53, 60.58, 60.63, 60.65, 60.84, 60.90, 60.91, 60.92, 60.120, 83.42 **3** 60.20 **4** 60.1
Hasselstrom, T. **1** 41.10
Hata, T. **1** 27.10, 28.20 **3** 50.20 **4** 27.5, 27.6, 64.35, 64.38
Hatfield, W.E. **3** 76.41, 81.18 **4** 82.9, 83.22, 83.23, 83.100
Hatsuda, Y. **4** 59.20
Hatton, J.R. **3** 42.3
Hattori, K. **1** 50.2
Hauge, S. **2** 70.13 **3** 8.12, 8.13 **4** 70.3
Haugen, L.P. **3** 86.31
Haupt, H.J. **4** 69.38
Hauptman, H. **1** 51.5 **4** 50.19, 51.15, 51.17, 51.24, 51.27, 51.33
Hauser, D. **4** 59.10
Hauser, F.M. **3** 58.19, 58.24, 58.50
Hauw, C. **1** 11.44, 24.21, 24.24, 24.25, 25.3, 25.4, 25.5, 25.6, 25.7, 25.13, 25.16, 25.17, 25.18, 25.21, 26.3, 26.37, 38.32, 38.35, 59.39 **3** 25.1, 25.2, 25.4, 25.5, 26.9, 60.32 **4** 25.1, 27.10, 35.9, 38.22
Hawkinson, S. **1** 19.43, 42.11, 42.12 **2** 83.150, 83.205 **3** 84.1, 84.7
Hawkinson, S.W. **2** 73.55, 83.143 **3** 47.4, 47.8, 49.14, 83.134, 83.135, 85.18
Hawley, D.M. **1** 53.12, 59.45 **2** 66.16, 66.17 **3** 39.38, 40.7, 53.24 **4** 50.13, 53.3
Hawthorne, M.F. **3** 62.15
Hayashi, M. **3** 53.20
Haynes, N.B. **1** 59.45
Hayter, R.G. **1** 3.64
Hazel, J.P. **3** 64.33
Hazell, A.C. **1** 41.8 **2** 62.21, 77.36 **3** 76.51 **4** 29.8
Hazell, R.G. **1** 3.53 **2** 85.6, 85.8, 85.15, 85.16 **3** 8.4 **4** 19.1, 31.27, 83.60
Healey, M.J. **2** 82.39
Healy, P.C. **4** 80.3, 80.9, 80.17
Heath, C. **2** 83.217 **3** 83.53, 83.129
Heath, C.E. **2** 83.123
Heath, G.A. **3** 85.24
Hecht, J.K. **3** 20.10
Hecht, S. **4** 47.12, 47.15
Hecht, S.M. **3** 83.25 **4** 47.16
Hechtfischer, S. **3** 31.4, 39.24, 39.25, 39.27, 39.33, 54.15
Hecker, E. **1** 54.24 **3** 54.8, 54.15 **4** 54.12
Hedberg, F.L. **4** 73.1
Hedberg, K. **2** 65.11
Heeres, G.J. **1** 39.53, 39.54, 39.55

Hefelfinger, D.T. **4** 30.5, 30.6
Hegenbarth, J.J. **1** 6.2
Hegstrom, R. **4** 40.9
Heidner, E.G. **1** 11.7
Heijkoop, E.van **1** 51.8
Heimbach, P. **4** 71.65
Heinrich, W. **2** 85.34
Heitsch, C.W. **2** 68.13
Helland-Madsen, G. **3** 39.26, 64.19
Helland, S. **1** 45.4, 45.5
Hellner, E. **1** 17.51, 36.25 **2** 64.68 **3** 63.9
Helm, D.van der **1** 2.78, 2.79, 2.80, 2.81, 41.7, 44.20, 53.10, 54.11 **2** 60.37, 60.95, 82.18, 82.22, 82.58, 84.44 **3** 20.5, 20.7, 31.13, 54.10, 82.6, 82.8, 82.12, 82.13, 82.14, 82.15, 82.28 **4** 4.4, 51.57, 51.58, 82.14, 82.15
Helmholdt, R.B. **2** 72.44 **3** 75.14
Hemingway, J.C. **1** 53.34
Hemmerich, P. **1** 39.27 **2** 83.54
Hencken, G. **3** 67.5
Hende, J.H.van den **1** 37.23, 50.22, 50.23, 50.28, 58.110, 59.15, 59.50 **2** 64.49, 64.50, 75.50
Hendricks, S.B. **1** 1.6, 1.7
Hendrickson, W.A. **3** 48.5
Henke, H. **4** 9.1
Henn, D.E. **1** 2.43 **2** 86.7, 86.67 **3** 60.36
Hennig, H. **4** 83.127
Henning, G. **3** 3.55
Hepburn, S.P. **2** 73.59
Hepworth, M.A. **3** 81.8 **4** 33.15, 41.1
Heraud, J. **1** 1.126
Herbstein, F.H. **1** 8.16, 19.6, 19.7, 24.1, 24.4, 30.16, 30.21, 36.10 **2** 60.69, 60.135, 60.154, 68.7, 68.8, 73.24 **3** 36.8, 60.45, 60.46 **4** 8.2, 29.1, 69.8, 69.13
Herceg, M. **2** 77.53 **4** 40.18, 67.12, 81.11
Herdklotz, J. **1** 41.11
Herdklotz, J.K. **3** 3.32, 7.6, 39.4
Hermann, C. **1** 1.9
Hermann, H.J.A. **3** 39.32
Hermodsson, Y. **1** 3.23 **3** 60.48
Herout, V. **1** 53.14
Herpin, P. **4** 21.13
Herrin, J. **3** 50.24
Herriott, J.R. **2** 64.12
Herz, W. **1** 53.3 **4** 53.1
Hesper, B. **1** 10.29, 51.8 **3** 51.37, 51.38
Hess, H. **2** 62.17, 62.20, 62.46, 63.23, 64.2 **3** 68.5 **4** 62.12
Hesse, M. **3** 58.67
Hesse, R. **2** 80.2, 80.37, 80.41 **3** 67.20, 68.4, 80.21, 80.22 **4** 80.20
Hessel, L.W. **3** 81.43
Hester Junior, J.B. **4** 36.20
Hettler, H. **3** 41.14
Hewertson, W. **3** 67.27, 72.30
Hewitt, T.G. **3** 72.22, 72.35, 72.50 **4** 77.4
Hewlins, M.J.E. **3** 64.50 **4** 44.11
Hewson, A.T. **4** 53.5
Hickel, D. **3** 50.26 **4** 50.14
Hickernell, G.L. **3** 59.32
Hidai, M. **4** 86.52
Higgs, M.A. **1** 1.29
High, D.F. **1** 51.6, 59.65
Higson, B.M. **4** 71.57, 78.13
Higuchi, T. **1** 36.26 **2** 63.9, 63.18, 73.94
Hill, N.J. **1** 3.38, 3.39
Hillery, S.S. **3** 9.19, 35.17
Hilton, J. **2** 75.47 **3** 69.3
Hinamoto, M. **4** 83.105
Hinazumi, H. **4** 48.14
Hinderer, A. **3** 68.5
Hine, R. **1** 48.34 **3** 9.7, 28.2
Hirabayashi, M. **1** 52.25
Hirai, K. **1** 50.35
Hirama, M. **4** 27.15
Hiramatsu, M. **1** 50.38 **3** 50.20
Hirata, Y. **1** 50.37, 54.9, 58.102, 58.114, 59.25 **3** 58.38 **4** 58.28, 58.29, 58.31, 58.39
Hirokawa, S. **1** 2.84, 3.13, 3.18, 3.42, 33.63, 33.65, 33.69, 48.32, 48.41 **3** 33.36
Hirotsu, K. **2** 73.94 **3** 5.3 **4** 13.14
Hirshfeld, F.L. **1** 18.24, 20.7, 23.16, 29.10, 29.24, 36.9 **3** 1.15
Hitchcock, P.B. **2** 72.16, 73.34, 86.84, 86.108 **4** 23.5, 71.43, 71.46, 71.49
Hitchman, M.A. **2** 76.21, 76.50, 83.215 **3** 77.24
Hjeds, H. **4** 40.3
Hjortaas, K.E. **1** 28.19 **2** 64.16
Hjortas, J. **4** 59.1
Hlavata, D. **4** 12.12
Ho, B.Y.K. **4** 86.49, 86.50
Ho, J.-Y. **1** 11.19
Hoard, J.L. **1** 3.16, 7.4, 31.1, 33.36, 49.18, 49.19, 49.24, 49.26, 49.29 **2** 76.59, 76.61, 76.63, 76.64, 76.65, 76.66, 76.69, 76.71, 76.72, 77.50, 78.45, 81.52, 81.70, 82.56, 83.137, 84.23 **3** 49.11, 49.12, 69.29, 69.30, 84.25 **4** 49.21, 49.22, 49.24, 76.34, 76.36, 81.9
Hoard, L.G. **2** 64.3
Hoare, R.J. **3** 71.53 **4** 71.14
Hobbs, J. **4** 47.1
Hobson, A.D. **1** 16.49
Hock, A.A. **2** 72.33, 73.74
Hodder, O.J.R. **2** 61.15 **3** 13.18, 49.3, 49.13, 54.14, 56.8, 56.11 **4** 2.24, 24.2, 31.11, 31.23, 40.8, 60.9, 60.10, 60.11
Hodges, R. **3** 59.28
Hodgkin, D.C. **1** 10.28, 10.30, 48.36, 49.32, 49.34, 50.13, 50.14, 50.16, 51.45, 51.50, 59.43 **3** 49.3, 50.31, 51.43, 58.14, 82.4 **4** 49.20, 49.23, 83.137, 83.138
Hodgson, D.J. **1** 2.16 **2** 86.75 **3** 71.9, 76.41, 86.69 **4** 76.13, 82.9, 83.22, 83.23, 83.100, 86.67
Hodgson, K.O. **4** 75.13, 75.14, 75.15, 75.25, 75.31
Hodgson, M. **3** 62.11
Hoekstra, A. **4** 12.10
Hoffman, A.U. **4** 49.24
Hoffmann, K. **3** 63.4, 63.5, 63.6
Hoffmann, R. **4** 12.9
Hoge, R. **1** 45.47 **3** 45.18, 45.20 **4** 59.24
Hohne, E. **1** 38.21, 38.51, 58.104, 58.105, 58.107, 58.109 **3** 32.17, 64.7 **4** 51.48, 51.56, 59.17, 85.20
Holah, D.G. **3** 80.12, 80.13, 80.15
Holan, G. **4** 17.15

Holden, J.R. **1** 12.6, 15.2, 15.18, 16.1, 24.6 **3** 16.5 **4** 32.14
Holian, B.L. **3** 79.13
Holland, R.V. **3** 82.17
Holm, R.H. **3** 83.71
Holmes, D.R. **1** 16.41
Holmesland, O. **2** 60.19, 60.20
Holst, J. **3** 85.5 **4** 4.1, 8.7
Holst, J.P.C. **2** 64.48
Holt Junior, S.L. **2** 79.34, 79.35
Holt, E.M. **3** 64.41, 79.21
Holt, S.L. **3** 62.16, 64.41, 79.21 **4** 78.9, 78.14
Holtedahl, B.H. **2** 64.31
Holtzberg, F. **1** 1.1
Holub, M. **1** 53.14 **3** 53.7
Hon, P. **2** 77.7, 77.40, 77.41, 77.42, 77.43
Hon, P.-K. **3** 77.24
Honzl, J. **2** 60.34
Hoogschagen, G. **1** 1.95
Hoogsteen, K. **1** 28.26, 44.50, 44.51
Hoogzand, C. **2** 72.61
Hope, H. **1** 8.49, 9.27, 9.29, 11.24, 33.71, 38.25, 40.5, 40.10, 51.26 **2** 60.35, 60.55, 60.56, 60.120, 83.87 **3** 1.19, 8.7, 11.11, 39.2, 39.30, 51.31, 53.8, 70.2 **4** 8.5, 31.17, 31.18, 63.13
Hopkins, T.E. **2** 62.33, 62.41, 65.39, 73.5
Hoppe, H. **1** 3.66 **2** 84.28
Hoppe, W. **1** 9.1, 28.24, 28.25, 31.6, 36.35, 49.16, 51.28, 51.49, 54.24 **2** 74.21, 75.61 **3** 31.4, 39.24, 39.25, 39.27, 39.33, 49.15, 54.6, 54.8, 54.15, 59.19, 59.25, 71.14 **4** 13.22, 27.19, 35.33, 36.37, 51.11, 54.8, 54.12, 54.16, 71.92
Hora, J. **4** 54.1
Hordvik, A. **1** 11.4, 39.6, 39.16, 39.44, 39.48, 39.59, 39.60, 41.1, 41.3, 41.4, 41.5, 45.9, 45.10 **3** 39.3, 39.16, 39.18, 39.42, 39.43, 39.47 **4** 39.2, 39.10, 39.13, 39.30, 39.32, 39.36, 39.44, 41.12, 41.23, 41.26, 42.1, 45.1
Horhammer, L. **3** 59.26
Horii, Z. **1** 58.9
Horn, P. **1** 47.19
Horner, L. **2** 64.48
Hornstra, J. **3** 21.20, 31.8, 51.8, 51.10, 51.17, 51.18, 51.21, 51.23, 51.24, 51.25
Horrocks Junior, W.DeW. **1** 3.80 **2** 65.42, 77.46 **3** 76.83, 77.27, 78.6, 78.19, 79.2 **4** 77.24, 80.8
Horsfield, E. **2** 83.131, 83.132
Hoshino, S. **1** 48.9
Hoskins, B.F. **1** 49.5 **2** 80.40, 83.90, 83.171, 83.213, 83.218, 85.63 **3** 49.5, 76.84, 80.3, 80.6, 83.78, 83.110, 83.151, 85.24 **4** 80.3, 83.88, 85.15
Hoskins, K. **4** 71.43, 71.46, 71.49
Hosokawa, T. **4** 71.51
Hosoo, T. **3** 81.17
Hosoya, S. **1** 39.68, 42.13, 42.14
Hospital, M. **1** 1.75, 1.82, 1.87, 1.94, 1.98, 1.105, 1.106, 1.111, 1.112, 1.124, 1.126, 1.129, 1.134, 26.10, 36.34, 51.4 **3** 1.26, 1.27, 1.28, 17.7, 17.9, 36.30, 51.5 **4** 8.12, 8.13, 13.8, 17.9, 17.10, 19.14, 51.4, 51.5, 51.8, 51.9, 51.12, 51.16, 51.18, 51.21
Hossain, M.B. **1** 33.10, 53.10, 54.11 **2** 82.22 **3** 54.10
Hosszu, J.L. **1** 44.72
Hota, N.K. **4** 72.29
Hotsyanova, T.L. **1** 16.33, 19.38
Hou, F.L. **4** 86.3, 86.4
Hough, E. **1** 59.48 **3** 58.34 **4** 38.23
House, D.A. **3** 76.5, 76.58 **4** 76.24
Housty, J. **1** 1.75, 1.82, 1.87, 1.94, 1.98, 1.105, 1.106, 1.111, 1.112, 1.124, 1.126, 1.129, 1.134, 1.140, 19.16, 41.45, 43.22, 43.26, 43.30 **3** 43.6, 43.9 **4** 25.1
Hovenkamp, S.G. **2** 64.4
Howard, J. **3** 45.45 **4** 69.26
Howard, W.J. **4** 83.21
Howatson, J. **3** 76.36, 76.37 **4** 33.40, 67.16
Howe, N.L. **3** 83.121, 83.122, 83.123, 83.124, 83.125, 83.126
Howell, P.A. **1** 39.11
Hoy, R.C. **2** 83.158
Hoy, T.G. **1** 13.49, 19.46, 19.47
Hoyano, J. **3** 69.27
Hoyano, J.K. **4** 73.13
Hseu, T.H. **4** 62.11
Hsiang-Ch'i, M. **4** 66.6
Hsieh, A.T.T. **4** 86.22
Hsiung, V. **4** 45.23
Hsu, I.-N. **3** 82.12, 82.13, 82.14, 82.15
Hsu, I.H.-C. **3** 58.51
Hsu, I.N. **4** 51.57, 51.58
Hubel, W. **2** 72.61, 73.116
Huber-Buser, E. **1** 23.6, 23.7
Huber, C.P. **1** 53.5, 58.78 **4** 37.7
Huber, C.S. **3** 58.37 **4** 35.12, 37.5, 40.21, 58.22
Huber, F. **4** 69.38
Huber, M. **1** 47.18 **2** 69.29 **4** 65.6, 66.5, 69.12
Huber, R. **1** 9.1, 28.24, 31.6, 51.28 **2** 75.61
Huddle, B.P. **3** 10.2 **4** 33.1
Hudson, D.W. **3** 31.11 **4** 38.12
Hudson, J.M. **4** 48.53
Hudson, P. **1** 38.15, 39.22, 39.24
Huebner, C.F. **3** 31.42
Huffman, J.C. **2** 83.190 **4** 62.6, 68.2
Hugel, R. **2** 84.66
Hughes, D.L. **3** 45.45 **4** 13.9
Hughes, D.O. **1** 1.24
Hughes, E.W. **1** 3.3, 8.17, 8.31, 33.8, 48.30, 48.83 **2** 60.8, 65.11 **3** 48.10
Hughes, R.E. **1** 7.2, 7.32, 37.2, 37.3 **3** 48.26, 50.17 **4** 50.5, 50.9, 71.9, 73.31
Huie, B.T. **4** 69.25
Huisgen, R. **3** 64.59
Huisman, H.O. **3** 51.15
Huitric, A.C. **3** 52.6 **4** 13.21
Hull, S.E. **3** 71.3, 85.7, 85.20 **4** 12.14, 28.11, 35.23
Hullen, A. **2** 76.9 **3** 3.55
Hulme, R. **2** 60.89, 60.114, 60.136, 60.137, 60.138, 69.16, 69.17, 71.20 **3** 60.19
Hulscher, J.B. **1** 38.47 **3** 69.40 **4** 64.7
Huml, K. **1** 12.10, 16.42, 31.21, 36.14, 36.15, 36.16, 36.17 **3** 37.11, 39.44
Hummelink, T.W. **4** 80.4
Hunt, A. **4** 65.5

Hunt, D.J. 1 47.26, 50.12 3 47.13

Hunt, G.W. 4 64.9

Hunter, F.D. 1 45.14

Hunter, S.H. 2 84.41

Huq, F. 3 83.79 4 71.52, 71.53, 83.82

Hursthouse, M. 4 71.27

Hursthouse, M.B. 1 2.92, 21.33, 50.27, 59.48 2 60.136, 60.137, 60.138, 63.24, 78.61, 83.123, 83.217, 84.65 3 75.16, 78.25, 82.25, 83.53, 83.102, 83.129 4 32.17, 38.21, 49.16, 54.1, 54.2, 59.15, 65.1, 65.2, 80.15, 83.26, 83.61, 83.85, 83.130, 83.136

Huse, G. 1 15.4 2 60.86

Huse, Y. 1 59.9, 59.21

Husebye, S. 2 64.27, 70.6, 70.7, 70.10, 70.14, 70.15, 70.17, 70.18 3 39.26, 40.9, 64.18, 64.19, 70.3 4 70.16, 70.17

Hussain, M.S. 2 83.87 3 67.25, 84.13, 84.18, 84.19

Hutcheon, W.L. 4 73.17

Huttner, G. 3 71.30 4 71.8, 71.30, 74.10, 75.1, 75.5, 75.11, 75.16

Hvoslef, J. 1 45.12, 45.17, 45.19, 45.20, 45.21 2 60.44, 60.45 4 45.17

Hwang, K. 3 40.11

Hwang, K.J. 3 60.8 4 37.8

Hybl, A. 1 45.76 3 1.39, 44.43 4 1.14, 5.8

Hyde, A.J. 1 33.58

Iball, J. 1 26.41, 28.16, 29.7, 29.11, 29.16, 29.21, 29.22, 30.2, 30.8, 30.11, 30.17, 30.20, 36.31, 36.32, 40.29, 40.33, 44.36, 47.2, 47.3, 47.4 2 77.23, 83.181 4 29.5, 45.16

Ibata, K. 4 27.5, 27.6

Ibekwe, S.D. 3 71.41, 71.42

Ibers, J.A. 1 1.13, 1.20, 28.15 2 64.52, 74.10, 75.51, 76.39, 76.40, 83.195, 83.200, 84.61, 85.30, 85.73, 85.97, 86.4, 86.16, 86.25, 86.39, 86.50, 86.58, 86.61, 86.66, 86.75, 86.76, 86.77, 86.79, 86.85, 86.87, 86.88, 86.98, 86.109, 86.112, 86.114 3 7.2, 7.3, 33.1, 65.7, 69.39, 71.20, 71.66, 73.2, 76.8, 76.16, 82.7, 83.22, 84.24, 85.36, 86.14, 86.21, 86.39, 86.47, 86.55, 86.56, 86.60, 86.63, 86.66, 86.68, 86.69, 86.73, 86.79, 86.89, 86.90, 86.93 4 71.48, 71.72, 71.93, 72.22, 72.23, 72.24, 72.30, 83.133, 86.20, 86.32, 86.33, 86.45, 86.51, 86.59

Ichikawa, M. 4 2.7

Ichikawa, T. 1 1.66, 1.67, 48.80

Ievins, A. 2 62.51, 85.80

Igarashi, T. 2 85.61

Ignat'ev, N.A. 3 27.10

Ihara, M. 1 58.45, 58.58 3 36.26, 58.58

Iida, K. 3 78.10

Iijima, I. 3 58.36

Iijima, K. 3 78.9, 81.23

Iitaka, Y. 1 1.66, 1.67, 21.22, 21.23, 24.33, 31.16, 41.31, 46.6, 47.13, 47.14, 48.3, 48.4, 48.42, 48.80, 50.9, 50.20, 50.21, 50.34, 50.35, 54.12, 55.1, 55.2, 55.3, 56.7, 58.90, 59.9, 59.21, 59.60 2 83.34 3 1.12, 1.17, 33.24, 36.20, 48.16, 48.30, 48.33, 50.18, 53.9, 56.3, 59.29, 71.21 4 22.3, 22.4, 22.5, 35.38, 48.28, 50.10, 51.54, 56.6, 59.23, 71.22, 72.28

Ikeda, T. 2 80.4

Ikehara, M. 3 47.32

Ikekawa, T. 1 50.9

Ikemoto, I. 2 60.142 3 60.53 4 60.33

Iljinsky, A.L. 3 77.11, 77.12, 77.28, 77.29

Illingworth, S.M. 3 69.27

Ilmaier, B. 3 86.97

Imado, S. 1 58.9

Immirzi, A. 1 21.32, 23.9, 29.20 2 61.17, 61.18, 61.19, 68.1, 72.20, 72.43, 72.45, 75.48 3 21.10, 58.28, 74.3, 74.4 4 58.26, 72.19

Inayama, S. 3 53.9 4 53.1

Inglis, F. 2 65.51

Ingram, L. 3 85.52

Ingri, N. 4 83.93

Innes, M. 1 16.34, 59.62 3 5.4

Inouye, H. 1 52.25

Interrante, L.V. 4 85.2, 85.6, 85.7

Inubushi, Y. 1 58.93

Iqbal, M.Z. 4 71.36

Ireland, P.R. 3 76.58, 77.1, 83.100 4 71.71, 71.73, 71.89, 76.24

Ireland, R.E. 3 27.11, 30.5 4 27.16

Irie, H. 1 1.108, 58.86 3 17.5, 58.43 4 58.43

Irie, T. 4 27.8

Irngartinger, H. 3 1.13, 29.3, 29.4, 31.37, 31.38, 31.40, 31.41

Irreverre, F. 4 48.20, 48.22

Irving, H.M.N.H. 1 41.35 4 85.29

Isaacs, N.W. 3 32.2, 47.25, 58.6, 69.15 4 45.18, 52.6, 56.2, 67.8

Isakov, I.V. 1 1.33, 16.25 2 83.19

Isaksson, G. 1 41.28

Ishida, M. 4 59.20

Ishihara, N. 4 81.33, 81.35

Ishii, H. 1 58.93

Ishimori, M. 4 22.7

Ishizuka, K. 4 58.43

Islam, K.M.S. 1 19.28, 29.8, 54.15, 59.54 2 73.102

Isobe, I. 4 77.3

Isono, Y. 1 50.4

Itai, A. 1 55.2 4 59.23

Ito, M. 3 76.54 4 76.20, 76.21, 76.40

Ito, S. 1 31.16, 58.111 4 22.3, 22.4, 22.5, 27.15

Ito, T. 2 85.61 3 60.13, 82.18 4 69.14, 76.2, 80.19

Ivanov, V.I. 2 77.33 3 77.12

Ivanova, N.S. 2 69.38, 80.21

Iwamoto, T. 3 61.4

Iwasaki, F. 2 60.117 3 21.9, 60.12 4 24.9

Iwasaki, F.F. 1 1.16, 1.17

Iwasaki, H. 1 1.17 2 60.123, 76.35, 76.58, 83.96 4 69.20

Iwashita, Y. 1 7.18

Iwata, M. 2 76.38

Iyer, K.N. **3** 58.39 **4** 58.38
Izatt, R.M. **4** 67.23

Jackobs, J. **1** 39.75 **3** 39.37, 45.40
Jackson, R.B. **2** 75.9 **4** 83.109
Jacob, L.M. **4** 47.4
Jacobi, P. **1** 54.24
Jacobs, T.L. **4** 20.11, 20.13
Jacobson, R.A. **1** 33.46, 38.14, 38.54, 45.65 **2** 62.45, 64.3, 73.36, 75.34, 77.69, 86.53 **3** 3.36, 33.8, 33.10, 33.11, 37.1, 45.4, 58.24, 58.50, 71.2, 83.73, 83.107 **4** 33.20, 64.31, 83.117, 83.135, 86.8, 86.27
Jaggi, H. **1** 1.92, 13.33 **2** 83.211
Jain, P.C. **2** 76.46, 78.10, 78.38, 78.39 **3** 76.47
Jain, S.C. **1** 2.72, 2.85 **4** 39.21, 39.29, 60.43, 60.44
Jakkal, V.S. **3** 41.1
Jakle, H. **4** 71.92
Jakovenko, V.I. **1** 11.30 **2** 62.29, 69.20
James, J.M. **3** 83.20
James, M. **3** 45.39
James, M.N.G. **1** 44.12, 50.14, 50.40 **4** 33.38, 47.2, 54.10, 65.4
James, T.A. **4** 83.131
James, V.H.T. **1** 51.38, 51.39, 51.40
James, V.J. **4** 21.9, 21.10, 21.14
Jameson, M.B. **1** 24.26
Jamet-Delcroix, S. **2** 60.27 **4** 3.2, 3.3
Jamieson, G. **4** 31.24
Jandacek, R.J. **1** 39.30 **3** 39.12, 48.31, 48.38 **4** 66.7
Janek, A. **2** 69.9
Jansing, J.A. **3** 38.4, 42.2
Janson, V. **4** 76.17, 76.18
Janusonis, G.A. **2** 83.65
Jaquier, R. **1** 9.47, 32.23 **3** 9.20
Jarchow, O.H. **1** 16.36
Jarski, M.A. **2** 78.8
Jarvis, B.B. **4** 11.9
Jarvis, J.A.J. **1** 11.10 **2** 83.3, 83.39, 83.168, 83.207, 86.8, 86.65, 86.104 **3** 71.57, 72.3, 80.1, 80.2, 86.28, 86.29, 86.41 **4** 67.27, 69.36, 71.62, 71.64, 72.2
Jayadavan, N.C. **2** 76.25, 77.68
Jayadevan, N.C. **3** 81.31, 81.44
Jayaweera, S.A.A. **3** 82.25
Jeannin, Y. **3** 83.46, 83.47, 83.48 **4** 32.8, 83.3
Jecny, J. **2** 60.34
Jefferey, J.W. **3** 84.28
Jefferies, P.R. **1** 59.41
Jeffrey, G.A. **1** 1.8, 1.54, 1.115, 1.145, 1.149, 2.28, 5.21, 16.5, 18.2, 19.8, 19.44, 19.51, 39.12, 39.29, 39.31, 40.27, 43.9, 43.12, 43.17, 44.17, 45.8, 45.15, 45.16, 45.18, 45.25, 45.37, 45.41, 45.44, 45.45, 45.67, 45.73, 52.18, 58.47, 58.82, 58.100, 58.101 **2** 61.1, 61.5, 61.6, 61.7, 61.8, 61.11, 61.12, 61.13, 61.14, 61.22 **3** 13.1, 45.3, 45.44, 60.23, 61.2, 61.5, 81.32 **4** 5.6, 21.1, 45.4, 45.6, 45.9, 45.10, 45.11, 45.12, 45.24, 60.22
Jeffreys, J.A.D. **1** 58.68 **3** 58.12, 58.54, 71.50, 71.51
Jeffs, P.W. **3** 58.20 **4** 58.19
Jelen, A. **3** 69.4
Jelenic, I. **2** 81.60
Jellinek, F. **1** 3.9, 38.38, 40.12 **2** 72.44, 74.11, 74.12
Jeminet, G. **3** 11.20, 11.21
Jenkins, P.N. **3** 58.34
Jenkins, R.N. **4** 64.32
Jenks, J.M. **3** 32.31 **4** 27.18
Jennings, P.W. **4** 86.11
Jennische, P. **4** 80.20
Jensen, B. **1** 38.16, 40.30 **3** 3.51, 39.11 **4** 1.13, 40.20, 40.25
Jensen, K.G. **1** 10.24, 10.27, 40.22 **3** 10.3, 76.2, 76.3
Jensen, L.H. **1** 1.135, 1.136, 1.138, 1.155, 9.25, 12.18, 13.21, 33.74, 35.2, 35.44, 37.20, 44.9, 44.10, 44.25, 44.28, 44.40, 44.47, 46.14, 47.10, 47.15, 47.27, 48.37, 48.67, 48.68, 59.26 **2** 60.121 **3** 24.4, 33.22, 44.10, 44.31, 48.29, 52.6 **4** 13.21, 33.17, 44.16, 45.19, 48.53
Jensovsky, L. **2** 79.29
Jepsen, H. **4** 41.32
Jeremic, M. **4** 75.21
Jerina, D.M. **4** 40.14
Jerome-Lerutte, S. **4** 3.10
Jerslev, B. **1** 8.7, 10.27, 19.21, 19.22, 19.26, 21.1, 21.2, 38.16, 40.22, 40.23, 59.28
Job, B.E. **2** 83.168
Joesten, M.D. **3** 67.25, 84.13, 84.18, 84.19
Johannsson, A.A. **2** 71.47
Johans, A.W. **4** 86.55
Johns, S.R. **4** 58.9, 58.10, 58.11, 58.35
Johnsen, J. **1** 8.36, 8.37
Johnsen, K. **1** 39.35, 39.36 **2** 70.3, 70.4
Johnson, B.F.G. **4** 75.3
Johnson, C.K. **1** 2.47, 2.77, 2.78, 2.79, 2.80, 2.81, 52.20, 52.21 **3** 1.6 **4** 81.4, 81.5, 81.6, 81.7, 81.8
Johnson, C.R. **2** 84.67 **3** 39.45
Johnson, D.A. **1** 50.26 **4** 83.29
Johnson, D.R. **3** 84.11, 84.17
Johnson, J.E. **3** 83.107 **4** 83.117, 83.135
Johnson, L.F. **4** 58.24
Johnson, L.N. **1** 45.51
Johnson, P.L. **1** 39.83 **3** 21.15, 21.16, 33.41, 39.23, 39.40, 39.41
Johnson, Q. **1** 3.37 **3** 3.28 **4** 21.21, 60.13
Johnson, S.M. **1** 31.7, 34.8, 50.33, 53.35 **3** 17.11, 23.8, 23.9, 37.10, 39.21, 39.46, 44.37, 50.24, 75.7
Johnson, W.S. **3** 31.10
Johnston, D.L. **3** 76.83, 77.27, 78.6 **4** 80.8
Johnston, J.P. **1** 37.28
Johnstone, R.M. **4** 60.27, 60.29
Jolly, P.W. **4** 71.65, 86.66
Jonas, K. **4** 86.66
Jones, C.J. **2** 71.9 **3** 85.7, 85.20
Jones, D.W. **1** 2.9 **4** 28.4
Jones, G. **4** 35.8, 58.4, 58.5
Jones, K. **3** 69.14
Jones, L.A. **4** 24.5
Jones, N.D. **1** 32.29, 36.21 **2** 60.71, 73.43 **3** 29.2, 38.20, 50.8 **4** 51.10, 58.42
Jones, P.E. **2** 85.33

Jones, R.A.Y. **4** 33.42, 33.45
Jones, R.D.G. **3** 11.12 **4** 11.5, 68.7, 86.9, 86.16, 86.18
Jones, R.E. **1** 1.26
Jones, R.O. **2** 72.35
Jongsma, C. **3** 65.6 **4** 20.7
Jonsson, P.-G. **1** 12.4 **3** 1.4, 12.4 **4** 48.1
Jordan, T.H. **1** 4.9 **2** 61.4, 61.5 **3** 60.21
Jose, P. **1** 48.12 **2** 82.19 **3** 77.15
Joshi, K.K. **2** 73.33 **3** 73.47, 73.48, 73.49
Jost, K.-H. **2** 67.2
Jotham, R.W. **3** 71.3
Joustra, A.H. **3** 51.1
Jovanovic, B. **4** 71.19, 71.21, 71.81
Joys, S. **1** 41.4
Jukes, D.E. **1** 44.43
Julshamn, K. **4** 39.2, 39.13, 41.12, 42.1
Jumas, J.C. **4** 64.3
Jungk, A.E. **3** 15.11, 15.12 **4** 13.23, 24.18

Kabalkina, S.S. **1** 24.7, 26.14
Kabuto, C. **1** 22.14 **3** 58.58
Kadonaga, M. **4** 72.27
Kadoya, S. **1** 21.22, 21.23
Kaesz, H.D. **2** 65.41
Kaftory, M. **4** 69.8
Kahrl, J. **4** 58.33
Kahrl, J.H. **3** 48.43
Kai, Y. **2** 68.28 **3** 35.12, 68.19 **4** 35.39, 68.9, 68.12, 71.28, 71.29, 71.58, 71.80
Kainosho, M. **3** 46.3
Kaiser, E.T. **1** 11.9, 42.1, 42.11, 42.12
Kajimoto, T. **4** 72.10
Kakati, K.K. **1** 9.38
Kakudo, M. **1** 2.3, 11.37, 13.3, 13.28, 13.29, 13.38, 20.1, 20.29, 20.30, 22.9, 27.13, 31.27, 33.42, 38.42, 48.18, 48.52, 48.56, 48.63, 48.84, 48.86, 48.88, 48.95, 56.12 **2** 63.2, 63.4, 63.6, 68.28, 69.31, 72.78, 78.1, 78.2, 82.13, 82.29, 86.71, 86.81 **3** 13.19, 13.20, 35.12, 36.9, 41.23, 59.15, 68.19, 69.11, 69.25, 72.10, 72.18, 73.33 **4** 44.14, 44.15, 48.33, 48.35, 48.57, 49.13, 54.7, 60.20, 67.9, 68.9, 68.12, 71.28, 71.29
Kalff, H.T. **1** 39.14, 39.15, 39.66
Kalish, R. **4** 39.39
Kalman, A. **1** 4.6, 4.7 **3** 11.28, 42.5 **4** 4.3, 11.13
Kalman, T.I. **1** 11.62 **3** 47.34
Kaluski, Z.L. **2** 73.85, 73.87, 73.104, 73.105, 73.107, 73.115 **3** 73.45
Kalyanaraman, A.R. **3** 78.11 **4** 44.24, 48.19, 77.20, 83.21, 84.14
Kalyani, V. **1** 13.56, 13.57, 38.48
Kamenar, B. **1** 49.5, 49.6 **2** 60.70, 60.74, 60.148, 69.15 **3** 65.10, 66.5, 83.9, 83.57
Kametani, T. **1** 58.45, 58.58 **3** 36.26, 58.58
Kamijo, N. **1** 58.115
Kaminskii, V.F. **4** 60.40
Kamiya, K. **1** 48.73, 50.5, 56.9, 58.35, 58.36, 58.57, 59.42, 59.52 **3** 44.42, 50.13, 50.22, 50.30 **4** 50.6, 54.3
Kamper, M.J. **1** 49.34
Kanamaru, F. **2** 83.156
Kanaoka, Y. **3** 31.14
Kanda, E. **1** 5.9
Kandler, H. **2** 63.25
Kane, A.R. **3** 86.10
Kane, J. **3** 3.28
Kaneda, M. **4** 51.54
Kaneko, H. **4** 58.43
Kaneko, T. **3** 59.1
Kannan, K.K. **1** 11.15
Kanters, J.A. **1** 1.32, 1.34, 38.47 **3** 1.8 **4** 1.1, 1.3, 1.7, 2.6, 2.15
Kapadi, A.H. **4** 58.44
Kapecki, J.A. **1** 37.24, 39.61 **3** 20.12 **4** 27.7
Kapicak, L.A. **4** 38.19
Kaplan, S.F. **2** 81.8, 81.10
Kapon, M. **3** 36.8
Kapovits, I. **1** 4.7 **3** 42.5
Karaoghlanian, B.V. **4** 39.14
Karim, A. **4** 58.40
Karle, I.L. **1** 1.119, 4.18, 5.41, 8.41, 16.60, 17.44, 17.46, 20.17, 20.18, 20.23, 20.25, 31.1, 31.18, 32.3, 33.73, 35.27, 35.28, 35.32, 35.43, 37.15, 37.17, 37.25, 38.40, 39.57, 41.39, 41.49, 42.4, 44.70, 46.20, 48.69, 48.89, 51.18, 52.27, 58.22, 58.31, 58.33, 58.51, 58.66, 58.113, 58.116 **3** 8.17, 11.3, 31.14, 31.34, 37.8, 47.27, 48.14, 48.47, 48.53, 51.28, 58.9, 58.61, 59.21, 60.52, 61.6, 61.7, 64.59, 67.7 **4** 9.12, 28.8, 37.16, 42.6, 44.17, 44.21, 44.26, 48.20, 48.22, 48.47, 54.5, 54.6, 54.9, 59.13
Karle, J. **1** 1.119, 13.51, 16.60, 20.17, 20.18, 31.1, 31.18, 35.32, 35.43, 37.17, 37.25, 38.40, 41.39, 41.49, 48.69, 48.89, 51.18, 58.22, 58.31, 58.33, 58.51, 58.66, 58.113, 58.116 **3** 37.8, 47.27, 48.5, 48.47, 59.21, 61.6, 61.7 **4** 31.5, 41.33, 48.20, 48.22, 54.5, 54.6, 59.13, 61.2
Karle, J.M. **4** 58.23
Karlsson, R. **4** 58.8
Karraker, D.G. **3** 77.32
Karrer, P. **3** 58.67
Kartha, G. **1** 2.66, 33.78, 48.31, 51.2, 51.3, 51.13, 56.10, 58.29, 59.68 **2** 60.151 **3** 44.14, 44.17, 44.19, 44.25, 54.9 **4** 48.51, 58.25
Kasai, N. **1** 31.27 **2** 63.4, 63.6, 68.28, 69.3, 69.31, 72.78, 86.71, 86.81 **3** 13.19, 13.20, 35.12, 36.9, 41.23, 68.19, 69.11, 69.25, 72.10, 72.18, 73.33 **4** 35.39, 60.20, 68.9, 68.12, 71.28, 71.29, 71.58, 71.80, 72.27
Kashino, S. **4** 33.10, 33.11
Kashiwaga, T. **2** 72.78
Kashiwagi, M. **3** 21.11 **4** 35.18, 72.10
Kashiwagi, T. **2** 86.71, 86.81
Kaski, B.A. **3** 58.30
Kasper, J.S. **4** 85.2, 85.6, 85.7
Kastalsky, V. **2** 85.32, 85.57
Katayama, C. **1** 13.36 **3** 53.20 **4** 54.18
Katayama, M. **1** 1.23
Kato, K. **3** 59.6 **4** 6.2, 11.16, 11.17
Kato, N. **4** 54.18
Kato, Y. **1** 13.17, 13.18, 20.29
Katritzky, A.R. **4** 32.22, 33.42, 33.45
Katsube, Y. **1** 13.28, 48.95 **4** 59.20

Katsumoto, K. 1 39.90 4 39.43
Katz, J.L. 1 1.38
Katz, L. 1 1.96, 38.7, 44.56, 44.58, 48.60 2 85.33
Katz, T.J. 2 75.35 3 75.34 4 31.12
Kaufman, H.W. 3 1.38
Kawada, I. 1 34.12 2 64.59, 86.63 3 39.34 4 68.6
Kawaguchi, H. 3 76.31, 76.52
Kawaguchi, S. 3 72.10, 76.31, 76.52, 84.10 4 71.58, 71.80, 77.3, 77.18
Kawaguchi, T. 4 5.1
Kawai, M. 3 59.30
Kawamoto, I. 4 64.35, 64.38
Kawasaki, T. 4 54.3
Kawase, T. 2 79.25
Kawashima, S. 4 58.43
Kawazu, K. 4 54.15
Kay, A. 3 82.16, 82.19
Kay, H.F. 1 23.17
Kay, J. 4 24.12, 81.46, 81.47
Kay, M.I. 1 1.57, 1.58, 1.96, 38.7 2 81.3, 81.4, 81.8, 81.10 3 28.3, 28.4 4 48.2
Kay, W.I. 1 50.11
Kayushina, R.L. 1 48.38
Kazitsyna, L.A. 1 9.19
Keats, S. 3 3.9
Keeler, B.T. 3 51.39
Keeton, M. 4 71.16, 71.17, 83.63
Keiderling, T.A. 3 3.36
Keller, C.E. 2 75.26
Kellerman, M. 2 76.20, 76.41
Kelley, B.P. 2 80.40 3 85.24
Kelley, J.A. 3 3.22, 77.14, 77.25, 83.65 4 78.10
Kelly, B.P. 3 80.3
Kelly, H.C. 4 3.5
Kelly, J.F. 3 42.2
Kelso, E. 4 39.6
Kemme, A.A. 3 27.12, 69.17
Kemmitt, R.D.W. 2 86.92 3 71.34 4 71.87, 85.34
Kende, A.S. 1 37.23
Kennard, C.H.L. 1 3.33, 3.35, 3.36 2 71.4, 71.5, 71.31, 76.66, 81.23, 81.85, 83.128, 83.171 3 16.6, 32.2, 60.40, 69.15, 75.9, 76.68, 83.68, 83.69, 83.136 4 17.15, 19.11, 45.18, 67.8, 68.4, 76.37, 83.73
Kennard, O. 1 8.32, 46.3, 51.35, 51.38, 51.39, 51.40, 56.4, 58.108 2 65.24, 83.208 3 20.17, 23.1, 46.5, 46.6, 47.25, 47.33, 51.42, 53.12, 53.13, 53.14, 54.13, 58.6, 58.47, 65.5 4 8.9, 23.2, 52.6, 53.5, 56.2
Kenny, N.C. 4 39.5
Kent, P.W. 3 45.31
Kerf, J.de 1 41.27
Kergomard, A. 1 16.55
Kerr, K.A. 1 19.62, 30.7, 51.35 2 65.24 3 20.17, 23.1, 24.5, 46.5, 46.6, 47.25, 58.47, 58.51 4 48.6, 58.18
Kersting, G.F. 4 12.14, 35.23
Kettle, S.F.A. 3 71.3
Keulen, E. 2 74.11 3 33.2, 33.3
Keve, E.T. 1 2.44
Khalil, M.F. 3 58.17
Khalilov, A.D. 3 77.12
Khan, M.Y. 1 13.24
Khandkarova, V.S. 3 74.9 4 74.5
Khare, G.P. 3 85.35 4 85.16, 85.39
Khodashova, T.S. 3 76.13
Khotsyanova, T.L. 1 5.2, 5.8, 16.3, 19.2, 24.8 2 66.9 3 16.1, 19.2
Khubchandani, P.G. 1 9.41
Khvatkina, A.N. 1 7.6, 11.30, 41.36 2 62.29, 69.20, 69.38, 80.21
Kido, M. 3 48.40
Kido, Y. 4 71.22
Kiechel, J.R. 3 50.19
Kierkegaard, P. 1 36.27 3 36.7, 36.10, 36.15, 58.4, 63.17, 80.5 4 45.13, 59.12
Kiers, C.T. 4 11.7
Kikkawa, S. 2 69.31
Kikuchi, T. 1 58.36, 58.45, 58.57
Kilbourn, B.T. 1 11.45, 50.44 2 69.48, 75.36, 83.168, 86.40, 86.41, 86.62 3 17.10, 67.27, 71.41, 71.42, 71.57, 72.3, 73.28, 73.47, 73.48, 73.49, 80.1, 80.2, 83.85, 86.12, 86.28, 86.29, 86.76, 86.88, 86.91 4 69.36, 71.62, 71.64, 72.2
Killean, R.C.G. 1 45.39 2 63.12, 68.2 3 64.23 4 45.7, 65.7
Kim, H.S. 1 45.15, 45.16, 45.45 3 13.1, 45.3, 45.44, 58.3, 60.23, 61.5 4 21.1, 60.22
Kim, N.E. 3 8.7
Kim, S.-H. 3 44.4
Kim, S.H. 1 8.8, 44.3, 45.8, 45.18, 45.38, 45.49 2 60.131 4 47.20
Kim, Y.-H. 3 51.35
Kime, N. 2 86.87
Kime, N.E. 3 73.50
Kimura, B.Y. 4 71.87
Kimura, E. 4 81.61
Kimura, S. 4 59.21
Kimura, T. 3 69.11, 69.25
Kindler, H. 1 22.7
King, G.S.D. 1 1.99, 33.38, 33.41 2 72.32, 72.69 3 23.8, 23.9, 32.9, 41.10, 42.4 4 32.24
King, M.G. 3 85.52
King, T.J. 1 54.1, 59.23 3 49.4, 59.31, 65.8 4 35.25, 69.37
Kinloch, M.E. 2 78.29
Kinoshita, T. 4 32.5
Kinoshita, Y. 1 44.17 2 83.114 3 21.11 4 35.18, 72.10
Kinson, P.L. 4 29.2
Kiosse, G.A. 2 66.11, 66.12
Kirby, A.J. 1 39.65 3 47.25
Kirchner, R.M. 3 83.121, 83.122, 83.123, 83.124, 83.125, 83.126 4 86.32, 86.33
Kirillova, N.I. 3 72.40
Kirkpatrick, J.L. 1 58.100, 58.101
Kirkwood, C.E. 3 84.12
Kirokawa, S. 1 1.85
Kirschke, K. 1 38.21
Kirson, I. 4 51.49
Kiryu, S. 2 83.144 4 60.31
Kishi, T. 1 50.5 3 50.22, 50.30
Kishida, E. 1 37.1
Kishida, Y. 1 39.88 4 64.35, 64.38
Kispert, L.D. 1 51.47
Kiss, B. 4 24.16

Kissing, W. **3** 37.9
Kistenmacher, T.J. **1** 3.67 **3** 40.4 **4** 35.24, 48.36, 82.12
Kitahara, Y. **1** 20.30, 22.14, 38.42 **3** 58.58
Kitaigorodskij, A.I. **1** 5.2, 5.8, 5.26, 24.7, 24.12, 26.11, 26.14, 29.28, 48.1 **2** 71.2, 71.26 **4** 33.32
Kitano, Y. **4** 35.18, 72.10
Kjaer, A. **3** 8.18
Kjoge, H.M. **1** 39.48, 39.60 **2** 70.11, 70.12 **3** 39.18, 64.8
Klaeren, A. **4** 39.7
Klanberg, F. **2** 71.40
Klanderman, K.A. **2** 72.23, 72.24, 72.31 **3** 72.1
Klasek, A. **4** 58.24
Kleinberg, R. **4** 48.2
Klem, E. **2** 83.184
Klement, U. **2** 84.4
Klewe, B. **1** 7.14, 7.15, 10.10, 12.13 **4** 7.6, 7.7, 7.8, 7.11, 9.4, 12.3, 12.8
Klimasenko, N.L. **3** 26.1, 26.2
Klimova, A.I. **2** 62.12
Kline, B.S. **1** 3.22
Klinga, M. **3** 76.32
Klohs, M.W. **3** 33.48, 38.17
Klug, A. **1** 19.12
Klug, H.P. **1** 16.28, 18.16 **2** 78.12, 80.24 **3** 16.3
Knappe, J. **1** 59.55
Knauer, K.H. **1** 39.27
Knazeva, A.N. **2** 78.30
Kneen, W.R. **4** 71.43, 71.46, 71.49
Knight, J. **4** 71.35, 86.22
Knobler, C. **1** 38.8, 38.10, 38.11, 38.12, 42.6 **2** 60.42, 79.13 **3** 39.2, 39.30, 70.2
Knobler, C.M. **4** 69.25
Knox, G.R. **3** 71.53, 73.56
Knox, J.R. **1** 53.34 **2** 71.27, 73.40, 73.75, 82.14, 83.38, 85.78
Knox, K. **1** 3.69 **2** 60.99 **3** 49.1, 49.7, 85.49 **4** 49.14
Knyazeva, A.N. **2** 77.60 **3** 77.26 **4** 77.2
Kobalkina, S.S. **2** 83.76
Kobayashi-Tamura, H. **4** 72.17
Kobayashi, A. **4** 83.78
Kobayashi, H. **3** 3.25, 35.4 **4** 60.24
Kobayashi, J. **2** 67.18, 67.19
Kobayashi, M. **4** 76.25
Kobayashi, N. **3** 59.29
Kobayashi, T. **4** 49.13, 49.18
Kobayashi, Y. **3** 36.20 **4** 71.22, 72.28
Kobelt, D. **2** 69.9 **4** 24.17, 40.5, 42.4, 44.13, 70.1, 70.5
Kobori, K. **3** 21.11
Koch, B. **1** 1.144 **4** 57.1
Koch, H.P. **1** 2.92
Koch, M. **4** 48.49
Koch, M.H.J. **3** 48.15 **4** 33.43, 33.44
Kocman, V. **4** 71.47
Koda, S. **2** 75.55 **3** 84.10 **4** 33.24, 77.3, 77.18
Kodama, T. **4** 86.52
Koenig, D.F. **1** 1.128, 28.15, 39.3, 49.17
Koetzle, T.F. **2** 62.25 **3** 62.5 **4** 48.5, 48.8, 48.10, 76.33
Koh, L.L. **4** 12.15
Koizumi, H. **2** 81.77
Kok, A.J.de **3** 1.14, 38.6
Koknat, F.W. **4** 3.7
Kokotailo, G.T. **3** 85.41, 85.42 **4** 69.24, 85.28
Kolakowski, B. **1** 1.88
Kolderup, M. **4** 59.3
Kolesnikov, S.P. **3** 60.9
Kolobova, N.E. **2** 71.47, 72.72, 73.68 **3** 71.56
Kolombos, A.J. **3** 67.24
Komiyama, Y. **2** 76.4, 76.13, 76.17 **3** 76.31, 78.9, 78.10, 81.23 **4** 78.1
Komoda, Y. **4** 58.40
Komori, T. **4** 54.3
Komson, R.C. **4** 81.48, 86.17
Kon, A.Yu. **2** 83.91
Kondrashev, Yu.D. **1** 9.30, 9.31, 9.32, 9.33, 9.42, 44.21
Kondrashev, Yu.D. **4** 9.10, 32.1
Konig, E. **4** 83.115, 83.116
Koning, R.E. **3** 28.8
Koningsveld, H.van **1** 5.14 **3** 58.10 **4** 21.7
Konishi, H. **1** 48.18
Konnert, J. **1** 7.13 **2** 71.18 **3** 47.27, 48.53 **4** 40.7, 44.17, 63.3, 69.6, 69.7
Konnert, J.H. **3** 11.4
Kono, Y. **4** 50.4
Konst, W.M.B. **3** 51.15
Koo, C.H. **1** 9.5, 9.18 **3** 3.43, 3.44, 58.3 **4** 72.25
Kopecky, W.J. **4** 37.9
Kopf, H. **3** 73.26 **4** 73.7, 73.8
Kopwillem, A. **3** 80.5
Korczynski, A. **2** 79.11, 79.12, 79.17, 79.21, 79.24
Koridze, A.A. **4** 73.10
Koritny, E.F. **3** 77.11, 77.29
Kornblau, M.J. **1** 7.32
Korotkevich, A.F. **4** 16.1
Korvenranta, J. **3** 76.57
Korytnyi, E.F. **2** 77.28 **3** 77.22 **4** 77.11
Korytnyk, W. **4** 33.22
Kosky, C.A. **4** 72.12
Kostansek, E.C. **3** 8.1
Koster, P.B. **3** 39.17
Kosturowicz, Z. **1** 11.43
Kostyanovskii, R.G. **1** 32.17 **3** 13.7
Kosuge, T. **4** 59.21
Koten, G.van **4** 71.82
Kotick, M.P. **1** 47.30
Kotsyanova, T.L. **1** 19.1, 19.3
Kovacs, A.L. **1** 37.10 **2** 60.97
Kovaleva, E.S. **2** 79.30
Kovama, H. **2** 81.74
Kowala, C. **3** 58.22
Koyama, G. **1** 41.31, 50.3, 50.20, 50.21, 50.34
Koyama, H. **1** 7.28, 27.12, 33.60, 36.30, 50.42, 51.20, 52.28, 58.18, 58.122 **2** 84.7, 85.18 **3** 31.12, 31.26, 38.13, 50.3, 50.4, 50.27, 51.20, 51.22, 51.33, 54.12, 56.7, 58.11, 58.16, 58.25 **4** 27.8, 39.37, 39.38, 48.24, 50.12
Koyama, Y. **3** 1.12
Koz'min, P.A. **3** 81.7 **4** 33.2
Koz'Min, P.A. **1** 33.27, 33.28, 33.30

Kozarek, W.J. **4** 85.13
Kozhin, V.M. **1** 48.1
Krahe, E. **2** 63.1, 63.17
Krakower, G.W. **3** 51.39
Krasnova, T.L. **4** 63.8
Krasochka, O.N. **3** 83.17
Krausse, J. **1** 13.9 **3** 71.47 **4** 71.12
Kraut, J. **1** 21.12, 31.39, 41.34, 46.1, 47.27, 48.27, 48.28, 51.6, 52.24, 52.30, 59.65
Krebs, B. **1** 1.128, 11.3, 39.1, 39.3
Kreisman, P. **2** 86.26
Kretsinger, R.H. **2** 82.49
Krieg, B. **4** 71.8, 75.5, 75.11
Krieger, M. **3** 35.24
Krigbaum, W.R. **1** 1.141 **3** 9.11 **4** 13.20, 19.5
Krishna Rao, K.V. **1** 31.25, 31.26 **4** 1.4
Krishnamachari, N. **4** 83.110
Kristensen, R. **4** 39.42
Kritskaya, I.I. **3** 71.49
Krivnov, V.Ya. **1** 7.6
Krogmann, K. **2** 72.15, 81.11, 81.17, 81.18, 81.19
Krogsgaard-Larsen, P. **4** 40.3
Krom, J.C. **1** 52.19
Kromer, D.T. **1** 3.22
Kronfeld, L.R. **1** 39.67
Kroon, J. **1** 1.32, 1.34, 2.63, 2.67 **3** 1.8, 69.40 **4** 1.1, 1.3, 2.6, 2.15, 64.7
Kroon, P.A. **3** 75.14
Kropf, R. **1** 30.27
Krueger, S.M. **3** 20.12
Kruger, C. **3** 72.17, 75.18 **4** 63.5, 71.65, 75.7, 86.46, 86.66
Kruger, G.J. **3** 12.7, 41.7 **4** 30.1, 41.9, 41.10
Kruiisink, C.A. **1** 21.5
Krujer, G.J. **1** 30.4
Krukonis, A.P. **1** 28.21, 28.22 **3** 28.7, 73.10 **4** 12.11, 73.12
Kruse, F.H. **1** 11.52 **2** 70.20
Kshnyakina, A.N. **1** 7.7
Ku, A.T. **4** 20.4
Kubina, L. **3** 51.11
Kubinyi, H. **3** 54.15
Kubota, T. **1** 7.28
Kucsman, A. **1** 4.7 **3** 11.28 **4** 4.3
Kuhn, H.J. **2** 64.51
Kukina, G.A. **2** 72.1, 72.3, 80.19, 81.44, 81.67 **3** 72.32 **4** 72.1, 81.43
Kuli-Zade, T.S. **3** 72.32
Kulishov, V.I. **3** 60.9, 73.22, 73.38
Kulpe, S. **1** 18.11, 18.13 **2** 83.145 **3** 32.29 **4** 18.2, 18.4
Kumakura, S. **2** 60.117
Kumashiro, I. **1** 7.18
Kumra, S.K. **1** 2.73, 58.123
Kunchur, N.R. **2** 71.25, 79.10, 83.44, 83.45, 83.61, 85.4, 85.23, 85.51, 85.95 **3** 71.15, 79.12
Kung-Du, C. **1** 3.54
Kunstmann, W. **4** 24.17
Kunz, V. **2** 74.13
Kuo, M.-C. **4** 60.28, 71.86
Kupchan, S.M. **1** 53.34 **3** 53.25, 58.57 **4** 53.10, 58.40, 59.29
Kupcik, V. **1** 53.36 **3** 83.37
Kupfer, D. **1** 33.39, 33.47, 33.48
Kupletskaya, N.B. **1** 9.19
Kupper, F.W. **2** 71.39, 71.41
Kurahashi, M. **1** 13.12 **3** 36.4
Kuribayashi, S. **1** 17.40, 48.32 **2** 62.40
Kuriyabashi, S. **1** 17.42
Kuroda, H. **2** 60.142 **3** 60.53 **4** 60.33
Kuroda, T. **1** 11.37
Kurokawa, F. **4** 49.18
Kuroya, H. **2** 76.2, 76.4, 76.8, 76.10, 76.11, 76.13, 76.36, 79.25 **3** 76.23, 76.28, 76.52, 76.66, 81.17, 83.109, 84.10 **4** 33.24, 76.3, 76.10, 77.3, 77.18, 83.105
Kursanov, D.N. **2** 63.28
Kusaka, T. **1** 50.5
Kushi, Y. **1** 32.33 **2** 76.2 **3** 32.32, 68.22, 81.17
Kutney, J.P. **4** 56.2
Kutoglu, A. **3** 73.26 **4** 41.32
Kutschabsky, L. **3** 52.2 **4** 19.7, 52.2, 52.3, 85.23
Kutsuma, T. **3** 36.20
Kuwada, Y. **1** 59.42
Kuz'mina, L.G. **4** 86.36
Kuzmina, L.G. **4** 83.128
Kuznetsov, S.I. **1** 16.3 **3** 16.1
Kuznetsov, V.G. **1** 33.27, 33.28, 33.30 **2** 79.30 **3** 81.7
Kvick, A. **3** 12.4, 33.6 **4** 48.1, 60.5
Kwan, T. **2** 83.34
Kyogoku, Y. **1** 46.6

La Placa, S.J. **1** 8.43, 8.44, 15.22, 28.15 **2** 64.61, 66.19, 66.21, 77.65, 86.76, 86.77 **3** 83.51 **4** 83.42
La Prade, M.D. **3** 64.4
Laan, J. **1** 58.1
Lacey, M.J. **4** 83.44
Lachmann, B. **4** 33.22
Ladd, M.F.C. **1** 51.37, 56.13
Ladell, J. **1** 1.2, 32.5, 32.35, 32.36, 32.37, 32.39, 32.40, 32.43, 32.44 **2** 64.60
Lahiri, B.N. **1** 36.19
Lai, T.F. **1** 13.27 **3** 47.22, 59.20, 86.40, 86.46, 86.71
Laine, R.M. **4** 72.32
Laing, K.R. **3** 86.61
Laing, M. **1** 37.16 **2** 83.131, 83.132 **3** 83.74, 85.46 **4** 31.32, 33.13, 33.14
Laing, M.B. **2** 73.72 **4** 36.2
Lajzerowicz-Bonneteau, J. **1** 12.8
Lajzerowicz, J. **3** 37.3 **4** 12.13
Lake, A.W. **4** 24.13, 24.14, 24.15
Lalancette, R. **4** 33.25
Lalancette, R.A. **2** 77.65
Lam, S.Y. **3** 86.71
Lam, T.W. **1** 11.69
Lamarque, A. **1** 17.11
Lambert, J.B. **3** 21.17
Lamberton, J.A. **4** 58.9, 58.10, 58.11, 58.35
Lamm, B. **1** 16.34
Landais, J. **1** 19.60
Lanfredi, A.M. **2** 81.35, 83.10

Lanfredi, A.M.M. **1** 9.2 **2** 83.4, 83.5, 83.7, 83.15 **3** 83.3, 83.4 **4** 67.1, 67.2, 79.4
Lange, J.J.de **1** 9.35
Lange, S. **4** 71.30, 74.10
Langer, E. **4** 71.92
Langer, R. **1** 9.1
Langford, P. **1** 42.11, 42.12
Langhoff, C.A. **3** 60.35
Langs, D.A. **2** 81.61, 81.62, 84.40, 85.31 **3** 11.2, 82.20, 82.29
Lapasset, J. **3** 27.5, 27.6 **4** 32.4, 32.13
LaPlaca, S.J. **2** 86.109, 86.114
Lappert, M.F. **3** 71.37
Lapperts, M.F. **4** 71.81
LaPrade, M.D. **2** 74.17, 75.58 **3** 86.57 **4** 75.3, 81.26, 81.28, 86.25
Larcheres, G. **3** 82.24
Lardy, I.A. **1** 39.56
Larking, I. **3** 76.4
Larrabee, R.B. **4** 69.35
Larsen, E. **3** 85.4
Larsen, F.K. **3** 32.8, 60.44
Larsen, I.K. **1** 8.7, 10.8, 10.24 **3** 10.7 **4** 13.12, 36.28, 36.29, 36.30
Larsen, M.L. **4** 3.1, 82.11
Larsen, S. **2** 82.2
Larson, A.C. **1** 15.19, 34.1, 34.2, 40.2 **3** 47.25, 47.33, 51.42, 83.19 **4** 23.2, 76.8, 83.11, 83.14, 84.8
Larsson, G. **1** 2.48, 5.32 **3** 60.3
Larsson, K. **1** 1.130, 1.142, 1.148, 1.154, 1.156, 1.157 **2** 63.10 **3** 1.35, 28.6
Lasheen, M.A. **1** 24.17
Lauberau, P.G. **4** 75.28
Laubereau, P.G. **3** 75.42
Laurent, A. **1** 10.9, 31.12, 33.40, 33.52 **2** 60.28, 64.38, 64.39 **3** 9.20, 11.20, 11.21 **4** 8.8
Lavallee, D. **1** 49.33
Laves, F. **1** 36.30
Lavie, D. **4** 51.49
Lawrence, J.L. **1** 28.2, 38.20 **2** 68.2 **3** 64.23 **4** 45.7
Lawson, A.M. **1** 38.58
Lawson, J. **3** 9.2
Lawson, M.B. **4** 82.14
Lawton, D. **2** 81.59
Lawton, S.L. **1** 33.46 **3** 33.11, 37.1, 64.28, 85.21, 85.41, 85.42 **4** 69.24, 85.28
Lazarev, I.I. **3** 77.28
Lazarini, F. **4** 60.4
Lazzaroni, R. **4** 72.11
Le Bihan, M.-T. **4** 83.4
Le Carpentier, J.-M. **4** 68.1
Le Carpentier, J.-M. **1** 12.5 **2** 66.10 **4** 12.1, 12.2, 12.5, 12.6, 33.8, 33.9, 66.2, 83.46, 84.3
Le Quesne, P.W. **4** 58.23
Leandri, G. **3** 16.12
Lebedda, J.D. **3** 68.12
LeBihan, M.-T. **2** 83.2 **4** 28.1, 28.2
Leclaire, A. **1** 41.14
Lecomte, C. **4** 73.26
Lectard, A. **4** 21.4
Ledesert, M. **3** 60.7 **4** 38.1
Lednicer, D. **4** 51.46
Lee, B. **2** 76.62, 76.71, 76.72
Lee, C. **4** 35.14
Lee, C.M. **3** 31.9
Lee, J.D. **1** 11.58, 11.66 **2** 60.88, 60.100, 64.24, 76.15, 84.47, 84.48 **3** 11.19, 11.24, 64.1, 64.10, 64.21 **4** 11.15, 42.3
Lee, K.S. **1** 38.23, 41.13 **4** 41.15
Lee, L.M. **3** 86.40
Lee, M. **2** 82.8
Lee, S. **4** 67.10
Lee, T. **2** 73.62, 73.65 **4** 67.10
Lee, T.-Y. **3** 67.9
Lee, Y. **2** 73.65
Leenhouts, J.I. **3** 21.20, 31.8, 51.8, 51.10, 51.17, 51.18, 51.21, 51.23, 51.24, 51.25 **4** 38.20
Leeuwestein, C.H. **4** 27.2
Lefebvre-Soubeyran, O. **1** 32.46, 32.47, 32.48 **4** 31.29
Legendre, J.-J. **4** 65.6, 66.5
Legg, J.I. **4** 82.11
Leguen, J.-L. **2** 84.14
Legzdins, P. **2** 75.66, 77.39, 77.61 **3** 73.20
Lehmann, M.S. **1** 36.4 **3** 32.8 **4** 19.1, 48.5, 48.8, 48.10
Lehr, R. **4** 4.4
Lehtonen, M. **3** 76.29 **4** 83.59
Leijonmarck, M. **3** 36.11, 36.15 **4** 36.17, 36.24
Leipoldt, J.G. **1** 33.67 **3** 33.44 **4** 33.33
Leiserowitz, L. **1** 1.80, 1.100, 1.117, 1.118, 7.22, 16.51, 16.52, 16.53 **3** 1.13, 29.3, 29.4, 31.37, 31.38, 31.40, 31.41, 38.2
Leligny, H. **4** 38.1
Lemmer, F. **3** 39.33
Lenhert, P.G. **1** 32.9, 49.35, 52.8 **2** 83.146 **3** 49.2, 67.25, 84.13, 84.18, 84.19
Lentz Junior, P.J. **1** 52.31
Leonard, A. **2** 64.18, 64.19
Leonard, N.J. **3** 44.37 **4** 37.6
Lepicard, G. **1** 32.23 **2** 64.38, 64.39
Lepore, U. **3** 75.36, 75.37 **4** 1.9, 83.90
Leppard, D.G. **1** 30.20
Lerbscher, J.A. **4** 1.4, 28.7, 35.30, 69.2
Leroy, F. **1** 43.30 **3** 43.6 **4** 17.10, 17.11, 51.5
Leroy, G. **1** 5.33
Leser, J. **4** 1.6, 13.4
Lessor, A.E. **1** 41.7
Lester, T.E. **2** 84.27
Letcher, R. **3** 58.34
Lethbridge, J.W. **3** 76.24
Leung, F. **1** 39.47 **3** 41.21 **4** 41.34
Leung, Y.C. **1** 48.92
Leutzinger, E.E. **4** 44.25
Levdik, V.F. **2** 72.11
Leverett, P. **4** 86.2
Levine, E.M. **4** 36.18
Levine, S.G. **4** 28.10, 51.20, 51.40, 51.41
Levinthal, C. **4** 58.13
Levy, H.A. **1** 2.53, 2.57, 8.4, 45.35, 45.69 **3** 1.6, 16.14, 16.15, 48.9 **4** 8.1, 13.2, 36.1, 44.7, 56.3
Lewin, A.H. **4** 83.90
Lewin, R. **2** 62.37
Lewis, D.F. **2** 81.46, 86.21 **3** 33.9, 86.92 **4** 71.69, 80.10, 85.22

Lewis, D.L. 4 82.9, 83.100
Lewis, E.L.V. 1 2.96
Lewis, J. 2 86.46 4 75.3
Lewis, J.C. 4 67.7
Lewis, R.M. 4 76.15
Li, C.-T. 1 46.16
Li, N.C. 2 83.97
Liebig, D.M. 2 83.29, 83.43
Liebling, G. 1 20.6
Liehr, G. 4 85.11
Lien, M.M. 4 39.25
Lightner, D.A. 4 30.5, 30.6
Liminga, R. 1 2.29, 2.30 2 60.13, 60.15, 60.24
Lin, B.T-S. 1 31.7
Lin, C.-F. 4 48.45, 48.48
Lin, C.-T. 2 77.19
Lin, G.H.-Y. 3 47.7, 47.12 4 8.5, 47.9, 47.10
Lin, H.Y. 4 44.19, 50.1
Lin, T.S. 3 59.13
Lind, M.D. 2 76.63, 76.64, 76.71, 76.72 3 3.19, 8.15
Lindberg, B. 4 45.13
Lindblom, L.A. 4 78.15
Linden, J.G.M.van der 3 80.11
Lindgren, J. 1 3.10, 3.17
Lindgren, O. 4 76.1
Lindh, B. 2 81.53
Lindley, P.F. 2 66.1 3 71.32, 75.11, 75.13, 75.25, 75.27, 77.4
Lindner, H.J. 2 71.39, 71.41, 84.53 3 28.11, 29.1, 35.9, 35.11 4 29.6
Lindoy, L.F. 4 85.25
Lindqvist, I. 1 11.42, 39.13 2 66.6, 82.3, 82.9 3 33.43
Lindqvist, O. 2 70.21 3 69.4
Lindsay, A.R. 1 46.2
Linek, A. 1 53.14, 53.36 2 79.29 3 53.7
Ling, N.C. 3 51.41, 58.35, 58.42
Lingafelter, E.C. 1 1.127, 1.143, 3.33, 3.34, 3.35, 3.36, 8.26, 8.27, 9.25, 10.18, 29.1 2 76.14, 76.17, 76.18, 76.32, 76.46, 77.8, 77.12, 77.13, 78.3, 78.8, 78.9, 78.10, 78.11, 78.19, 78.23, 78.36, 78.38, 78.39, 78.43, 78.47, 78.48, 78.51 3 8.6, 76.42, 76.46, 76.47, 78.18, 78.21, 81.25, 81.45, 81.46, 81.47, 83.29, 83.30, 83.44, 83.112, 83.118, 83.121, 83.122, 83.123, 83.124, 83.125, 83.126 4 8.3, 78.5, 83.56, 83.57
Lingier, W.R.F. 3 41.10
Linke, K.-H. 4 39.7
Linnros, B. 4 36.24
Lippard, S.J. 1 3.63 2 73.39, 77.39, 83.1, 83.183, 85.89, 86.69 3 72.34, 73.53, 85.31, 86.67, 86.92 4 71.69, 80.10, 85.22
Lippert, E.L. 2 77.9
Lipscomb, W.N. 1 3.1, 3.3, 4.9, 5.4, 5.11, 5.12, 7.19, 19.21, 19.22, 19.26, 22.17, 30.22, 39.11, 45.65, 52.14, 53.39, 54.26, 58.3, 58.8, 58.103, 58.125 2 62.1, 62.2, 62.3, 62.7, 62.9, 62.10, 62.11, 62.13, 62.15, 62.22, 62.23, 62.24, 62.25, 62.26, 62.31, 62.36, 62.37, 62.44, 64.14, 75.5, 75.19, 75.29, 85.45 3 3.26, 3.27, 46.1, 62.5, 64.26, 64.27 4 3.8
Liquori, A.M. 1 13.30, 33.1, 44.67, 59.8, 59.20 2 60.97, 60.110, 60.111, 60.112
Little, J.L. 2 62.39
Little, R.G. 2 85.31 3 60.44 4 7.5, 71.36, 81.60, 83.43
Littleton, C.D. 1 45.29
Liu, C.F. 3 76.8, 82.7
Liu, C.S. 4 63.2
Liu, S.J. 3 50.24
Livingstone, S.E. 2 85.92
Llaguno, E.C. 3 42.3 4 39.14, 39.26
Llewellyn, F.J. 1 1.35, 2.26, 35.11, 45.11, 45.33 2 76.22, 78.26
Lloyd, D. 4 65.13
Lloyd, D.J. 4 24.3
Loader, P.L. 2 80.32, 80.33
Lobacher, A.N. 1 8.3
Lobanova, G.M. 1 19.42
Locchi, S. 1 36.7, 36.8, 36.11, 36.29 4 32.27, 36.16, 36.25, 40.16
Lock, C.J.L. 2 76.25, 77.5, 77.68 3 81.31, 81.44 4 83.110
Locke, H.O. 1 45.40
Loehlin, J.H. 3 83.34
Loescher, B.R. 3 81.54
Logan, N. 3 65.8
Logiudice, F. 1 9.2
Logiudice, G. 1 29.19
Lohaus, G. 4 42.4
Lohn, J. 3 81.5
Lohr, L.L. 1 4.9
Lohs, K. 3 64.7
Loiseleur, H. 4 81.34, 81.38
Loliger, P. 3 30.5
Lomborg, J.G. 4 29.8
Lomer, T.R. 1 1.137, 2.96, 2.97, 21.13 4 81.44
Long Junior, L. 3 45.44
Long, G.J. 3 78.11
Long, H.A. 1 48.16 2 82.46
Long, L. 1 41.10
Long, R.E. 1 7.29, 20.3
Longworth, J.W. 1 44.72
Lonsdale, K. 1 3.77, 3.84, 19.59, 26.19, 26.20, 26.21, 26.22, 26.23, 31.25, 31.26, 33.11
Look, W. 2 65.35
Lopez-Castro, A. 1 8.46, 11.41 2 79.23 4 33.19
Loranova, G.M. 3 17.4
Losman, D. 4 58.8
Loub, J. 4 81.22
Love, W.E. 1 2.74, 2.80, 2.81, 44.20 2 64.12
Lovell, F.M. 1 32.28 3 21.13, 35.14, 48.51, 48.52, 50.2 4 25.7, 35.15, 35.16, 41.24
Low, B.W. 3 48.51, 48.52
Low, J.N. 1 30.20 4 45.16
Low, M. 1 21.28
Lowenstein, M.Z. 1 39.5
Lu, C.-S. 1 48.17 2 60.8
Lu, C.T. 1 51.2, 51.3, 51.7
Luber, J.R. 4 77.24
Ludi, A. 3 76.49
Ludin, A. 3 46.8
Ludwig, M.L. 4 31.2
Luger, P. 4 21.2, 21.3, 71.67
Lugli, G. 4 74.1
Luhan, P.A. 4 58.19, 64.13
Luk'yanova, N.A. 3 76.67

Lukas, J. **4** 75.33
Lumme, P. **3** 81.26
Lund, E.W. **1** 33.50
Lundberg, B.K.S. **2** 83.41 **4** 83.16, 83.93, 83.96
Lundell, S. **3** 76.35
Lundgren, G. **1** 5.13 **3** 81.26
Lundgren, J.-O. **4** 11.1, 11.2, 15.2
Lundin, P. **4** 11.2
Luss, H.R. **4** 28.3
Lussier, R. **3** 83.28
Luth, H. **1** 32.21 **2** 71.10, 72.66, 79.36, 85.19 **3** 68.10, 69.12, 74.11 **4** 10.3, 69.11, 74.11
Lutz, W.K. **4** 50.11
Luukkonen, E. **3** 76.29, 76.34, 76.56, 76.78
Luxmoore, A.R. **2** 72.13
Luzzati, V. **1** 40.6, 41.12, 42.7, 47.19
Lydon, J.E. **2** 72.37, 83.191
Lynch, T.R. **4** 39.31
Lynton, H. **1** 39.7, 39.70 **2** 83.161, 83.193, 84.35 **3** 56.1 **4** 31.26, 83.51

Ma, K.W. **3** 56.2
Maartmann-Moe, K. **1** 15.5, 15.6, 17.13, 17.18 **2** 70.3, 70.4 **4** 70.15
Mabbs, F.E. **2** 78.13, 78.14 **3** 78.3, 78.23 **4** 81.48
Mabis, A.J. **1** 1.155, 59.22
Macaulay, E.W. **2** 65.30, 65.31, 65.32 **3** 39.44, 51.26, 51.30, 65.11, 66.4 **4** 39.33, 62.14
Macchia, B. **4** 21.12
Macchia, F. **4** 21.12
MacConnell, J.G. **1** 51.1
Macdonald, A.C. **1** 31.17, 52.23, 58.69 **2** 73.86, 73.95
Macdonald, A.L. **4** 2.2, 2.3, 2.17, 2.18, 2.19
Macdonald, C.G. **4** 83.44
MacDonald, S.G.G. **1** 28.2, 30.8, 38.17, 38.20
MacGillavry, C.H. **1** 1.31, 1.83, 1.95, 1.144, 12.14, 18.18, 18.19, 38.4, 53.27, 57.1, 57.2, 58.1, 59.47 **2** 60.5, 64.4 **3** 1.37, 20.2, 37.4, 37.7
Machajdik, D. **2** 83.118
MacInnes, D. **3** 77.27
Macintyre, W.M. **1** 6.1, 35.1, 44.19, 44.71, 47.5 **2** 84.36 **3** 31.39
Mackay, I.R. **2** 60.156 **3** 51.44, 59.27, 60.42, 60.55
Mackay, K.J.H. **1** 26.41
Mackay, M. **3** 58.14
Mackay, M.F. **1** 58.16, 58.77, 58.118, 59.29 **3** 48.8 **4** 16.2, 16.4, 16.5, 41.6
MacKellar, F.A. **4** 34.3
MacLean, D.B. **4** 58.16
MacLean, H. **1** 59.61
MacNicol, D.D. **1** 22.6 **2** 61.16 **4** 61.1
Maconochie, G.H. **1** 45.30
Macrosson, W.D.K. **1** 31.38
MacSween, D.R. **3** 83.62
Madan, O.P. **1** 46.7
Madaule-Aubry, F. **2** 83.198, 83.199, 84.6
Madden, D.P. **3** 83.128, 83.137
Madden, J.J. **3** 47.20, 48.23
Maddox, H. **1** 20.3 **2** 60.43
Maeda, K. **1** 41.31, 50.3, 50.20, 50.21, 50.34
Maekawa, T. **4** 30.9
Magennis, I.M. **3** 64.23
Maglio, G. **3** 72.33
Magnuson, V.R. **3** 67.12, 68.15
Mague, J.T. **2** 86.59 **3** 71.68, 86.87 **4** 71.24, 71.86
Mahon, C. **4** 7.1, 7.2, 7.4
Maier, C.A. **3** 39.21, 50.23 **4** 27.7, 29.2
Main, P. **3** 54.2
Mair, G.A. **2** 86.7
Mais, R.H.B. **2** 72.28, 73.33, 75.36, 83.168, 86.6, 86.8, 86.24, 86.43, 86.52, 86.55, 86.62, 86.65, 86.104, 86.105, 86.106 **3** 67.27, 72.30, 80.1, 80.2, 86.41
Maisseu, J. **1** 19.25
Maitland, R. **1** 30.25
Maitlis, P.M. **3** 71.54 **4** 71.51
Majeste, R. **1** 3.44, 35.10, 35.29, 35.33, 35.37 **3** 51.2, 64.37
Majeste, R.J. **3** 83.108 **4** 13.19
Majid-Hamid, A. **4** 32.22
Majumdar, S.K. **1** 1.41
Mak, T.C.W. **1** 15.11, 16.37, 28.6, 28.11, 28.13 **2** 61.3, 61.4, 61.7, 61.10, 64.78, 71.3 **3** 60.21
Makarenko, G.V. **4** 12.4, 12.7
Maki, A.H. **3** 60.54
Malament, D.S. **3** 9.17
Malinovskii, T.I. **2** 81.32, 83.91, 83.134, 83.135, 83.136 **3** 81.30, 82.9, 83.26
Mallikarjunan, M. **1** 48.45, 48.94 **3** 45.15, 45.16, 48.22 **4** 48.26
Mallikarjunan, N. **1** 20.14
Mallinson, P.R. **2** 64.65 **3** 62.17 **4** 62.15, 64.37
Malmstrom, M.C. **4** 65.8
Malofsky, B.M. **1** 56.16
Malone, J.F. **2** 68.29 **3** 68.14, 68.16, 68.17, 71.12, 75.23
Malta, V. **3** 1.36
Mamedov, Kh.S. **2** 84.13
Mammi, M. **1** 1.51, 3.85, 4.16, 33.75, 34.4, 35.9, 38.2, 38.3, 39.8, 39.9, 39.39, 39.40, 39.45, 39.46, 39.62, 42.9 **2** 69.42 **3** 16.12, 32.12
Manassero, M. **2** 60.149 **3** 86.22, 86.99 **4** 86.57, 86.71
Mandava, N. **3** 37.6
Mandel, G. **4** 5.2
Mandel, G.S. **4** 24.1
Mandel, N. **3** 65.2 **4** 51.14, 52.4, 52.5
Mandyczewsky, R. **3** 86.8
Maner, R.J. **3** 11.13
Manfreddoti, A.G. **4** 79.2
Manfredotti, A. **1** 39.41 **2** 79.31
Manfredotti, A.G. **3** 85.12 **4** 8.6, 9.5, 9.6, 35.13, 76.38, 79.3, 81.1, 83.1, 83.2
Mangia, A. **1** 41.15 **2** 79.26, 83.46, 85.2 **3** 16.4, 41.2, 76.1, 79.4, 79.14, 83.10 **4** 28.5, 59.28, 76.22, 79.12, 83.5, 83.62, 83.74, 85.14
Mangini, A. **3** 39.39 **4** 39.27
Mani, N.V. **1** 13.56, 53.20 **2** 64.32, 64.64 **3** 64.29

Mann, F.G. **2** 65.7, 65.24, 86.99 **3** 65.5, 86.80
Mannan, K.A.I.F.M. **2** 86.90
Manning, A.R. **2** 73.110, 86.23, 86.60
Manohar, H. **1** 13.56, 58.75 **4** 84.6
Manojlovic-Muir, L. **2** 81.84 **3** 71.37, 73.2, 73.46, 73.52 **4** 71.19, 71.21, 71.81, 86.19, 86.26
Manojlovic, L. **1** 2.88, 14.6, 14.13
Manojlovic, L.M. **1** 17.39
Manor, P.C. **3** 75.21
Manotti Lanfredi, A.M. **2** 83.14
Mansell, C.M. **3** 71.50, 71.51
Manzini, G. **4** 77.14
March, F.C. **3** 83.144, 83.147 **4** 64.44, 65.12, 65.14, 66.3, 66.4, 66.9, 66.10, 85.1
Marchenko, E.I. **1** 3.29
Marchessault, R.H. **3** 45.36
Marconi, W. **4** 74.1
Marek, D. **1** 43.21
Margolis, N.V. **1** 7.3, 10.1, 10.4, 12.1 **3** 7.1, 12.3, 12.5, 12.6 **4** 12.4, 12.7
Margulis, T.N. **1** 16.30, 20.2, 20.9, 20.11, 20.21, 20.34, 28.3, 51.42 **2** 60.119, 64.23, 75.21 **3** 20.6, 20.13, 51.35 **4** 20.2, 20.3, 20.6
Mariani, C. **1** 5.28, 5.31, 9.13, 19.50 **2** 60.103, 80.17 **3** 19.11
Marignan, J. **4** 9.11
Mark, W. **3** 81.26, 84.16
Marks, T.J. **4** 75.6
Marlier, M. **4** 59.2
Marly, L. **3** 43.9
Marongiu, G. **1** 40.7, 40.9, 40.17, 40.18 **3** 40.6, 76.26, 76.46, 76.48 **4** 76.6, 78.5
Marongui, G. **1** 40.8
Maroy, K. **1** 4.14 **2** 70.3, 70.4, 70.7, 70.11, 70.12, 70.16, 76.12 **4** 70.6, 70.7
Marquez, R. **4** 33.19
Marr III, H.E. **3** 41.15
Marraud, M. **4** 48.23, 48.43
Marsau, P. **1** 19.39, 41.41, 41.42, 41.45 **3** 41.17, 41.18 **4** 41.28, 41.30, 41.31
Marsh, R. **1** 11.59 **2** 76.56, 86.26
Marsh, R.E. **1** 1.42, 11.52, 11.63, 11.64, 13.27, 13.39, 13.40, 13.41, 20.4, 20.6, 20.24, 33.12, 39.20, 39.21, 40.32, 44.15, 44.22, 44.23, 44.46, 44.62, 47.11, 48.5, 48.14, 48.23, 48.70, 48.91, 48.92 **2** 60.71, 70.20, 73.8, 73.43, 73.78, 76.81, 78.15, 82.10 **3** 2.6, 41.8, 44.22, 45.27, 47.14, 47.19, 47.23, 59.20, 76.17, 76.19, 77.9, 79.13, 86.77 **4** 24.8, 29.7, 30.7, 35.24, 44.8, 44.12, 48.36, 60.27, 76.28, 76.30, 78.15, 86.39, 86.40
Marsh, W.C. **3** 64.3, 64.9, 64.65, 83.127 **4** 35.8, 51.13, 51.22, 51.32, 58.4, 58.5, 85.21
Marshak, M.L. **1** 58.73
Marshall, G.E. **4** 46.3
Marsham, D.F. **1** 3.38, 3.39 **2** 83.215
Marstokk, K.-M. **2** 60.22
Martelli, A.F. **3** 1.36
Martelli, G. **3** 32.13
Martin Junior, L.L. **3** 63.16
Martin-Smith, M. **1** 39.50
Martin, G. **3** 73.53
Martin, H.A. **2** 72.44
Martin, J. **1** 31.30
Martin, J.C. **3** 17.11 **4** 11.18
Martin, J.L. **1** 3.83
Martin, J.O. **2** 83.79
Martin, L.L. **1** 38.26 **4** 50.2, 86.8
Martin, N.H. **4** 58.19
Martin, R.-P. **2** 82.39
Martin, R.L. **3** 49.5, 85.24
Martinez-Carrera, S. **1** 13.44, 32.11, 43.9, 59.46 **3** 13.13 **4** 13.5, 77.10
Martinez, J. **1** 41.45, 43.10
Martinez, N. **4** 73.6
Martuscelli, E. **1** 1.69, 1.78, 1.93, 1.101, 1.102, 1.104, 1.110, 1.120, 1.121, 38.18 **3** 1.20, 1.21, 1.32, 75.36
Martynenko, L.I. **2** 76.60, 81.43 **3** 68.9, 76.67, 76.70, 82.9 **4** 76.35
Martynov, V.P. **4** 72.5, 72.6
Marumo, F. **3** 1.23, 3.25, 35.4, 50.21, 76.54, 76.64, 76.71, 76.82, 82.18 **4** 20.1, 29.4, 50.4, 54.13, 60.32, 76.2, 76.20, 76.21, 76.25, 76.40, 81.31, 83.68, 83.78
Marusic, R. **4** 50.13
Maruyama, H. **2** 67.18, 67.19
Marx, G. **3** 71.47
Marzilli, L.G. **3** 76.74 **4** 76.17, 76.18
Marzilli, P.A. **2** 76.48, 76.49, 82.54
Marzin, C. **3** 9.18, 9.20
Masaki, N. **1** 52.25, 58.86 **3** 58.13, 58.43, 58.52 **4** 54.15, 58.14, 58.43, 59.21
Masakuni, M. **1** 3.18
Masamune, S. **4** 21.22, 28.12
Masamune, T. **4** 40.24
Mascarenhas, Y. **1** 43.6
Masironi, L.Di M. **1** 33.76
Maslen, E.N. **1** 16.11, 16.12, 16.13, 16.17, 16.20, 16.21, 40.3, 44.43, 50.13, 50.16, 54.5, 54.14, 56.11, 56.19, 58.78, 59.41, 59.44 **2** 72.35, 80.10, 80.14, 81.33 **3** 19.12, 19.13, 19.14, 60.40
Maslen, H.S. **1** 2.26
Mason, R. **1** 1.53, 23.11, 23.15, 26.31, 26.32, 28.18, 33.80, 36.33, 38.39, 39.28 **2** 71.48, 72.16, 72.21, 72.40, 72.42, 72.59, 73.26, 73.29, 73.34, 73.35, 73.46, 73.67, 74.19, 75.8, 75.17, 75.23, 77.4, 77.62, 81.59, 82.28, 83.86, 83.178, 83.182, 86.18, 86.19, 86.28, 86.46, 86.80, 86.84, 86.89, 86.108 **3** 71.28, 71.63, 71.70, 72.37, 72.38, 75.35, 82.5, 86.49, 86.70, 86.98 **4** 23.5, 71.16, 71.17, 71.43, 71.46, 71.49, 71.71, 71.73, 71.82, 71.89, 75.22, 83.63, 86.22
Mason, S.A. **1** 49.12 **3** 8.9, 33.12, 33.13, 49.5
Massaux, M. **2** 83.2 **4** 83.4
Massey, A.G. **3** 75.16
Masters, C. **4** 86.14
Masuko, A. **2** 83.105
Math, K.S. **3** 83.143 **4** 85.30
Math, V.B. **3** 49.13
Mathern, G. **4** 81.3, 81.19, 83.47, 83.48
Mathew, M. **1** 31.33, 38.55 **2** 71.25, 83.45, 86.70 **3** 39.9, 59.3, 71.15, 75.49, 78.5, 83.113, 83.137, 85.9 **4** 38.10, 38.19, 40.6, 60.2, 71.61, 72.29, 83.134, 84.13, 85.37
Mathews, F.S. **1** 44.30, 44.39, 44.55 **2** 62.44, 75.5

Mathieson, A.M.McL. **2** 82.42
Mathieson, A.McL. **1** 1.151, 11.57, 45.31, 48.46, 48.47, 48.62, 48.93, 50.7, 51.44, 52.15, 58.16, 58.23, 58.27, 58.32, 58.65, 58.77, 58.81, 58.94, 58.118, 59.16, 59.24, 59.29 **3** 26.3, 48.8 **4** 16.2, 16.5, 58.9, 58.10, 58.11, 58.35, 83.66
Mathieson, D.W. **1** 59.45
Mathisen, H. **1** 5.17, 5.18, 5.19, 5.22
Matkovic, B. **1** 3.7 **2** 77.47, 77.49, 84.49 **4** 81.11
Matsuda, H. **1** 17.37, 54.9, 58.35
Matsuda, S. **2** 69.31
Matsueda, S. **3** 53.18
Matsumaru, H. **1** 48.73 **3** 44.42
Matsumoto, K. **2** 76.10 **3** 76.23, 76.28, 76.52, 76.66 **4** 76.3, 76.10
Matsumoto, M. **1** 50.42
Matsumoto, T. **1** 58.120
Matsuura, T. **3** 59.30
Matsuura, Y. **3** 73.33
Matsuzaki, T. **3** 48.33
Mattes, R. **2** 72.15, 81.11, 81.18
Mattew, M. **3** 79.12
Matthews, B.W. **1** 41.24, 43.31 **3** 36.29, 37.13
Matthews, D.A. **4** 38.2, 38.3, 38.4
Matthews, J.D. **3** 73.42
Matthews, R.W. **3** 81.36 **4** 81.36
Matzat, E. **4** 35.3
Mauer, F.A. **2** 83.106, 83.107
Maurin, M. **4** 64.3
Mautner, H.G. **1** 44.11, 44.12 **3** 3.29, 3.30, 3.31, 3.34
Maver, F.A. **2** 62.35
Maverick, E. **4** 12.9, 20.11, 20.13
Maw Junior, C.E. **3** 83.93
Mawby, A. **4** 81.39, 85.29
Mawby, R.J. **2** 75.52
Maxwell, I.E. **2** 82.54 **3** 76.27, 76.38, 76.74, 76.75 **4** 76.23, 76.24, 83.83, 83.84
Mayer, E.F. **1** 49.2
Mayer, R.M. **1** 35.38
Mayers, R.F. **1** 2.44
Mays, M.J. **4** 71.35, 86.22
Mayweg, V.P. **2** 85.34
Maze, M. **1** 17.23, 17.26, 17.43, 17.45
Mazumdar, S.K. **1** 48.19, 48.71 **3** 48.21, 48.29
Mazurek, W. **3** 76.84
Mazwell, I.E. **4** 76.17, 76.18
Mazza, F. **3** 33.45, 44.14, 44.17, 44.19, 44.25 **4** 32.22, 33.42, 33.45, 35.26, 36.34
Mazzarella, L. **1** 1.102, 37.10 **2** 60.97 **3** 1.20, 1.21, 33.23
Mazzi, F. **1** 11.18 **3** 11.5
Mazzochi, P.H. **4** 37.9
Mazzone, G. **2** 80.20
McAdam, A. **4** 2.12, 2.13, 2.14, 72.22, 72.23, 72.24, 72.30
McBride, D.W. **3** 73.7
McBride, H.D. **2** 68.30
McBride, J.M. **3** 9.17
McCallum, G.H. **1** 48.21
McCammon, C.S. **1** 13.52
McCandlish, L.E. **1** 3.69
McCapra, F. **1** 54.3, 54.4
McCarley, R.E. **4** 3.7
McCarty, C.G. **3** 8.14
McClelland, B.W. **4** 45.2
McCleverty, J.A. **3** 85.7, 85.20 **4** 83.131
McClure, R.J. **1** 53.28 **3** 44.8, 53.28, 58.57 **4** 43.4
McConnell, J.F. **1** 9.44, 44.16, 52.15, 52.16 **2** 78.6, 83.157, 85.32, 85.57 **3** 82.1 **4** 21.9, 21.10, 83.44, 85.5
McCorkindale, N.J. **1** 59.53
McCormack, J.A. **4** 42.5
McCrae, L.G. **4** 65.5
McCredie, R.S. **3** 44.37
McCrindle, R. **1** 59.67
McCullough Junior, J.D. **4** 17.12, 17.13
McCullough, J.D. **1** 3.24, 11.20, 11.21, 11.22, 11.32, 11.35, 11.52, 11.53, 11.54, 11.63, 11.64, 39.21, 42.6 **2** 60.42, 60.43, 60.52, 60.55, 60.56, 60.57, 70.1, 70.19, 70.20, 70.22 **3** 39.2, 39.30, 70.2 **4** 70.10
McDaniel, M.C. **1** 41.38
McDonald, J. **4** 71.72
McDonald, T.R.R. **2** 81.20 **4** 68.14
McDonald, W.S. **1** 41.35 **2** 68.29, 68.32 **3** 11.10, 68.14, 68.16, 68.17, 75.23 **4** 3.17, 46.2, 68.14, 71.13, 71.33, 86.28
McDowell, J.J.H. **1** 41.40 **3** 41.20
McEachan, C.E. **1** 54.10
McEachen, C.E. **1** 53.29
McEwen, G.K. **4** 64.4
McEwen, R.S. **2** 79.8, 84.67, 85.48, 85.75 **3** 85.39
McEwen, W.E. **2** 66.19, 66.21
McGahren, W.J. **1** 59.15
McGandy, E.L. **2** 62.5 **3** 32.11, 32.23, 32.34, 38.10, 48.23, 58.23, 73.54 **4** 32.2, 48.21, 48.25
McGeachin, H.McD. **1** 45.34
McGeachin, S.G. **1** 50.43
McGinnety, J. **2** 86.87
McGinnety, J.A. **2** 86.79, 86.98, 86.112 **3** 64.46, 86.89, 86.90 **4** 84.9
McGregor, A. **3** 73.43
McGregor, D.R. **1** 11.40, 14.11
McIntosh, A.O. **1** 30.13
McIntosh, C.L. **3** 38.8
McKechnie, J. **2** 65.51
McKechnie, J.S. **1** 31.7, 32.45, 34.5, 51.24 **2** 65.46, 73.106, 75.15, 75.16, 75.20, 75.44, 75.78 **3** 12.11, 51.11, 59.24
McKenna, J. **1** 33.80
McKenzie, E.D. **1** 39.28 **2** 82.28, 83.85 **3** 78.7, 78.24, 83.82, 83.83 **4** 71.57, 78.13, 83.131
McKillop, T.F.W. **3** 53.24 **4** 53.3
McKinley, S.V. **3** 69.35
McKinnon, A.J. **2** 77.16, 78.4, 78.5, 78.17, 78.18, 78.46
McLachlan Junior, D. **1** 19.65
McLachlan, V.N. **3** 78.23
McLaughlin, G.M. **3** 50.19, 68.7, 68.8 **4** 54.11, 54.14
McLean, S. **3** 36.18, 36.25
McLean, W.J. **2** 61.6, 61.8
McMillan, J.A.S. **1** 50.25, 58.2

McMullan, R.K. **1** 13.55 **2** 61.1, 61.3, 61.5, 61.7, 61.11, 61.13, 61.14 **3** 61.2 **4** 44.19, 44.22
McPartlin, E.M. **2** 85.50, 85.66 **4** 83.25
McPartlin, M. **2** 72.40, 86.84, 86.108 **3** 71.28
McPhail, A.T. **1** 13.45, 16.4, 50.39, 51.48, 53.8, 53.11, 53.22, 53.29, 53.34, 53.37, 54.4, 54.10, 56.18, 58.67, 58.71, 58.87, 58.121, 59.70 **2** 60.118, 73.16, 73.44, 73.99, 74.6, 74.7, 74.14, 75.68, 84.30, 84.37, 85.12 **3** 1.29, 1.30, 3.38, 3.48, 45.48, 53.21, 53.25, 53.28, 53.29, 58.20, 59.7, 59.17, 59.34, 64.22, 68.12, 73.56, 78.3, 78.23, 86.1 **4** 28.10, 37.10, 51.40, 51.41, 58.19, 58.32, 59.19, 59.26, 64.13, 64.14, 64.24, 65.9, 81.48, 81.55, 86.17
McPherson, W.G. **1** 3.12, 33.35, 33.44, 33.45
McQuillan, G.P. **3** 85.52
Mealli, C. **3** 83.118, 83.121, 83.122, 83.123, 83.124, 83.125, 83.126 **4** 76.43, 76.44, 86.44
Mechlinski, W. **4** 50.17
Medlin, E.H. **3** 60.4, 60.5
Medrud, R.C. **1** 27.5 **2** 75.18
Meek, D.W. **3** 85.36 **4** 85.37, 86.49, 86.50
Meer, H.van der **1** 19.63 **3** 65.6 **4** 20.7, 27.4, 36.4
Meersche, M.van **4** 20.12
Meerssche, M.van **1** 5.27, 5.33, 10.14, 10.15, 10.16, 41.26, 41.27, 41.30, 41.46 **2** 60.23, 60.25, 60.26, 60.30, 60.32, 60.46, 60.64, 60.66, 60.109, 64.18, 64.19, 72.32, 72.54, 72.57, 73.38, 83.169 **3** 20.18, 60.17, 60.18, 67.3, 67.6, 71.17 **4** 20.10, 24.19, 32.21, 32.23, 40.17, 48.4, 48.9, 48.11, 48.27, 51.25, 51.30, 51.31, 51.34, 83.39
Meester, M.A.M. **3** 20.2 **4** 1.12
Meester, P.de **3** 83.43 **4** 83.108
Meier, W.M. **3** 3.14, 3.15
Meinnel, J. **1** 19.30, 19.31, 19.32
Meinwald, J. **4** 71.9
Mel'nikov, V.V. **1** 7.3, 10.1, 10.4, 12.1 **3** 7.1 **4** 12.4, 12.7
Mellor, I.P. **3** 3.43, 3.44 **4** 11.6, 39.24, 39.31
Melmed, K.M. **2** 86.69 **3** 86.67
Melsom, B.G. **3** 56.6
Melsom, B.G.A. **2** 76.15, 84.47, 84.48
Melson, G.A. **3** 73.3, 85.47
Melville, D. **1** 54.25
Melville, R.D. **1** 54.20
Menchetti, S. **4** 83.35
Menczel, G. **3** 9.4
Mendel, H. **1** 48.26, 48.36
Menegus, F. **2** 68.9
Meney, J. **3** 51.35
Mercer, M. **3** 83.62 **4** 67.24, 75.23
Merenyi, R. **3** 41.10
Meriani, S. **3** 85.32, 85.33, 85.40
Merlino, S. **1** 35.16, 35.18 **2** 80.7, 80.30 **4** 21.12, 72.11, 80.2
Merritt Junior, L.L. **1** 32.13, 33.61 **2** 83.162 **3** 35.2, 35.7
Merritt, L.L. **1** 41.7 **2** 83.197, 84.44
Messager, J.-C. **1** 19.30, 19.31 **4** 19.8, 19.9
Messager, J.C. **3** 32.30
Messmer, G.G. **2** 86.4, 86.10, 86.11 **3** 40.5, 84.23, 86.62
Mestvedt, B. **1** 33.62
Mesureur, C. **2** 60.14
Metras, F. **4** 21.4
Metras, J.C. **1** 25.2
Metz, B. **3** 67.19, 67.21, 67.22, 67.23 **4** 67.12, 67.20, 67.22
Meulemans, R. **3** 67.3, 67.6 **4** 48.9, 48.11, 48.27
Meunier-Piret, J. **2** 60.109, 72.32, 73.38 **3** 71.17 **4** 20.10, 32.23, 40.17
Meunier, J. **1** 41.46
Meuthen, B. **1** 17.10 **2** 78.20
Meyer Junior, E. **4** 49.19
Meyer Junior, E.F. **2** 83.189 **3** 49.9 **4** 49.2, 49.15
Meyer, E.F. **1** 34.11
Meyer, H.-J. **1** 3.19
Meyerhoffer, A. **1** 37.26 **3** 37.12
Meyers, E.A. **1** 3.12, 3.44, 33.21, 33.24, 33.25, 33.26, 33.35, 33.44, 33.45 **2** 60.150 **3** 8.8, 65.3, 83.108 **4** 13.19, 19.10, 65.3
Meyers, M. **1** 11.7, 11.68
Mez, H.-C. **1** 15.26
Mez, H.C. **2** 75.53 **3** 44.6
Michalov, J. **2** 81.78
Michel, P. **4** 17.1
Michl, R.J. **4** 83.90
Midgley, J.M. **3** 51.26, 51.30
Miesel, S.L. **2** 65.46
Migchelsen, T. **1** 3.61, 3.62 **4** 65.10
Migdal, S. **4** 15.6, 16.6
Mighell, A.D. **1** 38.14 **2** 83.102, 83.106, 83.107, 83.165 **3** 19.4, 83.6, 83.95 **4** 83.91, 83.92, 83.94, 83.95, 83.102, 84.11
Mikawa, H. **3** 36.9
Mikhailov, Yu. **2** 79.30
Miklos, D. **4** 50.13
Milberg, M.E. **1** 5.12
Milburn, G.H.W. **1** 39.27 **2** 60.145, 68.19 **3** 67.11, 67.24, 82.17, 82.25
Miler-Srenger, E. **3** 41.16
Miles, M.G. **2** 71.18 **4** 83.26
Milje, L. **4** 41.23
Millar, I.T. **2** 86.99 **3** 86.80
Milledge, H.J. **1** 3.77, 3.84, 19.9, 19.10, 19.11, 26.19, 26.20, 26.21, 26.22, 26.23, 31.25, 31.26 **3** 69.2
Miller, A. **1** 21.13
Miller, C.K. **1** 49.22, 49.23, 49.28, 49.31
Miller, D.P. **4** 67.28
Miller, J. **4** 71.20
Miller, R.A.L. **1** 41.10
Mills, H.H. **1** 2.42, 2.91, 8.25, 14.1, 50.43, 58.73 **2** 60.16, 61.16, 73.59
Mills, J.C. **2** 71.4, 71.5, 71.31
Mills, J.F.D. **1** 38.61, 38.62, 58.84
Mills, O.S. **2** 71.14, 71.24, 71.32, 71.46, 72.14, 72.33, 72.34, 72.52, 72.64, 73.42, 73.51, 73.54, 73.74, 73.79, 73.92, 74.20, 75.28, 75.31, 75.46, 75.53, 83.152, 83.170, 83.209, 83.212, 83.214, 86.47 **3** 31.11, 71.30, 71.32, 71.52, 71.53, 75.3, 75.11, 75.13, 75.25, 75.27, 75.28, 83.103 **4** 22.2, 38.12, 71.14, 72.15, 72.20, 75.1, 75.9, 75.16, 78.2
Milone, L. **4** 71.27
Mimum, A. **4** 31.21

Minacheva, L.K. 4 76.27
Minacheva, L.Kh. 3 76.15
Minacheva, L.Kh. 4 76.7
Minami, T. 3 41.23 4 35.39
Minasjanc, M.X. 4 72.14
Minasyants, M.Kh. 2 72.7, 72.19, 72.62, 72.63
Minato, H. 1 27.12, 50.42
Mingos, D.M.P. 3 71.66, 86.60, 86.63 4 86.59
Minichelli, D. 3 78.14 4 71.40
Minkin, J.A. 1 1.79, 2.76, 2.78, 2.79, 2.80, 2.81, 4.15, 44.20 2 85.46 3 2.22 4 2.21, 36.21
Minobe, M. 3 60.13
Miscovic, J.D. 1 16.48, 16.50
Mislow, K. 1 52.26 4 52.4, 52.5
Misra, R. 4 51.40, 51.41
Missavage, R.J. 3 81.18 4 37.6
Mitchell, D. 1 13.51
Mitchell, D.J. 1 11.39
Mitchell, J. 3 60.6
Mitchell, T.P. 3 76.77
Mitrofanova, N.D. 2 81.43
Mitscher, L.A. 1 59.15
Mitschler, A. 2 84.3 4 33.8, 33.9
Mitsui, T. 4 48.14, 54.7, 54.15
Mitsui, Y. 1 48.42
Miura, I. 4 35.12
Miyoshi, T. 3 61.4
Mizukami, F. 4 83.78
Mizuno, H. 3 44.12 4 44.14, 44.15
Mizuno, K. 1 48.73, 50.5 3 50.13
Mizushima, T. 2 83.114
Mizutani, I. 2 67.18, 67.19
Mladeck, M.H. 1 23.5
Mo, F. 1 5.39 3 38.12 4 56.1
Moats, W.L. 3 46.2
Mock, W.L. 4 31.8
Moeller, T. 3 11.8, 64.25
Moers, F.G. 4 85.3
Moise, C. 4 73.26
Mok, K.S. 3 86.46
Mokren, J.D. 4 45.15
Mokuolu, J.A.A. 2 64.76, 86.48, 86.56
Molin-Case, J. 4 81.61
Molin-Case, J.A. 3 32.14, 32.15 4 40.22
Mollier, Y. 4 39.26
Moncrief, J.W. 1 45.68, 58.125, 59.56 4 80.6, 80.7
Monier, J.-C. 3 60.7 4 84.5
Moniot, J.L. 4 58.21
Montagnoli, G. 4 72.11
Montenero, A. 4 83.64
Montgomery, H. 1 39.19 2 77.8, 77.12, 77.13, 77.26, 77.27
Mooney, R.C.L. 2 65.36
Moore, D.W. 3 33.5 4 38.17
Moore, E.B. 2 62.2
Moore, F.H. 2 78.25 4 2.5, 49.20, 67.8
Moore, J.W. 2 83.197 3 68.18 4 81.46, 81.47, 81.56, 81.58
Moore, R.D. 4 83.131
Moore, W.J. 3 48.10
Mootz, D. 1 5.35, 20.32, 21.31, 32.18, 43.17 2 64.62, 65.35 3 11.1, 13.2, 51.36, 63.8, 63.13 4 17.2, 20.9, 36.36, 63.7, 83.9
Moras, D. 2 83.124, 83.153 3 67.19, 67.21, 67.22, 67.23 4 67.12, 67.20, 67.22
Morassi, R. 3 86.52
Morehouse, S.M. 2 73.39, 81.25 3 72.34
Morelle, I. 3 48.25
Moreno, J. 1 16.14
Moret, C. 3 64.36
Moret, C.L. 1 32.34
Morgan, C.H. 1 47.2, 47.3, 47.4 2 77.23, 83.181
Moriarty, R.E. 4 72.32
Moriarty, R.M. 4 31.5, 41.33
Moriconi, E.J. 3 42.2
Morimoto, C.N. 3 8.6 4 8.3
Morimoto, Y. 1 50.2 4 32.28
Morisaki, M. 1 55.1
Morley, K.A. 1 10.26
Mornon, J.-P. 1 24.28, 24.29, 35.36 3 56.4 4 21.13, 41.8, 41.25, 60.26
Mornon, J.P. 3 13.16, 41.6
Morosin, B. 1 3.26, 3.34 2 67.9, 77.22, 77.24, 77.26, 77.27, 78.19 3 9.5, 76.36, 76.37 4 33.40, 67.11, 67.16
Morrell, D. 3 86.98
Morris Junior, E.D. 3 64.2
Morris, A. 3 65.8
Morris, D.G. 1 31.20 4 3.6, 3.20, 11.4, 64.34
Morris, D.M. 3 83.1
Morrison, J. 4 12.14, 35.23
Morrison, J.D. 1 1.55, 1.73, 1.84, 1.125
Morriss, R.H. 2 83.158
Morrow, J.C. 2 84.31 3 10.2, 77.2, 77.3 4 33.1
Mortillaro, L. 1 39.40, 42.9
Mosbo, J.A. 4 64.4
Moseley, P.T. 2 67.16, 71.6 3 76.72, 76.73 4 3.14
Moss, D.S. 2 83.174, 85.49
Moss, K.C. 4 83.8
Mossel, A. 1 21.7
Mostad, A. 1 9.12, 45.6, 45.63 3 48.42, 84.2 4 9.9, 48.39, 48.40, 59.3
Motherwell, W.D.S. 1 30.11, 36.31, 36.32, 40.33 3 47.25, 47.33, 51.42, 53.12, 53.13, 53.14, 58.6, 65.5 4 8.9, 23.2, 52.6, 56.2
Motohashi, M. 2 83.187
Mouvet, M. 3 2.9
Mowat, W. 4 71.52, 71.53
Muccini, U. 4 21.12
Mueller, M.H. 4 38.2, 38.3
Muetterties, E.L. 3 62.9, 86.10
Mugnoli, A. 1 5.28, 5.31, 9.13, 19.50, 22.12 2 60.103, 80.13 3 19.11, 39.35 4 31.20, 31.21, 31.22
Muhm, H. 4 71.92
Muir, K.M. 4 73.30
Muir, K.W. 1 31.40 2 74.22, 74.23 3 71.37, 73.2, 86.21, 86.68, 86.79 4 30.8, 64.10, 71.21, 71.81, 86.26
Muir, L.M. 3 64.69
Mukherjee, J. 1 39.58
Mukherjee, R. 4 41.33
Mullen, D. 3 60.19
Muller, E. 4 71.92
Muller, F.A. 4 47.13
Muller, H. 2 73.80

Mullins, J.R. **3** 83.82, 83.83
Mumme, W.G. **4** 69.15
Munakata, K. **4** 54.18
Mundy, B.W. **2** 83.162
Munnik, J. **1** 40.12
Munns, A.R.I. **3** 47.17, 47.18
Muntz, R.L. **4** 16.7
Murahahi, S. **1** 20.1
Murakami, Y. **3** 1.17
Muralidharan, K.V. **1** 41.9 **2** 76.24
Muraoka, Y. **1** 41.31
Murata, I. **1** 20.30, 38.42
Murata, T. **1** 56.9
Murayama, W. **1** 47.20 **3** 46.3
Murmann, R.K. **2** 83.24
Muroi, M. **1** 50.5
Murray Junior, R.K. **4** 52.5
Murray-Rust, P. **2** 60.146, 60.147, 81.27, 81.45, 81.48, 81.66, 81.81 **3** 33.38, 33.39, 33.40
Murty, B.V.R. **1** 26.18
Musatti, A. **1** 16.31, 35.8, 39.41, 41.21 **2** 79.14, 79.15, 79.26, 79.31, 81.80, 85.2, 85.37 **3** 81.19 **4** 83.55, 84.12
Muscio Junior, O.J. **4** 20.11, 20.13
Musco, A. **2** 75.37 **3** 72.33
Mutin, J.-C. **4** 2.9
Mutin, J.C. **3** 2.5
Muto, A. **3** 76.71
Muto, F. **3** 78.9, 81.23 **4** 78.1
Muxfeldt, H. **3** 50.17
Myasnikova, R.M. **1** 5.26 **2** 71.26
Myers, J.F. **4** 83.73

Nackert, M.L. **4** 33.20
Nadelli, M. **2** 80.1
Nagai, Y. **3** 36.34
Naganathan, P.S. **4** 48.13, 48.17, 48.32
Nagano, K. **3** 48.30
Nagao, R. **3** 1.23 **4** 81.31
Nagashima, N. **1** 7.18, 47.13, 47.14, 47.20
Nagel, A. **4** 39.5
Nahringbauer, G. **1** 11.42
Nahringbauer, I. **1** 2.2, 2.36, 2.37, 2.38, 2.39, 2.40, 2.41, 2.48 **3** 1.2, 1.3, 2.10, 2.11, 60.3
Nair, P.M. **1** 59.68
Nakahara, A. **2** 76.8 **3** 78.9, 78.10, 81.23 **4** 78.1
Nakai, H. **3** 31.14, 38.13, 56.7, 83.109 **4** 39.37, 39.38, 83.103
Nakai, Y. **1** 2.5 **2** 81.9
Nakajima, S. **3** 60.7
Nakamura, A. **1** 7.18 **2** 73.82
Nakamura, H. **1** 41.31 **3** 36.20
Nakamura, K. **3** 36.9
Nakamura, M. **1** 21.14
Nakamura, T. **1** 13.38
Nakamura, Y. **3** 84.10 **4** 77.3, 77.18
Nakanishi, H. **4** 33.41
Nakanishi, K. **4** 35.12
Nakanishi, N. **4** 44.14, 44.15
Nakanishi, T. **1** 56.20 **4** 56.5
Nakano, E. **3** 5.3
Nakano, T. **1** 58.115 **3** 58.59
Nakatsu, K. **1** 58.43 **2** 65.18, 65.20, 76.36, 76.37, 79.25 **4** 35.31, 35.32, 38.8
Nakatsu, S. **3** 36.34
Nakatzu, K. **2** 76.38
Naldini, L. **4** 86.71
Namba, Y. **1** 36.12, 36.26
Nancollas, G.H. **4** 76.15
Naqvi, R.R. **3** 16.16, 16.17 **4** 64.1, 64.2
Narasimhan, K. **3** 30.4
Narayanan, P. **3** 48.1, 48.2, 48.3, 54.6, 59.25 **4** 27.19, 51.11, 54.12
Nardelli, M. **1** 8.18, 9.3, 11.1, 32.15, 39.23, 39.25, 39.32 **2** 69.11, 79.5, 79.9, 79.16, 79.18, 79.20, 79.26, 79.31, 79.33, 81.80, 83.8, 83.9, 83.28, 83.46, 84.2, 84.5, 84.8, 84.9, 85.2, 85.3, 85.5, 85.14, 85.35 **3** 8.5, 8.11, 76.1, 79.14, 81.19, 83.10, 85.12 **4** 8.6, 28.5, 35.13, 41.21, 41.22, 59.28, 76.22, 76.38, 79.12, 83.1, 83.2, 83.5, 83.12, 83.55, 83.62, 83.64, 83.74, 84.12, 85.14
Nardin, G. **2** 78.59 **3** 76.81, 78.14, 78.26, 83.150, 85.32, 85.33, 85.34, 85.40 **4** 71.40, 77.14
Narita, H. **4** 59.21
Naruto, S. **4** 58.43
Nasakkala, M. **3** 76.32, 76.35
Nash, C.P. **3** 82.26
Nassimbeni, L.R. **4** 77.15, 86.6
Natarajan, S. **2** 79.6
Natile, G. **3** 86.25
Natsume, M. **1** 54.12
Natta, G. **2** 61.17, 61.19, 72.39, 74.5, 74.25
Nave, C. **4** 72.18
Nawata, Y. **1** 50.18 **2** 83.96 **3** 50.18 **4** 50.10
Naya, K. **1** 50.38 **3** 50.20, 53.20
Neegauz, M.S. **3** 17.4
Neel, J. **4** 48.23, 48.43
Neelakantan, L. **4** 40.22
Neely, S.C. **3** 31.13
Nefedov, O.M. **3** 60.9
Neguchi, T. **2** 82.33
Neidle, S. **1** 50.27, 59.48 **3** 45.14, 48.19 **4** 32.17, 38.21, 38.25, 49.16, 53.6, 59.15
Neigauz, M.G. **3** 40.2 **4** 24.7
Nelson, A.D. **1** 12.9 **2** 60.93, 60.94
Nelson, G.V. **4** 40.9
Nelson, N.J. **3** 73.50
Nelson, N.R. **1** 58.110, 59.50
Nemiroff, M. **3** 83.86
Nepomnyashchaya, V.N. **1** 3.29
Neri, C. **3** 71.25, 78.12, 78.16
Neronova, N.N. **1** 19.34
Nesmeyanov, A.N. **2** 72.72 **3** 71.49
Nesterova, Ya.M. **1** 9.19 **4** 9.7
Neuman, A. **4** 17.6, 17.7
Neuman, M.A. **4** 31.12, 38.13, 73.28
Newing, C.W. **3** 83.102 **4** 83.61, 83.85
Newlands, M.J. **3** 73.13
Newman, B.A. **1** 23.17
Newton, G. **1** 58.64
Newton, M.G. **1** 37.24, 41.38, 46.11 **2** 75.78 **3** 39.46, 54.5, 58.39, 58.40, 58.41, 58.48, 64.44 **4** 42.5, 58.38, 58.44, 64.11, 64.20, 64.26, 86.7
Ng, T.W. **2** 73.122

Nibbering, N. 1 38.4
Nice, J.P. 2 73.42, 73.54
Nice, R.van 3 84.26
Nicholas, A.F. 1 54.11 3 82.6
Nicholas, K. 3 72.41
Nicholson, C. 4 33.14
Nicholson, J.K. 3 71.12
Nickerk, J.N.van 2 81.21
Nicol, A.D.I. 1 11.33, 11.34
Niekerk, J.N.van 2 81.29, 81.34, 81.36, 81.37, 81.63, 81.64
Nielsen, A.T. 4 38.17
Nielsen, J. 2 73.116
Nielsen, K. 4 83.60
Nielson, D.O. 3 3.1, 3.12 4 82.11
Nigam, G.D. 1 26.28
Niggli, A. 1 25.20, 58.59, 58.122 2 83.188, 83.206 3 59.9
Niimura, N. 2 60.101
Nilson, L. 3 68.4, 80.21
Nilsson, A. 1 2.30
Nilsson, B. 1 30.36, 52.32, 53.16, 53.32, 58.21 3 5.4, 58.59
Nimrod, D.M. 1 46.8 2 64.17
Nishi, Y. 1 27.13, 56.12
Nishida, T. 3 47.32
Nishikawa, M. 1 48.73, 50.5, 56.9, 58.35, 58.36, 58.57, 59.52 3 44.42, 50.13, 50.22, 50.30 4 50.6
Nishikawa, Y. 4 77.18
Nishimura, K. 4 59.12
Nishio, K. 1 58.93
Nissen, H.M. 4 48.39
Nitta, I. 1 1.30, 1.89, 1.108, 2.3, 2.4, 2.6, 2.84, 5.36, 5.40, 9.5, 13.36, 16.15, 17.37, 22.3, 27.8, 35.17, 37.1, 48.32, 50.37, 50.38, 56.9, 58.35, 58.36, 58.57, 59.19, 59.25, 59.52 2 81.68, 82.4, 82.5, 83.156 3 17.5, 33.42, 36.4, 50.13, 50.20, 53.20 4 58.1, 58.2, 81.32, 81.33, 81.35
Nixon, J.F. 4 64.10
Nixon, R.E. 1 37.2, 37.3
Nobbs, C.L. 1 26.26
Nockolds, C.E. 1 39.27 3 83.20 4 49.20
Noda, T. 1 50.31, 50.38 3 50.20
Nofre, C. 1 44.41
Noguchi, M. 4 59.21
Noguchi, T. 1 3.18
Nojima, H. 4 32.28
Noland, W.E. 1 40.4 4 40.12
Nolte, M.J. 2 86.57
Noltes, J.G. 4 71.82
Nomura, T. 2 83.17, 83.105 3 76.64
Noordik, J.H. 1 11.47 4 80.4, 80.11, 80.12, 85.4
Nordman, C.E. 1 2.82, 30.3, 31.9, 31.10, 51.1, 58.43, 58.123 2 62.6, 68.4, 68.13 3 9.13, 30.2, 51.6, 64.2, 64.4 4 38.25, 59.24, 60.43
Norin, T. 4 59.12
Norman Junior, J.G. 3 83.132
Norman, J.G. 4 81.62
Norman, N. 1 5.17, 5.18, 5.19, 5.22, 20.16, 21.4
Norment, H.G. 1 17.44, 17.46, 31.28 2 62.19
Norrestam, R. 1 36.27, 45.27, 45.28 3 36.7, 36.10, 36.15 4 35.1, 36.13, 36.15, 36.19, 36.23, 60.30
North, P.P. 3 69.37, 69.38 4 27.1, 38.13, 38.14, 69.18
Northolt, M.G. 1 1.97, 34.9, 52.10 2 63.13 3 34.2, 34.5 4 31.8
Norton, D.A. 1 51.2, 51.3, 51.5, 51.7, 51.9, 51.10, 51.12, 51.13, 51.14, 51.17, 51.21, 51.22, 51.41 2 60.151, 82.53, 84.64 3 17.8, 48.41, 51.12, 51.16, 51.19, 60.14, 60.15, 60.43 4 11.12, 11.14, 51.2, 51.3, 51.15, 51.17, 51.24, 51.33, 51.42, 51.47, 51.50, 51.51, 60.39
Novak, C. 1 53.36 3 53.7 4 81.22
Novakovskaya, L.A. 3 17.4, 40.2
Novick, S. 2 63.15
Novikov, S.S. 3 40.2
Novozhilova, N.V. 2 76.60
Nowacki, M. 1 1.92
Nowacki, W. 1 13.33, 23.5, 44.5, 51.27, 58.28 2 74.13 3 58.26 4 58.42
Nowell, I.N. 2 73.3
Nowell, I.W. 2 86.92 3 86.5, 86.7 4 86.1
Nozakura, S. 1 20.1
Nozoe, S. 1 50.35, 55.1, 55.2 4 59.23
Nuno, M. 1 59.42
Nunzi, A. 4 70.4, 83.49, 85.8
Nuretdinov, I.A. 4 64.17
Nuttall, R.H. 3 83.77, 83.111 4 83.37, 83.38, 83.41, 83.52, 83.53
Nyberg, C.S. 3 81.54
Nyburg, S.C. 1 19.44, 38.61, 38.62, 39.47, 44.69, 44.73, 58.84 2 75.47 3 3.43, 3.44, 36.18, 36.25, 41.21, 60.34, 68.3, 83.72, 86.64 4 11.6, 39.24, 39.31, 39.46, 41.34, 63.2, 63.12, 69.1, 72.25, 72.26, 86.31
Nyholm, R.S. 2 86.46 3 67.24, 86.97 4 71.43, 71.46, 71.49, 71.71, 71.73, 71.89
Nyman, F. 2 73.33

O'Brien, E.J. 1 44.60, 44.61
O'Brien, J. 4 58.3, 58.6
O'Brien, R.J. 2 69.2 3 68.3 4 71.47
O'Brien, S. 2 72.28
O'Brien, T.A. 2 72.27, 73.112, 73.113 3 71.16, 71.43, 72.43
O'Connell, A.M. 1 1.139, 16.20, 16.21, 39.37, 54.14, 59.57
O'Connor, B.H. 1 16.17 2 80.14, 81.33 4 49.20
O'Connor, J.E. 2 73.48, 73.96, 79.28, 83.65 3 79.5
O'Donnell, E.A. 1 51.37
O'Donnell, M.M. 1 2.95
Oberhansli, W. 1 22.6 2 75.62
Oberhansli, W.E. 1 51.11 2 72.9, 75.80, 75.81 3 44.38, 58.32 4 31.28, 58.12, 58.17
Obodovskaya, A.E. 3 78.20
Ocenaskova, D. 1 12.10, 12.11
Ochiai, A. 4 59.21
Oda, K. 3 58.11, 58.16, 72.10, 72.18 4 48.24
Oda, T. 1 36.12, 36.26
Oftedal, P. 4 41.26

Ogawa, K. **1** 5.10 **2** 83.156 **3** 26.11 **4** 72.17, 82.6
Ogawa, Y. **3** 50.21
Ogihara, Y. **1** 59.60
Ogilvie, R.A. **3** 24.7
Ogura, T. **3** 72.10 **4** 71.58, 71.80
Oh, Y.L. **1** 54.5, 56.11
Ohashi, T. **1** 2.84
Ohashi, Y. **2** 60.101, 60.123 **3** 3.25
Ohloff, G. **3** 53.1, 53.2
Ohno, K. **3** 58.43
Ohrt, J. **3** 44.29
Ohrt, J.M. **1** 51.13, 51.14, 51.17 **3** 52.5
Ohsawa, T. **1** 56.7
Ohshiro, Y. **3** 41.23 **4** 35.39
Okabe, K. **1** 54.17
Okada, K. **2** 81.3, 81.4 **3** 31.26
Okada, Y. **4** 38.8
Okamoto, T. **3** 76.23, 76.66
Okamoto, Y. **1** 58.36, 58.57
Okamura, W.H. **3** 85.23
Okawara, R. **2** 69.3
Okaya, Y. **1** 1.57, 1.58, 4.4, 4.8, 8.30, 9.11, 13.15, 14.9, 14.10, 21.14, 31.4, 31.14, 33.65, 33.69, 40.11, 41.17, 45.54, 48.9, 48.11, 58.42, 59.2 **2** 63.21, 63.22, 64.7, 79.13, 82.8 **3** 13.4, 13.5, 28.3, 28.4, 69.28
Okeya, S. **3** 72.10
Okuda, S. **1** 50.35, 55.2
Okuda, T. **1** 58.111 **3** 45.24
Okunuki, M. **4** 60.7
Olcott, R.J. **3** 83.71
Oldham, G.C. **3** 77.2, 77.3
Oliver, J.D. **3** 72.42 **4** 71.84
Oliver, J.P. **3** 68.18
Ollendorf, W. **3** 81.11, 81.12, 81.13, 81.15
Olovsson, I. **1** 2.29, 2.30, 3.10, 3.17, 12.4 **3** 2.10, 12.4, 33.6 **4** 60.5
Olsen, B. **3** 85.4
Olsen, F.P. **4** 11.10
Olson, A.J. **1** 51.1
Olson, D.H. **2** 69.23
Omae, I. **2** 69.31
Omel'Chenko, Yu.A. **1** 9.31, 9.32, 9.33
Omel'chenko, Yu.A. **4** 32.1
Omura, S. **3** 50.20
Ono, K. **3** 72.20
Onuma, S. **1** 50.18 **3** 77.6
Onyszchuk, M. **4** 63.4, 69.16, 69.17
Ooi, S. **2** 76.4, 76.10, 76.11, 76.13, 79.25, 83.140, 85.70, 85.81 **3** 76.28, 77.15, 83.109, 84.10 **4** 33.24, 76.3, 76.10, 77.3, 77.18, 83.105
Oonishi, I. **3** 76.31, 78.9, 78.10, 81.23 **4** 78.1
Opedal, D. **3** 8.12, 8.13
Opferkuch, H.J. **3** 54.8
Oppolzer, W. **3** 40.12 **4** 83.112
Orehowsky Junior, W. **4** 2.20, 36.21
Orfanova, M.N. **2** 60.140
Organ, T.D. **3** 51.9
Orgel, L.E. **2** 73.22
Orii, S. **1** 13.38
Orio, A.A. **3** 86.77 **4** 86.39, 86.40
Orioli, P.L. **2** 76.76, 76.77, 76.78, 76.79, 78.3, 78.24, 78.32, 78.34, 78.36, 78.49, 78.53, 83.119, 83.120, 83.127, 86.49, 86.83, 86.95 **3** 76.62, 78.15, 83.18, 86.44 **4** 86.44
Osaki, K. **1** 2.4, 2.5, 16.51, 17.37, 52.25, 58.36, 58.57, 58.86, 58.115 **2** 61.21, 81.9, 81.77 **3** 2.1, 45.24, 45.25, 58.43, 59.30 **4** 58.43, 62.10, 67.3
Osawa, E. **1** 31.43
Osawa, Y. **4** 11.12, 11.14, 51.35, 51.36, 51.47, 51.50, 51.51
Osborn, J.A. **4** 86.70
Osborn, R.S. **4** 49.7
Oshio, H. **1** 59.52
Oskarsson, A. **2** 82.52 **4** 81.18
Osterberg, R. **3** 81.33, 81.51
Oswald, H.R. **4** 83.69, 83.70
Otaka, Y. **4** 5.4, 20.1, 29.4
Otake, N. **3** 50.21
Otsuka, S. **2** 73.82 **4** 72.17
Ott, H. **1** 58.87
Ottersen, T. **4** 48.40
Oughton, B.M. **1** 48.61
Overton, K.H. **1** 54.25 **3** 54.2
Owen, J.D. **3** 64.46 **4** 84.9
Owen, T.B. **1** 7.4, 31.1
Owston, P.G. **1** 35.24 **2** 72.5, 72.28, 73.33, 75.36, 80.5, 83.168, 86.6, 86.8, 86.13, 86.24, 86.27, 86.43, 86.52, 86.55, 86.65, 86.86, 86.104 **3** 17.10, 72.3, 72.30, 80.1, 80.2, 86.28, 86.29, 86.41 **4** 67.27, 72.2
Oyen, J.W.L.van **1** 18.21, 42.2
Oyum, P. **1** 11.56 **2** 70.24
Ozbirn, W.P. **4** 64.31
Ozeki, K. **1** 44.31
Ozerov, R.P. **1** 7.5, 7.7
Ozol, Ya. **2** 62.51
Ozola, A. **2** 85.80
Ozols, J. **2** 85.80

Pabst, A. **2** 81.13
Pace, A.de **1** 39.62
Pachler, K. **1** 5.20
Padanabhan, V.M. **2** 76.24
Paddock, N.L. **2** 64.45 **4** 64.27, 83.122
Padilla, J.O. **1** 53.40
Padmanabham, V.M. **1** 2.35, 9.41
Padmanabhan, V.M. **1** 2.24 **2** 67.8 **3** 77.8 **4** 48.3
Padwa, A. **1** 31.31 **4** 35.35, 39.8
Page, J.E. **1** 51.38, 51.39, 51.40
Page, S.W. **3** 58.40, 58.41 **4** 58.44
Paiaro, G. **2** 75.37
Paine, R.E. **3** 78.7
Pajunen, A. **2** 76.16, 83.50, 83.51 **3** 76.29, 76.33, 76.34, 76.43, 76.53, 76.56, 76.57, 76.78 **4** 83.59
Pakhomov, V.I. **1** 3.29 **2** 71.1, 71.2, 71.8, 71.19
Pal, B.C. **3** 85.18 **4** 47.4

Palenik, G.J. **1** 10.19, 15.12, 32.1, 32.30 **2** 64.11, 65.40, 68.12, 73.122, 75.11, 82.41, 83.155, 83.160, 86.68, 86.70 **3** 32.18, 32.19, 33.21, 33.27, 38.11, 39.9, 39.36, 40.5, 50.14, 59.3, 68.13, 73.8, 73.14, 75.49, 77.18, 77.31, 78.5, 83.113, 83.128, 83.137, 84.23, 85.9, 86.23, 86.42, 86.62, 86.83 **4** 15.1, 33.34, 38.10, 38.19, 40.6, 58.41, 60.2, 68.11, 71.61, 72.29, 77.13, 78.12, 83.134, 84.13, 85.37, 86.34
Palin, D.E. **2** 61.9
Pallister, E.T. **4** 45.22, 83.113
Palm, J.H. **1** 16.9, 21.17, 52.10 **2** 60.5
Palmer, H.T. **1** 11.70
Palmer, K.J. **1** 31.37, 38.23, 41.13 **4** 41.15, 67.7
Palmer, R.A. **1** 11.70, 46.12 **3** 1.33, 68.12
Palmieri, C. **1** 39.32 **3** 79.4 **4** 41.21, 41.22, 83.62, 83.64
Palumbo, R. **3** 1.20, 1.21, 72.33
Panattoni, C. **2** 68.9, 69.21, 73.19, 77.10, 78.22, 78.28, 78.31, 78.35, 78.44, 83.59, 83.60, 83.82, 83.83, 85.64, 86.38 **3** 39.39, 40.15, 64.42, 69.33, 72.49, 72.51, 81.55, 86.19, 86.25 **4** 19.2, 39.27, 40.19, 71.78
Pandraud, H. **1** 35.40, 38.53
Panetta, C.A. **1** 50.26
Pangborn, W.A. **4** 1.14
Panke, D. **2** 61.2 **3** 45.44, 61.1, 61.2
Pant, A.K. **1** 13.4, 13.5
Pant, L.M. **1** 2.52, 2.69, 2.89, 3.77, 3.84, 13.8, 13.10, 13.11, 19.9, 19.10, 19.11, 48.12 **2** 82.19 **4** 13.1, 13.6, 13.7, 15.4
Pantaleo, D.C. **4** 80.6, 80.7
Pantini, G. **2** 72.43, 72.45 **4** 72.19
Paoletti, P. **3** 76.46
Paquette, L.A. **1** 34.8 **4** 27.11, 27.12, 35.14, 36.22, 39.17
Park, D.Y. **2** 83.97
Park, J.J. **2** 76.61 **3** 69.29, 69.30
Park, Y.J. **3** 45.3 **4** 45.10, 45.11, 45.24
Parker Junior, E.B. **2** 77.25
Parker, L.M. **4** 68.5
Parker, R.G. **3** 33.48, 38.17 **4** 22.10
Parker, W. **3** 53.23 **4** 53.3
Parker, W.L. **1** 50.32
Parkes, A.S. **1** 7.2, 37.2, 37.3 **3** 24.7, 31.36, 75.21
Parry, G.S. **1** 1.61, 2.28
Parry, R.W. **2** 62.6, 68.13
Parthasarathy, R. **1** 45.52, 48.35, 50.19 **3** 2.13, 2.14, 32.33, 44.29, 44.32, 52.5 **4** 48.18, 48.30, 48.31, 48.44
Partridge, J.M. **2** 86.13
Partridge, M.W. **4** 35.25
Pascard-Billy, C. **1** 25.8, 25.9, 25.10, 58.10, 58.88, 58.89 **3** 43.1 **4** 19.13, 21.16, 45.14, 51.23, 53.8, 60.8
Pascard, C. **3** 53.30
Pascard, R. **4** 60.8
Pascher, I. **1** 46.13, 59.57
Pascoe, J.D. **4** 63.12
Pasternak, R.A. **1** 1.64, 8.35, 21.3, 32.24, 48.60, 48.90
Pasynskii, A.A. **2** 72.72
Patel, C.G. **3** 72.30
Patel, H.A. **3** 68.13 **4** 71.61, 72.29
Patel, M.B. **1** 58.73
Paternotte, C. **1** 41.26
Patil, H. **2** 83.154
Paton, M.G. **1** 59.44
Paton, P.N. **1** 51.38, 51.39, 51.40
Patrick, R. **3** 1.33
Pattabhi, V. **4** 48.15, 48.54
Pattabhiraman, T. **4** 51.57, 51.58
Patterman, S.P. **3** 67.7, 67.26
Patterson, A.L. **1** 1.79, 2.74, 2.76, 2.78, 2.79, 2.80, 2.81, 2.82, 4.15, 44.20 **2** 85.46
Paul-Roy, S. **1** 53.27
Paul, G.L. **3** 48.11
Paul, I.C. **1** 31.7, 32.45, 34.5, 34.8, 37.18, 37.24, 38.26, 39.61, 39.83, 41.19, 41.20, 41.29, 41.38, 50.33, 51.24, 53.35, 54.13, 59.51, 59.64 **2** 62.39, 62.48, 65.46, 73.71, 73.106, 75.15, 75.16, 75.20, 75.35, 75.44, 75.78, 81.49, 81.54 **3** 9.6, 11.8, 12.11, 17.11, 20.12, 23.8, 23.9, 33.41, 37.10, 39.21, 39.22, 39.23, 39.40, 39.41, 39.46, 42.3, 44.37, 50.23, 50.24, 51.11, 59.24, 64.25, 71.31, 75.7, 75.17, 75.29, 75.34, 77.24, 81.18, 81.52 **4** 11.18, 13.15, 13.16, 16.7, 17.12, 17.13, 27.7, 29.2, 32.30, 37.6, 39.14, 39.26, 50.18, 71.54, 71.55
Paul, S. **1** 1.56
Pauling, L. **1** 38.30
Pauling, P. **1** 58.24, 59.6, 59.13, 59.14 **2** 65.33, 71.52, 75.12, 77.62, 83.40, 84.68, 86.120 **3** 20.9, 38.3, 83.67 **4** 16.11
Pauling, P.J. **2** 77.4, 83.86 **3** 86.27, 86.97
Paulson, P.L. **3** 71.50
Paulus, E. **2** 75.61
Paulus, E.F. **2** 69.9, 73.79, 73.81, 73.92, 75.28, 85.65 **4** 24.17, 40.5, 42.4, 44.13, 70.1, 70.5
Paulus, H. **3** 35.5, 83.38
Pauson, P.L. **2** 83.214 **3** 71.53 **4** 72.15
Pautler, D. **4** 7.5
Pawley, G.S. **1** 22.17, 24.16, 27.7 **2** 62.4, 62.8, 62.27 **4** 19.1, 31.27
Pawson, B.A. **1** 52.29
Payne, D.S. **2** 64.76, 86.48, 86.56
Payne, N.C. **2** 82.37 **3** 76.63, 82.5, 86.73, 86.89, 86.90, 86.93 **4** 71.59, 76.19
Peach, J.M. **3** 40.19
Pechurova, N.I. **4** 76.35
Peck, D.N. **3** 60.14, 60.15
Pedersen, B. **1** 2.19 **2** 76.65
Pedersen, B.F. **1** 2.15, 2.17, 2.19, 2.20, 2.21, 2.27, 5.17, 5.22, 16.40 **2** 83.42 **3** 2.3 **4** 2.4
Pedone, C. **1** 1.93, 21.24, 21.25, 38.18, 44.53 **2** 64.13, 72.17, 72.18, 72.38 **3** 1.20, 1.21, 1.31, 20.4, 21.5, 21.6, 21.7, 21.8, 21.10, 33.16, 33.19, 39.1, 62.16, 72.19 **4** 1.8, 48.7, 71.25, 71.26, 72.8
Pedretti, U. **1** 29.19 **4** 74.1
Peerdeman, A.F. **1** 2.63, 2.65, 2.67, 48.25 **2** 64.48 **3** 59.12, 69.40 **4** 64.7
Pegel, K.H. **4** 31.32
Peiren, M.A. **3** 32.9
Pelizzi, C. **3** 85.12 **4** 9.5, 9.6, 41.22, 76.22, 83.64, 83.74, 85.14

Pelizzi, G. 1 41.15 2 83.46, 85.35 3 16.4, 41.2, 76.1, 79.14, 83.10 4 28.5, 35.13, 59.28, 76.22, 79.12, 83.1, 83.2, 83.5, 83.62, 83.74, 85.14
Pelletier, S.W. 1 58.64 3 54.5, 58.39, 58.40, 58.41, 58.48 4 58.38, 58.44
Pellinghelli, M.A. 4 8.4, 67.1, 67.2, 81.15, 81.16, 83.7, 83.50
Peneloux, A. 3 81.2, 81.3
Penfold, B.R. 1 1.22, 7.19, 13.23, 24.26, 26.35, 33.16, 33.17, 48.44 2 65.43, 69.24, 69.32, 83.139 3 71.35, 71.38, 71.40, 71.55, 71.59, 77.1, 83.146, 86.17 4 71.32, 71.39, 71.42, 71.76
Penna, G. 3 21.3, 21.4
Penrose, A.B. 4 53.3
Penzien, K. 4 5.5
Pepe, G. 4 41.5
Pepinsky, R. 1 8.30, 14.10, 33.29, 40.1, 48.9, 58.42 2 73.21, 82.8, 82.32
Perales, A. 1 59.46
Perazzo, P.K.de 4 38.6
Perdok, W.G. 1 11.31
Perego, G. 2 68.1, 68.20, 75.13 3 71.25, 78.12, 78.16
Peresie, H.J. 3 76.45
Perez-Rodriguez, M. 1 8.46
Perl'mutter, B.M. 3 60.9
Perloff, A. 2 62.16, 62.35, 83.92
Perotti, E. 3 71.25, 78.12
Perozzi, E.F. 4 11.18
Perrin, M. 4 17.1
Perrotti, E. 3 78.16
Perry, R.W. 4 64.9, 65.5
Perucaud, M.C. 4 28.1, 28.2
Perutz, M.F. 2 86.1
Petcher, T.J. 1 58.24 3 38.3 4 16.11
Peters, C. 1 1.9
Petersen, C.E.L. 4 31.27
Petersen, C.S. 1 45.57, 45.58, 45.59, 45.60, 45.64, 47.9 3 35.13 4 1.5, 44.1
Petersen, D.R. 1 20.8
Petersen, R.B. 4 73.31
Petersen, S.W. 4 13.2
Peterson, C.R. 2 62.39
Peterson, E.R. 3 3.1, 3.12
Peterson, J. 1 48.67, 48.68
Peterson, S.W. 1 2.57, 3.7, 3.8 3 3.16, 6.1, 48.9
Petraccone, V. 3 22.4
Petracek, F.J. 3 33.48, 38.17
Petrissans, J. 4 21.4
Petrukhin, O.M. 2 77.28 3 77.12, 77.22
Pettersen, R.C. 1 31.23, 49.14 2 73.25 4 49.10, 53.5
Petterson, R.C. 1 54.15
Pettit, G.R. 4 37.4
Pettit, L.D. 3 11.10
Pettit, R. 2 75.26 3 72.23, 72.25, 72.26, 72.41, 75.10 4 75.24
Peyronel, G. 2 80.26, 80.27, 80.28
Pflug, J.L. 3 62.11
Pfluger, C.E. 1 9.49, 19.23 2 77.7, 77.40, 77.41, 77.42, 77.43 3 77.24, 78.1 4 77.6
Phavanantha, P. 2 83.215
Philippot, E. 4 64.3
Philips, J.C. 4 39.17
Phillip, A.T. 3 76.84
Phillips, D.C. 1 36.22, 58.4 3 36.5
Phillips, L. 4 31.32
Phillips, R.K. 3 83.12, 83.33
Phillips, T. 2 77.30, 77.32 3 44.36, 44.39
Phipps, J.R. 4 35.8
Piacenza, L.P.L. 4 31.32
Piazzesi, A. 4 39.3, 39.4
Piazzesi, A.M. 1 4.16
Pickworth, J. 1 49.32
Pidcock, A. 3 86.1, 86.40
Pieret, A.F. 3 48.44 4 48.49
Piermarini, G.J. 3 19.4
Pierpont, C.G. 3 85.45, 86.96 4 73.15, 85.12, 86.29, 86.60, 86.63
Pierrot, M. 3 82.24 4 41.5
Pieterse, M.J. 4 54.17
Pignataro, E. 1 5.15 2 67.11
Pignedoli, A. 2 80.26, 80.27, 80.28
Pignolet, L.H. 3 78.19
Pijper, W.P. 3 1.5
Pilati, T. 4 31.22
Pilbrow, M.F. 4 83.63
Pilling, D. 1 49.34
Pilot, J.F. 4 64.25, 64.39
Pilotti, A.-M. 3 58.44 4 59.12, 71.37
Pinel, R. 4 39.26
Pinkerton, J.M.H. 1 58.14
Pinkerton, M. 1 50.45 3 50.25
Pinkus, J.L. 3 40.18
Pino, P. 2 72.39
Pioda, L.A.R. 1 50.44
Pioli, A.J.P. 3 71.57
Pipal, J.R. 3 83.133 4 72.9, 81.26, 81.28
Pipenberg, K.J. 1 50.11
Piret, J. 2 72.57
Piret, P. 1 10.14, 41.26, 41.27, 41.30 2 60.14, 60.23, 60.25, 60.26, 60.30, 60.32, 60.46, 60.66, 72.32, 72.54, 72.57, 73.38, 83.169 3 20.18, 60.17, 60.18, 67.3, 67.6, 71.17 4 20.12, 24.19, 32.21, 40.17, 48.4, 48.9, 48.11, 48.27
Pirkle, W.H. 4 16.7
Piron, J. 2 83.169 3 71.17
Pitman, D.T. 1 43.14
Pitman, I.H. 3 44.24
Pitt, G.J. 1 44.49
Pitts, W.D. 4 71.90
Plaksin, P.M. 4 83.134
Planinsek, F. 3 45.11
Plastas, H.J. 3 86.34, 86.35 4 86.38
Plastas, L.A. 1 32.32, 41.25 3 20.11 4 8.10, 33.31
Plato, K. 1 20.31
Pletcher, J. 3 3.47, 40.18, 41.4, 41.11, 59.5 4 13.13, 16.17, 41.17, 41.18, 46.6, 59.5
Plettinger, H.A. 2 81.42
Plieth, K. 1 5.34, 16.44, 20.31, 20.32 2 64.51, 76.7, 76.9 4 21.2, 21.3, 64.22, 64.23
Plymale, D.L. 2 86.73
Podberezskaya, N.V. 3 76.39 4 76.4, 76.5
Podlahova, J. 4 81.22
Pointer, D.J. 4 31.7, 31.11, 31.23, 41.27
Pokrywiecki, S. 4 11.12, 11.14, 51.35, 51.36, 60.39

Poling, M. 3 59.32
Poljak, R.J. 2 83.62
Pollak, M. 1 19.8
Pollard, D.R. 1 38.58, 59.45 3 60.42, 60.55, 64.56, 64.57 4 51.45, 64.8
Pollick, P.J. 2 85.71
Pollock, J.J.S. 1 40.33
Polynova, T.N. 1 9.22 2 66.14, 66.18, 76.60, 81.24 3 68.9, 76.67, 76.70 4 76.35
Pompa, F. 1 40.25, 40.26 2 86.74
Pond, J.S. 2 83.94
Poonia, N.S. 4 67.24
Poore, M.C. 3 66.2, 66.3
Popov, A.I. 2 60.93, 60.94 4 83.58
Popova, Z.V. 1 33.27, 33.28 4 33.2
Poppe, H. 1 36.35
Poppleton, B.J. 1 45.31, 59.24 3 45.21, 45.22, 45.23
Porai-Koshits, M.A. 1 8.6, 9.19, 9.22 2 66.14, 66.18, 72.11, 76.6, 76.27, 76.29, 76.60, 77.28, 77.33, 79.24, 80.19, 81.5, 81.14, 81.24, 81.43, 81.44, 83.78, 83.95, 83.138, 83.175, 83.176, 83.177, 83.186, 85.59 3 68.9, 72.32, 76.13, 76.15, 76.21, 76.67, 76.70, 77.11, 77.22, 77.28, 77.29, 77.30, 82.9 4 9.7, 76.7, 76.27, 76.35, 77.11, 77.12, 81.43, 81.45, 81.57
Porri, L. 2 72.43, 72.45, 72.67 3 74.3, 74.4 4 72.19
Porta, P. 2 75.52, 85.88 3 72.7, 85.22, 85.48, 85.50 4 83.89, 85.33
Porte, A.L. 1 1.71, 1.72
Porter, D.W. 3 86.27
Porter, E.J. 4 33.12
Porter, J.K. 4 81.48
Porter, S.K. 3 33.8, 33.10
Portheine, J.C. 3 51.32 4 51.43, 51.44
Post, B. 1 1.1, 1.2, 1.38, 1.48, 1.90, 5.15, 5.16, 8.1, 8.2, 20.15, 20.16, 39.2, 39.18, 39.71 2 63.7 3 21.7 4 78.9, 78.14
Potenza, J.A. 2 62.1, 62.3, 62.9, 62.10, 62.24
Potter, R. 1 33.68
Povet'eva, Z.P. 1 11.30 2 62.29, 69.20 4 41.4
Powell, H.M. 1 15.4, 16.41, 17.19, 45.2 2 60.86, 61.9, 61.15, 62.50, 68.22, 69.48, 71.16, 71.20, 71.21, 71.28, 73.12, 73.109, 75.52, 75.71, 86.7, 86.40, 86.41, 86.44, 86.45, 86.67, 86.99, 86.105, 86.106 3 86.32, 86.76, 86.81, 86.82, 86.86 4 71.68, 86.37, 86.42, 86.43, 86.56
Powell, J. 3 71.12
Power, L. 3 41.11, 59.5
Power, L.F. 3 11.12 4 2.5, 11.5, 83.66, 83.111
Powers, T.W. 4 30.5
Poyer, N. 3 19.16
Prabhu, K.V. 3 27.4
Prakash, A. 1 26.24, 26.25 4 58.16
Prasad, H.S. 4 66.9, 66.10
Prasad, S.M. 3 13.6 4 2.10, 2.16, 19.6
Pratt, C.S. 4 86.20
Pratt, M.R.A. 1 35.38
Pre, S.du 1 48.26
Precigoux, C. 4 51.18
Preer, J.R. 2 86.26
Preiss, H. 2 66.8 4 66.1
Prelesnik, B. 3 15.8
Prelog, V. 1 34.6 4 50.13
Presley, C.T. 3 21.2
Preston, H.S. 2 71.4, 71.5, 71.31, 83.128 3 15.1, 15.2, 35.21, 83.68, 83.69, 83.136 4 86.38
Preuss, J. 4 36.37
Preuss, L. 3 39.24, 39.25, 39.27
Preut, H. 4 69.38
Prewitt, C.T. 2 73.89
Price, D.A. 4 17.4
Price, K.A. 3 83.43
Prince, E. 4 83.94, 83.95
Prince, R.H. 2 83.208
Pringle, G.E. 4 81.39
Procter, I.M. 3 83.105
Prosen, R.J. 1 49.32 2 60.51
Protas, J. 3 2.5, 74.2, 74.5 4 2.9, 48.23, 48.43, 73.26, 74.2, 74.4, 74.6, 74.7, 75.8, 75.10
Prout, C.K. 1 49.5, 49.6 2 60.70, 60.73, 60.74, 60.75, 60.76, 60.107, 60.148, 71.27, 73.40, 73.75, 75.22, 81.27, 81.45, 81.48, 81.55, 81.66, 81.81, 82.14, 83.159, 83.167, 85.9, 85.10, 85.11, 85.36, 85.78 3 13.18, 25.6, 33.38, 33.39, 33.40, 35.18, 40.19, 40.20, 40.21, 45.31, 60.29, 64.43, 64.70, 65.10, 66.5, 71.5, 71.6, 73.47, 73.48, 73.49, 81.16, 81.22, 81.42, 82.23, 83.57 4 2.11, 2.24, 24.2, 36.35, 40.8, 60.9, 60.10, 60.11, 60.15, 60.16, 71.15, 73.19, 73.29, 81.12, 81.13, 81.14, 81.23, 81.41, 81.49, 81.54
Prusiner, P. 4 44.10
Przybylska, M. 1 17.29, 58.13, 58.15, 58.19, 58.50, 58.62, 58.80, 58.91, 58.92 3 54.11
Pucci, A. 4 86.63
Puckett, R.T. 1 9.49, 51.25 3 31.42, 58.55, 58.56 4 24.16, 33.47
Puliti, R. 1 59.8 2 60.110, 60.112
Pulsinelli, P.D. 3 41.4
Purdham, J.T. 4 21.22, 28.12
Putkey, E. 3 48.7
Putkey, E.F. 3 48.6 4 26.7
Putzeys, J.-P. 4 32.23, 40.17, 48.4
Puxley, D.C. 3 69.2
Pyronel, G. 2 63.5

Qasba, R. 1 28.24
Quick, A. 3 82.25
Quigley, G.J. 3 45.36
Quin, L.D. 3 64.22 4 64.13, 64.14, 86.17
Quinby, M.S. 3 76.40

Rabano, M. 4 24.12
Rabaud, H. 1 26.34
Rabinovich, D. 1 1.91, 18.22, 18.23, 18.24, 18.25, 18.27, 23.16 3 1.25, 17.6, 19.8, 19.9, 19.10, 36.8, 38.2 4 1.6, 13.4, 51.49
Rabinowitz, H.N. 3 71.31
Rabinowitz, I.N. 1 21.12
Racanelli, P. 2 72.43, 72.45 4 72.19

Racine-Weisbuch, M. **3** 19.15
Rademaker, W.J. **4** 62.1
Radlick, P. **4** 53.2, 59.8
Radonovich, L.J. **1** 3.83 **3** 30.4, 84.14 **4** 49.22, 83.77
Radoslovich, E.W. **4** 84.1
Rae, A.D. **2** 83.112, 83.194
Rae, A.I.M. **1** 16.13, 59.41 **2** 85.42 **3** 86.70
Raeburn, U.A. **3** 82.4, 86.12
Raghavan, V.K.V. **1** 59.68
Rague Schleyer, P.von **1** 31.43
Rahim, A.M.Abd.el **1** 56.15
Rahman, A. **3** 47.5
Rahn, R.O. **1** 44.72
Raimondi, M. **4** 31.20, 31.22
Ramachandran, G.N. **1** 45.43, 59.68 **3** 48.29
Ramage, R. **4** 58.30
Raman, S. **1** 58.8
Raman, S.V. **3** 48.1, 48.2, 48.3
Ramaseshan, S. **1** 19.41, 58.75
Ramaseshan, S.M. **1** 16.48, 16.50
Ramirez, F. **1** 46.7 **2** 64.61 **3** 64.69 **4** 64.25, 64.39, 64.45
Ramsay, W. **1** 59.6
Ramsoy, S. **4** 12.8
Randaccio, L. **1** 21.29 **2** 78.59, 84.50, 84.51 **3** 76.81, 78.14, 78.26, 83.150, 85.32, 85.33, 85.40 **4** 71.40, 77.14
Randall, E.W. **4** 71.27
Rannev, N.V. **1** 7.5, 7.7
Rao, G.S. **2** 81.47
Rao, J.K.M. **2** 82.1 **4** 82.4, 82.5
Rao, R.V.G.S. **2** 81.47
Rao, S.N. **4** 48.18
Rao, S.T. **1** 21.15, 48.94 **3** 45.35, 47.21, 47.30, 48.18, 48.22, 48.34 **4** 31.19, 47.8, 47.19
Raper, E.S. **3** 81.8 **4** 33.15, 41.1
Raper, G. **1** 41.35 **4** 71.13, 71.33, 86.28
Raphael, R.A. **1** 22.6, 39.56
Rasmussen, P.G. **3** 83.49
Rasmussen, S.E. **1** 3.52, 3.53, 36.4, 58.6 **2** 68.14, 76.47, 83.185, 84.25, 85.7 **3** 32.8, 85.11 **4** 83.60
Raspin, K.A. **2** 86.102, 86.117
Rassat, A. **1** 31.12
Raupach, M. **4** 84.1
Raveau, B. **3** 41.6
Ravn-Jonsen, E.J. **1** 59.28
Raymond, K.N. **2** 76.39, 76.40 **3** 76.65 **4** 71.50, 75.13, 75.14, 75.15, 75.25, 75.31, 76.14, 83.32
Rayner-Canham, G.W. **2** 83.104 **4** 71.85, 83.139
Read, L.K. **4** 27.11, 27.12, 36.22, 39.17
Reck, D. **1** 1.68
Reck, G. **3** 28.1, 52.2 **4** 52.2, 52.3
Reddy, B.S. **4** 47.2
Reddy, J.M. **1** 58.8 **2** 60.99
Redhouse, A.D. **2** 71.32, 71.46, 86.47 **3** 71.52
Reed, C. **4** 86.20
Reed, C.A. **4** 86.53
Reed, H.J. **1** 41.34
Reed, L.L. **3** 76.40 **4** 20.8, 22.8, 41.20, 51.52
Reed, T.B. **1** 5.11
Reedijk, J. **4** 83.81
Reeke Junior, G.N. **1** 44.62, 58.103
Reeke, G.N. **3** 69.37, 69.38
Rees, A.H. **1** 58.73
Rees, B. **1** 18.4, 18.5 **3** 18.1, 18.2, 18.3, 18.4, 33.29
Rees, G.V. **3** 73.47, 73.48, 73.49 **4** 73.29
Regev, H. **4** 29.1
Rehnberg, G. **4** 39.1
Reichstein, T. **1** 51.35
Reid, K.I.G. **3** 39.22, 39.23, 39.41, 75.17, 75.29
Reid, S.T. **1** 39.50
Reimann, C.W. **2** 83.92, 83.102, 83.106, 83.107, 83.165 **3** 83.6, 83.63, 83.95 **4** 83.94, 83.95, 83.102, 84.11
Reimlinger, H. **3** 41.10, 42.4 **4** 32.24
Reinheimer, H. **4** 71.51
Reis Junior, A.H. **3** 62.15, 83.90 **4** 62.7, 62.8, 83.86
Reiser, B. **4** 62.12
Reistad, T. **1** 39.35, 39.36
Reitveld, H.M. **3** 19.12, 19.13, 19.14
Remlinger, H. **3** 32.9
Remoortere, F.P.van **3** 20.8, 32.26, 38.18, 63.11, 69.37, 69.38 **4** 21.18, 21.19, 33.3, 33.18, 38.13, 69.18, 69.19
Renaud, M. **1** 5.1 **4** 19.3, 19.4
Rendall, I.F. **2** 78.61 **3** 78.25 **4** 80.15
Rendle, D.F. **4** 51.13, 51.22, 51.32, 64.44, 65.12, 65.13, 65.14
Replogle, L.L. **1** 39.90 **4** 39.43
Repmann, H. **1** 5.34
Rerat, B. **1** 16.43, 16.45, 16.54, 16.55, 16.56, 40.14, 43.20, 43.21 **2** 60.28 **3** 9.18, 9.20, 11.20, 11.21 **4** 32.18
Rerat, C. **1** 10.9, 10.21, 16.55, 17.43, 17.45, 31.12, 32.23, 33.14, 33.22, 33.34, 40.14, 43.20, 43.21, 52.6, 59.1 **2** 60.28, 64.38, 64.39 **3** 9.18, 9.20, 11.20, 11.21, 32.7 **4** 32.18
Restivo, R. **3** 33.21, 50.14, 73.55, 81.44 **4** 58.37, 68.11, 71.37, 73.32, 73.33
Restivo, R.J. **4** 59.29
Rettig, M.F. **3** 73.24
Rettig, S. **4** 48.38
Reynolds, C.T. **4** 3.14
Reynolds, W.F. **3** 60.34
Rhine, W. **3** 62.3
Ribi, M. **3** 9.8
Ricard, L. **4** 71.66
Ricci Junior, J.S. **4** 39.41
Ricci, G.M.B. **3** 86.94
Ricci, J. **1** 39.69
Ricci, J.S. **3** 11.14, 11.15 **4** 71.93, 86.45
Riccieri, F.M. **1** 35.12, 35.14
Rice, D.A. **3** 83.5 **4** 84.2
Rich, A. **1** 44.3, 44.30, 44.39, 44.55, 44.56, 44.58, 44.59 **2** 60.131, 73.22 **3** 44.4, 44.9 **4** 60.34
Richards, A. **3** 78.3
Richards, G.F. **2** 72.47, 73.17, 75.3, 75.6 **4** 52.1
Richards, J.H. **2** 73.43
Richards, J.P.G. **1** 6.7, 13.32, 17.21 **4** 17.4
Richards, R.L. **2** 71.38

Richards, S. **2** 76.65
Richardson, D.C. **2** 76.44
Richardson, M.F. **2** 77.29 **3** 77.23 **4** 83.45
Riche, C. **1** 20.28 **4** 19.13, 38.18
Rickard, C.E.F. **2** 76.26 **3** 80.12, 80.13, 80.15
Ridley, D.R. **4** 66.3, 66.4, 66.9, 66.10
Riedel, E.F. **4** 86.27
Riedl, M.J. **4** 83.75
Rietveld, H.M. **2** 80.10
Riganti, V. **1** 36.7, 36.8, 36.11, 36.13
Rigatti, G. **1** 33.75
Rigau, J.J. **3** 39.45
Riley, P.E. **3** 85.30
Rimmer, B. **1** 53.8
Rimmer, B.M. **1** 51.45
Rimsky, A. **2** 60.7 **4** 8.8
Rinehart Junior, K.L. **1** 50.33 **4** 50.18
Ringertz, H. **1** 2.71, 8.39, 8.40, 44.24 **3** 13.9, 62.1
Ripamonti, A. **1** 33.81, 59.8, 59.20 **2** 60.110, 60.111, 60.112, 64.6, 84.50, 84.51 **3** 78.26, 85.32, 85.33, 85.34, 85.40
Ripperger, H. **1** 58.107, 58.109
Ristey, W.J. **1** 4.10
Ritter, H.L. **1** 33.58
Riva di Sanseverino, L. **1** 5.37, 5.38, 51.38, 51.39, 51.40, 56.4, 58.108 **3** 36.21, 47.25, 54.13, 58.47 **4** 9.13, 9.14
Rivest, R. **4** 83.124
Rix, C.J. **3** 74.17
Rizvi, S.H. **1** 16.46, 19.36
Ro, G. **4** 33.6, 33.7
Robert, J.-B. **4** 24.8
Roberts, J.C. **3** 59.31
Roberts, J.D. **3** 33.48, 38.17 **4** 24.8
Roberts, J.S. **3** 31.18, 53.23
Roberts, P.J. **3** 86.5, 86.7, 86.9, 86.17, 86.18, 86.37 **4** 52.6
Robertson, B.E. **4** 83.30, 84.7
Robertson, B.K. **1** 33.26, 33.44, 33.45
Robertson, C.G. **3** 73.56
Robertson, G.B. **1** 19.37, 39.28 **2** 75.17, 77.2, 77.3, 77.4, 81.75, 82.28, 82.37, 83.40, 83.86, 84.68, 86.46, 86.80, 86.89 **3** 72.38, 72.52, 74.17, 75.35, 76.63, 82.5, 86.27, 86.98 **4** 71.43, 71.46, 71.49, 71.56, 75.34, 75.35, 86.21
Robertson, I. **2** 77.14
Robertson, I.C. **3** 71.50, 71.51
Robertson, J.H. **1** 2.23, 39.22, 39.24, 45.48, 49.32, 58.54 **2** 83.29, 83.43, 85.22
Robertson, J.M. **1** 1.11, 1.43, 1.55, 1.71, 1.72, 1.73, 1.77, 1.84, 1.125, 3.49, 3.50, 5.25, 9.34, 9.35, 11.26, 11.57, 13.20, 17.14, 17.16, 17.29, 19.40, 19.49, 19.53, 19.62, 20.19, 24.13, 24.30, 27.6, 29.12, 29.13, 30.12, 30.13, 30.14, 30.15, 30.19, 30.22, 30.24, 30.28, 30.29, 30.33, 30.35, 35.6, 41.10, 48.21, 49.8, 49.9, 49.13, 51.11, 52.7, 52.13, 53.8, 53.12, 53.15, 53.17, 54.13, 54.22, 54.25, 56.1, 56.2, 56.3, 58.48, 58.49, 58.70, 58.71, 58.74, 58.76, 59.45, 59.58, 59.64, 59.70 **2** 60.156, 84.36 **3** 24.5, 26.3, 28.5, 31.18, 51.26, 51.30, 53.17, 53.23, 53.24, 58.67, 59.16, 60.42, 60.55 **4** 30.8, 53.3
Robin, M.B. **2** 60.99 **3** 83.50
Robinson, A.G. **4** 83.26
Robinson, B. **1** 24.20
Robinson, B.H. **3** 71.38 **4** 71.39
Robinson, D.J. **3** 60.40, 75.9, 76.68
Robinson, G. **2** 72.14, 72.52, 82.15
Robinson, W.R. **2** 71.48, 81.88 **4** 60.35, 81.37
Robinson, W.T. **1** 26.35, 52.17 **2** 65.43, 69.24, 79.34, 79.35, 86.25 **3** 33.30, 33.31, 33.32, 71.38, 71.55, 71.66, 76.5, 76.58, 77.1, 83.100, 83.144, 83.147 **4** 76.17, 76.18, 76.24, 78.4, 78.6, 78.7
Robsin, R. **4** 85.15
Robson, A. **2** 75.64, 77.37
Rodesiler, P.F. **2** 63.24, 84.65 **4** 41.2, 69.11, 69.22, 69.27, 74.3, 74.8, 74.11, 75.19
Rodgers, J. **3** 64.12, 83.73 **4** 64.33, 83.75, 83.76
Rodig, O.R. **4** 59.6
Rodley, G.A. **2** 84.41, 84.68, 86.32
Rodrigo, S. **1** 54.1
Rodrigues, M.M. **1** 43.23, 43.24, 43.25
Rodrique, L. **2** 60.23, 72.54
Roe, A.M. **3** 3.48
Roe, R.-J. **1** 1.141
Roelofsen, G. **3** 1.8 **4** 1.7
Rogers, D. **1** 2.92, 50.12, 50.27, 53.2, 53.13, 53.23, 58.30, 59.48 **2** 76.21, 76.31, 76.50, 77.1, 82.34, 83.215, 86.78 **3** 10.5, 45.14, 48.19, 50.15, 52.1, 53.15, 58.34, 59.8, 75.4 **4** 36.27, 38.25, 49.7, 53.6, 59.7, 59.15
Rogers, J.B. **1** 17.30
Rogido, R. **3** 32.20
Rohrbaugh, W.J. **4** 85.28
Rohrbaugh, W.L. **3** 76.83 **4** 80.8
Rohrer, D.C. **3** 44.11, 47.11, 47.26 **4** 45.21, 45.25
Rohrl, M. **1** 54.24 **3** 54.6, 54.15, 59.19, 59.25 **4** 54.8, 54.12, 54.16
Rolfe, N. **3** 83.5 **4** 83.8
Rollett, J.S. **1** 1.107, 5.21, 46.21, 56.4, 58.108, 59.43
Romain, P. **1** 32.31
Romers, C. **1** 1.70, 10.29, 18.12, 18.17, 18.18, 18.19, 18.20, 18.21, 18.26, 21.7, 35.45, 36.36, 38.4, 38.8, 38.9, 38.10, 38.11, 38.12, 38.13, 39.14, 39.15, 39.66, 42.2, 42.3, 42.5, 51.8, 51.30, 51.31, 51.32, 51.33, 51.34 **2** 77.59 **3** 1.14, 22.3, 38.6, 51.32, 51.37, 51.38, 81.43 **4** 21.6, 21.17, 27.2, 51.43, 51.44
Romming, C. **1** 7.10, 9.12, 9.14, 9.15, 9.16, 9.20, 9.23, 20.13, 21.10, 47.9 **2** 60.17, 60.18, 60.19, 60.20, 60.49, 60.62, 60.63, 60.82, 60.90, 60.139 **3** 48.42 **4** 9.2, 9.4, 9.9, 48.39, 48.40, 59.3
Ronnquist, O. **1** 45.28 **3** 36.7, 36.22
Roos, I.A.G. **3** 85.24
Roper, W.R. **3** 86.61 **4** 86.35, 86.53
Ros, R. **3** 85.19
Rose, I.A. **1** 2.47
Rosen, L. **4** 5.8
Rosen, L.S. **3** 44.43
Rosenberg, E. **4** 71.27
Rosenberg, H. **4** 73.1
Rosenblum, M. **2** 60.119, 75.21

Rosenfield Junior, R.E. **4** 48.30, 48.31, 48.44
Rosenqvist, E. **3** 84.2 **4** 38.7
Rosenstein, R. **2** 82.3, 82.9
Rosenstein, R.D. **1** 32.22, 45.14, 45.15, 45.18, 45.25, 45.38, 45.44, 45.45, 45.46, 58.82 **3** 32.22, 45.2, 45.5, 45.6, 45.11, 45.12, 45.13, 45.34, 47.10, 58.23, 59.26, 60.23 **4** 37.4, 44.9, 44.18, 44.25, 48.21, 60.14
Rosini, G. **4** 9.13, 9.14
Ross, F.K. **3** 3.39, 37.2, 64.69, 83.21 **4** 33.4, 64.45
Rossi, G. **4** 83.35
Rossi, U. **2** 61.17, 61.19
Rossmann, M.G. **1** 11.69, 17.47, 30.22, 30.32, 52.12, 52.31, 53.39 **4** 21.5
Rossotti, F.J.C. **2** 81.27, 81.45, 81.48, 81.66, 81.81 **3** 81.16, 81.22, 81.42 **4** 2.11, 2.24, 81.12, 81.13, 81.14, 81.23, 81.41, 81.49, 81.54
Roth, R.J. **4** 31.12
Roth, W.L. **2** 63.8
Roth, W.R. **1** 23.1
Roti, I. **4** 70.8, 70.12
Rousseaux, M.-P. **4** 32.23
Roux, J.P. **2** 60.40, 60.41 **3** 67.17
Rowe, I. **1** 39.71
Rowe, J.M. **1** 35.24 **2** 72.5, 72.10, 80.5, 86.13, 86.86
Roy, P.K. **3** 2.17
Roy, P.N. **1** 1.41
Rozenberg, L.P. **2** 74.16 **4** 60.41
Ruban, G. **1** 16.44 **2** 76.9 **4** 21.2, 21.3, 71.67
Ruben, H. **4** 39.15
Rubezhov, A.Z. **3** 71.4
Ruble, J.R. **4** 82.3
Rucklidge, J.C. **4** 71.47
Rudko, A.D. **3** 48.51, 48.52
Rudman, R. **1** 5.16 **2** 63.15 **3** 13.3 **4** 5.3, 64.5
Rudolph, R.W. **3** 62.11
Ruegg, M. **3** 76.49
Ruff, J.K. **3** 3.37, 64.62, 64.63, 64.64 **4** 3.15, 3.16, 64.41, 64.42
Rule, L. **2** 69.17
Rumball, S.S. **3** 76.5
Rundle, R.E. **1** 11.11, 11.12, 11.13, 19.54, 19.55, 38.19, 45.76 **2** 67.1, 67.10, 67.14, 67.17, 68.3, 69.23, 71.29, 74.1, 75.24, 83.20, 83.21, 83.57, 83.58, 84.19
Runsink, J. **4** 65.10
Rupainwar, D.C. **1** 41.35
Rusholme, G.A. **1** 39.28 **4** 71.10
Russ, B.J. **2** 83.49 **4** 83.19, 83.20
Russell, D.R. **2** 72.59, 73.3, 83.178, 86.92 **3** 66.2, 66.3, 71.34, 71.41, 71.42, 72.13, 72.16 **4** 71.16, 71.17, 71.83, 71.87, 83.63, 85.34
Russell, J.H. **1** 33.66 **4** 33.28, 33.29, 33.30, 33.46
Russell, S.E. **4** 54.2
Russell, S.W. **1** 21.33
Russo, M. **1** 39.40, 42.9
Rutherford, J.S. **1** 32.22 **3** 8.2, 8.3, 32.22
Rutherford, K.W. **2** 85.53
Rutten, E.W.M. **1** 1.70, 38.4, 51.31, 51.34 **4** 21.6, 51.43, 51.44
Ryan, R.R. **1** 48.15 **3** 83.85
Ryan, T.J. **3** 9.8
Rybin, L.V. **4** 86.36
Rybinskaya, M.I. **4** 86.36
Rychnovsky, V. **1** 39.49
Ryderstedt-Nahringbauer, I. **1** 1.146

Sabacky, M.J. **3** 17.11
Sabelli, C. **1** 35.5, 40.19, 40.20 **4** 35.4, 35.10
Sabesan, M.N. **3** 32.25, 48.49 **4** 48.52
Sabine, T.M. **1** 1.12, 1.18, 1.19, 43.15 **3** 33.1, 48.11
Sacconi, L. **2** 78.22, 78.24, 78.31, 78.32, 78.34, 78.36, 78.44, 78.53, 86.49, 86.95 **3** 76.87, 78.15, 86.43, 86.48, 86.52, 86.84, 86.85 **4** 76.43, 76.44, 83.126, 86.44
Sachdev, H.S. **3** 40.13
Sackman, P. **3** 3.40, 3.41, 60.30, 60.31 **4** 35.35, 36.18, 60.17
Sadanaga, R. **1** 2.62, 19.45
Sadikov, G.G. **2** 81.44, 81.67 **4** 81.43
Sadler, P.J. **4** 60.9, 60.10, 60.11
Saenger, W. **1** 17.30 **3** 40.13, 41.14, 44.21, 47.1, 47.9 **4** 20.5, 35.37, 47.1, 47.3, 47.5, 47.7, 47.17, 47.18
Saethre, L.J. **3** 39.47 **4** 39.32, 39.36
Sager, R.S. **2** 84.32, 84.34, 84.38, 84.39 **3** 84.8, 84.9, 84.22
Saha, N.N. **1** 1.41, 48.19 **3** 48.21, 82.21 **4** 40.11
Sahu, B.N. **3** 2.15 **4** 2.16
Sahu, M. **4** 13.10
Sahu, R.G. **3** 2.16, 2.18 **4** 2.16
Saint Clair, D. **3** 62.8, 62.12 **4** 62.9, 71.5
Saint-Giniez-Liebig, D.de **2** 64.38, 64.39
Saint-Giniez, D.de **1** 32.23
Saito, N. **3** 45.43
Saito, Y. **1** 1.16, 1.17, 2.3, 2.6, 50.18, 59.4 **2** 60.101, 60.117, 60.123, 76.4, 76.8, 76.35, 76.36, 76.38, 76.58, 81.74, 83.17, 83.96, 83.105 **3** 1.23, 3.25, 35.4, 50.21, 60.12, 76.54, 76.64, 76.71, 76.82, 82.18 **4** 20.1, 29.4, 50.4, 54.13, 60.24, 60.32, 76.2, 76.20, 76.21, 76.25, 76.40, 81.31, 83.68, 83.78
Sakabe, N. **1** 15.23, 16.39, 44.31, 50.37, 54.17, 58.35, 58.102, 58.114, 59.25 **3** 58.36, 58.38
Sakakibara, H. **3** 59.1
Sakanoue, S. **3** 35.12 **4** 60.20
Sakata, K. **1** 13.17, 13.18 **4** 54.15
Sakata, Y. **1** 1.30
Sakore, T.D. **1** 13.10, 13.11 **3** 39.13, 39.14, 44.13, 44.18, 44.19, 44.25, 44.26, 44.27, 44.28 **4** 60.43
Sakurai, H. **1** 58.102
Sakurai, K. **1** 3.15, 13.16, 13.19
Sakurai, T. **1** 16.5, 17.1, 17.2, 17.17, 17.35, 17.36, 18.2, 50.4 **3** 60.13 **4** 60.7, 60.12
Salah, O.M.A. **4** 71.94
Salamon, R. **3** 39.15
Salas-Cimingo, G. **1** 13.26, 14.7, 14.8
Sales, D.L. **2** 85.26, 85.27, 85.52
Salmon, O.N. **1** 7.4
Salomone, R.A. **3** 42.2
Samarskaya, V.D. **1** 5.26

Samuel, E.B. **4** 21.21
Samuel, G. **3** 40.10, 64.24, 64.38, 64.40
Samus', I.D. **2** 83.80
Sanders, D.A. **3** 68.18
Sanders, W.W. **1** 1.70
Sandler, S. **1** 29.10, 29.24
Sandmark, C. **2** 83.166
Sands, D.E. **1** 11.16 **2** 62.38, 77.29, 77.30, 77.32, 77.34 **3** 77.23 **4** 84.16
Sandstrom, J. **1** 41.28
Sanseverino, L.Riva di **3** 28.10, 32.13
Sansoni, M. **2** 71.33 **3** 3.54, 86.22, 86.99 **4** 86.58, 86.71
Santavy, F. **4** 58.24, 58.30
Santis, P.de **1** 44.67 **2** 60.97, 60.110, 60.112
Santo, W. **4** 71.47
Santoro, A. **2** 83.102, 83.165 **3** 83.6, 83.95 **4** 83.91, 83.92, 83.94, 83.95, 83.102, 84.11
Sanz-Ruiz, F. **4** 77.10
Sanz, F. **3** 64.49, 65.1, 71.65
Saperas, B. **4** 83.124
Sargeson, A.M. **2** 76.34, 76.48, 76.49, 82.2, 82.24, 82.54 **4** 76.11, 76.17, 76.18
Sarko, A. **3** 45.36
Sarma, V.R. **1** 16.7, 16.8, 33.37, 48.94
Sartain, D. **2** 65.29, 85.91
Sartori, F. **4** 80.2
Sasada, Y. **1** 5.7, 11.37, 13.3, 13.28, 13.29, 13.38, 20.1, 20.29, 20.30, 22.3, 22.4, 22.5, 22.9, 22.14, 27.8, 27.10, 27.13, 28.20, 31.27, 33.18, 33.42, 35.17, 38.33, 38.34, 38.42, 48.52, 48.56, 48.84, 48.86, 48.88, 48.95, 56.12, 58.7, 58.12, 58.45, 58.58 **2** 78.1, 78.2, 82.13, 82.29 **3** 22.1, 31.21, 36.26, 45.43, 53.22, 58.46, 58.58, 59.15 **4** 5.7, 27.5, 27.6, 33.41, 40.24
Sasaki, K. **3** 58.38 **4** 58.28, 58.29, 58.31
Sasaki, Y. **3** 61.3, 61.4 **4** 58.39
Sass, R.L. **1** 1.29, 1.36, 1.37, 1.76, 7.16, 7.26, 10.11, 11.28, 23.1, 33.55, 39.67, 41.11 **3** 3.32, 3.49, 7.6, 9.2, 21.2, 39.4
Sassmannshausen, G. **2** 64.62
Sasvari, K. **1** 21.28 **3** 45.7, 45.8 **4** 4.3, 9.8, 11.13, 21.15
Satake, S. **1** 42.14
Sato, N. **4** 40.24
Sato, S. **1** 39.88, 44.76, 48.7, 48.8 **4** 32.5
Sato, T. **1** 27.12, 40.21, 51.20, 52.28, 58.18 **2** 84.7, 85.18 **3** 31.12, 50.4, 51.20, 51.22, 51.33, 58.25
Sato, Y. **3** 51.48
Saucy, G. **1** 52.29 **4** 27.16
Saunderson, C.P. **1** 51.46
Savage, D.S. **3** 51.44
Sax, M. **1** 1.145, 13.13, 13.55, 35.26, 35.30, 39.74 **3** 3.47, 40.18, 41.4, 41.11, 59.5 **4** 13.13, 16.17, 41.17, 41.18, 46.6, 59.5
Sayed, K.E. **1** 26.19, 26.20, 26.21, 26.22, 26.23
Sayre, D. **1** 29.9, 29.23 **3** 51.43
Scala, A. **3** 59.33
Scales, C.G. **3** 13.18
Scane, J.G. **2** 65.38
Scaramuzza, L. **2** 65.27, 80.16, 85.36 **3** 68.20, 85.48 **4** 79.11, 85.17, 85.18, 85.19, 85.31
Scarbrough, F.E. **2** 62.25 **4** 3.8
Scatturin, V. **1** 40.25, 40.26 **2** 86.9, 86.72, 86.74, 86.100, 86.118, 86.119
Scavnicar, S. **2** 77.53, 84.49
Schaap, H. **4** 85.15
Schade, G. **3** 28.5, 31.18
Schaefer, J.P. **1** 1.122, 5.23, 39.92, 50.15 **3** 21.15, 21.16, 34.4 **4** 20.8, 22.8, 31.9, 41.14, 41.20, 51.52
Schaefer, W.P. **2** 78.15 **3** 78.30, 82.26 **4** 78.15
Schaeffer, E. **4** 81.59
Schaeffer, R. **1** 3.37 **3** 3.28
Schafer, H.L. **2** 84.31
Schaffner, C.P. **4** 50.17
Schaffrin, R. **3** 32.18, 32.19, 48.20
Schaffrin, R.M. **3** 39.29
Schairer, H.U. **1** 54.24
Schalkwyk, T.G.D.van **1** 19.18
Schapiro, P.J. **2** 62.30
Scheel, H.-J. **3** 26.11
Scheidt, W.R. **3** 83.49 **4** 76.34, 76.36, 81.9
Scheie, A. **2** 70.8, 70.9
Schein, B.J.B. **1** 8.26, 8.27
Scheit, K.H. **3** 47.9
Schelle, S. **3** 71.30
Schenk, H. **1** 53.27, 58.1 **3** 3.11, 20.2, 20.16, 37.4, 39.5, 40.14, 51.1, 51.15 **4** 1.12, 21.8, 27.17, 30.4, 37.2, 37.3, 51.19
Scheringer, C. **1** 17.9 **2** 78.19
Scherr, P.A. **3** 68.18
Scheuerman, R.F. **1** 1.76
Scheurman, R.F. **1** 1.36, 1.37
Schevitz, R.W. **1** 52.12
Schibilla, H. **2** 67.3
Schieltz, N.C. **1** 50.11
Schiff, L. **2** 75.21
Schiffer, M. **1** 1.52
Schilling, J.W. **1** 30.3 **3** 9.13
Schipperijn, A.J. **4** 75.33
Schlemper, E.O. **1** 7.24 **2** 65.3, 69.1, 69.6, 69.8, 69.37, 83.24, 83.84 **3** 83.51, 83.52 **4** 69.31, 83.42
Schlessinger, R.H. **3** 39.13, 39.14
Schlueter, A.W. **2** 75.45 **4** 73.3, 73.4
Schlupp, R. **4** 83.46, 84.3
Schlupp, R.L. **3** 22.2
Schmid, H. **3** 58.67, 59.9
Schmidt, G.M.J. **1** 1.80, 1.91, 1.100, 1.117, 1.118, 7.22, 15.20, 16.52, 16.53, 17.5, 17.6, 18.23, 18.25, 18.27, 23.16, 28.28, 29.10, 29.24, 30.21, 36.9, 36.10, 38.29, 39.42, 39.43 **3** 1.13, 1.38, 15.11, 15.12, 17.6, 19.9, 19.10, 29.3, 29.4, 31.37, 31.38, 31.40, 31.41 **4** 13.23, 16.12, 24.18, 26.1, 26.2, 26.4
Schmidt, W.H. **1** 39.85
Schmitkons, D.J. **1** 31.9, 31.10
Schmitt, R.D. **3** 60.54
Schmitz, F.J. **4** 51.57, 51.58
Schneider, G. **4** 59.17
Schneider, M.L. **3** 73.43 **4** 49.17, 86.41
Schnell, H.W. **4** 5.5
Schnering, H.G.von **2** 63.17 **4** 83.10
Schnoes, H.K. **3** 53.25
Schodl, G. **3** 71.47 **4** 71.12
Schoemaker, D.P. **4** 67.14
Schoenborn, B.P. **1** 52.15, 52.16

Schoening, F.R.L. 2 81.21, 81.29, 81.34, 81.36, 81.37, 81.63, 81.64
Scholer, F.R. 3 71.60, 75.1 4 75.2
Scholtz, D. 3 40.18
Schomaker, V. 1 20.24, 48.33 2 72.34, 72.74, 72.75, 75.79 4 62.11
Schomeker, V. 3 48.9
Schoone, J.C. 1 1.60, 1.62, 1.63, 58.55 2 82.15, 82.47
Schorno, K.S. 1 53.28
Schrauber, H. 4 19.7
Schrauzer, G.N. 2 85.34 3 71.31, 75.17 4 71.54, 71.55
Schreiber, G. 4 86.34
Schreiber, G.H. 3 77.18 4 77.13
Schreiber, K. 1 58.104, 58.105, 58.107, 58.109 4 51.48, 51.56, 59.17
Schrieke, R.R. 4 73.24
Schroeder, E.D. 1 33.61
Schubert, W.M. 2 60.121
Schugar, H. 2 83.183
Schuijff, A. 1 20.26, 29.3
Schulte-Elte, K.-H. 3 53.1, 53.2
Schultz, A.J. 4 85.16
Schultz, R.J. 3 83.8
Schulz, M. 1 38.21
Schunn, R.A. 2 69.35, 73.63, 73.89
Schurr, J.M. 4 48.53
Schuster, D.I. 3 27.4
Schvoerer, M. 3 25.1, 25.2 4 25.1, 25.2
Schwalbe, C.H. 3 3.26, 3.27 4 20.5
Schwartz, A. 4 85.5
Schwartz, S.L. 3 31.42
Schwarzenbach, D. 1 31.5, 33.13 3 81.38
Schwarzhans, K.E. 2 85.65
Schweikert, W.W. 2 60.150
Scoffer, M.D. 1 40.11
Scordamaglia, R. 4 71.63
Scott, A.I. 1 50.26, 54.3, 54.4, 54.7
Scott, C.C. 1 39.50, 58.49
Scott, J.W. 4 31.28
Scott, W.E. 4 52.1
Scouloudi, H. 2 76.19
Scruton, J.C. 2 60.89 3 60.19
Sears, M.C. 4 83.51
Seccombe, R.C. 3 16.6
Sedmera, P. 4 58.24
Seeden, R.P.A. 2 84.46
Seely Junior, O. 1 48.72, 48.79
Seely, O. 3 48.28, 48.48
Seeman, N.C. 4 45.3, 47.20, 48.21
Seff, K. 1 11.7, 33.64 2 64.66 3 85.30 4 67.14, 77.23, 82.3
Segrott, G.R. 3 85.53, 85.55 4 85.40
Seidel, I. 4 51.48
Seigler, D.S. 3 20.5, 20.7
Seki, M. 1 3.18
Seki, S. 1 37.1
Sekita, T. 3 53.9
Sekizaki, M. 4 83.68
Seleborg, M. 4 78.9
Selig, W. 4 60.13
Seligmann, O. 3 59.26
Selinger, B.K. 1 31.42
Semeniuk, G.M. 4 64.24, 65.9
Semin, G.K. 1 19.1, 19.2, 19.3 2 71.1 3 16.1
Semion, V.A. 2 71.47, 73.53 3 71.4, 71.10, 71.61, 71.62
Semmingsen, D. 4 6.1
Sen, D.K. 3 36.6
Sendo, Y. 3 58.36
Senko, M.E. 1 8.1, 8.2, 32.7, 59.3
Senma, M. 4 67.3
Senoff, C.V. 4 85.27
Senol, A. 4 53.11
Sequeira, A. 1 14.14 3 2.2, 85.16
Serator, M. 2 83.117
Serewicz, A.J. 1 33.26
Sergeeva, G.N. 3 71.56
Sergienko, V.S. 3 76.13
Serpone, N. 4 77.1
Sevastyanov, N.G. 1 24.10
Sevcik, J. 3 73.54
Seyferth, D. 3 71.60
Sgamellotti, A. 2 85.88 3 85.22, 85.50
Sgarabotto, P. 2 85.37 3 8.5, 61.8 4 26.5, 35.2, 39.11, 76.12, 83.7, 83.12, 83.33, 83.125, 83.129
Shafer, J.L. 4 83.32
Shaffner, T.J. 1 52.8 3 49.2
Shahat, M. 1 1.49
Shaked, Z. 3 19.9, 19.10
Shakked, Z. 4 51.49
Shallcross, F.V. 1 7.11
Shamala, N. 4 48.16
Shamma, M. 4 58.21
Shankar, J. 1 9.41 2 67.8
Shannon, J.S. 4 83.44
Shapiro, B.L. 4 21.11
Sharma, B.D. 1 44.16, 48.30 3 32.1
Sharma, G.M. 3 58.7
Sharma, V.C. 4 45.7
Sharp, D.W.A. 3 83.62
Shaver, A. 4 75.21
Shaw, B. 4 48.53
Shaw, B.L. 2 71.9 3 71.12, 86.49 4 71.83, 86.14
Shaw, B.W. 2 72.64
Shaw, C.M. 4 28.9
Shaw, R.A. 4 64.28, 64.29, 64.30
Shchegolev, I.F. 4 60.40
Shearer, H.M.M. 1 1.55, 5.42, 23.10, 27.6, 39.10 2 67.5, 67.16, 71.6, 71.17, 71.23, 71.50, 71.54, 72.29, 72.30, 75.53 3 73.43, 76.72, 76.73 4 68.13
Sheat-Rumball, S.V. 2 78.58
Sheat, S.V. 2 78.21
Sheel, H. 1 36.30
Shefter, E. 1 11.62, 31.31, 39.52, 42.8, 44.11, 44.12, 47.7, 47.16, 47.30, 47.31 2 60.104, 60.132 3 3.29, 3.30, 3.31, 3.34, 3.39, 3.40, 3.41, 44.24, 44.33, 47.34, 60.24, 60.30, 60.31 4 27.13, 35.35, 36.18, 40.2, 41.3, 47.19, 60.17
Sheldrick, B. 1 45.48
Sheldrick, G.M. 3 3.10, 63.1, 63.12, 68.2, 69.8, 69.9, 69.13, 69.14, 69.16 4 10.1, 33.12, 63.10
Sheldrick, W.S. 3 3.10, 63.1, 63.12, 68.2, 69.13, 69.14

Shelestov, V.M. **3** 12.1
Shelton, C.A. **3** 61.1
Shelton, G. **3** 49.4
Shen, K. **2** 66.19, 66.21
Shen, M.S. **4** 5.6, 45.9, 45.12
Sherer, H. **2** 69.9
Sherfinski, J.S. **3** 47.14
Shetty, P.S. **2** 85.60 **3** 83.87, 83.94, 85.28
Shiba, T. **3** 59.1
Shibaeva, R.P. **1** 13.43, 32.17 **2** 60.140, 74.16 **3** 13.7, 27.12, 69.17 **4** 60.40, 60.41
Shibata, S. **1** 56.7, 59.60 **3** 59.29, 77.6
Shibayama, S. **4** 54.18
Shields, K.G. **4** 67.8
Shimada, A. **1** 13.12, 21.14, 22.4, 22.5, 36.26, 45.1 **2** 63.9, 63.18, 65.12, 65.14, 65.18, 65.20, 73.94 **3** 5.3, 36.4 **4** 13.14, 38.8
Shimakawa, Y. **3** 42.2
Shimanouchi, H. **1** 20.30, 22.9, 22.14, 27.10, 28.20, 38.42, 58.45, 58.58 **3** 22.1, 31.21, 36.26, 45.43, 53.22, 58.46, 58.58 **4** 5.7, 27.5, 27.6, 33.41, 40.24
Shimanouchi, T. **3** 1.12
Shimizu, K. **2** 83.34
Shimizu, Y. **1** 47.20
Shintani, R. **1** 9.9
Shiomi, M.T. **4** 72.25
Shiono, R. **1** 22.2, 39.12, 45.13 **3** 47.10, 59.26
Shipley, G.G. **2** 60.78, 60.79, 60.80, 60.81
Shiro, M. **1** 7.28, 27.12, 51.20, 52.28, 58.9, 58.18 **2** 76.36, 84.7, 85.18 **3** 31.12, 38.13, 50.4, 50.27, 51.20, 51.22, 51.33, 54.12, 83.141, 85.29 **4** 59.6, 68.10, 83.104
Shive, W. **1** 27.15
Shklover, V.E. **4** 63.8, 73.10
Shkol'nikova, L.M. **2** 77.20, 77.60, 78.30, 84.33, 85.92 **3** 77.26, 78.20 **4** 77.2, 78.3
Shmueli, U. **4** 38.5, 38.9, 39.18, 39.19, 60.23, 64.18
Shoemaker, C.B. **1** 11.14
Shoemaker, D.P. **1** 22.7, 48.17, 48.33 **2** 73.66 **3** 75.21
Shohedrin, B.M. **3** 83.26
Shok, L.N. **4** 24.4, 24.6, 24.7
Shokhor, I.N. **1** 7.3, 10.1, 10.4, 12.1 **3** 7.1, 12.5
Shokhor, N.I. **3** 12.6
Short, M.R. **4** 36.22
Shortland, A. **4** 71.3, 71.53
Shrivastava, H.N. **1** 14.2, 14.3, 30.34
Shriver, D.F. **2** 68.4 **3** 7.2, 7.3, 73.50
Shugam, E.A. **2** 77.20, 77.60, 80.11, 80.36, 84.33, 85.92 **3** 77.5, 77.26, 78.20 **4** 69.28, 77.2, 78.3
Shungu, T. **4** 58.43
Shuto, Y. **3** 71.21
Shvelashvili, A.E. **3** 76.21 **4** 76.7
Siddall, J.B. **1** 51.25
Sieber, A.A.F. **3** 9.8
Siegel, L.A. **2** 64.49, 64.50
Sieker, L.C. **4** 45.19
Sieler, J. **4** 83.127, 85.20
Sievers, R.E. **3** 77.23 **4** 83.45
Sigel, C.W. **4** 59.29
Sigg, H.P. **4** 59.10
Sigwart, C. **2** 83.54
Sikka, S.K. **1** 17.3
Silver, L. **4** 5.3, 64.5
Silverman, J. **1** 26.13, 26.16, 26.38, 28.21, 28.22, 44.65 **3** 28.7, 73.10 **4** 12.11, 18.3, 28.6, 73.12
Silvers, S.J. **1** 49.25, 58.72
Silverton, E.W. **2** 81.70
Silverton, J.V. **1** 11.29, 59.70 **2** 76.65, 77.50, 81.52, 81.70 **3** 11.2, 51.48 **4** 31.25, 40.14, 76.9
Sim, G.A. **1** 1.132, 3.57, 3.58, 3.59, 13.1, 13.2, 13.6, 13.20, 13.45, 16.4, 19.62, 20.20, 27.6, 31.30, 31.40, 37.21, 39.50, 39.56, 41.10, 48.21, 50.17, 50.39, 51.25, 51.36, 51.48, 52.7, 52.13, 53.6, 53.7, 53.8, 53.11, 53.22, 53.25, 53.29, 53.34, 53.37, 54.3, 54.4, 54.7, 54.10, 54.13, 54.25, 56.1, 56.2, 56.3, 56.17, 56.18, 58.48, 58.49, 58.67, 58.68, 58.71, 58.74, 58.76, 58.87, 58.121, 59.33, 59.34, 59.35, 59.51, 59.58, 59.64, 59.70 **2** 60.118, 71.38, 73.16, 73.44, 73.99, 74.6, 74.7, 74.14, 74.22, 75.68, 79.8, 84.30, 84.37, 84.67, 85.12, 85.44, 85.48, 85.75 **3** 1.29, 1.30, 3.38, 8.18, 11.6, 50.19, 51.9, 53.5, 53.21, 53.23, 53.25, 53.28, 53.29, 54.7, 58.51, 58.53, 58.55, 58.56, 58.57, 58.67, 59.17, 68.7, 68.8, 73.1, 73.16, 73.17, 73.46, 73.52, 73.56, 75.6, 85.39 **4** 4.2, 4.5, 24.13, 24.14, 24.15, 37.12, 53.4, 53.7, 53.9, 54.11, 54.14, 58.32, 73.2, 73.5
Sime, J.G. **1** 11.50, 54.15, 59.53 **2** 60.156, 82.47 **3** 2.13, 2.14, 71.50, 71.51
Sime, R.J. **3** 76.25
Simmons, G.L. **1** 29.1 **2** 78.19
Simon, K. **3** 45.7, 45.8
Simon, M.S. **1** 17.30
Simon, W. **1** 50.44
Simonetta, M. **1** 5.28, 5.31, 9.13, 17.25, 17.31, 17.32, 19.50, 52.1 **2** 60.103 **3** 15.10, 19.11, 31.1, 31.25 **4** 31.4, 31.6, 31.10, 31.20, 31.21, 31.22
Simonov, Yu.A. **2** 81.32, 81.76, 81.82 **3** 81.30, 83.26
Simonsen, O. **4** 40.13
Simonsen, S.H. **1** 19.23, 27.15, 39.30 **2** 83.189 **3** 23.10, 31.2, 31.5, 32.31, 39.12, 39.19, 39.20, 48.31, 83.23, 83.120, 83.131 **4** 27.18, 38.24, 39.12, 83.99
Simpson, H.D. **3** 72.14, 75.15 **4** 75.24
Simpson, H.J. **1** 48.14
Simpson, K.A. **4** 63.6
Simpson, P.G. **1** 58.63 **2** 62.15, 62.37 **3** 31.10, 48.36, 48.43, 58.42
Simpson, R. **4** 45.21
Simpson, T.D. **1** 12.20 **3** 11.13
Simpson, T.H. **1** 59.45
Simpson, W.S. **1** 1.22
Sims, J.J. **3** 20.5, 20.7 **4** 53.2, 59.8
Sims, M.J. **4** 35.5
Sims, S.P. **1** 45.68
Simundza, G. **3** 44.27, 44.28
Sinclair, R.L. **2** 82.27, 82.30, 82.31, 83.31
Sinclair, V.C. **1** 29.13 **3** 26.3
Singer, J. **1** 44.68
Singer, M.I.C. **4** 65.13
Singer, N. **3** 73.42, 83.145

Singer, S.J. 4 71.90
Singh, C. 1 43.16, 43.18
Singh, I. 3 59.13
Singh, N. 3 58.48 4 58.38
Singh, P. 1 16.61, 35.1, 38.57 2 64.81 3 83.139 4 35.5, 83.22, 83.23, 86.67
Singh, S. 4 59.27
Singleton, E. 2 71.9
Sinha, S.P. 2 83.174
Sinha, U.C. 3 9.9
Sinn, E. 4 78.4, 78.6, 78.7
Sinnreich, D. 3 1.38
Sintes, A. 1 1.129
Sipe III, J.P. 4 77.24
Sirigu, A. 1 1.86 2 64.13, 72.17, 72.18, 73.111 3 72.33 4 48.7, 72.4
Siskova, J. 2 79.29
Sivertsen, B.K. 3 38.12
Sixma, F.L.J. 1 1.95
Sjoberg, B. 3 81.33, 81.51
Sjolset, O. 4 39.32
Sjostrom, L. 3 1.35
Skakke, P.E. 3 85.11
Skapski, A.C. 1 2.44, 3.38, 3.39 2 65.17, 71.7, 71.55, 83.104, 86.78, 86.110, 86.115, 86.116 3 73.36, 81.29, 83.43, 83.79, 86.74, 86.95 4 71.3, 71.52, 71.53, 80.5, 80.14, 83.82, 83.108, 85.38
Skelton, B.W. 4 86.53
Skinner, C.G. 1 27.15
Skinner, J.M. 1 14.4
Sklar, N. 1 8.1, 8.2
Skolnik, S. 1 58.1
Skripkin, V.V. 2 73.68
Sky, K. 2 60.65
Slade, P.G. 4 84.1
Sletten, E. 1 35.2, 39.44, 39.59, 44.25 2 83.73 3 33.15, 33.18, 39.42, 39.43, 83.104 4 83.54
Sletten, J. 1 35.2, 39.16, 41.5, 44.25, 44.28 3 39.42, 39.43, 39.49, 44.1 4 39.23, 39.28, 39.42
Sljukic, M. 4 81.11
Sloan, G.J. 4 36.12
Slomp, G. 4 34.3
Slone, C.J. 4 48.12
Sly, W.G. 1 22.7, 27.16, 57.3 2 72.56
Small, R.W.H. 1 1.24, 1.39, 1.40, 2.18, 2.33, 2.51 3 10.1 4 1.2
Smallwood, R.J. 4 83.130, 83.136
Smart, M.L. 1 3.38, 3.39 3 50.15, 75.4, 81.29
Smelyanskaya, E.M. 4 24.7
Smiley, I.E. 1 17.47
Smirnova, V.I. 3 19.2
Smissman, E.E. 3 3.29, 3.30, 3.31, 3.34, 3.40, 3.41 4 27.13
Smith, A.E. 1 36.2 2 72.8, 85.34 3 72.36, 85.2, 85.3 4 39.39, 72.13, 72.31
Smith, A.J. 2 81.51 3 78.7
Smith, A.W. 3 77.4
Smith, C.H. 3 9.12, 9.15
Smith, C.P. 1 46.7 2 64.61 4 64.25
Smith, D.E. 4 67.23
Smith, D.L. 1 35.23, 41.37 2 72.46, 73.58, 75.75 3 28.13, 36.33 4 28.3, 32.26, 32.29
Smith, G.D. 3 77.21 4 31.13, 64.16, 64.19, 79.5
Smith, G.S. 1 3.37 2 63.11, 76.66, 76.69
Smith, G.W. 3 56.5, 56.6
Smith, H.E. 1 52.8
Smith, H.G. 2 74.1
Smith, H.M. 2 84.22, 84.31
Smith, H.W. 1 4.9
Smith, J.A.S. 3 19.3
Smith, J.D. 3 68.7, 68.8 4 73.24
Smith, J.E.W.L. 2 83.23, 84.10
Smith, M.R. 3 65.3 4 39.6, 65.3
Smith, N.E. 4 80.6, 80.7
Smith, N.O. 2 71.21
Smith, P. 3 13.12 4 13.11
Smith, P.J. 3 69.2
Smith, P.M. 4 76.9
Smith, R.A. 1 37.18 3 83.128
Smith, R.M. 4 58.36, 58.40
Smits, D.W. 1 48.87
Smolin, Yu.I. 2 76.43
Smutny, E.J. 4 39.39
Snaauw, G.J. 1 5.6
Snader, K.M. 4 59.27
Snaith, R. 4 68.13
Snare, D.L. 1 16.47
Snatzke, G. 4 58.24
Sneath, T.C. 3 36.3, 53.19, 53.27
Sneeden, R.P.A. 3 71.65
Snow, A.I. 2 67.1
Snow, M.R. 2 76.48, 76.49, 82.5, 83.173 3 76.55, 83.142 4 71.72, 76.31, 76.32, 81.24, 83.24
Snyder, D.A. 3 76.6
Snyder, R. 3 33.7
Snyder, R.G. 2 75.51
Snyder, R.L. 3 32.34 4 60.14
Snyman, J.A. 3 60.45, 60.46
Sobell, H.M. 1 44.38, 44.52, 44.57, 44.59, 47.22, 47.23 3 39.13, 39.14, 44.13, 44.14, 44.17, 44.18, 44.19, 44.25, 44.26, 44.27, 44.28 4 39.21, 39.29, 60.43, 60.44
Soboleva, S.V. 3 78.22
Sobti, R.R. 4 51.20
Soderberg, B. 1 2.59
Soderberg, E. 3 58.4 4 35.1, 35.6
Soderberg, R.H. 2 60.155, 83.99, 84.21
Soderquist, R. 3 81.33, 81.51
Soest, T.C.van 3 59.10, 59.11, 59.12
Sokolova, G.V. 2 81.24
Sokolova, Y.A. 4 81.10, 84.10
Sokolova, Yu.A. 2 83.78
Solbakk, J. 3 44.7, 45.33 4 44.4
Soleimani, E. 1 52.2
Solheim, J.O. 4 70.15
Solheimsnes, S. 2 70.8, 70.9
Soling, H. 3 76.2, 76.3
Solodova, O.P. 2 69.39
Soltwisch, M. 3 75.19
Soltzberg, L. 4 20.2
Soltzberg, L.J. 3 20.13 4 12.11
Sommerville, P. 3 85.46 4 31.32, 33.13
Sondheimer, F. 4 75.22
Song Loong, W. 1 8.22, 8.23
Sonin, A.S. 1 3.29
Sorensen, A.M. 1 21.1, 21.2 2 60.15 3 5.2

Sorm, F. **3** 53.7
Sorum, H. **1** 5.39, 11.17, 59.7 **4** 33.6, 33.7, 59.1
Sotofte, I. **2** 83.185 **3** 32.8 **4** 71.82
Soule, F.B. **3** 2.22
Souron, C. **1** 24.25
Southwick, J. **3** 45.2
Southwick, P.L. **1** 30.27
Sowerby, J.D. **4** 68.13
Sozzi, L. **2** 83.46
Spagna, R. **2** 85.69 **3** 32.36, 72.5, 72.6, 85.1 **4** 70.9, 72.3, 86.64
Sparks, R.A. **1** 7.29, 36.18, 47.31, 49.32 **2** 70.1 **3** 24.1, 26.6, 26.7 **4** 36.2
Sparrow, N. **4** 33.13
Speakman, J.C. **1** 2.10, 2.11, 2.32, 2.42, 2.45, 2.88, 2.91, 11.40, 14.1, 14.2, 14.3, 14.4, 14.5, 14.11, 14.13, 28.4, 30.34, 46.17, 49.3 **2** 64.76, 86.48, 86.56 **3** 2.12, 2.13, 2.14, 35.8, 35.22 **4** 2.2, 2.3, 2.12, 2.13, 2.14, 2.17, 2.18, 2.19
Speckamp, W.N. **3** 51.15
Spencer, A. **4** 81.62
Spencer, C.B. **2** 71.17, 71.50 **3** 76.72, 76.73
Spencer, C.J. **2** 64.14
Spencer, J.L. **3** 71.35, 71.38, 71.40 **4** 71.39
Spencer, R.C. **2** 83.208
Spink, J.M. **2** 81.20
Spiro, T.C. **2** 71.30
Spiro, T.G. **3** 76.65
Spivack, B. **4** 82.13
Spoelder, T. **4** 12.10
Spofford III, W.A. **3** 79.8, 79.24, 79.25 **4** 85.10, 85.27
Spofford, W.A. **2** 85.19, 86.51 **3** 79.6, 79.7, 79.9, 79.22
Spohn, R.J. **3** 71.60
Spraggins, R.L. **4** 51.57, 51.58
Spratley, R. **3** 72.1
Spratley, R.D. **2** 64.60
Sprecher, C.M. **4** 31.8
Sprenkels, A.J.J. **1** 2.64
Springer, J.P. **4** 59.16
Sproul, G.D. **4** 36.3
Spruijt, A.M. **4** 85.4
Srenger, E.M. **4** 41.29
Srikanta, S. **3** 2.2
Srikantha, S. **1** 2.24
Srikrishnan, T. **4** 48.42
Srinivasan, L. **3** 83.14, 83.15 **4** 83.79, 83.80
Srinivasan, R. **1** 44.45, 48.24, 48.66, 48.71, 50.33 **3** 48.29 **4** 11.11, 44.24, 48.19, 48.34, 48.37, 48.42
Srinivasan, S. **1** 9.17
Srivastava, P. **1** 13.24, 25.12
Srivastava, R.C. **1** 3.33, 3.35, 3.36, 22.7 **2** 78.10 **4** 83.36, 83.67
Srivastava, S.N. **1** 26.33, 58.15 **3** 54.11
Srivastava, T.S. **4** 49.19, 49.24
Stackelberg, M.v. **1** 5.20, 17.10 **2** 78.20
Stadler, H.P. **1** 30.25, 30.30, 36.37 **3** 3.35, 30.3
Stahlin, W. **4** 83.69, 83.70
Stalhandske, C.-I. **3** 66.1
Stalick, J.K. **2** 86.88 **3** 86.14, 86.56 **4** 86.49, 86.50
Stalik, J.K. **3** 69.39, 71.20, 86.39
Stam-Thole, I. **4** 18.3
Stam, C.H. **1** 5.37, 5.38, 8.34, 19.63, 28.1, 59.47 **3** 33.46, 37.7 **4** 18.3, 84.4
Stanfield, R.A. **1** 30.27
Stanford Junior, R.H. **1** 1.81, 41.2 **3** 24.6, 28.12, 30.5 **4** 22.10
Stanford, R.H. **4** 59.4
Stanko, J.A. **3** 3.53, 76.45 **4** 58.21, 64.40
Stanko, V.I. **2** 62.12, 62.14 **3** 62.2
Stanley, E. **1** 2.90, 46.23 **4** 83.30, 84.7
Starikova, Z.A. **2** 78.56, 78.57 **3** 77.5
Starovskii, O.V. **2** 73.14, 73.108, 74.15
Steadman, J.P. **3** 3.3
Steeple, H. **1** 3.6, 41.32
Steer, I.A. **4** 65.1, 65.2
Steer, R.J. **3** 19.1
Steggarda, J.J. **4** 79.10
Steggerda, J.J. **2** 80.8, 80.9, 80.34
Steigemann, W. **3** 31.4
Stein, M.T. **4** 64.40
Steinfink, H. **2** 63.7 **3** 81.4
Steinhauser, S. **3** 68.5
Steinrauf, L.K. **1** 48.58, 48.67, 48.68, 48.76, 50.45, 58.14, 59.55 **2** 77.55 **3** 48.27, 48.28, 48.45, 48.48, 50.25 **4** 50.16, 58.34
Stelling, G.D. **3** 31.10
Stempelova, D. **2** 81.86
Stemple, N.R. **1** 1.57, 1.58 **3** 84.9, 84.22
Sten Rouwelaar, L.N.A. **2** 81.79
Stensland, B. **1** 36.27 **3** 63.17 **4** 35.1, 36.13
Stenson, P. **3** 41.13, 83.130
Stepanovich, V.M. **3** 76.13
Stephen Junior, W.F. **3** 3.40, 3.41
Stephens, D.N. **3** 36.3
Stephens, F.A. **2** 86.116
Stephens, F.S. **2** 64.72, 64.73, 76.51, 83.113 **3** 1.34, 60.37, 72.29, 73.39, 76.20, 76.22, 76.59, 76.69, 81.40, 81.49, 83.36, 83.105 **4** 33.23, 81.40, 83.107
Stephensen, T.A. **4** 85.36
Stephenson, N.C. **1** 3.65 **2** 76.34, 77.54, 83.33, 83.100, 83.101, 83.154, 85.50, 85.66, 86.31, 86.33, 86.34, 86.35 **3** 45.37, 45.41, 45.42, 76.30, 82.1, 83.76, 86.13, 86.20, 86.26 **4** 45.22, 83.113, 86.5
Sterling, C. **1** 2.13, 2.22, 57.4, 57.5 **2** 69.30
Stermitz, F.R. **1** 58.52
Stern, F. **1** 1.59
Sterns, M. **1** 31.42 **3** 36.16, 83.119 **4** 59.9, 83.71
Steuck, M.J. **3** 76.40
Stevens, J.D. **3** 45.37, 45.41, 45.42 **4** 45.22
Stevens, K.L. **1** 56.14
Stevenson, D.L. **2** 73.27, 86.20
Stevenson, R. **1** 51.42 **3** 51.35
Stewart, G. **1** 14.4
Stewart, J.M. **1** 8.26, 8.27, 16.1, 32.32, 41.25 **2** 72.83, 78.9, 78.11 **3** 11.18, 15.1, 15.2, 32.28, 35.21, 36.32, 39.6, 39.28, 41.15, 86.34, 86.35 **4** 16.17, 32.3, 32.14, 39.43, 67.11, 86.38
Stewart, R.F. **1** 44.9, 44.10, 44.47 **3** 83.60 **4** 31.8
Stezowiski, J.J. **4** 50.5
Stezowski, J.J. **3** 73.21 **4** 73.31, 77.8

Stiefel, E.I. **1** 3.40 **3** 85.17, 85.27
Stock, D. **2** 72.12
Stockel, K. **1** 51.35 **4** 75.22
Stocklin, W. **1** 51.35
Stocks, J. **4** 85.34
Stoeckli-Evans, H. **4** 49.23
Stogsdill, R.M. **2** 78.51
Stokely, P.F. **3** 11.16, 11.17, 73.3 **4** 82.7
Stokes, J. **2** 85.26, 85.27, 85.52
Stomberg, R. **2** 76.1, 83.35, 83.71, 83.72, 83.93 **3** 76.4, 76.11, 76.12, 81.1
Stone, A. **1** 42.11, 42.12, 49.21, 49.27
Stone, A.L. **1** 49.30 **2** 75.26
Stone, F.G.A. **4** 71.36, 71.70, 71.84, 71.86
Stone, M.E. **4** 83.30, 84.7
Stora, C. **3** 19.16, 45.17 **4** 17.14
Stork-Blaisse, B.A. **4** 83.81
Storr, A. **4** 68.3
Stosick, A.J. **1** 48.2
Stoufer, R.C. **4** 83.134
Stout, C.D. **3** 83.25
Stout, G.H. **1** 56.14, 56.16, 59.26, 59.27, 59.40, 59.65 **3** 59.13, 59.32
Stout, V.F. **1** 56.16, 59.40
Strahs, G. **1** 33.51
Strandberg, B. **2** 82.3
Strandberg, R. **4** 83.96
Streib, W.E. **2** 62.7, 62.22, 62.28, 75.9, 83.190 **3** 62.19 **4** 62.6, 68.2, 83.109
Strel'tsova, I.N. **1** 19.4, 19.24
Streurman, H.J. **3** 39.5
Strickler, P. **1** 21.16, 21.18 **2** 85.67, 85.94
Strid, K. **1** 7.1
Strieter, F.J. **1** 1.36, 1.37, 1.65 **2** 60.1
Strijk, B. **1** 21.5
Strom-Gundersen, E.K. **3** 36.25
Strom, E.T. **4** 31.30
Stromme, K.O. **1** 16.2 **2** 60.22, 60.29, 60.36, 60.84, 60.85
Stroud, R.M. **4** 39.9
Struchkov, Yu.T. **1** 5.2, 5.8, 11.43, 15.1, 16.33, 19.4, 19.24, 19.38, 24.2, 24.3, 24.5, 24.8, 24.11, 24.12, 28.8, 28.9, 28.10, 29.28 **2** 62.12, 62.14, 63.28, 65.13, 66.9, 69.4, 69.5, 69.18, 69.39, 69.45, 71.34, 71.47, 72.7, 72.19, 72.62, 72.63, 72.72, 72.73, 72.84, 73.14, 73.41, 73.49, 73.53, 73.68, 73.69, 73.85, 73.87, 73.101, 73.105, 73.107, 73.108, 73.119, 73.120, 74.15 **3** 40.2, 60.9, 62.2, 64.30, 69.31, 71.4, 71.10, 71.49, 71.56, 71.61, 71.62, 72.40, 72.48, 72.53, 73.22, 73.29, 73.38, 73.45, 74.8, 74.9, 74.16, 75.38, 75.46 **4** 63.8, 64.17, 74.5, 83.128
Struchkov, Yu.T. **3** 74.10 **4** 16.1, 69.29, 71.38, 72.5, 72.6, 72.14, 73.10, 73.21, 86.36
Stryukov, V.B. **4** 60.40
Stucky, G. **2** 67.14, 67.17
Stucky, G.D. **1** 3.21, 3.27, 3.67 **2** 67.12, 67.15, 68.18 **3** 37.2, 40.4, 62.3, 67.7, 67.8, 67.12, 67.13, 67.18, 67.26, 68.1, 68.11, 68.15, 68.21, 83.21 **4** 38.2, 38.3, 38.4, 67.26, 86.10
Studd, B.F. **2** 83.196 **3** 83.145
Stulen, G. **1** 39.91 **3** 39.48
Sturdivant, J.H. **2** 71.29, 83.115
Suares, H. **4** 58.10, 58.11, 58.35
Subramanian, E. **1** 16.32, 31.19, 47.26, 48.64, 48.81 **3** 1.11, 44.22, 47.13, 59.2, 64.11 **4** 36.31
Subrtova, V. **1** 37.11 **4** 62.5
Suchy, M. **3** 53.7
Suck, D. **3** 44.21 **4** 47.1, 47.3, 47.7, 47.18
Sudarikov, V.S. **3** 73.22
Suddath Junior, F.L. **4** 32.10
Suffness, M.I. **3** 58.57
Suga, H. **1** 37.1
Sugawara, I.T. **1** 5.9
Sugawara, T. **1** 2.3
Sugihara, A. **2** 82.29
Sugino, T. **3** 50.30
Suginome, H. **4** 40.24
Sugisaka, N. **3** 38.17
Sugiyama, K. **4** 5.7
Suito, E. **4** 49.13, 49.18
Sukhanova, L.S. **3** 82.9
Sullivan, R.A.L. **1** 8.9, 16.57
Sullivan, S. **2** 60.119
Sumida, Y. **4** 33.10, 33.11
Sumner, G.G. **2** 78.12
Sumner, R.H. **2** 78.41
Sumsion, H.T. **1** 19.65
Sundaralingam, M. **1** 1.136, 12.18, 13.21, 16.5, 21.15, 25.22, 27.14, 39.75, 40.27, 46.14, 47.10, 47.28 **3** 39.37, 44.11, 45.35, 45.40, 47.7, 47.11, 47.12, 47.21, 47.26, 47.30, 48.6, 48.7, 48.22, 48.35, 50.16, 83.25, 84.6 **4** 11.3, 11.8, 20.4, 26.7, 31.15, 31.19, 44.2, 44.3, 44.10, 44.19, 44.22, 47.8, 47.9, 47.10, 47.12, 47.15, 47.16, 47.19, 50.1, 83.28, 83.40
Sundarama, K. **2** 81.47
Sundaresan, T. **4** 40.4, 60.25
Sung, N. **1** 19.43
Surazhakaya, M.D. **3** 81.7
Sussman, J.L. **4** 47.20, 58.13
Susz, B.P. **2** 60.50
Sutherland, H.H. **1** 13.49, 13.50, 19.46, 19.47, 19.48, 48.10 **3** 13.14
Sutherland, S.A. **1** 50.26, 54.7, 56.3, 59.61
Sutor, D.J. **1** 1.35, 2.26, 44.54, 44.64, 44.66, 44.67, 47.21, 58.32 **2** 65.7, 65.25, 65.26
Sutton, D. **3** 83.1 **4** 71.85, 83.139
Sutton, P.W. **2** 71.22
Suzuki, K. **3** 36.34
Suzuki, S. **1** 50.4
Suzuki, Y. **4** 35.38
Svaeren, S.E. **4** 70.16, 70.17
Svedung, D.H. **3** 83.7
Svelasvili, A.E. **2** 76.27, 76.70, 76.74
Svetich, G.W. **1** 46.9, 46.24 **4** 82.10, 84.18
Swallow, A.G. **2** 77.58, 83.164, 83.196 **3** 73.42, 83.145
Swaminatha Reddy, B. **4** 47.9, 47.10
Swaminathan, S. **1** 9.17, 26.28 **2** 79.6 **3** 84.4 **4** 12.9
Swan-Walstra, S. **4** 65.10
Swan, E.P. **1** 59.61
Swank, D. **3** 84.1, 84.7
Swank, D.D. **1** 46.7 **2** 64.25 **4** 64.39
Swann, D.A. **4** 83.101
Swanson, B. **3** 7.2, 7.3
Swanson, J. **4** 38.2, 38.3
Swanson, J.S. **3** 3.17

Swanwick, M.G. **3** 71.5, 71.6
Swardstrom, J.W. **4** 67.28
Sweet, R.M. **1** 44.22, 44.23 **2** 69.35 **3** 50.7, 50.9, 50.11
Swen-Walstra, S.C. **3** 31.33
Swift, D.R. **3** 85.49 **4** 83.121
Swink, L.N. **2** 76.42
Sydow, E.von **1** 1.148, 1.150, 5.32
Szabo, W. **4** 38.10
Szmuszkovicz, J. **4** 34.3, 35.28, 35.29, 40.23
Szymanski, J.T. **2** 60.114, 82.25, 82.35 **3** 83.72 **4** 63.2, 63.12, 69.1

Tadini, C. **1** 11.18
Taga, T. **3** 45.24, 45.25, 58.43, 59.30 **4** 67.3
Tagawa, H. **4** 60.12
Taguchi, I. **1** 16.15
Tai, C.-C. **3** 40.11
Taira, Z. **4** 62.10
Tait, A.M. **4** 83.111
Takabatake, A. **1** 59.52
Takada, S. **1** 54.17 **4** 21.22, 28.12
Takagi, S. **3** 45.13, 45.34, 59.26
Takahashi, R. **4** 56.6
Takahashi, S. **2** 86.71, 86.81
Takaki, Y. **1** 13.16, 13.17, 13.18, 13.19, 13.38, 27.8, 33.18
Takanashi, S. **3** 33.24
Takano, T. **1** 13.29, 33.42 **2** 63.6
Takashina, K. **4** 5.1
Takats, J. **2** 75.39, 75.57 **3** 73.44 **4** 73.23
Take, T. **1** 50.31 **3** 50.20, 59.1
Takeda, T. **3** 58.46
Takei, W.J. **1** 43.7
Takemoto, K. **3** 36.34
Takenaka, A. **1** 1.89, 50.31 **2** 75.55 **3** 33.42 **4** 81.32, 81.33, 81.35
Takeshita, H. **4** 27.15
Takeuchi, S. **4** 50.4
Takeuchi, Y. **1** 33.29, 59.4
Takigawa, T. **1** 48.84, 48.86
Takita, T. **1** 41.31
Takusagawa, F. **4** 13.14
Takwale, M.G. **4** 13.6
Talaty, E.R. **4** 32.30
Talbot, J.H. **2** 81.29
Tamburin, H.J. **4** 37.9
Tamm, C. **3** 50.19
Tamura-Kobayashi, H. **4** 82.6
Tamura, C. **1** 35.17, 37.21, 39.88, 50.1, 51.36, 58.7, 58.12 **3** 58.53 **4** 32.5, 64.35, 64.38
Tamura, G. **1** 44.76 **3** 50.18
Tamura, S. **4** 54.13
Tamura, Z. **3** 33.24
Tanaka, J. **1** 15.23, 16.39, 30.5, 44.31
Tanaka, K. **3** 72.20
Tanaka, N. **1** 13.3, 31.27, 56.7 **3** 59.15 **4** 59.20
Tanaka, O. **1** 56.7 **3** 56.3
Tanaka, T. **4** 5.1
Tanako, T. **1** 20.1
Tang, Y.-C. **1** 11.19 **2** 83.115
Tangocci, P. **1** 35.5
Tani, H. **2** 68.28 **3** 68.19 **4** 68.9
Tani, M.E.V. **3** 8.11 **4** 8.6, 28.5
Taniguchi, T. **1** 13.16, 13.17, 13.18, 13.19
Tanimura, M. **2** 83.114 **3** 21.11
Tapscott, R.E. **2** 81.54
Tarantelli, T. **4** 85.33
Tarkhova, T.N. **2** 81.76, 81.82, 81.87
Tarnopol'skii, B.L. **3** 27.12, 33.28, 64.30
Tartakovskii, V.A. **3** 40.2
Tashpulatov, Yu. **1** 41.18
Tasker, P.A. **3** 83.117, 83.138
Tatsch, C.E. **3** 82.28
Tauer, K.J. **1** 5.4
Tavale, S.S. **1** 2.52, 2.69, 2.72, 2.85, 2.89, 13.11 **3** 44.13, 44.18 **4** 13.1
Taylor Junior, I.F. **3** 74.11, 74.12, 74.13, 79.24, 79.25 **4** 79.6, 86.24
Taylor, C.A. **1** 10.26
Taylor, D.A.H. **3** 56.11
Taylor, D.W. **3** 83.77, 83.111 **4** 83.37, 83.38, 83.41, 83.52, 83.53
Taylor, E.C. **4** 40.2, 41.3
Taylor, H.F.W. **3** 76.24
Taylor, H.L. **4** 59.19, 59.26
Taylor, H.P. **3** 71.39
Taylor, I.F. **2** 79.3 **3** 74.7, 74.14
Taylor, J.C. **1** 8.12, 8.13, 8.14; 8.15, 59.16 **2** 83.23
Taylor, J.H. **2** 60.83
Taylor, K.A. **2** 86.104 **3** 86.91
Taylor, M.R. **1** 2.46, 2.47, 4.15 **2** 82.26, 85.46 **3** 83.14, 83.15 **4** 83.79, 83.80
Taylor, N. **4** 83.137, 83.138
Taylor, N.E. **1** 35.39, 59.43
Taylor, P.J. **3** 33.35
Taylor, R.C. **3** 64.44 **4** 62.2, 62.3, 62.4, 86.7
Taylor, R.M. **2** 85.24
Taylor, R.P. **2** 65.42 **3** 78.19
Taylor, S.R. **3** 71.55
Taylor, T.G. **1** 31.39, 52.24, 52.30
Taylor, W.I. **3** 58.51
Tazzoli, V. **3** 11.5 **4** 83.35
Teare, P.W. **1** 5.43
Tebbe, F.N. **4** 71.44
Tedder, J.M. **1** 40.33
Teitel, S. **4** 58.3, 58.6
Temple, D.M. **3** 60.40
Templeton, D.H. **1** 1.26, 1.36, 1.37, 1.65, 3.31, 3.80, 27.1, 27.2, 27.3, 31.37, 32.7, 49.36, 54.8, 59.3 **2** 60.1, 62.33, 62.41, 64.23, 65.39, 65.42, 71.30, 73.5, 73.28, 77.6, 77.46, 82.59, 83.210, 83.219, 85.39, 85.40 **3** 48.17, 49.6, 62.8, 62.12, 76.25, 77.32 **4** 39.15, 49.8, 49.12, 62.9, 71.5
Templeton, L.K. **1** 27.1, 27.2, 27.3
Temussi, P.A. **1** 1.86, 1.97, 21.24, 21.25
Tench, A.H. **3** 31.39 **4** 47.11
Teo, B.K. **4** 22.7
Terao, Y. **1** 58.115
Terhaar, G. **2** 62.6
Ternay Junior, A.L. **1** 39.74
Terpstra, P. **1** 11.31
Terry III, N.W. **4** 86.48
Terry, J.C. **4** 33.16, 36.1
Terzis, A. **3** 76.65
Tewari, R. **4** 83.36

Thewalt, U. **2** 76.56, 83.11, 83.12, 85.38 **3** 44.20, 47.16, 47.19, 47.23, 64.54, 76.17, 76.19 **4** 35.11, 44.8, 44.20, 47.14, 64.36, 68.6, 76.28, 76.29, 76.30, 76.39, 83.6
Thiele, G. **4** 85.11
Thierry, J.-C. **3** 53.10, 53.11, 53.12, 53.13, 53.14, 53.16
Thiessen, W.E. **1** 40.5, 40.10 **3** 45.19, 51.47, 53.3, 53.8 **4** 8.1, 44.7, 56.3
Thill, R.J. **1** 51.1
Thoai, N. **3** 1.7, 30.6
Thom, E. **1** 7.15 **3** 27.7, 51.40, 51.46
Thomas, C.K. **3** 3.53
Thomas, G.J. **4** 58.40
Thomas, J. **3** 47.15
Thomas, J.T. **2** 85.22
Thomas, P. **4** 85.20
Thomas, P.C. **1** 41.29 **2** 62.48
Thomas, R. **1** 11.14 **3** 59.8 **4** 22.9
Thompson, B. **4** 83.137, 83.138
Thompson, B.P. **1** 30.11
Thompson, C.M. **1** 19.23
Thompson, D.J. **3** 59.31
Thompson, D.T. **2** 86.6, 86.8, 86.43, 86.55 **3** 86.12, 86.41
Thompson, D.W. **4** 77.1
Thompson, G.L. **4** 39.17
Thompson, J.C. **4** 63.2
Thompson, L.C. **1** 3.83
Thornley, A.S. **3** 78.3
Thorup, N. **3** 76.2, 76.3 **4** 41.11
Thuijl, J.van **1** 35.45, 36.36 **2** 77.59
Thurmann-Moe, T. **2** 60.61
Thyagaraja Rao, S. **1** 48.24, 48.45
Tichy, K. **1** 1.113 **3** 15.8
Tickle, I.J. **4** 36.35
Tieghi, G. **4** 5.9, 35.21, 71.60, 76.26
Tillack, J.V. **4** 67.8
Tillberg, O. **4** 36.19, 60.30
Timkovich, R. **3** 49.10
Timms, R.E. **2** 69.41, 69.47
Ting, H.-Y. **4** 3.5, 56.4, 59.14
Tiripicchio Camellini, M. **1** 8.28
Tiripicchio, A. **1** 8.28, 9.2 **2** 83.4, 83.5, 83.7, 83.14, 83.15 **3** 83.2, 83.3, 83.4 **4** 8.4, 67.1, 67.2, 79.1, 79.4, 81.2, 81.15, 81.16, 83.50
Tirouflet, J. **3** 74.2, 74.5 **4** 73.26, 75.10
Tishchenko, G.N. **2** 83.95
Tisley, D.G. **3** 86.75 **4** 80.13
Titova, N.S. **3** 62.2
Titus, D.D. **3** 86.77 **4** 86.39, 86.40
Titze, H. **3** 77.19, 77.20
Tjomsland, O. **1** 41.6
Tjornhom, T. **2** 60.82
Tkatchenko, I. **4** 71.65
Toan, T. **4** 73.25
Tobe, M.L. **2** 83.86
Tobias, R.S. **2** 71.18
Todd, G. **1** 53.15
Todd, L.J. **2** 62.39
Todd, M.J. **3** 9.8
Todd, P.F. **2** 83.168
Toft, L. **1** 10.24
Tokuma, Y. **4** 32.28
Tokunan, H. **4** 71.58, 71.80
Tokuyama, T. **3** 58.9
Tollin, P. **1** 19.48, 47.21, 47.25 **3** 47.17, 47.18
Tomalia, D.A. **1** 11.9, 42.1
Toman, K. **1** 12.10, 12.11 **2** 60.34
Tomiie, Y. **1** 9.5, 37.1, 50.37, 50.38, 56.9, 58.35, 58.36, 58.57, 59.19, 59.25, 59.52 **2** 67.19 **3** 50.5, 50.13, 50.20, 72.20, 81.24 **4** 58.1, 58.2
Tomile, Y. **1** 54.9 **2** 67.18
Tomita, K. **1** 44.42, 44.48, 44.52, 44.56, 44.58, 44.59, 56.20, 58.93 **2** 82.4, 82.5, 82.21, 83.62 **3** 33.26, 35.10, 44.12, 47.32, 48.40, 56.13 **4** 44.14, 44.15, 56.5
Tomita, K.I. **1** 44.38
Tomita, M. **1** 58.36, 58.57
Tomkins, I.B. **3** 72.52 **4** 86.55
Tomko, J. **1** 58.104
Tomlinson, A.A.G. **3** 85.22 **4** 81.1
Tomlinson, A.J. **3** 75.16
Tomoko, J. **1** 58.105
Tomuka, Y. **1** 48.63
Tondello, E. **3** 80.19, 80.20, 83.45 **4** 39.27
Toney, J. **2** 67.15 **3** 67.18 **4** 67.26
Tong, Y.C. **1** 11.9, 42.1
Toniolo, C. **1** 33.75
Toome, V. **3** 35.23
Torbjornsson, L. **1** 36.27 **3** 36.15
Torelli, L. **2** 80.35
Torii, A. **4** 72.17
Torii, K. **3** 48.16 **4** 48.28
Torre, L.P. **3** 83.112, 83.121, 83.122, 83.123, 83.124, 83.125, 83.126
Torri, A. **4** 82.6
Torti, E. **1** 21.32, 29.17, 29.18
Toube, T.P. **4** 38.21, 54.1
Tougard, P. **3** 47.3 **4** 47.6
Toussaint, J. **1** 11.46, 11.48, 11.49, 13.7, 17.33, 38.24
Towl, A.D.C. **2** 86.46 **3** 71.63, 78.24 **4** 71.43, 71.46, 71.49
Towns, R. **1** 35.25, 35.37
Towns, R.L. **3** 32.34, 34.6, 64.37 **4** 32.19
Towns, R.L.R. **3** 39.19, 39.20 **4** 21.11, 37.13, 37.14, 51.7
Townsend, L.B. **4** 44.25
Tranter, T.C. **1** 48.49
Traverso, G. **1** 39.45
Trefonas, L.M. **1** 3.44, 3.55, 32.34, 35.10, 35.25, 35.29, 35.31, 35.33, 35.37, 40.21 **2** 62.44 **3** 32.34, 34.6, 36.19, 51.2, 64.36, 64.37 **4** 32.19, 35.7, 35.17, 37.13, 37.14, 40.7, 51.1, 51.7, 81.59
Treichel, P.M. **1** 3.64 **2** 73.90
Trinderup, P. **3** 85.4
Trinh-Toan, K **4** 73.27
Trinh-Toan, M **4** 73.28
Trommel, J. **1** 48.65
Trost, B.M. **4** 29.2

Trotter, J. **1** 3.60, 13.52, 15.9, 15.11, 15.15, 15.16, 15.17, 15.24, 16.6, 16.37, 18.14, 18.15, 19.29, 19.33, 19.35, 24.9, 24.22, 24.23, 26.12, 26.15, 26.17, 26.30, 26.36, 28.6, 28.7, 28.11, 28.13, 28.17, 29.5, 29.12, 29.13, 29.14, 30.6, 30.10, 30.14, 30.15, 30.18, 30.23, 30.29, 30.33, 30.35, 31.17, 31.44, 32.21, 37.27, 45.3, 45.47, 45.55, 45.61, 45.74, 47.6, 51.43, 52.23, 53.38, 58.46, 58.69, 58.79 **2** 64.45, 64.78, 65.1, 65.2, 65.4, 65.8, 65.10, 65.19, 65.21, 65.22, 65.23, 65.34, 65.48, 65.49, 69.2, 69.36, 70.5, 71.3, 71.10, 73.57, 73.86, 73.95, 73.98, 75.33, 75.60 **3** 1.10, 1.11, 10.5, 27.2, 27.3, 32.18, 32.19, 35.15, 39.29, 45.18, 45.20, 45.45, 48.20, 51.49, 56.10, 58.64, 59.2, 63.15, 64.3, 64.9, 64.11, 64.20, 64.65, 69.32, 73.41, 83.32, 83.127, 86.5, 86.7, 86.9, 86.11, 86.17, 86.18, 86.37, 86.45 **4** 1.4, 13.9, 28.7, 31.1, 31.3, 35.30, 36.31, 48.38, 64.27, 68.3, 69.2, 71.18, 73.9, 81.17, 83.122, 85.21, 86.1
Trotz, U.O. **3** 36.18
Troughton, P.G.H. **1** 3.38, 3.39 **2** 71.7, 71.55, 86.78, 86.110, 86.115 **3** 73.36, 86.74, 86.95 **4** 36.27
Trueblood, K.N. **1** 7.21, 7.29, 11.7, 11.68, 15.12, 16.10, 20.3, 22.8, 31.22, 31.35, 33.64, 36.18, 47.7, 47.19, 47.31, 48.40, 49.32, 50.29, 51.45 **2** 60.51, 64.66, 67.6, 67.7, 73.72 **4** 12.9, 20.11, 20.13, 30.5, 30.6, 31.17, 31.18, 36.2, 60.37
Truelock, M.M. **4** 71.81
Trus, B.L. **1** 36.6 **2** 60.130 **3** 36.2, 41.8, 77.9 **4** 29.7, 30.7, 44.12, 60.27
Truter, M.R. **1** 4.3, 8.11, 8.42, 11.2, 11.6, 11.41 **2** 60.145, 65.29, 68.19, 72.13, 72.37, 72.66, 75.64, 76.80, 77.9, 77.14, 77.36, 77.37, 77.58, 78.33, 79.10, 79.23, 79.36, 83.29, 83.164, 83.191, 85.53, 85.91 **3** 38.15, 38.21, 38.22, 44.15, 44.16, 67.11, 67.24, 68.10, 71.46, 71.48 **4** 10.2, 10.3, 38.26, 44.5, 44.6, 60.6, 67.15, 67.17, 67.21, 67.24, 67.25, 72.18, 77.7
Tsai, C. **2** 62.28 **3** 48.26, 62.19 **4** 37.8, 50.5, 81.9
Tsay, Y.-H. **3** 51.48 **4** 75.7, 86.46, 86.66
Tschesche, R. **1** 51.49
Tsein, L.Y. **1** 1.123
Tselinski, I.V. **1** 7.3, 10.1, 10.4, 12.1
Tselinskii, I.V. **3** 7.1, 12.3, 12.5, 12.6 **4** 12.4, 12.7
Tsernoglou, D. **1** 44.7, 44.8
Tsintsadze, G.V. **2** 83.138
Tsoucaris, G. **1** 3.56, 33.33, 33.39, 33.47, 33.48 **3** 58.1
Tsuboi, M. **1** 48.42
Tsuchida, Y. **4** 40.24
Tsuchiya, H. **4** 60.32
Tsuda, K. **1** 58.7, 58.12
Tsukihara, T. **4** 44.14, 44.15, 48.33, 48.35
Tsukuda, Y. **1** 50.42, 51.20 **2** 84.7 **3** 31.12, 50.3, 50.4, 51.20, 51.22, 51.33 **4** 50.12
Tsuneno, T. **4** 40.24
Tsutsui, M. **4** 49.19
Tuck, B. **4** 59.18
Tuck, D.G. **1** 3.60, 3.68 **3** 68.6
Tucker, P.A. **3** 3.52, 64.51
Tufte, T. **2** 60.90
Tuggle, R.M. **3** 71.58 **4** 75.27, 75.30
Tulinsky, A. **1** 39.85, 49.25, 50.28, 58.72, 59.38 **2** 60.93, 60.94, 67.11 **3** 49.10 **4** 49.1, 49.9
Turco, A. **2** 86.9, 86.38
Turley, J.W. **1** 3.32, 10.17, 40.1 **2** 63.16, 63.26 **3** 9.1, 63.2, 63.3, 76.85 **4** 32.12, 33.3, 33.18, 40.1
Turnbull, K.R. **2** 82.2, 82.24 **4** 76.17, 76.18
Turner, J.D. **1** 1.143
Turner, R.B. **1** 23.1
Turner, R.W. **2** 74.2, 74.3
Turney, T.W. **4** 86.21
Tustin, G.C. **3** 85.23
Tvedten, O. **1** 8.36, 8.37
Tyrrell, J. **1** 19.27
Tysseland, M. **4** 70.3
Tyurk de Garcia Banos, D. **3** 68.9, 76.67

Ubbelohde, A.R. **1** 17.16
Ubell, E. **1** 1.42, 18.27
Uchida, T. **2** 60.128 **4** 86.52
Uchida, Y. **4** 86.52
Uchtman, V.A. **2** 73.64 **3** 73.27, 73.37
Ucko, D.A. **4** 81.26, 81.28
Ueda, H. **1** 15.23
Ueda, I. **1** 18.1, 18.3
Ueki, T. **1** 48.56, 48.95 **2** 69.31, 78.1, 78.2, 82.13, 83.210, 86.81 **3** 69.25, 72.18, 73.33 **4** 48.33, 48.35, 48.57
Uesu, T. **3** 11.20
Uesu, Y. **2** 67.18, 67.19 **3** 9.18, 9.20 **4** 32.18
Uggla, R. **3** 76.32, 76.35
Ugo, R. **2** 77.2, 77.3, 86.80, 86.89 **3** 86.98
Ugro, J.V. **2** 84.20, 86.36
Uhlemann, E. **4** 85.20
Ukaji, T. **1** 1.103
Ulku, D. **3** 10.2, 77.2, 77.3
Ulrich, L. **3** 59.9
Umans, A.J.H. **1** 18.20
Umanskij, M.M. **1** 15.3, 24.10
Umezawa, H. **1** 41.31, 50.3, 50.9, 50.20, 50.21, 50.34
Umland, F. **4** 83.10
Ungaretti, L. **3** 11.5
Uno, T. **2** 83.187
Unsworth, W.D. **4** 83.25
Uramoto, M. **3** 50.21
Urch, D.S. **3** 75.16
Urdy, C.E. **2** 84.57
Urry, D.W. **3** 32.14, 32.15
Uskokovic, M. **1** 41.29 **3** 35.23
Utsumi, H. **1** 1.108 **3** 17.5, 33.42 **4** 81.32, 81.33, 81.35
Utsumi, Y. **3** 50.5, 76.82
Uttech, R. **2** 72.22
Uyeda, N. **4** 49.13, 49.18
Uyeo, S. **1** 58.86 **3** 58.43 **4** 58.43

Vaccaro, A.M.T. **3** 59.33

Vaciago, A. 1 3.70, 4.12, 33.70, 35.3, 35.4, 35.7, 35.12, 35.14, 35.15, 50.10, 50.48 2 65.27, 80.13, 80.16, 80.17, 80.20, 80.32, 80.33, 80.35, 85.9, 85.10, 85.11, 85.36, 85.69 3 9.10, 33.45, 50.28, 58.15, 83.24, 85.1 4 32.22, 33.42, 33.45, 35.26, 54.19, 70.9, 80.15
Vagg, R.S. 3 83.92
Vahrenkamp, H. 2 73.91 4 73.14, 85.24, 86.15
Vaidya, S.N. 2 62.47
Vainshtein, B.K. 1 8.3, 8.10, 48.38
Vaira, M.di 1 3.33, 3.35, 3.36 2 76.76, 76.77, 76.78, 76.79, 78.24, 78.32, 78.49, 78.53, 83.119, 83.120, 83.127 3 76.62, 76.86, 78.15, 83.88, 86.84
Valambal, V. 1 48.24
Valle, G. 1 39.8, 39.9, 39.39, 39.40, 42.9 3 16.12, 32.12 4 39.3, 39.4
Valle, M. 4 71.27
Valle, S. 2 71.45
Vallino, M. 3 67.14
Valverde-Lopez, S. 3 58.13, 58.52
Van Soest, T.C. 4 59.11
Vand, V. 1 5.42, 30.13
Vanderauwera, J. 1 41.46
Vandewalle, J.J.M. 3 41.10
VanEtten, R.L. 3 32.11, 32.23, 32.34
Vannerberg, N.-G. 1 7.1 2 81.71 3 31.16
Varenkamp, H. 2 73.64
Varshavskii, Y.S. 4 83.128
Vaska, L. 2 86.83 4 86.48
Vassian, E.G. 3 83.65
Vassova, A. 1 58.104
Vaughan, P.A. 1 29.2
Veal, J.T. 3 71.9 4 76.13, 82.9, 83.100
Veen, J.van der 3 27.4
Vega, E. 4 35.35
Veidis, M.V. 2 86.68 3 32.18, 32.19, 77.18 4 71.74, 71.75, 73.16, 73.18
Vekris, J. 2 83.47
Veldis, M.V. 2 82.41
Velichko, F.K. 2 71.1
Ven, C.F.W.van de 4 27.17, 51.19
Venanzi, L.M. 2 75.52, 75.71, 80.32, 80.33, 86.67, 86.105 3 72.5, 72.6 4 86.64
Venkataraman, K. 1 59.68
Venkatesan, K. 1 19.41, 23.8, 31.19, 48.94 3 32.25, 48.49 4 48.13, 48.15, 48.16, 48.17, 48.32, 48.52, 48.54, 49.20
Verbist, J. 2 60.64 3 67.3 4 48.4
Verbist, J.F. 4 48.10
Vergamini, P.J. 4 73.14, 85.24
Verhoef, L.H.W. 2 60.11, 60.98
Verhoeven, A.M. 4 80.11
Verkade, J.G. 1 46.8 3 64.12 4 64.4
Verma, H.K.L. 3 26.10
Veronese, A.C. 1 33.75
Verschoor, G.C. 1 42.2 3 33.2, 33.3 4 83.81
Vesela, L. 1 53.36
Vetter, W. 4 31.28
Vialard, A.-M. 1 38.60
Vickery, B.L. 2 60.145 3 44.15, 44.16 4 44.5, 44.6, 60.6, 67.17, 77.7
Victor, D. 1 9.29
Vidal, J.-P. 4 63.9
Vidali, M. 3 85.19 4 78.11, 83.65
Vigato, P.A. 4 71.41, 78.11, 83.65
Vihovde, E.H. 1 21.9 2 70.2
Vijayalakshmi, B.K. 4 11.11
Vijayan, M. 1 13.57, 38.48 2 67.20, 67.21, 69.49 4 32.15
Villa, A. 2 80.1
Villa, A.C. 1 11.1 3 8.11, 79.4, 85.12 4 8.6, 35.13, 76.38, 79.2, 79.3, 81.1, 83.1, 83.2, 83.5
Villa, J.F. 3 81.18
Villiers, J.P.R.de 3 77.34 4 77.19, 77.21
Vimala, T.M. 3 84.4
Vimba, S. 2 62.51
Vincent, A.T. 4 16.16, 64.43
Vincent, R.L. 1 58.103
Vinciguerra, N. 2 85.88 3 85.50
Viossat, C. 1 17.27
Visser, G.J. 1 3.25, 35.42, 39.26, 39.53, 39.54, 39.55, 39.91 3 11.23, 28.8, 31.33, 39.17, 39.48 4 18.1, 32.16, 35.22
Visser, J.P. 4 75.33
Visti, J. 3 76.32, 76.35
Viswamitra, M. 4 83.137, 83.138
Viswamitra, M.A. 1 11.15 2 62.47, 67.20, 67.21, 69.49, 81.15 3 50.31 4 47.2, 47.9, 47.10, 82.4, 82.5
Viswanathan, K.S. 2 83.61
Viterbo, D. 1 10.13, 13.53, 40.15, 40.16 2 83.25, 83.26 3 40.17, 59.4 4 24.2, 40.8
Vitulli, G. 2 72.67 3 74.3, 74.4
Vizzini, E.A. 1 43.23, 43.24, 43.25, 43.28, 43.29 2 79.3 3 43.4, 43.5 4 43.6
Vlahov, R. 1 53.14
Vlasse, M. 1 45.25
Vliegenthart, J.A. 1 1.34 3 1.8
Vloten, G.W.van 1 21.5
Vmanskij, M.M. 1 7.23
Voblikova, V.A. 4 78.3
Voecks, G.E. 4 86.11
Voet, D. 2 62.26, 62.31 3 44.9, 60.25 4 60.18, 60.34
Voge, A.A. 4 82.10, 84.18
Vogt Junior, L.H. 4 33.27
Vogt, L.H. 1 49.36
Voigt, D. 4 51.48
Vol'pin, M.E. 2 63.28
Volponi, L. 3 86.19, 86.25
VonDreele, R.B. 4 77.8
Vooren, M. 3 37.4
Vorontsova, L.G. 1 4.1, 11.38, 39.73 3 33.28
Vos, A. 1 3.25, 3.61, 3.62, 11.47, 16.42, 35.42, 39.26, 39.53, 39.54, 39.55, 40.12 2 60.122, 64.70, 72.44 3 28.8, 33.1, 33.4 4 11.7, 12.10, 16.8, 16.9, 27.14, 32.16, 35.22
Vozzennikov, V.M. 1 11.30 2 62.29, 69.20
Vranka, R.G. 2 68.10, 79.1, 83.16 3 86.72 4 79.9
Vries, A.de 1 48.31
Vukosavovich, M. 3 3.13
Vuuren, P.J.van 4 71.9

Wada, T. 1 37.1
Wada, Y. 3 50.22, 50.30 4 50.6, 54.3

Wade, K. **4** 62.16, 68.13
Wadley, L.G.B. **4** 79.8
Waerstad, K. **1** 9.15
Wagner, A.J. **2** 64.70 **3** 64.29 **4** 64.15
Wagner, H. **3** 59.26
Wagner, W.F. **2** 77.29, 77.30, 77.32, 77.34 **4** 84.16
Wahl Junior, G.H. **4** 31.31
Waight, E.S. **4** 31.32
Waite, M.G. **3** 11.6, 58.55, 58.56, 75.6 **4** 4.2, 4.5
Wakahara, A. **3** 35.10, 48.40
Wakamiya, T. **3** 59.1
Wakayama, N.I. **4** 41.19
Wakisaka, K. **1** 58.45 **3** 36.26
Walker, A. **2** 86.70 **4** 69.1
Walker, C. **3** 81.16, 81.42
Walker, G. **1** 2.9
Walker, J. **1** 8.32
Walker, L.A. **1** 32.13
Walker, W.R. **2** 82.23, 82.36, 82.38
Wall, M.E. **4** 59.19, 59.26
Wallace, J.C. **4** 45.8
Wallace, J.E. **3** 78.29
Wallbridge, M.G.H. **2** 65.6, 68.5, 68.6
Walling, C. **2** 83.183
Wallwork, S.C. **1** 5.29, 17.19, 33.66, 54.1, 59.23 **2** 60.72, 60.78, 60.79, 60.80, 60.81, 60.83, 60.88, 60.100, 60.113, 60.116, 60.125, 60.126, 60.133, 60.134, 65.37, 84.1 **3** 59.34, 65.8, 69.3 **4** 33.28, 33.29, 33.30, 33.46, 40.4, 60.25
Walter, R. **1** 1.128
Walter, W. **1** 4.16 **3** 85.5 **4** 4.1, 8.7
Walthers, K.K. **3** 76.40 **4** 31.9
Walton, R.A. **2** 83.103, 85.54, 86.21 **3** 33.9, 81.36, 86.75 **4** 80.13, 81.36, 81.37
Wampler, D.L. **2** 83.32 **3** 47.25 **4** 8.9, 23.2
Wan, C. **2** 77.5 **4** 73.31
Wan, K.Y. **4** 72.26, 86.31
Wang Lund, E. **1** 21.6
Wang, A.H.-J. **4** 32.30, 37.6, 50.18, 71.54, 71.55
Wang, B.-C. **3** 33.34, 78.30
Wang, B.C. **3** 83.115, 83.116 **4** 83.118, 83.119
Wang, C.-H. **2** 82.25
Wang, F.E. **2** 62.15
Wani, M.C. **4** 59.19, 59.26
Ward, D.L. **4** 79.5, 84.15, 86.11
Ward, J.E.H. **3** 86.5, 86.7
Wardle, R. **4** 75.23
Warr, W.A. **3** 35.18
Warren, R.C. **3** 82.1
Warren, S. **3** 47.25
Warrener, R.N. **3** 36.16 **4** 44.16
Warwicker, J.O. **1** 43.12
Waser, J. **2** 65.9, 65.11 **3** 45.27
Wasson, J.R. **3** 76.77
Wasson, S.J. **4** 84.16
Watanabe, I. **1** 55.3
Watanabe, K. **2** 76.2
Watanabe, T. **1** 2.5, 16.15, 22.4, 22.5, 33.18, 36.12, 50.31, 50.31, 58.120 **2** 61.21, 75.55, 81.9, 81.77, 82.32 **3** 50.20 **4** 5.1, 72.17, 82.6
Watanabe, Y. **2** 69.22, 69.27 **3** 11.9, 80.4
Watase, H. **1** 59.17, 59.19 **2** 81.68
Watase, T. **2** 63.2
Watelle-Marion, G. **4** 2.9
Watelle, G. **3** 2.5
Watenpaugh, K. **1** 47.15 **2** 77.48, 84.22, 84.54
Waterhouse, C.R. **3** 86.1
Waters, J.M. **1** 49.6 **2** 76.22, 78.46, 82.57 **3** 59.28, 60.47, 78.2, 78.4, 78.27, 78.29, 86.61, 86.65 **4** 86.35
Waters, T.N. **1** 49.6, 54.2 **2** 76.26, 77.16, 77.17, 78.4, 78.5, 78.7, 78.16, 78.17, 78.18, 78.21, 78.26, 78.27, 78.29, 78.40, 78.41, 78.46, 78.50, 78.58, 82.55, 82.57, 83.109, 83.112, 83.192, 83.194 **3** 54.1, 54.4, 59.28, 60.1, 60.16, 60.39, 60.47, 78.2, 78.4, 78.8, 78.27, 78.29 **4** 76.42, 83.101, 86.53
Watkin, D. **2** 86.99
Watkin, D.J. **1** 42.10 **2** 84.52 **3** 86.81, 86.82, 86.86 **4** 22.1
Watkins, D.J. **4** 42.2
Watkins, E.D. **2** 77.30
Watkins, S.F. **3** 19.1, 69.26, 71.64, 86.2 **4** 71.90
Watling, R.C. **2** 78.33
Watson, D.G. **1** 27.6, 44.22, 44.23, 46.3, 47.21, 51.35, 56.2 **2** 65.24 **3** 20.17, 23.1, 46.5, 46.6, 47.25, 47.33, 51.42, 53.12, 53.13, 53.14, 54.13, 58.47, 65.5 **4** 8.9, 23.2
Watson, H.C. **1** 24.19
Watson, K.J. **1** 15.7, 15.8, 50.40, 53.5, 59.44 **2** 82.2 **3** 79.21, 85.4 **4** 83.115, 83.116
Watson, W.H. **2** 77.19, 84.32, 84.34, 84.38, 84.39, 85.28 **3** 84.8, 84.9, 84.11, 84.17, 84.22 **4** 3.5, 31.30, 32.11, 33.5, 51.53, 56.4, 59.14, 84.8
Watt, R. **4** 71.56
Watton, E.C. **2** 83.101 **3** 83.92
Watts Junior, P.H. **4** 1.14, 37.9, 39.43
Watts, P.H. **3** 39.6, 39.28
Watts, W.E. **2** 73.106
Weakliem, H.A. **2** 76.59
Weaver, D.L. **2** 84.24, 84.29, 85.29, 85.55 **3** 71.44, 71.58, 76.6, 76.76 **4** 75.27, 75.30
Weaver, J. **4** 71.31
Webb, L.E. **1** 49.1, 49.22, 49.23, 49.28, 49.31 **2** 81.30 **3** 79.16 **4** 48.45, 48.48
Webb, N.C. **1** 4.5, 31.3, 59.22 **2** 73.78
Weber, A.H. **3** 33.47
Weber, H.P. **1** 23.4, 34.7, 53.40 **2** 69.40, 73.31, 73.121, 85.83 **3** 40.12 **4** 39.35, 59.10, 83.112
Webster, M. **2** 63.12 **3** 3.9, 69.6 **4** 69.3
Webster, M.S. **1** 10.25, 50.29
Wedegaertner, D.K. **4** 21.21
Weedon, B.C.L. **1** 21.33 **4** 54.1, 54.2
Weeks, C. **4** 47.19
Weeks, C.M. **2** 82.53 **3** 17.8, 51.12, 51.19 **4** 11.12, 11.14, 51.15, 51.17, 51.24, 51.26, 51.27, 51.33
Wegner, P.A. **3** 62.9
Wei, C.H. **1** 44.72 **2** 72.48, 73.90, 85.62, 85.74, 85.82, 85.84 **3** 44.34, 83.31, 83.97, 83.98, 83.99 **4** 30.2, 30.3, 39.20, 48.55, 48.56, 83.98

Wei, L. 2 78.51
Weichselgartner, H. 1 49.16
Weickhardt, P.L. 3 85.24
Weidenborner, J.E. 4 69.35
Weigel, D. 4 73.20
Weigel, F. 3 81.11, 81.12, 81.13, 81.15
Weimann, J. 3 1.7, 30.6
Weinblum, D. 1 44.73
Weinheimer, A.J. 4 51.57, 51.58
Weiniger, M.S. 2 79.28
Weininger, M.S. 3 79.24, 79.25 4 69.11, 69.22, 69.27, 74.11, 86.24
Weir, C.E. 3 19.4
Weisbach, J.A. 1 58.100, 58.101
Weiss, E. 2 73.20 3 63.4, 63.5, 63.6, 67.5
Weiss, G.S. 1 37.2, 37.3
Weiss, J. 2 64.1, 64.8, 64.20, 65.5, 65.15, 83.11, 83.12, 85.1, 85.38 3 64.5, 71.45 4 69.4
Weiss, R. 1 12.5, 18.4, 18.5 2 66.10, 83.124, 83.153, 84.3 3 33.29, 40.10, 53.10, 53.11, 53.12, 53.13, 53.14, 53.16, 60.51, 64.24, 64.38, 64.40, 67.19, 67.21, 67.22, 67.23, 72.21, 72.31 4 12.1, 12.2, 12.5, 12.6, 33.8, 33.9, 40.18, 66.2, 67.12, 67.20, 67.22, 68.1, 71.66, 72.7, 77.5, 77.16, 81.3, 81.19, 81.50, 83.46, 83.47, 83.48, 84.3
Weiss, U. 4 28.8
Weisz, O. 2 86.1
Weitz, G. 1 58.39
Welberry, T.R. 3 31.35
Welch, A.J. 4 83.61, 83.85, 83.130, 83.136
Welcker, P.S. 2 62.39 3 86.92
Weldon, A.S. 1 2.82
Wells, A.F. 2 66.3, 66.4, 66.5, 83.39, 86.2, 86.14
Wells, J.L. 1 32.41 4 60.27
Wells, R.B. 4 38.21
Welsh, H.K. 1 1.151 2 82.42
Welsh, M.J. 1 59.40
Wengelin, R.F. 1 14.12
Wentworth, R.A.D. 2 83.190
Weperen, K.J.van 4 18.1
Werbin, H. 4 31.30
Werkema, M.S. 1 6.1, 35.1
Werner, P.-E. 1 45.27, 45.28 3 36.7, 36.11, 36.15, 36.22 4 36.17, 36.24
Wessels, G.F.S. 4 33.33
West, B.O. 4 86.2
West, R. 4 22.7
Westcott, N.D. 4 56.2
Westerdahl, A. 1 1.131, 41.28
Westley, J.W. 4 50.15
Wet, J.F.de 2 81.63
Wetzel, J. 2 66.15
Weyenberg, A.M. 4 60.3
Whalley, W.B. 1 2.92, 59.45 3 51.26, 51.30 4 28.8
Wheatley, P.J. 1 4.2, 4.11, 4.17, 5.23, 13.42, 20.22, 22.16, 32.14, 33.6, 33.9, 41.22, 41.23, 41.47, 41.48, 44.6, 50.15 2 63.14, 63.30, 64.10, 64.36, 64.43, 64.57, 64.75, 64.79, 66.22, 68.21, 69.34, 85.47 3 16.16, 16.17 4 16.16, 64.1, 64.2, 64.43, 83.123
Wheeler, A.G. 2 60.107, 72.21, 72.42, 83.159
Wheeler, D.H. 4 67.8
Wheeler, G.L. 4 33.26, 33.39, 35.19, 35.27
Whillans, F.D. 3 76.84, 83.78, 83.110
Whimp, P.O. 3 72.37, 72.52, 74.17, 75.35, 81.48 4 71.56, 75.34, 75.35, 86.21
Whinnery, J.E. 4 31.30, 32.11, 51.53
White Junior, R.P. 4 3.15, 3.16, 64.41, 64.42
White, A.H. 3 80.6 4 80.3, 80.9, 80.17
White, B.F. 1 49.12
White, B.M. 2 73.9
White, D.N.J. 3 51.9 4 50.13, 53.4, 53.9, 58.32
White, D.W. 3 64.12
White, E.H. 3 9.8
White, J.C.B. 1 13.23, 49.12 3 8.9, 33.12, 33.13 4 78.8
White, J.D. 4 59.27
White, J.G. 1 3.20, 24.13, 30.9, 30.12, 30.19 3 38.4, 42.2
White, N.E. 1 44.14
Whitfield, H.J. 3 85.10
Whiting, D.A. 3 56.12, 59.23 4 20.14, 59.18
Whitla, A. 2 73.12
Whitla, W.A. 2 75.71
Whitlow, S.H. 1 16.6 2 64.45 3 64.20, 83.32 4 2.23, 67.4
Whitt, C.D. 4 67.18, 68.5
Whittaker, E.J.W. 1 17.38
Whittle, K.R. 3 86.65 4 86.35
Wickramasinghe, J.A.F. 4 51.55
Widdowson, D.A. 3 58.34
Wiebenga, E.H. 1 5.6, 19.61, 48.87, 52.19
Wieland, D.M. 3 8.14
Wiesner, J.R. 1 3.33, 3.35, 3.36 2 76.14
Wiest, R. 4 77.5, 77.16
Wiewiorowski, M. 1 59.32
Wijnhoven, J.G. 4 80.16
Wilcox, H.E. 1 1.42
Wildman, G.C. 4 19.5
Wildman, W.C. 3 58.19, 58.24, 58.50
Wiley, P.F. 4 45.23
Wilford, J.B. 2 71.21, 73.12, 73.109, 86.44 3 86.81, 86.82, 86.86 4 31.7, 31.11, 31.23, 41.27
Wilke, G. 4 71.65
Wilkes, C.E. 2 76.54, 77.69
Wilkes, G.R. 2 73.90
Wilkinson, G. 4 71.3, 71.52, 71.53, 81.62
Will, G. 1 49.16 3 32.6
Willett, R.D. 1 3.5, 3.11 2 83.20, 83.21, 84.19 3 3.1, 3.2, 3.3, 3.12, 3.13, 84.1, 84.3, 84.7 4 3.1, 3.4, 82.11
Williams, D.A. 2 84.11 3 7.4 4 7.13
Williams, D.E. 1 12.12, 12.17, 19.54, 19.56, 38.19, 45.76 2 83.57, 83.58 3 12.10
Williams, D.G. 1 6.2, 6.3, 6.4 3 45.38
Williams, D.J. 1 58.30 2 77.1 3 59.8
Williams, G. 1 39.72
Williams, G.J.B. 4 33.38, 47.2
Williams, J.M. 1 3.7, 3.8 3 3.16, 3.17, 6.1, 76.7 4 13.2
Williams, J.R. 3 36.29, 37.13
Williams, P.P. 1 52.11
Williams, R.J. 2 84.32, 85.28 3 84.8 4 33.5, 76.8, 84.8
Williams, R.M. 2 60.72, 60.113, 60.125, 60.126

Williams, T. 1 41.29
Willis, B.T.M. 2 73.23
Willis, J.S. 3 15.1, 15.2
Williston, C.S. 1 15.15 2 70.5, 73.98 3 73.41
Willoughby, T.V. 2 60.95
Wills, D.L. 4 80.13
Willstadter, E. 2 82.56
Wilson Junior, E.W. 4 82.7
Wilson Junior, G.E. 3 39.1
Wilson, A. 2 60.133, 60.134
Wilson, F.B. 2 61.16 4 61.1
Wilson, F.C. 2 73.66
Wilson, H.R. 1 44.36, 47.2, 47.3, 47.4, 47.25 3 47.5, 47.18
Wilson, J.N. 1 48.30
Wilson, S.A. 1 19.19, 19.20
Wilson, V.A. 3 73.47, 73.48, 73.49
Wilson, W.S. 3 36.16
Wing, R.M. 2 62.34, 64.56 3 22.2, 37.5, 60.54, 73.24, 83.148, 85.23 4 53.2, 59.8
Wingard Junior, R.E. 4 39.17
Wingfield, J.N. 4 64.27, 83.122
Winkler, F.K. 4 50.11
Winkler, H. 2 64.48
Winquist, B.H.C. 2 86.97 3 77.33
Winter, G. 4 69.15
Wirth, J.G. 4 33.27
Wise, J.J. 2 77.56
Wiseman, T.J. 2 83.167
Witkop, B. 3 31.14, 58.9 4 48.20, 48.22
Witmore, W.G. 1 7.20
Witt, J.R. 4 7.1, 7.2, 7.3, 7.4
Witters, R.D. 2 84.58
Witters, R.W. 1 53.3
Wodak, S. 4 58.13, 58.15
Wohlauer, G. 2 83.57, 83.58
Wojcicki, A. 2 85.71
Wojtkunski, J. 4 33.47
Wolf, A.P. 2 66.19, 66.21
Wolf, G.-U. 2 67.2
Wolf, N.de 1 42.5
Wolf, W. 1 42.8
Wolfe, G.A. 1 39.4 3 53.1, 53.2
Wolfel, E. 1 58.39 2 63.25
Wolfram, W. 1 8.6
Wolodina, G.F. 2 60.87
Wolters, A.P. 4 3.12, 86.55
Wolters, J. 1 39.53, 39.54, 39.55
Wolthius-Spuy, M. 3 1.37
Wong, C. 4 67.10
Wong, C.-H. 3 67.9
Wong, C.H. 1 20.24 2 73.62, 73.65
Wong, C.S. 4 81.61
Wong, P.-S. 4 37.10
Wong, R.Y. 4 41.15, 67.7
Wong, W. 4 71.90
Wood, A.M. 2 73.33, 82.23, 82.36, 82.38, 86.55
Wood, J.S. 2 65.45, 71.44, 72.25, 74.9, 77.57, 80.23, 83.47, 83.49, 83.172, 84.27 3 64.48, 65.9, 79.19 4 78.17, 83.19, 83.20, 84.17
Wood, M.K. 3 31.5
Wood, R.G. 1 39.72
Woodlock, A. 3 33.12, 33.13
Woods, J.D. 1 1.141
Woods, M. 4 64.28, 64.29, 64.30
Woodward, I. 1 5.25, 9.35, 19.49, 49.8, 49.9
Woodward, P. 2 66.1, 73.47, 85.25, 85.26, 85.27, 85.52, 86.17 3 3.52, 19.1, 71.39, 71.64 4 69.23, 69.26, 71.31, 71.45, 71.77
Woolfson, M.M. 2 64.26
Worch, H. 1 58.107
Worch, H.-H. 4 51.56
Wormald, J. 3 71.8, 71.24, 72.9, 73.35, 73.40, 75.22, 75.31, 85.14 4 71.35, 75.2, 85.9, 86.70
Worsham Junior, J.E. 1 8.4, 8.21
Worthington, C.R. 2 67.11
Worthington, J.M. 4 83.131
Worwald, J. 2 73.77
Wright, D.A. 1 48.70 2 84.11 3 7.4 4 7.13, 67.14
Wright, G.F. 1 9.7 2 71.11, 71.12
Wright, H. 3 54.5
Wright, H.E. 1 58.64
Wright, J.D. 2 60.70, 60.146, 60.147, 60.148, 75.26
Wright, L.H. 3 54.5, 58.48 4 58.38, 58.44
Wright, W.B. 1 33.38, 33.41, 48.82 4 19.10
Wu, V.Y. 1 17.4
Wudl, F. 4 39.5
Wulf, H. 1 58.59
Wulff, G. 1 51.49
Wunderlich, H. 3 11.1 4 17.2, 20.9
Wunderlich, J.A. 1 45.65, 52.14, 58.17, 58.38 3 35.1, 58.22
Wynberg, H. 1 35.42 3 39.48

Yadav, B. 4 2.10
Yadav, V.S. 4 48.3
Yagen, B. 4 51.55
Yagubskii, E.B. 4 60.40
Yaji, H. 1 36.26
Yakel Junior, H.L. 1 48.83
Yakel, H.L. 1 8.31
Yakovenko, V.I. 2 80.25
Yakushi, K. 3 60.53 4 60.33
Yamada, N. 2 67.18, 67.19
Yamada, Y. 1 7.18
Yamahata, K. 3 80.4
Yamakawa, M. 1 7.28
Yamamoto, T. 1 1.89 4 81.32
Yamamura, S. 1 54.9
Yamasaki, K. 4 83.68
Yamashita, S. 2 69.12, 69.22
Yamataka, K. 4 35.39
Yamauchi, H. 3 56.13 4 56.5
Yamazaki, H. 4 72.28
Yamazaki, S. 4 54.13
Yamazaki, Y. 1 1.89 2 67.18, 67.19
Yanagiya, M. 1 58.120
Yanai, H.S. 1 58.3
Yannoni, N.F. 1 26.13, 26.16, 26.38, 28.21, 28.22, 44.65 3 28.7, 73.10 4 12.11, 28.6, 73.12
Yao, S.Y. 4 58.21
Yasaki, E. 3 76.31
Yasin, Y.M.G. 2 61.15
Yasucka, N. 4 72.27

Yasuda, H. 2 68.28 3 68.19 4 68.9
Yasuda, K. 2 69.3 4 51.42
Yasuda, N. 4 71.28, 71.29
Yasuoka, N. 1 17.40, 17.42 2 68.28, 69.31, 72.78, 86.71, 86.81 3 13.19, 13.20, 35.12, 36.9, 41.23, 68.19, 69.11, 69.25, 72.10, 72.18, 73.33 4 35.39, 60.20, 68.9, 68.12, 71.28, 71.29
Yates, P. 4 39.46
Yawney, D.B.W. 3 79.11, 81.35 4 81.60
Yeats, E.A. 1 24.16
Yen, T.-M. 2 73.62
Yerkess, J. 4 28.4
Yonehara, H. 3 50.21 4 50.4
Yonemitsu, O. 3 31.14
Yonezawa, M. 3 76.52
Yoo, C.S. 4 13.13, 16.17
Yoshida, M. 2 69.31
Yoshida, N. 2 69.27
Yoshida, T. 2 73.82 4 72.17, 75.35
Yoshino, A. 1 58.90 3 33.24, 71.21
Yoshioka, H. 3 36.34 4 35.31, 35.32
Yoshioka, T. 1 44.76
You-Chi, T. 1 3.54
Young, A.E. 1 11.9, 42.1
Young, D.W. 1 19.48, 30.2, 30.17, 45.39, 46.2, 47.25, 48.10, 54.3, 54.4, 54.7 3 47.18
Young, J.M. 3 3.43, 3.44
Yu, A.P. 1 23.1
Yumal', E.M. 4 78.3

Zabkiewicz, J.A. 1 22.6
Zaborsky, O.R. 1 11.9, 42.1
Zacharias, D.E. 1 58.82, 58.85, 58.100 3 58.65
Zachariasen, W.H. 1 2.1 2 81.42
Zacharis, H. 3 36.19
Zacharis, H.M. 1 35.31 4 35.7, 35.17
Zahrobsky, R.F. 1 44.71 2 84.36
Zakharova, G.N. 2 69.18 4 69.29
Zakrzewski, G. 3 83.44
Zakrzewski, G.A. 3 81.45, 81.46, 81.47 4 83.56, 83.57
Zalkin, A. 1 3.31, 3.75, 3.80, 49.36, 54.8 2 62.33, 62.38, 62.41, 65.39, 71.30, 73.5, 77.6, 77.46, 82.59, 83.210, 83.219, 85.39, 85.40 3 48.17, 62.8, 62.12, 76.25, 77.32 4 39.15, 49.8, 49.12, 62.9, 71.5, 75.14, 75.15
Zalkow, L.H. 1 53.28
Zalukaev, L.P. 3 27.10
Zambonelli, L. 1 33.70, 35.13 2 80.13, 80.16, 80.17, 80.20, 80.32, 80.33, 80.35 3 32.36, 72.4, 72.5, 72.6, 83.24 4 70.9, 72.3, 86.64
Zanari, A.R. 4 85.8
Zanazzi, P.F. 1 3.4, 35.5, 38.43, 40.19, 40.20 4 35.4, 35.10, 70.4, 83.49, 85.8
Zand, R. 1 20.14 4 48.26, 48.34, 48.37, 48.42
Zanella, A. 4 76.11
Zanker, V. 4 36.37
Zannetti, R. 1 39.46 2 83.59, 83.60, 83.74 3 1.36
Zanzari, A.R. 4 70.4
Zarli, B. 3 80.19, 80.20, 81.55 4 85.26, 85.32
Zaslow, B. 3 3.6 4 3.11
Zavalishin, E.I. 3 27.10
Zazzetta, A. 3 71.25, 78.12, 78.16 4 74.1
Zechmeister, K. 3 39.24, 39.25, 39.27, 54.6, 54.8, 54.15, 59.25 4 27.19, 36.37, 47.7, 51.11, 54.8, 54.12, 54.16
Zedler, A. 3 32.29
Zee, J.J.van der 4 67.8
Zeiger, W. 3 30.4
Zeiss, H.H. 3 71.65
Zenda, H. 4 59.21
Zeppezauer, M. 3 33.43
Zhdanov, G.S. 1 7.23, 15.3, 24.10, 39.77, 41.18 3 26.1, 35.20
Ziegler, M. 2 72.51, 85.1 4 69.4
Ziegler, M.L. 2 64.8, 64.9 3 64.5, 71.45
Ziffer, H. 3 36.29, 37.13
Zimmer, J.-C. 2 69.29
Zimmerman, I.C. 1 11.20
Zimmermann-Barlow, I. 1 39.17
Zingaro, R.A. 3 65.3 4 65.3
Zinnius, A. 3 63.8, 63.13 4 63.7
Ziolkovska, B. 2 71.26
Ziolo, R.F. 3 86.96 4 86.61, 86.68
Zirakzadeh, M. 1 44.19
Ziser, M. 3 44.24
Zobel, T. 1 16.6 2 65.2
Zocchi, M. 2 72.67, 83.165 3 83.63 4 5.9, 35.21, 71.60, 76.26, 83.102
Zoer, H. 4 64.15
Zolnikova, G.P. 3 71.49
Zubieta, J. 4 86.22
Zubieta, J.A. 3 85.31 4 80.10, 85.22
Zuccaro, D.E. 1 11.22
Zuckerman, J.J. 4 86.49, 86.50
Zurr, Z. 4 64.18
Zussman, J. 1 47.29, 48.39
Zvarykina, A.V. 4 60.40
Zvonkova, Z.V. 1 7.6, 11.30, 16.25, 41.18, 41.36 2 62.29, 62.42, 69.20, 69.38, 80.21, 80.25 4 41.4